NANOTUBES
AND NANOSHEETS

NANOTUBES AND NANOSHEETS

Functionalization and Applications of Boron Nitride and Other Nanomaterials

Edited by
Ying (Ian) Chen

CRC Press
Taylor & Francis Group
Boca Raton   London   New York

CRC Press is an imprint of the
Taylor & Francis Group, an **informa** business

CRC Press
Taylor & Francis Group
6000 Broken Sound Parkway NW, Suite 300
Boca Raton, FL 33487-2742

First issued in paperback 2021

© 2015 by Taylor & Francis Group, LLC
CRC Press is an imprint of Taylor & Francis Group, an Informa business

No claim to original U.S. Government works

Version Date: 20141002

ISBN 13: 978-0-367-78359-4 (pbk)
ISBN 13: 978-1-4665-9809-6 (hbk)

Library of Congress Cataloging-in-Publication Data

Nanotubes and nanosheets : functionalization and applications of boron nitride and other
    nanomaterials / editor Ying (Ian) Chen.
        pages cm
    Includes bibliographical references and index.
    ISBN 978-1-4665-9809-6 (alk. paper)
    1. Nanotubes. 2. Nanostructured materials. I. Chen, Ying (Professor), editor of compilation.

    TA418.9.N35N35773 2015
    620'.5--dc23   2014030808

Visit the Taylor & Francis Web site at
http://www.taylorandfrancis.com

and the CRC Press Web site at
http://www.crcpress.com

Contents

SECTION III   Computation and Modeling

SECTION IV   Functionalization

SECTION V   Applications

Preface

There are many books about carbon nanotubes and graphene, but this is the first book dedicated to nanotubes and nanosheets made of boron nitride (BN). BN is isostructural to carbon (C) and exists in various crystalline forms. The hexagonal form (h-BN) is analogous to graphite, and the cubic form (c-BN) is similar to diamond. BN nanotubes have exactly the same structure as C nanotubes. Because B and N are neighbors to C on the periodic table, BN nanotubes have a similar density and the same excellent mechanical strength as C nanotubes. Other one-dimensional BN nanomaterials such as nanowires, nanoribbons, nanofibers, and nanorods have also been developed. Recently, inspired by graphene, BN nanosheets (sometimes called white graphene) have been intensively investigated. Although these BN nanomaterials have the same structures and many similar properties to their C counterparts, they cannot replace but only complement each other in many applications.

As this book shows, BN nanotubes and nanosheets are almost electrically insulating, chemically inert, resistant to oxidation at high temperatures, radiation shielding, and biologically safe. These properties have led to many exciting applications where C materials cannot be used, including high-temperature, metal- and ceramic-based composites, substrates for graphene and other semiconducting layers in electronic devices, reusable absorbents for oil and other contaminants, dry solid lubricants, and biomedical applications.

Although BN is akin to a sibling of C in the nanomaterial family, it has received much less attention. One main reason is that the synthesis of BN nanotubes and nanosheets is so complicated that many popular processes used for C nanomaterial production do not work efficiently for BN. Many authors of this book attended the 13th International Conference on the Science and Application of Nanotubes held in Brisbane, Australia, June 25–29, 2012. A symposium dedicated to BN nanotubes and nanosheets was an integral part of the conference for the first time. All participants realized the need to promote research in BN nanomaterials, and this book is one of the outcomes. As exciting new research results in BN nanomaterials continue to appear, many more books will follow.

This book is a reference work that reveals the innovative research work on BN nanotubes and nanosheets. The contributors include many active researchers working in different areas of BN nanomaterials—from synthesis and characterization to computer simulation and applications. An important focus of the book is the applications of BN nanotubes and nanosheets, which are also the focus of current nanotechnology research. Without practical applications, new materials will not have a long-term future. The book describes various applications, including BN nanotube–reinforced, metal- and ceramic-based composites, field emission, desalination, cleanup of oil spillages, biosensing and bioimaging, drug delivery, and biomedical applications, as well as energy storage using BCN and TiO_2 nanorods and nanosheets as electrode materials. The book also includes chapters on C and other nanotubes and nanosheets to give readers a broad view of current nanomaterials research.

Finally, I thank my family for their support; my staff, students, collaborators, and colleagues for their important contributions; and the staff at Taylor & Francis Group for their assistance in publishing this book.

Editor

Professor Ying (Ian) Chen is chair of nanotechnology at the newly established Institute for Frontier Materials at Deakin University and node head of the ARC Centre of Excellence for Functional Nanomaterials. Professor Chen invented the ball-milling and annealing method, making his team a world leader in nanomaterials synthesis and commercialization. His research at Deakin University focuses on fundamental research in nanomaterials for energy storage (batteries and capacitors), environmental protection, and medical applications.

Professor Chen earned his BS from Tsinghua University in Beijing, People's Republic of China, and his PhD from the University of Paris-Sud, France. He is listed by the Web of Knowledge as the top author on the two subjects of nanotubes and ball milling. He has contributed to three best-selling books on nanotechnology published by CRC Press. His publications have been cited more than 3000 times over the past 10 years. Professor Chen is a fellow of the Institute of Physics and member of the American Physics Institute, the Materials Research Society, the Australian Materials Union, and the International Mechanochemical Association. He has been honored with several prestigious awards, including Australian Research fellowships from the Australian Research Council and the "1000 Talents" professorship in 2011.

Contributors

Arvind Agarwal
Department of Mechanical and Materials
  Engineering
Florida International University
Miami, Florida

Afsana Ahmed
Faculty of Engineering and Industrial Science
and
Centre for Molecular Simulation
Swinburne University of Technology
Melbourne, Victoria, Australia

Anjana Asthana
Department of Physics
Michigan Technological University
Houghton, Michigan

Xuedong Bai
Beijing National Laboratory for Condensed
  Matter Physics
Institute of Physics
Chinese Academy of Sciences
Beijing, People's Republic of China

Yoshio Bando
World Premier International Center for
  Materials Nanoarchitectonics
National Institute for Materials Science
Tsukuba, Japan

Colin J. Barrow
School of Life and Environmental Sciences
Deakin University
Geelong Campus at Waurn Ponds, Victoria,
  Australia

Hua Chen
Centre for Advanced Microscopy
The Australian National University
Canberra, Australian Capital Territory,
  Australia

Ying (Ian) Chen
Institute for Frontier Materials
Deakin University
Geelong Campus at Waurn Ponds, Victoria,
  Australia

Zhi-Gang Chen
Division of Materials Engineering
School of Mechanical and Mining Engineering
The University of Queensland
Brisbane, Queensland, Australia

Zhiqiang Chen
Institute for Frontier Materials
Deakin University
Geelong Campus at Waurn Ponds, Victoria,
  Australia

Zhongfang Chen
Department of Chemistry
Institute for Functional Nanomaterials
University of Puerto Rico
San Juan, Puerto Rico

Hui-Ming Cheng
Shenyang National Laboratory for Materials
  Science
Institute of Metal Research
Chinese Academy of Sciences
Shenyang, People's Republic of China

Shin-Ho Chung
Research School of Biology
The Australian National University
Canberra, Australian Capital Territory,
  Australia

Gianni Ciofani
Center for Micro-BioRobotics @SSSA
Italian Institute of Technology
Pontedera (Pisa), Italy

John W. Connell
Langley Research Center
National Aeronautics and Space
  Administration
Hampton, Virginia

Ben Corry
Research School of Chemistry
The Australian National University
Canberra, Australian Capital Territory,
  Australia

Jiabin Dai
School of Engineering
University of South Australia
Adelaide, South Australia, Australia

Xiujuan J. Dai
Institute for Frontier Materials
Deakin University
Geelong Campus at Waurn Ponds, Victoria,
  Australia

Aijun Du
Faculty of Science and Engineering
School of Chemistry, Physics and Mechanical
  Engineering
Queensland University of Technology
Brisbane, Queensland, Australia

Daniel Fox
Centre for Research on Adaptive
  Nanostructures and Nanodevices
and
School of Physics
Trinity College Dublin
Dublin, Ireland

Lei Ge
School of Chemical Engineering
The University of Queensland
Brisbane, Queensland, Australia

Dmitri Golberg
World Premier International Center for
  Materials Nanoarchitectonics
National Institute for Materials Science
Tsukuba, Japan

Boyi Hao
Department of Physics
Michigan Technological University
Houghton, Michigan

Tamsyn A. Hilder
Research School of Biology
The Australian National University
Canberra, Australian Capital Territory,
  Australia

Anita J. Hill
Division of Process Science and Engineering
Commonwealth Scientific and Industrial
  Research Organisation
Melbourne, Victoria, Australia

Debrupa Lahiri
Department of Metallurgical and Materials
  Engineering
Indian Institute of Technology
Roorkee, Uttarakhand, India

Chee Huei Lee
Department of Physics
Michigan Technological University
Houghton, Michigan

Weiwei Lei
Institute for Frontier Materials
Deakin University
Geelong Campus at Waurn Ponds, Victoria,
  Australia

Lu Hua Li
Institute for Frontier Materials
Deakin University
Geelong Campus at Waurn Ponds, Victoria,
  Australia

Yunlong Liao
National Institute of Aerospace
Hampton, Virginia

and

Department of Physics
University of Puerto Rico
San Juan, Puerto Rico

Yi Lin
National Institute of Aerospace
Hampton, Virginia

and

Department of Applied Science
The College of William and Mary
Williamsburg, Virginia

Dan Liu
Institute for Frontier Materials
Deakin University
Geelong Campus at Waurn Ponds, Victoria,
    Australia

Yun Liu
Research School of Chemistry
The Australian National University
Canberra, Australian Capital Territory,
    Australia

Jun Ma
School of Engineering
and
Mawson Institute
University of South Australia
Adelaide, South Australia, Australia

Renzhi Ma
World Premier International Center for
    Materials Nanoarchitectonics
National Institute for Materials Science
Tsukuba, Japan

Majumder Mainak
Department of Mechanical and Aerospace
    Engineering
Monash University
Melbourne, Victoria, Australia

Motilal Mathesh
School of Life and Environmental Sciences
Deakin University
Geelong Campus at Waurn Ponds, Victoria,
    Australia

Virgilio Mattoli
Center for Micro-BioRobotics @SSSA
Italian Institute of Technology
Pontedera (Pisa), Italy

Barbara Mazzolai
Center for Micro-BioRobotics @SSSA
Italian Institute of Technology
Pontedera (Pisa), Italy

Qingshi Meng
School of Engineering
University of South Australia
Adelaide, South Australia, Australia

Wenjun Meng
Department of Physics and Materials Science
City University of Hong Kong
Kowloon, Hong Kong

Amir Pakdel
World Premier International Center for
    Materials Nanoarchitectonics
National Institute for Materials Science
Tsukuba, Japan

Ho Bum Park
Department of Energy Engineering
Hanyang University
Seoul, Republic of Korea

David Portehault
Chimie de la Matière Condensée de Paris
Université Pierre et Marie Curie (Paris VI)
Paris, France

Si Qin
Institute for Frontier Materials
Deakin University
Geelong Campus at Waurn Ponds, Victoria,
    Australia

Thomas Rufford
School of Chemical Engineering
The University of Queensland
Brisbane, Queensland, Australia

Nasser Saber
School of Engineering
University of South Australia
Adelaide, South Australia, Australia

Takayoshi Sasaki
World Premier International Center for
    Materials Nanoarchitectonics
National Institute for Materials Science
Tsukuba, Japan

Tao Tao
Institute for Frontier Materials
Deakin University
Geelong Campus at Waurn Ponds, Victoria,
    Australia

Michael Thomas
VLSCI Life Sciences Computation Centre
La Trobe Institute for Molecular Science
La Trobe University
Melbourne, Victoria, Australia

Aaron W. Thornton
Division of Materials Science and Engineering
Commonwealth Scientific and Industrial
    Research Organisation
Melbourne, Victoria, Australia

Hongbin Wang
School of Life and Environmental Sciences
Deakin University
Geelong Campus at Waurn Ponds, Victoria,
    Australia

Jiesheng Wang
Department of Physics
Michigan Technological University
Houghton, Michigan

Li Wang
School of Chemical Engineering
The University of Queensland
Brisbane, Queensland, Australia

Lianzhou Wang
School of Chemical Engineering
The University of Queensland
Brisbane, Queensland, Australia

Wenlong Wang
Beijing National Laboratory for Condensed
    Matter Physics
Institute of Physics
Chinese Academy of Sciences
Beijing, People's Republic of China

Xungai Wang
Institute for Frontier Materials
Deakin University
Geelong Campus at Waurn Ponds, Victoria,
    Australia

Dustin Winslow
Nano Innovations, LLC
Houghton, Michigan

Zhi Xu
Beijing National Laboratory for Condensed
    Matter Physics
Institute of Physics
Chinese Academy of Sciences
Beijing, People's Republic of China

Wenrong Yang
School of Life and Environmental Sciences
Deakin University
Geelong Campus at Waurn Ponds, Victoria,
    Australia

Yoke Khin Yap
Department of Physics
Michigan Technological University
Houghton, Michigan

Delai Ye
School of Chemical Engineering
The University of Queensland
Brisbane, Queensland, Australia

Hua Yu
School of Chemical Engineering
The University of Queensland
Brisbane, Queensland, Australia

Yuanlie Yu
Centre for Advanced Microscopy
The Australian National University
Canberra, Australian Capital Territory,
    Australia

Dongyan Zhang
Department of Physics
Michigan Technological University
Houghton, Michigan

Hongzhou Zhang
Centre for Research on Adaptive
    Nanostructures and Nanodevices
and
School of Physics
Trinity College Dublin
Dublin, Ireland

Chunyi Zhi
Department of Physics and Materials Science
City University of Hong Kong
Kowloon, Hong Kong

Yangbo Zhou
Centre for Research on Adaptive
    Nanostructures and Nanodevices
and
School of Physics
Trinity College Dublin
Dublin, Ireland

Zhonghua Zhu
School of Chemical Engineering
The University of Queensland
Brisbane, Queensland, Australia

Jin Zou
Division of Materials Engineering
School of Mechanical and Mining Engineering
The University of Queensland
Brisbane, Queensland, Australia

Synthesis and Fabrication

Chapter 1

Nanoboron Nitrides

Amir Pakdel, Yoshio Bando, and Dmitri Golberg

CONTENTS

1.1   INTRODUCTION

This chapter provides an overview of the past and current state of research on boron nitride (BN) nanomaterials in terms of synthesis, structure, properties, and applications. Eighteen years after the initial synthesis of BN nanotubes, research on BN nanomaterials has developed far enough to establish them as one of the most promising inorganic nanosystems. In this regard, unique properties of BN nanomaterials, such as superb mechanical stiffness, high thermal conductivity, strong ultraviolet emission, excellent thermal stability, and chemical inertness, are envisaged to play the key role for prospective developments (Pakdel et al. 2012c).

BN is a chemical compound, consisting of equal numbers of boron (B) and nitrogen (N) atoms, which is not found in nature and is therefore produced synthetically. BN is isostructural to carbon (C) and exists in various crystalline forms. The hexagonal form (h-BN) is analogous to graphite with a layered structure. This can also be stacked in a rhombohedral form. The other common structure of BN is the cubic form (c-BN), which is similar to diamond and is the second hardest material known so far. There is also a rare wurtzite form, which is similar to lonsdaleite (Pakdel et al. 2014).

Four years after the identification of C nanotubes (Iijima 1991), BN nanotubes were successfully synthesized (Chopra et al. 1995). Subsequently, other one-dimensional (1D) BN nanomaterials such as nanowires, nanoribbons, nanofibers, and nanorods were synthesized (Yong et al. 2006; Chen et al. 2008; Qiu et al. 2009; Zhang et al. 2006). Moreover, inspired by C_{60} buckyballs (Kroto et al. 1985), 0D nested and single-layered octahedral BN fullerenes were produced in 1998 (Stephen et al. 1998; Golberg et al. 1998). In addition, after the rise of graphene (Novoselov et al. 2004) and the research progress on layered 2D nanostructures, free-standing 2D BN flakes were peeled off from a BN crystal (Novoselov et al. 2005). However, the initial growth of BN nanosheets in the form of nanomeshes on metallic substrates had been reported a year earlier (Corso et al. 2004). Structural models of 0D, 1D, and 2D BN nanomaterials are illustrated in Figure 1.1.

1.2   SYNTHESIS

BN is a synthetic material and has to be produced from other natural or refined raw materials. The first synthesis of BN was performed in 1842 by Balmain (Balmain et al. 1842a,b), using the reaction between molten boric acid (H_3BO_3) on potassium cyanide (KCN). It was not possible to

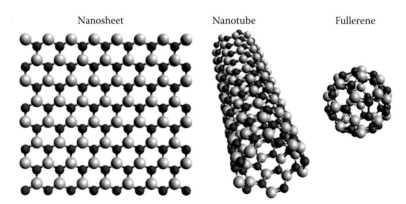

Nanosheet   Nanotube   Fullerene

Figure 1.1   Structural models of low-dimensional BN nanostructures: 2D nanosheet: 1D nanotube, and 0D fullerene. (Reprinted from *Mater. Today*, 15(6), Pakdel, A., Zhi, C.Y., Bando, Y., and Golberg, D., Low-dimensional boron nitride nanomaterials, 15(6), 256–265, Copyright 2012, with permission from Elsevier.)

stabilize this material in the form of powders until the early 1960s, when advances in technology made BN an economically affordable material. In this section, the synthetic routes to produce BN nanostructures are concisely reviewed.

1.2.1   Synthesis of 1D BN Nanostructures

Most techniques known for the growth of C nanotubes have been modified to synthesize BN nanotubes as well. These methods can be classified into two broad categories: (1) high-temperature synthesis (>2000°C), such as arc discharge and laser ablation and (2) medium-/low-temperature synthesis (<2000°C), such as carbothermal methods, ball milling, and chemical vapor deposition (CVD).

1.2.1.1   Arc Discharge
The first successful synthesis of BN nanotubes was reported in 1995 by Chopra et al. They used a tungsten electrode filled with h-BN powder as anode and a copper (Cu) electrode as cathode to produce multiwalled BN nanotubes, which included metallic particles at their tips. Subsequently, single- and double-walled BN nanotubes were synthesized by arcing Hafnium diboride (HfB_2) electrodes in an inert atmosphere (Loiseau et al. 1996). In another approach, double-walled BN nanotubes were mass-produced by arcing B electrodes containing nickel (Ni) or cobalt (Co) in a nitrogen (N_2) atmosphere (Cumings et al. 2000). Later in 2006, a continuous process using an arc-jet technique at high temperature (5,000–20,000 K) was developed as a promising method for the large-scale production of single- and multiwalled BN nanotubes (Lee et al. 2006). A mixture of h-BN and catalyst metal powders was used as the precursor, and a combination of argon (Ar) and N_2 formed the plasma gas.

1.2.1.2   Laser Ablation
The second synthesis method of BN nanotubes production was a laser-assisted technique used by Golberg et al. (1996) to generate multiwalled BN nanotubes. Single-crystal samples of c-BN or h-BN were laser heated for a short time in a diamond anvil cell under high pressures of N_2. However, a large quantity of BN flakes and particles were also included in the product. Figure 1.2 shows typical BN nanotubes produced by this method. Laude et al. (2000) and Chen et al. (2002) also produced multiwalled BN nanotubes by this method. Further development of the laser technique resulted in the synthesis of single-, double-, and triplewalled BN nanotubes by using a mixture of BN, Ni, and Co powders in an inert atmosphere (Yu et al. 1998). Later, a rotating catalyst-free BN target in a continuous laser ablation reactor under N_2 was used to produce bulk quantities of BN nanotubes (Lee et al. 2001). Subsequently, high yields of single- and multiwalled BN nanotubes were obtained based on the vaporization of h-BN targets via a continuous CO_2 laser under N_2 flow at ~3500 K (Arenal et al. 2007). However, the product contained unwanted morphologies besides tubular structures.

1.2.1.3   Carbothermal Methods
An alternative method to synthesize BN nanotubes was developed by Han et al. (1998). It was based on the idea of turning C nanotubes to BN nanotubes through the substitution of C atoms with B and N atoms. Boron trioxide (B_2O_3) powders covered by C nanotubes were under N_2 flow at 1500°C inside an induction furnace. While the C nanotubes were oxidized by B_2O_3, B and N atoms filled the generated vacancies in the hexagonal structure. As a result, multiwalled BN nanotubes were produced with diameters similar to those of the initial C nanotubes (Golberg et al. 2000a). This technique was further optimized to produce single-walled BN nanotube bundles by using pure single-walled C nanotubes as the precursor (Golberg et al. 1999a). However, the product

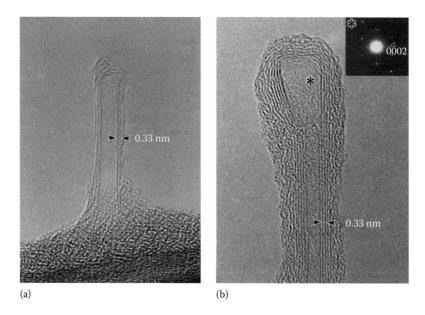

(a)   (b)

Figure 1.2   HRTEM images of BN nanotubes produced from laser-heated BN at N_2 pressures of (a) 8.4 and (b) 5.4 GPa. (Reprinted with permission from Golberg, D., Bando, Y., Eremets, M., Takemura, K., Kurashima, K., and Yusa, H., Nanotubes in boron nitride laser heated at high pressure, *Appl. Phys. Lett.*, 69(14), 2045–2047, 1996. Copyright 1996, American Institute of Physics.)

included B–C and B–C–N compound single-walled nanotubes as well. Further work demonstrated that C content in the product can be reduced by adding an extra oxidizing agent to the system (Golberg et al. 2000b). This technique could also be applied to generate BN nanotubes with smaller diameters than those of the starting C nanotubes (Han et al. 2006). Transmission electron microscope (TEM) images of BN nanotubes with one to six shells and a schematic of the substitution process are illustrated in Figure 1.3.

In another approach, Bartnitskaya et al. (1999) employed boron carbide (B_4C) as the C source instead of C nanotubes. The carbothermal reduction of B_xO_y and the simultaneous nitriding at 1100°C–1450°C resulted in the production of BN tubes with large diameters and a bamboo-like structure. They suggested that the formation of the BN nanotubes involved both solid and gas phases in line with the following reactions:

$$B_2O_3 \text{ (s)} + 3B_4C \text{ (s)} + 7N_2 \text{ (g)} \rightarrow 14BN \text{ (s)} + 3CO \text{ (g)}$$

$$B_2O_3 \text{ (s)} + 3CO \text{ (g)} + N_2 \text{ (g)} \rightarrow 2BN \text{ (s)} + 3CO_2 \text{ (g)}$$

Also, synthesis of multiwalled BN nanotubes by using boric acid (H_3BO_3) and active carbon or C nanotubes in the presence of iron compounds under ammonia (NH_3) atmosphere has been reported (Deepak et al. 2002).

1.2.1.4   Mechanothermal Methods

The initial mechanothermal method for producing BN nanotubes was pioneered by Chen et al. (1999b). It included ball milling of B powder in NH_3 gas, followed by thermal annealing at 1000°C–1200°C under N_2 or Ar. The BN products had a bamboo-like structure with 50–75 nm external diameters. They also applied this process to h-BN powder, obtained bamboo-like nanotubes with

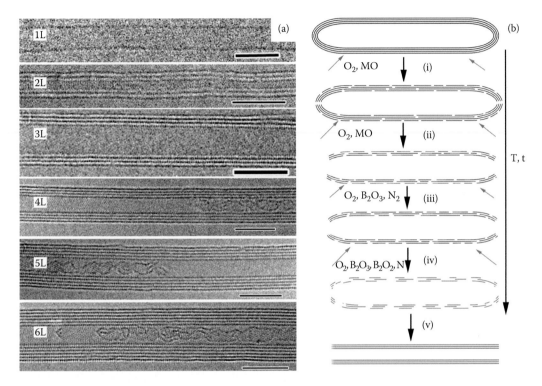

Figure 1.3   (a) HRTEM images of nanotubes with one to six walls. (b) Schematic of the substitutional reactions in C nanotubes, developed with temperature (T) and time (t). (i) Void formation in C layers due to oxidation of C with O_2 or metal oxides (MOs). (ii) More voids formation and outer layers peeling off. (iii) Reaction of B_2O_3 and N_2 with C to form BN domains. (iv) End of the substitution reactions. (v) Shrinkage of the BN layers by the rearrangement of B and N atoms to eliminate large voids. (Reprinted with permission from Han, W.Q., Todd, P.J., and Strongin, M., Formation and growth mechanism of ^{10}BN nanotubes via a carbon nanotube-substitution reaction, *Appl. Phys. Lett.*, 89(17), 173103, 2006. Copyright 2006, American Institute of Physics.)

diameters ranging from ~11 to ~280 nm, and suggested that ball milling of h-BN powder could generate highly disordered or amorphous nanostructures, which could promote nucleation and growth of BN nanotubes after annealing at high temperatures (Chen et al. 1999a). It has also been proposed that surface diffusion in milled powder could be the key factor for growing BN nanotubes through a tip-growth model, due to the presence of Fe particles at the tip of the BN tubes (Chadderton et al. 1999), as shown in Figure 1.4. Further development of this method by other research groups resulted in the synthesis of BN nanotubes with diameters of 5–30 nm (Tang et al. 2002b) and 40–60 nm (Bae et al. 2003), and lengths of several μm, depending on the chemical composition of initial powders and synthesis temperature. High yields of multiwalled BN nanotubes with diameters <10 nm were achieved by Yu et al (2005). using a ball milling–annealing process in which amorphous B was ball-milled for 150 h and annealed at 1200°C under NH_3. The produced nanotubes had no trace of metallic particles, which may suggest a root-growth mechanism for their growth.

1.2.1.5 Chemical Vapor Deposition

The pioneering work on the CVD synthesis of BN nanotubes, by using borazine ($B_3H_6N_3$) as precursor and Ni_2B particles as catalyst at 1000°C–1110°C, was carried out by Lourie et al. (2000).

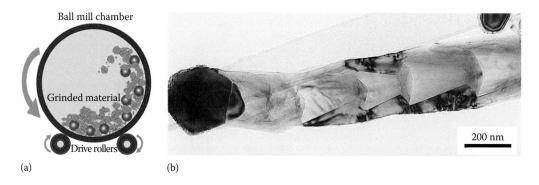

Ball mill chamber

Grinded material

Drive rollers

(a)

200 nm

(b)

Figure 1.4   (a) A schematic of a ball milling apparatus. (b) TEM image of a bamboo-like BN nanotube, containing a metallic tip, synthesized by ball milling. (Reprinted from *Phys. Lett. A*, 263(4–6), Chadderton, L.T. and Chen, Y., Nanotube growth by surface diffusion, 263(4–6), 401–405. Copyright 1999, with permission from Elsevier Ltd.)

The nanotubes often possessed bulbous, flag-like, and/or club-like tips. Later, an efficient CVD route was developed via heating a mixture of B, MgO, and FeO powders to 1300°C to synthesize significant amounts of BN nanotubes. The growth vapors were transported by an Ar flow to the reaction chamber, where a substrate was placed and heated to 1550°C under NH_3 flow. BN nanotubes (both cylindrical and bamboo like) with diameters of 20–100 nm were produced by this method (Tang, 2002a). Recently, this technique has successfully been applied to generate high yields of thin BN nanotubes with an average diameter of 10 nm by using a mixture of Li_2O and B powders as precursor (Huang et al. 2011). There are several reports on the development of various CVD techniques and using different precursor materials to obtain BN nanotubes at 450°C–1200°C in different shapes and sizes (Kim et al. 2008; Xu et al. 2003; Chen et al. 2005; Dai et al. 2007).

An interesting feature of the CVD process is the growth of thin films and coatings made of nanostructured materials. Fe-particle-functionalized substrates were used by Yap's group for growing BN nanotube bundles at 600°C (Wang et al. 2005). They used a plasma-enhanced pulsed-laser deposition technique with a negative substrate bias voltage to generate the reactive condition for tubes' growth perpendicular to the substrates. In another approach, a plasma technique was used to produce highly ordered multiwalled BN nanotube arrays (Wang et al. 2007). Later, conventional tube furnaces were used to synthesize BN nanotube films by thermal CVD (Lee et al. 2008), and recently, Pakdel et al. (2012b) systematically studied the effect of process variables on the CVD growth of BN nanotube films with different sizes and morphology, as shown in Figures 1.5 and 1.6.

It was demonstrated that raising the growth temperature from 1200°C increased the tubes' diameter; however, the secondary growth of very thin tubes could be observed at 1400°C. The BN nanotube films grown at 1200°C and 1300°C consisted of straight uniformly distributed tubes with high aspect ratios, whereas the ones grown at 1400°C contained tube bundles with a flower-like structure. Moreover, the BN nanotubes synthesized with low metal oxide content had a curly morphology; however, the very thick ones synthesized with high-metal-oxide content showed a randomly tilted flower-like morphology. Detailed electron microscopy investigations revealed that the thin nanotubes grown at 1200°C and 1400°C grow by a base-growth mechanism, whereas the thick ones grown at 1300°C and 1400°C grow by a tip-growth mechanism (Pakdel et al. 2012b). This CVD technique has been further modified to grow other novel 1D BN nanostructures, such as nanofunnels and nanomikes, on Si/SiO_2 substrates as a uniform film (Figure 1.7) (Pakdel et al. 2013a).

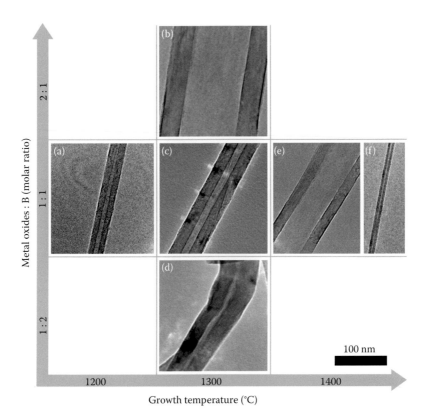

Figure 1.5 Effect of growth temperature (a, c, e, f) and catalyst content (b, c, d) on the size of BN nano-tubes, as revealed by TEM. (From Pakdel, A., Zhi, C., Bando, Y., Nakayama, T., and Golberg, D., A comprehensive analysis of the CVD growth of boron nitride nanotubes, *Nanotechnology*, 23(21), 215601, 2012b. Copyright IOP Publishing. With permission.)

1.2.2 Synthesis of 2D BN Nanostructures

A variety of methods have been used to synthesize 2D BN nanostructures. Most of them are similar to the well-known techniques utilized for the growth of graphene sheets and ribbons with slight modifications. In this section, these methods will be discussed in detail.

1.2.2.1 Self-Assembly

The first 2D BN nanostructure was prepared by the decomposition of borazine ($B_3H_6N_3$) in the form of a nanomesh on a rhodium (Rh) single-crystalline surface by Corso et al. (2004). The preparation procedure consisted of exposing the atomically clean Rh(111) surface at 800°C to $B_3H_6N_3$ vapor inside an ultrahigh vacuum chamber and consecutive cooling to room temperature. The regular mesh structure was observed via scanning tunneling microscopy (STM), and it was suggested that the hole formation was driven by the lattice mismatch of the BN film and the Rh substrate. BN nanomeshes were later grown on Ru(0001) substrates by the same process, as shown in Figure 1.8a. Both nanomeshes grown on Rh(111) and Ru(0001) are similar in many aspects, for example, they show a highly regular 12 × 12 superstructure, comprising 2 nm wide apertures with a depth of about 0.1 nm, as depicted in Figure 1.8b and c (Goriachko et al. 2007). Further investigation on

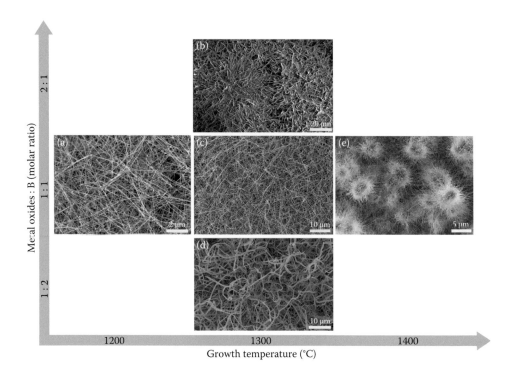

Figure 1.6 Effect of growth temperature (a, c, e) and catalyst content (b, c, d) on the morphology of BN nanotube films, as revealed by SEM. (From Pakdel, A., Zhi, C., Bando, Y., Nakayama, T., and Golberg, D., A comprehensive analysis of the CVD growth of boron nitride nanotubes, *Nanotechnology*, 23(21), 215601, 2012b. Copyright IOP Publishing. With permission.)

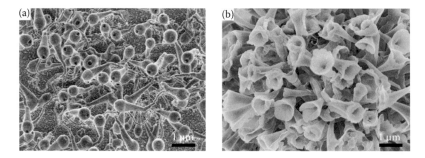

Figure 1.7 SEM images of conical BN nanostructures: (a) nanomikes and (b) nanofunnels. (Reprinted with permission from Pakdel, A., Bando, Y., and Golberg, D., Morphology-driven nonwettability of nano-structured BN surfaces, *Langmuir*, 29(24), 7529–7533, 2013a. Copyright 2013, American Chemical Society.)

BN nanomesh on Rh(111) surfaces demonstrated that it is a single but corrugated monolayer of BN. The 2 nm–sized pores are formed by regions where the layer binds strongly to the underlying metal, while the regular hexagonal network of mesh wires represents regions where the layer is not bonded to the Rh surface but it is stable through strong cohesive forces within the film itself (Berner et al. 2007).

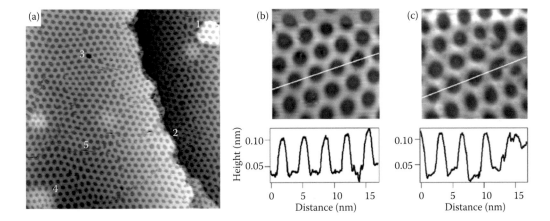

Figure 1.8 (a) STM image of an h-BN nanomesh on Ru(0001): scan size 86 × 86 nm², sample voltage U_{sample} = 1.3 V, tunneling current IT = 1 nA. The numbered areas indicate peculiar features of the h-BN nanomesh: 1—single-layer step edges; 2—double-layer step edges; 3—dark pits, which are deeper than typical nanomesh apertures; 4—subsurface argon bubbles in the form of protrusions of round shape originating from the Ar^+ sputtering; and 5—BN-nanomesh domain boundaries. (b, c) Higher magnification STM images of the BN nanomeshes on Rh(111) and Ru(0001), respectively. Their cross-sectional profiles along the white lines show the similarity of the two as they reveal the same apparent buckling of 0.07 ± 0.02 nm between the apertures and the wires and the same aperture size (2 nm) and wire thickness (1 nm). (Reprinted with permission from Goriachko, A., He, Y.B., Knapp, M. et al., Self-assembly of a hexagonal boron nitride nanomesh on Ru(0001), *Langmuir*, 23(6), 2928–2931, 2007. Copyright 2007, American Chemical Society.)

1.2.2.2 Micromechanical Cleavage

The pioneering procedure to obtain atomic sheets of h-BN was micromechanical cleavage, reported by Novoselov et al. (2004). In this method, layers of h-BN are peeled off with adhesive tapes and attached to a substrate. Atomic force microscopy and TEM investigations of BN nanosheets prepared by this method in Zettl's group revealed the clean well-ordered nanosheets, with thicknesses between 3.5 and 80 nm (Pacile et al. 2008).

Another approach to the mechanical exfoliation of BN is the utilization of shear forces instead of direct pulling forces via peeling. In fact, during mechanical peeling, the pulling force easily breaks the weak van der Waals bonding between adjacent BN layers and leaves the strongly sp^2 bonded in-plane structure intact. A shear force can have a similar effect. In 2011, a mild wet ball milling process was used to produce BN nanosheets from BN powder precursor through gentle shear forces, under an N_2 atmosphere. In this process, benzyl benzoate ($C_{14}H_{12}O_2$) was used as the milling agent to reduce the ball impacts and milling contamination. Figure 1.8 shows SEM images of the nanosheets produced by this method and their corresponding peeling mechanisms (Li et al. 2011).

1.2.2.3 Chemical Exfoliation (Sonication)

Preparation of mono- and few-layered nanosheets from a single-crystalline h-BN by a chemical solution–derived method was first accomplished by Han et al. (2008). The h-BN crystal was sonicated in an organic solution to break up into few-layered h-BN sheets. Figure 1.9 displays typical TEM images of the obtained BN nanosheets. Later, the exfoliation of nanosheets from BN particles dispersed in a strong polar solvent was performed via vigorous sonication–centrifugation (Zhi et al. 2009). The solvent facilitated the exfoliation, due to strong interactions between its

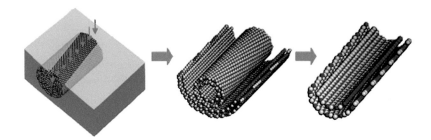

Figure 1.9   Schematic of a BN nanotube unzipping processes induced by plasma etching. The stepwise opening/unzipping, removing, and exfoliating of tube walls to form nanoribbons are sketched. (Reprinted with permission from Zeng, H.B., Zhi, C.Y., Zhang, Z.H. et al., "White graphenes": Boron nitride nanoribbons via boron nitride nanotube unwrapping, *Nano Lett.*, 10(12), 5049–5055, 2010. Copyright 2010, American Chemical Society.)

polar molecules and the BN surface. Milligram levels of pure BN nanosheets with 2–10 nm thicknesses were achieved. Different solvents have been employed in chemical exfoliation methods to promote BN nanosheet formation (Warner et al. 2010; Wang et al. 2011c).

An alternative method capable of high-throughput and large-scale production of few-layered BN nanosheets is the high-pressure microfluidization process. In this method, BN powder and a combination of DMF and chloroform (as solvent) are inserted inside a microfluidizer processor and are kept at a constant pressure. Then, the product is accelerated into the interaction chamber within which the product stream separates into micro channels of various geometries. The stream is then forced to collide upon itself, creating incredible forces of impact and shear, which are several orders of magnitude greater than the case of sonicators. This results in exfoliation of BN with a reported yield efficiency of 45% (Yurdakul et al. 2012).

1.2.2.4   High-Energy Electron Irradiation

Fabrication of freestanding single-layered BN nanosheets by controlled electron irradiation through a layer-by-layer sputtering process in situ inside a TEM was reported by Jin et al. (2009) and Meyer et al. (2009). First, h-BN nanoflakes and powders underwent mechanical cleavage to obtain h-BN sheets with a reduced number of layers. Then, the h-BN nanoflakes were further thinned down to monolayers by focusing an intensive electron beam onto the specimens. By scanning the electron beam under a manual control, h-BN nanoflakes were burned layer-by-layer until h-BN monolayers were obtained.

1.2.2.5   Unzipping of BN Nanotubes

BN nanotubes have been unwrapped via Ar plasma etching (Zeng et al. 2010). In this process, multiwalled BN nanotubes were first deposited on a Si substrate and spin coated with a poly(methyl methacrylate) (PMMA) film. The polymer film with embedded nanotubes was peeled off and turned over. Then, the composite PMMA–BN nanotube film was subjected to Ar plasma etching for 100 s. While the bottom surface of the BN nanotube was covered and protected by the polymer matrix, the ablation of the top surface of the nanotubes led to the formation of half-opened BN nanoribbons. PMMA was removed by acetone vapor, and the produced BN nanoribbons were heated at 600°C for 6 h to remove the PMMA residue and oxidize possible carbonaceous contaminants. The nanoribbons were as narrow as 15 nm with lengths up to a few μm. Figure 1.9 illustrates the unzipping process.

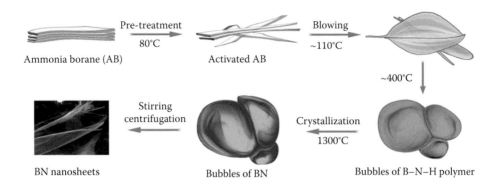

Figure 1.10 A scheme of BN nanosheet synthesis via a chemical blowing method. (From Wang, X.B., Pakdel, A., Zhi, C.Y. et al., High-yield boron nitride nanosheets from 'Chemical Blowing': Towards practical applications in polymer composites, *J. Phys. Condens. Matter*, 24(31), 314205, 2012. Copyright IOP Publishing. With permission.)

1.2.2.6 Solid-State Reactions

Flower-like BN nanoflakes were produced by a template-free solid-phase reaction. $NaBF_4$, NH_4Cl, and NaN_3 powders were mixed, pressed into pellets at room temperature, and then heated in an autoclave at 300°C for 20 h. The BN production reaction is as follows:

$$NaBF_4 + 3NH_4Cl + 3NaN_3 \rightarrow BN + NaF + 3NaCl + 4N_2 + 3NH_3 + 3HF$$

The as-prepared BN nanoflowers were composed of vertically standing BN nanoflakes. In order to prepare few-layered BN nanosheets, the nanoflakes were sonicated afterward (Lian et al. 2011).

Figure 1.10 displays a chemical blowing method by which few-layered BN nanosheets with large lateral dimensions (over 100 μm) were produced. This technique does not require a catalyst or substrate, instead uses moderate heating of NH_3BH_3 in atmospheric pressure. The precursor is preheated at 80°C, and then the temperature increases to ~110°C to begin the blowing process due to dehydrogenization and finally to ~400°C to dehydrogenate, where hydrogen is rapidly released from the soft swollen B–N–H compound. Further heating to 1400°C for 3 h crystallizes the material into a desired BN product. Ultrasonic stirring and centrifugation remove bulky portions from the BN suspension and result in few-layered polycrystalline BN nanosheets (Wang et al. 2011, 2012).

1.2.2.7 Chemical Vapor Deposition

CVD is one of the most common techniques to prepare BN nanosheets. Relatively large quantities of thick h-BN sheets were synthesized via a catalyst-free CVD process at 1100°C–1300°C by Gao et al. (2009). B_2O_3 and melamine powders were mechanically mixed and placed in an induction furnace. The temperature was raised to 1000°C–1300°C under N_2 flow, and after ~1 h, BN sheets with thicknesses of 25–50 nm (depending on the synthesis temperature) were obtained. Later, a multistep thermal catalytic CVD method was employed to fabricate h-BN films, consisting of two to five atomic layers, under ammonia borane (NH_3–BH_3) flow on a Cu foil as the substrate (Song et al. 2010). Few-layered h-BN films were also synthesized by ambient pressure CVD on polycrystalline Ni films with thicknesses of ~5–50 nm, depending on the growth conditions (Shi et al. 2010). The h-BN grew continuously on the entire Ni surface, and the regions with uniform thicknesses were up to 20 μm in lateral size. In another set of experiments, reaction of polyborane decaborane ($B_{10}H_{14}$) with NH_3 formed BN nanosheets on either polycrystalline Ni or Cu foils at 1000°C (Chatterjee et al. 2011). Recently, monolayer h-BN was grown on Cu foils by using NH_3–BH_3 via low-pressure CVD with two heating zones (Kim et al. 2012).

Vertically aligned BN nanosheets on a Si substrate were grown at 800°C from a gas mixture of BF_3–N_2–H_2 via a microwave plasma CVD technique (Yu et al. 2010). The growth of such protruding BN nanosheets rather than a uniform granular film was attributed to the strong etching effect of fluorine and the generated electrical field in plasma sheath. Alternatively, a thermal CVD technique has been developed in which solid precursors (B, MgO, and FeO powders) were heated up to 1000°C–1300°C in a horizontal tube furnace under NH_3 flow (Pakdel et al. 2011, 2013b). As a result, vertically standing BN nanosheets were grown on Si/SiO_2 substrates with different size and morphology, depending on the synthesis temperature. As the temperature increased from 1000°C to 1200°C, the nanosheets grew higher in lateral size, and at 1300°C, branching of subnanosheets on the surface of the main nanosheets resulted in a peculiar 3D nanostructure (Figure 1.11a and b). The thicknesses of the nanosheets were mostly less than 4 nm, and the nanosheets grown at higher temperatures displayed better crystallinity. In another set of experiments, B and B_2O_3 powders were used as precursors at 1200°C under NH_3 flow to produce BN nanosheets with thicknesses smaller than 5 nm protruding from the Si/SiO_2 substrate. This suggests that the BN nanosheet growth is not strongly dependent on specific precursor materials, as long as enough reactive B and N are provided in the growth atmosphere. Figure 1.11c illustrates the proposed model for the two-stage growth of such BN nanosheets. That is, before the onset of the vertical growth of the nanosheets, there is a planar growth stage, in which the base layers are flat and parallel to the substrate. After the development of sufficient levels of force at grain boundaries, the edges of the top layers curl upward, and the vertical growth of the nanosheets begins. BN species at the synthesis temperature

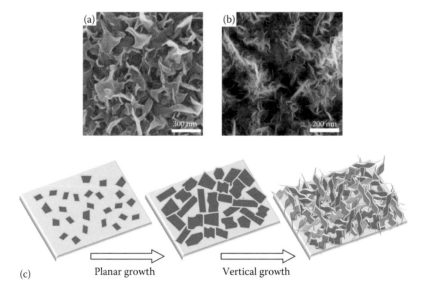

Figure 1.11 Partially vertically aligned BN nanosheets grown on a Si/SiO_2 substrate at (a) 1200°C and (b) 1300°C. (Reprinted with permission from Pakdel, A., Zhi, C.Y., Bando, Y., Nakayama, T., and Golberg, D., Boron nitride nanosheet coatings with controllable water repellency, *ACS Nano*, 5(8), 2011, 6507–6515. Copyright 2011, American Chemical Society; From Pakdel, A. et al., *Surf. Innov.*, 1(1), 32, 2013b. Copyright Thomas Telford Ltd. With permission.) (c) A model illustrating the nucleation and two-stage growth of the BN nanosheets on the Si/SiO_2 substrate. (Reprinted from *Acta Mater.*, 61(4), Pakdel, A., Wang, X.B., Bando, Y., and Golberg, D., Nonwetting "white graphene" films, 1266–1273. Copyright 2013, with permission from Elsevier.)

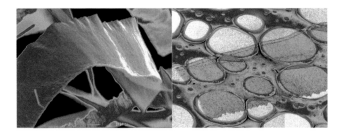

Figure 1.12 Electron microscope images of hollow BN ribbons produced by a CVD-templating method. (Reprinted with permission from Chen, Z.G., Zou, J., Liu, G. et al., Novel boron nitride hollow nanoribbons, *ACS Nano*, 2(10), 2008, 2183–2191. Copyright 2008, American Chemical Society.)

($\geq 1000°C$) can be assumed to have very high mobility; therefore, upon landing on the surface of a growing nanosheet, they quickly move along the surface toward the edge of the nanosheet and covalently bond to the edge atoms before being re-evaporated. Thus, the BN nanosheets tended to grow higher rather than thicker (Pakdel et al. 2011, 2013c).

Hollow nanoribbons of BN have been fabricated by a combined CVD-templating method using ZnS powder and a B–N–O–Fe mixture (produced by ball milling of B, B_2O_3, and Fe_2O_3 under NH_3). The precursor materials were mixed and placed inside a furnace with Ar, NH_3, and H_2 flow as the carrier and reaction gases, respectively. The furnace temperature was raised to 1220°C for 2 h to form ZnS/BN core/shell heterostructures. Subsequently, the temperature was increased to 1350°C and maintained for 4 h in vacuum at 50 Torr to remove ZnS template. The synthesized BN product is displayed in Figure 1.12.

1.3 STRUCTURE

Generally, BN nanomorphologies can be considered as h-BN layers formed in different ways. In this section, the atomic structure of BN nanomaterials is discussed in detail. Theoretical calculations and experimental results based on high-resolution TEM (HRTEM) investigations of different BN nanomorphologies are presented. For clarification, some structural features of BN nanotubes and nanosheets are compared with those of C nanotubes and graphenes.

1.3.1 Atomic Structure of BN Nanotubes

BN nanotubes can be defined as cylinders of concentric h-BN layers (from one to many layers), with diameters in the nanometer scale. The structure of multiwalled BN tubes can be described by two different models. In the *Russian Doll* model, h-BN sheets are arranged in separate concentric cylinders (e.g., a single-walled nanotube within a larger one), and in the *Parchment* model, a single sheet of h-BN is rolled around itself, resembling a scroll of parchment or a rolled newspaper (McNeish et al. 2008). However, multiwalled nanotubes are usually believed to be made of concentric tubes, rather than a spiraling layer, due to energy considerations.

1.3.1.1 Theoretical Aspects

Ab initio calculations have demonstrated that folding of an isolated h-BN sheet onto a tubular BN is slightly more favorable than that of graphene onto a C nanotube of the same radius, due to a *buckling* effect that stabilizes the BN tubular structure. That is, the elastic energy contributed to h-BN sheet bending onto a tube is smaller than that of graphene, because in h-BN tubes, N atoms are located closer to the tube axis (i.e., relaxed inward), compared to B atoms. First-principle calculations indicate that in the minimum energy structure, all B atoms are arranged in one cylinder and all N atoms in a larger concentric one, and due to charge transfer from B to N, the buckled tubular structure forms a dipolar shell. As a result, each B atom is virtually located on a plane formed by its three neighboring N atoms, so that the sp^2 environment for the B atom in the planar hexagonal structure is restored. This buckling effect significantly reduces the occupied band energy in the case of BN compounds; thus, folding an h-BN sheet into a tube is energetically more favorable than in the case of graphene. Also, the calculated energy of dangling bonds associated with opened BN strips is much larger than that of tubes; thus, it is energetically more favorable to preserve the cylindrical geometry despite the elastic energy resulting from the curvature (Blase et al. 1994).

Considering rolled-up h-BN sheets along a chosen axis, boundary limits constrain the number of choices for the helicity of the hexagonal layer relative to the tube axis. That is, in order to maintain the continuity of hexagons on the cylinder, the number of possible choices for the helicity becomes limited. As in the case of C nanotubes, BN nanotubes can be labeled with two indices (n,m) that define uniquely the chiral vector $\mathbf{C}_h = n\mathbf{a}_1 + m\mathbf{a}_2$, which is perpendicular to the tube axis direction and defines the tube folding direction. The angle between the chiral vector ($\mathbf{C}_h$) and the lattice vectors ($\mathbf{a}_1$ and $\mathbf{a}_2$) is known as the chiral angle or helicity (θ). Based on the value of the chiral angle, BN nanotubes can be classified as *zigzag* ($\theta = 0°$), *armchair* ($\theta = 30°$), and *chiral* ($0° < \theta < 30°$), as illustrated in Figure 1.13 (Arenal et al. 2010). Although both armchair and helical structures have been observed in BN nanotubes, the majority of tubes studied by several research groups have displayed zigzag or near-zigzag configurations, independent of the synthesis method and morphology of the BN nanotubes (Golberg et al. 1996, 1999, 2000a,b, 2003; Lee et al. 2001; Terauchi et al. 2000; Ma et al. 2001). This is a distinguishing feature in BN nanotubes, compared to C nanotubes in which all helicities are statistically equally probable. It is because crystallization of BN layers is governed by the strong tendency to have the atomically perfect B–N stacked consecutive layers; therefore, the shells within an individual BN nanotube tend to have the same layouts. In contrast, in C nanotubes, the relative freedom in rotational disorders between consecutive shells can lead to a wide variety of helicities (Golberg et al. 2010).

1.3.1.2 Experimental Aspects

HRTEM instruments are widely utilized to obtain direct images of the atomic arrangements in the hexagonal shells of the BN nanotubes under particular imaging conditions (Scherzer focus) (Wang et al. 2003). This imaging not only determines the helicity of the tubes but also reveals the defects present in tube walls. Nevertheless, the HRTEM technique is very sensitive to the orientation of the tubes with respect to the electron beam and to the defocus conditions, which make it difficult to set the aforementioned particular conditions. Thus, there are limitations for the use of this technique in such studies. Some other techniques, such as STM, which have been used to determine the atomic structure in C nanotubes, are not applicable to BN nanotubes due to their wide bandgap.

Electron diffraction is another means to investigate the atomic structure of nanotubes. This technique does not include the HRTEM limitations and has been considered as a reliable method to obtain the chiral indices of nanotubes. To interpret the electron diffraction patterns (EDPs), it is ideally considered that a single-walled BN nanotube is formed by wrapping up a sheet of h-BN along a tubule axis. Therefore, the tube can be seen as two parallel sheets rotated with respect to each other by 2Φ, that is, the EDPs are the superposition of two rotated trigonal networks, as shown in Figure 1.14 (Arenal et al. 2006a). Moreover, because of the finite size of the BN nanotubes in the radial direction, the intensity distribution of the EDP elongates perpendicular to the tubule axis and forms a set of layer lines normal to the tubule axis. The practical

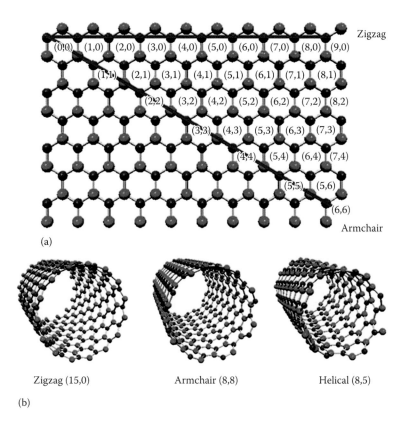

(a)

Zigzag (15,0)   Armchair (8,8)   Helical (8,5)

(b)

Figure 1.13   (a) A BN sheet with possible wrapping structures. (b) Structural models of three configurations of single-walled BN nanotubes made of wrapped BN sheets. (Reprinted with permission from Golberg, D., Bando, Y., Huang, Y. et al., Boron nitride nanotubes and nanosheets, *ACS Nano*, 4(6), 2010, 2979–2993. Copyright 2010, American Chemical Society.)

examples to calculate the tube diameter and the chiral indices are discussed by Zhang et al. (1993), Amelinckx et al. (1999), and Gao et al. (2003).

Having examined EDPs of multiwalled BN nanotubes, Arenal et al. (2006b) reported that 25% of the tubes had helicities from 0° to 5°, and the distribution of other helicities was uniform. However, contrary to the case of large multiwalled BN nanotubes, a strong correlation between the chiral angles of the constituent concentric tubes was not observed in double-walled nanotubes, which may indicate a lower interaction between the layers of a double-walled nanotube compared to that of a multiwalled tube (Lim et al. 2007).

1.3.1.3   Polygonization

EDP analysis by Celik-Atkas et al. (2005a,b) and Golberg et al. (2007c) showed the presence of a coaxial polygonized cylindrical structure on multiwalled BN nanotubes. Celik-Atkas et al. suggested the presence of a double-helix structural model in their nanotubes; one of them was polygonal in cross section and highly crystalline and the other was circular and less ordered, as shown in Figure 1.15. The double-helix structure of BN nanotubes may correspond to a stronger wall–wall interactions in them (compared to C nanotubes), due to ionic bonding.

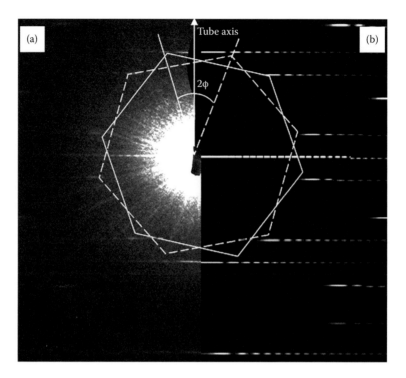

Figure 1.14   (a) Experimental and (b) simulated electron diffraction patterns of a single-walled BN nano-tube. (Reprinted with permission from Arenal, R., Kociak, M., Loiseau, A., and Miller, D.J., Determination of chiral indices of individual single- and double-walled boron nitride nanotubes by electron diffraction, *Appl. Phys. Lett.*, 89(7), 73104, 2006b. Copyright 2006, American Institute of Physics.)

Golberg et al.'s results showed that the whole range of chiral angles, from zigzag to armchair, could coexist in a polygonized or tubular multiwalled BN nanotube. Laude and Matsui (2004) recorded EDPs on multiwalled BN nanotubes and observed that both zigzag and armchair helicities were dominant, without any preference.

Polygonization of cross sections is favorable in the multiwalled BN nanotubes since the B–N–B–N stacking order across a BN tube could be better preserved within the multiple polygonized shells with flat facets (Golberg et al. 2007b). Such stacking can be broken in cylindrical multiwalled nanotubes due to the existence of different circumferences in consecutive shells. In fact, a multiwalled nanotube with a polygonal cross section can have a lower energy than a nanotube with a circular cross section. The energy of a nanotube, relative to the bulk material, can be divided into four components: strain energy, interfacial energy, defect energy, and surface energy. The total energy of the nanotube can therefore be calculated according to the following equation:

$$E_{Total} = N \cdot \Omega_{bulk} + E_{strain} + E_{interface} + E_{defect} + E_{surface}$$

where N is the number of atoms, Ω_{bulk} is the bulk energy per atom, and all other energy terms represent the excess energy of the nanotube due to various components. The surface energy per unit area will be approximately identical for both aforementioned structures (polygonal and cylindrical tubes). But defects can lower the energy of polygonal nanotubes and increase the likelihood of their formation. Moreover, the polygonal cross section will result in higher strain

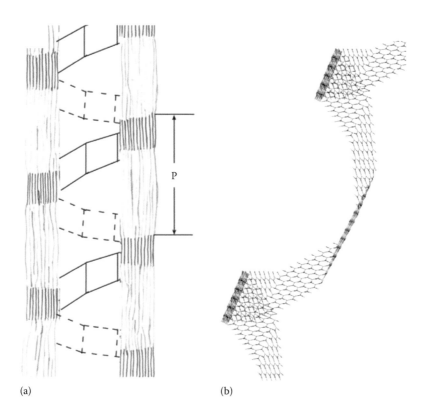

(a)   (b)

Figure 1.15   (a) Schematic of a structure model for the multiwalled BN nanotubes. (b) Atomic structure model of a hexagonal helix in the case of a zigzag tube. (From Celik-Aktas, A. et al., *Acta Crystallogr. A*, 61, 533, 2005b. Copyright International Union of Crystallography. With permission.)

energy due to the small bending radii at the corners, but this can be easily compensated by their lower interfacial energy. This energy reduction occurs because the flat sections of the nanotube can have a coherent interface with no tensile strain. But when the cross section is circular, it is not possible to maintain a coherent interface without an excessive amount of tensile strain, due to the difference in the circumference of consecutive shells. When the interface is coherent, the same number of unit cells has to be spread out over a length that increases with distance from the center of the nanotube. This is unlikely for multiwalled nanotubes due to the large amount of strain required (Tibbetts et al. 2009).

Figure 1.16 is a schematic of two layers of BN in two different conditions. Figure 1.16a illustrates two flat layers, analogous to the bulk BN, in which B atoms in one layer project directly above N atoms in a preceding layer and vice versa. In such a coherent interface, no strain is required to maintain alignment throughout the layers. Figure 1.16b depicts two bent BN layers undergoing a large amount of strain to maintain the bulk alignment. This tensile strain energy grows rapidly with the number of layers, thus favoring the formation of polygonized cross sections.

Figure 1.17 schematically shows that the bending strain energy is localized at the corners of the polygon, resulting in increased strain energy. However, the flat sides of the polygon provide a coherent interface leading to a large reduction in the interfacial energy. Because the interfacial energy is much larger than the bending strain energy, the polygonal model can often result in lower overall energy than that for a nanotube with a circular cross section (Tibbetts et al. 2009).

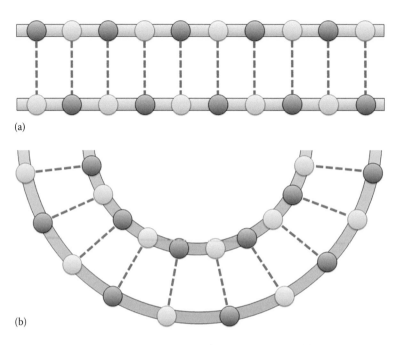

Figure 1.16   Schematic diagrams of bulk alignment in (a) flat BN layers and (b) curved BN layers under stress.

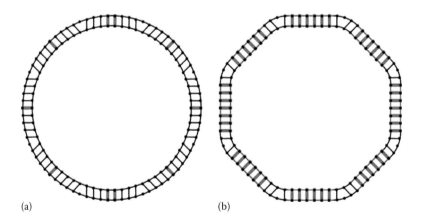

Figure 1.17   Coherency at interface in (a) multiwalled cylindrical nanotubes and (b) polygonal nanotubes. Thick and thin lines represent coherent and incoherent interfaces, respectively. (Reprinted with permission from Tibbetts, K., Doe, R., and Ceder, G., Polygonal model for layered inorganic nanotubes, *Phys. Rev. B*, 80(1), 014102, 2009. Copyright 2009 by the American Physical Society.)

The diverse cross sections of BN nanotubes are due to different growth conditions, which also determine the wall thicknesses and tube diameters. Representative TEM images of several possible BN nanotube cross sections are depicted in Figure 1.18. The polygonal tubes in Figure 1.18a and b with sharp facets were observed most commonly in multiwalled BN nanotubes, while some tubes have more or less round cross sections, as seen in Figure 1.18c (yet, some faceted domains are clearly visible).

Figure 1.18 TEM images of different cross sections in polygonal BN nanotubes; sharp (a, b) and round (c) facets. (Reprinted with permission from Golberg, D., Costa, P.M.F.J., Lourie, O. et al., Direct force measurements and kinking under elastic deformation of individual multiwalled boron nitride nanotubes, *Nano Lett.*, 7(7), 2007a, 2146–2151. Copyright 2007, American Chemical Society.)

In multiwalled tubes, it is common to observe some dark (or bright) spots on both wall sides in bright- (or dark-) field imaging modes, as shown in Figure 1.19. Kim et al. (2005) attributed the dark spots and the contrast difference to compositional changes along the multiwalled BN nanotubes. Celik-Aktas et al. (2005b) did not observe any significant impurity in their BN nanotubes, but their EELS study revealed a significant difference in the ratio of σ^* and π^* peaks measured at the dark spots and at other locations. This difference could be due to the change in the relative orientation of the atomic planes with respect to the electron beam across the tube axis. TEM images in Figure 1.19b and c demonstrate that dark regions in the bright-field image appear bright in the dark-field image. These regions contribute to a strong 002 peak, since the dark-field images were formed by using the 002 spot along the equatorial line of the nanotube diffraction pattern. The 002 peak is formed by interference between the different walls of the nanotube. For a circular multiwalled tube, the intensity of the 002 peak depends on the regularity of the wall spacing and the number of walls. If the multiwalled tube is faceted, the peak intensity also depends on the relative orientation

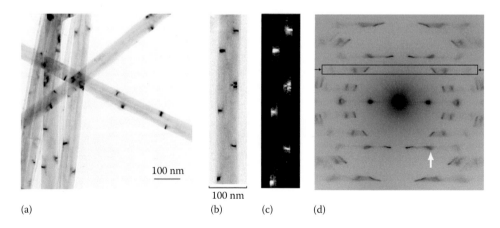

Figure 1.19 (a) TEM image of the CVD-grown multiwalled BN nanotubes showing dark spots. (b) Bright-field image of a BN nanotube. (c) Central dark-field image of the same area as (b). (d) Typical EDP of a CVD-grown BN nanotube. (From Celik-Aktas, A. et al., *Acta Crystallogr. A*, 61, 533, 2005b. Copyright International Union of Crystallography. With permission.)

between the electron beam and the facet (Celik-Aktas et al. 2005b). In short, such contrasts in wall domains may reflect the tube areas where stress/strain fields exist due to the presence of various cross-sectional and/or chirality shell packages within the multiwalled structure. The corresponding EPDs also show additional spots that are not allowed in a perfectly cylindrical tube but can appear in a faceted polygonal cross-sectional multilayer nanostructure (Golberg et al. 2010).

1.3.2   Atomic Structure of BN Nanosheets

BN nanosheets may be depicted as graphene layers, in which C atoms have been fully substituted by alternating B and N atoms. Within each h-BN layer, atoms are bound together by strong covalent bonds, while there are weak van der Waals forces between different layers. The crystallographic parameters of h-BN and graphite are almost identical, as summarized in Table 1.1 (Pauling 1966).

1.3.2.1   Theoretical Aspects

Unlike the popular monolayer graphenes and single-walled C nanotubes, their corresponding BN sister systems have rarely been observed (Lin et al. 2011; Arenal et al. 2006a), due to the peculiar B–N stacking characteristics. The hexagons of neighboring planes in h-BN are superposed, that is, B and N atoms are in succession along the c axis, while in graphite, they are shifted by half a hexagon. Moreover, due to the difference in the electronegativity of B and N, the B–N bonds in BN materials are partially ionic, in contrast with the purely covalent C–C bonds in graphitic structures. These can lead to the so-called lip–lip interactions between neighboring layers/ shells in BN nanostructures, that is, chemical bonds form as bridges or *spot-welds* between the atoms of adjacent layers/shells. This phenomenon contributes to a metastable energy minimum through decreasing the number of dangling bonds at the edges/tips and reducing the *frustration* effect (i.e., when B–B and N–N bonds form instead of the energetically more favorable B–N bonds). Therefore, formation of multilayers/shells stabilizes the whole structure (Blase et al. 1998; Charlier et al. 1999).

1.3.2.2   Experimental Aspects

TEM investigations have revealed that in typical multi-layered nanosheets and multiwalled nanotubes of BN, layers and walls are ordered with an interlayer distance of ~0.33–0.34 nm, characteristic of the interplanar spacing in h-BN (002) planes. Some researchers have reported that the interlayer spacing of BN nanotubes may be slightly larger than that of bulk h-BN (i.e., ~0.333 nm), which might result from the inner stresses within the bent walls. Figure 1.20 shows a low-magnification TEM image of the BN nanosheets prepared by a thermal CVD method. It indicates the compact BN network consisting of very thin nanosheets that are almost transparent to the electron beam. In addition, intrinsic bending and scrolling of the nanosheets can be noticed in Figure 1.20a. The dark parts are generally the cross sections of the nanosheets folded back. Figure 1.20b is an HRTEM image of the BN nanosheets, in which lattice fringes can be observed with a 0.33 nm interlayer spacing. The hexagonal structure of BN can be observed

TABLE 1.1 CRYSTALLOGRAPHIC INFORMATION OF h-BN AND GRAPHITE				
Material	Crystal Structure	Nearest Neighbor Distance (nm)	Lattice Parameters (nm)	Interlayer Spacing (nm)
h-BN	Hexagonal	0.144	a: 0.250, c: 0.666	0.333
Graphite	Hexagonal	0.142	a: 0.246, c: 0.670	0.335

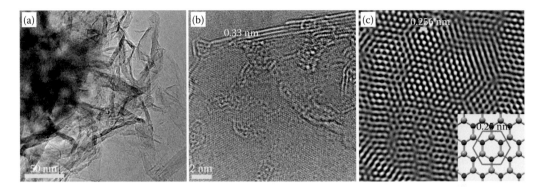

Figure 1.20 (a) Low-magnification TEM and (b) HRTEM images of BN nanosheets. (From Pakdel, A. et al., *Surf. Innov.*, 1(1), 32, 2013b. Copyright Thomas Telford Ltd. With permission.) (c) Lattice image of the hexagonal structure of BN nanosheets. (From Wang, X.B., Pakdel, A., Zhi, C.Y. et al., High-yield boron nitride nanosheets from 'Chemical Blowing': Towards practical applications in polymer composites, *J. Phys. Condens. Matter*, 24(31), 314205, 2012. Copyright IOP Publishing. With permission.)

in the lattice image in Figure 1.20c. The spacing between adjacent white dots is about 0.25 nm, corresponding to the (100) lattice constant of h-BN and the B–B or N–N atom separations in h-BN. Therefore, the white dots correspond to the hexagonal rings rather than individual atoms.

1.4 PROPERTIES AND APPLICATIONS

Bulk h-BN has been a matter of interest for a long time due to its low density, high thermal conductivity, electrical insulation, superb oxidation resistance, excellent inertness (i.e., passivity to reactions with acids, alkalis, and melts), and low friction coefficient. BN nanotubes and nanosheets, as inheritors of h-BN, also possess these advantageous properties.

1.4.1 Electronic and Optical Properties

BN nanostructures are generally recognized as insulators with wide bandgaps of 5.0–6.0 eV. The theoretical analysis of electronic properties of BN nanotubes and nanosheets reviewed by Arenal et al. (2010) showed their bandgap to be first order independent of the details of the atomic structure. This explains the so-called stability of the bandgap in BN nanotubes, which indeed hardly depends on curvature and helicity (Blase et al. 1994), except in the limit of very small diameters. Recent experimental studies by Lee et al. (2010) and Pakdel et al. (2012a) have pointed out optical bandgaps of ~6 and ~5.7 eV for BN nanotubes and BN nanosheets, respectively, which approach that of h-BN single crystals. This insulating behavior of BN nanostructures encourages their applications as protective shield encapsulating nanomaterials. In this regard, considerable research has been performed to fill BN nanotubes during or after the synthesis with fullerenes or with crystalline compounds, such as cobalt, iron–nickel, and magnesium oxide (Mickelson et al. 2003; Bando et al. 2001; Golberg et al. 2003). Terrones et al. (2008) have theoretically and experimentally demonstrated that BN nanoribbons with zigzag edges can show metallic behavior. Their porous BN nanospheres showed stable field emission properties at low turn-on voltages (e.g., 1–1.3 V/μm) due to the presence of these ribbons protruding from the surface of the spheres. Such BN nanostructures may find applications as catalysts or field emitters in the future.

Another approach to modify the wide bandgap of 2D BN nanomaterials is the addition of a third element into their structure (Zeng et al. 2010; Pakdel et al. 2012a, Ci et al. 2010; Qin et al. 2012). Bearing in mind the similar lattice parameters and crystal structure of BN and graphene, it can be expected that C addition into the BN network forms stable structures. In fact, combination of the semimetallic properties of graphene with the electrical insulation of BN has drawn particular attention to the ternary B–C–N system. Boron carbonitride ($B_xC_yN_z$) nanostructures are especially attractive for applications in electronic and luminescent devices owing to their semiconducting properties with variable bandgap energies.

BN and BN–C nanostructures also exhibit distinctive violet or ultraviolet (UV) luminescence emissions, which promote their application in downsized UV lasing devices for sterilization, surgery, photocatalysis, and optical storage (Golberg et al. 2007a).

1.4.2   Thermal Properties

BN possesses remarkable thermal conductivity and high specific heat. Theoretical calculations by Xiao et al. (2004b) confirm the high specific heat of BN nanotubes and predict their thermal conductivity to be higher than that of C nanotubes. They also showed that thermal conductance of single-walled BN nanotubes at low temperatures is independent of the tube diameter and chirality (Xiao et al. 2004a). A recent study indicates that the strong phonon–phonon scattering in h-BN is the cause for its lower thermal conductivity compared to graphite; however, reduction in such scattering in a single-layer BN sheet leads to a substantial increase in its conductivity (>600 W/m/K at room temperature) (Lindsay et al. 2011). Experimentally, Chang et al. (2006) measured the thermal conductivity values for BN nanotubes to be ~350 W/m/K at room temperature and demonstrated that if they were mass-loaded externally and heterogeneously with heavy molecules (e.g., $C_9H_{16}Pt$), they would possess asymmetric axial thermal conductance properties. Thus, BN nanotube thermal rectifiers were suggested to have substantial implications for diverse nanoscale calorimeters, microelectronic processors, macroscopic refrigerators, and energy-saving buildings.

Thermal stability experiments on multiwalled BN nanotubes were performed by Golberg et al. (2001) and indicated excellent oxidation resistance up to 830°C in air. Moreover, thermogravimetric analysis showed that the onset temperature for the oxidation of BN nanotubes (800°C) was much higher than that of C nanotubes (400°C) under the same conditions. Later, it was shown that thin BN nanotubes with diameters smaller than 20 nm can resist oxidation up to 900°C (Chen et al. 2014).

1.4.3   Mechanical Properties

Theoretical studies by Hernandez et al. (1998) revealed the elastic modulus of BN nanotubes to vary between ~0.84 and ~0.91 TPa with diameters ranging from 0.81 to 2.08 nm. Chopra and Zettl (1998) estimated the elastic modulus of a cantilevered individual double-walled BN nanotube by measuring the amplitude of the thermal-induced vibrations in a TEM at room temperature. They found a value of 1.22 ± 0.24 TPa, which is similar to the elastic modulus of C nanotubes. Subsequently, an electric-field-induced resonance method in TEM was utilized to calculate elastic modulus values between 0.51 and 1.03 TPa (Suryavanshi et al. 2004). More recently, Golberg's group performed extensive in situ TEM bending and tensile experiments on individual multiwalled tubes with diameters of 40–100 nm and estimated their elastic modulus as 0.5–0.6 TPa or up to 1.3 TPa, and their tensile strength as 33 GPa. However, in the case of bamboo-like BN nanotubes, the elastic modulus and tensile strength were estimated as 225 and 8 GPa, respectively (Wei et al. 2010; Tang et al. 2011).

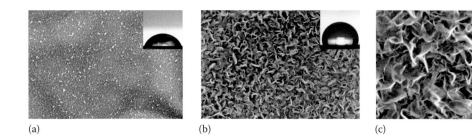

(a)   (b)   (c)

Figure 1.21   SEM images of BN nanosheet coatings with different levels of surface roughness. As a result, the water repellency of the coating is different: (a) hydrophilic, (b) hydrophobic, and (c) superhydrophobic. (Reprinted with permission from Pakdel, A., Zhi, C.Y., Bando, Y., Nakayama, T., and Golberg, D., Boron nitride nanosheet coatings with controllable water repellency, *ACS Nano*, 5(8), 2011, 6507–6515. Copyright 2011, American Chemical Society.)

Superior mechanical properties and thermal conductivity of BN nanostructures make them attractive as nanofillers in composite materials. For instance, BN nanotubes can reinforce polymer matrices and improve their thermal conductivity, while transparency of the polymers is preserved. Zhi et al. (2010) initiated research on polymer matrix composites reinforced by BN nanotubes and showed an effective improvement in the elastic modulus of polystyrene and PMMA by ~20% with an addition of only 1 wt.% of nanotubes. A recent work on the nanomechanical characterization of single-walled BN nanotubes revealed their axially strong, but radially supple, characteristics and suggested that they might be superior to single-walled C nanotubes as reinforcing additives for nanocomposite applications (Zheng et al. 2012). BN nanotubes can be ideal reinforcing agents in lightweight metal matrix composites. Aluminum (Al) ribbons reinforced with BN nanotubes have been recently fabricated via melt spinning in an Ar atmosphere. BN nanotubes could be randomly dispersed within the microcrystalline Al matrix and improved the room-temperature ultimate tensile strength of the composite more than twice that of pure Al ribbons produced in similar conditions (Yamaguchi et al. 2013).

1.4.4   Wetting Properties

While h-BN films are partially wetted by water with a contact angle of ~50°, BN nanostructure films can achieve superhydrophobic state with water contact angle exceeding 150° Pakdel et al. 2011; Pakdel et al. 2013a,b,c; Lee et al. 2009; Li et al. 2010. In a systematic approach, Pakdel et al. succeeded in growing partially vertically aligned BN nanosheets on Si substrates with controllable water-repellency levels, as shown in Figure 1.21. As a result, water contact angles from ~51° to ~159° were obtained, indicating a significant change from hydrophilicity to superhydrophobicity. Due to outstanding chemical intentness of BN, the pH value of water did not affect the wetting characteristics of BN nanosheet and BN nanotube films. Therefore, BN nanostructure films are anticipated to find industrial applications in water-repelling, antifouling, self-cleaning, and anticorrosion systems.

1.5   PERSPECTIVES

BN has been considered as one of the best substrates to maintain the excellent transport behavior of graphene, because it provides an extraordinarily flat surface for graphene layers. This significantly reduces electron–hole puddles as compared to SiO_2 (Xue et al. 2011). By reducing the charge fluctuations, the low-density regime and the Dirac point can be more readily accessed.

Graphene devices on BN substrates have demonstrated mobilities and carrier inhomogeneities that are almost an order of magnitude better than devices on SiO_2 (Dean et al. 2010). A hybrid structure consisting of h-BN and graphene would enable tailoring of physical properties in graphene-based structures. (Ci et al. 2010; Lin et al. 2012). It can find various applications, such as field-effect transistors for radio frequency apparatus (Wang et al. 2011a).

Further applications of BN nanomaterials can be in nanomedical fields due to their good biocompatibility, scanning probe microscopy as ultrathin stiff tips, new generations of semi-conducting materials with adjustable bandgap through doping of other elements in BN structure, and hybrid nanostructures through surface functionalization of BN nanomaterials with nanoparticles of other materials. Recently, functionalized BN nanotubes have been tested in vitro on fibroblast cells and demonstrated optimal cytocompatibility even at high concentrations in the culture medium (Ciofani et al. 2012). Also, BN nanotube/mesoporous silica composites with controllable surface zeta potential have been tested for intracellular delivery of doxorubicin and showed satisfactory doxorubicin intracellular endocytosis efficiency and LNcap prostate cancer cell killing ability.

BN and other layered materials not only have proven to be of technological importance in a variety of industrial applications but also are critical to a number of environmental issues involving the fate of contaminants and groundwater quality. Very recently, porous BN nanosheets with high specific surface areas exhibited excellent sorption performances for a wide range of oils, solvents, and dyes and were proved effective for the removal of such contaminations from water (Lei et al. 2013).

Moreover, considering the relatively flexible interlayer space in layered nanomaterials, BN nanostructures may be able to accommodate various-sized molecules such as proteins, viruses, and pesticides. Therefore, BN nanomaterials may find ecological and biological applications, provided that extensive research on the biocompatibility and toxicity levels of these layered nanomaterials is performed. A recent study suggests that if layered materials (including BN) are functionalized with bioactive molecules, they can be stabilized as uniformly dispersed suspensions with biocompatibility factored in (Zhang et al. 2013). Such functionalized hybrid nanosheets do offer scope for applications in biomedical fields and beyond.

ACKNOWLEDGMENT

The authors acknowledge the financial support of the World Premier International Center for Materials Nanoarchitectonics (WPI-MANA) of the National Institute for Materials Science (NIMS), Tsukuba, Japan.

REFERENCES

Amelinckx, S., Lucas, A., and Lambin, P. Electron diffraction and microscopy of nanotubes. *Reports on Progress in Physics* 62(11) (1999): 1471–1524.

Arenal, R., Blase, X., and Loiseau, A. Boron-nitride and boron-carbonitride nanotubes: Synthesis, characterization and theory. *Advances in Physics* 59(2) (2010): 101–179.

Arenal, R., Ferrari, A.C., Reich, S. et al. Raman spectroscopy of single-wall boron nitride nanotubes. *Nano Letters* 6(8) (2006a): 1812–1816.

Arenal, R., Kociak, M., Loiseau, A., and Miller, D.J. Determination of chiral indices of individual single- and double-walled boron nitride nanotubes by electron diffraction. *Applied Physics Letters* 89(7) (2006b): 73104.

Arenal, R., Stephan, O., Cochon, J.L., and Loiseau, A. Root-growth mechanism for single-walled boron nitride nanotubes in laser vaporization technique. *Journal of the American Chemical Society* 129(51) (2007): 16183–16189.

Bae, S.Y., Seo, H.W., Park, J., Choi, Y.S., Park, J.C., and Lee, S.Y. Boron nitride nanotubes synthesized in the temperature range 1000–1200°C. *Chemical Physics Letters* 374(5–6) (2003): 534–541.

Balmain, W.H. XLVI. Observations on the formation of compounds of boron and silicon with nitrogen and certain metals. *Philosophical Magazine Series 3* 21(138) (1842a): 270–277.

Balmain, W.H. Bemerkungen über die Bildung von Verbindungen des Bors und Siliciums mit Stickstoff und gewissen Metallen. *Journal für Praktische Chemie* 27(1) (1842b): 422–430.

Bando, Y., Ogawa, K., and Golberg, D. Insulating 'Nanocables': Invar Fe-Ni alloy nanorods inside BN nanotubes. *Chemical Physics Letters* 347(4–6) (2001): 349–354.

Bartnitskaya, T.S., Oleinik, G.S., Pokropivnyi, A.V., and Pokropivnyi, V.V. Synthesis, structure, and formation mechanism of boron nitride nanotubes. *JETP Letters* 69(2) (1999): 163–168.

Berner, S., Corso, M., Widmer, R. et al. Boron nitride nanomesh: Functionality from a corrugated monolayer. *Angewandte Chemie-International Edition* 46(27) (2007): 5115–5119.

Blase, X., De Vita, A., Charlier, J.C., and Car, R. Frustration effects and microscopic growth mechanisms for Bn nanotubes. *Physical Review Letters* 80(8) (1998): 1666–1669.

Blase, X., Rubio, A., Louie, S.G., and Cohen, M.L. Stability and band-gap constancy of boron-nitride nanotubes. *Europhysics Letters* 28(5) (1994): 335–340.

Celik-Aktas, A., Zuo, J.M., Stubbins, J.F., Tang, C., and Bando, Y. Structure and chirality distribution of multi-walled boron nitride nanotubes. *Applied Physics Letters* 86(13) (2005a): 133110.

Celik-Aktas, A., Zuo, J.M., Stubbins, J.F., Tang, C.C., and Bando, Y. Double-helix structure in multiwall boron nitride nanotubes. *Acta Crystallographica Section A* 61 (2005b): 533–541.

Chadderton, L.T. and Chen, Y. Nanotube growth by surface diffusion. *Physics Letters A* 263(4–6) (1999): 401–405.

Chang, C.W., Fennimore, A.M., Afanasiev, A. et al. Isotope effect on the thermal conductivity of boron nitride nanotubes. *Physical Review Letters* 97(8) (2006): 085901.

Charlier, J.C., Blase, X., De Vita, A., and Car, R. Microscopic growth mechanisms for carbon and boron-nitride nanotubes. *Applied Physics A: Materials Science and Processing* 68(3) (1999): 267–273.

Chatterjee, S., Luo, Z., Acerce, M., Yates, D.M., Johnson, A.T.C., and Sneddon, L.G. Chemical vapor deposition of boron nitride nanosheets on metallic substrates via decaborane/ammonia reactions. *Chemistry of Materials* 23(20) (2011): 4414–4416.

Chen, X., Wang, X., Liu, J., Wang, Z., and Qian, Y. A reduction-nitridation route to boron nitride nanotubes. *Applied Physics A: Materials Science and Processing* 81(5) (2005): 1035–1037.

Chen, Y., Chadderton, L.T., FitzGerald, J., and Williams, J.S. A solid-state process for formation of boron nitride nanotubes. *Applied Physics Letters* 74(20) (1999a): 2960–2962.

Chen, Y., Conway, M., Williams, J.S., and Zou, J. Large-quantity production of high-yield boron nitride nanotubes. *Journal of Materials Research* 17(8) (2002): 1896–1899.

Chen, Y., Fitz Gerald, J.D., Williams, J.S., and Bulcock, S. Synthesis of boron nitride nanotubes at low temperatures using reactive ball milling. *Chemical Physics Letters* 299(3–4) (1999b): 260–264.

Chen, Y., Zou, J., Campbell, S.J., and Le Caer, G. Boron nitride nanotubes: Pronounced resistance to oxidation. *Applied Physics Letters* 84(13) (2004): 2430–2432.

Chen, Z.G., Zou, J., Liu, G. et al. Novel boron nitride hollow nanoribbons. *ACS Nano* 2(10) (2008): 2183–2191.

Chopra, N.G., Luyken, R.J., Cherrey, K. et al. Boron-nitride nanotubes. *Science* 269(5226) (1995): 966–967.

Chopra, N.G. and Zettl, A. Measurement of the elastic modulus of a multi-wall boron nitride nanotube. *Solid State Communications* 105(5) (1998): 297–300.

Ci, L., Song, L., Jin, C.H. et al. Atomic layers of hybridized boron nitride and graphene domains. *Nature Materials* 9(5) (2010): 430–435.

Ciofani, G., Genchi, G.G., Liakos, I. et al. A simple approach to covalent functionalization of boron nitride nanotubes. *Journal of Colloid and Interface Science* 374(1) (2012): 308–314.

Corso, M., Auwarter, W., Muntwiler, M., Tamai, A., Greber, T., and Osterwalder, J. Boron nitride nanomesh. *Science* 303(5655) (2004): 217–220.

Cumings, J. and Zettl, A. Mass-production of boron nitride double-wall nanotubes and nanococoons. *Chemical Physics Letters* 316(3–4) (2000): 211–216.

Dai, J., Xu, L.Q., Fang, Z. et al. A convenient catalytic approach to synthesize straight boron nitride nanotubes using synergic nitrogen source. *Chemical Physics Letters* 440(4–6) (2007): 253–258.

Dean, C.R., Young, A.F., Meric, I. et al. Boron nitride substrates for high-quality graphene electronics. *Nature Nanotechnology* 5(10) (2010): 722–726.

Deepak, F.L., Vinod, C.P., Mukhopadhyay, K., Govindaraj, A., and Rao, C.N.R. Boron nitride nanotubes and nanowires. *Chemical Physics Letters* 353(5–6) (2002): 345–352.

Gao, M., Zuo, J.M., Twesten, R.D., Petrov, I., Nagahara, L.A., and Zhang, R. Structure determination of individual single-wall carbon nanotubes by nanoarea electron diffraction. *Applied Physics Letters* 82(16) (2003): 2703–2705.

Gao, R., Yin, L.W., Wang, C.X. et al. High-yield synthesis of boron nitride nanosheets with strong ultraviolet cathodoluminescence emission. *Journal of Physical Chemistry C* 113(34) (2009): 15160–15165.

Golberg, D., Bando, Y., Eremets, M., Takemura, K., Kurashima, K., and Yusa, H. Nanotubes in boron nitride laser heated at high pressure. *Applied Physics Letters* 69(14) (1996): 2045–2047.

Golberg, D., Bando, Y., Fushimi, K., Mitome, M., Bourgeois, L., and Tang, C.-C. Nanoscale oxygen generators: MgO$_2$-based fillings of BN nanotubes. *Journal of Physical Chemistry B* 107(34) (2003): 8726–8729.

Golberg, D., Bando, Y., Han, W., Kurashima, K., and Sato, T. Single-walled B-doped carbon, B/N-doped carbon and Bn nanotubes synthesized from single-walled carbon nanotubes through a substitution reaction. *Chemical Physics Letters* 308(3–4) (1999a): 337–342.

Golberg, D., Bando, Y., Huang, Y. et al. Boron nitride nanotubes and nanosheets. *ACS Nano* 4(6) (2010): 2979–2993.

Golberg, D., Bando, Y., Kurashima, K., and Sato, T. Ropes of BN multi-walled nanotubes. *Solid State Communications* 116(1) (2000a): 1–6.

Golberg, D., Bando, Y., Kurashima, K., and Sato, T. MoO$_3$-promoted synthesis of multi-walled BN nanotubes from C nanotube templates. *Chemical Physics Letters* 323(1–2) (2000b): 185–191.

Golberg, D., Bando, Y., Kurashima, K., and Sato, T. Synthesis and characterization of ropes made of BN multi-walled nanotubes. *Scripta Materialia* 44(8–9) (2001): 1561–1565.

Golberg, D., Bando, Y., Stephan, O., and Kurashima, K. Octahedral boron nitride fullerenes formed by electron beam irradiation. *Applied Physics Letters* 73(17) (1998): 2441–2443.

Golberg, D., Bando, Y., Tang, C.C., and Zhi, C.Y. Boron nitride nanotubes. *Advanced Materials* 19(18) (2007a): 2413–2432.

Golberg, D., Costa, P.M.F.J., Lourie, O. et al. Direct force measurements and kinking under elastic deformation of individual multiwalled boron nitride nanotubes. *Nano Letters* 7(7) (2007b): 2146–2151.

Golberg, D., Han, W., Bando, Y., Bourgeois, L., Kurashima, K., and Sato, T. Fine structure of boron nitride nanotubes produced from carbon nanotubes by a substitution reaction. *Journal of Applied Physics* 86(4) (1999b): 2364–2366.

Golberg, D., Mitome, M., Bando, Y., Tang, C.C., and Zhi, C.Y. Multi-walled boron nitride nanotubes composed of diverse cross-section and helix type shells. *Applied Physics A: Materials Science and Processing* 88(2) (2007c): 347–352.

Golberg, D., Rode, A., Bando, Y., Mitome, M., Gamaly, E., and Luther-Davies, B. Boron nitride nanostructures formed by ultra-high-repetition rate laser ablation. *Diamond and Related Materials* 12(8) (2003): 1269–1274.

Goriachko, A., He, Y.B., Knapp, M. et al. Self-assembly of a hexagonal boron nitride nanomesh on Ru(0001). *Langmuir* 23(6) (2007): 2928–2931.

Han, W.Q., Bando, Y., Kurashima, K., and Sato, T. Synthesis of boron nitride nanotubes from carbon nanotubes by a substitution reaction. *Applied Physics Letters* 73(21) (1998): 3085–3087.

Han, W.Q., Todd, P.J., and Strongin, M. Formation and growth mechanism of ^{10}BN nanotubes via a carbon nanotube-substitution reaction. *Applied Physics Letters* 89(17) (2006): 173103.

Han, W.Q., Wu, L.J., Zhu, Y.M., Watanabe, K., and Taniguchi, T. Structure of chemically derived mono- and few-atomic-layer boron nitride sheets. *Applied Physics Letters* 93(22) (2008): 223103.

Hernandez, E., Goze, C., Bernier, P., and Rubio, A. Elastic properties of C and B$_x$C$_y$N$_z$ composite nanotubes. *Physical Review Letters* 80(20) (1998): 4502–4505.

Huang, Y., Lin, J., Tang, C.C. et al. Bulk synthesis, growth mechanism and properties of highly pure ultrafine boron nitride nanotubes with diameters of sub-10 nm. *Nanotechnology* 22(14) (2011): 145602.

Iijima, S. Helical microtubules of graphitic carbon. *Nature* 354(6348) (1991): 56–58.

Jin, C.H., Lin, F., Suenaga, K., and Iijima, S. Fabrication of a freestanding boron nitride single layer and its defect assignments. *Physical Review Letters* 102(19) (2009): 195505.

Kim, K.K., Hsu, A., Jia, X.T. et al. Synthesis of monolayer hexagonal boron nitride on Cu foil using chemical vapor deposition. *Nano Letters* 12(1) (2012): 161–166.

Kim, M.J., Chatterjee, S., Kim, S.M. et al. Double-walled boron nitride nanotubes grown by floating catalyst chemical vapor deposition. *Nano Letters* 8(10) (2008): 3298–3302.

Kim, Y.I., Jung, J.K., Ryu, K.S., Nahm, S.H., and Gregory, D.H. Quantitative phase analysis of boron nitride nanotubes using Rietveld refinement. *Journal of Physics D: Applied Physics* 38(8) (2005): 1127–1131.

Kroto, H.W., Heath, J.R., O'Brien, S.C., Curl, R.F., and Smalley, R.E. C_{60}: Buckminsterfullerene. *Nature* 318(6042) (1985): 162–163.

Laude, T. and Matsui, Y. Fine modulations in the diffraction pattern of boron nitride nanotubes synthesised by non-ablative laser heating. *European Physical Journal: Applied Physics* 28(3) (2004): 293–300.

Laude, T., Matsui, Y., Marraud, A., and Jouffrey, B. Long ropes of boron nitride nanotubes grown by a continuous laser heating. *Applied Physics Letters* 76(22) (2000): 3239–3241.

Lee, C.H., Drelich, J., and Yap, Y.K. Superhydrophobicity of boron nitride nanotubes grown on silicon substrates. *Langmuir* 25(9) (2009): 4853–4860.

Lee, C.H., Wang, J.S., Kayatsha, V.K., Huang, J.Y., and Yap, Y.K. Effective growth of boron nitride nanotubes by thermal chemical vapor deposition. *Nanotechnology* 19(45) (2008): 455605.

Lee, C.H., Xie, M., Kayastha, V., Wang, J.S., and Yap, Y.K. Patterned growth of boron nitride nanotubes by catalytic chemical vapor deposition. *Chemistry of Materials* 22(5) (2010): 1782–1787.

Lee, C.M., Choi, S.I., Choi, S.S., and Hong, S.H. Synthesis of boron nitride nanotubes by Arc-Jet plasma. *Current Applied Physics* 6(2) (2006): 166–170.

Lee, R.S., Gavillet, J., de la Chapelle, M.L. et al. Catalyst-free synthesis of boron nitride single-wall nanotubes with a preferred zig-zag configuration. *Physical Review B* 64(12) (2001): 121405(R).

Lei, W., Portehault, D., Liu, D., Qin, S., and Chen, Y. Porous boron nitride nanosheets for effective water cleaning. *Nature Communications* 4 (2013): 1777.

Li, L.H. and Chen, Y. Superhydrophobic properties of nonaligned boron nitride nanotube films. *Langmuir* 26(7) (2010): 5135–5140.

Li, L.H., Chen, Y., Behan, G., Zhang, H.Z., Petravic, M., and Glushenkov, A.M. Large-scale mechanical peeling of boron nitride nanosheets by low-energy ball milling. *Journal of Materials Chemistry* 21(32) (2011): 11862–11866.

Li, X., Zhi, C., Hanagata, N., Yamaguchi, M., Bando, Y., and Golberg, D. Boron nitride nanotubes functionalized with mesoporous silica for intracellular delivery of chemotherapy drugs. *Chemical Communications* 49 (2013): 7337–7339. doi:10.1039/C3CC42743A.

Lian, G., Zhang, X., Tan, M., Zhang, S.J., Cui, D.L., and Wang, Q.L. Facile synthesis of 3D boron nitride nanoflowers composed of vertically aligned nanoflakes and fabrication of graphene-like BN by exfoliation. *Journal of Materials Chemistry* 21(25) (2011): 9201–9207.

Lim, S.H., Luo, J.Z., Ji, W., and Lin, J. Synthesis of boron nitride nanotubes and its hydrogen uptake. *Catalysis Today* 120(3–4) (2007): 346–350.

Lin, T.Q., Wang, Y.M., Bi, H. et al. Hydrogen flame synthesis of few-layer graphene from a solid carbon source on hexagonal boron nitride. *Journal of Materials Chemistry* 22(7) (2012): 2859–2862.

Lin, Y., Williams, T.V., Xu, T.B., Cao, W., Elsayed-Ali, H.E., and Connell, J.W. Aqueous dispersions of few-layered and monolayered hexagonal boron nitride nanosheets from sonication-assisted hydrolysis: Critical role of water. *Journal of Physical Chemistry C* 115(6) (2011): 2679–2685.

Lindsay, L. and Broido, D.A. Enhanced thermal conductivity and isotope effect in single-layer hexagonal boron nitride. *Physical Review B* 84(15) (2011): 155421.

Loiseau, A., Willaime, F., Demoncy, N., Hug, G., and Pascard, H. Boron nitride nanotubes with reduced numbers of layers synthesized by Arc discharge. *Physical Review Letters* 76(25) (1996): 4737–4740.

Lourie, O.R., Jones, C.R., Bartlett, B.M., Gibbons, P.C., Ruoff, R.S., and Buhro, W.E. CVD growth of boron nitride nanotubes. *Chemistry of Materials* 12(7) (2000): 1808–1810.

Ma, R., Bando, Y., and Sato, T. CVD synthesis of boron nitride nanotubes without metal catalysts. *Chemical Physics Letters* 337(1–3) (2001): 61–64.

McNeish, T., Gumbs, G., and Balassis, A. Model of plasmon excitations in a bundle and a two-dimensional array of nanotubes. *Physical Review B* 77(23) (2008): 235440.

Meyer, J.C., Chuvilin, A., Algara-Siller, G., Biskupek, J., and Kaiser, U. Selective sputtering and atomic resolution imaging of atomically thin boron nitride membranes. *Nano Letters* 9(7) (2009): 2683–2689.

Mickelson, W., Aloni, S., Han, W.Q., Cumings, J., and Zettl, A. Packing C_{60} in boron nitride nanotubes. *Science* 300(5618) (2003): 467–469.

Novoselov, K.S., Geim, A.K., Morozov, S.V. et al. Electric field effect in atomically thin carbon films. *Science* 306(5696) (2004): 666–669.

Novoselov, K.S., Jiang, D., Schedin, F. et al. Two-dimensional atomic crystals. *Proceedings of the National Academy of Sciences of the United States of America* 102(30) (2005): 10451–10453.

Pacile, D., Meyer, J.C., Girit, C.O., and Zettl, A. The two-dimensional phase of boron nitride: Few-atomic-layer sheets and suspended membranes. *Applied Physics Letters* 92(13) (2008): 133107.

Pakdel, A., Bando, Y., and Golberg D. Nano boron nitride flatland. *Chemical Society Reviews* 43(3) (2014): 934–959.

Pakdel, A., Bando, Y., and Golberg, D. Morphology-driven nonwettability of nanostructured BN surfaces. *Langmuir* 29(24) (2013a): 7529–7533.

Pakdel, A., Bando, Y., Shtansky, D., and Golberg, D. Nonwetting and optical properties of BN nanosheet films. *Surface Innovations* 1(1) (2013b): 32–39.

Pakdel, A., Wang, X.B., Bando, Y., and Golberg, D. Nonwetting "white graphene" films. *Acta Materialia* 61(4) (2013c): 1266–1273.

Pakdel, A., Wang, X.B., Zhi, C.Y. et al. Facile synthesis of vertically aligned hexagonal boron nitride nanosheets hybridized with graphitic domains. *Journal of Materials Chemistry* 22(11) (2012a): 4818–4824.

Pakdel, A., Zhi, C., Bando, Y., Nakayama, T., and Golberg, D. A comprehensive analysis of the CVD growth of boron nitride nanotubes. *Nanotechnology* 23(21) (2012b): 215601.

Pakdel, A., Zhi, C.Y., Bando, Y., and Golberg, D. Low-dimensional boron nitride nanomaterials. *Materials Today* 15(6) (2012c): 256–265.

Pakdel, A., Zhi, C.Y., Bando, Y., Nakayama, T., and Golberg, D. Boron nitride nanosheet coatings with controllable water repellency. *ACS Nano* 5(8) (2011): 6507–6515.

Pauling, L. The structure and properties of graphite and boron nitride. *Proceedings of the National Academy of Sciences of the United States of America* 56(6) (1966): 1646–1652.

Qin, L., Yu, J., Kuang, S.Y., Xiao, C., and Bai, X.D. Few-atomic-layered boron carbonitride nanosheets prepared by chemical vapor deposition. *Nanoscale* 4(1) (2012): 120–123.

Qiu, Y.J., Yu, J., Yin, J. et al. Synthesis of continuous boron nitride nanofibers by solution coating electrospun template fibers. *Nanotechnology* 20(34) (2009): 345603.

Shi, Y.M., Hamsen, C., Jia, X.T. et al. Synthesis of few-layer hexagonal boron nitride thin film by chemical vapor deposition. *Nano Letters* 10(10) (2010): 4134–4139.

Song, L., Ci, L.J., Lu, H. et al. Large scale growth and characterization of atomic hexagonal boron nitride layers. *Nano Letters* 10(8) (2010): 3209–3215.

Stephan, O., Bando, Y., Loiseau, A. et al. Formation of small single-layer and nested BN cages under electron irradiation of nanotubes and bulk material. *Applied Physics A: Materials Science and Processing* 67(1) (1998): 107–111.

Suryavanshi, A.P., Yu, M.F., Wen, J.G., Tang, C.C., and Bando, Y. Elastic modulus and resonance behavior of boron nitride nanotubes. *Applied Physics Letters* 84(14) (2004): 2527–2529.

Tang, C., Bando, Y., Sato, T., and Kurashima, K. A novel precursor for synthesis of pure boron nitride nanotubes. *Chemical Communications* (12) (2002a): 1290–1291.

Tang, C.C., Bando, Y., and Sato, T. Catalytic growth of boron nitride nanotubes. *Chemical Physics Letters* 362(3–4) (2002b): 185–189.

Tang, D.M., Ren, C.L., Wei, X.L. et al. Mechanical properties of bamboo-like boron nitride nanotubes by in situ TEM and MD simulations: Strengthening effect of interlocked joint interfaces. *ACS Nano* 5(9) (2011): 7362–7368.

Terauchi, M., Tanaka, M., Suzuki, K., Ogino, A., and Kimura, K. Production of zigzag-type BN nanotubes and BN cones by thermal annealing. *Chemical Physics Letters* 324(5–6) (2000): 359–364.

Terrones, M., Charlier, J.C., Gloter, A. et al. Experimental and theoretical studies suggesting the possibility of metallic boron nitride edges in porous nanourchins. *Nano Letters* 8(4) (2008): 1026–1032.

Tibbetts, K., Doe, R., and Ceder, G. Polygonal model for layered inorganic nanotubes. *Physical Review B* 80(1) (2009): 014102.

Wang, H., Taychatanapat, T., Hsu, A. et al. BN/graphene/BN transistors for RF applications. *IEEE Electron Device Letters* 32(9) (2011a): 1209–1211.

Wang, J.S., Kayastha, V.K., Yap, Y.K. et al. Low temperature growth of boron nitride nanotubes on substrates. *Nano Letters* 5(12) (2005): 2528–2532.

Wang, X.B., Pakdel, A., Zhi, C.Y. et al. High-yield boron nitride nanosheets from 'chemical blowing': Towards practical applications in polymer composites. *Journal of Physics-Condensed Matter* 24(31) (2012): 314205.

Wang, X.B., Zhi, C.Y., Li, L. et al. "Chemical blowing" of thin-walled bubbles: High-throughput fabrication of large-area, few-layered BN and C_X-BN nanosheets. *Advanced Materials* 23(35) (2011b): 4072–4076.

Wang, X.Z., Wu, Q., Hu, Z., and Chen, Y. Template-directed synthesis of boron nitride nanotube arrays by microwave plasma chemical reaction. *Electrochimica Acta* 52(8) (2007): 2841–2844.

Wang, Y., Shi, Z., and Yin, J. Boron nitride nanosheets: Large-scale exfoliation in methanesulfonic acid and their composites with polybenzimidazole. *Journal of Materials Chemistry* 21(30) (2011c): 11371–11377.

Wang, Z.L. and Hui, C. *Electron Microscopy of Nanotubes*. Boston, MA: Kluwer Academic Publishers, 2003.

Warner, J.H., Rummeli, M.H., Bachmatiuk, A., and Buchner, B. Atomic resolution imaging and topography of boron nitride sheets produced by chemical exfoliation. *ACS Nano* 4(3) (2010): 1299–1304.

Wei, X.L., Wang, M.S., Bando, Y., and Golberg, D. Tensile tests on individual multi-walled boron nitride nanotubes. *Advanced Materials* 22(43) (2010): 4895–4899.

Xiao, Y., Yan, X.H., Cao, J.X., Ding, J.W., Mao, Y.L., and Xiang, J. Specific heat and quantized thermal conductance of single-walled boron nitride nanotubes. *Physical Review B* 69(20) (2004a): 205415.

Xiao, Y., Yan, X.H., Xiang, J. et al. Specific heat of single-walled boron nitride nanotubes. *Applied Physics Letters* 84(23) (2004b): 4626–4628.

Xu, L.Q., Peng, Y.Y., Meng, Z.Y. et al. A Co-pyrolysis method to boron nitride nanotubes at relative low temperature. *Chemistry of Materials* 15(13) (2003): 2675–2680.

Xue, J.M., Sanchez-Yamagishi, J., Bulmash, D. et al. Scanning tunnelling microscopy and spectroscopy of ultra-flat graphene on hexagonal boron nitride. *Nature Materials* 10(4) (2011): 282–285.

Yamaguchi, M., Pakdel, A., Zhi, C.Y. et al. Utilization of multiwalled boron nitride nanotubes for the reinforcement of lightweight aluminum ribbons. *Nanoscale Research Letters* 8 (2013): 3.

Yong, J.C., Hong, Z.Z., and Ying, C. Pure boron nitride nanowires produced from boron triiodide. *Nanotechnology* 17(3) (2006): 786–789.

Yu, D.P., Sun, X.S., Lee, C.S. et al. Synthesis of boron nitride nanotubes by means of excimer laser ablation at high temperature. *Applied Physics Letters* 72(16) (1998): 1966–1968.

Yu, J., Chen, Y., Wuhrer, R., Liu, Z.W., and Ringer, S.P. In situ formation of BN nanotubes during nitriding reactions. *Chemistry of Materials* 17(20) (2005): 5172–5176.

Yu, J., Qin, L., Hao, Y.F. et al. Vertically aligned boron nitride nanosheets: Chemical vapor synthesis, ultraviolet light emission, and superhydrophobicity. *ACS Nano* 4(1) (2010): 414–422.

Yurdakul, H., Göncü, Y., Durukan, O. et al. Nanoscopic characterization of two-dimensional (2D) boron nitride nanosheets (BNNSs) produced by microfluidization. *Ceramics International* 38(3) (2012): 2187–2193.

Zeng, H.B., Zhi, C.Y., Zhang, Z.H. et al. "White graphenes": Boron nitride nanoribbons via boron nitride nanotube unwrapping. *Nano Letters* 10(12) (2010): 5049–5055.

Zhang, F., Chen, X., Boulos, R.A. et al. Pyrene-conjugated hyaluronan facilitated exfoliation and stabilisation of low dimensional nanomaterials in water. *Chemical Communications* 49(42) (2013): 4845–4847.

Zhang, H.Z., Yu, J., Chen, Y., and FitzGerald, J. Conical boron nitride nanorods synthesized via the ball-milling and annealing method. *Journal of the American Ceramic Society* 89(2) (2006): 675–679.

Zhang, X.F., Zhang, X.B., Vantendeloo, G., Amelinckx, S., Debeeck, M.O., and Vanlanduyt, J. Carbon nanotubes—Their formation process and observation by electron-microscopy. *Journal of Crystal Growth* 130(3–4) (1993): 368–382.

Zheng, M., Chen, X.M., Bae, I.T. et al. Radial mechanical properties of single-walled boron nitride nanotubes. *Small* 8(1) (2012): 116–121.

Zhi, C., Bando, Y., Tang, C., and Golberg, D. Boron nitride nanotubes. *Materials Science and Engineering: R: Reports* 70(3–6) (2010): 92–111.

Zhi, C.Y., Bando, Y., Tang, C.C., Kuwahara, H., and Golberg, D. Large-scale fabrication of boron nitride nanosheets and their utilization in polymeric composites with improved thermal and mechanical properties. *Advanced Materials* 21(28) (2009): 2889–2893.

Chapter 2

Boron Nitride Nanotubes and Nanoribbons Produced by Ball Milling Method

Lu Hua Li and Ying (Ian) Chen

CONTENTS

2.1 INTRODUCTION

In the wake of the discovery of carbon nanotubes (CNTs) in 1991 (Iijima 1991), the exploitation of other materials of a similar cylinder structure began. Though hexagonal boron nitride (hBN) has the closest structure to graphite, the first non-CNTs discovered were not boron nitride nanotubes (BNNTs) but metal dichalcogenide tungsten disulfide (WS_2) nanotubes, which were produced in 1992 (Tenne et al. 1992). The BNNT structure was theoretically predicted in 1994 (Blase et al. 1994; Rubio et al. 1994), and the successful experimental synthesis of cylindrical BNNTs was reported by a Berkley laboratory 1 year later (Chopra et al. 1995). Since then, BNNTs have become the most important non-CNTs and aroused great research attention, as evidenced by the increasing number of publications and patents (Golberg et al. 2007a).

Similar to CNTs, BNNTs can be viewed as hBN sheets rolled into tubes seamlessly with diameters less than 100 nm. BNNTs thus have superior mechanical strength and high thermal conductivity. A thermal vibration method was first used to determine the elastic property of individual BNNTs, and the calculated Young's modulus was 1.22 ± 0.24 TPa (Chopra and Zettl 1998). Electric-field-induced resonance and direct force measurements were also used to show Young's modulus of BNNTs with diameters of 34–100 nm, and the resultant values were in the range of 0.5–1.1 TPa (Golberg et al. 2007b; Suryavanshi et al. 2004). Recently, exceptional compressive/bending strength of 1.2 GPa was observed from BNNTs with diameters less than 10 nm (Huang et al. 2013). Because of the anisotropic deformation of BNNTs under axial compression and the higher interlayer interaction, the formation energy of dislocation in BNNTs and the thermodynamic yield limit of BNNTs under constant strain are larger than those of CNTs (Dumitrica et al. 2003; Srivastava et al. 2001). In spite of hBN's lower thermal conductivity than graphite, BNNTs have a thermal conductivity in the same order of that of CNTs (Xiao et al. 2004). This can be attributed to the dramatically reduced phonon scattering due to the phonon confinement in the nanotubes. BNNTs with a high isotopic content can have drastic increased thermal conductivity excelling most of CNTs (Chang et al. 2006; Stewart et al. 2009).

BNNTs also have many intriguing properties unavailable in CNTs. In contrast to the zero or small bandgap of CNTs, BNNTs have a bandgap of ~6 eV, which is almost independent of their chirality and diameter (Blase et al. 1994). Because of the wide bandgap, BNNTs are white and have been proposed as insulating cover on CNTs for electronics. In addition, the wide bandgap of BNNTs makes them promising deep ultraviolet (DUV) light emitters, especially both the electro-optical and photoluminescence quantum yield of BNNTs are predicted to be higher than hBN and CNTs (Chen et al. 2007b; Jaffrennou et al. 2007a,b; Li et al. 2010c). BNNTs are more thermally stable than CNTs. CNTs start to oxidize at 400°C and burn out at 700°C when heated in air (Ajayan et al. 1993; Golberg et al. 2001; Tsang et al. 1993); BNNTs can sustain 600°C–950°C (dependent on the nanotube structure) and the burnout happens at above 1200°C (Chen et al. 2004; Golberg et al. 2001). So, BNNTs are much more favorable for high-temperature applications. Thanks to the noncentrosymmetric structure, BNNTs are intrinsic piezoelectric (Mele et al. 2002; Nakhmanson et al. 2003), potentially useful in nanoelectromechanics. BNNTs can act as effective ion channel for selective permeation of certain ionic species and harness osmotic energy (Hilder et al. 2009; Siria et al. 2013).

Compared to CNTs, BNNTs are much more difficult to produce, which makes the synthesis an important part of BNNT research. BNNTs have been produced by various methods, including plasma arc discharge, laser heating, chemical vapor deposition (CVD), substitution reaction, thermal reaction, and ball milling and annealing. BNNTs were first prepared by the plasma arc discharge method, in which copper and hBN inserted tungsten were used as electrodes (Chopra et al. 1995). Later, different starting materials, anodes/cathodes, and environmental gases were experimented for improving the length, yield, and purity of the BNNT product (Cumings and Zettl 2000; Loiseau et al. 1996; Terrones et al. 1996b). It should be mentioned that plasma arc discharge was also the first method that synthesized single-walled BNNTs (Loiseau et al. 1996). Laser ablation was the second method that successfully produced BNNTs. This method requires extremely high temperature so that cubic boron nitride or hBN melts to form BNNTs (Golberg et al. 1996; Lee et al. 2001). Laser ablation was then modified to the so-called oven-laser ablation method that a furnace was included in the experimental setting so that extra heat up to 1200°C could help the growth of BNNTs (Yu et al. 1998). More recently, laser with higher repetition and short pulse was employed to increase the yield of BNNTs. Though the oven-laser ablation method is still the only method that can produce predominant single-walled BNNTs, the BNNTs produced by this way normally have a wide distribution of diameters (Golberg et al. 2003). Because both plasma arc discharge and laser ablation methods involve a high-temperature and high-energy process, the BNNT products are normally of a high crystallinity. However, the yield of these two methods is relatively low. In 2009, a new laser ablation method called the pressurized vapor/condenser method was proposed

(Smith et al. 2009). In this new method, boron was first vaporized at over 4000°C and then condensed on a cooled metal wire for nucleation before confronting N_2 gas to form BNNTs. The BNNTs have a natural alignment on the macroscopic scale similar to cotton fiber and can be easily spun to yarn.

CVD is the most popular method in CNT synthesis; however, this method is far less efficient in the case of BNNTs. The early CVD methods mainly produced BNNTs with a bamboo-like structure (Gleize et al. 1994; Lourie et al. 2000). High-temperature, high-pressure novel catalysts or other assisted conditions are required for the formation of cylindrical BNNTs by CVD (Kim et al. 2008; Ma et al. 2001; Xu et al. 2003). Another problem is that the common boron sources in the CVD such as borazine ($B_3N_3H_6$) and borane (BH_3) are highly toxic. The advantage of CVD method is that BCN nanotubes can be prepared when carbon-containing precursors are used. For example, the pyrolysis of acetonitrile (CH_3CN) and boron trichloride (BCl_3) or BH_3 with N,N-dimethylmethanamine ($N(CH_3)_3$) and pyridine (C_5H_5N) at 900°C–1000°C produced BCN nanotubes of bamboo structure (Sen et al. 1998; Terrones et al. 1996a). Also, CVD is one of the few methods that can grow BNNTs on substrates.

The substitution reaction method can produce BNNTs via chemical conversion of CNTs whose synthesis technique is much more sophisticated. CNTs are normally oxidized first to maximize the substitution reaction (Han et al. 1998), but unexpectedly, the BNNTs produced by substitution normally have a relatively good crystallization and uniform chiralities, thanks to the high-temperature recrystallization process after substitution (Golberg et al. 1999). The addition of metal catalysts to the substitution reaction can result in well-aligned BNNT ropes. Carbon residues are commonly found in the BNNTs produced by substitution, though the carbon contamination can be sometimes minimized (Han et al. 1998).

BNNTs of relatively high quantities can be produced by thermal reaction. For example, the heating of iron oxide (FeO), magnesium oxide (MgO) and boron (B) powder in ammonia (NH_3) gas could produce 0.2 g BNNTs each time (Lee et al. 2008; Zhi et al. 2005). This method can also grow BNNTs on alumina (Al_2O_3)-coated silicon (Si) wafer (Lee et al. 2010). Recently, lithium oxide (Li_2O) was added to B powder to produce high-quality BNNTs with diameters smaller than 10 nm (Huang et al. 2009).

Ball milling and annealing method, first proposed in 1999, combines mechanical and chemical processes to synthesize BNNTs (Chen et al. 1999a,b). In the method, B powder is ball milled to form active nanosized particles, and the heating of the particles in nitrogen-containing gases produces BNNTs. The large capacity of ball milling process makes BNNTs available for commercial applications for the first time. Now the ball milling and annealing method can produce BNNTs of larger quantities and higher purity, or grow BNNT films on different substrates and in situ unzip BNNTs to high-quality BN nanoribbons (BNNRs). This chapter focuses on the synthesis, property, and application of BNNTs and other BN nanomaterials produced by the ball milling and annealing method.

2.2   LARGE-QUANTITY SYNTHESIS OF BNNTs

Ball milling is a technique that utilizes mechanical impacts between milling balls and vial to break or mix materials. This technique was originally invented for the production of nickel and iron superalloys and is now applied to produce a variety of materials, such as nonequilibrium alloys, amorphous powders, and nanoparticles (Suryanarayana 2001). In 1999, Chen et al. (1999a,b,c) found that ball milling could dramatically reduce the formation temperature of BNNTs. In the early attempts, hBN powder was used as the starting material and milled in a rolling jar with several steel balls in nitrogen (N_2) gas. The milled hBN powder was then heated in N_2 atmosphere at up to 1300°C. Both cylindrical and bamboo-like BNNTs were found in the product. Though the yield of BNNTs was not high in the early attempts, the importance is that a novel solid-state route was established. This is of particular interest by considering the well-known high melting temperature of both hBN and B. To improve the yield of BNNTs, B powder was then adopted as the starting material (Chen et al. 1999b,c; Ji et al. 2006). B powder also has a high melting temperature of ~2000°C, but nitriding reactions could be introduced by heating the milled B in N_2 or NH_3 atmosphere so that higher yields of BNNTs could be obtained. There are several key factors that make the ball milling and annealing method efficient in BNNT production. During the ball milling, B particles were repeatedly ground under mechanical impacts to form nanosized particles full of structural defects. The highly deformed B nanoparticles become much more chemically reactive: they are self-ignited if exposed to air. Thanks to their high chemical reactivity, the milled B nanoparticles can react with nitrogen-containing gases more efficiently at lower temperatures. Besides, small BN structures are formed during the ball milling in NH_3, which can act as nucleation sites for BNNT growth during the following annealing treatment. Metal contaminations from milling balls and vial are inevitable in the ball milled B particles. These metal nanoparticles can act as catalysts for nanotube growth (Chen et al. 1999b; Velazquez-Salazar et al. 2005).

Although the steel nanoparticles from the ball milling process can play catalytic roles during BNNT growth, it is found that the addition of catalyst can greatly improve the production yield. One way to add additional catalyst is through mechanical mixing during ball milling (Li et al. 2012). For example, B powder with 10 wt.% ferric nitrate ($Fe(NO_3)_3 \cdot 9H_2O$) was ball milled in a stainless steel vial with several steel balls. The milling atmosphere was NH_3 gas at a pressure of 250 kPa. The subsequent heating of the milled powder in N_2 + 15% H_2 atmosphere at 1100°C produced a high density of BNNTs with diameters of 60–100 nm and lengths of 50–200 μm (see Figure 2.1). Both bamboo and cylindrical BNNTs were found in the product. Most of the BNNTs had metal catalysts at the tips, indicating a growth mechanism of vapor–liquid–solid. The energy dispersive x-ray (EDX) spectroscopy results confirmed the chemical composition of the product. When the milled powder was heated in NH_3 atmosphere at 1300°C, cylindrical BNNTs with diameters of 10 nm and lengths of 5–8 μm were synthesized (see Figure 2.2). The different nanotube structures grown in different nitrogen gases are due to the different nitriding reaction rates (Yu et al. 2007). In another example, nickel boride (NiB_x) powder was added to B particles during ball milling (Lim et al. 2007). The NiB_x powder was produced by the reduction of nickel chloride with potassium tetrahydroborate (KBH_4). The heating of the mixture of NiB_x and B from the ball milling treatment produced BNNTs with diameters of 20–40 nm and length of >250 nm.

Compared to the ball milling and annealing method without the addition of catalyst, the mechanical mixing of additional metal catalysts with B powder can greatly improve the yield of BNNTs (Li et al. 2012). As revealed by the thermogravimetric analyses (TGA) in Figure 2.3, the B powder milled with $Fe(NO_3)_3 \cdot 9H_2O$ (curve ii) has much higher mass increases than the B powder milled alone (curve i) during the heating up to 1300°C in N_2 atmosphere, representing a much higher rate of nitriding and higher yield of BNNT formation. The mechanism of the yield improvement was disclosed by the x-ray photoelectron spectroscopy (XPS) investigation of the two milled B powders. The B powder milled with $Fe(NO_3)_3$ had a 71% more N content than the B powder milled alone. The N was from the NH_3 milling atmosphere and

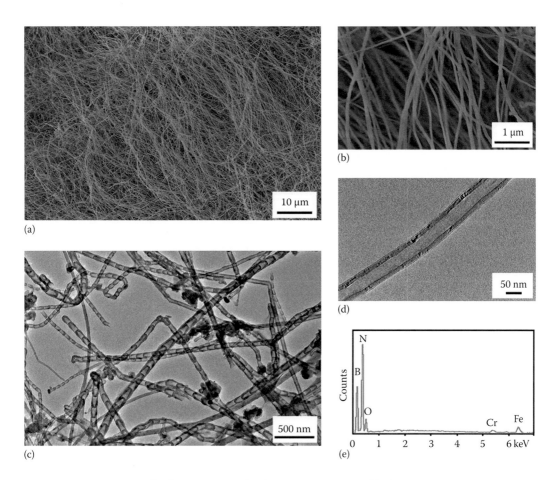

Figure 2.1 SEM images (a, b), TEM images (c, d), and EDX spectrum (e) of BNNTs. The BNNTs were produced by heating the $Fe(NO_3)_3$ milled powder in N_2 + 15% H_2 gas at 1100°C.

mainly in the form of B–N bonds. The higher content of N with the presence of $Fe(NO_3)_3$ can be attributed to the Fe decomposed from the $Fe(NO_3)_3$, because Fe is well known for its catalytic effect in nitriding as well as NH_3 decomposition. The decomposition of the $Fe(NO_3)_3$ to Fe was confirmed by the absence of NO_3 peak in the N 1s XPS spectrum and the appearance of Fe 2p XPS peaks. The XRD results also supported this proposition. The Fe that is more reactive than the steel particles from the milling balls and vial can provide additional and more efficient catalyst for BNNT growth. Besides, the higher amount of B–N bonds in the B powder milled with $Fe(NO_3)_3$ can create more nucleation sites for BNNT growth at lower temperatures.

In addition to metal nitrate, Li_2O can also be used in the ball milling–assisted catalyst mixing to enhance BNNT growth (Li et al. 2013a). Long cylindrical BNNTs with diameters of 10–50 nm were produced from the heating of B + Li_2O milled powder at 1200°C in NH_3 atmosphere (see Figure 2.4). The XPS studies of the product showed 47.94% B and 49.16% N, along with small amounts of Li and O, suggesting a high chemical purity. Similar to the case of $Fe(NO_3)_3$, the B + Li_2O powder after ball milling contained a large quantity of N and showed an increased rate of nitriding reaction during heating in nitrogen-containing gas.

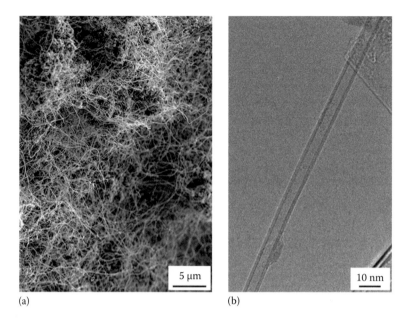

(a)   (b)

Figure 2.2   (a) SEM and (b) TEM images of the BNNTs produced in NH$_3$ gas at 1300°C.

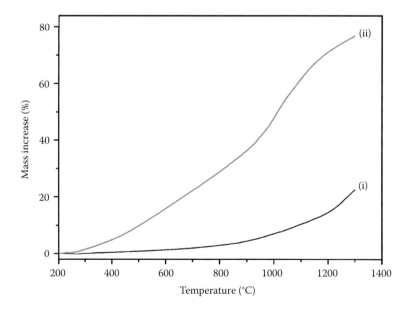

Figure 2.3   TGA curves of (i) B powder ball milled alone and (ii) B powder ball milled with 0.04 M Fe(NO$_3$)$_3$, heated up to 1300°C in N$_2$ gas with a temperature increasing rate of 10°C/min.

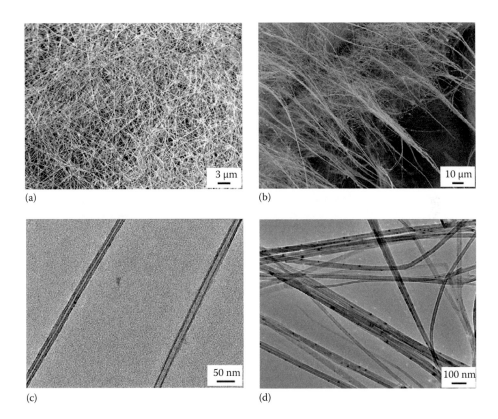

(a)   (b)

(c)   (d)

Figure 2.4   (a, b) SEM images of high-yield BNNTs and (c, d) TEM images of the multiwalled cylindrical BNNTs produced by heating B powder milled with Li_2O in NH_3 atmosphere. (Reprinted from *Microelectron. Eng.*, 110, Li, L., Liu, X., Li, L., and Chen, Y., High yield BNNTs synthesis by promotion effect of milling-assisted precursor, 256–259, Copyright 2013, with permission from Elsevier.)

2.3   BORON INK METHOD

Another way to add additional catalyst to the ball-milled B powder is via ethanol solution of metal nitrate (Li et al. 2008, 2010b). In the process, the B powder, milled in a similar way to the traditional ball milling and annealing method, is added to $Fe(NO_3)_3$ and $Co(NO_3)_2$ ethanol solution with the help of bath sonication to form a dark ink-like solution, therefore named B ink. The ink can be poured in a crucible and heated in nitrogen-containing gases for large-quantity production of BNNTs. The BNNTs produced from the B ink method are normally white. The color change from the dark brown of the B ink to the white suggests a high conversion rate of the B particles to BNNTs. Similar to the traditional ball milling and annealing method, the structure and size of the BNNT product can be controlled by atmosphere and temperature of heating. When $N_2 + 5\%$ H_2 gas was used during the heating of the B ink at 1000°C–1100°C, majority of the BNNTs have a bamboo structure, diameters of 50–80 nm, and lengths of 100–200 μm (see Figure 2.5). Sometimes, well-aligned BNNTs could be found at the edge of the crucible, which was believed to be caused by the gas flow. When the B ink was heated in NH_3 gas at 1300°C, high-quality cylindrical BNNTs with diameters of 2–10 nm (including double-walled BNNTs) and lengths of 3–5 μm were harvested (see Figure 2.6).

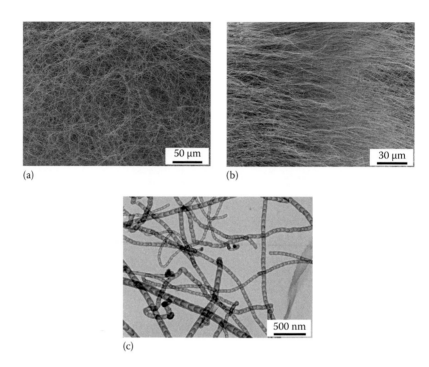

(a)   (b)

(c)

Figure 2.5   (a) SEM image showing high-purity BNNTs produced with Fe(NO$_3$)$_3$ in N$_2$ + 5% H$_2$ gas at 1100°C, (b) SEM image showing the aligned BNNTs found near the crucible edge, and (c) TEM image showing most of these nanotubes having a bamboo-like structure with a metal catalyst at the tip. (Li, L.H., Chen, Y., and Glushenkov, A.M., Synthesis of boron nitride nanotubes by boron ink annealing, *Nanotechnology*, 21, 105601, 2010b. Reproduced by permission of IOP Publishing. All rights reserved.)

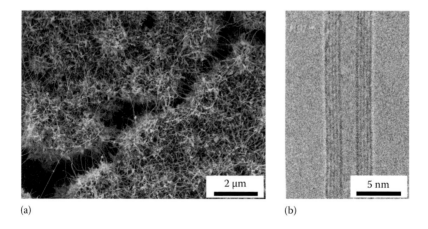

(a)   (b)

Figure 2.6   (a) SEM image of the BNNTs produced with Co(NO$_3$)$_2$ in NH$_3$ gas at 1300°C and (b) TEM image revealing the small diameter and cylindrical structure of the nanotubes. (Li, L.H., Chen, Y., and Glushenkov, A.M., Synthesis of boron nitride nanotubes by boron ink annealing, *Nanotechnology*, 21, 105601, 2010b. Reproduced by permission of IOP Publishing. All rights reserved.)

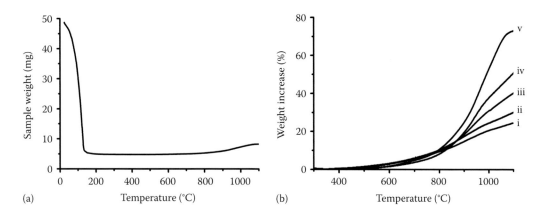

Figure 2.7 (a) TGA curve showing the B ink weight change during annealing to 1100°C in N_2 + 5% H_2 gas, (b) five TGA curves comparing the nitriding reaction rates of (i) ball-milled dry B particles, (ii) ball-milled B particles with pure ethanol, (iii) B ink with 0.013 M $Fe(NO_3)_3$ ethanol solution, (iv) B ink with 0.04 M $Fe(NO_3)_3$ ethanol solution, and (v) B ink with 0.1 M $Fe(NO_3)_3$ ethanol solution. All B inks were prepared by adding 100 mg of ball-milled B particles to 1 mL of solution. (Li, L.H., Chen, Y., and Glushenkov, A.M., Synthesis of boron nitride nanotubes by boron ink annealing, *Nanotechnology*, 21, 105601, 2010b. Reproduced by permission of IOP Publishing. All rights reserved.)

The heating process of the B ink involves three different stages, as identified by the TGA (see Figure 2.7a). The first stage is from room temperature to 145°C, in which the weight of the B ink plummets almost 90% due to the evaporation of the ethanol. No chemical change is present in this stage. The second stage is from 145°C to 400°C, in which the sample weight shows little variation. However, the metal nitrate decomposes to nanosized metal particles that can promote the growth of BNNTs. The third stage is from 400°C to 1100°C, in which the sample weight regains due to the nitriding reaction between the B particles and N_2 gas with the formation of BNNTs. The TGA also show that the nitriding rate highly depends on the amount of metal nitrate added (see Figure 2.7b). The B powder alone only has a 24.7% weight gain after 1100°C heating. The mixture of pure ethanol and the ball-milled B powder shows a weight gain of 30.2%, slightly higher compared to that of the dry B powder. The B ink with 0.013, 0.04, and 0.1 M $Fe(NO_3)_3$ shows much higher weight increases of 40.2%, 50.9%, and 73.2%, respectively. So, the $Fe(NO_3)_3$ in the B ink can dramatically enhance the nitriding reaction, giving rise to higher yield of BNNTs. This also indicates that the steel particles from the ball milling process are not numerous or efficient enough to transform all B particles into BNNTs. The ethanol in the B ink assists the higher yield of BNNTs via the protection of the active B particles from oxidation and the breakup of the B particle agglomeration via sonication. The ethanol seems not to leave any detectable C residues in the BNNT product if no excessive volume is used. Other metal nitrates, such as copper nitrate ($Cu(NO_3)_2$), magnesium nitrate ($Mg(NO_3)_2$), and aluminum nitrate ($Al(NO_3)_3$), can result in a similar enhancement in BNNT growth.

The large amount of high-purity BNNTs produced by the B ink method makes it possible to prepare BNNT-reinforced composite in bulk sizes. For example, BNNT-reinforced polyurethane (PU) composites in the form of cylinder with a diameter of 31.5 mm and heights of 16–20 mm were produced from around 1 g of BNNTs produced from B ink (see Figure 2.8) (Li et al. 2013c). Comparing to the neat PU, the 0.5 vol.% BNNT-reinforced composite shows a 38.2% enhancement of compressive modulus. The SEM investigations on the fracture surfaces of the composite revealed that the 0.5 vol.% BNNT composite had a good dispersion of BNNTs and relatively

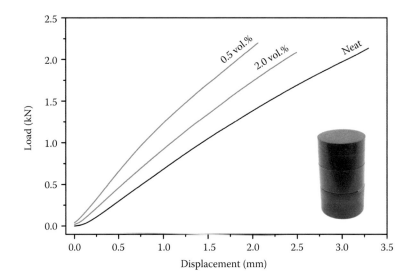

Figure 2.8   Compressive stress–strain curves of neat PU, 0.5 and 2.0 vol.% BNNT-reinforced PU composites. The inset shows a digital photo of the neat PU (bottom), 0.5 vol.% PU composite (middle), and 2.0 vol.% PU composite (top). (Reprinted from *Prog. Nat. Sci. Mater. Int.*, 23, Li, L.H., Chen, Y., and Stachurski, Z.H., Boron nitride nanotube reinforced polyurethane composites, 170–173, Copyright 2013, with permission from Elsevier.)

strong adhesion between BNNTs and the PU matrix. Most BNNTs tended to break rather than pull out so that the load could be transferred from the matrix to the reinforcing nanotubes.

2.4   GROWTH OF BNNT FILMS

The growth of nanotube films on various substrates and surfaces is vital for many applications, that is, nanoscale functional devices, antifouling, heat transfer, and biological and biomedical tests. The B ink can be used to grow high-density BNNT films in any desired pattern (Li et al. 2010a). The B ink–coating method normally consists of four steps (see Figure 2.9). First, ball milling is used to produce chemically active B nanoparticles in NH_3 gas. Second, B ink is prepared by mixing the ball-milled B nanoparticles with metal nitrate ethanol solution, as described earlier. Third, the B ink can be painted, sprayed, or inkjet printed on substrate with any desired pattern. The last step is to heat the substrate covered by the thin layer or pattern of the B ink at 950°C–1300°C in N_2, N_2 + 15% H_2, or NH_3 gas to grow BNNTs. The B ink–coating method is flexible and easy to use, as demonstrated by the BNNT coating in a complicated Chinese calligraphic pattern created by the B ink painting (see Figure 2.10). The BNNTs are found to grow only on the B ink–painted area. Regular BNNT patterns can be produced by mask coating. The B ink is compatible with commercial inkjet printer to prepare repeated BNNT patterns on smaller scales, because the B particles have an average size of 45 nm. Figure 2.11 shows an optical microscopy image and an SEM image of the BNNT film patterns in one group of letters that has a size of 1.5 mm × 2.0 mm. It is believed that BNNT film patterns with micrometer resolution are possible by using a high-resolution inkjet printer.

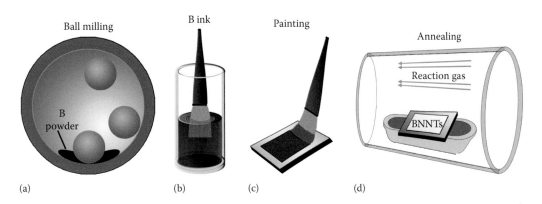

Figure 2.9   Schematic diagram showing the four steps involved in the B ink–painting method. (a) Ball milling of B powder in NH_3 to produce nanosized B particles. (b) Mixing ball-milled B particles, metal nitrate, and ethanol to form B ink. (c) Painting the B ink on the substrate. (d) Annealing of the painted substrate in a nitrogen-containing atmosphere to grow the BNNT film. (Li, L.H., Chen, Y., and Glushenkov, A.M., Boron nitride nanotube films grown from boron ink painting, *J. Mater. Chem.*, 20, 9679–9683, 2010a. Reproduced by permission of The Royal Society of Chemistry.)

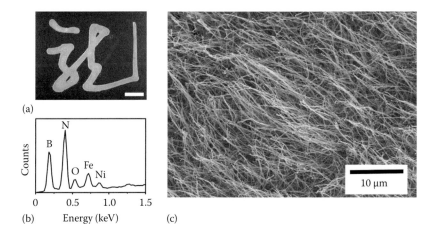

Figure 2.10   Patterned BNNT film produced by B ink painting. (a) Optical microscope photo of a white BNNT film with a complicated calligraphic pattern, scale bar 5 mm; (b) EDS spectra from the BNNT layer; (c) SEM image showing high-density BNNTs in the film. (Li, L.H., Chen, Y., and Glushenkov, A.M., Boron nitride nanotube films grown from boron ink painting, *J. Mater. Chem.*, 20, 9679–9683, 2010a. Reproduced by permission of The Royal Society of Chemistry.)

Using this method, BNNT films can be grown on objects of various and complex shapes for the first time. Three examples are shown in Figure 2.12. The steel mesh is homogeneously coated by high-density BNNTs, and its square openings less than 100 μm in size are not blocked; both the internal and external surfaces of the steel syringe are covered by BNNTs; BNNTs grow on the threads and the gaps of the tiny screw. Besides steel, silicon oxide–covered silicon (SiO_2/Si) and alumina (Al_2O_3) substrates were also successfully coated by BNNTs using the B ink method.

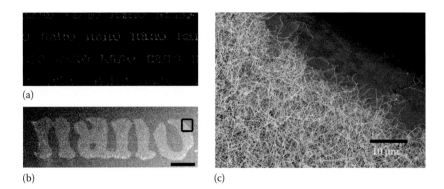

(a)

(b)   (c)

Figure 2.11   BNNT films produced by B ink printing. (a) Optical microscope photo of inkjet-printed arrays of BNNTs forming the word "nano"; (b) SEM image of one word in "Times New Roman" font, scale bar 500 mm; (c) higher-magnification SEM image of the top right of the letter "o," indicated by the square. (Li, L.H., Chen, Y., and Glushenkov, A.M., Boron nitride nanotube films grown from boron ink painting, *J. Mater. Chem.*, 20, 9679–9683, 2010a. Reproduced by permission of The Royal Society of Chemistry.)

The BNNTs in the film are bond to the substrate (see Figure 2.13). The adhesion force between the BNNT film and the steel substrate was measured by a dynamic mechanical analyser (DMA). In the measurement, a double-sized tape of 1 cm × 1 cm was glued to the BNNT film on one side (but not in contact with the steel substrate behind the BNNT film) and to a steel strip on the other side. The steel strip and the BNNT film substrate were pulled away by a DMA. A typical stress–strain curve is shown in Figure 2.13, and the average failure shear stress based on five BNNTs films was 4.76 ± 0.35 kPa.

2.4.1   Single-Band Deep Ultraviolet Light Emission

New properties were discovered from the high-density BNNT films grown by the B ink method. Single DUV light emission was recorded from a film of high-quality cylindrical multiwalled BNNTs with diameters less than 10 nm (Li et al. 2010c). The BNNT film was produced by heating the $Co(NO_3)_2$ containing B ink–coated SiO_2/Si substrate in NH_3 gas at 1300°C for 3 h (see Figure 2.14). The film shows strong light emission at 225 nm at both 10 K and room temperature after the surface adsorption is cleaned at an elevated temperature. The asymmetry of the DUV band suggests the existence of several excitonic peaks that have been observed from high-quality hBN crystals (Museur et al. 2008; Watanabe et al. 2004, 2009). The observed strong DUV emission can be credited to the good crystallinity and high purity of the BNNTs produced by the B ink method. In contrast, the films of bamboo-structured BNNTs that normally contain more defects and impurities showed a strong UV peak at around 300 nm and relatively weak DUV emission.

2.4.2   Surface Wetting Properties

The BNNT films produced by the B ink method are highly nonwettable to water. Among the eight BNNT films tested, four give equilibrium contact angles (CAs) around 170° and CA hysteresis values less than 10°; the other four had CAs around 160° and hysteresis larger than 10°. So, half of the BNNT films were superhydrophobic, and the other half were hydrophobic. The observed

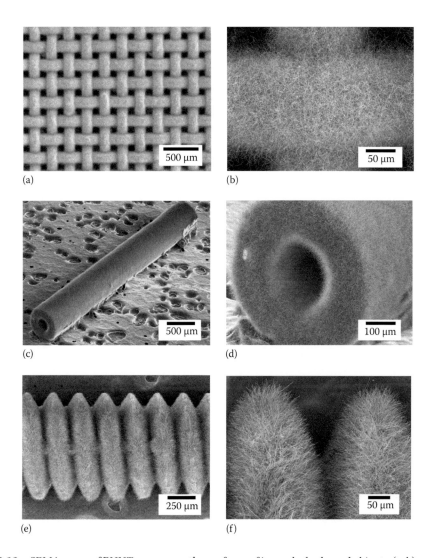

Figure 2.12 SEM images of BNNTs grown on the surfaces of irregularly shaped objects: (a, b) woven steel mesh of wires of 150 μm diameter, (c, d) steel syringe needle with outer diameter 510 μm and inner diameter 260 μm, and (e, f) steel screw of 0.9 μm diameter and the thread spacing of 200 μm. (Li, L.H., Chen, Y., and Glushenkov, A.M., Boron nitride nanotube films grown from boron ink painting, *J. Mater. Chem.*, 20, 9679–9683, 2010a. Reproduced by permission of The Royal Society of Chemistry.)

different wettability from the eight BNNT films was due to the different length of the BNNTs, which gave rise to different arrangements of the BNNTs on surface, that is, semierect or prostrate. The strong water repellence of the BNNT films suggests a Cassie state, in which water drops sit mostly on air. The wettability study gave many insights into the nature of BNNTs. Though hBN is hydrophilic (CA much less than 90°), BNNTs have a calculated Young's CA 85.6°–89.3° and hence almost hydrophobic. The hydrophobicity of BNNTs is mainly because of the positive Laplace pressure introduced by the very small nanotube diameter that opposes the disjoining pressure and prevents spreading of a water film on the nanotube surface. The apolar and polar components

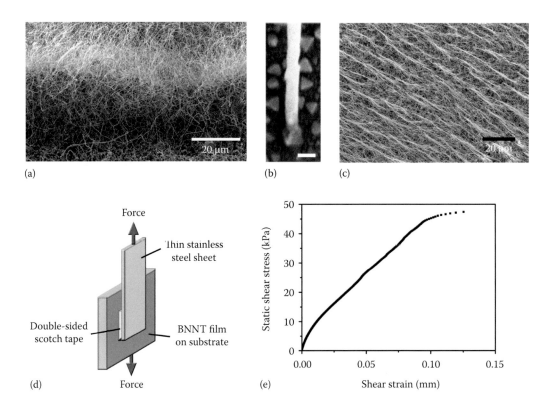

Figure 2.13 (a) Cross-sectional view of partly erect BNNTs in the film, (b) a BNNT securely attached to the steel substrate via a catalyst particle, (c) SEM image of BNNT film after high-pressure air blowing at a short distance, (d) schematic diagram showing the setup of the adhesion measurement, and (e) the stress–strain curve of a 1 cm² BNNT film under controlled shear force. (Li, L.H., Chen, Y., and Glushenkov, A.M., Boron nitride nanotube films grown from boron ink painting, *J. Mater. Chem.*, 20, 9679–9683, 2010a. Reproduced by permission of The Royal Society of Chemistry.)

(S_d and S_p) of the initial spreading coefficient (S_i) and the apolar components ($\gamma_S^d, \gamma_{SL}^d, \gamma_L^d$) of the surface tensions of BNNTs are also revealed and used for the calculation of the initial CA (θ_i). It was found that the BNNT films showed a 5° difference between the initial and equilibrium CAs. This indicates that when a dry BNNT had contact with a drop, water spontaneously spreads on its wall until the equilibrium CA is reached. In spite of the very slight water spreading on BNNTs before equilibrium state, no catastrophic seeping was found, and the water repellence was quite stable at <4 kPa for the prostrate BNNT films and at <9 kPa for the semierect BNNT films. In addition, the anti-wetting properties of BNNT films were not much affected by strong acids and bases (pHs). The examination of the water drop impact behavior on the BNNT films (see Figure 2.15) disclosed that the semierect BNNT films had an estimated first bounce-off restitution coefficient (ε) of 0.79. This means that water drops could rebound for several times on the surface of the semierect BNNT films and, in most cases, bounced off the films. Much less or no drop bounce was observed from the prostrate BNNTs. This suggests that the semierect BNNT films are highly efficient in self-cleaning. To confirm this, irregular-shaped copper particles with sizes of 2–10 μm were sprayed on both the semierect superhydrophobic and the prostrate hydrophobic BNNT films. The copper

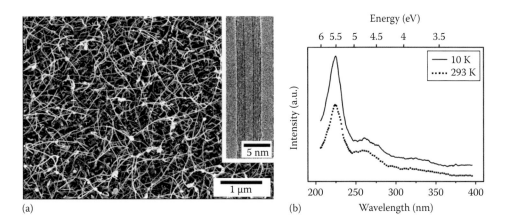

(a) (b)

Figure 2.14 (a) SEM image of the film of BNNTs with small diameters and TEM image (the inset) showing a well-crystallized multiwall structure and (b) PL spectra from the BNNT film at 10 and 293 K after vacuum annealing. (Reprinted with permission from Li, L.H., Chen, Y., Lin, M.Y., Glushenkov, A.M., Cheng, B.M., and Yu, J., Single deep ultraviolet light emission from boron nitride nanotube film, *Appl. Phys. Lett.*, 97, 141104, 2010. Copyright 2010, American Institute of Physics.)

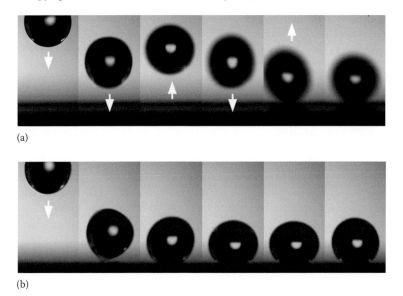

(a)

(b)

Figure 2.15 Dynamic behavior of a free-falling drop on (a) the semierect BNNT film and (b) the prostrate BNNT film with the arrows indicating the direction of movement of the drop. (Reprinted with permission from Li, L.H. and Chen, Y., Superhydrophobic properties of nonaligned boron nitride nanotube films, *Langmuir*, 26, 5135–5140, 2010. Copyright 2009, American Chemical Society.)

particles could be almost completely cleaned off the semierect BNNT films by using both static drop sitting on the film and fast moving drops rolling off the film (particles remained <2.5%); in contrast, the static drop cleaning was effective to remove about 50% of the copper particles from the prostrate BNNT films, though the fast rolling-off drops could have a good cleaning (see Figure 2.16). The different self-cleaning property was attributed to the different wettability as well

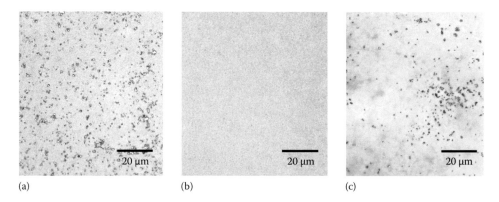

(a)   (b)   (c)

Figure 2.16   (a) Optical microscope image of copper particles smaller than 10 μm (red dots) spread on BNNT film for the self-cleaning test. (b) Semierect BNNT film thoroughly cleaned after a static drop rolled off. (c) Still many particles left on the prostrate BNNT film after the same cleaning. (Reprinted with permission from Li, L.H. and Chen, Y., Superhydrophobic properties of nonaligned boron nitride nanotube films, *Langmuir*, 26, 5135–5140, 2010. Copyright 2009, American Chemical Society.)

as particle adhesion between the copper particles and the BNNT films. So, BNNTs are potentially useful in many applications, such as water-repellent and self-cleaning coatings, microfluidic lab-on-a-chip systems for chemical, biological, and medical analyses, and low water friction surface for ocean freighters. It should be noted that BNNTs are superoleophilic or superwettable to oil (Li et al. 2010a). The CA of the BNNT film to paraffin oil was ~5°. The combination of the super-hydrophobicity and superoleophilicity of the BNNT films is useful in separating water and oil on microliter and liter volume scales. For example, the BNNT-coated mesh produced by the B ink method can support a water drop with a close-to-round shape but allow paraffin or any other oil to penetrate in a fraction of second.

2.5   ISOTOPICALLY ENRICHED ^{10}BNNTS

Compared to ^{11}B, ^{10}B is a more efficient neutron absorber with a neutron-capture cross section of 3800 barn. To reduce the secondary radiation, ^{10}BN rather than pure ^{10}B is used as neutron-cap-turing materials in nuclear reactor, spacecraft, and neutron therapy. ^{10}BNNTs also have a dramat-ically improved thermal conductivity than the natural BNNTs (Chang et al. 2006; Stewart et al. 2009). Natural hBN or BNNTs contain ~80.1% ^{11}B and less than 19.9% ^{10}B. Isotopically enriched ^{10}BNNTs are first produced by the ball milling and annealing method in which the enriched ^{10}B powder was ball milled for 150 h in NH_3 gas and then heated in NH_3 at 1100°C for 6 h (Yu et al. 2006). The ^{10}BNNTs have a cylindrical structure, diameters of around 10 nm, and lengths of several micrometers (see Figure 2.17). Rutherford backscattering spectrometry (RBS) revealed the pres-ence of ^{10}B, O, N, and Fe (see Figure 2.18b). The Fe was the steel contamination from the milling vial and balls. The secondary ion mass spectrometry (SIMS) spectrum confirmed the dominant presence of ^{10}B over ^{11}B with an intensity ratio of about 14:1 (see Figure 2.18c). The growth mecha-nism of ^{10}BNNTs is the same as that of natural BNNTs by the ball milling and annealing method. The ^{10}BNNTs exhibit a strong resistance to oxidation at high temperatures, as shown by the TGA that a weight increase caused by the oxidation of the BN nanotubes with the formation of boron oxides occurring at ~900°C in a pure oxygen atmosphere. This suggests that the ^{10}BNNTs have

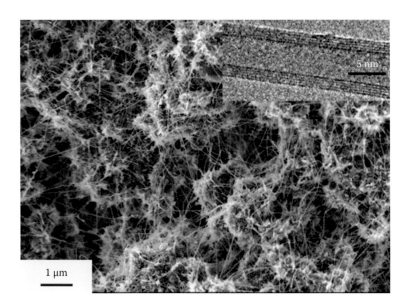

5 nm

1 µm

Figure 2.17   SEM image of ^{10}BNNTs produced by annealing of the milled ^{10}B powder at 1100°C for 6 h in NH$_3$. The inset shows a typical TEM image showing parallel fringes, indicating a well-crystallized, multi-walled, cylindrical structure. (Yu, J., Chen, Y., Elliman, R. G., and Petravic M., Isotopically Enriched ^{10}BN Nanotubes, *Adv. Mater.* 18, 2157, 2006. Copyright 2006, Wiley-VCH Verlag GmbH & Co. KGaA.)

a well-crystallized structure and few defects. The density of the ^{10}BNNTs was 1.85 g/cm^3, determined from pressed pellets. So, the ^{10}BNNTs that are lightweight, have excellent resistance to oxidation, are mechanically strong and efficient in neutron capturing are promising spacecraft materials.

2.6   DOPING OF BNNTS

Ball milling and annealing method can produce doped BNNTs via an in situ process. For example, in situ doping of europium (Eu) in BNNTs can be achieved by ball milling B powder with 1.0 at.% Eu (Chen et al. 2007a). During the ball milling, the Eu particles are broken to nanoparticles and homogeneously mixed with B. The mixture of B and Eu from the milling was heated in N$_2$ + 5% H$_2$ at 1200°C for 2 h to grow Eu-doped BNNTs (see Figure 2.19). The XRD analyses reveal that EuB$_6$ as an intermediate phase was formed at the temperature range of 800°C–1100°C first and further heating up to 1200°C generating the reaction between the EuB$_6$ phase and nitrogen with the formation of BNNTs and pure Eu. Eu was thought to precipitate mainly between the (002) BN basal planes rather than in the bamboo central voids or on the external surface. The estimated Eu concentration was 0.5 at.% and adjustable by varying the starting Eu content. Wide-bandgap semiconductors doped by rare-earth ions are an important type of luminescent materials, and the light emission properties of the Eu-doped BNNTs were tested by cathodoluminescence (CL). Compared to the undoped BNNTs, the Eu doping introduced a strong light emission peaked at 490 nm, which is attributed to the 4f^6 5d^1–4f^7 band emission of the Eu^{2+} ions in the BNNTs. The intensity of the visible-light emission peak can be controlled by changing the

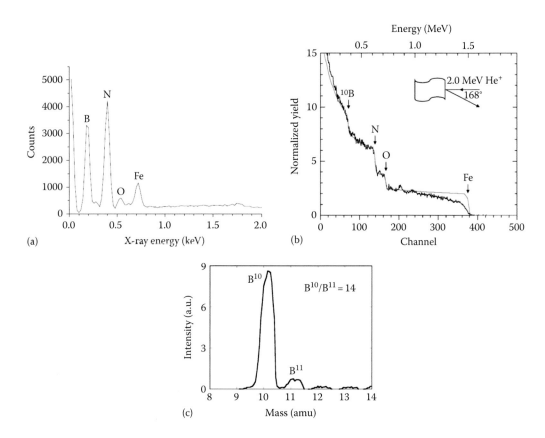

Figure 2.18   (a) EDX spectrum of the BNNTs; (b) Rutherford backscattering spectra taken from a ^{10}BNNT pellet using a 2.0 MeV He beam; (c) secondary ion mass spectrometry of the same ^{10}BN nanotube sample, showing the dominant presence of ^{10}B over ^{11}B. (Chen, H., Chen, Y., Li, C. P., Zhang, H., Williams, J. S., Liu, Y., Liu Z., Ringer S. P., Eu-doped Boron Nitride Nanotubes as a Nanometer-Sized Visible-Light Source, *Adv. Mater.* 19, 1845, 2007. Copyright 2006, Wiley-VCH Verlag GmbH & Co. KGaA.)

excitation voltage, and the light emission can be changed by varying the Eu content. The strong light emission in the visible range from the entire Eu-doped BNNTs may be useful in nanosized lighting sources and nanospectroscopy.

2.7   HIGH-QUALITY BN NANORIBBONS

BNNRs are strips of BN sheets with a very small width, which gives BNNRs many interesting properties. Zigzag-edged BNNRs can show a tunable bandgap and strong magnetism under transverse electric fields (Barone et al. 2008; Zhang et al. 2008). The electronic and magnetic properties of BNNRs can also be controlled by doping or edge terminations (Du et al. 2007; Park et al. 2008; Samarakoon et al. 2012; Tang et al. 2010). Therefore, BNNRs are of great interest to both fundamental research and practical application. However, the synthesis of nanoribbons is more challenging than that of nanotubes. Unzipping of nanotubes is a popular approach for nanoribbon production, in which two separate steps are normally required: nanotube synthesis and posttreatments of the as-grown nanotubes for unzipping. BNNRs have been previously

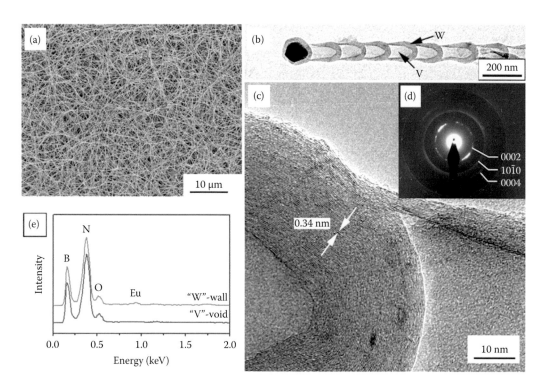

Figure 2.19 (a) SEM image of the doped BNNTs, indicating a high formation yield; (b) TEM image, showing a BN nanotube with a bamboo-type structure; (c) high-resolution lattice image of a bamboo tube, showing the cup-shaped BN layers and central void; (d) electron diffraction pattern; and (e) EDX spectra taken from different areas of a doped bamboo tube, indicating Eu doping in the tube-wall area. (Chen, H., Chen, Y., Li, C. P., Zhang, H., Williams, J. S., Liu, Y., Liu, Z., and Ringer, S. P., Eu-doped Boron Nitride Nanotubes as a Nanometer-Sized Visible-Light Source, *Adv. Mater.* 19, 1845, 2007. Copyright 2007, Wiley-VCH Verlag GmbH & Co. KGaA.)

produced by plasma etching and alkali metal intercalation, but the yields in both methods are low. Ball milling and annealing method was found capable of in situ unzipping BNNTs during their growth process (Li et al. 2013), in which the growth and unzipping of BNNTs happen simultaneously in one step: BNNTs are longitudinally unzipped during the BNNT synthesis so that BNNRs can be produced without the need of any posttreatment.

In the synthesis, amorphous B powder was ball milled at room temperature in NH_3 gas at a pressure 300 kPa. The milling lasted for 150 h at a speed of 110 rpm. The milled B powder was then placed at the bottom of an alumina crucible and covered by Li_2O powder at a molar ratio of 0.15–0.30:1. After the two powders are slightly mixed, the crucible with the powder was heated to 1200°C in NH_3 gas. After heating, the product is a white fluffy material. The SEM investigations showed that it contained a high density of needle-like structures, and the XRD of the product revealed a domination of hBN phases, along with a small amount of rhombohedral boron nitride and steel as the milling contamination.

The TEM investigations showed ~40% completely or partially unzipped BNNRs in the product (see Figure 2.20). The BNNRs are normally tens of nanometers in thickness and several micrometers in length. The BNNRs had well-defined stacking with little misalignment. Different from CNTs that normally show no preferential helicity, it is found that BNNTs are more likely to have a single

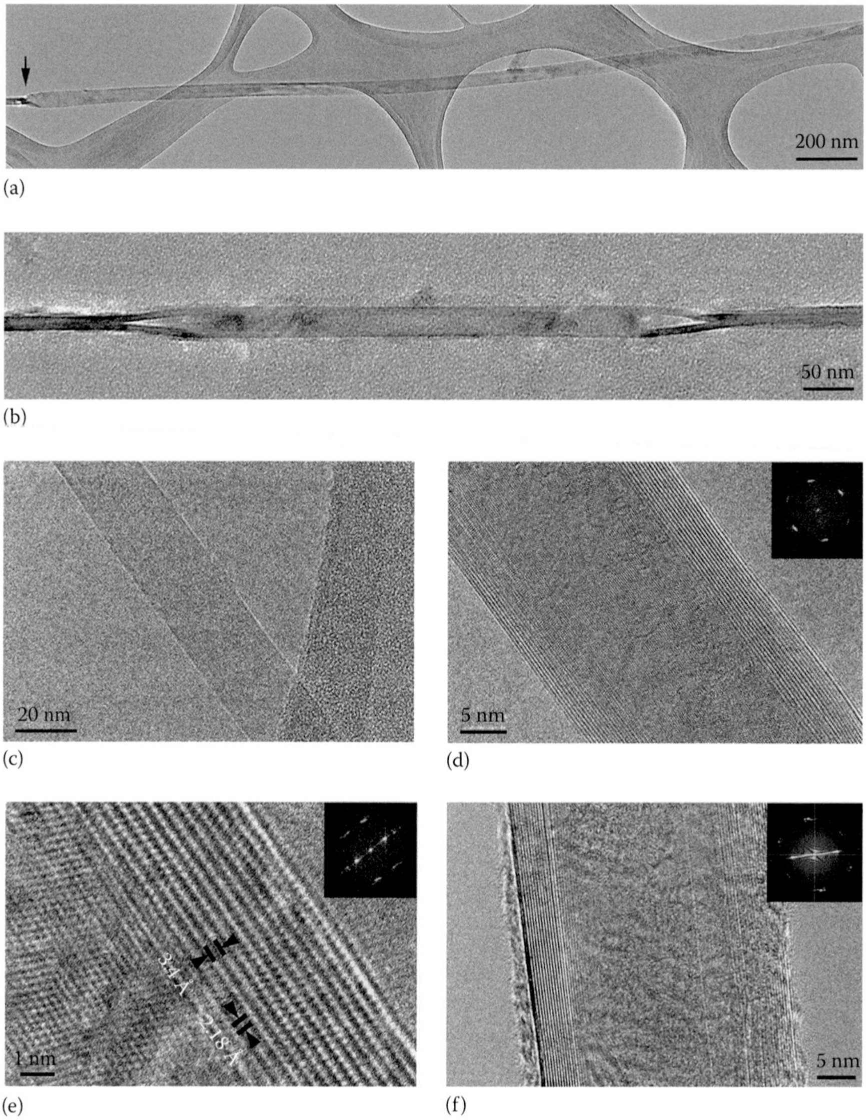

Figure 2.20 TEM images of (a) a partially unzipped BNNR (the unzipping stop site is marked with an arrow); (b) a partially unzipped BNNR with two unzipping stop sites; (c–f) two BNNRs at higher magnifications, with the FFTs inserted. (Li, L., Li, L. H., Chen, Y., Dai, X. J., Lamb, P. R., Cheng, B.-M., Lin, M.-Y., and Liu, X., 2013. High-Quality Boron Nitride Nanoribbons: Unzipping during Nanotube Synthesis, 52, 4212, 2013. Copyright 2013, Wiley-VCH Verlag GmbH & Co. KGaA.)

orientation in each tube (Golberg et al. 2007). It was found that the produced BNNRs also had a uniform helicity with a majority of zigzag orientation and the rest of armchair orientation. Because BNNRs are much more flexible than BNNTs, twisting and deformation can be easily observed. Sometimes, the twisting enables the distinction between BNNRs and BNNTs under SEM. Some of the BNNRs have narrowed and incompletely opened ends. This is because they are unzipped from

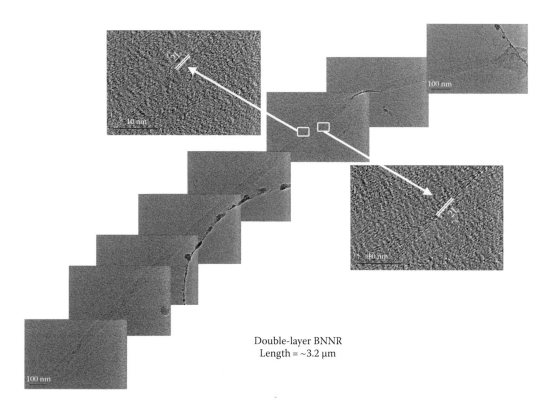

Double-layer BNNR
Length = ~3.2 μm

Figure 2.21 TEM images of a double-layer BNNR of ~3.2 μm long from bath sonication and centrifugation. The two insets show the edges of the BNNR.

BNNTs with flat caps, where the unzipping is more difficult to happen than the body of BNNTs. Bath sonication and centrifugation could increase the unzipping ratio to ~60% and produce few-layer BNNRs (see Figure 2.21). It was proposed that BNNTs were formed first during heating and the Li–NH$_3$ species expanded the BN interplanar spacing and achieved the in situ unzipping.

The BNNRs had a high chemical purity and good crystallinity, as revealed by the near-edge x-ray absorption fine structure (NEXAFS) spectroscopy. Figure 2.22 shows the comparison of the NEXAFS spectra of the BNNR product and a single-crystal hBN. Both samples show a strong and sharp π^* resonance at 192.0 eV and broad σ^* resonances above 197 eV, corresponding to core-level electron transitions to the unoccupied antibonding π^* and σ^* orbitals according to the dipole selection rule. The full widths at half maximum of the π^* resonances were quite similar for the two samples, and almost no oxygen-decorated nitrogen vacancies were present in the NEXAFS spectra of the BNNR-containing sample. No carbon- or lithium-related bonds were observable by the NEXAFS.

The photoluminescence excitation (PLE) measurements using a synchrotron vacuum ultraviolet spectroscopy showed different optical properties of the BNNRs to BNNTs (see Figure 2.23). In the case of BNNTs, the 5.98 eV (207.3 nm) peak can be attributed to the Frenkel excitons with lattice interaction, and the onset of the intensity increase at 6.20 eV (200 nm) can be assigned to free excitons and the bandgap of BNNTs. The BNNR plus BNNT sample showed a similar bandgap value but with a Frenkel exciton peak located at a lower energy of 5.91 eV (209.8 nm). The 0.07 eV difference indicates that the BNNRs have a larger exciton binding energy than the BNNTs. Because of the more localized excitons, BNNRs should be more efficient in light emission than BNNTs.

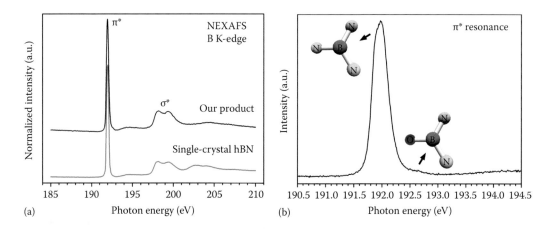

Figure 2.22   (a) Comparison of NEXAFS spectra of our product (black) and an hBN single crystal (gray) around the B K-edge region; (b) a high-resolution scan through the π* resonance of the product, showing only a very weak satellite peak at 192.6 eV representing one O atom occupying an N vacancy site in the planar ring. (Li, L., Li, L. H., Chen, Y., Dai, X. J., Lamb, P. R., Cheng, B.-M., Lin, M.-Y., and Liu, X., 2013. High-Quality Boron Nitride Nanoribbons: Unzipping during Nanotube Synthesis, 52, 4212, 2013. Copyright 2013, Wiley-VCH Verlag GmbH & Co. KGaA.)

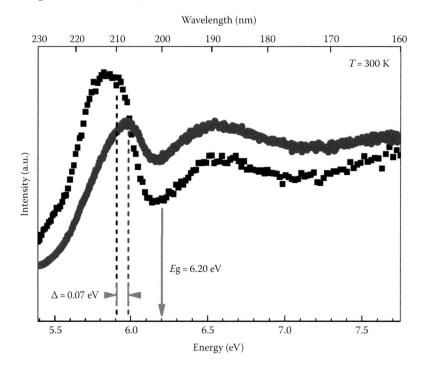

Figure 2.23   PLE spectra from a mixture of BNNRs and BNNTs (black) as well as BNNTs only (gray) at room temperature. (Li, L., Li, L. H., Chen, Y., Dai, X. J., Lamb, P. R., Cheng, B.-M., Lin, M.-Y., and Liu, X., 2013. High-Quality Boron Nitride Nanoribbons: Unzipping during Nanotube Synthesis, 52, 4212, 2013. Copyright 2013, Wiley-VCH Verlag GmbH & Co. KGaA.)

2.8   CONCLUSION AND CHALLENGES

With the progress in the last decade or so, the ball milling and annealing method is efficient for the production of a larger quantity of BNNTs in various forms with a high purity. Doped and isotopic BNNTs can be achieved by choosing different ball milling precursors. In addition to the powder-like form, BNNT films can grow on different surfaces of even complicated contours by the ink method, a modified ball milling and annealing method, which dramatically broadens the application of BNNTs such as in optoelectronics, self-cleaning, and biology. The BNNRs from the unzipping of BNNTs in one step will definitely contribute to the future research and application on these intriguing new nanomaterials. However, there remain some challenges. For example, the diameter of the BNNTs produced by the ball milling and annealing method is still nonuniform, which may relate to the nonuniform-sized steel nanoparticles from the milling process. Also, different from the high temperature and high-energy method for BNNT production, there has been no report on the synthesis of single-walled BNNTs from ball milling method. In addition, vertically aligned CNT films have been around for a long time, but BNNT film of a similar structure has not been achieved yet. These challenges should be the focuses of the future research on BNNTs by ball milling and annealing method.

REFERENCES

Ajayan, P. M., Ebbesen, T. W., Ichihashi, T., Iijima, S., Tanigaki, K., and Hiura, H. (1993), Opening carbon nanotubes with oxygen and implications for filling. *Nature*, *362*(6420), 522–525.

Barone, V. and Peralta, J. E. (2008), Magnetic boron nitride nanoribbons with tunable electronic properties. *Nano Letters*, *8*(8), 2210–2214.

Blase, X., Rubio, A., Louie, S. G., and Cohen, M. L. (1994), Stability and band-gap constancy of boron-nitride nanotubes. *Europhysics Letters*, *28*(5), 335–340.

Chang, C. W., Fennimore, A. M., Afanasiev, A., Okawa, D., Ikuno, T., Garcia, H., Li, D. Y., Majumdar, A., and Zettl, A. (2006), Isotope effect on the thermal conductivity of boron nitride nanotubes. *Physical Review Letters*, *97*(8), 085901.

Chen, H., Chen, Y., Li, C. P., Zhang, H. Z., Williams, J. S., Liu, Y., Liu, Z. W., and Ringer, S. P. (2007a), Eu-doped boron nitride nanotubes as a nanometer-sized visible-light source. *Advanced Materials*, *19*(14), 1845–1848.

Chen, H., Chen, Y., Liu, Y., Xu, C. N., and Williams, J. S. (2007b), Light emission and excitonic effect of boron nitride nanotubes observed by photoluminescent spectra. *Optical Materials*, *29*(11), 1295–1298.

Chen, Y., Chadderton, L. T., FitzGerald, J., and Williams, J. S. (1999a), A solid-state process for formation of boron nitride nanotubes. *Applied Physics Letters*, *74*(20), 2960–2962.

Chen, Y., Fitz Gerald, J. D., Williams, J. S., and Bulcock, S. (1999b), Synthesis of boron nitride nanotubes at low temperatures using reactive ball milling. *Chemical Physics Letters*, *299*(3–4), 260–264.

Chen, Y., FitzGerald, J., Williams, J. S., and Willis, P. (1999c), Mechanochemical synthesis of boron nitride nanotubes. *Journal of Metastable and Nanocrystalline Materials*, *2–6*, 173–178.

Chen, Y., Zou, J., Campbell, S. J., and Le Caer, G. (2004), Boron nitride nanotubes: Pronounced resistance to oxidation. *Applied Physics Letters*, *84*(13), 2430–2432.

Chopra, N. G., Luyken, R. J., Cherrey, K., Crespi, V. H., Cohen, M. L., Louie, S. G., and Zettl, A. (1995), Boron-nitride nanotubes. *Science*, *269*(5226), 966–967.

Chopra, N. G. and Zettl, A. (1998), Measurement of the elastic modulus of a multi-wall boron nitride nanotube. *Solid State Communications*, *105*(5), 297–300.

Cumings, J. and Zettl, A. (2000), Mass-production of boron nitride double-wall nanotubes and nanococoons. *Chemical Physics Letters*, *316*(3–4), 211–216.

Du, A. J., Smith, S. C., and Lu, G. Q. (2007), First-principle studies of electronic structure and C-doping effect in boron nitride nanoribbon. *Chemical Physics Letters*, *447*(4–6), 181–186.

Dumitrica, T., Bettinger, H. F., Scuseria, G. E., and Yakobson, B. I. (2003), Thermodynamics of yield in boron nitride nanotubes. *Physical Review B*, *68*(8), 085412.

Gleize, P., Schouler, M. C., Gadelle, P., and Caillet, M. (1994), Growth of tubular boron-nitride filaments. *Journal of Materials Science*, *29*(6), 1575–1580.

Golberg, D., Bando, Y., Eremets, M., Takemura, K., Kurashima, K., and Yusa, H. (1996), Nanotubes in boron nitride laser heated at high pressure. *Applied Physics Letters*, *69*(14), 2045–2047.

Golberg, D., Bando, Y., Kurashima, K., and Sato, T. (2001), Synthesis and characterization of ropes made of Bn multiwalled nanotubes. *Scripta Materialia*, *44*(8–9), 1561–1565.

Golberg, D., Bando, Y., Tang, C. C., and Zhi, C. Y. (2007a), Boron nitride nanotubes. *Advanced Materials*, *19*(18), 2413–2432.

Golberg, D., Costa, P. M. F. J., Lourie, O., Mitome, M., Bai, X. D., Kurashima, K., Zhi, C. Y., Tang, C. C., and Bando, Y. (2007b), Direct force measurements and kinking under elastic deformation of individual multiwalled boron nitride nanotubes. *Nano Letters*, *7*(7), 2146–2151.

Golberg, D., Han, W., Bando, Y., Bourgeois, L., Kurashima, K., and Sato, T. (1999), Fine structure of boron nitride nanotubes produced from carbon nanotubes by a substitution reaction. *Journal of Applied Physics*, *86*(4), 2364–2366.

Golberg, D., Rode, A., Bando, Y., Mitome, M., Gamaly, E., and Luther-Davies, B. (2003), Boron nitride nano-structures formed by ultra-high-repetition rate laser ablation. *Diamond and Related Materials*, *12*(8), 1269–1274.

Han, W. Q., Bando, Y., Kurashima, K., and Sato, T. (1998), Synthesis of boron nitride nanotubes from carbon nanotubes by a substitution reaction. *Applied Physics Letters*, *73*(21), 3085–3087.

Hilder, T. A., Gordon, D., and Chung, S. H. (2009), Boron nitride nanotubes selectively permeable to cations or anions. *Small*, *5*(24), 2870–2875.

Huang, Y., Bando, Y., Tang, C., Zhi, C., Terao, T., Dierre, B., Sekiguchi, T., and Golberg, D. (2009), Thin-walled boron nitride microtubes exhibiting intense band-edge UV emission at room temperature. *Nanotechnology*, *20*(8), 085705.

Huang, Y., Lin, J., Zou, J., Wang, M. S., Faerstein, K., Tang, C. C., Bando, Y., and Golberg, D. (2013), Thin boron nitride nanotubes with exceptionally high strength and toughness. *Nanoscale*, *5*(11), 4840–4846.

Iijima, S. (1991), Helical microtubules of graphitic carbon. *Nature*, *354*(6348), 56–58.

Jaffrennou, P., Barjon, J., Lauret, J. S., Maguer, A., Golberg, D., Attal-Tretout, B., Ducastelle, F., and Loiseau, A. (2007a), Optical properties of multiwall boron nitride nanotubes. *Physica Status Solidi B*, *244*(11), 4147–4151.

Jaffrennou, P., Donatini, F., Barjon, J., Lauret, J. S., Maguer, A., Attal-Tretout, B., Ducastelle, F., and Loiseau, A. (2007b), Cathodoluminescence imaging and spectroscopy on a single multiwall boron nitride nanotube. *Chemical Physics Letters*, *442*(4–6), 372–375.

Ji, F. Q., Cao, C. B., Xu, H., and Yang, Z. G. (2006), Mechanosynthesis of boron nitride nanotubes. *Chinese Journal of Chemical Engineering*, *14*(3), 389–393.

Kim, M. J., Chatterjee, S., Kim, S. M., Stach, E. A., Bradley, M. G., Pender, M. J., Sneddon, L. G., and Maruyama, B. (2008), Double-walled boron nitride nanotubes grown by floating catalyst chemical vapor deposition. *Nano Letters*, *8*(10), 3298–3302.

Lee, C. H., Wang, J. S., Kayatsha, V. K., Huang, J. Y., and Yap, Y. K. (2008), Effective growth of boron nitride nanotubes by thermal chemical vapor deposition. *Nanotechnology*, *19*(45), 455605.

Lee, C. H., Xie, M., Kayastha, V., Wang, J. S., and Yap, Y. K. (2010), Patterned growth of boron nitride nanotubes by catalytic chemical vapor deposition. *Chemistry of Materials*, *22*(5), 1782–1787.

Lee, R. S., Gavillet, J., de la Chapelle, M. L., Loiseau, A., Cochon, J. L., Pigache, D., Thibault, J., and Willaime, F. (2001), Catalyst-free synthesis of boron nitride single-wall nanotubes with a preferred zig-zag configuration. *Physical Review B*, *64*(12), 121405.

Li, L., Li, L. H., Chen, Y., Dai, X. J., Lamb, P. R., Cheng, B. M., Lin, M. Y., and Liu, X. W. (2013a), High-quality boron nitride nanoribbons: Unzipping during nanotube synthesis. *Angewandte Chemie International Edition*, *52*(15), 4212–4216.

Li, L., Li, L. H., Chen, Y., Dai, X. J. J., Xing, T., Petravic, M., and Liu, X. W. (2012), Mechanically activated catalyst mixing for high-yield boron nitride nanotube growth. *Nanoscale Research Letters*, *7*(1), 417.

Li, L., Liu, X., Li, L., and Chen, Y. (2013b), High yield BNNTS synthesis by promotion effect of milling-assisted precursor. *Microelectronic Engineering*, *110*(5226), 256–259.

Li, L. H. and Chen, Y. (2010), Superhydrophobic properties of nonaligned boron nitride nanotube films. *Langmuir*, *26*(7), 5135–5140.

Li, L. H., Chen, Y., and Glushenkov, A. M. (2010a), Boron nitride nanotube films grown from boron ink painting. *Journal of Materials Chemistry*, *20*(43), 9679–9683.

Li, L. H., Chen, Y., and Glushenkov, A. M. (2010b), Synthesis of boron nitride nanotubes by boron ink annealing. *Nanotechnology*, *21*(10), 105601.

Li, L. H., Chen, Y., Lin, M. Y., Glushenkov, A. M., Cheng, B. M., and Yu, J. (2010c), Single deep ultraviolet light emission from boron nitride nanotube film. *Applied Physics Letters*, *97*(14), 141104.

Li, L. H., Chen, Y., and Stachurski, Z. H. (2013c), Boron nitride nanotube reinforced polyurethane composites. *Progress in Natural Science-Materials International*, *23*(2), 170–173.

Li, L. H., Li, C. P., and Chen, Y. (2008), Synthesis of boron nitride nanotubes, bamboos and nanowires. *Physica E: Low-Dimensional Systems and Nanostructures*, *40*(7), 2513–2516.

Lim, S. H., Luo, J. Z., Ji, W., and Lin, J. (2007), Synthesis of boron nitride nanotubes and its hydrogen uptake. *Catalysis Today*, *120*(3–4), 346–350.

Loiseau, A., Willaime, F., Demoncy, N., Hug, G., and Pascard, H. (1996), Boron nitride nanotubes with reduced numbers of layers synthesized by arc discharge. *Physical Review Letters*, *76*(25), 4737–4740.

Lourie, O. R., Jones, C. R., Bartlett, B. M., Gibbons, P. C., Ruoff, R. S., and Buhro, W. E. (2000), CVD growth of boron nitride nanotubes. *Chemistry of Materials*, *12*(7), 1808–1810.

Ma, R., Bando, Y., and Sato, T. (2001), CVD synthesis of boron nitride nanotubes without metal catalysts. *Chemical Physics Letters*, *337*(1–3), 61–64.

Mele, E. J. and Kral, P. (2002), Electric polarization of heteropolar nanotubes as a geometric phase. *Physical Review Letters*, *88*(5), 056803.

Museur, L. and Kanaev, A. (2008), Near band-gap photoluminescence properties of hexagonal boron nitride. *Journal of Applied Physics*, *103*(10), 103520.

Nakhmanson, S. M., Calzolari, A., Meunier, V., Bernholc, J., and Nardelli, M. B. (2003), Spontaneous polarization and piezoelectricity in boron nitride nanotubes. *Physical Review B*, *67*(23), 235406.

Park, C. H. and Louie, S. G. (2008), Energy gaps and stark effect in boron nitride nanoribbons. *Nano Letters*, *8*(8), 2200–2203.

Rubio, A., Corkill, J. L., and Cohen, M. L. (1994), Theory of graphitic boron-nitride nanotubes. *Physical Review B*, *49*(7), 5081–5084.

Samarakoon, D. K. and Wang, X.-Q. (2012), Intrinsic half-metallicity in hydrogenated boron-nitride nanoribbons. *Applied Physics Letters*, *100*(10), 103107.

Sen, R., Satishkumar, B. C., Govindaraj, A., Harikumar, K. R., Raina, G., Zhang, J. P., Cheetham, A. K., and Rao, C. N. R. (1998), B-C-N, C-N and B-N nanotubes produced by the pyrolysis of precursor molecules over Co catalysts. *Chemical Physics Letters*, *287*(5–6), 671–676.

Siria, A., Poncharal, P., Biance, A. L., Fulcrand, R., Blase, X., Purcell, S. T., and Bocquet, L. (2013), Giant osmotic energy conversion measured in a single transmembrane boron nitride nanotube. *Nature*, *494*(7438), 455–458.

Smith, M. W., Jordan, K. C., Park, C., Kim, J. W., Lillehei, P. T., Crooks, R., and Harrison, J. S. (2009), Very long single- and few-walled boron nitride nanotubes via the pressurized vapor/condenser method. *Nanotechnology*, *20*(50), 505604.

Srivastava, D., Menon, M., and Cho, K. (2001), Anisotropic nanomechanics of boron nitride nanotubes: Nanostructured "skin" effect. *Physical Review B*, *63*(19), 195413.

Stewart, D. A., Savic, I., and Mingo, N. (2009), First-principles calculation of the isotope effect on boron nitride nanotube thermal conductivity. *Nano Letters*, *9*(1), 81–84.

Suryanarayana, C. (2001), Mechanical alloying and milling. *Progress in Materials Science*, *46*(1–2), 1–184.

Suryavanshi, A. P., Yu, M. F., Wen, J. G., Tang, C. C., and Bando, Y. (2004), Elastic modulus and resonance behavior of boron nitride nanotubes. *Applied Physics Letters*, *84*(14), 2527–2529.

Tang, S. B. and Cao, Z. X. (2010), Carbon-doped zigzag boron nitride nanoribbons with widely tunable electronic and magnetic properties: Insight from density functional calculations. *Physical Chemistry Chemical Physics*, *12*(10), 2313–2320.

Tenne, R., Margulis, L., Genut, M., and Hodes, G. (1992), Polyhedral and cylindrical structures of tungsten disulfide. *Nature*, *360*(6403), 444–446.

Terrones, M., Benito, A. M., MantecaDiego, C., Hsu, W. K., Osman, O. I., Hare, J. P., Reid, D. G. et al. (1996a), Pyrolytically grown BxCyNz nanomaterials: Nanofibres and nanotubes. *Chemical Physics Letters*, *257*(5–6), 576–582.

Terrones, M., Hsu, W. K., Terrones, H., Zhang, J. P., Ramos, S., Hare, J. P., Castillo, R. et al. (1996b), Metal particle catalysed production of nanoscale Bn structures. *Chemical Physics Letters*, *259*(5–6), 568–573.

Tsang, S. C., Harris, P. J. F., and Green, M. L. H. (1993), Thinning and opening of carbon nanotubes by oxidation using carbon-dioxide. *Nature*, *362*(6420), 520–522.

Velazquez-Salazar, J. J., Munoz-Sandoval, E., Romo-Herrera, J. M., Lupo, F., Ruhle, M., Terrones, H., and Terrones, M. (2005), Synthesis and state of art characterization of Bn bamboo-like nanotubes: Evidence of a root growth mechanism catalyzed by Fe. *Chemical Physics Letters*, *416*(4–6), 342–348.

Watanabe, K. and Taniguchi, T. (2009), Jahn-Teller effect on exciton states in hexagonal boron nitride single crystal. *Physical Review B*, *79*(19), 193104.

Watanabe, K., Taniguchi, T., and Kanda, H. (2004), Direct-bandgap properties and evidence for ultraviolet lasing of hexagonal boron nitride single crystal. *Nature Materials*, *3*(6), 404–409.

Xiao, Y., Yan, X. H., Cao, J. X., Ding, J. W., Mao, Y. L., and Xiang, J. (2004), Specific heat and quantized thermal conductance of single-walled boron nitride nanotubes. *Physical Review B*, *69*(20), 205415.

Xu, L. Q., Peng, Y. Y., Meng, Z. Y., Yu, W. C., Zhang, S. Y., Liu, X. M., and Qian, Y. T. (2003), A co-pyrolysis method to boron nitride nanotubes at relative low temperature. *Chemistry of Materials*, *15*(13), 2675–2680.

Yu, D. P., Sun, X. S., Lee, C. S., Bello, I., Lee, S. T., Gu, H. D., Leung, K. M., Zhou, G. W., Dong, Z. F., and Zhang, Z. (1998), Synthesis of boron nitride nanotubes by means of excimer laser ablation at high temperature. *Applied Physics Letters*, *72*(16), 1966–1968.

Yu, J., Chen, Y., Elliman, R. G., and Petravic, M. (2006), Isotopically enriched 10bn nanotubes. *Advanced Materials*, *18*(16), 2157–2160.

Yu, J., Li, B. C. P., Zou, J., and Chen, Y. (2007), Influence of nitriding gases on the growth of boron nitride nanotubes. *Journal of Materials Science*, *42*(11), 4025–4030.

Zhang, Z. H. and Guo, W. L. (2008), Energy-gap modulation of Bn ribbons by transverse electric fields: First-principles calculations. *Physical Review B*, *77*(7), 075403.

Zhi, C. Y., Bando, Y., Tan, C. C., and Golberg, D. (2005), Effective precursor for high yield synthesis of pure Bn nanotubes. *Solid State Communications*, *135*(1–2), 67–70.

Chapter 3

Boron (Carbon) Nitride Nanomaterials

Synthesis, Porosity, and Related Applications

Weiwei Lei, Dan Liu, Si Qin, David Portehault, and Ying (Ian) Chen

CONTENTS

3.1 INTRODUCTION

Similarities between carbon and boron nitride (BN) are striking. Indeed, BN is isoelectronic to carbon. It exhibits different polymorphs, including a wurtzite structure (w-BN), a sphalerite (cubic) structure (c-BN), and a hexagonal structure (h-BN). The two latter forms are strongly related to the two common carbon allotropes diamond and graphite, respectively. h-BN is the most stable phase at ambient conditions. Despite strong structural relationships between h-BN and graphite, they are not strictly speaking isomorphs because they slightly differ in the stacking between (001) layers—they exhibit strongly different properties. The reason of this discrepancy lies in the bonding scheme of BN: the strong polarity or partial ionicity of B–N bonds reduces the electron mean free path drastically, so that h-BN is a wide band gap semiconductor, exhibiting related UV luminescence. h-BN also possesses very good thermal conductivity, high mechanical strength, outstanding chemical inertness, thermal stability, and resistance to oxidation. Because of all these peculiar properties and its ease of fabrication, h-BN is by far the most studied polymorph of BN, and all the studied BN nanostructures are somewhat related to this structure. The development of BN nanomaterials has been strongly linked to the advent of *carbon nanoscience*. Indeed, BN nanotubes were discovered soon after carbon nanotubes (Chopra et al. 1995); BN fullerenes were identified a bit more than 10 years after C_{60} (Golberg et al. 1998, Stephan et al. 1998), and BN nanosheets, containing one (so-called white graphene) or a few layers of h-BN, were obtained at the same time as graphene (Corso et al. 2004, Novoselov et al. 2005). Still now, the increasing research effort on BN nanostructures is most often strongly related to research on carbon nanostructures. This current *way of thinking* BN research has two consequences.

First, BN is primarily considered as a compound that could modify, extend, or tune the properties of carbon nanomaterials, essentially nanotubes and graphene or nanosheets. The electrical properties and band gap opening in intimate mixes of graphene and BN sheets in sandwich structures are one example that evidences the impact that BN nanostructures can have on carbon ones (Giovannetti et al. 2007, Kim et al. 2013). The evolution from these *simple* structures, using Van der Waals interactions between carbon and h-BN domains, toward the development of complex objects with strong covalent bonds between carbon and h-BN nano-areas, is an evident extension of this concept of *hybridization*. This paradigm is now at the origin of a growing number of works on boron carbon nitrides (BCNs), where the content and the (nano)size of carbon and BN regions are of outmost importance for the control of the properties of the resulting material (Rocha et al. 2011).

Second, most studies target the properties that are already highly relevant for carbon nanostructures, among all electric and thermal transport behaviors, for implementation in various types of optoelectronic devices. However, from the carbon side, the range of properties and potential applications has quickly expanded during the last 10 years, when materials chemists took over carbon nanomaterials for energy storage (Simon and Gogotsi 2008) or environmental remediation (Sohn et al. 2012), for instance. B(C)N nanostructures are comparatively in their infancy and confined to few topics as optoelectronics. Even if those are of utmost importance, one could wonder if, like for carbon nanomaterials, there are other fields worth of investigation.

The aim of this short review is to provide the reader with some elements to answer this apparently simple question from the point of view of materials chemists: "what's new with BN nanomaterials compared to carbon nanostructures?" For this purpose, we will focus on original properties with BN and BCN nanomaterials, which have been investigated in fields such as environmental remediation, visible light fluorescence, energy storage, and gas storage, while summarizing briefly the electronic properties of these nanostructures, which are extensively discussed in other chapters of the current book and other comprehensive reviews (Terrones et al. 2002, Arenal et al. 2010). Because the properties are strongly related to the synthesis processes used for the design of the materials, a big part of this chapter is dedicated to the synthesis of BN and BCN nanostructures. Indeed, a strong limitation to scope widening in terms of properties in the field of B(C)N materials is the development of novel synthetic approaches that will allow exquisite control over the composition, the local structure, the nanostructure (shape, size), and assembly of B(C)N nano-objects. Although not directly in the scope of the present book dedicated to 1D and 2D nanostructures, it appears that 0D and 3D nano-objects have been mostly overlooked in recent reviews on B(C)N materials, even in the few feature articles describing parts of the activity in the field of boron (carbon) nitrides (Kumar et al. 2013). Yet, the properties of 3D B(C)N nanostructured materials are appealing more and more research teams who are investigating electrical and storage behaviors, strongly related to those of the 1D and 2D nanostructures. Therefore, discussing B(C)N properties cannot be done without taking into account all the known nanostructures. For this reason, we present the synthesis and properties of nanoparticles and porous micro-/mesostructures, together with those of more common 1D and 2D materials. Since other chapters of the present book are especially dedicated to BN nanotubes and nanosheets, we briefly describe their most important features to shed light on the whole range of available nanostructured B(C)N materials.

3.2   BORON NITRIDE NANOMATERIALS

3.2.1   Nanostructures of Boron Nitride

h-BN attracts more and more attention because of its unique properties including a wide energy band gap, electrical insulation, UV photoluminescence (PL), high thermal conductivity and stability, high resistance to oxidation, and chemical inertness (Paine and Narula 1990, Rudolph 1994). h-BN exhibits various nanostructures, such as nanotubes (Chen et al. 1999), nanocapsules (Oku et al. 2001), hollow spheres (Chen et al. 2004), nanocages (Pan et al. 2005, Bernard et al. 2011), hollow nanoribbons (Chen et al. 2008), nanomesh structures (Corso et al. 2004, Goriachko et al. 2007), and porous structures (Lei et al. 2013a). The present section describes the synthesis and some properties of the most common pure h-BN nanostructures, although other exotic nano-objects were reported, such as fullerenes (Golberg et al. 1998, Stephan et al. 1998) and nanocages (Chen et al. 1999, Oku et al. 2001, Chen et al. 2004, Pan et al. 2005, Wood et al. 2006).

3.2.1.1   Synthesis of BN Nanotubes and Nanoribbons

Arenal et al. (2010) and Pakdel et al. (2012b), among others, reviewed extensively the structures and synthesis of BN nanotubes. Readers are directed to these detailed reports for further information on these objects. Briefly, BN nanotubes can be synthesized by high-temperature processes, above 2000°C, including arc discharge and laser ablation (Chopra et al. 1995). A second category of procedures relies on lower temperatures, using carbothermal synthesis (Han et al. 1998), ball milling followed by heat treatment (Chen et al. 1999), and chemical vapor deposition (CVD; Lourie et al. 2000, Dai et al. 2007).

At the frontier between 1D and 2D nanostructures, BN nanoribbons are also attracting interest due to modified optoelectrical properties compared to nanotubes and nanosheets (Zeng et al. 2010, Yu et al. 2011). Chen et al. (2008) proposed to couple the CVD growth of h-BN with nanoribbons of ZnS acting as templates. After the deposition of BN layers over ZnS nanostructures at 1220°C under Ar/NH$_3$/H$_2$ atmosphere, a temperature increase up to 1350°C under vacuum led to the evaporation of the template and the recovery of BN hollow nanoribbons. Later, Zeng et al. (2010) obtained BN nanoribbons by the encapsulation in PMMA of a longitudinal section of BN nanotubes, followed by etching of the other *hemisphere* to *open* the nanotubes, and then removal of the PMMA protecting coating. Nanoribbons were then released. A close alternative to this process was proposed by nanotube *unzipping* provoked by potassium or lithium intercalation (Erickson et al. 2011, Li et al. 2013).

3.2.1.2 Synthesis of BN Nanosheets

Synthesis methods still remain a central focus in 2D BN nanomaterial, since different synthesis routes may result in products with very different physical structure and chemical composition, thus affecting the morphology and properties of the products. To achieve the desired 2D layered nanostructure—known as nanosheet, nanoflake, or nanolamina—routes distinct from the traditional ways must be introduced. Depending on the synthesis processes, the methods can be categorized into three groups.

One group of methods produces nanosheets by *breaking the van der Waals interactions* between layers of bulk or multilayer h-BN. Recently, few-layered BN nanosheets have been prepared by ultrasonication, mechanical cleavage, and chemical exfoliation from bulk BN crystals (Han et al. 2008, Pacile et al. 2008, Zhi et al. 2009) by using polar organic solvents (Zhi et al. 2009), water (Du et al. 2013), or molten hydroxides (Li et al. 2013).

Another group of methods produces nanosheets by *CVD* with the use of various chemical precursors such as BCl_3/NH_3 (Rozenberg et al. 1993), $B_3N_3H_6/N_2$ (Shi et al. 2010), or NH_3–BH_3/Ar–H_2 (Song et al. 2010). Typical temperatures involved in CVD processes for BN nanostructures are in the range 900°C–1300°C. The as-grown BN nanosheets consist in few layers with a large band gap and have unique mechanical properties (Shi et al. 2010, Song et al. 2010).

The third group of methods produces nanosheets by the *bottom-up chemical processes*, which seem to be the most suitable ways due to the ease for scale-up, ability to tune product's structure and morphology, and after all, the simplicity and versatility. Lei et al. (2011) demonstrated a simple, high-yield chemical synthesis process by using a molten salt as a reaction medium. The LiCl and KCl eutectic mixture melts at a low temperature (~350°C) compared to those where the reaction between B and N precursors usually occurs (~900°C for CVD processes). The molten salt then provides a liquid medium where the reaction can take place, thus enhancing the reaction kinetics and enabling the formation of h-BN at lower temperature than 900°C. It also remains stable at the temperatures of 600°C–800°C required in the BN synthesis. Sodium borohydride ($NaBH_4$) as boron source and urea (CH_4N_2O) as nitrogen source with a molar ratio of 1:1 were finely mixed with the LiCl/KCl salt. Then, the mixture was heated at 700°C for 2 h under nitrogen flow. After cooling down to room temperature, the LiCl/KCl salt could be easily removed by washing the powder with water. Samples produced by this method contain a few layers of BN nanosheets confirmed by atomic force microscopy (AFM), transmission electron microscopy (TEM), and HRTEM results (see Figure 3.1).

3.2.1.3 Synthesis of Porous BN Nanosheets

Perfect BN nanosheets with high crystallinity and high purity always seem to be the major interest in this area. However, by introducing *defects* or *holes*, BN nanosheets with a porous nanostructure can be fabricated and possess some desirable properties that the defect-free nanosheets lack, such as super high surface area, more edges as chemical reactive sites, higher superhydrophobicity, and more fluid or gas diffusion pathways. Porous BN nanosheets with these unique properties may shed light on many innovative application fields. Wang et al. (2009) prepared h-BN micromeshes by using B_2O_3, Mg, and NaN_3 as starting materials. In addition, h-BN nanomeshes were also formed by self-assembly through thermal decomposition of borazine $(HBNH)_3$ at about 800°C on a transition-metal surface, such as Rh(111) or Ru(0001) (Corso et al. 2004, Goriachko et al. 2007). Meng et al. (2010) prepared meshy BN at 550°C by using $CS(NH_2)_2$ and $NaBH_4$ as raw materials. More recently, Lei et al. (2013a) showed a method to fabricate BN nanosheets with high-porosity structure. Boron oxide and guanidine chloride were used as starting materials with a molar ration of 1:5. After mixing in methanol and drying the resulting suspension, white powders containing a salt of boron oxide and guanidine chloride were recovered and heated at 1100°C for

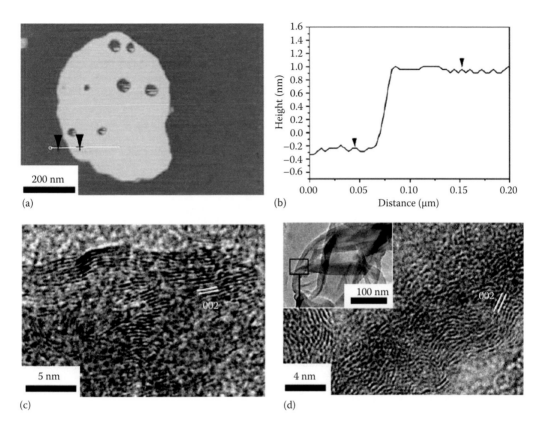

Figure 3.1 AFM (a) picture, corresponding line-scan profile (b), and HRTEM and TEM (c, d) images of BN nanosheets. (From Lei, W.W. et al., *J. Am. Chem. Soc.*, 133, 7121, 2011.)

2 h under nitrogen/hydrogen gas flow. BN nanosheets produced by this method have a layered porous BN structure with many holes of diameters ranging from 20 to 100 nm, which is confirmed by SEM, TEM, and AFM images (see Figure 3.2). The nanosheets also show a high surface area reaching 1427 m^2 g^{-1}. The porous structure of the nanosheets might arise from gas bubbles, so-called dynamic templates, generated by the decomposition of guanidine chloride.

3.2.1.4 Synthesis of BN Mesostructures

Recently, BN mesostructures have also gathered a lot of attention due to their porous structure and high specific surface area, particularly for the design of catalyst supports with high thermal stability (Postole et al. 2005, Lin et al. 2009). Dibandjo et al. (2005a) and Schlienger et al. (2012) reported the fabrication of h-BN mesostructures by *hard templating* (see Figure 3.3a, b). In a typical synthesis process, borazines (tri(methylamino)borazine, tri(chloro)borazine, polyborazylene, etc.) are used as boron and nitrogen sources. Different porous materials (silica, zeolites, and carbonaceous foams) can be impregnated by the molecular precursors. The mixture is usually heated above 1200°C under nitrogen gas flow. When silica or aluminosilicate templates are used, borazines decompose into BN wrapping up the surface of the template material. The BN mesostructures, recovered as inverse replica of the templates, are collected upon further removal of the template material. When carbon templates are used, depending on the nature of precursors, boron and nitrogen can substitute carbon, so that direct replica can be obtained, as described later in

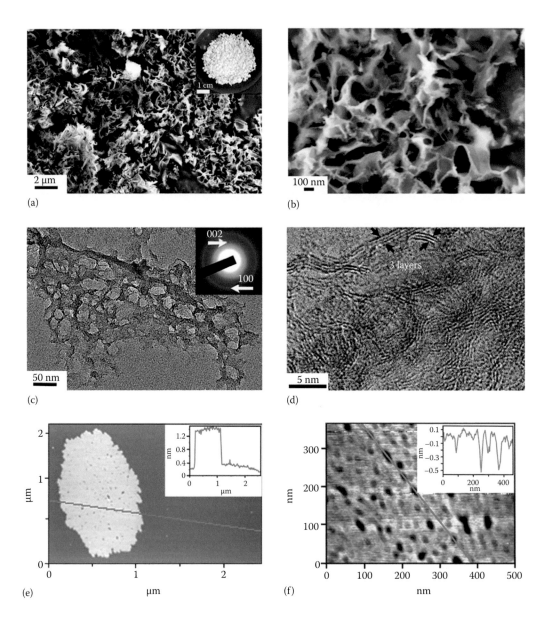

Figure 3.2 (a) Low-magnification SEM image of the porous BN nanosheets. The inset shows the typical white powder obtained after synthesis. (b) High-magnification SEM image revealing the porous nanosheet structure. (c) TEM image of a single nanosheet showing holes on the nanosheet, inserted SAED pattern indicating a layered BN structure. (d) HRTEM image of the edge folding of a nanosheet with three BN layers. (e) AFM image of a nanosheet and the inserted height profiles showing typical size and thickness of a single nanosheet. (f) High-magnification AFM image and the corresponding height profiles inserted. The porous structure can be seen clearly. (From Lei, W.W. et al., *Nat. Commun.*, 4, 1777, 2013a.)

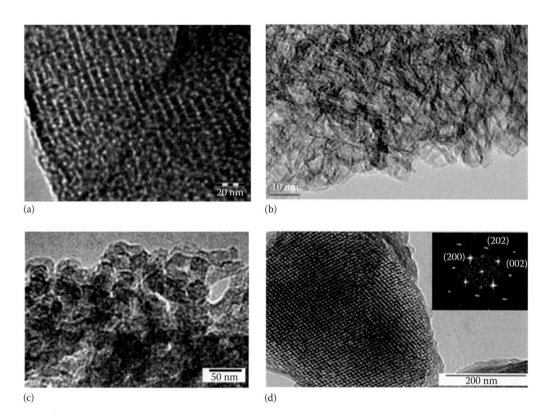

(a)

(b)

(c)

(d)

Figure 3.3   Representative TEM images of the BN mesostructures synthesized by two synthesis methods: hard templating (a, b), (From Dibandjo, P. et al., *Adv. Mater.*, 17(5), 571, 2005a; and (From Schlienger, S. et al., *Chem. Mater.*, 24(1), 88, 2012.); and soft templating (c, d), (From Dibandjo, P. et al., *J. Eur. Ceramic Soc.*, 27, 313, 2007a.) and (From Dibandjo, P. et al., *Microporous Mesoporous Mater.*, 92, 286–291, 2006.)

Section 3.3.1.4 (Vinu et al. 2005). Han et al. (2004) used this mechanism for fabricating a BCN replica of an activated carbon. After calcination under air to remove remaining carbon, a porous h-BN material was obtained, with surface area close to 170 $m^2\ g^{-1}$. Interestingly, borazine-derived (macro) molecules are precondensed BN precursors. Upon heating, polymerization of the BN network readily occurs, so that the B, N molecular source does not have time to react with the carbon template. As a result, inverse replica of carbon templates is recovered from borazine-derived precursors after carbon removal by calcination in air. By using various templates, the BN porous structure could be varied extensively in terms of specific surface area, pore morphology and size, and mesoporous structure: surface areas ranging from 53 to 865 $m^2\ g^{-1}$, pore volume ranging from 0.24 to 0.8 $cm^3\ g^{-1}$, pore size from 2 to 50 nm, hexagonal, cubic mesostructures, zeolite-derived microporous structures were obtained (Dibandjo et al. 2005a, 2006, 2007a–b, Alauzun et al. 2011, Schlienger et al. 2012). Interestingly, Sneddon's group used the frustules of diatom—bio-silica—as templates for the bio-inspired design of porous BN nano- and mesostructures (Kusari et al. 2007). Foams could also be designed by using carbon foam templates derived from integrative chemistry (Alauzun et al. 2011).

Soft templating was also reported to yield BN mesostructures using the sol-gel process (Figure 3.3c, d). Porosity can then be created by surfactant micelles (Dibandjo et al. 2007a) or self-assembled copolymers (Malenfant et al. 2007).

3.2.2   Related Applications

BN nanostructures show the typical semiconductor and luminescence properties, good thermal and chemical inertness of the corresponding bulk compounds. They also possess outstanding mechanical properties, especially studied for BN nanotubes. After a short summary of the luminescence properties, we focus in the following subsections on emergent properties and applications of BN nanomaterials in the field of environmental remediation.

3.2.2.1   Luminescence in BN Materials

Recent significant scientific attention has been paid to hexagonal BN (h-BN) as one promising material, because h-BN has strong luminescence in the UV range, which makes it very different from carbon materials and attractive for different kinds of applications as blue light and UV emitters (Watanabe et al. 2004). Recently, theories and experiments on the optical properties of BN nanostructures have made significant progress. Some theoretical works predict a band gap between 5.5 and 6 eV for BN single-wall nanotubes, using the tight binding method or ab initio calculations (Rubio et al. 1994, Marinopoulos et al. 2004, Ng and Zhang 2004). Experimentally, the optical properties of BN nanotubes were investigated by electron energy loss spectroscopy (Misewich et al. 2003, Arenal et al. 2005). Lauret et al. (2005) directly measured the optical properties of single-walled BN nanotubes that showed different band gaps of 4.45, 5.50, and 6.15 eV using optical absorption spectroscopy. Recently, the studies of BN nanotubes were carried out in the time-resolved PL spectra, which show a broad emission peak between 3.5 and 4.2 eV (Wu et al. 2004). In addition, Zhi et al. (2005) also observed in BN nanotubes a strong cathodoluminescence (CL) peak around 3.3 eV with a shoulder at 4.1 eV. For the other morphologies of BN nanostructures, Chen et al. (2008) synthesized novel BN hollow nanoribbons that exhibit an ultraviolet CL emission at 5.33 eV. Gao et al. (2009) fabricated BN nanosheets with strong ultraviolet CL emission between 3.79 and 3.96 eV. Meng et al. (2010) reported that meshy BN material exhibits a strong peak at 3.50 eV in the ultraviolet range by CL spectra. The optical properties of BN whiskers also were investigated by PL and CL measurements (Zhu et al. 2004). In short, BN materials are wide band gap semiconductors with optical transitions in the UV range above 5 eV. This property makes them very different from carbon materials and attractive for applications in nanoelectronics or optoelectronics such as blue-light emitters and excitonic lasers.

3.2.2.2   Water Cleaning in Porous BN Nanosheets

Effective removal of oils, organic solvents, and dyes from water is of significant importance for environmental and water source protections on a global scale. Advanced sorbent materials with excellent sorption capacity need to be developed. We review here the excellent sorption performances for a wide range of oils, solvents, and dyes of porous BN nanosheets.

3.2.2.2.1   Removal of Oil and Organic Solvent from Water

In recent years, oil spillage and organic solvents have been primary pollutants in water sources (Aurell and Gullet 2010, Dalton and Jin 2010). Among the existing techniques used for cleaning up these species, sorption is an efficient way due to its simplicity and high efficiency. However, the common absorbents, including activated carbon (Bayat et al. 2005), zeolites (Adebajo et al. 2003), and natural fibers (Deschamps et al. 2003), suffer from low separation selectivity and low absorption capacity. Therefore, new effective absorbent materials with a higher absorption capacity, higher separation efficiency, and recyclability need to be explored. In addition, as absorption materials, a porous structure with high surface area is an essential requirement. Therefore, BN nanostructures combining porosity and a layered structure will provide an opportunity as novel absorbents and are expected to possess superior performances.

Lei et al. (2013a) have reported that porous BN nanosheets exhibit excellent sorption performances for a wide range of oils and solvents. Figure 3.4a shows the absorbencies for toluene, pump oil, used engine oil, ethanol, and ethylene glycol. The values are ranging from 2000 to 3300 wt.%. The porous BN nanosheets absorb up to 33 times their own weight of ethylene glycol and 29 times of used engine oil. These uptakes are much higher than those of commercial activated carbon and BN particles (see Figure 3.4b), as well as the nonporous BN nanosheets, indicating essential performance enhancement of porosity. The different steps of the used engine oil absorption are shown in Figure 3.4c and d. The starting BN nanosheets have a white color, and the used oil is brown. After oil uptake, the oil-filled BN nanosheets become black and flow on the water surface due to the superhydrophobicity of porous BN nanosheets, which facilitates their collection. Within just 20 s, nearby oil has been sucked into the BN nanosheets (see the inset in Figure 3.4d). After only 2 min, all oil has been taken by the sample, and it becomes black (see Figure 3.4d).

Importantly, the absorbed oil can be removed from the nanosheets by directly burning them in air and then reused for several times (Figure 3.4e, f). The nanosheets can be cleaned completely by heating in a furnace at 600°C for 2 h in air to almost regenerate all original properties (see Figure 3.5a, b). Additional TEM pictures also confirm that the porosity of porous BN nanosheets is maintained (see Figure 3.5c). Moreover, solvent extraction can be used as an alternative cleaning method to avoid the evolution of potentially harmful gases from burning oil; washing with organic solvents like ethanol or petroleum enables the recovery of the initial absorption capacity of porous BN nanosheets (see Figure 3.5d).

A clear shift of the (002) diffraction peak to low angles is observed by XRD for oil-saturated (e.g., used engine oil) porous and nonporous BN nanosheets (see Figure 3.6a). This indicates that the interdistance between the (002) basal planes can increase by 37% from 0.35 to 0.48 nm through intercalation of the organic molecules in the interlayer space. In contrast, the (002) peak of commercial bulk BN particles does not change indicating that the bulk BN particles cannot stand such high strains (see Figure 3.6b). Therefore, the intercalation of molecules in the interlayer space contributes to the uptake increase in the nanosheets. The sorption properties of porous BN nanosheets for oils and organic solvents may result from the following factors: (1) surface effect of molecules on the hydrophobized BN nanosheet surface, (2) capillarity effects for filling of the holes inside the nanosheets and the space between the sheets, and (3) intercalation of molecules in the interlayer space of the BN nanosheets.

3.2.2.2.2 Removal of Dye from Water

Congo red (CR), a textile dye (anionic), is considered as a primary toxic pollutant in water resources (Wang et al. 2012a). UV–vis spectroscopy on the CR solution after adding BN porous nanosheets was used to monitor the amount of remaining dye and the kinetics of the adsorption process (see Figure 3.7a). After only 60 min, the characteristic band of CR became too weak to be observed, suggesting the high efficiency for CR removal. Figure 3.7a and b shows that about 99% of the CR can be removed from the water within 180 min without using any additives at room temperature. The adsorption isotherm (Figure 3.7c) fitted by the Langmuir model gives a maximum adsorption capacity Q_m of 782 mg g^{-1}, corresponding to complete monolayer coverage (Wu et al. 2005). To investigate the versatility of porous BN nanosheets, further experiments have been carried out by testing other dyes with opposite charges commonly used in

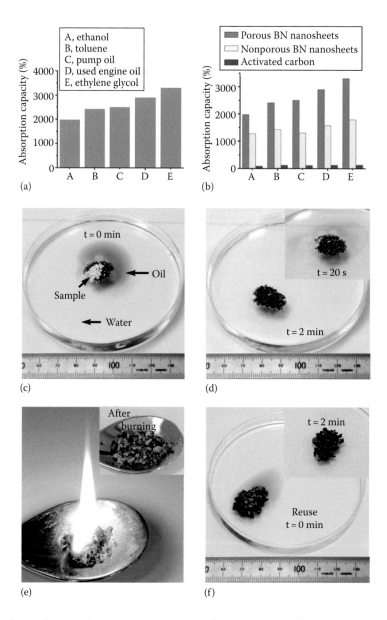

Figure 3.4 (a) Gravimetric absorption capacities of the porous nanosheets for five organic solvents and oils. (b) Comparison of the absorption capacities of the porous BN nanosheets with nonporous BN nanosheets, commercial bulk BN particles, and activated carbon. (c) Photograph of the setup for oil absorption tests with white porous BN nanosheets. (d) Photograph of porous BN nanosheets saturated with oil after 2 min of absorption, inset showing the absorption process after 20 s. (e) Photograph of burning oil-saturated porous BN nanosheets in air for cleaning purpose, inset showing the color change after burning. (f) Photograph of the cleaned nanosheets for second oil absorption test, inset showing the absorption result after 2 min. (From Lei, W.W. et al., *Nat. Commun.*, 4, 1777, 2013a.)

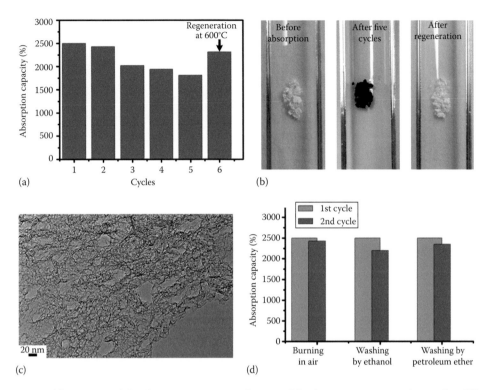

Figure 3.5 (a) Variation of the absorption capacity of pump oil for the porous BN nanosheets after different cycles and regeneration. (b) Photographs of the porous BN nanosheets: before absorption, after five absorption cycles, and after regeneration at 600°C for 2 h in air. (c) TEM picture of the porous BN nanosheets after regeneration at 600°C in air. (d) Variation of the absorption capacity of pump oil by the porous BN nanosheets between the first and second cycles of sorption cleaning by burning in air, ethanol washing, and petroleum ether washing. (From Lei, W.W. et al., *Nat. Commun.*, 4, 1777, 2013a.)

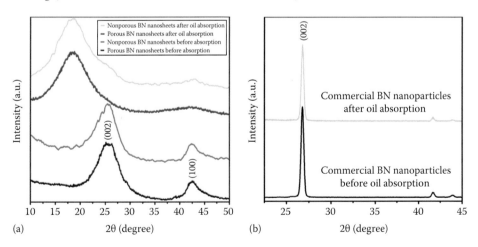

Figure 3.6 (a) XRD patterns of porous and nonporous BN nanosheets before and after absorption of oils. (b) XRD patterns of commercial BN particles before and after oil absorption. (From Lei, W.W. et al., *Nat. Commun.*, 4, 1777, 2013a.)

the textile industry, such as the cationic dyes methylene blue (MB) (see Figure 3.7d through f) and basic yellow 1 (BY) (see Figure 3.7g through i). Figure 3.7f and i shows that the maximum adsorption capacities of MB and BY are 313 and 556 mg g^{-1}, respectively. More importantly, the adsorbed dyes can be easily separated from porous BN nanosheets by simple heating of the powder at 400°C in air for 2 h (see insets in Figure 3.7c, f, and i). The BN adsorbents are highly stable and can be reused for several cycles without losing activity. Here, dye sorption relies on the surface effect (adsorption of molecules on the hydrophobized BN nanosheet surface) of porous BN nanosheets. In addition, a large number of unsaturated atoms along the edges of the nanosheets and pores are chemically more reactive and might be stronger adsorption sites for organic molecules such as dyes (Petravic et al. 2010).

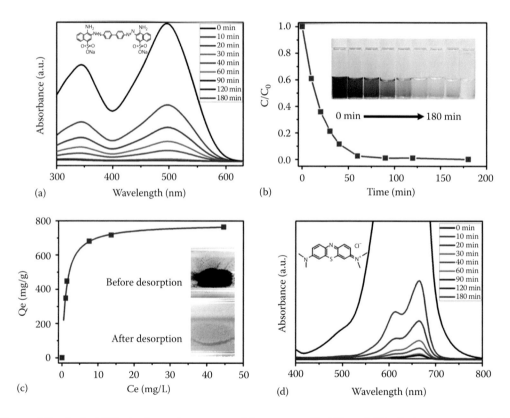

Figure 3.7 (a, d, g) UV–vis absorption spectra of the aqueous CR, MB, and BY solutions, respectively, in the presence of porous BN nanosheets at different time intervals. The insets show the molecular structures of CR, MB, and BY, respectively. (b, e, h) Adsorption rates of CR, MB, and BY on porous BN nanosheets, respectively. The insets show the corresponding photographs. (c, f, and i) Adsorption isotherms of CR, MB, and BY on porous BN nanosheets, respectively. Qe (mg g^{-1}) is the amount of dyes adsorbed at equilibrium, and Ce (mg L^{-1}) is the equilibrium solute concentration. The insets present the photographs of the BN porous nanosheets with CR and MB composites before (upper) and after (bottom) recovery by heating at 400°C for 2 h, respectively. (From Lei, W.W. et al., *Nat. Commun.*, 4, 1777, 2013a.)   (*Continued*)

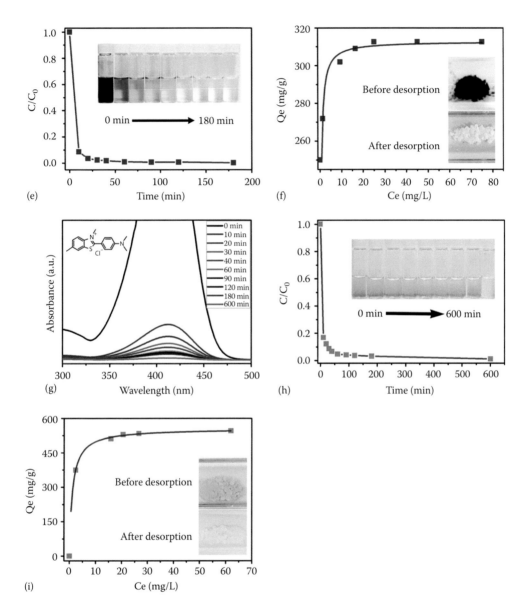

Figure 3.7 (Continued)   (a, d, g) UV–vis absorption spectra of the aqueous CR, MB, and BY solutions, respectively, in the presence of porous BN nanosheets at different time intervals. The insets show the molecular structures of CR, MB, and BY, respectively. (b, e, h) Adsorption rates of CR, MB, and BY on porous BN nanosheets, respectively. The insets show the corresponding photographs. (c, f, and i) Adsorption isotherms of CR, MB, and BY on porous BN nanosheets, respectively. Qe (mg g^{-1}) is the amount of dyes adsorbed at equilibrium, and Ce (mg L^{-1}) is the equilibrium solute concentration. The insets present the photographs of the BN porous nanosheets with CR and MB composites before (upper) and after (bottom) recovery by heating at 400°C for 2 h, respectively. (From Lei, W.W. et al., *Nat. Commun.*, 4, 1777, 2013a.)

3.3 BORON CARBON NITRIDE NANOMATERIALS

A big community of materials scientist has turned to BCN nanostructures besides pure carbon and BN, because the ternary B–C–N compositions allow for extensive adjustment of the properties, especially transport properties. The flagships of BCN nanostructures are BCN nanotubes, which are intensively studied since the 1990s, and BCN nanosheets, which are evident extensions of graphene. But other nanostructures emerge since few years and attract interest because of peculiar properties. Those are nanoparticles and meso- and microporous materials. The following subsection describes the main synthesis pathways for the fabrication of nanostructured BCN materials with 0D (nanoparticles), 1D (nanotubes), 2D (nanosheets), and 3D (porous materials) features. Then, the most striking corresponding properties are presented.

3.3.1 Nanostructures of Boron Carbon Nitride: Syntheses and Features

3.3.1.1 BCN(O) Nanoparticles

At present, there are mainly four methods to prepare BCN nanoparticles. One is ball milling. Torres et al. (2007) have prepared BCN nanoparticles with a diameter of 60 nm by low-energy ball milling method, using as precursors h-BN, graphite, and polypropylene micrometric powders. The modified O'Connor's method is the second way, in which boric acid, urea, and PEG/TEG (polyethylene/tetraethylene glycol) are usually used as the boron, nitrogen, and carbon sources, respectively (Liu et al. 2009a–c, Wang et al. 2011a). Liu et al. (2009c) employed boric acid, urea, and PEG combustion at 800°C and then soaked the resulting mixture in distilled water to remove residual B_2O_3, leaving the nanosized BCNO nanocrystals in the solution. The third method consists of preparing a solid-state BCNO precursor by mixing melamine, formaldehyde, and trimethyl borate in liquid phase (Zhang et al. 2013). The resulting precursor was then sintered at low temperature of approximately 550°C–750°C. The fourth method yields nanoparticles with a diameter of 5 nm (Lei et al. 2011). It relies on the use of a simple eutectic LiCl/KCl salt melt, in which sodium borohydride, urea, or guanidine hydrochloride is mixed as the boron, carbon, and nitrogen sources, respectively. This environmentally friendly solution route provides easy access to BCN(O) nanostructures with high yield, controlled composition, and aqueous dispersion properties. Figure 3.8 shows the AFM and TEM pictures of BCN(O) demonstrating that the samples are composed of approximately 5 nm nanoparticles. A few other 0D nanostructures were reported, such as hollow BCN nanoparticles made by reactive magnetron sputtering (Johansson et al. 2000).

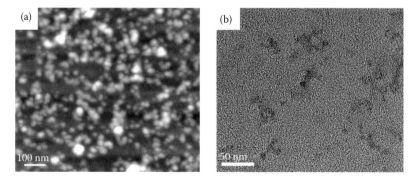

Figure 3.8   AFM (a) and TEM (b) pictures of BCN(O) nanoparticles. (From Lei, W.W. et al., *J. Am. Chem. Soc.*, 133, 7121, 2011.)

3.3.1.2 BCN Nanotubes

BCN nanotubes have been widely reported and studied during the last years. Some reviews and chapters of the present book deal extensively with the fascinating topic of their synthesis and properties, so that readers interested in this field can refer to more dedicated reports (Terrones et al. 2002, Arenal et al. 2010). We might, however, quote few relevant synthesis pathways and structural features for these specific nanostructures. Arc discharge and laser ablation/vaporization are commonly used high-temperature processes (Stephan et al. 1994, Arenal et al. 2010). Lower-temperature methods, between 500°C and 1700°C, often rely on (metal-catalyzed) CVD (Yin et al. 2005, Wang et al. 2006, Liao et al. 2007, Arenal et al. 2010).

As for BCN nanosheets, phase separation in BCN nanotubes between carbon and h-BN domains often occurs (Suenaga et al. 1997, Terrones et al. 2002), although some reports highlight homogeneous BCN distribution, depending on the carbon content (Yin et al. 2005, Kim et al. 2007).

3.3.1.3 BCN Nanosheets

The ternary composition of BCN nanosheets is of particular interest because there is scope for tuning the electronic properties by adjusting the content of each hetero-element and its localization into the structure. Especially, calculations and experimental data show that boron preferentially binds to nitrogen atoms while carbon forms C–C and C–N single or multiple bonds (Rocha et al. 2011), so that most often, nanosegregation between h-BN and carbon domains is expected (Ci et al. 2010, Pakdel et al. 2012a, Chang et al. 2013). B and N atoms are usually incorporated with similar contents. In this case, first-principle calculations and experiments like XPS measurements show that the h-BN/C interface is made of N–C bonds (Chang et al. 2013), while B–C bonds are rarely encountered (Pakdel et al. 2012a). h-BN/graphene hetero-junctions can be produced at selected areas by posttreatment of h-BN or graphene films (Chang et al. 2013, Kim et al. 2013, Liu et al. 2013a,b). Spontaneous phase segregation can also occur in situ during the synthesis of BCN nanosheets. Actually, calculations tend to suggest that this separation is thermodynamically favorable (Rocha et al. 2011).

Experimentally, the composition and the extent of the h-BN and carbon domains in the 2D nanostructures can be conveniently adjusted by modifying the ratio between the molecular precursors during the synthesis process (Ci et al. 2010, Chang et al. 2013). Three main approaches toward BCN nanosheets are listed in the following:

1. *Chemical vapor deposition*: Ci et al. (2010) have used the thermal catalytic CVD method to synthesize 2D BCN nanosheets on Cu substrate from a mixture of ammonia borane and methane (see Figure 3.9). The sheets possess a nanostructure consisting of randomly distributed BN (h-BN) and graphene nanodomains, which provides a novel route to engineer the band gap of graphene. Few-atomic-layered boron carbonitride nanosheets have been grown on Si substrate by microwave plasma CVD from a gas mixture of $CH_4-N_2-H_2-BF_3$ (Pakdel et al. 2012a, Qin et al. 2012).
2. *Combustion reactions*: Combustion reactions are the second route, in which boric acid/boric oxide/boron trichloride, urea/activated charcoal/graphite, and urea/N_2/NH_3 gases are usually used as the boron, carbon, and nitrogen sources, respectively. Lim and coworkers (2007) have prepared BCN nanosheets via the partial substitution of carbon atoms of GO nanosheets with boron and nitrogen atoms, where B_2O_3 powder and NH_3 gas are used as the B and N sources, respectively (see Figure 3.9) (Yin et al. 2005, Raidongia et al. 2010, Han et al. 2011, Lin et al. 2012).

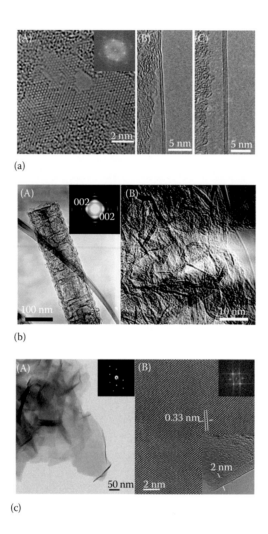

(a)

(b)

(c)

Figure 3.9   TEM images of BCN nanosheets. (a) An HRTEM image of (A) one-, (B) two-, or (C) three-atomic-layer thick sheets obtained by chemical vapor deposition method. (From Ci, L. et al., *Nat. Mater.*, 9, 430, 2010.) (b) (A) TEM and (B) HRTEM images of a porous BCN tubular fiber synthesized by combustion reaction method. (From Yin, L.-W. et al., *J. Am. Chem. Soc.*, 127, 16354, 2005.) (c) (A) TEM and (B) HRTEM images of BCN nanosheets fabricated by the molten salt method. (From Lei, W.W. et al., *Chem. Commun.*, 49, 352, 2013b.)

3. *Colloidal synthesis in molten salts*: Lei et al. (2011, 2013b) have used the salt melt approach for the synthesis of few-layered BCN nanosheets. Sodium borohydride was used as the B precursor and urea (CH_4N_2O) as C and N sources. A molten eutectic mixture of LiCl and KCl (45/55 wt.%) was used as a sustainable solvent that is cheap, water soluble, and recyclable. The BCN nanosheets were obtained after heating (750°C) under nitrogen flow and washing in water. Most of the BCN sheets are composed of two to six atomic layers, as shown in Figure 3.9.

Other miscellaneous processes could be mentioned, such as the decomposition of ammonia–borane in the presence of ethanol (Wang et al. 2011b). In this case, thermal dehydrogenation of

NH_3–BH_3 yields microbubbles of hydrogen, which template the growth of a porous network. The latter collapses upon heating, releasing BCN nanosheets that were constituting the walls of the network.

3.3.1.4 Mesoporous BCN

Implementation of nanostructures into many devices obviously calls for the management of charge percolation, ionic, fluid, and/or gas diffusion pathways. Combining these features can be done by carefully processing nanoparticles as primary nanobuilding blocks. Another fruitful alternative is to design nanostructured porous materials: the outer dimensions of the single object are not nano, but the solid incorporates nanoscale features, especially micro- (diameter < 2 nm), meso- (2 nm < diameter < 50 nm), and macropores (diameter > 50 nm), as well as nanoscale walls between pores. During the last 5 years, some efforts were devoted to the design of porous nanostructured BCN materials. The most significant results are reported in this section and summarized in Figure 3.10, according to the number of steps involved in the production of the material: two-step approaches or one-step approaches.

First two-step approach—*Hydrothermal synthesis coupled to freeze-drying*: A first approach (see Figure 3.10), nicely demonstrated by Feng and coworkers (Wu et al. 2012), consists in the production of a BCN aerogel. The treatment of graphene oxide nanosheets under hydrothermal conditions (180°C) in the presence of ammonia boron trifluoride yielded a hydrogel made of a percolated network of B- and N-doped graphene sheets. After freeze-drying to ensure the absence of collapse during solvent removal, a B- and N-doped carbon aerogel was obtained, with a surface area of about 250 $m^2 \cdot g^{-1}$. According to XPS measurements, the B and N contents were rather low of 0.6 and 3.0 at.%, respectively. Also, at such low synthesis temperatures, strong amounts of oxygen-containing functional groups are still present. Fellinger et al. (2012) reported a similar approach combining the hydrothermal treatment of sugar and borax solutions with freeze-drying to yield carbon aerogels, but the boron content was not assessed. Interestingly, post-carbonization up to 1000°C led to a strong increase in the electron conductivity, presumably through the elimination of oxidized functional groups and the enhancement of graphitic order.

Second two-step approach—*Salt templating*: Fechler et al. (2013) recently reported on an elegant route toward porous BCNs by the so-called salt templating (see Figure 3.10). B- and N-containing ionic liquids were used as precursors, yielding BCNs upon calcination at 1000°C. Small amounts of low-melting-point eutectic mixtures of alkali chlorides and zinc chloride were mixed with the ionic liquids (inorganic salt/liquid ratio of ca. 1 wt.%). Upon calcination under nitrogen, carbonization of the organic precursor occurs together with nanophase segregation of the salt, yielding salt inclusions into the final BCN compounds. After washing out the salt with water, highly porous BCNs (B, C, N, O contents according to XPS of 2, 90, 2, 6 at.%, respectively) could be obtained, some of them reaching surface areas of 2000 $m^2 \cdot g^{-1}$, among the highest reported. Indeed, although other authors reported values up to 2900 $m^2 \cdot g^{-1}$ (Raidongia et al. 2010), the corresponding nitrogen sorption isotherms and their interpretation are still questionable. Possible recycling of the porogen salt was envisioned, but it could not be applied in a straightforward approach, since the saturated vapor pressure of the inorganic salts at 1000°C is significant, so that some of the porogen should be lost while the initial eutectic composition could also change. Nevertheless, it is worth to note the ability to modify the porous structure by only changing the nature of the inorganic salt (see Figure 3.11). The exact origin of this effect is not clear yet but might be related to solubility effects of the inorganic salt into the organic liquid.

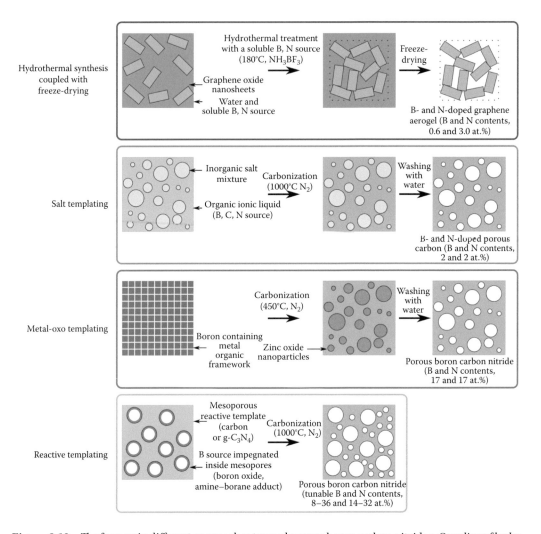

Figure 3.10   The four main different approaches toward porous boron carbon nitrides. *Coupling of hydrothermal treatment with freeze-drying* (Wu et al., 2012), *salt templating* (Jayarmulu et al., 2013), and *metal oxo-templating* (Fechler et al., 2013) are two-step procedures, while *reactive templating* (Portehault et al., 2010) involves one single step.

Third two-step approach—*metal-oxo templating*: Inspired by works using the decomposition of metal-organic frameworks to yield microporous carbons, Jayaramulu et al. reported recently on the use of a boron-containing zinc imidazolate framework as precursor for BCN (Jiang et al. 2011, Jayaramulu et al. 2013). After calcination at 450°C, a BCN matrix incorporating ZnO nanoparticles was obtained. Washing with acidic water resulted in a porous BCN, with a surface area of 988 m$^2 \cdot$g^{-1}. Although the mechanism for porosity creation was not discussed, one should consider the potential templating by ZnO nanoparticles and other Zn–O-based species resulting from the decomposition pathway. We then name this route *metal-oxo templating*. The authors described B, C, and N contents of 17, 66, and 17 at.%, respectively, but did not account for the oxygen content, though the low-temperature process would yield a sample poorly resistant to oxidation upon air exposure.

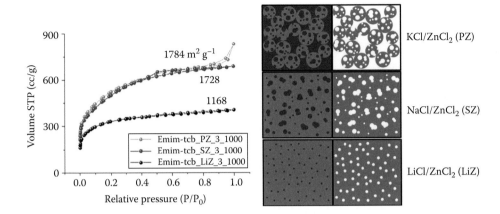

Figure 3.11 Salt templating: By adjusting the nature of the inorganic salt eutectic mixtures added to the ionic liquid (1-ethyl-3-methyl-imidazolium tetracyanoborate Emim-tcb), the surface area and microstructure can be modified. (From Fechler, N. et al., *Adv. Mater.*, 25, 75, 2013.)

One-step approach—*Reactive templating*: To our knowledge, the only *one-pot* synthesis of porous BCNs reported up to now was proposed by Golberg's group and further developed by some of us and others (see Figure 3.10) (Vinu et al. 2005, Portehault et al. 2010). The basis of the approach is the use of a reactive template, that is to say a reactant (C or C and N sources) acting also as a template. Vinu et al. (2005) first described the conversion of mesoporous carbon by its reaction with boron oxide (B source) and nitrogen (N source) at approximately 1500°C. The B, C, and N contents of the resulting materials were 21%, 38%, and 23%, respectively, while the hexagonal mesoporous structure was maintained, yielding a surface area of approximately 700 $m^2 \cdot g^{-1}$. Later on, we proposed graphitic carbon nitride precursor g-C_3N_4 as a C and N precursor, obtained as a mesoporous solid through the use of silica nanoparticles as templates (Portehault et al. 2010). The mesopores of the carbon nitride were infiltrated by liquid amine–borane adducts (e.g., tert-butylamine-borane). Upon calcination, boron atoms substitute carbon, while some C atoms segregate into nanodomains (see Figure 3.12). This results in a direct replica of the initial template. Further heating leads to carbon elimination and the formation of micropores, thus resulting in a BCN with hierarchical porous structure. Specific surface areas close to 1600 $m^2 \cdot g^{-1}$ were reached. Noteworthy, in this case, the B and N contents can be adjusted and belong to the high B and N concentration regions of the ternary B–C–N phase diagram, contrary to the two-step methods described earlier: B and N contents can be adjusted here within the 8–36 and 14–32 at.% ranges, respectively.

Alternatively, Xue and coworkers (2013) reported recently the fabrication of BCN foams by coupling CVD with hard templating: BCN nanosheets first formed by CVD on a Ni foam. After nickel etching, a BCN foam was recovered, exhibiting a macroporous structure.

3.3.2 Properties of BCN Nanostructures

Most of the properties and potential applications of BCN nanostructures rely on the distinct transport properties of these materials, compared to wide band gap h-BN and conductive graphene. In the following section, we highlight few relevant fields. However, one should keep in my mind that a comprehensive survey of the fascinating properties of nanostructured BCNs exceeds

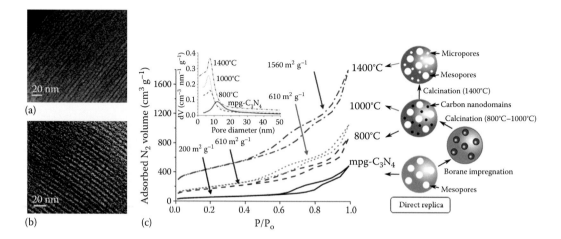

Figure 3.12 Reactive templating: The conversion of a porous reactant (carbon [a] [Vinu et al., 2005] or g-C$_3$N$_4$ [c]) into a direct replica of boron carbon nitride can be assessed by TEM (original carbon template [a] and its BCN relative [b]) and nitrogen sorption (c). The latter highlights the conservation of the initial g-C$_3$N$_4$ mesoporous structure, together with the creation of micropores. (From Portehault, D. et al., *Adv. Funct. Mater.*, 20, 1827, 2010.)

the scope of this chapter. Indeed, outstanding mechanical properties, chemical and thermal inertness, and good thermal conductivity are already known, while intriguing behaviors have appeared recently, related to information technology (magnetism) (Kumar et al. 2013), energy harnessing (Iyyamperumal et al. 2012), and catalysis (Wang et al. 2012b, Xue et al. 2013, Zheng et al. 2013).

3.3.2.1 Electrical Properties of BCN Nanostructures

The study of the electrical and luminescence properties of BCN nanostructures experienced intensive research efforts during two periods. The 1990s and 2000s were the *golden age* of nanotubes and related 1D systems (Terrones et al. 2002, Arenal et al. 2010), while the 2000s and 2010s saw—and still see—an increasing number of works targeting the control of the electrical properties of graphene, and more generally band structure engineering in BCN 2D nanostructures, especially through B- and N-doped graphene. The large literature available on this topic, including reviews and other chapters in this book, should be referred to by interested readers (Terrones et al. 2002, Arenal et al. 2010, Jariwala et al. 2013, Kumar et al. 2013, Song et al. 2013, Xu et al. 2013). Briefly, simulations predict the possibility to adjust the band gap through the control of B and N contents and of the atomic distribution into BCN nanostructures, as in bulk compounds (Watanabe et al. 1999, Rocha et al. 2011). As explained in Section 3.3.1.2, nanosegregation between h-BN and carbon domains usually occurs in BCN nanotubes and nanosheets. Such a phase separation drives changes in the electrical properties. Especially, BN doping in graphene leads to band gap opening. The semiconducting properties of BCN nanotubes (Bai et al. 2000a–b) and nanosheets (Ci et al. 2010, Pakdel et al. 2012a) were demonstrated in a series of experimental reports, dealing usually with B and N contents above 2 at.% (Chang et al. 2013). The possibility to tune the band gap of graphene by B and N incorporation was experimentally ascertained, especially by Ci et al. (2010) and Chang et al. (2013) through the adjustment of the BN content (ranging from 10 to 100 at.% [Ci et al. 2010], from 0 to 27 at.% [Chang et al. 2013]) and the extent of h-BN domains by in situ doping during CVD growth.

3.3.2.2   Photoluminescence of BCN Nanoparticles

Theoretical studies suggested that it should be possible to use BCN materials to tune the wavelength of emitted light across the visible light spectrum by varying the composition of BCN compounds (Rubio et al. 1994, Blasé et al. 1997, Mazzoni et al. 2006). Therefore, much effort has been devoted to the development of tunable light emission properties of BCN(O) compounds by controlling the carbon and oxygen contents. Ogi et al. (2008) synthesized a non-RE-doped BCNO with a wide range of emission spectra with luminescence varying from violet to near-red regions. In addition, Liu et al. (2009) developed a facile strategy based on the reaction of urea and boric acid for the synthesis of BCNO nanocrystals containing carbon impurities and exhibiting efficient multicolor fluorescence under both single-photon and two-photon excitations. Okuyama and coworkers prepared BCNO phosphor nanofibers with intense green and yellow emissions by electrospinning (Suryamas et al. 2011). Very recently, Tang et al. reported a two-step method to prepare BCNO phosphors (Zhang et al. 2013). The emission spectra of BCNO phosphors can be easily tuned across the visible light spectrum by changing the sintering temperatures, formaldehyde volumes, and the molar ratio of trimethyl borate precursor. In this regard, Lei et al. (2011) have reported the facile synthesis of BCN(O) compounds in a simple eutectic salt melt. The nanoparticles are readily dispersed in water and exhibit tunable PL that covers a spectral region from the UV to the visible regions (see Figure 3.13), which is governed by the elemental composition. The efficiency of the fluorescence is reflected by the high absolute quantum yield (QY) measured. QY values as high as 26% for B/G:1/5 are achieved without further specific optimization of the as-obtained materials. The origin of the

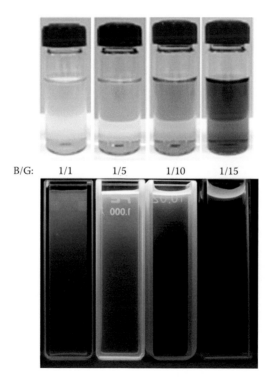

Figure 3.13   Digital photographs of aqueous colloidal dispersions of BCN(0) nanoparticles made in molten salts: increase in the carbon content from left to right, in the absence (top) and presence (bottom) of Hg-365 nm UV irradiation. (From Lei, W.W. et al., *J. Am. Chem. Soc.*, 133, 7121, 2011.)

fluorescence for the BCN(O) nanoparticles should be from a defect-related nature, rather than an inter band transition as has been found for bulk BN crystals as well as nanotubes. Some paramagnetic centers appear, originating from electrons trapped at nitrogen vacancies. These defects are responsible for the observed PL behavior of the BCN(O) nanoparticles (Liu et al. 2009).

3.3.2.3 Lithium Storage in BCN Nanosheets

Few-layered BCN nanosheets, as analogues of graphene, are expected to be a candidate for high-performance electrode material of lithium-ion batteries. It has been reported that the Li-intercalation into bulk BCN materials occurred with smaller interlayer expansion ratio than graphite, thus suggesting better cyclability (Ishikawa et al. 1995, Kawaguchi et al. 2006). In this regard, Lei et al. (2013b) reported the electrochemical behavior of few-layered BCN nanosheets with a surface area of 173 m^2 g^{-1}. Figure 3.14a shows the charge and discharge curves of the BCN nanosheets cycled between 3.0 and 0.0 V at a current density of 30 mA g^{-1} in the 1st, 2nd, 5th, and 50th cycles. A distinctive discharge plateau is observed at about 0.9 V with another less marked event at 0.6 V, in agreement with faradaic events associated with Li insertion into bulk BCNs (Kawaguchi et al. 2006). The BCN nanosheet electrode delivers initial discharge/charge capacities of 832 and 424 mAh g^{-1},

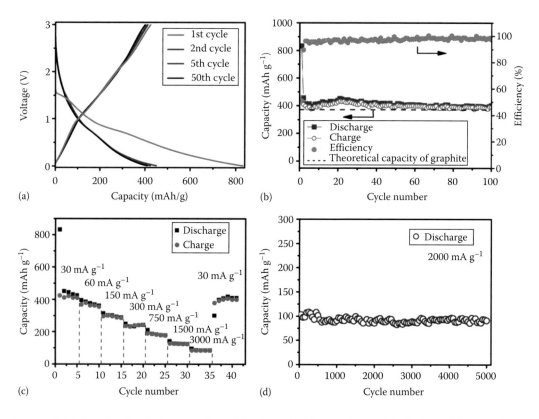

Figure 3.14 Electrochemical properties of few-layer BCN nanosheets: (a) discharge/charge voltage curves under a current density of 30 mA g^{-1}, (b) cycling performance and Coulombic efficiency at 30 mA g^{-1} for 100 cycles, (c) rate performance of few-layer BCN nanosheets at rates from 30 to 3000 mA g^{-1}, and (d) cycling performance of few-layer BCN nanosheets electrode at the high rate of 2 A g^{-1}. (From Lei, W.W. et al., *Chem. Commun.*, 49, 352, 2013b.)

respectively, with a low Coulombic efficiency of 51%, which may be caused by irreversible lithium loss during the formation of the solid electrolyte interphase (Kawaguchi et al. 2006). The charge and discharge capacities stabilize at about 390 mAh g^{-1} after 100 cycles (see Figure 3.14b) with a Coulombic efficiency of ~98%. This reversible capacity is higher than the theoretical capacity of graphite (372 mAh g^{-1}) (Noel and Suryanarayan 2002). This is further confirmed by the excellent charge and discharge performances at high current densities. The rate capability of the sample was evaluated by charging/discharging at various current rates from 30 to 3000 mA g^{-1} (see Figure 3.14c). The BCN nanosheets show excellent capacity retention at each rate. Notably, a reversible capacity of 100 mAh g^{-1} is sustained at the highest current rate of 3 A g^{-1}. More importantly, a stable capacity of ~420 mAh g^{-1} can be recovered at 30 mA g^{-1} after charging–discharging at a high current rate, demonstrating very good reversibility and structural stability of the BCN nanosheets.

Excellent cycling stability is also obtained at a very high current rate. Figure 3.14d shows the cycling performance of a BCN nanosheet electrode at 2 A g^{-1} for 5000 cycles after 5 cycles at 30 mA g^{-1}. Ninety percent of the starting capacity (102 mAh g^{-1}) is maintained after 5000 cycles.

The improved performances in terms of cyclability and cycling rate may be attributed to three aspects. First, the high stability of the 2D layer structure enables the BCN nanosheet electrode to be used at a very high current rate with superior cycling performance. Second, the high surface area with accessible pores provides a large electrode/electrolyte contact area and short path length for Li$^+$ transport (Wu et al. 2011). Third, the high density of >B–N< and ≡C–N< groups ensures a high concentration of redox centers for reaction with Li$^+$ ions (Reddy et al. 2010), thus providing high capacity.

3.3.2.4   Hydrogen Storage in B(C)N Nanostructures

Hydrogen storage has been the focus of works dedicated to boron (carbon) nitrides since the beginning of the twenty-first century. Because nanostructured carbons possess interesting uptake capacities, it has been quickly suggested that structurally similar-layered BN could provide enhanced H$_2$ adsorption enthalpy because of the polarity of B–N bonds (Wang et al. 2002, Ma et al. 2002). Furthermore, incorporation of carbon into pure BN could lead to BCNs with further ability to tune the sorption properties (Pattanayak et al. 2002, Jhi and Kwon 2004). As on carbons, hydrogen interaction on B(C)Ns involves mainly physisorption, without H$_2$ dissociation and with relatively low adsorption enthalpy. Then, an important limitation of these materials is the decrease in uptake values when the temperature is increased close to room temperature, so that current realistic scenarios involving B(C)Ns for hydrogen storage rely on cryogenic and pressurized tanks. Although few reports deal with storage of other gases, such as CO$_2$ (Raidongia et al. 2010) and CH$_4$ (Kumar et al. 2013), the majority of works is dedicated to the topic of H$_2$ uptake, which is detailed in the following paragraphs.

Hydrogen storage in BN materials: Most of the works reported during the 2010s have dealt with BN nanostructures, especially those related to nanotubes. In 2002, Wang et al. (2002) obtained a 2.6 wt.% hydrogen concentration after ball milling bulk h-BN under hydrogen atmosphere at 10 bars. The same year, Ma et al. (2002) demonstrated H$_2$ uptake of 2.6 wt.% at 100 bars and 293 K on CVD-grown bamboo-like nanotubes (see Figure 3.15b), while Tang et al. (2002) reported on further capacity increase up to 4.2 wt.% at 100 bar and 298 K after Pt-catalyzed collapse of BN nanotubes at 1500°C (see Figure 3.15d–g). This structural evolution was accompanied by an increase in the specific surface area from 254 to 789 m$^2 \cdot$ g^{-1}, which could explain the uptake enhancement. Later, similar capacities were reported for BN nanotubes (e.g., 2.2 wt.% at 60 bars and 293 K) (Lim et al. 2007). Interestingly, research on hydrogen storage onto BN nanostructures only recently evolved toward more complex architectures than nanotubes. For instance, ultrathin-shell BN hollow spheres exhibited 4.1 wt.% at 100 bars (Lian et al. 2012), while BN porous microbelts

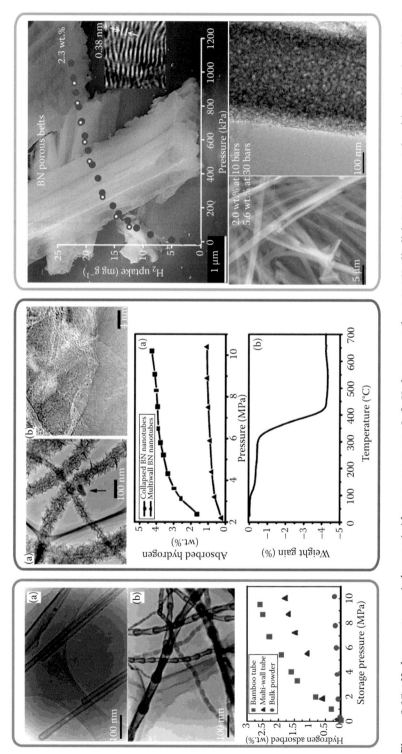

Figure 3.15 Hydrogen storage in boron nitride nanostructures. Left: Hydrogen uptake at 293 K of bulk h-BN, BN nanotubes (a) and bamboo-like BN nanotubes (b). (From Ma, R. et al., *J. Am. Chem. Soc.*, 124, 7672, 2002.) Middle: Hydrogen uptake at 298 K of BN nanotubes (d) and collapsed nanotubes (e) and desorption after storage at 102 bars in the collapsed nanotubes. (From Tang, C. et al., *J. Am. Chem.*, 124, 14550, 2002.) Right: Storage capacity of BN porous microbelts at 77 K (g) (From Weng, Q. et al., *ACS Nano.*, 7, 1558, 2013.) and of porous nanowhiskers at 298 K (h). (From Li, J. et al., *Nanotechnology*, 24, 155603, 2013a.)

obtained by Golberg and coworkers showed interesting uptake values up to 2.3 wt.% at 10 bars and 77 K (see Figure 3.15g–h) (Weng et al. 2013). Li et al. (2013) seemed to hold one of the highest H_2 uptake performances at 298 K on these materials with 2.0 wt.% at 10 bars and 5.6 wt.% at 30 bars with porous BN nanowhiskers (see Figure 3.15, right). Noteworthy, the porous microbelts and nanowhiskers originate from very similar procedures involving the reaction between boric acid and melamine, but different heating profile (final temperature between 1100°C and 1600°C) and atmospheres (ammonia for the belts, nitrogen for the whiskers). The direct comparison of the properties is, however, hindered by the different temperatures for uptake studies, but also because Li et al. (2013) used the BET method on nitrogen sorption isotherms to evaluate the specific surface area over the relative pressure range 0.01–0.30, which is not suitable for highly microporous solids. However, it is worth to note that Weng et al. (2013) obtained different samples after calcination at 900°C, 1000°C, and 1100°C, exhibiting surface areas of, respectively, 1161, 1488, and 1144 $m^2 \cdot g^{-1}$ but a monotonous increase in the H_2 uptake with 1.6, 1.8, and 2.3 wt.%, respectively. The nontrivial relation between the total surface area and the storage capacity shows that the chemical state of the surface might also play a crucial role on the H_2 adsorption properties.

Hydrogen storage in BCN materials: Although the earlier-mentioned surface effect on hydrogen sorption is not clear yet, it has already been observed on BCNs. Indeed, the hierarchical porous BCNs obtained by reactive templating and described earlier were used to investigate the impact of surface states (Portehault et al. 2010). Studies were performed at 77 K at a max. H_2 pressure of 1 bar. Although relatively low uptake values were observed (up to 1.1 wt.% with the 1560 $m^2 \cdot g^{-1}$ BCN obtained at 1400°C), evidences of the important role of the surface could be obtained. Indeed, samples calcined at 800°C and 1000°C show different uptakes of 0.82 and 0.55 wt.% (see Figure 3.16), even if they possess identical surface areas (610 $m^2 \cdot g^{-1}$) and microporous volume. This discrepancy clearly points at the impact of surface state, especially structural disorder, for uptake enhancement. Later on, Jayaramulu et al. (2013) reported on a porous BCN (B, C, N contents of 17, 66, and 17 at.%, respectively) with a specific surface area of 988 $m^2 \cdot g^{-1}$, with uptake capacities similar to those of the reactive templating-derived BCNs. The authors showed that further pressure increase at 77 K led to H_2 uptake enhancement to 2.6 wt.% at 10 bars and 3.3 wt.% at 40 bars.

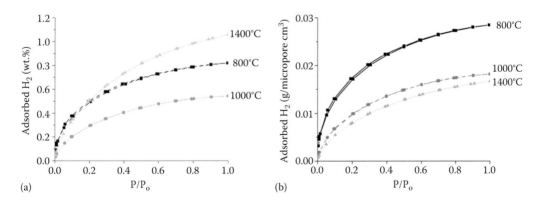

Figure 3.16 H_2 sorption isotherms at 77 K of boron carbon nitrides obtained after treatment at different temperatures. (a) H_2 weight uptake and (b) normalization versus the microporous volume. The samples calcined at 800°C and 1000°C have the same surface area and microporous volume. Therefore, the discrepancies between the values for both samples highlight the impact of surface chemical states, including local disorder. (From Portehault, D. et al., *Adv. Funct. Mater.*, 20, 1827, 2010.)

3.4 CONCLUSIONS

The various nanostructures of BN (hexagonal h-BN) and BCNs have motivated a renewal of interest in BN and BCN materials, due to their wide energy band gap, electrical-insulating property, UV PL, high thermal conductivity and stability, high resistance to oxidation, and good chemical inertness. These properties for BCN 1D and 2D nanomaterials are of particular interest in combination with the transport properties of carbon nanotubes and nanosheets and enable to envision, among all, full tuning of electrical and thermal conductivities. However, now additional properties are emerging, which rely mainly on the high surface-to-volume ratio of these nanostructures. Among these new upcoming properties and potential applications, one may cite not only the whole environment and health domains but also energy storage and harnessing. Because high surfaces is of utmost importance in these fields for enhanced mass transfer at the interface between the nano-objects and liquid or gas media, nanostructured porous materials are becoming increasingly important to investigate. Clearly, 2D and 3D porous BN and BCN materials are expected to show novel behaviors in areas such as water cleaning, lithium ion batteries, and hydrogen storage.

Still, synthesis is one of the narrowest bottlenecks for detailed investigations in the aforementioned fields. Indeed, while there are many ongoing research projects on the properties and applications of 1D and 2D (porous) BN and BCN nanomaterials, new synthesis processes are still highly sought for providing easy, reliable access to nanostructured systems with high yield, controlled composition, and morphology. Deep understanding of the unique combination of 2D and porous structure as well as surface characteristics of BN and BCN nanomaterials is critical for taking their full advantage in a wide range of practical applications such as electronics, energy storage, clean energy, and environment protection.

REFERENCES

Adebajo, M. O., Frost, R. L., Kloprogge, J. T., Carmody, O., and Kokot, S. (2003) Porous materials for oil spill cleanup: A review of synthesis and absorbing properties, *J. Porous Mater. 10*, 159–170.

Alauzun, J. G., Ungureanu, S., Brun, N., Bernard, S., Miele, P., Backov, R., and Sanchez, C. (2011) Novel monolith-type boron nitride hierarchical foams obtained through integrative chemistry, *J. Mater. Chem. 21*, 14025–14030.

Arenal, R., Blase, X., and Loiseau, A. (2010) Boron-nitride and boron-carbonitride nanotubes: Synthesis, characterization and theory, *Adv. Phys. 59*, 101–179.

Arenal, R., Stephan, O., Kociak, M., Taverna, D., Loiseau, A., and Colliex, C. (2005) Electron energy loss spectroscopy measurement of the optical gaps on individual boron nitride single-walled and multiwalled nanotubes, *Phys. Rev. Lett. 95*, 127601.

Aurell, J. and Gullet, K. B. (2010) Aerostat sampling of PCDD/PCDF emissions from the gulf oil spill in situ burns, *Environ. Sci. Technol. 44*, 9431–9437.

Bai, X. D., Guo, J. D., Yu, J., Wang, E. G., Yuan, J., and Zhou, W. (2000a) Synthesis and field-emission behavior of highly oriented boron carbonitride nanofibers, *Appl. Phys. Lett. 76*, 2624–2626.

Bai, X. D., Wang, E. G., Yu, J., and Yang, H. (2000b) Blue–violet photoluminescence from large-scale highly aligned boron carbonitride nanofibers, *Appl. Phys. Lett. 77*, 67–69.

Bayat, A., Aghamiri, S. F., Moheb, A., and Vakili-Nezhaad, G. R. (2005) Oil spill cleanup from sea water by sorbent materials, *Chem. Eng. Technol. 28*, 1525–1528.

Bernard, S., Salles, V., Li, J., Brioude, A., Bechelany, M., Demirci, U. B., and Miele, P. (2011) High-yield synthesis of hollow boron nitride nano-polyhedrons, *J. Mater. Chem. 21*, 8694–8699.

Blase, X., Charlier, J. C., DeVita, A., and Car, R. (1997) Theory of composite BCN nanotube heterojunctions, *Appl. Phys. Lett. 70*(2), 197.

Chang, C.-K., Kataria, S., Kuo, C.-C., Ganguly, A., Wang, B.-Y., Hwang, J.-Y., Huang, K.-J. et al. (2013) Band gap engineering of chemical vapor deposited graphene by in situ BN doping, *ACS Nano 7*, 1333–1341.

Chen, L. Y., Gu, Y. L., Shi, L., Yang, Z. H., Ma, J. H., and Qian, Y. T. (2004) A room-temperature approach to boron nitride hollow spheres, *Solid State Commun. 130*, 537–540.

Chen, Y., Gerald, J. F., Williams, J. S., and Bulcock, S. (1999) Synthesis of boron nitride nanotubes at low temperatures using reactive ball milling, *Chem. Phys. Lett. 299*, 260–264.

Chen, Z. G., Li, F., Wang, Y., Wang, L. Z., Yuan, X. L., Sekiguchi, T., Cheng, H. M., and Lu, G. Q. (2008) Novel boron nitride hollow nanoribbons, *ACS Nano 2*(10), 2183–2191.

Chopra, N. G., Luyken, R. J., Cherrey, K., Crespi, V. H., Cohen, M. L., Louie, S. G., and Zettl, A. (1995), Boron nitride nanotubes, *Science 269*, 966–967.

Ci, L., Song, L., Jin, C., Jariwala, D., Wu, D., Li, Y., Srivastava, A. et al. (2010) Atomic layers of hybridized boron nitride and graphene domains, *Nat. Mater. 9*, 430–435.

Corso, M., Auwärter, W., Muntwiler, M., Tamai, A., Greber, T., and Osterwalder, J. (2004) Boron nitride nanomesh, *Science 303*, 217–220.

Dai, J., Xu, L., Fang, Z., Sheng, D., Guo, Q., Ren, Z., Wang, K., and Qian, Y. (2007) A convenient catalytic approach to synthesize straight boron nitride nanotubes using synergic nitrogen source, *Chem. Phys. Lett. 440*, 253–258.

Dalton, T. and Jin, D. (2010) Extent and frequency of vessel oil spills in US marine protected areas, *Mar. Pollut. Bull. 60*, 1939–1945.

Deschamps, G., Caruel, H., Borredon, M., Bonnin, C., and Vignoles, C. (2003) Oil removal from water by selective sorption on hydrophobic cotton fibers. 1. Study of sorption properties and comparison with other cotton fiber-based sorbents, *Environ. Sci. Technol. 37*, 1013–1015.

Dibandjo, P., Bois, L., Chassagneux, F., Cornu, D., Letoffe, J.-M., Toury, B., Babonneau, F., and Miele, P. (2005a) Synthesis of boron nitride with ordered mesostructure, *Adv. Mater. 17*(5), 571–574.

Dibandjo, P., Bois, L., Chassagneux, F., and Miele, P. (2007a) Thermal stability of mesoporous boron nitride templated with a cationic surfactant, *J. Eur. Ceramic Soc. 27*, 313–317.

Dibandjo, P., Chassagneux, F., Bois, L., Sigala, C., and Miele, P. (2005b) Comparison between SBA-15 silica and CMK-3 carbon nanocasting for mesoporous boron nitride synthesis, *J. Mater. Chem. 15*, 1917–1923.

Dibandjo, P., Chassagneux, F., Bois, L., Sigala, C., and Miele, P. (2006) Synthesis of boron nitride with a cubic mesostructure, *Microporous Mesoporous Mater. 92*, 286–291.

Dibandjo, P., Chassagneux, F., Bois, L., Sigala, C., and Miele, P. (2007b) Condensation of borazinic precursors for mesoporous boron nitride synthesis by carbon nanocasting, *J. Mater. Res. 22*, 26–34.

Du, M., Wu, Y., and Hao, X. (2013) A facile chemical exfoliation method to obtain large size boron nitride nanosheets, *CrystEngComm 15*, 1782–1786.

Erickson, K. J., Gibb, A. L., Sinitskii, A., Rousseas, M., Alem, N., Tour, J. M., and Zettl, A. K. (2011) Longitudinal splitting of boron nitride nanotubes for the facile synthesis of high quality boron nitride nanoribbons, *Nano Lett. 11*, 3221–3226.

Fechler, N., Fellinger, T.-P., and Antonietti, M. (2013) "Salt Templating": A simple and sustainable pathway toward highly porous functional carbons from ionic liquids, *Adv. Mater. 25*, 75–79.

Fellinger, T.-P., White, R. J., Titirici, M.-M., and Antonietti, M. (2012) Borax-mediated formation of carbon aerogels from glucose, *Adv. Funct. Mater. 22*, 3254–3260.

Gao, R., Yin, L. W., Wang, C. X., Qi, Y. X., Lun, N., Zhang, L. Y., Liu, Y. X., Kang, L., and Wang, X. F. (2009) High-yield synthesis of boron nitride nanosheets with strong ultraviolet cathodoluminescence emission, *J. Phys. Chem. C 113*, 15160–15165.

Giovannetti, G., Khomyakov, P., Brocks, G., Kelly, P., and van den Brink, J. (2007) Substrate-induced band gap in graphene on hexagonal boron nitride: Ab initio density functional calculations, *Phys. Rev. B 76*, 073103.

Golberg, D., Bando, Y., Stéphan, O., and Kurashima, K. (1998) Octahedral boron nitride fullerenes formed by electron beam irradiation, *Appl. Phys. Lett. 73*, 2441–2443.

Goriachko, A., He, Y., Knapp, M., and Over, H. (2007) Self-assembly of a hexagonal boron nitride nanomesh on Ru (0001), *Langmuir 23*, 2928–2931.

Han, W., Bando, Y., Kurashima, K., and Sato, T. (1998) Synthesis of boron nitride nanotubes from carbon nanotubes by a substitution reaction, *Appl. Phys. Lett. 73*, 3085–3087.

Han, W. Q., Brutchey, R., Tilley, T. D., and Zettl, A. (2004) Activated boron nitride derived from activated carbon, *Nano Lett. 4*, 173–176.

Han, W. Q., Wu, L. J., Zhu, Y. M., Watanabe, K., and Taniguchi, T. (2008) Structure of chemically derived mono- and few-atomic-layer boron nitride sheets. *Appl. Phys. Lett. 93*, 223103-1–223103-3.

Han, W. Q., Yu, H. G., and Liu, Z. X. (2011) Convert graphene sheets to boron nitride and boron nitride–carbon sheets via a carbon-substitution reaction. *Appl. Phys. Lett. 98*, 203112.

Ishikawa, M., Nakamura, T., Morita, M., Matsuda, Y., Tsujioka, S., and Kawashima, T. (1995) Boron-carbon-nitrogen compounds as negative electrode matrices for rechargeable lithium battery systems, *J. Power Sources 55*, 127–130.

Iyyamperumal, E., Wang, S., and Dai, L. (2012) Vertically aligned BCN nanotubes with high capacitance, *ACS Nano 6*, 5259–5265.

Jariwala, D., Sangwan, V. K., Lauhon, L. J., Marks, T. J., and Hersam, M. C. (2013) Carbon nanomaterials for electronics, optoelectronics, photovoltaics, and sensing, *Chem. Soc. Rev. 42*, 2824–2860.

Jayaramulu, K., Kumar, N., Hazra, A., Maji, T. K., and Rao, C. N. R. (2013) A nanoporous borocarbonitride (BC4 N) with novel properties derived from a boron-imidazolate-based metal-organic framework, *Chem. Euro. J. 19*, 6966–6970.

Jhi, S.-H. and Kwon, Y.-K. (2004) Hydrogen adsorption on boron nitride nanotubes: A path to room-temperature hydrogen storage, *Phys. Rev. B 69*, 245407.

Jiang, H. L., Liu, B., Lan, Y. Q., Kuratani, K., Akita, T., Shioyama, H., Zong, F., and Xu, Q. (2011) From metal–organic framework to nanoporous carbon: Toward a very high surface area and hydrogen uptake, *J. Am. Chem. Soc. 133*, 11854.

Johansson, M. P., Suenaga, K., Hellgren, N., Colliex, C., Sundgren, J.-E., and Hultman, L. (2000) Template-synthesized BN:C nanoboxes, *Appl. Phys. Lett. 76*(7), 825–827.

Kim, S. M., Hsu, A., Araujo, P. T., Lee, Y.-H., Palacios, T., Dresselhaus, M., Idrobo, J.-C., Kim, K. K., and Kong, J. (2013) Synthesis of patched or stacked graphene and hBN flakes: A route to hybrid structure discovery, *Nano Lett. 13*, 933–941.

Kim, S. Y., Park, J., Choi, H. C., Ahn, J. P., Hou, J. Q., and Kang, H. S. (2007) X-ray photoelectron spectroscopy and first principles calculation of BCN nanotubes, *J. Am. Chem. Soc. 129*, 1705–1716.

Kumar, N., Moses, K., Pramoda, K., Shirodkar, S. N., Mishra, A. K., Waghmare, U. V., Sundaresan, A., and Rao, C. N. R. (2013) Borocarbonitrides, BxCyNz, *J. Mater. Chem. A 1*, 5806–5821.

Kusari, U., Bao, Z., Cai, Y., Ahmad, G., Sandhage, K. H., and Sneddon, L. G. (2007) Formation of nanostructured, nanocrystalline boron nitride microparticles with diatom-derived 3-D shapes, *Chem. Commun. 11*, 1177–1179.

Lauret, J. S., Arenal, R., Ducastelle, F., and Loiseau, A. (2005) Optical transitions in single-wall boron nitride nanotubes, *Phy. Rev. Lett. 94*, 037405.

Lei, W. W., Portehault, D., Dimova, R., and Antonietti, M. (2011) Boron carbon nitride nanostructures from salt melts: Tunable water-soluble phosphors, *J. Am. Chem. Soc. 133*, 7121–7127.

Lei, W. W., Portehault, D., Liu, D., Qin, S., and Chen, Y. (2013a) Porous boron nitride nanosheets for effective water cleaning, *Nat. Commun. 4*, 1777.

Lei, W. W., Qin, S., Liu, D., Portehault, D., Liu, Z. W., and Chen, Y. (2013b) Large scale boron carbon nitride nanosheets with enhanced lithium storage capabilities, *Chem. Commun. 49*, 352–354.

Li, J., Lin, J., Xu, X., Zhang, X., Xue, Y., Mi, J., Mo, Z. et al. (2013a) Porous boron nitride with a high surface area: Hydrogen storage and water treatment, *Nanotechnology 24*, 155603.

Li, L., Li, L. H., Chen, Y., Dai, X. J., Lamb, P. R., Cheng, B.-M., Lin, M.-Y., and Liu, X. (2013b) High-quality boron nitride nanoribbons: Unzipping during nanotube synthesis, *Angew. Chem. Int. Edn. 125*, 4306–4310.

Li, X., Hao, X., Zhao, M., Wu, Y., Yang, J., Tian, Y., and Qian, G. (2013c) Exfoliation of hexagonal boron nitride by molten hydroxides, *Adv. Mater. 25*, 2200–2204.

Lian, G., Zhang, X., Zhang, S. J., Liu, D., Cui, D. L., and Wang, Q. L. (2012) Controlled fabrication of ultrathin-shell BN hollow spheres with excellent performance in hydrogen storage and wastewater treatment, *Energy Environ. Sci. 5*, 7072–7080.

Liao, L., Liu, K., Wang, W., Bai, X., Wang, E., Liu, Y., Li, J., and Liu, C. (2007) Multiwall boron and carbonitride/carbon nanotube junction and its rectification behavior, *J. Am. Chem. Soc. 129*, 9562–9563.

Lim, S. H., Luo, J., Ji, W., and Lin, J. (2007) Synthesis of boron nitride nanotubes and its hydrogen uptake, *Catal. Today 120*, 346–350.

Lin, L., Li, Z., Zheng, Y., and Wei, K. (2009) Synthesis and application in the CO oxidation conversion reaction of hexagonal boron nitride with high surface area, *J. Am. Chem. Soc. 92*, 1347–1349.

Lin, T. W., Su, C. Y., Zhang, X. Q., Zhang, W. J., Lee, Y. H., Chu, C. W., Lin, H. Y., Chang, M. T., Chen, F. R., and Li, L. J. (2012) Converting graphene oxide monolayers into boron carbonitride nanosheets by substitutional doping, *Small 8*(9), 1384–1391.

Liu, L., Sham, T.-K., and Han, W. (2013) Investigation on the electronic structure of BN nanosheets synthesized via carbon-substitution reaction: The arrangement of B, N, C and O atoms, *Phys. Chem. Chem. Phys. 15*, 6929–6934.

Liu, X. F., Qiao, Y. B., Dong, G. P., Ye, S., Zhu, B., Zhuang, Y. X., and Qiu, J. R. (2009a) BCNO-based long-persistent phosphor, *J. Electrochem. Soc. 156*, 81–84.

Liu, X. F., Ye, S., Dong, G. P., Qiao, Y. B., Ruan, J., Zhuang, Y. X., Zhang, Q., Lin, G., Chen, D. P., and Qiu, J. R. (2009b) Spectroscopic investigation on BCNO-based phosphor: Photoluminescence and long persistent phosphorescence, *J. Phys. D Appl. Phys. 42*, 215409.

Liu, X. F., Ye, S., Qiao, Y. B., Dong, G. P., Zhang, Q., and Qiu, J. R. (2009c) Facile synthetic strategy for efficient and multi-color fluorescent BCNO nanocrystals, *Chem. Commun. 27*, 4073–4075.

Liu, Z., Ma, L., Shi, G., Zhou, W., Gong, Y., Lei, S., Yang, X., Zhang, J., Yu, J., and Hackenberg, K. P. (2013) In-plane heterostructures of graphene and hexagonal boron nitride with controlled domain sizes, *Nat. Nanotechnol. 8*, 119–124.

Lourie, O., Jones, C., Bartlett, B. M., Gibbons, P. C., Ruoff, R. S., and Buhro, W. E. (2000) CVD growth of boron nitride nanotubes, *Chem. Mater. 12*, 1808–1810.

Ma, R., Bando, Y., Zhu, H., Sato, T., Xu, C., and Wu, D. (2002) Hydrogen uptake in boron nitride nanotubes at room temperature, *J. Am. Chem. Soc. 124*, 7672–7673.

Malenfant, P. R. L., Wan, J., Taylor, S. T., and Manoharan, M. (2007) Self-assembly of an organic-inorganic block copolymer for nano-ordered ceramics, *Nat. Nanotechnol. 2*, 43–46.

Marinopoulos, A. G., Wirtz, L., Marini, A., Olevano, V., Rubio, A., and Reining, L. (2004) Optical absorption and electron energy loss spectra of carbon and boron nitride nanotubes: A first-principles approach, *Appl. Phys. A 78*, 1157–1167.

Mazzoni, M. S. C., Nunes, R. W., Azevedo, S., and Chacham, H. (2006) Electronic structure and energetics of B {x} C {y} N{z} layered structures, *Phys. Rev. B Condens. Mater. Phys. 73*, 073108.

Meng, X. L., Lun, N., Qi, Y. Q., Bi, J. Q., Qi, Y. X., Zhu, H. L., Han, F. D. et al. (2010) Low-temperature synthesis of meshy boron nitride with a large surface area, *Eur. J. Inorg. Chem. 2010*(20), 3174–3178.

Misewich, J. A., Martel, R., Avouris, P. H., Tsang, J. C., Heinze, S., and Tersoff, J. (2003) Electrically induced optical emission from a carbon nanotube FET, *Science 300*, 783–786.

Ng, M. F. and Zhang, R. Q. (2004) Optical spectra of single-walled boron nitride nanotubes, *Phys. Rev. B 69*, 115417.

Noel, M. and Suryanarayanan, V. (2002) Role of carbon host lattices in Li-ion intercalation/de-intercalation processes, *J. Power Sources 111*, 193–209.

Novoselov, K. S., Jiang, D., Schedin, F., Booth, T. J., Khotkevich, V. V., Morozov, S. V., and Geim, A. K. (2005) Two-dimensional atomic crystals, *Proc. Natl. Acad. Sci. USA 102*, 10451–10453.

Ogi, T., Kaihatsu, Y., Iskandar, F., Wang, W., and Okuyama, K. (2008) Facile synthesis of new full-color-emitting BCNO phosphors with high quantum efficiency, *Adv. Mater. 20*, 3235–3238.

Oku, T., Kuno, M., Kitahara, H., and Narita, I. (2001) Formation, atomic structures and properties of boron nitride and carbon nanocage fullerene materials, *Int. J. Inorg. Mater. 3*, 597–612.

Pacile, D., Meyer, J. C., Girit, C. O., and Zettl, A. (2008) The two-dimensional phase of boron nitride: Few-atomic-layer sheets and suspended membranes, *Appl. Phys. Lett. 92*, 133107-1–133107-3.

Paine, R. T. and Narula, C. K. (1990) Synthetic routes to boron nitride, *Chem. Rev. 90*, 73–91.

Pakdel, A., Wang, X., Zhi, C., Bando, Y., Watanabe, K., Sekiguchi, T., Nakayama, T., and Golberg, D. (2012a) Facile synthesis of vertically aligned hexagonal boron nitride nanosheets hybridized with graphitic domains, *J. Mater. Chem. 22*, 4818–4824.

Pakdel, A., Zhi, C., Bando, Y., and Golberg, D. (2012b) Low-dimensional boron nitride nanomaterials, *Mater. Today 15*, 256–265.

Pan, Y., Huo, K. F., Hu, Y. M., Fu, J. J., Lu, Y. N., Dai, Z. D., Hu, Z., and Chen, Y. (2005) Boron nitride nanocages synthesized by a moderate thermochemical approach, *Small 1*, 1199–1203.

Pattanayak, J., Kar, T., and Scheiner, S. (2002) Boron-nitrogen (BN) substitution of fullerenes: C_{60} to $C_{12}B_{24}N_{24}$ CBN ball, *J. Phys. Chem. A 106*, 2970–2978.

Petravic, M., Petravic, M., Peter, R., Kavre, I., Li, L. H., Chen, Y., Fan, L. J., and Yang, Y. W. (2010) Decoration of nitrogen vacancies by oxygen atoms in boron nitride nanotubes, *Phys. Chem. Chem. Phys. 12*, 15349–15353.

Portehault, D., Giordano, C., Gervais, C., Senkovska, I., Kaskel, S., Sanchez, C., and Antonietti, M. (2010) High-surface-area nanoporous boron carbon nitrides for hydrogen storage, *Adv. Funct. Mater. 20*, 1827–1833.

Postole, G., Caldararu, M., Ionescu, N. I., Bonnetot, B., Auroux, A., and Guimon, C. (2005) Boron nitride: A high potential support for combustion catalysts, *Thermochim. Acta 434*, 150–157.

Qin, L., Yu, J., Kuang, S. Y., Xiao, C., and Bai, X. D. (2012) Few-atomic-layered boron carbonitride nanosheets prepared by chemical vapor deposition, *Nanoscale 4*, 120–123.

Raidongia, K., Nag, A., Hembram, K. P. S. S., Waghmare, U. V., Datta, R., and Rao, C. N. R. (2010) BCN: A graphene analogue with remarkable adsorptive properties, *Chem. Eur. J. 16*, 149–157.

Reddy, A. L. M., Srivastava, A., Gowda, S. R., Gullapalli, H., Dubey, M., and Ajayan, P. M. (2010) Synthesis of nitrogen-doped graphene films for lithium battery application, *ACS Nano 4*, 6337–6343.

Rocha Martins, J. da. and Chacham, H. (2011) Disorder and segregation in B-C-N graphene-type layers and nanotubes: Tuning the band gap, *ACS Nano 5*, 385–393.

Rozenberg, A. S., Sinenko, Y. A., and Chukanov, N. V. (1993) Regularities of pyrolytic boron nitride coating formation on a graphite matrix, *J. Mater. Sci. 28*(20), 5528–5533.

Rubio, A., Corkill, J. L., and Cohen, M. L. (1994) Theory of graphitic boron nitride nanotubes, *Phys. Rev. B 49*, 5081–5084.

Rudolph, S. (1994) Boron-nitride, *Am. Ceram. Soc. Bull. 73*(6), 89–90.

Schlienger, S., Alauzun, J., Michaux, F., Vidal, L., Parmentier, J., Gervais, C., Babonneau, F., Bernard, S., Miele, P., and Parra, J. B. (2012) Micro-, mesoporous boron nitride-based materials templated from zeolites, *Chem. Mater. 24*(1), 88–96.

Shi, Y., Hamsen, C., Jia, X., Kim, K. K., Reina, A., Hofmann, M., Hsu, A. L. et al. (2010) Synthesis of few-layer hexagonal boron nitride thin film by chemical vapor deposition, *Nano Lett. 10*, 4134–4139.

Simon, P. and Gogotsi, Y. (2008) Materials for electrochemical capacitors, *Nat. Mater. 7*, 845–854.

Sohn, K., Joo Na, Y., Chang, H., Roh, K.-M., Dong Jang, H., and Huang, J. (2012) Oil absorbing graphene capsules by capillary molding, *Chem. Commun. 48*, 5968–5970.

Song, L., Ci, L. J., Lu, H., Sorokin, P. B., Jin, C. H., Ni, J., Kvashnin, A. G. et al. (2010) Direct growth of graphene/hexagonal boron nitride stacked layers, *Nano Lett. 10*, 3209–3215.

Song, X., Hu, J., and Zeng, H. (2013) Two-dimensional semiconductors: Recent progress and future perspectives, *J. Mater. Chem. C 1*, 2952–2969.

Stéphan, O., Ajayan, P. M., Colliex, C., Redlich, P., Lambert, J. M., Bernier, P., and Lefin, P. (1994) Doping graphitic and carbon nanotube structures with boron and nitrogen, *Science 266*, 1683–1685.

Stéphan, O., Bando, Y., Loiseau, A., Willaime, F., Shramchenko, N., Tamiya, T., and Sato, T. (1998) Formation of small single-layer and nested BN cages under electron irradiation of nanotubes and bulk material, *Appl. Phys. A Mater. Sci. Process. 67*, 107–111.

Suenaga, K., Colliex, C., and Demoncy, N. (1997) Synthesis of nanoparticles and nanotubes with well-separated layers of boron nitride and carbon, *Science 278*, 653–655.

Suryamas, A. B., Munir, M. M., Ogi, T., Khairurrijal, and Okuyama, K. (2011) Intense green and yellow emissions from electrospun BCNO phosphor nanofibers, *J. Mater. Chem. 21*, 12629–12631.

Tang, C., Bando, Y., Ding, X., Qi, S., and Golberg, D. (2002) Catalyzed collapse and enhanced hydrogen storage of BN nanotubes, *J. Am. Chem. Soc. 124*, 14550–14551.

Terrones, M., Grobert, N., and Terrones, H. (2002) Synthetic routes to nanoscale BxCyNz architectures, *Carbon 40*, 1665–1684.

Torres, R., Carettia, I., Gagoa, R., Martínd, Z., and Jiméneza, I. (2007) Bonding structure of BCN nanopowders prepared by ball milling, *Diam. Relat. Mater.* 16, 1450–1454.

Vinu, A., Terrones, M., Golberg, D., Hishita, S., Ariga, K., and Mori, T. (2005) Synthesis of mesoporous BN and BCN exhibiting large surface areas via templating methods, *Chem. Mater.* 17, 5887–5890.

Wang, B., Wu, H., Yu, L., Xu, R., and Lim, T. T. (2012a) Template-free formation of uniform urchin-like α-FeOOH hollow spheres with superior capability for water treatment, *Adv. Mater.* 24, 1111–1116.

Wang, L. C., Xu, L. Q., Sun, C. H., and Qian, Y. T. (2009) A general route for the convenient synthesis of crystalline hexagonal boron nitride micromesh at mild temperature, *J. Mater. Chem.* 19, 1989–1994.

Wang, P., Orimo, S., Matsushima, T., Fujii, H., and Majer, G. (2002) Hydrogen in mechanically prepared nanostructured h-BN: A critical comparison with that in nanostructured graphite, *Appl. Phys. Lett. 80*, 318–320.

Wang, S., Zhang, L., Xia, Z., Roy, A., Chang, D. W., Baek, J.-B., and Dai, L. (2012b) BCN graphene as efficient metal-free electrocatalyst for the oxygen reduction reaction, *Angew. Chem. Int. Edn. 51*, 4209–4212.

Wang, W. L., Bai, X. D., Liu, K. H., Xu, Z., Golberg, D., Bando, Y., and Wang, E. G. (2006) Direct synthesis of B-C-N single-walled nanotubes by bias-assisted hot filament chemical vapor deposition, *J. Am. Chem. Soc. 128*, 6530–6531.

Wang, W. N., Ogi, T., Kaihatsu, Y., Iskandar, F., and Okuyama, K. (2011a) Novel rare-earth-free tunable-color-emitting BCNO phosphors, *J. Mater. Chem. 21*, 5183–5189.

Wang, X., Zhi, C., Li, L., Zeng, H., Li, C., Mitome, M., Golberg, D., and Bando, Y. (2011b) "Chemical blowing" of thin-walled bubbles: High-throughput fabrication of large-area, few-layered BN and C(x)-BN nanosheets, *Adv. Mater. 23*, 4072–4076.

Watanabe, K., Taniguchi, T., and Kanda, H. (2004) Direct-bandgap properties and evidence for ultraviolet lasing of hexagonal boron nitride single crystal, *Nat. Mater. 3*, 404–409.

Watanabe, M., Mizushima, K., Itoh, S., and Mashita, M. (1999) Semiconductor device using semiconductor BCN compounds, US Patent 5747118.

Weng, Q., Wang, X., Zhi, C., Bando, Y., and Golberg, D. (2013) Boron nitride porous microbelts for hydrogen storage, *ACS Nano 7*, 1558–1565.

Wood, G. L., Paine, R. T., February, R. V., Re, V., Recei, M., and August, V. (2006) Aerosol synthesis of hollow spherical morphology boron nitride particles, *Chem. Mater. 18*(20), 4716–4718.

Wu, J., Han, W. Q., Walukiewicz, W., Ager, J. W., Shan, W., Haller, E. E., and Zettl, A. (2004) Raman spectroscopy and time-resolved photoluminescence of BN and BxCyNz nanotubes, *Nano Lett. 4*(4), 647–650.

Wu, R. C., Qu, J. H., and Chen, Y. S. (2005) Magnetic powder MnO–Fe$_2$O$_3$ composite—A novel material for the removal of Azo-dye from water, *Water Res. 39*, 630–638.

Wu, Z. S., Ren, W. C., Xu, L., Li, F., and Cheng, H. M. (2011) Doped graphene sheets as anode materials with superhigh rate and large capacity for lithium ion batteries, *ACS Nano 5*, 5463–5471.

Wu, Z.-S., Winter, A., Chen, L., Sun, Y., Turchanin, A., Feng, X., and Müllen, K. (2012) Three-dimensional nitrogen and boron co-doped graphene for high-performance all-solid-state supercapacitors, *Adv. Mater. 24*, 5130–5135.

Xu, M., Liang, T., Shi, M., and Chen, H. (2013) Graphene-like two-dimensional materials, *Chem. Rev. 113*, 3766–3798.

Xue, Y., Yu, D., Dai, L., Wang, R., Li, D., Roy, A., Lu, F., Chen, H., Liu, Y., and Qu, J. (2013) Three-dimensional B,N-doped graphene foam as a metal-free catalyst for oxygen reduction reaction, *Phys. Chem. Chem. Phys. 15*, 12220–12226.

Yin, L.-W., Bando, Y., Golberg, D., Gloter, A., Li, M.-S., Yuan, X., and Sekiguchi, T. (2005) Porous BCN nanotubular fibers: Growth and spatially resolved cathodoluminescence, *J. Am. Chem. Soc. 127*, 16354–16355.

Yu, Z., Hu, M. L., Zhang, C. X., He, C. Y., Sun, L. Z., and Zhong, J. (2011) Transport properties of hybrid zigzag graphene and boron nitride nanoribbons, *J. Phys. Chem. C 115*, 10836–10841.

Zeng, H., Zhi, C., Zhang, Z., Wei, X., Wang, X., Guo, W., Bando, Y., and Golberg, D. (2010) "White graphenes": Boron nitride nanoribbons via boron nitride nanotube unwrapping, *Nano Lett. 10*(12), 5049–5055.

Zhang, X. H., Lu, Z. M., Lin, J., Fan, Y., Li, L. L., Xu, X. W., Hu, L., Meng, F. B., Zhao, J. L., and Tang, C. C. (2013) Spectra properties of BCNO phosphor prepared by a two-step method at low sintering temperature, *ECS J. Solid State Sci. Technol. 2*(3), R39–R43.

Zheng, Y., Jiao, Y., Ge, L., Jaroniec, M., and Qiao, S. Z. (2013) Two-step boron and nitrogen doping in graphene for enhanced synergistic catalysis, *Angew. Chem. Int. Edn. 52*, 3110–3116.

Zhi, C., Bando, Y., Tang, C., Kuwahara, H., and Golberg, D. (2009) Large-scale fabrication of boron nitride nanosheets and their utilization in polymeric composites with improved thermal and mechanical properties, *Adv. Mater. 21*, 2889–2893.

Zhi, C. Y., Bando, Y., Tang, C. C., Golberg, D., Xie, R. G., and Sekigushi, T. (2005) Phonon characteristics and cathodolumininescence of boron nitride nanotubes, *Appl. Phys. Lett. 86*(21), 213110.

Zhu, Y. C., Bando, Y., Xue, D. F., Sekiguchi, T., Golberg, D., Xu, F. F., and Liu, Q. L. (2004) New boron nitride whiskers: Showing strong ultraviolet and visible light luminescence, *J. Phys. Chem. B 108*, 6193–6196.

Chapter 4

Fabrication, Characterization, and Application of Boron Nitride Nanomaterials

Zhi-Gang Chen, Jin Zou, and Hui-Ming Cheng

CONTENTS

4.1 INTRODUCTION

Hexagonal boron nitride (h-BN) is a layered material analogous to graphite that exhibits a hexagonal crystal structure with the lattice parameters of $a = 2.50$ Å, $c = 6.66$ Å. In such an h-BN crystal, the hexagonal layers stack each other, and the B_3N_3 hexagons overlap and alternate with N_3B_3 hexagons (Figure 4.1). h-BN crystals are insulators with a band gap of ~5.8 eV. Individual BN hexagonal sheets roll ideally into nanotube (NT) (similar to carbon nanotubes [CNTs]), firstly proposed in 1994. BNNTs are believed to be a wide-gap semiconductor with a band gap independent of their morphologies and/or geometries (tube diameter, chirality, and number of the wall). Due to their excellent optical and mechanical properties, high thermal conductivity, good oxidation

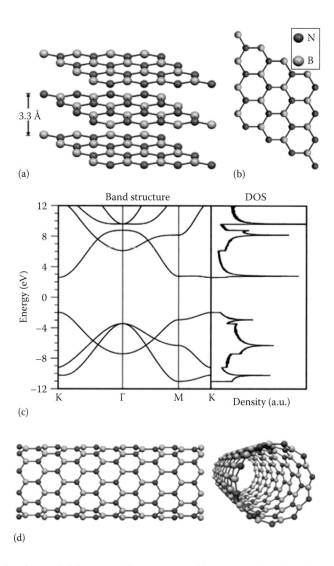

(a)

(b)

(c)

(d)

Figure 4.1   (a) Molecular model depicting the structure of h-BN; note the three layers and the interlayer spacing of 0.33 nm; (b) top view of h-BN showing alternating B and N atoms; (c) calculated band structure and density of states for a single sheet of h-BN showing that this material is an insulator with a band gap of ~4.5 eV (E_f located at zero) (note that this local density approximation calculation underestimates the band gap; other studies considering the GW approximation reveal a band gap of 5.4 eV for bulk h-BN and 6.0 eV for a single h-BN layer); and (d) molecular model of a BNNT exhibiting a (10,0) zigzag chirality (two views are depicted). (Reprinted from *Mater. Today*, 10(5), Terrones, M., Romo-Herrera, J.M., Cruz-Silva, E. et al., Pure and doped boron nitride nanotubes, 10 (5), 30–38, Copyright 2007, with permission from Elsevier.)

resistivity, and chemical inertness, BNNTs show great potential for applications as unique electromechanical and optoelectronic components for laser, light-emitting diode, and medical diagnosis. Inspired by BNNTs, various BN nanostructures, such as NTs, nanobamboos, nanohorns (NHs), nanowires (NWs), and yard-glass BNNTs, have been synthesized. A wide range of catalysts, such as Fe, Ni, Co, Mg, and metal oxides, have been used to synthesize BN nanostructures by different methods including arc discharge, laser heating/ablation, ball milling–annealing, CNT substitution, soft chemical method, and chemical vapor deposition (CVD).

In this chapter, we will summarize the latest progress in synthesis and applications of boron nitride nanomaterials. Especially, we will focus on our findings in this area. At the same time, their doping and their functionalization will be systematically reviewed as well as their various applications.

4.2 BORON NITRIDE NANOTUBES

4.2.1 Fabrications

From the first observation of BNNTs by a carbon-free plasma discharge, extensive methods, including arc discharge, laser heating/ablation, ball milling–annealing, CVD, CNT substitution, and soft chemical method, have been employed for the large-scale synthesis of BNNTs with different layers.

4.2.1.1 Arc Discharge

Inspired by CNTs, multiwalled BNNTs were firstly discovered by using a carbon-free arc discharge between a BN-packed tungsten rod and a cooled copper electrode. Consequently, few-layer BNNTs including single-walled BNNTs were observed through an improved arc discharge using HfB_2 or ZnB_2 electrodes. Such BNNT ends were closed by flat layers or encapsulated metal particles or amorphous BN materials. Double-walled BNNTs were mass fabricated by Ni- and Co-improved boron-rich electrodes in arc discharge methods.

4.2.1.2 Laser Ablation

Pioneering work on laser ablation to fabricate BNNTs was reported by Golberg et al. In this method, single-crystal cubic BN (c-BN) targets are laser heated for 1 min in a diamond anvil cell under N_2 pressures of 5–15 GPa. A temperature of approximately 5000°C was estimated on the target surface, where short and pure BNNTs were observed. A main disadvantage of this technique is that the yield is not high and the quantity of impurities, such as unwanted BN particles, and flakes, is high. Significant developments of the laser technique have been reported by Yu et al. The laser ablation of BN powder mixed with nanosized Ni and Co powder in an inert atmosphere (e.g., Ar, N_2, or He) and heated at 1200°C resulted in the production of single-, double-, and triple-walled BNNTs. Lee et al. reported on a catalyst-free continuous laser ablation of BN targets to generate bulk quantities of samples including single-walled BNNTs. A root-growth model was proposed based on the observation of boron particles at the tube ends. By contrast, though applying a nearly similar experimental setup, Laude et al. did not observe single-walled BNNTs, but observed long rope multiwalled BNNTs. Wang et al. reported high-yield synthesis of multiwalled BNNTs deposited on Fe film–coated oxidized silicon substrates using laser ablation at 600°C, which indicated the feasibility of laser ablation to prepare BNNTs in large scale.

4.2.1.3 Chemical Vapor Deposition

CVD synthesis of BNNTs was initially reported by Lourie et al. using nickel boride catalyst and borazine as a precursor at 1000°C–1100°C. The fabricated BNNTs exhibited lengths up to several micrometers and often processed bulbous, flag-like, and or club-like tips. Presently, CVD is the widely used method for synthesizing BNNTs, which usually needs catalytic growth by metal particles following typical vapor-liquid-solid (VLS) mechanism. However, available boron source suitable for CVD synthesis is severely restricted. Normally, frequently used gaseous boron-contained compounds are BCl_3 and B_2H_6. Su et al. reported large-scale fabrication of BNNTs by using B_2H_6-based gas precursors catalyzed by containing Ni or Fe particles at low temperature (below 1000°C) in a plasma-assisted CVD system.

Ma et al. used a specially designed home made $B_4N_3O_2H$ precursor prepared from boric acid and melamine to synthesize BNNTs under a nitrogen gas atmosphere at 1700°C. Structure analysis showed that the obtained BNNTs exhibited zigzag arrangement and rhombohedral stacking order. Growth observation suggested that B oxynitride nanoclusters doped with Si, Al, and Ca served effective promoters for the growth. Subsequently, controlled synthesis of BNNTs, nanobamboos, and nanocables was implemented using the $B_4N_3O_2H$ precursors under tunable different conditions. Tang et al. developed a new synthetic system based on a novel precursor composed of B_2O_2 and Mg that was in situ generated by reacting B and MgO at 1300°C to synthesize bulk amounts of pure BNNTs in an ammonia atmosphere. Zhi et al. developed this method and added FeO into starting precursors as promoters to fabricate BNNTs in large scale. Up to now, this route remains one of the most effective carbon-free routes to fabricate BNNTs in large scale, which advanced further functionalization and performance research of BNNTs. Huang et al. developed a modified B_2O_2 and Li_2O precursors to fabricate multiwalled BNNTs. The prepared BNNTs show a specific characteristic, which have average external diameters of sub-10 nm and lengths of up to tens of micrometers.

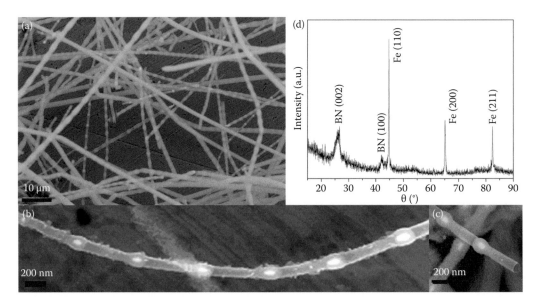

Figure 4.2 SEM images of YG-BNNTs: (a) a typical low-magnification image; (b, c) typical high-magnification images showing uniform units connection and an open end of the tube, respectively. (d) XRD patterns showing two crystalline phases—hexagonal BN and α-Fe. (From Chen, Z.G., Zou, J., Li, F. et al.: Growth of magnetic yard-glass shaped boron nitride nanotubes with periodic iron nanoparticles. *Adv. Funct. Mater.* 2007, 17(16), 3371–3376. Copyright Wiley-VCH Verlag GmbH & Co. KGaA. Reproduced with permission.)

Chen et al. developed a new floating catalyst CVD method and used a homemade B–O–Fe precursor and ferrocenes as floating catalysts to fabricate high-purity yard-glass BNNTs, as shown in Figure 4.2. Such yard-glass BNNTs have Fe periodically filling and show strong ferromagnetic property at the room temperature. Kim et al. used nickelocene as the floating catalyst to fabricate double-walled BNNTs with ~2 nm diameters from borazine as B precursors. Such floating CVD method could fabricate different morphologies of BNNTs in large scale, which are similar to the growth of CNTs.

4.2.1.4 Ball Milling

Ball milling and annealing method was firstly introduced for the synthesis of BNNTs by Chen et al. (1999). Such method involved the ball milling of B in NH_3 followed by thermal annealing at 1000°C–1300°C under N_2 or Ar. In such method, the high-energy ball milling induced a nitration reaction between the B powder and the NH_3 gas and then a metastable material formed comprising disordered BN and nanocrystalline B. The BN fibers grew from the metastable and chemically activated phase during the nitrogen heat treatment. Especially, over 1.0 mm BNNTs with huge resistivity could be synthesized by an optimized ball milling and annealing method. The authors believed that the annealing temperature of 1100°C is crucial for the growth of the long BNNTs because at this temperature, there is a fast nitrogen dissolution rate in Fe and the B/N ratio in Fe is 1. Li et al. developed an improved ball milling method for high-yield BNNT synthesis, in which metal nitrate, such as $Fe(NO_3)_3$, and amorphous boron powder are milled together to prepare a more effective precursor. This method provides an effective mechanothermal route for the growth of BNNTs from solid phase, which is suitable for the mass production of BNNTs.

4.2.1.5 Other Chemical Methods

Due to the structural similarities between CNTs and BNNTs and high reaction activity of carbon at high temperature, CNTs could be ideal templates for the substitution growth of BNNTs. Han et al. initially developed pyrolytically grown CNTs as templates to prepare BNNTs through substituting C with B and N by reacting B_2O_3 with nitrogen gas in the presence of CNTs at 1500°C. In such substituted reaction, BNNTs of similar diameters to the starting CNT materials are obtained. The CNTs are removed by the intense oxidation condition caused by B_2O_3. B and N atoms (supplied during the reaction) may be simultaneously positioned in the vacancies generated with the graphitic network, which result in the formation of BN domains. The major drawback of the synthesis was a significant fraction of ternary BCN NTs possessing significant amounts (2–10 at%) of residual C at the expense of pure BNNTs. A dramatic increase in the yield of pure BNNTs was achieved by Golberg et al. through adding the substitution reaction promoters. Low-sublimation-temperature metal oxides, for example, MoO_3, V_2O_5, PbO, CuO, and Cu_2O, were found to be useful as oxidization promoters in such substitution reaction system. The sublimated oxide vapors caused the starting CNT template opening at the initial reaction stages and at rather moderate temperatures (400°C–700°C) via intense oxidation, which resulted in effective substitution of C for BN on both sides of multiwalled CNT templates and crystallization of BN domains from inside-out and outside-in on the tubular bodies. The substitution reaction route was also utilized for the first time by Golberg et al. to produce single-walled B-doped, B–C–N and BNNTs from bundles of pure single-wall CNTs in large scale. Other templates, such as porous anodic aluminum oxide, were utilized for the fabrication of highly ordered BNNTs. Solution method also developed to fabricate multiwalled BNNTs by co-pyrolyzing NH_4BF_4, KBH_4, and NaN_3 in a temperature range from 450 to 600.

4.2.2   Doped Boron Nitride Nanotubes

Impurity doping can significantly affect the properties of BNNTs and result in novel properties including mainly modified band gaps and magnetic properties. Owing to their similarity in structure, carbon-doped BNNTs were firstly and intensively studied from both theoretical predictions and experiments. It is believed that C doping could be possible to reduce the electronic band gap. It is anticipated that as the C content is decreased, the energy gap increases from 0.15 to 4.2 eV. Main types of C-doped BNNTs, homogeneous B–C–N NTs, and BN/C heterostructure NTs are well considered and fabricated. Especially, the substitution synthesis method could well yield homogeneous C-doped BNNTs. The precise controlling C content is still a big problem. Guo et al. described the synthesis of BN/C heterostructure NTs using a two-stage hot filament CVD process with varying flows of C, B, and N sources under a polycrystalline Ni substrate. This suggests that this kind of heterostructures could be fabricated in a controlled way.

Fluorine-doped (F-doped) BNNTs were reported by Tang et al., and semiconducting behaviors were also observed after being doped with highly electronegative fluorine. Subsequently, theoretical studies on fluorine-doped BNNTs anticipated that F-doped BNNTs exhibited long-ranged ferromagnetic spin ordering along the tube as well as strong spin polarization around the Fermi level, which enables the F-doped BNNTs to function as piezomagnetic NTs. Si-doped bamboo-shaped BNNTs were synthesized via catalyst-assisted pyrolysis of a silicon-containing polymeric precursor under N_2 atmosphere. Intense visible PL emission was observed from the Si-BNNTs in the range between 500 and 800 nm. Very recently, a similar structure with Si dopant concentration of 5% was produced using B pieces, BN powder, and a piece of Si wafer as reactants. The valence band analysis showed that the band gap of BNNTs was reduced by ~1.7 eV after Si doping. All these results indicated that Si could be more stable than C in such doping system.

Different elements, such as V, Ga, In, Al, Be, K, O, P, S, Ni, Pd, and Pt, were theoretically studied to be doped into BNNTs, and expect to modify the optical, electrical, transport, and magnetic properties, but need further experimental investigation. Heavy-element-doped BNNTs, such as Eu and Au, were demonstrated by annealing of ball-milled B-containing mixtures.

Especially, Chen reported that Cu-doped BNNTs were synthesized using nanoscale Cu powders as catalysts. Such Cu-doped BNNTs have large inner diameter and flat tips and self-assembled into microbelts as shown in Figure 4.3. These experimental results prove that Cu can act as a catalyst for the fabrication of BNNTs and self-assembling BNNTs to form microbelts. Cu doping and Cu filling in the BNNTs could greatly benefit their electronic and optoelectronic properties.

4.3   BORON NITRIDE NANOWIRES

Apart from NTs, boron nitride nanowires (BNNWs) are another important morphology of 1D BN nanomaterials, which can act as insulating components in nanoscale devices. Compared with BNNTs, BNNWs always served as the by-product or modified products from the fabrication of BNNTs. Presently, the reports on the BNNWs are not very much.

In 2002, Deepak et al. observed BNNWs obtained by reacting a mixture of H_3BO_3 and activated carbon with NH_3 at 1300°C. Huo et al. fabricated well crystalline BNNWs through the reaction of homemade B-rich FeB nanoparticles with the mixed nitrogen and ammonia at 1100°C. While using the same preparation procedure except for the dramatically reduced boron content in the FeB precursor, bamboo-like BNNTs were obtained.

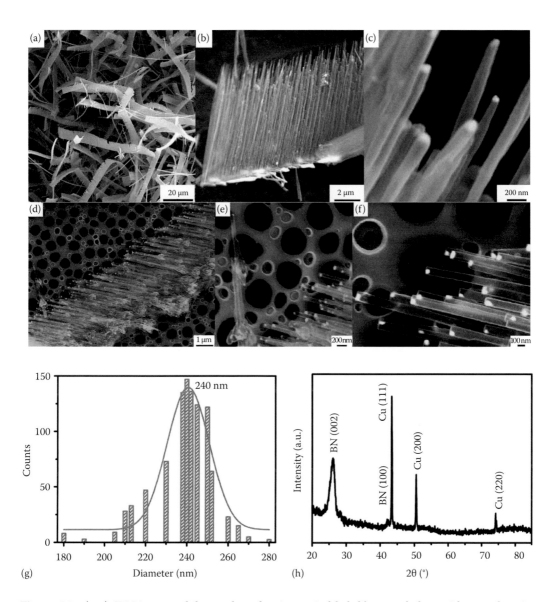

Figure 4.3 (a–c) SEM images of the product showing typical belt-like morphology with a catalyst tip. (d–f) SEM images of the product showing microbelts that consist of arrays of numerous NTs with a flat tip. (g) Outer diameter distribution and (h) XRD spectrum of the product. (Reprinted with permission from Chen, Z.G., Zou, J., Liu, Q.L. et al., Self-assembly and cathodoluminescence of microbelts from Cu-doped boron nitride nanotube. *ACS Nano* 2008, 2(8), 1523–1532, 2008b. Copyright 2008, American Chemical Society.)

Tang et al. developed a floating catalyst CVD method through different catalysts and reaction temperature to selective synthesis of BNNTs, bamboo-shaped BNNTs, and platelet BNNWs. Especially, high-purity and large-scale platelet BNNWs can be fabricated via a floating catalyst-enhanced thermal reaction method by using amorphous boron powder and boron oxide powder as boron source, NH_3 as nitrogen source, nickelocene as catalyst precursor, and iron (II) sulfide as a growth promoter, as shown in Figure 4.4. Such BNNWs are composed of

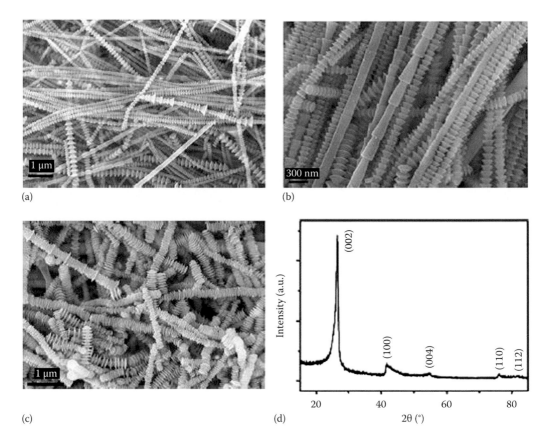

Figure 4.4   (a) Low- and (b) high-magnification SEM images of the as-prepared BN nanowires. (c) SEM image of the nanowires decorated with protruding nanosheets. (d) XRD pattern of the as-prepared BN nanowires. (Reprinted with permission from Tang D. M., Liu C., Cheng H. M., Platelet boron nitride nanowires, *Nano* 2006, 1(1), 65–71. Copyright 2006, World Scientific Publication Co. PTE Ltd.)

periodic cup-like single-crystalline h-BN sections growing along <0001> direction. Crystalline BNNWs were also prepared by heating milled mixture of ZnO and B powder on stainless-steel foils under N_2/H_2 flow.

Turbostratic BN structure (t-BN) NWs, not the hexagonal BN (h-BN) structure as previously reported, were successfully synthesized via a nitriding reaction between boron triiodide (BI_3) and ammonia (NH_3) at 1100°C. Such turbostratic structure may be due to the low deposition temperature (1100°C). Cubic BNNWs also have been predicted to be excited in Si NT, but have not been demonstrated in experiments.

4.4   BORON NITRIDE NANORIBBONS

Nanoribbons (NBs) are one of important morphology shows great application in important applications as optoelectronic devices and/or nanoscale reactors for drug delivery, separation, optoelectronics, and electronics.

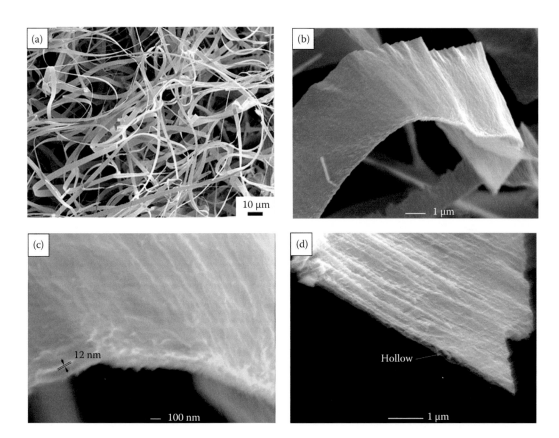

Figure 4.5 (a) Low-magnification SEM image showing high density of synthesized BN products. (b–d) High-magnification SEM images of a synthesized BN product showing their typical thickness and hollow characteristic. (Reprinted with permission from Chen, Z.G., Zou, J., Liu, G. et al., Novel boron nitride hollow nanoribbons, *ACS Nano,* 2008, 2(10), 2183–2191. Copyright 2008, American Chemical Society.)

Zeng et al. reported few- and single-layered BNNRs, mostly terminated with zigzag edges, by unwrapping multiwalled BNNTs through delicate plasma etching. Such BNNRs show metallic characteristics, which is matched with the theoretical predicts. BNNRs also observed co-existed in BN fibers, which fabricated by ZnS NWs as the templates. Chen et al. fabricated high-purity hollow BNNRs using ZnS as template. Such BNNRs duplicated the ZnS NR morphology and have hollow structure. C-doped hollow BNNRs also can be fabricated in such ZnS template methods (Figure 4.5).

Lin et al. reported a facile chemical method to exfoliate h-BNs using amine molecules, resulting in exfoliated BNNRs that could be homogeneously dispersed in organic solvents and/or water. Such wet chemistry method strongly implicates the rich chemistry of these BNNRs, paving the way for future exploitation in their wet processing for a variety of electronics and nanocomposite-related applications. Zhi et al. also reported that BNNRs can peel from the BN particles for large-scale applications.

Single-layer BNNRs can be fabricated by controlled energetic electron irradiation through a layer-by-layer sputtering process. Boron monovacancies are found to be preferably formed, and the dominating zigzag-type edges are proved to be nitrogen terminated in such single-layer BNNRs.

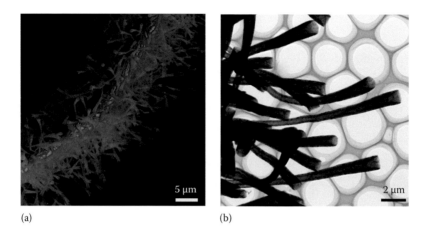

(a)   (b)

Figure 4.6   (a) Typical SEM image of microworms assembled by BNNHs. (b) High-magnification SEM image. (Reprinted from *Chem. Eng. J.*, 165(2), Chen, Z.-G., Cheng, L., Lu, G., and Zou, J., Microworms self-assembled by boron nitride horns for optoelectronic applications, 165(2), 714–719, Copyright 2010, with permission from Elsevier.)

4.5 BORON NITRIDE NANOHORNS

Among them, a BN cone-shaped structure, called BN nanohorns (NHs), is of special interest since the deviations from a flat B–N hexagonal sheet surface are accompanied by the appearance of topological defects located at its apex. This nanostructure was often observed as a by-product of the BNNT synthesis, and its yield was marginally low. A large-scale fabrication of BNNHs was achieved at high temperature (1700°C) by using a mixture of MgO and B powder. The synthesized product exhibited specific aggregation behavior, and Mg catalysts always can be observed at the end of the tips. Therefore, it is still a great significance to search for more efficient methods to fabricate high-purity BNNHs (Figure 4.6).

A lower-temperature and large-scale fabrication of BNNHs was developed at a temperature of 1240°C by a floating catalyst method in a multizone horizontal tube furnace. A key characteristic is that all the BNNHs self-assembled into microworms. Such BNNHs having a conical hollow structure and tens of micrometers in length are self-assembled to form microworms with tens of micrometers in diameter and hundreds of micrometers in length. From detailed characterizations of intermediate products, the infancy BN horns are initiated by the boron particles, then self-intertwisted together to form worm-like structures and finally to form microworms. Such growth is governed by a root-based growth mechanism, which is different from that at 1700°C.

4.6 PROPERTIES AND APPLICATIONS

4.6.1 Physical Properties

The partly ionic structure of BN layers in h-BN reduces the covalency and electrical conductivity, whereas the interlayer interaction increases resulting in higher hardness of h-BN relative to graphite. The significantly different bonding characteristics—strong covalent within the basal planes (planes where boron and nitrogen atoms are covalently bonded) and weak between them—cause high anisotropy of most properties of h-BN. For example, the hardness, electrical conductivity, and thermal conductivity are much higher within the planes than perpendicular to them. On the contrary, the properties of amorphous BN (a-BN), c-BN, and wurtzite (w-BN) are more homogeneous. In fact, in some case, BN nanomaterials conserve or express better physical properties.

The oxidation experiment of BNNTs starts at 950°C, while that of CNTs is at 550°C, which demonstrated that BNNTs display far better thermal and chemical stabilities than CNTs. Therefore, the BNNTs are preferred to serve as protective shells for the device applications at a high-temperature and in a chemically active or hazardous environment. The oxidation property of the synthesized BN/ZnS core–shell heterostructure NWs and ZnS NWs was evaluated through TG analysis in air, and the results are given in Figure 4.7. It is of interest to note that the weight loss of BN/ZnS core–shell heterostructure occurs at a temperature 150°C higher than that for ZnS NWs, indicating that BN/ZnS core–shell heterostructures have a better oxidation resistance. It is believed that the improved oxidation property of the BN/ZnS core–shell heterostructures is attributed to the high chemical inertness of BN shells. Thermal stability and chemical toughness of BNNTs may also be of high importance for the decent performance of NT-based field emitters in flat panel displays and in field-emission (FE) tips for scanning tunneling (STM) and atom force (AFM) microscopes.

Measurements of the thermal conductivity (k) of BNNTs produced with B in its natural abundance and ^{11}B isotopically enriched BNNTs have demonstrated that isotopically pure BNs in fact rival CNTs. Values on the order of 350 W/(m·K) were obtained at room temperature for BNNTs having an outer diameter of 30–40 nm. A dramatic dependence of k on the isotopic disorder was found with a room temperature enhancement in k of 50%, the largest for any material known. The theoretical k of hexagonal BNNRs can approach 1700–2000 W/(m·K), which has the same order of magnitude as the experimentally measured value for graphene, and can be compared to the theoretical calculations for graphene NBs. Moreover, the thermal transport in the BNNRs is anisotropic. The thermal conductivity of zigzag-edged BNNRs is about 20% larger than that of armchair-edged NBs at room temperature.

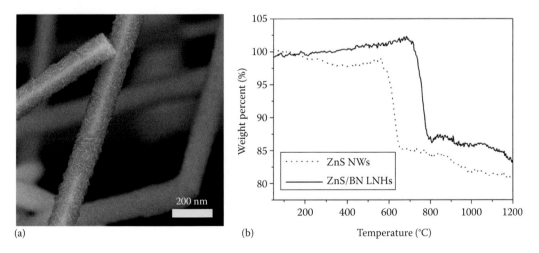

Figure 4.7 (a) Typical SEM image of BN/ZnS core–shell heterostructures; (b) TG analyses of BN/ZnS core–shell heterostructures and ZnS nanowires. (Reprinted with permission from Chen, Z. G., Zou, J., Lu, G. Q. et al., ZnS nanowires and their coaxial lateral nanowire heterostructures with BN, *Applied Physics Letters* 2007, 90(7), 103117–103119. Copyright 2007, American Institute of Physics.)

The mechanical stiffness of BNNTs is assumed to be comparable to that of CNTs. Chopra reported that the yield stress of BNNTs was measured to be 1.1–1.3 TPa. Wei et al. found out Young's modulus of MWBNNT of about 895 GPa from linear fitting of the stress strain plot bamboo-shaped BNNTs show a high tensile fracture strength and Young's modulus up to 8.0 and 225 GPa, respectively. Very recently, by using ab initio density functional theory and the Perdew–Burke–Ernzerhof generalized gradient approximation, Young's modulus of double-wall BNNTs is estimated to be 764–821 GPa. Theoretical value of 4.45 eV was obtained for the band gap energy of double-wall BNNTs. The calculated Young's modulus for armchair NTs is compared to the existing experimental and theoretical data.

4.6.2   Cathodoluminescent Properties

Cathodoluminescence is a strong method to investigate the optoelectronic properties of BN nanostructures. CL spectrum of the yard-glass BNNTs and bamboo-shaped BNNTs is shown in Figure 4.8a. Compared with a single emission at ~352 nm from the bamboo-shaped BNNTs, the spectrum of yard-glass BNNTs is composed of three relatively strong emissions centered at 352, 440, and 685 nm. The peaks centered at 352 (corresponding to 3.52 eV) and 440 nm (corresponding to 2.82 eV) are typical for BN nanostructures or multiwalled cylindrical BNNTs, which are excited by high-energy electrons. It is of interest to note that the emission at 685 nm (corresponding to 1.80 eV) is the strongest one among the three emissions. The band structures of BNNTs were investigated both theoretically and experimentally. It was reported that the electronic structures of BNNTs can be tuned between 1 and 5.8 eV by the direct band gap, radiative transitions, excitonic effects, and doping/replacement. The broad low-energy emissions in the 500–600 nm range are often considered to be due to defects and the B and/or N vacancies. Moreover, visible emissions at 490 and 550 nm were observed due to Si doping and Eu doping in BS-BNNTs,[7] respectively. Taking all these analyses into account, the emission at 685 nm may be attributed to the specific structures of yard-glass BNNTs, although there is a possibility that the BNNTs were doped by Fe. In fact, no Fe was detected in YG-BNNTs from EELS analysis, and the

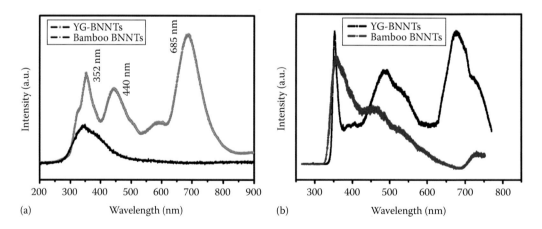

Figure 4.8   (a) CL spectra of a single YG-BNNT and bamboo-shaped BNNT. (b) PL spectra of YG-BNNTs and bamboo BNNTs. T = 300 K, excitation at 325 nm. (Reprinted with permission from Chen, Z.G., Zou, J., Liu, G. et al., Long wavelength emissions of periodic yard-glass shaped boron nitride nanotubes, *Applied Physics Letters* 2007, 94(2), 023105–023107, 2009. Copyright 2009, American Institute of Physics.)

intensity of the 685 nm emission is significantly strong; the possibility that this long-wavelength emission is caused by Fe doping can be ruled out. It is believed that the YG-BNNTs have the inserting connection mode that gestates lattice defects to induce this emission at 685 nm. Such long-wavelength emission could be important for the optoelectronic applications in scatheless biological essay and medial analysis, and also indicates that the light emission from near-UV to near-red light range can be tuned by structural engineering (introducing defects), not by doping.

The PL spectra, excited at 325 nm (~3.81 eV) and 300 K, of a concentrated ensemble of the YG-BNNTs and bamboo BNNTs, are shown in Figure 4.7b. The PL spectrum of YG-BNNTs is composed of three individual luminescent emissions centered at ~352, ~470, and ~680 nm, which are well consistent with the CL observations. On the contrary, only one peak centered at ~352 nm with a shoulder at ~450 nm is observed in bamboo-shaped BNNTs. From these comparisons, it can be concluded that the PL regions at ~350–470 nm can be attributed to the intrinsic emission from BS-BNNTs, which have a large number of bent BN layers and associated defects similar to the cup-shaped BN layers in the bamboo-shaped NTs. The near-red emission centered at ~680 nm can be attributed to lattice defects in the periodical structures and their inserting connection mode.

Two CL peaks, respectively, centered at 341.1 and 415.77 nm, were observed in BNNHs. The peak centered at 341.1 nm (3.64 eV) is a typical CL emission from hexagonal BN structure or multiwalled cylindrical BNNTs, excited by high-energy electron. The peak (centered at 415.77 nm, corresponding 2.98 eV) is most likely associated with the large-diameter conical-structured BN nanostructures as the emission at the same energy range was observed only from BNNHs, suggesting that they are attributed to deep-level emissions associated with defect-related centers (B or N vacancy-type defect-trapped states). Visible emission peaks around 532.85 (2.33 eV) and 602.95 nm (2.57 eV) also were discovered in BNNHs, which could be attributed to large curve in the assembly processing. In fact, the electronic structure of BNNH has been investigated in a few theoretical studies. The gap of $B_{31}N_{31}$ nanohorn was given to be 0.8 eV by molecular orbital calculation, while a more reasonable band gap at around 3.0 eV is given by density function theory calculation. Moreover, the band structures of associated single-walled or multiwalled BNNTs and BN structures have been investigated both theoretically and experimentally, and the electronic structures are varied between 1 and 5.8 eV by the direct band gap and radiative transitions. Therefore, the variable CL peak positions (ultraviolet and visible emissions) for BN nanomaterials could be induced by intrinsic structure characteristics of variable impurities, defects, or B and N vacancies.

4.6.3 Field Emitters

High-quality field emitters with a low turn-on field, high emission current, and stability are strongly desirable for practical applications in a wide range of FE-based devices, such as high-efficiency flat-panel displays, vacuum microwave amplifiers, parallel-electron-beam microscopes, and x-ray sources. Extensive investigations have indicated that controlling geometrical configurations, band-engineering, and doping modification can greatly enhance the performances of 1D nanostructure. As a wide band gap semiconductor, BN (5.5 eV for hexagonal-BN and ~6.4 eV for cubic-BN) and its associated low-dimensional nanostructures have continuously attracted wide attentions due to their structurally independent electric properties. In addition, BN nanomaterials have excellent optical and mechanical properties, high thermal conductivity, resistance to oxidation, chemical inertness, and negative electron affinity. These unique features make BN a promising cold electron emission material. For this reason, exploring material fabrications is indispensable for the realization of high-efficiency field emitters, and extensive efforts have been devoted to manufacturing BN cathodes and to understanding their potential superior FE properties.

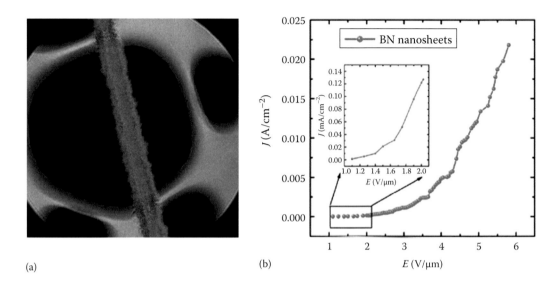

(a)   (b) E (V/μm)

Figure 4.9 (a) Typical TEM image of BN fibers with ultrathin BN sheets and (b) their corresponding filed-emission properties. (Reprinted with permission from Chen, Z.G. and Zou, J., Field emitters: Ultrathin BN nanosheets protruded from BN fibers, *Journal of Material Chemistr*, 2011, 21(4), 1191–1195. Copyright 2011, The Royal Chemical Society.)

The form ultrathin BN sheets protruded from BNNTs templated by ZnS NWs showed excellent FE properties, as shown in Figure 4.9. The turn-on field for the film made of the ultrathin BN nanosheets protruding from BN fiber is 1.9 V·μm^{-1}, compared with that of random-oriented NWs (4.0 V·μm^{-1}) of ZnS NWs, both measured at a current density of 10 μA·cm^{-2}. The threshold field of the ultrathin BN nanosheets can be determined to be 4.65 V·μm^{-1} at a current density of 10 mA·cm^{-2}. The emission current density reached 21.8 mA·cm^{-2} at a macroscopic field of 5.8 V·μm^{-1} (not saturated) for ultrathin BN nanosheets protruding from BN fiber, which is close to the CNTs. This value is more than 10 times higher compared to that of randomly oriented ZnS NWs (~2.2 mA·cm^{-2} under the same macroscopic field). Cumings et al. studied the FE properties of BNNTs using an in situ manipulation stage inside a transmission electron microscope (TEM), and the emission current turned to be detectable at an electric field of 25 V·μm^{-1}. Tang et al. reported a turn-on electric field of 4.85 V·μm^{-1} for platelet BN NWs with rough surfaces. Sugino et al. investigated the FE properties of BN nanofilms, and an electric field of 8.3 V·μm^{-1} was required to obtain an emission current of 1×10^{-11} A. Tang et al. reported a turn-on electric field of 14 V·μm^{-1} for bulky BNNTs. Yin et al. claimed a turn-on electric field of 4.02 V·μm^{-1} for BN nanosheets protruding from Si$_3$N$_4$ NWs. The results of the turn-on and threshold at 1.9 and 4.65 V·μm^{-1}, respectively, from ultrathin BN nanosheets, much lower than those of BN nanostructures and close to the BN porous nanourchins with metallic BN edges. The unique structural characteristics of the as-prepared BN nanostructures can be responsible for their superior FE performance. Therefore, it is believed that the ultrathin BN nanosheets are necessary for securing these extraordinary FE properties, as summarized in Table 4.1.

Although the work function of BN is comparable to that of AlN (3.7 eV), SiC (4.0 eV), GaN (6 eV), Si (3.6 eV), CNTs (5 eV), TiO$_2$ (4.5 eV), and ZnO (5.3 eV), the FE properties of BN nanostructures are similar or even better than those of many other materials because of their thermal and chemical

TABLE 4.1   KEY FE PERFORMANCE PARAMETERS OF THE BN AND OTHER NANOSTRUCTURES REPORTED IN THE LITERATURE

Samples	Turn-On Field[a] $(V \mu m^{-1})$	Field Enhancement Factor (β)	Stability: Testing Time and Fluctuation
BN sheets	1.9	1600	1.0%
BN porous nanourchins	1–1.3	—	—
BN nanosheets	4.02	—	2%
BNNTs	14	—	—
BN nanofilms	8.3 at 1×10^{-11} A	—	—
BNNTs	25 (in situ)	—	—
BNNWs	4.85	—	4%
ZnS pyramids	2.81	3000	8 h, <0.9%
CNTs	2.7	—	10%
TiO_2 nanotubes	11.3	—	3%
AlN nanotips	4.7	1175–1888	0.74%

[a] We define the turn-on field at a field producing emission current density of 10 $\mu A \ cm^{-2}$. If the other values are used, then this is mentioned separately.

stabilities. Furthermore, it can be anticipated that structural engineering can greatly improve the FE properties of BN nanomaterials and result in using as candidates for applications as field emitters or other device systems.

4.7   PERSPECTIVES AND CONCLUSIONS

By taking into account all of the earlier-mentioned and preexisting restrictions/limitations of the BNNT/nanosheet syntheses and analyses, their modifications, and practical utilizations, the scientists should focus more on a design of the desired diameter and thickness with controllable syntheses and targeted functionalizations of BN nanostructures. The final goal of these efforts is the detailed evaluation of their advanced properties required for future high-performance composites, nano-, bio-, electromechanical, and medical devices.

ACKNOWLEDGMENTS

This work was supported by the Australian Research Council. ZGC thanks QLD government for a smart state future fellowship and a UQ research foundation excellent award.

REFERENCES

Baierle, R. J., P. Piquini, T. M. Schmidt, and A. Fazzio. Hydrogen adsorption on carbon-doped boron nitride nanotube. *Journal of Physical Chemistry B* 110(42) (2006): 21184–21188.

Blase, X., A. Rubio, S. G. Louie, and M. L. Cohen. Stability and band-gap constancy of boron-nitride nanotubes. *Europhysics Letters* 28(5) (1994): 335–340.

Chang, C. W., A. M. Fennimore, A. Afanasiev et al. Isotope effect on the thermal conductivity of boron nitride nanotubes. *Physical Review Letters* 97(8) (2006): 085901.

Chen, H., Y. Chen, C. P. Li et al. Eu-doped boron nitride nanotubes as a nanometer-sized visible-light source. *Advanced Materials* 19(14) (2007): 1845–1848.

Chen, H., Y. Chen, Y. Liu et al. Rare-earth doped boron nitride nanotubes. *Materials Science and Engineering B-Solid State Materials for Advanced Technology* 146(1–3) (2008a): 189–192.

Chen, H., Y. Chen, Y. Liu, L. Fu, C. Huang, and D. Llewellyn. Over 1.0 mm-long boron nitride nanotubes. *Chemical Physics Letters* 463(1–3) (2008): 130–133.

Chen, Y., L. T. Chadderton, J. FitzGerald, and J. S. Williams. A solid-state process for formation of boron nitride nanotubes. *Applied Physics Letters* 74(20) (1999): 2960–2962.

Chen, Y., J. D. Fitz Gerald, J. S. Williams, and S. Bulcock. Synthesis of boron nitride nanotubes at low temperatures using reactive ball milling. *Chemical Physics Letters* 299(3–4) (1999b): 260–264.

Chen, Y., J. Zou, S. J. Campbell, and G. Le Caer. Boron nitride nanotubes: Pronounced resistance to oxidation. *Applied Physics Letters* 84(13) (2004): 2430–2432.

Chen, Y. J., B. Chi, D. C. Mahon, and Y. Chen. An effective approach to grow boron nitride nanowires directly on stainless-steel substrates. *Nanotechnology* 17(12) (2006): 2942–2946.

Chen, Y. J., L. Fu, Y. Chen, J. Zou, J. Li, and W. H. Duan. Tunable electric conductivities of Au-doped boron nitride nanotubes. *Nano* 2(6) (2007): 367–372.

Chen, Z. G., J. Zou, F. Li et al. Growth of magnetic yard-glass shaped boron nitride nanotubes with periodic iron nanoparticles. *Advanced Functional Materials* 17(16) (2007): 3371–3376.

Chen, Z. G. and J. Zou. Field emitters: Ultrathin BN nanosheets protruded from BN fibers. *Journal of Materials Chemistry* 21(4) (2011): 1191–1195.

Chen, Z. G., J. Zou, G. Liu et al. Novel boron nitride hollow nanoribbons. *ACS Nano* 2(10) (2008): 2183–2191.

Chen, Z. G., J. Zou, Q. L. Liu et al. Self-assembly and cathodoluminescence of microbelts from Cu-doped boron nitride nanotubes. *ACS Nano* 2(8) (2008): 1523–1532.

Chen, Z. G., J. Zou, G. Liu et al. Long wavelength emissions of periodic yard-glass shaped boron nitride nanotubes. *Applied Physics Letters* 94(2) (2009): 023105–023107.

Chen, Z. G., J. Zou, G. Q. Lu, G. Liu, F. Li, and H. M. Cheng. ZnS nanowires and their coaxial lateral nanowire heterostructures with BN. *Applied Physics Letters* 90(10) (2007): 103117–103119.

Chen, Z. G., J. Zou, D. W. Wang et al. Field emission and cathodoluminescence of ZnS hexagonal pyramids of zinc blende structured single crystals. *Advanced Functional Materials* 19(3) (2009): 484–490.

Chen, Z.-G., L. Cheng, G. Lu, and J. Zou. Microworms self-assembled by boron nitride horns for optoelectronic applications. *Chemical Engineering Journal* 165(2) (2010): 714–719.

Cho, Y. J., C. H. Kim, H. S. Kim et al. Electronic structure of Si-doped BN nanotubes using x-ray photoelectron spectroscopy and first-principles calculation. *Chemistry of Materials* 21(1) (2009): 136–143.

Chopra, N. G., R. J. Luyken, K. Cherrey et al. Boron-nitride nanotubes. *Science* 269(5226) (1995): 966–967.

Chopra, N. G. and A. Zettl. Measurement of the elastic modulus of a multi-wall boron nitride nanotube. *Solid State Communications* 105(5) (1998): 297–300.

Ci, L., L. Song, C. H. Jin et al. Atomic layers of hybridized boron nitride and graphene domains. *Nature Materials* 9(5) (2010): 430–435.

Cumings, J. and A. Zettl. Mass-production of boron nitride double-wall nanotubes and nanococoons. *Chemical Physics Letters* 316(3–4) (2000): 211–216.

Cumings, J. and A. Zettl. Field emission and current-voltage properties of boron nitride nanotubes. *Solid State Communications* 129(10) (2004): 661–664.

Deepak, F. L., C. P. Vinod, K. Mukhopadhyay, A. Govindaraj, and C. N. R. Rao. Boron nitride nanotubes and nanowires. *Chemical Physics Letters* 353(5–6) (2002): 345–352.

Deheer, W. A., A. Chatelain, and D. Ugarte. A carbon nanotube field-emission electron source. *Science* 270(5239) (1995): 1179–1180.

Duan, X. F., C. M. Niu, V. Sahi et al. High-performance thin-film transistors using semiconductor nanowires and nanoribbons. *Nature* 425(6955) (2003): 274–278.

Fakhrabad, D. V. and N. Shahtahmassebi. First-principles calculations of the Young's modulus of double wall boron-nitride nanotubes. *Materials Chemistry and Physics* 138(2–3) (2013): 963–966.

Fan, S. S., M. G. Chapline, N. R. Franklin, T. W. Tombler, A. M. Cassell, and H. J. Dai. Self-oriented regular arrays of carbon nanotubes and their field emission properties. *Science* 283(5401) (1999): 512–514.

Golberg, D., Y. Bando, L. Bourgeois, K. Kurashima, and T. Sato. Insights into the structure of BN nanotubes. *Applied Physics Letters* 77(13) (2000): 1979–1981.

Golberg, D., Y. Bando, M. Eremets, K. Takemura, K. Kurashima, and H. Yusa. Nanotubes in boron nitride laser heated at high pressure. *Applied Physics Letters* 69(14) (1996): 2045–2047.

Golberg, D., Y. Bando, W. Han, K. Kurashima, and T. Sato. Single-walled B-doped carbon, B/N-doped carbon and BN nanotubes synthesized from single-walled carbon nanotubes through a substitution reaction. *Chemical Physics Letters* 308(3–4) (1999): 337–342.

Golberg, D., Y. Bando, Y. Huang et al. Boron nitride nanotubes and nanosheets. *ACS Nano* 4(6) (2010): 2979–2993.

Golberg, D., Y. Bando, K. Kurashima, and T. Sato. MoO$_3$-promoted synthesis of multi-walled BN nanotubes from C nanotube templates. *Chemical Physics Letters* 323(1–2) (2000): 185–191.

Golberg, D., Y. Bando, K. Kurashima, and T. Sato. Ropes of BN multi-walled nanotubes. *Solid State Communications* 116(1) (2000): 1–6.

Golberg, D., Y. Bando, K. Kurashima, and T. Sato. Synthesis and characterization of ropes made of BN multi-walled nanotubes. *Scripta Materialia* 44(8–9) (2001): 1561–1565.

Golberg, D., Y. Bando, K. Kurashima, and T. Sato. Synthesis, HRTEM and electron diffraction studies of B/N-doped C and BN nanotubes. *Diamond and Related Materials* 10(1) (2001): 63–67.

Golberg, D., Y. Bando, C. C. Tang, and C. Y. Zhi. Boron nitride nanotubes. *Advanced Materials* 19(18) (2007): 2413–2432.

Golberg, D., P. Costa, O. Lourie et al. Direct force measurements and kinking under elastic deformation of individual multiwalled boron nitride nanotubes. *Nano Letters* 7(7) (2007): 2146–2151.

Gou, G. Y., B. C. Pan, and L. Shi. The nature of radiative transitions in o-doped boron nitride nanotubes. *Journal of the American Chemical Society* 131(13) (2009): 4839–4845.

Guo, J. D., C. Y. Zhi, X. D. Bai, and E. G. Wang. Boron carbonitride nanojunctions. *Applied Physics Letters* 80(1) (2002): 124–126.

Han, W. Q., Y. Bando, K. Kurashima, and T. Sato. Synthesis of boron nitride nanotubes from carbon nanotubes by a substitution reaction. *Applied Physics Letters* 73(21) (1998): 3085–3087.

Han, W. Q., W. Mickelson, J. Cumings, and A. Zettl. Transformation of B$_x$C$_y$N$_z$ nanotubes to pure BN nanotubes. *Applied Physics Letters* 81(6) (2002): 1110–1112.

Han, W. Q., H. G. Yu, C. Zhi et al. Isotope effect on band gap and radiative transitions properties of boron nitride nanotubes. *Nano Letters* 8(2) (2008): 491–494.

Hernandez, E., C. Goze, P. Bernier, and A. Rubio. Elastic properties of C and B$_x$C$_y$N$_z$ composite nanotubes. *Physical Review Letters* 80(20) (1998): 4502–4505.

Hu, S., X. Lu, J. Yang, W. Liu, Y. Dong, and S. Cao. Prediction of formation of cubic boron nitride nanowires inside silicon nanotubes. *Journal of Physical Chemistry C* 114(47) (2010): 19941–19945.

Huang, Y., J. Lin, C. Tang et al. Bulk synthesis, growth mechanism and properties of highly pure ultrafine boron nitride nanotubes with diameters of sub-10 nm. *Nanotechnology* 22(14) (2011): 145602.

Huo, K. F., Z. Hu, F. Chen et al. Synthesis of boron nitride nanowires. *Applied Physics Letters* 80(19) (2002): 3611–3613.

Huo, K. F., Z. Hu, J. J. Fu, H. Xu, X. Z. Wang, and Y. N. Lu. Microstructure and growth model of periodic spindle-unit BN nanotubes by nitriding Fe-B nanoparticles with nitrogen/ammonia mixture. *Journal of Physical Chemistry B* 107(41) (2003): 11316–11320.

Iyyamperumal, E., S. Y. Wang, and L. M. Dai. Vertically aligned BCN nanotubes with high capacitance. *ACS Nano* 6(6) (2012): 5259–5265.

Jaffrennou, P., J. Barjon, T. Schmid et al. Near-band-edge recombinations in multiwalled boron nitride nanotubes: Cathodoluminescence and photoluminescence spectroscopy measurements. *Physical Review B* 77(23) (2008): 235422-1–235422-7.

Jalili, S. and R. Vaziri. Curvature effect on the electronic properties of BN nanoribbons. *Molecular Physics* 108(24) (2010): 3365–3371.

Jin, C. H., F. Lin, K. Suenaga, and S. Iijima. Fabrication of a freestanding boron nitride single layer and its defect assignments. *Physical Review Letters* 102(19) (2009): 195505.

Kahaly, M. U. and U. V. Waghmare. Electronic structure of carbon doped boron nitride nanotubes: A first-principles study. *Journal of Nanoscience and Nanotechnology* 8(8) (2008): 4041–4048.

Kan, E. J., F. Wu, H. J. Xiang, J. L. Yang, and M. H. Whangbo. Half-metallic dirac point in b-edge hydrogenated BN nanoribbons. *Journal of Physical Chemistry C* 115(35) (2011): 17252–17254.

Kaur, J. and N. Goel. Influence of carbon-doping by boron/nitrogen substitution in boron nitride nanotube, a density functional theory study of nuclear quadrupole resonance parameters. *Journal of Computational and Theoretical Nanoscience* 10(1) (2013): 48–53.

Kim, M. J., S. Chatterjee, S. M. Kim et al. Double-walled boron nitride nanotubes grown by floating catalyst chemical vapor deposition. *Nano Letters* 8(10) (2008): 3298–3302.

Kim, S. Y., J. Park, H. C. Choi, J. P. Ahn, J. Q. Hou, and H. S. Kang. X-ray photoelectron spectroscopy and first principles calculation of BCN nanotubes. *Journal of the American Chemical Society* 129(6) (2007): 1705–1716.

Kim, Y. H., K. J. Chang, and S. G. Louie. Electronic structure of radially deformed BN and bc3 nanotubes. *Physical Review B* 63(20) (2001): 205408.

Lan, J. H., J. S. Wang, C. K. Gan, and S. K. Chin. Edge effects on quantum thermal transport in graphene nanoribbons: Tight-binding calculations. *Physical Review B* 79(11) (2009): 115401.

Laude, T., Y. Matsui, A. Marraud, and B. Jouffrey. Long ropes of boron nitride nanotubes grown by a continuous laser heating. *Applied Physics Letters* 76(22) (2000): 3239–3241.

Lauret, J. S., R. Arenal, F. Ducastelle et al. Optical transitions in single-wall boron nitride nanotubes. *Physical Review Letters* 94(3) (2005): 037405–037407.

Lee, R. S., J. Gavillet, M. L. de la Chapelle et al. Catalyst-free synthesis of boron nitride single-wall nanotubes with a preferred zig-zag configuration. *Physical Review B* 64(12) (2001): 121405.

Li, L., L. H. Li, Y. Chen et al. Mechanically activated catalyst mixing for high-yield boron nitride nanotube growth. *Nanoscale Research Letters* 7(1) (2012): 417.

Lim, Y. B., E. Lee, Y. R. Yoon, M. S. Lee, and M. Lee. Filamentous artificial virus from a self-assembled discrete nanoribbon. *Angewandte Chemie-International Edition* 47(24) (2008): 4525–4528.

Lin, Y., T. V. Williams, and J. W. Connell. Soluble, exfoliated hexagonal boron nitride nanosheets. *Journal of Physical Chemistry Letters* 1(1) (2010): 277–283.

Liu, C., Y. Tong, H. M. Cheng, D. Golberg, and Y. Bando. Field emission properties of macroscopic single-walled carbon nanotube strands. *Applied Physics Letters* 86(22) (2005): 223114–223115.

Liu, G., F. Li, D. W. Wang et al. Electron field emission of a nitrogen-doped TiO_2 nanotube array. *Nanotechnology* 19(2) (2008): 025606.

Loh, K. P., I. Sakaguchi, M. N. Gamo, S. Tagawa, T. Sugino, and T. Ando. Surface conditioning of chemical vapor deposited hexagonal boron nitride film for negative electron affinity. *Applied Physics Letters* 74(1) (1999): 28–30.

Loiseau, A., F. Willaime, N. Demoncy, G. Hug, and H. Pascard. Boron nitride nanotubes with reduced numbers of layers synthesized by arc discharge. *Physical Review Letters* 76(25) (1996): 4737–4740.

Lourie, O. R., C. R. Jones, B. M. Bartlett, P. C. Gibbons, R. S. Ruoff, and W. E. Buhro. CVD growth of boron nitride nanotubes. *Chemistry of Materials* 12(7) (2000): 1808–1810.

Ma, R. Z., Y. Bando, and T. Sato. Controlled synthesis of BN nanotubes, nanobamboos, and nanocables. *Advanced Materials* 14(5) (2002): 366–368.

Ma, R. Z., Y. Bando, T. Sato, and K. Kurashima. Growth, morphology, and structure of boron nitride nanotubes. *Chemistry of Materials* 13(9) (2001): 2965–2971.

Machado, M., P. Piquini, and R. Mota. Electronic properties of selected BN nanocones. *Materials Characterization* 50(2–3) (2003): 179–182.

Mirzaei, M. and M. Giahi. Computations of the quadrupole coupling constants in aluminum doped boron nitride nanotubes. *Physica B-Condensed Matter* 405(18) (2010): 3991–3994.

Nishiwaki, A., T. Oku, and I. Narita. Formation and atomic structures of boron nitride nanohoms. *Science and Technology of Advanced Materials* 5(5–6) (2004): 629–634.

Pan, Y. F. and Z. Q. Yang. Electronic structures and spin gapless semiconductors in BN nanoribbons with vacancies. *Physical Review B* 82(19) (2010): 195308.

Peyghan, A. A., M. T. Baei, M. Moghimi, and S. Hashemian. Phenol adsorption study on pristine, Ga-, and In-doped (4,4) armchair single-walled boron nitride nanotubes. *Computational and Theoretical Chemistry* 997(2012): 63–69.

Rahman, G. and S. C. Hong. Possible magnetism of Be-doped boron nitride nanotubes. *Journal of Nanoscience and Nanotechnology* 8(9) (2008): 4711–4713.

Rubio, A., J. L. Corkill, and M. L. Cohen. Theory of graphitic boron-nitride nanotubes. *Physical Review B* 49(7) (1994): 5081–5084.

Saito, Y., M. Maida, and T. Matsumoto. Structures of boron nitride nanotubes with single-layer and multilayers produced by arc discharge. *Japanese Journal of Applied Physics Part 1-Regular Papers Short Notes & Review Papers* 38(1A) (1999): 159–163.

Sanchez-Portal, D. and E. Hernandez. Vibrational properties of single-wall nanotubes and monolayers of hexagonal BN. *Physical Review B* 66(23) (2002): 235415.

Sharma, S., P. Rani, A. S. Verma, and V. K. Jindal. Structural and electronic properties of sulphur-doped boron nitride nanotubes. *Solid State Communications* 152(9) (2012): 802–805.

Shelimov, K. B. and M. Moskovits. Composite nanostructures based on template-crown boron nitride nanotubules. *Chemistry of Materials* 12(1) (2000): 250–254.

Shen, G. Z., Y. Bando, J. Q. Hu, and D. Golberg. High-symmetry ZnS hepta- and tetrapods composed of assembled ZnS nanowire arrays. *Applied Physics Letters* 90(12) (2007): 123101–123103.

Soltani, A., S. G. Raz, V. J. Rezaei, A. D. Khalaji, and M. Savar. Ab initio investigation of Al- and Ga-doped single-walled boron nitride nanotubes as ammonia sensor. *Applied Surface Science* 263 (2012): 619–625.

Su, C.-Y., W.-Y. Chu, Z.-Y. Juang et al. Large-scale synthesis of boron nitride nanotubes with iron-supported catalysts. *Journal of Physical Chemistry C* 113(33) (2009): 14732–14738.

Su, C.-Y., Z.-Y. Juang, K.-F. Chen et al. Selective growth of boron nitride nanotubes by the plasma-assisted and iron-catalytic CVD methods. *Journal of Physical Chemistry C* 113(33) (2009): 14681–14688.

Sugino, T., C. Kimura, and T. Yamamoto. Electron field emission from boron-nitride nanofilms. *Applied Physics Letters* 80(19) (2002): 3602–3604.

Sun, M., G. L. Wu, T. Ye, H. Zhang, Z. D. Yang, and Z. S. Li. 2012. The electronic properties of al-, p-doped and al, p co-doped boron nitride nanotubes. In *New Materials, Applications and Processes, Pts 1–3*, J. M. Zeng, Y. H. Kim, and Y. F. Chen (eds.).

Sun, Y. G., W. M. Choi, H. Q. Jiang, Y. G. Y. Huang, and J. A. Rogers. Controlled buckling of semiconductor nanoribbons for stretchable electronics. *Nature Nanotechnology* 1(3) (2006): 201–207.

Tang, C., Y. Bando, T. Sato, and K. Kurashima. A novel precursor for synthesis of pure boron nitride nanotubes. *Chemical Communications* 21(12) (2002): 1290–1291.

Tang, C. C. and Y. Bando. Effect of BN coatings on oxidation resistance and field emission of SiC nanowires. *Applied Physics Letters* 83(4) (2003): 659–661.

Tang, C. C., Y. Bando, D. Golberg, and R. Z. Ma. Cerium phosphate nanotubes: Synthesis, valence state, and optical properties. *Angewandte Chemie-International Edition* 44(4) (2005): 576–579.

Tang, D. M., C. Liu, and H. M. Cheng. Platelet boron nitride nanowires. *Nano* 1(1) (2006): 65–71.

Tang, D. M., C. Liu, and H. M. Cheng. Controlled synthesis of quasi-one-dimensional boron nitride nanostructures. *Journal of Materials Research* 22(10) (2007): 2809–2816.

Tang, D. M., C. L. Ren, X. L. Wei et al. Mechanical properties of bamboo-like boron nitride nanotubes by in situ TEM and MD simulations: Strengthening effect of interlocked joint interfaces. *ACS Nano* 5(9) (2011): 7362–7368.

Tao, O. Y., Y. P. Chen, Y. E. Xie, K. K. Yang, Z. G. Bao, and J. X. Zhong. Thermal transport in hexagonal boron nitride nanoribbons. *Nanotechnology* 21(24) (2010): 245701.

Terrones, M., J. C. Charlier, A. Gloter et al. Experimental and theoretical studies suggesting the possibility of metallic boron nitride edges in porous nanourchins. *Nano Letters* 8(4) (2008): 1026–1032.

Terrones, M., D. Golberg, N. Grobert et al. Production and state-of-the-art characterization of aligned nanotubes with homogeneous BCxN (1 < = x < = 5) compositions. *Advanced Materials* 15(22) (2003): 1899–1903.

Terrones, M., J. M. Romo-Herrera, E. Cruz-Silva et al. Pure and doped boron nitride nanotubes. *Materials Today* 10(5) (2007): 30–38.

Tontapha, S., N. Morakot, V. Ruangpornvisuti, and B. Wanno. Geometries and stabilities of transition metals doped perfect and Stone-Wales defective armchair (5,5) boron nitride nanotubes. *Structural Chemistry* 23(6) (2012): 1819–1830.

Vaccarini, L., C. Goze, L. Henrard, E. Hernandez, P. Bernier, and A. Rubio. Mechanical and electronic properties of carbon and boron-nitride nanotubes. *Carbon* 38(11–12) (2000): 1681–1690.

Wang, J. S., V. K. Kayastha, Y. K. Yap et al. Low temperature growth of boron nitride nanotubes on substrates. *Nano Letters* 5(12) (2005): 2528–2532.

Wang, R. X. and D. J. Zhang. Theoretical study of the adsorption of carbon monoxide on pristine and silicon-doped boron nitride nanotubes. *Australian Journal of Chemistry* 61(12) (2008): 941–945.

Wang, S. Y., E. Iyyamperumal, A. Roy, Y. H. Xue, D. S. Yu, and L. M. Dai. Vertically aligned bcn nanotubes as efficient metal-free electrocatalysts for the oxygen reduction reaction: A synergetic effect by co-doping with boron and nitrogen. *Angewandte Chemie-International Edition* 50(49) (2011): 11756–11760.

Wang, W. L., Y. Bando, C. Y. Zhi, W. Y. Fu, E. G. Wang, and D. Golberg. Aqueous noncovalent functionalization and controlled near-surface carbon doping of multiwalled boron nitride nanotubes. *Journal of the American Chemical Society* 130(26) (2008): 8144–8145.

Wang, X. Z., Q. Wu, Z. Hu, and Y. Chen. Template-directed synthesis of boron nitride nanotube arrays by microwave plasma chemical reaction. *Electrochimica Acta* 52(8) (2007): 2841–2844.

Watanabe, K., T. Taniguchi, and H. Kanda. Direct-bandgap properties and evidence for ultraviolet lasing of hexagonal boron nitride single crystal. *Nature Materials* 3(6) (2004): 404–409.

Wei, X. L., M. S. Wang, Y. Bando, and D. Golberg. Tensile tests on individual multi-walled boron nitride nanotubes. *Advanced Materials* 22(43) (2010): 4895–4899.

Wei, X. L., M. S. Wang, Y. Bando, and D. Golberg. Electron-beam-induced substitutional carbon doping of boron nitride nanosheets, nanoribbons, and nanotubes. *ACS Nano* 5(4) (2011): 2916–2922.

Wu, J., W. Q. Han, W. Walukiewicz et al. Raman spectroscopy and time-resolved photoluminescence of BN and $B_xC_yN_z$ nanotubes. *Nano Letters* 4(4) (2004): 647–650.

Wu, J. B. and W. Y. Zhang. Tuning the magnetic and transport properties of boron-nitride nanotubes via oxygen-doping. *Solid State Communications* 149(11–12) (2009): 486–490.

Xu, L. Q., Y. Y. Peng, Z. Y. Meng et al. A co-pyrolysis method to boron nitride nanotubes at relative low temperature. *Chemistry of Materials* 15(13) (2003): 2675–2680.

Xu, S. F., Y. Fan, J. S. Luo et al. Phonon characteristics and photoluminescence of bamboo structured silicon-doped boron nitride multiwall nanotubes. *Appl. Phys. Lett.* 90(1) (2007): 13115–13117.

Yan, B., C. Park, J. Ihm, G. Zhou, W. Duan, and N. Park. Electron emission originated from free-electron-like states of alkali-doped boron-nitride nanotubes. *Journal of the American Chemical Society* 130(50) (2008): 17012–17015.

Yang, K. K., Y. P. Chen, Y. E. Xie, X. L. Wei, T. Ouyang, and J. X. Zhong. Effect of triangle vacancy on thermal transport in boron nitride nanoribbons. *Solid State Communications* 151(6) (2011): 460–464.

Yao, K. L., Y. Min, Z. L. Liu, H. G. Cheng, S. C. Zhu, and G. Y. Gao. First-principles study of transport of V doped boron nitride nanotube. *Physics Letters A* 372(34) (2008): 5609–5613.

Yong, J. C., Z. Z. Hong, and C. Ying. Pure boron nitride nanowires produced from boron triiodide. *Nanotechnology* 17(3) (2006): 786–789.

Yu, D. P., X. S. Sun, C. S. Lee et al. Synthesis of boron nitride nanotubes by means of excimer laser ablation at high temperature. *Applied Physics Letters* 72(16) (1998): 1966–1968.

Zeng, H. B., C. Y. Zhi, Z. H. Zhang et al. "White graphenes": Boron nitride nanoribbons via boron nitride nanotube unwrapping. *Nano Letters* 10(12) (2010): 5049–5055.

Zhang, H. R., E. J. Liang, P. Ding, Z. L. Du, and X. Y. Guo. Production and growth mechanisms of BCN nanotubes. *Acta Physica Sinica* 51(12) (2002): 2901–2905.

Zhang, J., K. P. Loh, J. W. Zheng, M. B. Sullivan, and P. Wu. Adsorption of molecular oxygen on the walls of pristine and carbon-doped (5,5) boron nitride nanotubes: Spin-polarized density functional study. *Physical Review B* 75(24) (2007): 245301.

Zhang, Z. H. and W. L. Guo. Tunable ferromagnetic spin ordering in boron nitride nanotubes with topological fluorine adsorption. *Journal of the American Chemical Society* 131(19) (2009): 6874–6879.

Zhao, J. X. and B. Q. Dai. DFT studies of electro-conductivity of carbon-doped boron nitride nanotube. *Materials Chemistry and Physics* 88(2–3) (2004): 244–249.

Zheng, F. L., Y. Zhang, J. M. Zhang, and K. W. Xu. Effects of the period vacancy on the structure, electronic and magnetic properties of the zigzag BN nanoribbon. *Journal of Molecular Structure* 984(1–3) (2010): 344–349.

Zhi, C. Y., Y. Bando, C. C. Tang, D. Golberg, R. G. Xie, and T. Sekiguchi. Large-scale fabrication of boron nitride nanohorn. *Applied Physics Letters* 87(6) (2005): 63107–63109.

Zhi, C. Y., Y. Bando, C. C. Tang, H. Kuwahara, and D. Golberg. Large-scale fabrication of boron nitride nanosheets and their utilization in polymeric composites with improved thermal and mechanical properties. *Advanced Materials* 21(28) (2009): 2889–2893.

Zhi, C. Y., Y. Bando, C. C. Tang, R. G. Xie, T. Sekiguchi, and D. Golberg. Perfectly dissolved boron nitride nanotubes due to polymer wrapping. *Journal of the American Chemical Society* 127(46) (2005): 15996–15997.

Zhi, C. Y., J. D. Guo, X. D. Bai, and E. G. Wang. Adjustable boron carbonitride nanotubes. *Journal of Applied Physics* 91(8) (2002): 5325–5333.

Zhou, G. and W. H. Duan. Spin-polarized electron current from carbon-doped open armchair boron nitride nanotubes: Implication for nano-spintronic devices. *Chemical Physics Letters* 437(1–3) (2007): 83–86.

Zhu, Y. C., Y. Bando, L. W. Yin, and D. Golberg. Field nanoemitters: Ultrathin BN nanosheets protruding from si3n4 nanowires. *Nano Letters* 6(12) (2006): 2982–2986.

Chapter 5

Nanosheets and Nanocones Derived from Layered Hydroxides

Renzhi Ma and Takayoshi Sasaki

CONTENTS

5.1 INTRODUCTION

Layered metal hydroxides (Braterman 2004, Rives 2006, Carillo and Griego 2012, Wang and O'Hare 2012), consisting of metal-hydroxyl host layers (sheets, slabs) and/or charge-balancing anions in the interlayer galleries, afford a large variety of functionality and hybrid possibility for potential applications as anion exchangers and adsorbents (Constantino and Pinnavaia 1995, Nijs et al. 1998, Rives and Ulibarri 1999, Millange et al. 2000, Aisawa et al. 2001), catalysts (Cavani et al. 1991, Kumbhar et al. 1998), active electrode materials (Mousty et al. 1994, Robins and Dutta 1996, Lee et al. 2011, Zhao et al. 2011, Liu et al. 2012), and drug delivery systems (Choy et al. 1999, 2000, Khan et al. 2001, Del Arco et al. 2004, Gordijo et al. 2005).

The most basic and simplest form of layered metal hydroxides is based on magnesium hydroxide or so-called brucite structure ($M^{2+}(OH)_2$), also known as β-type hydroxide, in which neutral host layers are composed of divalent metal cation (M^{2+})–centered octahedra coordinated with six hydroxyl groups (OH^-) locating in the vertices, whereas each OH^- is surrounded by three metal cations. The neutral layers are held together by van der Waals force and hydrogen bonding. Other closely related phases of layered hydroxides may be perceived as derivates from compositional changes and corresponding structural modifications to the brucite prototype. For example, if trivalent cations (M^{3+}) isomorphically replace some of the M^{2+} sites in the brucite structure, inducing excessive positive electric charge to the host sheets, the structure needs to intercalate chemical species bearing compatible counter charges, that is, anions, resulting in a substantial expansion in the dimension along layer-stacking direction. Such an evolved structure is well known as layered double hydroxide (LDH) (Allmann 1968a, Braterman et al. 2004, Duan and Evans 2006, Rives 2006), which may be expressed by a generic formula of $[M^{2+}_{1-x}M^{3+}_{x}(OH)_2]^{x+}[A^{n-}_{x/n}]^{x-} \cdot mH_2O$ (M^{2+} and M^{3+} represent di- and trivalent metal cations, M^{2+} = Mg^{2+}, Fe^{2+}, Co^{2+}, Ni^{2+}, Zn^{2+}, etc.; M^{3+} = Al^{3+}, Ga^{3+}, Cr^{3+}, Co^{3+}, Fe^{3+}, etc., A^{n-} represents charge-balancing n-valent anion, $0.2 \leq x \leq 0.33$). Another variation to the brucite structure is the so-called α-type hydroxide (Allmann 1968, Oliva et al. 1982, Delahaye-Vidal and Figlarz 1987, Rajamathi et al. 1997, 2000, Ma 2006b), in which approximately one-sixth to one-fifth of the octahedral M^{2+} sites are replaced by pairs of tetrahedrally coordinated M^{2+} on each side of the host layer. Anions are intercalated in the gallery near the middle plane, directly coordinating with M^{2+} to form tetrahedral apex. The α-type hydroxide structure may be represented by a typical formula of $[M_{octa}{}^{2+}_{1-x}M_{tetra}{}^{2+}_{2x}(OH)_2]^{2x+}[A^{n-}_{2x/n}]^{2x-} \cdot mH_2O$ ($M_{octa}{}^{2+}$ and $M_{tetra}{}^{2+}$ represent octahedrally and tetrahedrally coordinated divalent metal cations, respectively, M^{2+} = Co^{2+}, Ni^{2+}, Zn^{2+}, etc.; A^{n-} represents charge-balancing n-valent anion, $0.166 \leq x \leq 0.2$). Figure 5.1 shows the schematic illustrations of these closely related structures: brucite (β-type hydroxide), LDH, as well as α-type hydroxide.

The morphology of layered metal hydroxides is determined by the crystallization process, which depends not only on the chemical nature of the metal cations and anionic species but also on the synthetic conditions, for example, reaction temperature, time, and posttreatment. Generally, euhedral hydroxide crystals adopt a lamellar or plate-like morphology (Cai et al. 1994, Liu et al. 2005).

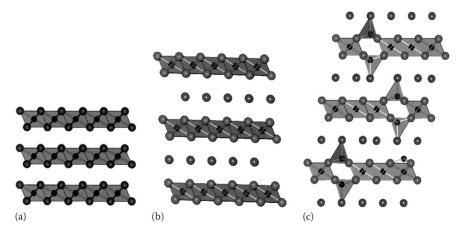

(a)   (b)   (c)

Figure 5.1   Typical structure models of layered metal hydroxides. (a) β-Type: $M^{2+}(OH)_2$, (b) LDH: $M^{2+}_{1-x}M^{3+}_{x}(OH)_2]^{x+}[A^{n-}_{x/n}]^{x-} \cdot mH_2O$, and (c) α-type: $[M_{octa}{}^{2+}_{1-x}M_{tetra}{}^{2+}_{2x}(OH)_2]^{2x+}[A^{n-}_{2x/n}]^{2x-} \cdot mH_2O$. Atoms for M, OH, and guests (interlayer anion, H_2O) are represented by small, large, and medium balls, respectively. (Reprinted with permission from Ma, R., Liu, Z., Takada, K. et al., Tetrahedral Co(II) coordination in α-type cobalt hydroxide: Rietveld refinement and x-ray absorption spectroscopy, *Inorg. Chem.*, 45, 3964–3969, 2006. Copyright 2006, American Chemical Society.)

Recently, new synthetic processes have been developed to prepare versatile morphologies (Xu and Braterman 2004), even 1D nanotubes/nanocones (Liu et al. 2010, 2012).

Due to the structural anisotropy of layered hydroxides, that is, covalent bonding in the host layers is much stronger than layer-to-layer interaction, the structural characteristics endow layered hydroxides with the capacity of incorporating anions and/or solvent species into the interlayer region, which may induce substantial gallery expansion or osmotic swelling. Concomitant with the swelling, layered hydroxide can be exfoliated into unilamellar host slabs, that is, 2D nanosheets, under appropriate conditions (Ma et al. 2006a, Ma and Sasaki 2010, Wang and O'Hare 2012).

In this chapter, we are going to introduce some recent progress, together with basic principles, on the preparation, characterization, and some application prospects of nanomaterials derived from layered metal hydroxides, with a particular focus on nanocones and nanosheets.

5.2 SYNTHESIS OF LAYERED HYDROXIDES

The most commonly used method to precipitate a metal hydroxide is to raise the pH value of a solution containing a single or multiple metal cation(s). The routine procedure to adjust the pH of the solution is using caustic alkali, such as sodium hydroxide (NaOH) or sodium hydrogen carbonate ($NaHCO_3$). The precipitation may be affected by several parameters, but the most important factor is the pH value of the solution. Metal hydroxides are generally amphoteric, that is, they may become soluble at both low and high pH, and the point of minimum solubility (optimum pH for precipitation) occurs at a different pH value for every metal cation.

5.2.1 Platelets

When the pH of a solution containing metal cations is raised by using caustic alkali, metal hydroxide usually forms as aggregates of poor crystallinity and irregular shape, but these can be improved into more regular shaped platelets through prolonged aging treatments. These treatments cause smaller and imperfect crystallites to evolve into larger platelets through Ostwald ripening by a dissolution/recrystallization process (Kahlweit 1975). For example, LDH crystallites can be easily prepared by a constant pH co-precipitation method, in which NaOH or $NaHCO_3$ is added into a solution of mixed metal salts (M^{2+}–M^{3+}) (Cavani et al. 1991, Braterman et al. 2004, Rives and Ulibarri 2006). One of the main advantages of co-precipitation is that a great number of LDH materials with versatile M^{2+}–M^{3+} compositions and different anion combinations can be routinely synthesized. However, the size of the crystallites obtained is typically in the range of several tens to hundreds of nanometers. The shape is neither perfect nor uniform even after longtime aging treatment.

On the other hand, hydroxide platelets with uniform shape and monodispersed size can be synthesized via a method called as homogeneous precipitation. The homogeneous synthesis of large metal hydroxide platelets in micrometer size has been demonstrated by typically using urea ($CO(NH_2)_2$) or hexamethylenetetramine (HMT, $C_6H_{12}N_4$) as hydrolysis agents under mild solvothermal or hydrothermal conditions (Costantino et al. 1998, Ogawa and Kaiho 2002, Adachi-Pagano et al. 2003, Iyi et al. 2004a, Liu et al. 2005). Progressive hydrolysis of urea or HMT produces ammonia, which slowly makes the solution containing metal cations alkaline and induces homogeneous nucleation and crystallization of hydroxides. A slow nucleation rate, resulting from a low degree of supersaturation, is favorable for the growth of large crystallites.

For example, highly crystalline hexagonal platelets of brucite (β) hydroxides such as $Co(OH)_2$, $Co^{2+}_{1-x}Fe^{2+}_x(OH)_2$, and $Co^{2+}_{1-x}Ni^{2+}_x(OH)_2$ can be synthesized via refluxing a dilute aqueous solution of divalent cobalt and/or ferrous (nickel) cations with HMT as a hydrolysis agent under nitrogen gas protection (Liu et al. 2005, Ma et al. 2007, 2008, 2011, 2012, Liang et al. 2010). Figure 5.2 displays scanning electron microscopic (SEM) images of hexagonal crystallites of β-$Co(OH)_2$ and $Co_{3/4}Fe_{1/4}(OH)_2$ prepared from homogeneous precipitation. The products typically consist of uniform hexagonal platelets with a mean lateral size of a few to 10 μm and a thickness of approximately several tens of nanometers. In bimetallic hydroxide platelets such as $Co^{2+}_{1-x}Fe^{2+}_x(OH)_2$ and $Co^{2+}_{1-x}Ni^{2+}_x(OH)_2$, the ratio of metallic contents (Co/Fe, Co/Ni) in the brucite-like products was usually found consistent with the designed molar ratio in the

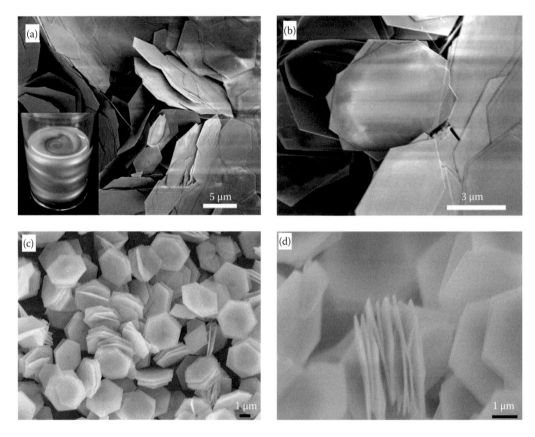

Figure 5.2 SEM images of as-synthesized uniform hexagonal platelets of brucite hydroxide. (a, b) β-$Co(OH)_2$. (Reprinted with permission from Liu, Z., Ma, R., Osada, M., Takada, K., and Sasaki, T., Selective and controlled synthesis of alpha- and beta-cobalt hydroxides in highly developed hexagonal platelets, *J. Am. Chem. Soc.*, 127, 13869–13874, 2005. Copyright 2005, American Chemical Society.) The inset in panel (a) shows a vial of the suspension obtained by dispersing β-$Co(OH)_2$ platelets in ethanol. (c, d) $Co_{3/4}Fe_{1/4}(OH)_2$. (Reprinted with permission from Ma, R., Liang, J., Liu, X., and Sasaki, T., General insights into structural evolution of layered double hydroxide: Underlying aspects in topochemical transformation from brucite to layered double hydroxide, *J. Am. Chem. Soc.*, 134, 19915–19921. Copyright 2012, American Chemical Society.) Some platelets in the images are tilted, showing nanometer thickness of the platelets.

starting solution, demonstrating a comparable range of optimum pH for the precipitation of mixed $Fe^{2+}/Co^{2+}/Ni^{2+}$ salts.

During the homogeneous precipitation of brucite (β) hydroxides under dilute salt concentration, a transformation proceeds with the dissolution of initially formed α-type hydroxide, and subsequent conversion into brucite (β) nucleus was observed. This $\alpha \to \beta$ transformation could be delayed or even prevented by adding an excess of salts (anions) into the reaction solution. For example, the addition of a certain amount of NaCl in dilute $CoCl_2$ solutions during the homogeneous precipitation using HMT could guarantee the yield of a pure phase of chloride-intercalated α-type cobalt hydroxide (Liu et al. 2005). Figure 5.3 shows SEM images of synthesized α-$Co(OH)_2$ with a composition of $[Co_{octa}{}^{2+}{}_{0.83}Co_{tetra}{}^{2+}{}_{0.34}(OH)_2]^{0.34+}Cl_{0.34}\cdot0.46H_2O$, in which approximately one-sixth of the Co^{2+} at octahedral sites is replaced by pairs of tetrahedrally coordinated Co^{2+} on each side of the hydroxide plane (Ma et al. 2006b). The existence of tetrahedral Co^{2+} gives rise to a typical green color appearance, which can be easily distinguished from the pink color of β-$Co(OH)_2$ in Figure 5.2a.

The homogeneous precipitation method can also be employed for the synthesis of well-developed LDH crystallites. For example, Mg–Al LDH, prepared by precipitating Mg^{2+} and Al^{3+} salts through hydrolyzing urea in a sealed vessel at 75°C –80°C for a month, yielded large LDH platelets with lateral sizes that could reach 20 μm (Cai et al. 1994). Similarly, monodispersed hexagonal Mg–Al LDH crystallites, with a typical lateral size of ~10 μm, were readily produced by a simple hydrothermal treatment of Mg^{2+} and Al^{3+} salts with HMT, as shown in Figure 5.4a (Iyi et al. 2004a, Li et al. 2005). On the other hand, hydrothermal synthesis may bring about oxidation to some divalent transition metal cations such as Co^{2+} and Fe^{2+}. Refluxing under a protective atmosphere (e.g., N_2) was proven effective to avoid such oxidation. In this way, uniform hexagonal platelets of Co–Al, Zn–Al, and Ni–Al LDHs (Liu et al. 2006, 2007), with a typical lateral size of ~4 μm, could be reproducibly obtained (see Figure 5.4).

All the homogeneously precipitated platelets generally exhibit a clear anisotropic stream when dispersing in water or ethanol (see insets in Figures 5.2a and 5.3a), implying a large aspect

Figure 5.3 SEM images of α-$Co(OH)_2$. The inset in panel (a) shows a vial of the suspension obtained by dispersing the platelets in ethanol. (Reprinted with permission from Liu, Z., Ma, R., Osada, M., Takada, K., and Sasaki, T., Selective and controlled synthesis of alpha- and beta-cobalt hydroxides in highly developed hexagonal platelets, *J. Am. Chem. Soc.*, 127, 13869–13874, 2005. Copyright 2005, American Chemical Society.)

Figure 5.4   SEM images of LDH hexagonal platelets prepared by homogeneous precipitation. (a) Mg–Al and (b) Co–Al. (Reprinted with permission from Liu, Z., Ma, R., Osada, M. et al., Synthesis, anion exchange, and delamination of Co-Al layered double hydroxide: Assembly of the exfoliated nanosheet/polyanion composite films and magneto-optical studies, *J. Am. Chem. Soc.*, 128, 4872–4880, 2006. Copyright 2006, American Chemical Society.) (c) Zn–Al and (d) Ni–Al. (Reprinted with permission from Liu, Z., Ma, R., Ebina, Y., Iyi, N., Takada, K., and Sasaki, T., General synthesis and delamination of highly crystalline transition-metal-bearing layered double hydroxides, *Langmuir*, 23, 861–867, 2007. Copyright 2007, American Chemical Society.)

ratio. It is estimated that the aspect ratio is typically much higher than 100, indicating the in-plane growth is greatly preferred. All the hydroxides are crystallized in hexagonal or rhombohedral symmetry, with in-plane lattice constant of $a \approx 0.3$ nm whereas interlayer spacing c usually ≥ 0.48 nm depending on the anion intercalation. There are more OH groups per unit area in the host plane than those in other crystallographic planes across the gallery. To maximize the exposure of OH groups to the aqueous phase, that is, more hydrogen bonds between water and OH groups in the growing crystallites or less surface tension, crystallite growth along the in-plane axes is more energy favorable (Xu and Braterman 2004). In other words, nucleation of a new layer will be less facile than continual growth at the edge of one layer that already exists, a prerequisite for forming 2D anisotropy. Moreover, the isotropic in-plane growth along six equivalent a axes in hexagonal symmetry finally leads to the formation of uniform hexagonal platelets.

5.2.2 Nanocones

The introduction of organic surfactant anions during the precipitation of hydroxides may dramatically change their morphology. For example, the surface in inorganic hydroxide is usually hydrophilic. It may become hydrophobic due to surfactant modification. These surface modifications, related to the structure feature of surfactant and/or its arrangement between hydroxide layers, remarkably affect the structure, growth habit, and morphology of final products. Because of the hydrophobic interactions and multiple packing modes of surfactants in the gallery, it has reported that bar-like or ribbon-like crystallites with micrometer or submicrometer dimensions could be synthesized with the addition of surfactants (Xu and Braterman 2004).

Recently, a unique synthetic process for preparing nanocones of layered cobalt hydroxides has been reported by using HMT or urea as alkaline reagents and sodium dodecyl sulfate (SDS, $C_{12}H_{25}OSO_3Na$) as both surfactant and structure-directing agent (Liu et al. 2010, 2012). The products, several micrometers in length, display a conical structure with an apparent hollow interior (see Figure 5.5). The nanocones also exhibit a green color, ascribed to the existence of tetrahedral Co^{2+} coordination. They were therefore identified as DS^--intercalated α-type cobalt hydroxide.

In addition to cobalt hydroxide, this synthetic strategy may be extended to prepare layered hydroxide nanocones with other transition metal or bimetallic compositions (Ni, Co–Ni, Co–Zn, Co–Cu). For example, typical x-ray diffraction (XRD) patterns of layered Co–Ni hydroxides at varied Co/Ni ratios are shown in Figure 5.6. The initial value of 2.4 nm for bimetallic Co–Ni hydroxides with low nickel content is close to that of cobalt hydroxide nanocones. A proposed model for the DS-intercalated α-type hydroxide nanocones is shown in Figure 5.6b. The tetrahedral fourth apex pointing into the interlayer space may coordinate with the sulfate group of a DS anion. The thickness of hydroxide layers consisting of octahedral and tetrahedral coordination is estimated at 0.80 nm, whereas the length of DS ion is 1.78 nm. Therefore, the interlayer spacing of 2.4 might correspond to antiparallel grafting of DS at a slant angle between the host layers. However, the interlayer distance is slightly expanded to about 2.7 nm

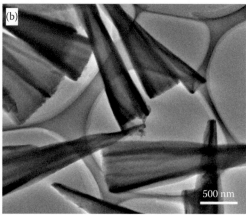

Figure 5.5   SEM (a) and transmission electron microscopic (TEM) (b) images of as-prepared nanocones of layered cobalt hydroxide. (Liu, X., Ma, R., Bando, Y., and Sasaki, T.: A general strategy to layered transition-metal hydroxide nanocones: Tuning the composition for high electrochemical performance. *Adv. Mater.* 2012. 24. 2148–2153. Copyright Wiley-VCH Verlag GmbH & Co. KGaA. Reproduced with permission.)

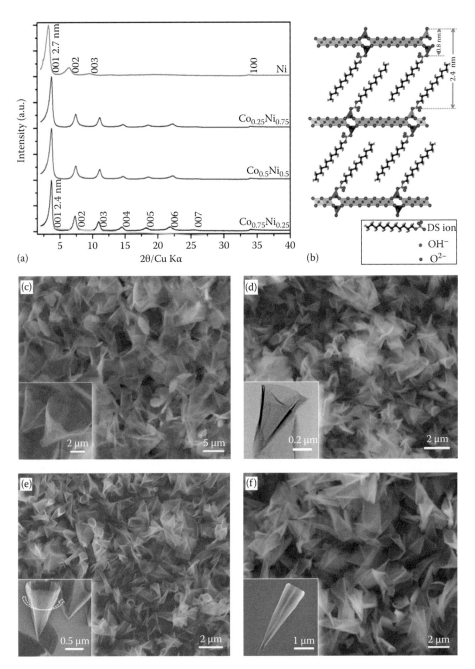

Figure 5.6 (a) XRD patterns of layered Co–Ni hydroxide nanocones at varied Co/Ni ratios, showing the gallery expansion with increasing nickel content. (b) Structure model showing antiparallel arrangement of DS anions in the gallery. (c through f) SEM images of layered hydroxide nanocones with different Co–Ni compositions. (c) Ni, (d) $Co_{0.25}$–$Ni_{0.75}$, (e) $Co_{0.5}$–$Ni_{0.5}$, and (f) $Co_{0.75}$–$Ni_{0.25}$. Corresponding high-magnification images of individual nanocones are displayed as insets. (Liu, X., Ma, R., Bando, Y., and Sasaki, T.: A general strategy to layered transition-metal hydroxide nanocones: Tuning the composition for high electrochemical performance. *Adv. Mater.* 2012. 24. 2148–2153. Copyright Wiley-VCH Verlag GmbH & Co. KGaA. Reproduced with permission.)

for nanocones of nickel hydroxide. The reason for the gallery expansion might be related to some change in the coordination or orientation of DS anions in the gallery. By increasing the amounts of SDS during the synthesis, the interlayer distance of the nanocones can be further expanded to 3.7 nm (Liu et al. 2012). SEM images of Ni, $Co_{0.25}-Ni_{0.75}$, $Co_{0.5}-Ni_{0.5}$, and $Co_{0.75}-Ni_{0.25}$ hydroxide nanocones are given in Figure 5.6c through f, respectively. As one can clearly see in the insets, the conical feature becomes more obvious, and the aspect ratio also increases with higher cobalt content.

The elemental line-scan profiling and mapping results of an individual DS-intercalated $Co_{0.5}-Ni_{0.5}$ hydroxide nanocone are given in Figure 5.7. The line-scan analysis in Figure 5.7a reveals the profile of Co and Ni contents across the radial direction. A low plateau in the center area indicates a hollow core. A homogeneous distribution of Co, Ni, and S elements confirms that Co and Ni were indeed incorporated into the host layer of individual nanocones without segregation.

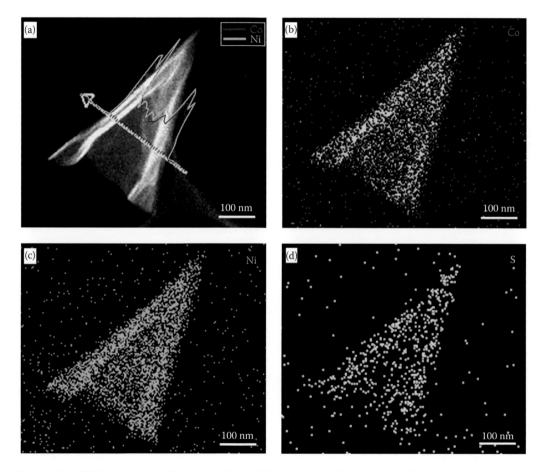

Figure 5.7 (a) Line-scan profile of an individual DS-intercalated $Co_{0.5}-Ni_{0.5}$ hydroxide nanocone across the radial direction. Elemental maps of (b) Co, (c) Ni, and (d) S. (Liu, X., Ma, R., Bando, Y., and Sasaki, T.: A general strategy to layered transition-metal hydroxide nanocones: Tuning the composition for high electrochemical performance. *Adv. Mater.* 2012. 24. 2148–2153. Copyright Wiley-VCH Verlag GmbH & Co. KGaA. Reproduced with permission.)

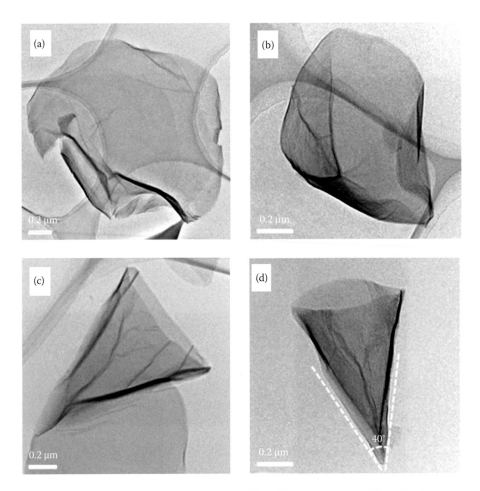

Figure 5.8 (a through d) TEM images exhibit a possible rolling-up scenario of lamellae into nanocones. (Liu, X., Ma, R., Bando, Y., and Sasaki, T.: A general strategy to layered transition-metal hydroxide nanocones: Tuning the composition for high electrochemical performance. *Adv. Mater.* 2012. 24. 2148–2153. Copyright Wiley-VCH Verlag GmbH & Co. KGaA. Reproduced with permission.)

The morphological evolution of layered hydroxides during the synthesis offers a possible rolling-up mechanism for the formation of nanocones. Figure 5.8 summarizes a series of morphological observations supposed to be in different stages of forming nanocones. Firstly, layered hydroxide flake or lamella intercalated with dodecyl sulfate (e.g., thickness in the ranges of a few nanometers, width in the ranges of several hundred nanometers to several micrometers) was initially formed (Figure 5.8a). The flake tends to form a conical angle due to the curve of the edge (Figure 5.8b). Subsequently, it gradually rolls up and naturally grows following the conical angle to form a cone-shaped object (Figure 5.8c). Finally, the flake can fold up onto itself to form a perfect and regular nanocone, as shown in Figure 5.8d.

The growth of peculiar 1D nanocones rather than conventional 2D platelets may be attributed to the introduction of hydrophobic interactions from surfactant anions (dodecyl sulfate). The hydroxide crystallite surface becomes hydrophobic, and thus, there is no preferred growth along any particular axis. Specifically, the surfactant anions are grafted to the hydroxide host layer and thus direct their hydrophobic tails toward the aqueous phase. The hydrophobic tails attract additional surfactants to form an antiparallel packing (see Figure 5.6b), which will facilitate nucleation of new hydroxide layers for crystallite growth. Therefore, growth along layer stacking direction (*c* axis) may become kinetically favorable in surfactant-containing environment than that in pure inorganic solution. Thus, the final nanocone shapes might be an outcome of the balance among the nature of surfactant anions (type, chain length, chain structure, etc.) as well as preparative conditions (temperature, concentration, aging time, etc.).

5.3 EXFOLIATION OF LAYERED HYDROXIDES INTO NANOSHEETS

Over the last decade, there has been a widespread interest in the attempt to exfoliate layered metal hydroxides, typically anion-containing LDHs and α-type hydroxides, into single-layer host units, that is, unilamellar nanosheets with ultimate 2D features. As a direct consequence of the high charge density of host slabs, there is a strong electrostatic attraction among hydroxide host layers, and an integrated hydrogen bonding network developed in the interlayer gallery of LDHs, or directly coordinated anions in α-type hydroxides, hindering their exfoliation. Recently, there have been some success in tackling this issue and made the exfoliation possible (Ma 2006, Ma and Sasaki 2010, Nie and Hou 2011, Wang and O'Hare 2012).

5.3.1 Nanosheets in Organic Solvents

Earlier reports on the exfoliation of LDHs were performed in short-chain alcohols. The interlayer gallery of LDHs was substantially expanded after the insertion of anionic surfactants, and then the exfoliation was promoted via heating or mechanical treatments. For example, refluxing Zn–Al LDH intercalating dodecyl sulfate in butanol led to a translucent colloidal suspension, indicating the exfoliation into nanosheets (Adachi-Pagano et al. 2000). Other solvents such as amyl alcohol and hexanol exhibit similar effects to those of butanol (Singh et al. 2004). The study showed that a successful exfoliation might depend on whether or not interlayer water could be replaced by the short-chain alcohol. When alcohol molecules can quickly replace the interlayer water, for example, under refluxing conditions, the exfoliation is initiated. Furthermore, it seems that the exfoliating process was also dependent on the chain length and head size of surfactants used to modify the layers. Some experimental data indicated that larger benzene sulfonates may promote exfoliation, whereas smaller sulfates seem to be less effective (Venugopal et al. 2006).

In general, the intercalation or modification with anionic surfactants presents the advantage of effectively separating the hydroxide layers, resulting in a large basal spacing typically ≥2.5 nm and a weakened electrostatic attraction. Exfoliation is anticipated if alcohols, assisted by heating or mechanical treatment, are able to solvate the hydrophobic tails of pre-inserted surfactants. Although LDH nanosheets can be obtained using such a combination of surfactants and alcohols, the conditions for exfoliation appear to be complex (e.g., under refluxing or sonication, long reaction time, low yield, etc.).

A new approach for the exfoliation of Mg-Al LDH in a polar solvent, formamide ($HCONH_2$), was revealed to occur instantly and spontaneously at room temperature (Hibino and Jones 2001).

It was later revealed that formamide might act as a general exfoliating agent for a large number of LDHs including Co–Al, Zn–Al, and Ni–Al (Hibino 2004). The mechanism was proposed that the interlayer environment modified with amino acid or other suitable counter anions (dodecyl sulfate, nitrate, etc.) become attractive to formamide driven by the formation of new hydrogen bonding in the gallery and the penetration of a large amount of formamide molecules leads to the swelling and subsequent exfoliation assisted with a mechanical force (Guo et al. 2005, Wu et al. 2005, 2007).

However, LDH crystallites containing carbonate (CO_3^{2-}) as counter anions are not able to be exfoliated directly in formamide because binding force of carbonate is much stronger than other counter anions such as nitrate. For example, LDH platelets prepared through homogeneous precipitation using urea or HMT always contain carbonate due to the fact that induced hydrolysis of urea releases carbonate, while the disproportionation of formaldehyde formed during hydrolysis of HMT may also produce carbonate. The high affinity of carbonate anions to LDH host layers poses a major impediment for the exfoliation of these well-crystallized platelets. To overcome this, de-intercalating carbonate anions and converting them into other anion-containing LDHs, also called as decarbonation, were required. Generally, decarbonation of LDHs is performed using a mixed solution of a very dilute acid (HCl, HNO_3) and a concentrated salt (NaCl, $NaNO_3$) solution (Iyi et al. 2004b). This so-called acid–salt treatment is capable of exchanging carbonate anions into other weaker counter anions (Cl^-, NO_3^-). A precise control of the acid concentration necessary for removing carbonate anions while not causing any apparent morphological changes to the LDH platelets is necessary. Once LDHs are decarbonated, the exfoliation could be achieved by dispersing and mechanically shaking them in formamide for 1–2 days (Li et al. 2005, Liu et al. 2006, 2007).

Typical colloidal suspensions of exfoliated nanosheets from corresponding platelets are shown in Figure 5.9. Tyndall scattering, a light scattering effect by small particles in a colloid, was clearly observed for all suspensions by using a side-incident light beam. Figure 5.10 displays atomic force microscopy (AFM) images of obtained nanosheets on Si substrates, respectively. Well-defined nanosheets, most of them exhibiting a few micrometers in lateral size and only ~0.8 nm in thickness, were clearly observed. The thickness is consistent with the crystallographic thickness of LDH host slabs (0.48 nm). These findings indicate that a combination of homogeneous precipitation of well-crystallized LDH platelets and exfoliation in formamide after decarbonation is an optimum protocol for preparing well-defined LDH nanosheets.

Similarly, through dispersing DS-intercalated nanocones (α-type hydroxide) in formamide and agitating in a mechanical shaker, a translucent colloidal suspension could also be obtained, suggesting successful exfoliation (Liu et al. 2012). Interestingly, nanocones intercalating dodecyl sulfate with an interlayer spacing of 3.7 nm could be more quickly exfoliated than those with a value of 2.4 nm. The reason might be attributed to the advantage of further separating the hydroxide layers and a more hydrophobic interlayer environment for the penetration of formamide. AFM images of as-prepared nanosheets exfoliated from nanocones with different compositions are shown in Figure 5.11. The thickness of the nanosheets was measured to be approximately 1.0 nm, which is close to the crystallographic thickness of α-type hydroxide host layer (0.8 nm). As mentioned earlier (see Figure 5.8 and related descriptions), a nanocone may be formed by the rolling-up of a lamella. The successful exfoliation instead indicates that a nanocone can also be unwrapped/exfoliated into unilamellar nanosheets. Such observations are fundamentally

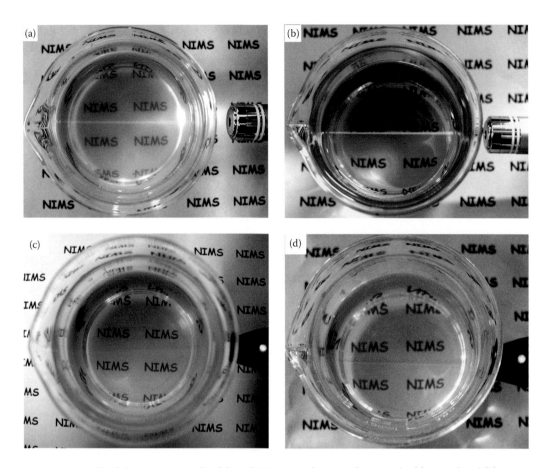

Figure 5.9   Colloidal suspensions of exfoliated LDH nanosheets in formamide. (a) Mg–Al and (b) Co–Al. (Reprinted with permission from Liu, Z., Ma, R., Osada, M. et al., Synthesis, anion exchange, and delamination of Co–Al layered double hydroxide: Assembly of the exfoliated nanosheet/polyanion composite films and magneto-optical studies, *J. Am. Chem. Soc.*, 128, 4872–4880, 2006. Copyright 2006, American Chemical Society.) (c) Zn–Al and (d) Ni–Al. (Reprinted with permission from Liu, Z., Ma, R., Ebina, Y., Iyi, N., Takada, K., and Sasaki, T., General synthesis and delamination of highly crystalline transition-metal-bearing layered double hydroxides, *Langmuir*, 23, 861–867, 2007. Copyright 2007, American Chemical Society.)

important for revealing the formation mechanism of nanotubes/nanocones as well as the energy balance between layered hydroxides and their exfoliated form.

In short, exfoliating LDHs and α-type hydroxides with suitable anions in formamide is a direct reaction occurring without the need for heating or refluxing treatment. Presumably, polar solvent such as formamide may readily penetrate into the gallery and instantly replace water molecules, causing substantial swelling. Assisted by mechanical treatment imposing a transverse sliding force on the swollen phase, the host slabs eventually come apart, that is, exfoliation. Nevertheless, formamide gradually dissolves hydroxide nanosheets over an extended stocking time. Thus, it is considered necessary to achieve exfoliation in more gentle and environmentally friendly solvents such as water.

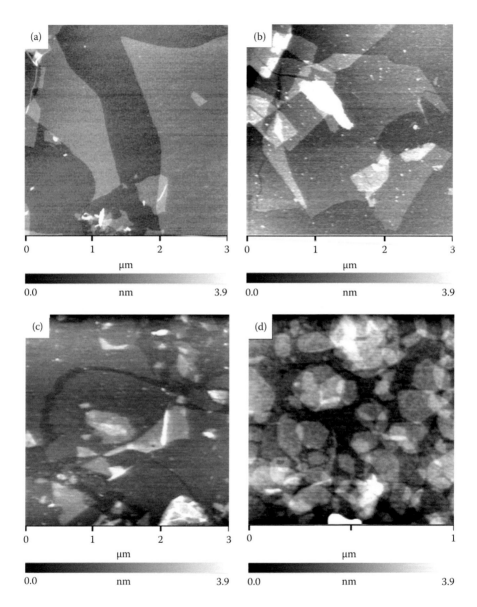

Figure 5.10   AFM images of exfoliated LDH nanosheets. (a) Mg–Al. (Reprinted with permission from Li, L., Ma, R., Ebina, Y., Iyi, N., and Sasaki, T., Positively charged nanosheets derived via total delamination of layered double hydroxides, *Chem. Mater.*, 17, 4386–4391, 2005. Copyright 2005, American Chemical Society.) (b) Co–Al. (Reprinted with permission from Liu, Z., Ma, R., Osada, M. et al., Synthesis, anion exchange, and delamination of Co-Al layered double hydroxide: Assembly of the exfoliated nanosheet/ polyanion composite films and magneto-optical studies, *J. Am. Chem. Soc.*, 128, 4872–4880, 2006. Copyright 2006, American Chemical Society.) (c) Zn–Al and (d) Ni–Al. (Reprinted with permission from Liu, Z., Ma, R., Ebina, Y., Iyi, N., Takada, K., and Sasaki, T., General synthesis and delamination of highly crystalline transition-metal-bearing layered double hydroxides, *Langmuir*, 23, 861–867, 2007. Copyright 2007, American Chemical Society.)

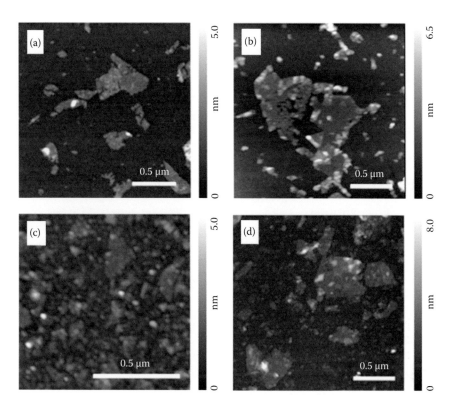

Figure 5.11   AFM images of the exfoliated nanosheets derived from nanocones. (a) Co hydroxide nano-cones with interlayer spacing of 2.4 nm, (b) Co hydroxide nanocones with interlayer spacing of 3.7 nm, (c) Ni hydroxide nanocones, and (d) $Co_{0.5}$–$Ni_{0.5}$ hydroxide nanocones. (Liu, X., Ma, R., Bando, Y., and Sasaki, T.: A general strategy to layered transition-metal hydroxide nanocones: Tuning the composition for high electrochemical performance. *Adv. Mater.* 2012. 24. 2148–2153. Copyright Wiley-VCH Verlag GmbH & Co. KGaA. Reproduced with permission.)

5.3.2   Nanosheets in Water

It was first reported that LDH containing lactate anions swelled and became exfoliated in water (Hibino and Kobayashi 2005). It thus becomes a common sense that intercalating organic anions may be effective for the exfoliation of layered hydroxides in water. Again, balancing the hydrophilicity and hydrophobicity of the intercalant is important. LDHs containing short-chain carboxylates such as acetate and propionate seemed to satisfy this requirement, undergoing exfoliation in water (Iyi et al. 2008). However, short-chain carboxylates tend to degrade into LDH carbonate when exposed to ambient air for a long time, one of the notorious drawbacks for LDHs in contact with water.

A recent report demonstrated that Mg–Al LDHs containing various short-chain organic sulfonate anions, such as isethionate ($HO(CH_2)_2SO_3$), exhibited swelling behavior in water (Iyi et al. 2011). Upon adding water, these isethionate-intercalating LDHs immediately swelled to form a viscous gel, and further addition of water led to the formation of semitransparent colloidal suspensions. AFM observations revealed the existence of nanosheets with a thickness of 1.5–2 nm having the same morphological features as the initial LDHs. Other Ni–Al and Co–Al LDHs with different M^{2+}/M^{3+} ratios also showed water-swelling properties and formed viscous gels when isethionate was similarly incorporated.

5.4 APPLICATION PROSPECTS

Layered metal hydroxides are the most commonly used family of halogen-free flame retardants. These mineral compounds are used in polyolefins, thermoplastic elastomers, polyvinylchloride (PVC), rubbers, and some engineering polymers (such as polyamide). Aluminum hydroxide is usually used when processing temperature is under 200°C. When processing temperature exceeds 200°C, magnesium hydroxide is instead required.

Mg–Al LDHs are often used as polymer additives especially for PVC. As little as 0.05%–1% well-crystallized LDH could improve the thermal stability and weather resistance of thermoplastic resins containing halogen and/or acidic components from the manufacturing process (Miyata and Kuroda 1981). Chlorine-containing polymers commonly undergo degradation under heat and UV light, giving off gaseous HCl. This may catalyze further degradation, leading to brittleness and discoloration. In that aspect, LDHs are regarded as a very good acid absorbent. Though inorganic LDHs are immiscible with polymer because of the surface tension, coating or intercalating surfactant anions bring about a hydrophobic surface that can well combine LDH and polymer (van der Ven et al. 2000, Fisher and Gielgens 2002a,b, Costa et al. 2008). LDH partially exchanged with anionic surfactants was mixed with monomeric materials, for example, caprolactam, to make well-blended polymer–LDH composites.

On the other hand, due to electrochemical and redox activity, transition metal hydroxides (Co, Ni, etc.) are frequently used in electrochemical cells. In particular, as a high-performance active electrode material, they have already been used or proposed for use in various electrochemical energy storage and conversion devices such as rechargeable batteries and next-generation supercapacitors (Gao and Yang 2010).

5.4.1 Nanosheet Assembly

As new elementary building blocks or inorganic *macromolecules* bearing a positive charge, hydroxide nanosheets may be self-assembled via electrostatic interaction into novel nanocomposites, ultrathin films, and other nanoarchitectures (Ma and Sasaki 2010, 2012).

5.4.1.1 Flocculation

In using as a polymer additive, a newly emerging technique is to incorporate LDH nanosheets into polymer matrix (Leroux and Besse 2001, Leroux et al. 2001). The nanometer-scale interaction between polymer and LDH nanosheets may improve properties such as mechanical strength and hardness, thermal stability, flexibility and durability, and impermeability. For example, LDH nanosheets were dispersed into styrene, followed by in situ polymerization. The presence of individually dispersed LDH single layers was observed by high-resolution TEM images. It was also found that the almost complete exfoliation of Mg–Al LDH in polar acrylate monomer solution led to the incorporation of the exfoliated nanosheets into the polymer matrix after in situ polymerization (O'Leary et al. 2002). This polymer/LDH nanocomposite showed a much higher thermal stability than the pure polymer itself.

By simply spreading a colloidal formamide suspension of LDH onto a glass substrate, well-oriented films of an organic LDH composite (*p*-toluene sulfonate/LDH) were fabricated (Okamoto et al. 2006). Composites of carboxymethyl cellulose (CMC) and Mg–Al LDHs were also prepared by exfoliation-assembly method (Kang et al. 2009). LDH was first exfoliated in formamide. Then, CMC dissolved in formamide was added into the exfoliated solution. After centrifugation and water washing to remove the formamide, strong electrostatic interaction between the CMC and LDH nanosheets caused the wrapping of CMC onto LDH nanosheets to form a stable aqueous dispersion. After drying, a restacked layered structure of CMC–LDH was obtained. The thermal decomposition temperature of CMC in the interlayer was increased by 160°C in comparison with that of the pure form. Similarly, swelling reassembly of water-soluble macromolecular thiacalix[4]arene tetrasulfonate (TCAS) and exfoliated Mg–Al LDH nanosheets was also reported (Huang et al. 2010). The hybrid composite was obtained by swelling LDH in formamide and mixed with an appropriate amount of TCAS. When the TCAS concentration was low, a monolayer TCAS was intercalated with a gallery spacing of 1.3 nm. With increasing TCAS concentration, intermolecular π–π interactions led to the formation of an antiparallel arrangement of TCAS in the gallery, yielding a larger spacing of 1.54 nm. TCAS molecules in the interlayer have a higher thermal stability, with the decomposition temperature increased by 100°C, than that of the pure form.

Mg–Al LDH nanosheets were used to assemble with DNA molecules (Park et al. 2010). Electrostatic interaction enabled the adsorption of LDH nanosheets onto the spherical surface of DNA molecules. After stirring and freeze-drying, DNA–LDH core–shell structure was prepared. Elemental line-scan profile proved that the outer shell was covered by LDH (Mg, Al), whereas the inner core was the characteristic of DNA (P, N) composition. This type of core–shell structure of nano-hybrid materials can be used for advanced gene delivery systems and biomedical diagnostics, etc.

Direct flocculation of hydroxide nanosheets (positively charged) and their counterpart nanosheets with opposite charge, for example, oxide nanosheets or dichalcogenide nanosheets (negatively charged), has also been achieved (Li et al. 2007, Corrodano et al. 2010). A lamellar superlattice with two kinds of nanosheets sandwiching each other resulted in a basal spacing compatible with the sum of the crystallographic thickness of the constituent nanosheets. As shown in Figure 5.12, flocculation of Mg–Al LDH and $Ti_{0.91}O_2$ or $Ca_2Nb_3O_{10}$ nanosheets produced restacked products with a basal spacing of 1.2 and 2.0 nm, respectively. These values agree well with the sum of the crystallographic thickness of the LDH nanosheets (0.48 nm) and corresponding oxide nanosheets ($Ti_{0.91}O_2$: 0.73 nm and $Ca_2Nb_3O_{10}$: 1.44 nm). TEM characterization of flocculated samples clearly showed two types of nanosheets with different thicknesses stacked in alternating sequence. On the other hand, the flocculation of Ni–Al LDH nanosheets with TaS_2 nanosheets realized the coexistence of superconductivity and magnetism at 4 K (Corrodano et al. 2010). In that work, it is further demonstrated that the magnetic ordering in Ni–Fe LDH/TaS_2 superlattice is shifted from 4 to 16 K. This self-assembly technique directly using nanosheet as building blocks provides a breakthrough in the rational design of functionalities by constructing artificial superlattice-like hybrid materials in a highly controllable way.

5.4.1.2 Layer-by-Layer Assembly

Bearing a positive charge at a 2D molecular scale, hydroxide nanosheets can be used as a model system for electrostatic layer-by-layer (LbL) assembly in combination with polyanionic counterparts such as poly(sodium 4-styrene sulfonate) (PSS). The film deposition is a well-established protocol, by alternately dipping a solid substrate in a colloidal suspension of LDH nanosheets and PSS solution. Multilayer films can be routinely obtained by repeating the deposition multiple times under optimized conditions. The growth can be monitored by UV–vis absorption spectra

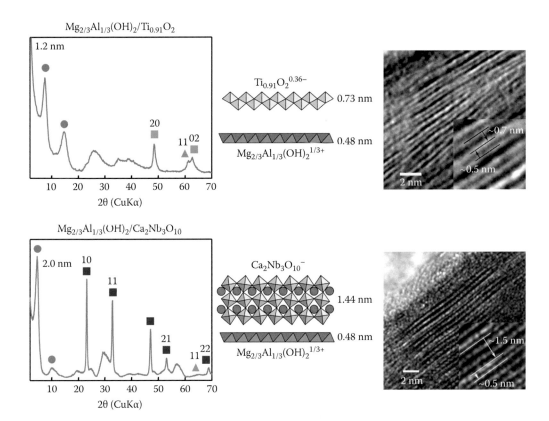

Figure 5.12 Direct flocculation of positively charged Mg–Al LDH nanosheets and negatively charged oxide nanosheets ($Ti_{0.91}O_2$ or $Ca_2Nb_3O_{10}$). The basal spacings of flocculated products are consistent with the thickness sum of two types of nanosheets. TEM observations show that hydroxide and oxide nanosheets are stacked in alternating sequence. (Reprinted with permission from Li, L., Ma, R., Ebina, Y., Fukuda, K., Takada, K., and Sasaki, T., Layer-by-layer assembly and spontaneous flocculation of oppositely charged oxide and hydroxide nanosheets into inorganic sandwiched layered materials, *J. Am. Chem. Soc.*, 129, 8000–8007, 2007. Copyright 2007, American Chemical Society.)

Li et al. 2005, Huang et al. 2009. As shown in Figure 5.13, the monitoring of $(Co–Al\ LDH/PSS)_n$ multilayer film exhibited a linear increase in the absorption band of PSS at around 200 nm as a function of the number of deposition cycles n, evidencing the stepwise and regular film growth of LDH nanosheets. In addition, a repeating periodicity of 2.0 nm in the XRD patterns of the multilayer films was consistent with the thickness sum of LDH nanosheets (0.48 nm) and PSS (1.5 nm). The peak intensity also increased progressively with the increase in the number of layer pairs, a clear sign for the successful multilayer buildup (Liu et al. 2006).

The magneto-optical effects of Co–Al LDH nanosheet multilayer films were investigated using magnetic circular dichroism (MCD) spectroscopy (see Figure 5.13c). It can be seen that these multilayer films exhibit notable positive magneto-optical signals at wavelengths of >300 nm. The magnetic field dependence of the MCD signal clearly indicates a room-temperature ferromagnetic behavior, which can be ascribed to the spin-orbit coupling in Co^{2+} in octahedral coordination. In more detail, the signal at 390 nm may be attributed to the charge transfer of O2p to Co π^*, and the peaks at 470 and 615 nm may originate from the d–d* transitions ($A_{2g} \rightarrow T_{2g}$ and $A_{2g} \rightarrow T_{1g}$,

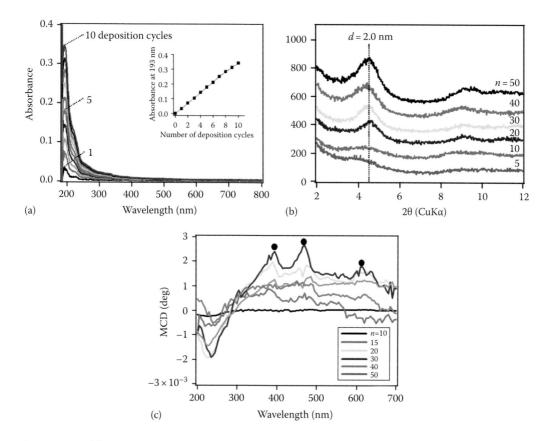

Figure 5.13 (a) UV–vis absorbance spectra of multilayer film of (Co–Al LDH/PSS)$_n$ (n = 1–10) assembled on a quartz glass. The inset shows the observed absorbance at 193 nm plotted against the number of deposition cycles. (b) XRD patterns of the multilayer films of (Co–Al LDH/PSS)$_n$ (n = 5, 10, 20, 30, 40, 50). (c) MCD spectra for the multilayer films of (Co–Al LDH/PSS)$_n$ (n = 10, 15, 20, 30, 40, 50) measured in 10 kOe at 300 K. (Reprinted with permission from Liu, Z., Ma, R., Osada, M. et al., Synthesis, anion exchange, and delamination of Co-Al layered double hydroxide: Assembly of the exfoliated nanosheet/polyanion composite films and magneto-optical studies, *J. Am. Chem. Soc.*, 128, 4872–4880, 2006. Copyright 2006, American Chemical Society.)

respectively) of Co–Co. The increment of MCD signal with increasing number of LDH layers hints strong interlayer coupling between the electronically isolated LDH nanosheets. These results indicate the potential of employing LDH nanosheets as an important testing ground for 2D ferromagnetism.

LbL deposition of photosensitive ultrathin films employing LDH nanosheets and luminescent components was interesting in photoluminescent exploration (Yan et al. 2009, 2010a,b). The films usually show longer fluorescence lifetime and fluorescence polarization than a luminescent entity alone. For example, Mg–Al LDH nanosheets were paired with sulfonated poly(p-phenylene) anionic derivatives (APPP) for electrostatic deposition. A periodic density function calculation indicates that the film is a type of multiple quantum-well structure that confines the valence electrons of the π-conjugated polymer in the energy wells formed by the inert LDH nanosheets. The film has a longer fluorescence lifetime and higher photostability for UV irradiation than the comparison film containing polymer and PDDA pairs, in which PDDA fail to impose the same isolation effect on APPP as effectively as with the LDH nanosheets.

Multilayer films using LDH nanosheets and negatively charged polymer of azobenzene polymer (PAZO) show reversible *cis–trans* isomerization, under alternating ultraviolet and visible irradiation, exhibiting light-induced switching performance (Han et al. 2010). In another work, Co–Al LDH nanosheets were assembled with porphyrin, and the film showed interesting electrocatalytic properties (Shao et al. 2010). A reversible thermochromic effect of the fluorescent thin film (BSB-LDHs)$_n$ exhibits high stability against light and heat, implying potential application in light sensors and molecular thermometers (Yan et al. 2011).

More remarkably, sequential deposition of oxide and LDH nanosheets can be used to produce multilayer films entirely composed of nanosheets without any polymer component (Li et al. 2007). UV–vis absorption spectra exhibit progressive enhancement of optical absorption due to oxide nanosheets as a function of deposition cycles, providing strong evidence for the regular growth of nanosheet pairs. As in the case of flocculation, the combination of Mg–Al LDH/Ti$_{0.91}$O$_2$ and Mg–Al LDH/Ca$_2$Nb$_3$O$_{10}$ yields a basal spacing of 1.2 and 2.0 nm, respectively, clearly validating the alternating stacking of two types of nanosheets.

Hollow nanoshells can also be produced by adopting a sacrificial template, on which nanosheets are sequentially deposited in an LbL manner (Li et al. 2006). For example, polystyrene (PS) beads were dispersed in formamide suspension containing Mg–Al LDH nanosheets, and the adsorption of LDH nanosheets onto the PS surface was promoted by ultrasonical agitation. After centrifugation and washing with water, the sample was then dispersed in an aqueous solution of PSS. The product was recovered by further centrifugation. Core–shell composites coated with multilayer shells of (PSS/LDH)$_n$ were synthesized by repeating the earlier procedures n times. Similar to the results on flat substrate, a repeating nanostructure of LDH/PSS with a periodicity of about 2.0 nm was obtained. Figure 5.14 shows SEM images of the obtained core–shell composite. The homogeneous curvature of the spherical PS beads was preserved after the deposition of the LDH/PSS shell.

The obtained core–shell sample was carefully calcined to combust the PS core and PSS, losing 91% of its weight accompanied by huge exotherms. Hollow nanocapsules of Mg–Al mixed oxides with a nanometer shell thickness were produced. Finally, the calcined capsules were exposed to humid air, in which the layered structure of LDH was recovered due to a reconstruction mechanism (Li et al. 2006). Because of the low density, high surface area, and high stability, such inorganic hollow nanocapsules with well-defined architecture are interesting candidates for a range of applications including catalysis, adsorption, and drug delivery.

5.4.2 Nanocones

Layered transition metal (Co, Ni, etc.) hydroxides with large surface area and large interlayer spacing are promising for desirable electrochemical activity derived from their redox reaction (Gao et al. 2010, Lee et al. 2011, Zhao et al. 2011). If different metal cations are co-incorporated in the host layer, that is, bimetallic Co–Ni hydroxides, the capacity and cycling stability may be further improved in comparison with monometallic hydroxides, which therefore offer an effective way for achieving high electrochemical performance (Chen et al. 2008, Gupta et al. 2008, Hu et al. 2009).

The electrochemical behavior of DS-intercalated Co$_{1-x}$Ni$_x$ (x = 0, 0.25, 0.5, 0.75, 1.0) hydroxide nanocones deposited on graphite substrate was investigated by cyclic voltammetry (CV) and galvanic charge–discharge in 1 M KOH aqueous electrolyte (Liu et al. 2012). Figure 5.15 shows typical CV curves. The redox current peaks were explicitly observed, which can be related to a combined effect of electrochemical redox reactions derived from Co and Ni hydroxides in alkaline

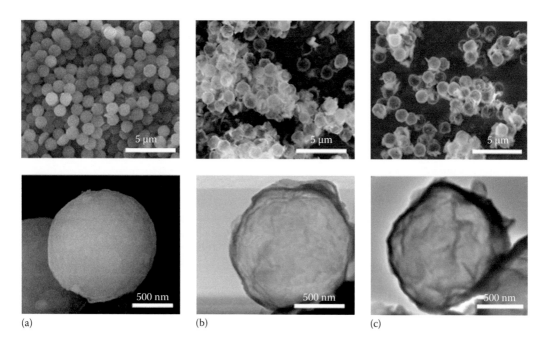

Figure 5.14 Hollow nanocapsules fabricated using Mg–Al LDH nanosheets and PSS on PS beads. (a) SEM images of $(LDH/PSS)_{20}$ deposited on PS beads. (b) SEM and TEM images of hollow nanoshells formed after thermal decomposition of PS beads. (c) Hollow nanoshells with LDH structure recovered after exposure to humid air. (Li, L., Ma, R., Iyi, N., Ebina, Y., Takada, K., and Sasaki, T., Hollow nanoshell of layered double hydroxide, *Chem. Commun.*, 29, 3125–3126, 2006. Reproduced by permission of The Royal Society of Chemistry.)

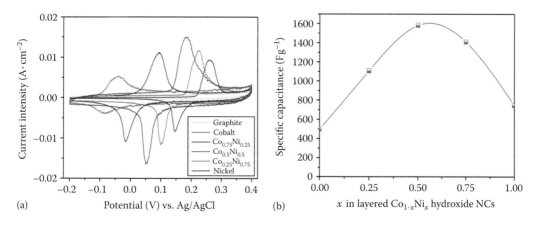

Figure 5.15 Electrochemical characterizations of layered hydroxide nanocones. (a) CV curves for nanocones with different compositions of $Co_{1-x}Ni_x$ (x = 0, 0.25, 0.5, 0.75, 1.0) deposited on graphite substrates at a scan rate of 10 mV s^{-1}. (b) Specific capacitance of nanocones dependent on metallic composition at a charge and discharge current density of 10 A g^{-1}. (Liu, X., Ma, R., Bando, Y., and Sasaki, T.: A general strategy to layered transition-metal hydroxide nanocones: Tuning the composition for high electrochemical performance. *Adv. Mater.* 2012. 24. 2148–2153. Copyright Wiley-VCH Verlag GmbH & Co. KGaA. Reproduced with permission.)

electrolyte: $Co(OH)_2 + OH^- \leftrightarrow CoOOH + H_2O + e^-$ and $Ni(OH)_2 + OH^- \leftrightarrow NiOOH + H_2O + e^-$. In contrast, the background signal originated from graphite substrates was almost negligible. Furthermore, the oxidation and reduction peaks at −40 and −85 mV for monometallic Co hydroxide nanocones shift to 260 and 145 mV for monometallic Ni hydroxide nanocones, respectively, due to the higher redox potential of $Ni(OH)_2/NiOOH$ couple. It is interesting to note that the oxidation and reduction peaks are gradually enhanced up to a Ni/Co ratio of 0.5 in the bimetallic $Co_{1-x}Ni_x$ hydroxide nanocones. After that ratio, the peak currents gradually decrease though the peak positions remain shifted to a higher potential with the increase in Ni content. In other words, the highest redox current was observed for $Co_{0.5}Ni_{0.5}$ hydroxide nanocones. This offers the feasibility of rationally tuning the transition metal composition for optimum electrochemical performance.

Figure 5.15b shows the specific capacitances of hydroxide nanocones with different Co–Ni compositions. A clear trend was identified that the specific capacitances of bimetallic hydroxide nanocones were generally superior to those of monometallic hydroxides. Consistent with the CV profiles, $Co_{0.5}Ni_{0.5}$ hydroxide nanocones possess the highest specific capacitance, which may be attributed to the enhancement of the electroactive sites participated in the redox reaction due to possible valence interchange or charge hopping between Co and Ni cations (Gupta et al. 2008). For example, in traditional nickel–cadmium (Ni–Cd) and nickel–metal hydride (Ni–MH) alkaline rechargeable batteries, Co hydroxide has been widely used as an additive to enhance the conductivity and stability of $Ni(OH)_2$ electrodes (Zhao et al. 2012). The present Co–Ni hydroxide nanocones, taking advantage of their tunable bimetallic composition, peculiar hollow feature, and large interlayer spacing, may be particularly advantageous for potential uses in next-generation redox supercapacitors.

5.5 SUMMARY

Based on a rational design on the structure, morphology, and composition, a wide variety of layered metal hydroxides, brucite (β-type), LDH, and α-type can be custom synthesized. The morphology may vary from platelets to nanocones depending on the preparative parameters. By exfoliating these well-crystallized platelets or peculiar nanocones in formamide, well-defined and positively charged hydroxide nanosheets may be obtained in a highly reproducible way. Some approaches to fabricate nanocomposites or functional assemblies using hydroxide nanosheets as building blocks and to explore their properties and applications have been developed. Especially, transition metal (Co, Ni)-bearing hydroxides are redox and magnetic active. They might be used in the development of sophisticated optical and magnetic nanodevices, and as ideal active electrode material for high-performance energy storage and conversion applications.

REFERENCES

Adachi-Pagano, M., Forano, C., and Besse, J. P. (2000), Delamination of layered double hydroxides by use of surfactants. *Chemical Communications* 1: 91–92.

Adachi-Pagano, M., Forano, C., and Besse, J. P. (2003), Synthesis of Al-rich hydrotalcite-like compounds by using the urea hydrolysis reaction-control of size and morphology. *Journal of Materials Chemistry* 13: 1988–1993.

Allmann, R. (1968a), The crystal structure of pyroaurite. *Acta Crystallographica Section B* 24: 972–977.

Allmann, R. (1968b), Verfeinerung der Struktur des Zink Hydrochlorides II. *Zeitschrift für Kristallographie* 126: 417–426.

Aisawa, S., Takahashi, S., Ogasawara, W., Umetsu, Y., and Narita, E. (2001), Direct intercalation of amino acids into layered double hydroxides by coprecipitation. *Journal of Solid State Chemistry* 162: 52–62.

Braterman, P. S., Xu, Z. P., and Yarberry, F. (2004), Layered double hydroxides (LDHs). In *Handbook of Layered Materials*, Auerbach, S. M., Carrado, K. A., and Dutta, P. K. (Eds.), pp. 373–474. New York: Marcel Dekker, Inc.

Cai, H., Hiller, A. C., Franklin, K. R., Nunn, C. C., and Ward, M. D. (1994), Nanoscale imaging of molecular adsorption. *Science* 266: 1551–1555.

Carillo, A. C. and Griego, D. A. (2012), *Hydroxides: Synthesis, Types and Applications.* New York: Nova Science Publishers, Inc.

Cavani, F., Trifiro, F., and Vaccari, A. (1991), Hydrotalcite-type anionic clays: Preparation, properties and applications. *Catalysis Today* 11: 173–301.

Chen, W., Yang, Y., Shao, H., and Fan, J. (2008), Tunable electrochemical properties brought about by partial cation exchange in hydrotalcite-like Ni–Co/Co–Ni hydroxide nanosheets. *Journal of Physical Chemistry C* 112: 17471–17477.

Choy, J. H., Kwak, S. Y., Park, J. S., Jeong, Y. J., and Portier, J. (1999), Intercalative nanohybrids of nucleoside monophosphates and DNA in layered metal hydroxide. *Journal of American Chemical Society* 121: 1399–1400.

Choy, J. H., Kwak, S. Y., Jeong, Y. J., and Park, J. S. (2000), Inorganic layered double hydroxides as nonviral vectors. *Angewandte Chemie International Edition* 39: 4042–4045.

Corodano, E., Martí-Gastaldo, C., Navarro-Moratalla, E., Ribera, A., Blundell, S. J., and Baker, P. J. (2010), Coexistence of superconductivity and magnetism by chemical design. *Nature Chemistry* 2: 1031–1036.

Costa, F. R., Saphiannikova, M., Wagenknecht, U., and Heinrich, G. (2008), Layered double hydroxide based polymer nanocomposites. *Advances in Polymer Science*, 210: 101–168.

Constantino, V. R. L. and Pinnavaia, T. J. (1995), Basic properties of $Mg^{2+}_{1-x}Al^{3+}_x$ layered double hydroxides intercalated by carbonate, hydroxide, chloride, and sulfate anions. *Inorganic Chemistry* 34: 883–892.

Costantino, U., Marmottini, F., Nocchetti, M., and Vivani, R. (1998), New synthetic routes to hydrotalcite-like compounds: Characterisation and properties of the obtained materials. *European Journal of Inorganic Chemistry* 10: 1439–1446.

Del Arco, M., Cebadera, E., Gutiérrez, S. et al. (2004), Mg, Al layered double hydroxides with intercalated indomethacin: Synthesis, characterization, and pharmacological study. *Journal of Pharmaceutical Science* 93: 1649–1658.

Delahaye-Vidal, A. and Figlarz, M. (1987), Textural and structural studies on nickel hydroxide electrodes. II. Turbostratic nickel (II) hydroxide submitted to electrochemical redox cycling. *Journal of Applied Electrochemistry* 17: 589–599.

Duan, X. and Evans, D. G. (2006), *Layered Double Hydroxides, Structure and Boding Series.* Berlin, Germany: Springer-Verlag.

Fisher, H. R. and Gielgens, L. H. (2002a), Nanocomposite material. US Patent 6,372,837.

Fisher, H. R. and Gielgens, L. H. (2002b), Nanocomposite material. US Patent 6,365,661.

Gao, X. P. and Yang, H. X. (2010), Multi-electron reaction materials for high energy density batteries. *Energy & Environmental Science* 3: 174–189.

Gordijo, C. R., Barbosa, C. A. S., Ferreira, A. M. D. C., Constantino, V. R. L., and Silva, D. O. (2005), Immobilization of ibuprofen and copper-ibuprofen drugs on layered double hydroxides. *Journal of Pharmaceutical Science* 94: 1135–1148.

Guo, Y., Zhang, H., Zhao, L., Li, G. D., Chen, J. S., and Xu, L. (2005), Synthesis and characterization of Cd-Cr and Zn-Cd-Cr layered double hydroxides intercalated with dodecyl sulfate. *Journal of Solid State Chemistry* 178: 1830–1836.

Gupta, V., Gupta, S., and Miura, N. (2008), Potentiostatically deposited nanostructured Co_xNi_{1-x} layered double hydroxides as electrode materials for redox-supercapacitors. *Journal of Power Sources* 175: 680–685.

Han, J. B., Yan, D. P., Shi, W. Y. et al. (2010), Layer-by-layer ultrathin films of azobenzene-containing polymer/layered double hydroxides with reversible photoresponsive behavior. *The Journal of Physical Chemistry B* 114: 5678–5685.

Hibino, T. (2004), Delamination of layered double hydroxides containing amino acids. *Chemistry of Materials* 16: 5482–5488.

Hibino, T. and Jones, W. (2001), New approach to the delamination of layered double hydroxides. *Journal of Materials Chemistry* 11: 1321–1323.

Hibino, T. and Kobayashi, M. (2005), Delamination of layered double hydroxides in water. *Journal of Materials Chemistry* 15: 653–656.

Hu, Z.-A., Xie, Y.-L., Wang, Y.-X., Wu, H.-Y., Yang, Y.-Y., and Zhang, Z.-Y. (2009), Synthesis and electrochemical characterization of mesoporous Co_xNi_{1-x} layered double hydroxides as electrode materials for supercapacitors. *Electrochimica Acta* 54: 2737–2741.

Huang, G. L., Ma, S. L., Zhao, X. H., Yang, X. J., and Ooi, K. (2010), Intercalation of bulk guest into LDH via osmotic swelling/restoration reaction: Control of the arrangements of thiacalix[4]arene anion intercalates. *Chemistry of Materials* 22: 1870–1877.

Huang, S., Cen, X., Peng, H. D., Guo, S. Z., Wang, W. Z., and Liu, T. X. (2009), Heterogeneous ultrathin films of poly(vinyl alcohol)/layered double hydroxide and montmorillonite nanosheets via layer-by-layer assembly. *The Journal of Physical Chemistry B* 113: 15225–15230.

Iyi, N., Ebina, Y., and Sasaki, T. (2008), Water-swellable MgAl-LDH (layered double hydroxide) hybrids: Synthesis, characterization, and film preparation. *Langmuir* 24: 5591–5598.

Iyi, N., Ebina, Y., and Sasaki, T. (2011), Synthesis and characterization of water-swellable LDH (layered double hydroxide) hybrids containing sulfonate-type intercalant. *Journal of Materials Chemistry* 21: 8085–8095.

Iyi, N., Matsumoto, T., Kaneko, Y., and Kitamura, K. (2004a), A novel synthetic route to layered double hydroxides using hexamethylenetetramine. *Chemistry Letters* 33: 1122–1123.

Iyi, N., Matsumoto, T., Kaneko, Y., and Kitamura, K. (2004b), Deintercalation of carbonate ions from a hydrotalcite-like compound: Enhanced decarbonation using acid-salt mixed solution. *Chemistry of Materials* 16: 2926–2932.

Kahlweit, M. (1975), Ostwald ripening of precipitates. *Advances in Colloid and Interface Science* 5: 1–35.

Kang, H. L., Huang, G. L., Ma, S. L. et al. (2009), Coassembly of inorganic macromolecule of exfoliated LDH nanosheets with cellulose. *The Journal of Physical Chemistry C* 113: 9157–9163.

Khan, A. I., Lei, L. X., Norquist, A. J., and O'Hare, D. (2001), Intercalation and controlled release of pharmaceutically active compounds from a layered double hydroxide. *Chemical Communications* 22: 2342–2343.

Kumbhar, P. S., Sanchez-Valente, J., and Figueras, F. (1998), Modified Mg-Al hydrotalcite: A highly active heterogeneous base catalyst for cyanoethylation of alcohols. *Chemical Communications* 10: 1091–1092.

Lee, J. W., Ko, J. M., and Kim, J.-D. (2011), Hierarchical microspheres based on α-$Ni(OH)_2$ nanosheets intercalated with different anions: Synthesis, anion exchange, and effect of intercalated anions on electrochemical capacitance. *The Journal of Physical Chemistry C* 115: 19445–19454.

Leroux, F., Adachi-Pagano, M., Intissar, M., Chauvière, S., Forano, C., and Besse, J. P. (2001), Delamination and restacking of layered double hydroxides. *Journal of Materials Chemistry* 11: 105–112.

Leroux, F. and Besse, J. P. (2001), Polymer interleaved layered double hydroxide: A new emerging class of nanocomposites. *Chemistry of Materials* 13: 3507–3515.

Liu, X., Ma, R., Bando, Y., and Sasaki, T. (2010), Layered cobalt hydroxide nanocones: Microwave-assisted synthesis, exfoliation, and structural modification. *Angewandte Chemie International Edition* 49: 8253–8256.

Liu, X., Ma, R., Bando, Y., and Sasaki, T. (2012), A general strategy to layered transition-metal hydroxide nanocones: Tuning the composition for high electrochemical performance. *Advanced Materials* 24: 2148–2153.

Li, L., Ma, R., Ebina, Y., Iyi, N., and Sasaki, T. (2005), Positively charged nanosheets derived via total delamination of layered double hydroxides. *Chemistry of Materials* 17: 4386–4391.

Li, L., Ma, R., Ebina, Y., Fukuda, K., Takada, K., and Sasaki, T. (2007), Layer-by-layer assembly and spontaneous flocculation of oppositely charged oxide and hydroxide nanosheets into inorganic sandwich layered materials. *Journal of the American Chemical Society* 129: 8000–8007.

Li, L., Ma, R., Iyi, N., Ebina, Y., Takada, K., and Sasaki, T. (2006), Hollow nanoshell of layered double hydroxide. *Chemical Communications* 29: 3125–3126.

Liang, J., Ma, R., Iyi, N., Ebina, Y., Takada, K., and Sasaki, T. (2010), Topochemical synthesis, anion exchange, and exfoliaiton of Co–Ni layered double hydroxides: A route to positively charged Co–Ni hydroxides nanosheets with tunable composition. *Chemistry of Materials* 22: 371–378.

Liu, Z., Ma, R., Osada, M., Takada, K., and Sasaki, T. (2005), Selective and controlled synthesis of alpha- and beta-cobalt hydroxides in highly developed hexagonal platelets. *Journal of the American Chemical Society* 127: 13869–13874.

Liu, Z., Ma, R., Osada, M. et al. (2006), Synthesis, anion exchange, and delamination of Co-Al layered double hydroxide: Assembly of the exfoliated nanosheet/polyanion composite films and magneto-optical studies. *Journal of the American Chemical Society* 128: 4872–4880.

Liu, Z., Ma, R., Ebina, Y., Iyi, N., Takada, K., and Sasaki, T. (2007), General synthesis and delamination of highly crystalline transition-metal-bearing layered double hydroxides. *Langmuir* 23: 861–867.

Ma, R., Liang, J., Takada, K., and Sasaki, T. (2011), Topochemical synthesis of Co-Fe layered double hydroxides at varied Fe/Co ratios: Unique intercalation of triiodide and its profound effect. *Journal of the American Chemical Society* 133: 613–620.

Ma, R., Liang, J., Liu, X., and Sasaki, T. (2012), General insights into structural evolution of layered double hydroxide: Underlying aspects in topochemical transformation from brucite to layered double hydroxide. *Journal of the American Chemical Society* 134: 19915–19921.

Ma, R., Liu, Z., Li, L., Iyi, N., and Sasaki, T. (2006a), Exfoliating layered double hydroxides in formamide: A method to obtain positively charged nanosheets. *Journal of Materials Chemistry* 16: 3809–3813.

Ma, R., Liu, Z., Takada, K. et al. (2006b), Tetrahedral Co(II) coordination in α-type cobalt hydroxide: Rietveld refinement and X-ray absorption spectroscopy. *Inorganic Chemistry* 45: 3964–3969.

Ma, R., Liu, Z., Takada, K., Iyi, N., Bando, Y., and Sasaki, T. (2007), Synthesis and exfoliation of Co^{2+}-Fe^{3+} layered double hydroxides: An innovative topochemical approach. *Journal of the American Chemical Society* 129: 5257–5263.

Ma, R. and Sasaki, T. (2010), Nanosheets of oxides and hydroxides: Ultimate 2D charge-bearing functional crystallites. *Advanced Materials* 22: 5082–5104.

Ma, R. and Sasaki, T. (2012), Synthesis of LDH nanosheets and their layer-by-layer assembly. *Recent Patents on Nanotechnology* 6: 159–168.

Ma, R., Takada, K., Fukuda, K., Iyi, N., Bando, Y., and Sasaki, T. (2008), Topochemical synthesis of monometal-lic (Co^{2+}-Co^{3+}) layered double hydroxide and its exfoliation into positively charged $Co(OH)_2$ nanosheets. *Angewandte Chemie International Edition* 47: 86–89.

Millange, F., Walton, R. I., Lei, L. X., and O'Hare, D. (2000), Efficient separation of terephthalate and phthalate anions by selective ion-exchange intercalation in the layered double hydroxide $Ca_2Al(OH)_6NO_3 \cdot 2H_2O$. *Chemistry of Materials* 12: 1990–1994.

Miyata, S. and Kuroda, M. (1981), Method for inhibiting the thermal or ultraviolet degradation of thermoplastic resin and thermoplastic resin composition having stability to thermal or ultraviolet degradation. US Patent 4,299,759.

Mousty, C., Therias, S., Forano, C., and Besse, J. P. (1994), Anion-exchanging clay-modified electrodes: Synthetic layered double hydroxides intercalated with electroactive organic anions. *Journal of Electroanalytical Chemistry* 374: 63–69.

Nie, H. Q. and Hou, W. G. (2011), Methods and applications for delamination of layered double hydroxides. *Acta Physico-Chimica Sinica* 27: 1783–1796.

Nijs, H., Clearfield, A., and Vansant, E. F. (1998), The intercalation of phenylphosphonic acid in layered double hydroxides. *Microporous and Mesoporous Materials* 23: 97–108.

Ogawa, M. and Kaiho, H. (2002), Homogeneous precipitation of uniform hydrotalcite particles. *Langmuir* 18: 4240–4242.

Okamoto, K., Sasaki, T., Fujita T., and Iyi, N. (2006), Preparation of highly oriented organic-LDH hybrid films by combining the decarbonation, anion-exchange, and delamination processes. *Journal of Materials Chemistry* 16: 1608–1616.

O'Leary, S., O'Hare, D., and Seeley, G. (2002), Delamination of layered double hydroxides in polar monomers: New LDH-acrylate nanocomposites. *Chemical Communications* 14: 1506–1507.

Oliva, P., Leonardi, J., Laurent, J. F. et al. (1982), Review of the structure and the electrochemistry of nickel hydroxides and oxy-hydroxides. *Journal of Power Sources* 8: 229–255.

Park, D. H., Kim, J. E., Oh, J. M., Shul, Y. G., and Choy, J. H. (2010), DNA core@inorganic shell. *Journal of the American Chemical Society* 132: 16735–16736.

Rajamathi, M., Kamath, P. V., and Seshadri, R. (2000), Chemical synthesis of α-cobalt hydroxide. *Materials Research Bulletin* 35: 271–278.

Rajamathi, M., Subbanna, G. N., and Kamath, P. V. (1997), On the existence of a nickel hydroxide phase which is neither α nor β. *Journal of Materials Chemistry* 7: 2293–2296.

Rives, V. (2006), *Layered Double Hydroxides: Present and Future*. New York: Nova Science Publishers, Inc.

Rives, V. and Ulibarri, M. A. (1999), Layered double hydroxides (LDH) intercalated with metal coordination compounds and oxometalates. *Coordination Chemistry Reviews* 181: 61–120.

Robins, D. S. and Dutta, P. K. (1996), Examination of fatty acid exchanged layered double hydroxides as supports for photochemical assemblies. *Langmuir* 12: 402–408.

Shao, M. F., Han, J. B., Shi, W. Y., Wei, M., and Duan, X. (2010), Layer-by-layer assembly of porphyrin/layered double hydroxide ultrathin film and its electrocatalytic behavior for H_2O_2. *Electrochemistry Communications* 12: 1077–1080.

Singh, M., Ogden, M. I., Parkinson, G. M., Buckley, C. E., and Connolly, J. (2004), Delamination and re-assembly of surfactant-containing Li/Al layered double hydroxides. *Journal of Materials Chemistry* 14: 871–874.

van der Ven, L., van Gemert, M. L. M., Batenburg, L. F. et al. (2000), On the action of hydrotalcite-like clay materials as stabilizers in polyvinylchloride. *Applied Clay Science* 17: 25–34.

Venugopal, B. R., Shivakumara, C., and Rajamathi, M. (2006), Effect of various factors influencing the delamination behavior of surfactant intercalated layered double hydroxides. *Journal of Colloid and Interface Science* 294: 234–239.

Wang, Q. and O'Hare, D. (2012), Recent advances in the synthesis and application of layered double hydroxide (LDH) nanosheets. *Chemical Reviews* 112: 4124–4155.

Wu, Q. L., Olafsen, A., Vistad, Ø. B., Roots, J., and Norby, P. (2005), Delamination and restacking of a layered double hydroxide with nitrate as counter anion. *Journal of Materials Chemistry* 15: 4695–4700.

Wu, Q. L., Sjåstad, A. O., Vistad, Ø. B. et al. (2007), Characterization of exfoliated layered double hydroxide (LDH, Mg/Al = 3) nanosheets at high concentrations in formamide. *Journal of Materials Chemistry* 17: 965–971.

Xu, Z. P. and Braterman, P. S. (2004), Self-assembly and multiple phases of layered double hydroxides. In *Encyclopedia of Nanoscience and Nanotechnology*, Schwarz, J. A. Contescu, C. I., and Putyera, K. (eds.), pp. 3387–3398. New York: Marcel Dekker, Inc.

Yan, D. P., Lu, J., Chen, L. et al. (2010b), A strategy to the ordered assembly of functional small cations with layered double hydroxides for luminescent ultra-thin films. *Chemical Communications* 46: 5912–5914.

Yan, D. P., Lu, J., Ma, J., Wei, M., Evans, D. G., and Duan, X. (2011), Reversibly thermochromic, fluorescent ultra-thin films with a supramolecular architecture. *Angewandte Chemie International Edition* 50: 720–723.

Yan, D. P., Lu, J., Wei, M. et al. (2009), Ordered poly(p-phenylene)/layered double hydroxide ultrathin films with blue luminescence by layer-by-layer assembly. *Angewandte Chemie International Edition* 48: 3073–3076.

Yan, D. P., Qin, S. H., Chen, L. et al. (2010a), Thin film of sulfonated zinc phthalocyanine/layered double hydroxide for achieving multiple quantum well structure and polarized luminescence. *Chemical Communications* 46: 8654–8656.

Zhao, T., Jiang, H., and Ma, J. (2011), Surfactant-assisted electrochemical deposition of α-cobalt hydroxide for supercapacitors. *Journal of Power Sources* 196: 860–864.

Zhao, X. Y., Ma, L. Q., and Shen, X. D. (2012), Co-based anode materials for alkaline rechargeable Ni/Co batteries: A review. *Journal of Materials Chemistry* 22: 277–285.

Structures and Properties

Chapter 6

Synthesis and Properties of Boron Carbon Nitride Nanosheets and Nanotubes

Zhi Xu, Wenlong Wang, and Xuedong Bai

CONTENTS

6.1 INTRODUCTION

Carbon nanomaterials are nevertheless the hottest nanomaterial system in the last two decades, starting with fullerene, followed by carbon nanotubes (CNTs), and recently, graphene, thus covering all the low dimensions. Tremendous progress has been achieved in fundamental research on carbon nanomaterials and their applications in various fields. Soon after the discovery of CNTs, substitutional doping of CNTs with B and N heteroatoms has been found to be possible and feasible (Rubio et al. 1994). Given that B and N are the nearest neighbors of C in the periodic table and that their atomic radii are very similar to that of C, B, and N atoms, they have a reasonable probability of entering the CNT lattices through atomic substitutional doping. The co-doping of C nanotubes/nanosheets with both B and N leads to the formation of a ternary system, which is also known as boron carbonitride (BCN) nanotube/nanosheet. The band gap of CNT depends on chirality and diameter. For graphene, its two linear bands cross at Dirac point, resulting in a zero-band gap semiconductor or semimetal, whereas for hexagonal boron nitride

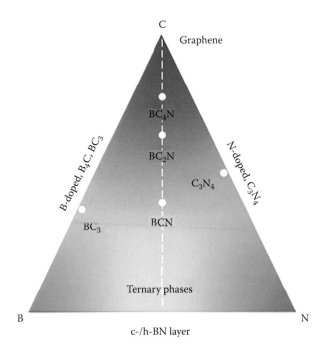

Figure 6.1   Phase segregation in BCN ternary phase diagram. In this diagram, layered architectures could be built from graphene (C), h-BN (BN), doped graphene (BC and CN), and BCN composite. The C–B and C–N binary chemical compounds are possibly formed with certain stoichiometry, such as BC_3, B_2C, and CN_3, while BCN, BC_2N, and BC_4N are located in the ternary phase diagram (dashed line). All these stoichiometries have specific electronic structure and properties.

(h-BN), it is an insulator with an approximately 5.5 eV band gap. The band gaps of BN nanotubes are similar to those of h-BN, which are independent of chiralities and morphologies. B, C, and N can be atomically mixed together to form various semiconducting hexagonal-layered structures with varying stoichiometries (Figure 6.1). Theoretical calculation shows that their band gaps are primarily determined by their chemical compositions and independent of their geometry, which could be further tailored to smartly build interesting nano-heterojunctions. BCN nanostructures have elicited much attention because these ternary systems provide the most feasible method to control the semiconducting properties of graphene or BN.

In this chapter, we will present an up-to-date overview of the synthesis strategy on B- and/or N-doped nanotubes and nanosheets, with discussions covering the binary systems of CB and CN, the ternary BCN system, and their physical properties.

6.2   SYNTHESES

Two different strategies may be used to produce BCN nanotubes and nanosheets: direct synthesis and postsynthetic substitution reaction. Direct synthesis methods are derived from techniques that have been well established for pure CNT synthesis, including arc discharge, laser ablation, and chemical vapor deposition (CVD). The postsynthetic substitution reaction approach refers to the use of pre-synthesized pure CNTs or graphene as starting materials, which are then reacted with B- and/or N-containing compounds at high temperature; this reaction process induces partial substitution of the carbon lattice by foreign B and/or N atoms. Substitution can also start from pure BN nanotubes or nanosheets, doping them with carbon-containing sources.

6.2.1 Postsynthetic Substitution Reaction Route

The postsynthetic substitution reaction approach was initially developed by Han et al. (1998, 1999) for the synthesis of multiwalled BN and BCN nanotubes. Later, this method was applied by Golberg et al. (1999, 2000) for single-walled nanotube (SWNT) doping. In Golberg's experiments, the laser ablation–produced C-SWNT bundles were employed as starting materials; they were mixed with B_2O_3 powder and heated in a flowing N_2 atmosphere. The maximum yield of doped SWNTs was obtained under synthesis at 1553 K over 30 min, with an accumulating B dopant concentration up to ~10 at.% and N concentration up to ~2 at.%. Although the B and N contents could increase to higher levels by increasing synthesis time and temperature, the majority of SWNT bundles had collapsed and transformed to large multiwalled nanotubes (MWNTs), nanorods, and polygonal particles. This result indicates that the high-temperature substitution reaction process of SWNTs with corrosive B_2O_3 vapor deteriorates their single-shelled tube wall structure. Within the doped SWNTs via substitution reactions, the B/C and B/N ratios were found to vary irregularly, and a few thin bundles consisting of pure BN-SWNTs were also accidentally detected.*

Post-doping of graphene was first implemented by Wang et al. (2009). Individual graphene nanoribbons were covalently functionalized by nitrogen species through high-power electrical joule heating in ammonia gas, leading to n-type electronic doping. The formation of the C–N bond occurs mostly at the edges of graphene where chemical reactivity is high. Li, who was from the same group as Wang, developed a simple chemical method to obtain bulk quantities of N-doped, reduced graphene oxide (GO) sheets through thermal annealing of GO in ammonia (Li et al. 2009b). Starting from GO, N-doping occurs at a temperature as low as 300°C, whereas the highest doping level close to 5% N was achieved at 500°C. Wang (2009) subsequently reported a facile strategy to prepare N-doped graphene by using nitrogen plasma treatment of graphene synthesized via a chemical method. By controlling the exposure time, the N content in host graphene can be regulated, ranging from 0.11% to 1.35%. Pure graphene has a perfect honeycomb structure, so introducing heteroatoms into the lattice is difficult. Substitutions usually occur at defect sites. With this point of view, Guo et al. (2010) first introduced defects into a graphene plane by N-ion irradiation and then restored them with ammonia. Their approach provided a physical mechanism for the introduction of defects and subsequent hetero dopant atoms into the graphene material in a controllable fashion.

Wang et al. (2012) produced nanostructured BCN by thermal annealing of GO in the presence of boric acid and ammonia. Three samples of different chemical compositions ($B_{0.38}C_{0.28}N_{0.34}$, $B_{0.12}C_{0.77}N_{0.11}$, and $B_{0.07}C_{0.87}N_{0.06}$) were obtained. The products were not layered; instead, they had a porous structure, but their detailed crystal structures were unclear. However, the concentrations of boron and nitrogen were always very close, and the ratio of carbon/BN varied in a very large range, so a hybrid structure of BN and graphene is reasonably expected.

To avoid the formation of such hybrid structure, Zheng et al. (2013) developed a two-step method. Synthesis was also started from GO. First, N was incorporated by annealing with NH_3 at an intermediate temperature (e.g., 500°C), and then B was introduced by pyrolysis of

---

* Excerpt and modified with permission from Wang et al. (2007). Copyright 2007, World Scientific Publishing Company.

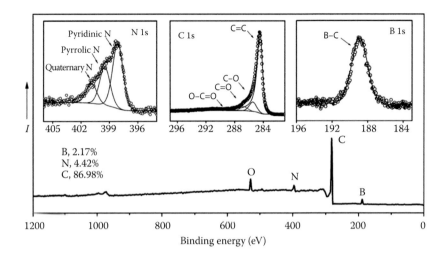

Figure 6.2   XPS survey and high-resolution spectra of N, C, and B 1s core levels in B, N-graphene.

the intermediate material (N-graphene) with H_3BO_3 at a higher temperature (e.g., 900°C). The nanosheet morphology was preserved after the two-step doping procedure. Detailed XPS analysis (Figure 6.2) shows that all N heteroatoms in the resulting B–N graphene are bonded to the surrounding C atoms in pyridinic N (~2.03 at.%), pyrrolic N (~1.29 at.%), and graphitic N forms (~0.90 at.%) without any N–B configuration, whereas all B heteroatoms are bonded to C atoms only in the form of a BC_3 structure (~2.17 at.%), but not coupled with N. This BC_3 bonding type indicates that all B heteroatoms have replaced C atoms in the framework of graphene without the formation of the undesired h-BN hybrid.

6.2.2   Direct Synthesis: Arc Discharge and Laser Ablation

Both arc discharge and laser ablation are high-temperature vaporization methods that have demonstrated remarkable success in the synthesis of standard C-SWNTs. The arc discharge method was the first to be applied for doped CNT synthesis when Stephan et al. (1994) initiated the direct synthesis of B- and N-doped MWNTs using this method. A year later, the successful arc discharge synthesis of BC_2N and BC_3 MWNTs was reported by Wengsieh et al. (1995) in Zettl's group. However, the direct synthesis of doped SWNTs by arc discharge was not achieved until 2002. Droppa et al. (2002) doped SWNTs with N by the arc discharge evaporation of graphite in a He–N_2 atmosphere in the presence of Fe–Co–Ni catalysts. The use of laser ablation technique for the B-doped SWNT synthesis was first reported by Gai (2004). In their growth experiments, the ablation targets were prepared by mixing different amounts of elemental B with C paste and the Co–Ni catalyst. More high-yield doped SWNTs with B contents up to 1.8 at.% were successfully produced by Blackburn (2006), wherein nickel boride alloy acted as both the B precursor and the catalyst.[*]

Because of the very successful application of CVD in graphene synthesis, arc discharge and laser ablation methods are not widely used in producing doped graphene. Very few studies are available in the literature. Panchakarla et al. (2009) carried out arc discharge using graphite electrodes in

---

[*] Excerpt and modified with permission from Wang et al. (2007). Copyright 2007, World Scientific Publishing Company.

the presence of $H_2 + B_2H_6$ (BG1) or using boron-stuffed graphite electrodes (BG2). They prepared nitrogen-doped graphene by carrying out arc discharge in the presence of H_2 + pyridine (NG1) or H_2 + ammonia (NG2). XPS analysis showed that BG1 and BG2 contained 1.2 and 3.1 at.% of boron, respectively, whereas NG1 and NG2 contained 0.6 and 1.0 at.% of nitrogen, respectively.

6.2.3 Direct Synthesis: CVD Growth

CVD is a well-established and reliable synthetic route to produce standard C-MWNTs and C-SWNTs. Over the past years, the direct CVD synthesis of B- and/or N-doped MWNTs has been extensively explored, and a number of studies about the production of binary CN- and ternary BCN-MWNTs and nanofibers have been reported (Terrones et al. 2002, Ma et al. 2004a, Zhi et al. 2004, Ewels and Glerup 2005). More recently, a breakthrough in the CVD growth of ternary BCN-SWNTs has been achieved by the present authors through a bias-assisted hot filament CVD (HFCVD) growth process (Wang et al. 2006). Prior to our work, ternary SWNTs were made available only through the postsynthetic substitution route. Therefore, our work represents the first example of the direct synthesis of ternary BCN-SWNTs; in fact, during the past few years, the same HFCVD method has been employed by the present authors to grow aligned BCN-MWNTs, cactus-like BCN nanofibers, and C/B−C−N nanotubular heterojunctions (Bai et al. 2000a, Guo et al. 2002, Zhi et al. 2002). Thus, the work on ternary SWNT synthesis is a continuation of the author's ongoing efforts toward the rational CVD synthesis of BCN nanotubular materials.*

BCN-SWNT growth by HFCVD was achieved over a high-activity powdery catalyst, namely, MgO-supported Fe–Mo bimetallic catalyst (denoted as Fe–Mo/MgO), by using $CH_4B_2H_6$ and ethylenediamine vapor as the reactant gases and inert He as the diluting gas. TEM observation revealed that the as-grown SWNTs have clean and smooth surfaces, as shown in Figure 6.3a. A typical EELS spectrum taken from the SWNT bundles is displayed in Figure 6.3b, where the core loss of the K-edges of B, C, and N located at 188, 284, and 401 eV, respectively, can be clearly identified. In Figure 6.3c, the zero loss and energy-filtered TEM images of two adjacent BCN-SWNT bundles are shown, where all components of B, C, and N elements are homogeneously distributed within the SWNT tube shells.*

Besides the authors' work on HFCVD growth of ternary BCN-SWNTs, a few recent studies concerning the synthesis of N-doped SWNTs by conventional thermal CVD have also emerged. Keskar et al. (2005) initially studied the growth of isolated CN-SWNTs on quartz and SiO_2/Si substrate surfaces through a liquid injection CVD process by using the mixture of xylene and acetonitrile (CH_3CN, N precursor) as feeding source. Liu et al. (2006) published a previous example of the thermal CVD growth of N-doped SWNTs by pyrolysis of pyridine (C_5H_5N) over powdery Fe–Mo/MgO catalyst (the same kind of catalyst used in our experiment on HFCVD growth for BCN-SWNTs). For the N-doping levels within the SWNTs, the doping levels should be lower than 1 at.%, which can hardly be detected by EELS and even XPS.*

Graphite segregation at surfaces and grain boundaries of metals has been studied for a long time. Previous research has shown that thin graphite with well-controlled thickness, including monolayer and low defect density, can be segregated from metals and metal carbides (Oshima and Nagashima 1997, Rosei et al. 1983). However, growing graphene by CVD method has not drawn any interest until a large-area and high-quality graphene was synthesized by Kim et al. (2009) and Li et al. (2009a).

---

* Excerpt and modified with permission from Wang et al. (2007). Copyright 2007, World Scientific Publishing Company.

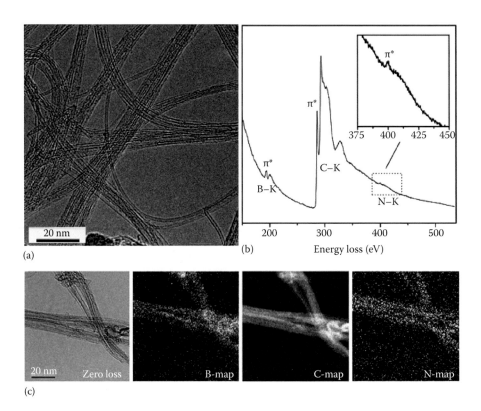

(a)

(b)

(c)

Figure 6.3 (a) TEM image of the as-grown BCN-SWNTs and (b) EELS spectrum taken from a SWNT bundle. (c) Zero loss and energy-filtered images of two adjacent BCN-SWNT bundles. (Reprinted with permission from Wang, W.L., Bai, X.D., Liu, K.H. et al., Direct synthesis of B-C-N single-walled nanotubes by bias-assisted hot filament chemical vapor deposition, *J. Am. Chem. Soc.*, 128, 6530–6531, 2006. Copyright 2013, American Chemical Society.)

The direct synthesis of graphitic boron carbon nitrogen ternary composite was first prepared by thermally decomposing B-, C-, and N-containing precursors by CVD in 1978, with thickness ranging from 100 nm to a few micrometers (Kaner et al. 1987). Few-layer BCN ternary films were synthesized by Ci et al. (2010) using CVD technology similar to graphene growth. Cu foils were used as substrate, whereas methane and ammonia borane (NH_3–BH_3) were used as precursors for carbon and BN, respectively. The atomic ratio of B, C, and N can be tuned by controlling the experimental parameters, keeping the B/N ratio unity. Thus far, the atomic percentage of C could be tuned from about 10% to ~100%. High-resolution TEM observations indicate that these h-BCN films are about 1 nm thick, consisting of mainly two to three layers. The B, N peaks in the XPS spectrum are very similar to those of h-BN, indicating that h-BN domains exist in the film (Figure 6.4). The C 1s peak is located at 284.4 eV, which is close to the value observed in graphite (284.9 eV). This finding suggests that the C–C bonds stay together and form graphene domains. The existence of C–N bonds and C–B bonds is also noticed as small shoulders shown at B 1s, N 1s, and C 1s appear, indicating that the obtained h-BCN ternary films are composed of hybridized h-BN and graphene domains. Ci et al. (2010) argued that BN segregation occurred for BN concentrations higher than 10%.

BN segregation is a general phenomenon during the direct synthesis of ternary BCN materials. To systematically study its crystal and electronic structures, Chang et al. (2013) synthesized

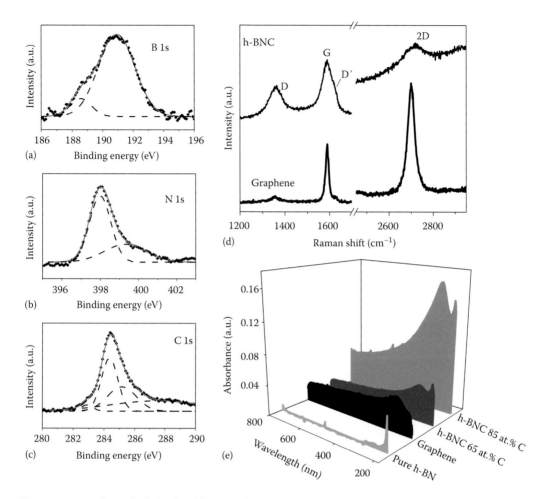

Figure 6.4 Evidence for hybridized h-BN and graphene domain-like structure of h-BNC. (a–c) XPS spectra of B, N, and C 1s core levels, respectively. The spectrum curves (filled diamonds) are deconvoluted (black dashed curves) by Gaussian fitting (solid curves), indicating possible multibonding information. (d) Raman spectrum of an h-BNC and a CVD-grown graphene film. (e) Ultraviolet–visible absorption spectra of different graphene films. (Reprinted by permission from Macmillan Publishers Ltd. *Nat. Mater.*, Ci, L., Song, L., Jin, C. et al., Atomic layers of hybridized boron nitride and graphene domains, 9, 430–435, Copyright 2010.)

few-layer BCN nanosheets with BN concentration from 2 to 56 at.%. The resulting films exhibited structural evolution from homogeneously dispersed small BN clusters to large-sized BN domains with embedded diminutive graphene domains. Their experimental setup was similar to that of large-area graphene synthesis (using CH_4 gas as the carbon source and Cu foils as substrates), except that ammonia borane was heated to 70°C to ~110°C and introduced as vapor to obtain different BN contents. From the XPS spectra of 2% BN concentration (2BNG) to 27% BN concentration (27BNG), one should clearly notice that the N_B peak is not detected in the case of 2BNG and 6BNG films, whereas the peak N_B at 398 eV increases and the peak N_C at 399.5 eV diminishes as BN concentration is increased, which can be directly related to the increasing number of N–B bonds over N–C bonds (Figure 6.5). This result indicates the preferential bonding between N and B atoms, leading to an increased number of BN domains.

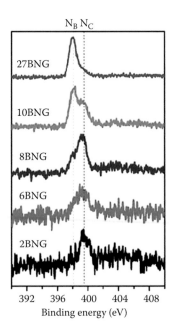

Figure 6.5 N 1s spectra for different BNG films as a function of increasing BN concentration. Dashed lines are guide to eyes. (Reprinted with permission from Chang, C., Kataria, S., Kuo, C. et al., Band gap engineering of chemical vapor deposited graphene by in situ BN doping, *ACS Nano*, 7, 1333–1341, 2013. Copyright 2013, American Chemical Society.)

Graphene and h-BN nanosheets can be both grown on copper or nickel foils by CVD method, so Levendorf et al. (2012) and Liu et al. (2013) showed that in-plane heterostructures of graphene and h-BN can be obtained via two-step growth (Figure 6.6). After growing the first film of graphene/h-BN, the film was patterned using lithography and etching. Next, the h-BN/graphene film was grown on the etched place in the second-step growth. This approach can create periodic arrangements of domains with sizes ranging from tens of nanometers to millimeters, which shows potential applications in property control and functional devices.

6.3 PHYSICAL PROPERTIES

6.3.1 Electronic Property

A SWNT can be considered as a roll-up of graphene, and understanding the former requires knowledge about the latter. However, research history shows that the discovery and study of nanotubes preceded that of graphene. To evaluate the doping effect of B/N on carbon nanostructures, we will start our review by discussing B- and/or N-doped graphene.

6.3.1.1 B-Doped Graphene

B-doped graphene is expected to show p-type conductivity. Wu et al. (2010) studied the electronic structure of B-doped graphene using VASP code, with PAW formalism for the electron–ion interactions and PBE for the exchange–correlation functional (Figure 6.7). The linear energy

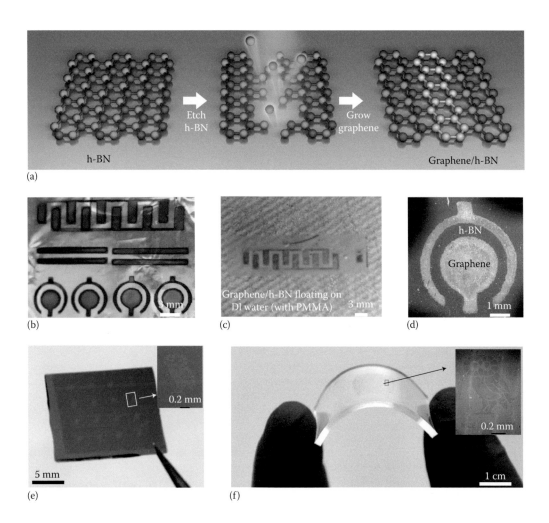

Figure 6.6 Creation of millimeter-sized graphene/h-BN in-plane heterostructures. (a) Illustration of the fabrication procedure for in-plane graphene/h-BN heterostructures. Steps: Preparation of h-BN films using the CVD method; partial etching of h-BN by argon ions to give predesigned patterns; and subsequent CVD growth of graphene on the etched regions. (b) Optical image of the as-grown graphene/h-BN patterned layers (shaped as combs, bars, and rings) on a copper foil. Light areas are h-BN and dark areas are graphene. (c) Optical image of a graphene/h-BN film separated from copper, on water, after coating with PMMA and etching the copper foil. (d) SEM image showing an h-BN ring surrounded by graphene. (e, f) Graphene/h-BN owl patterns that have been transferred on silica and PDMS, respectively. Insets: Optical images of individual owls. (Reprinted by permission from Macmillan Publishers Ltd. *Nat. Nanotechnol.*, Liu, Z., Ma, L., Shi, G. et al., In-plane heterostructures of graphene and hexagonal boron nitride with controlled domain sizes, 8, 119–124, Copyright 2013.)

dispersion near the Dirac point is almost maintained because the planar honeycomb graphene lattice structure is well preserved after boron doping. However, because of the lowering of the symmetry upon doping, the shape of all bands is altered, leading to an opening of a small gap that arises at K between the π and π^* bands. The projected density of states (PDOS) indicates that the boron atom p_σ orbitals (the three orbitals after sp^2 hybridization of the original p$_x$ and p$_y$ orbitals) strongly hybridize with the carbon atom p_σ orbitals over a wide range from

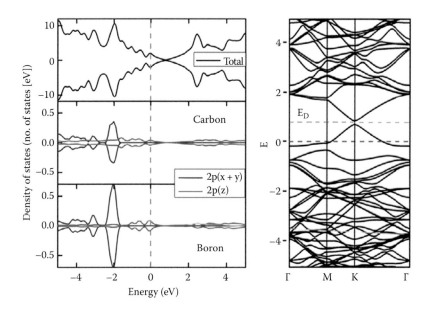

Figure 6.7 The DOS and PDOS of B-doped graphene. All the Fermi levels are adjusted to zero. The black solid line in the DOS plot represents the total DOS of the system, while the dark gray and light gray solid lines represent the PDOS of $p_{x,y}$ and p_z states, respectively. The PDOS of the carbon atom comes from the one bonded to the dopant. We show the band structure as a reference, although in realistic doped systems, the dopant cannot be periodic as in the model system. The upper dashed line represents the Dirac energy level. (Reproduced from Wu, M., Cao, C., and Jiang, J. 2010. Light non-metallic atom (B, N, O and F)-doped graphene: A first-principles study. *Nanotechnology* 21, 505202. With permission of IOP Publishing. All rights reserved.)

~E_F–1.7 to ~E_F–5 eV. The carbon π orbitals dominate the states around the Dirac point. Bader charge analysis shows that 1.84 electrons transfer from the boron atom to the carbon atom because of the higher electronegativity of the carbon atom than that of the boron atom.

The main features of theoretical prediction on B-doped graphene have been confirmed by Cattelan et al. (2013) using angle-resolved valence band spectra. Figure 6.8a and b shows images of the π-band near the K-point, with the dispersion direction perpendicular to the Γ–K direction of pure and B-doped graphene. Despite the relatively high level of doping (about 1% in their CVD-grown sample), the band structure of B-doped graphene is similar to that of pure graphene with a linear dispersion of the π-band close to the Dirac point, which is the typical fingerprint of massless fermions. For pure graphene on copper, the π-band intersects the K-point at 0.3 eV below the Fermi level, indicating the n-doping result induced by the Cu metal substrate. For B-doped graphene, the n-doping operated by the metal contact counterbalances the intrinsic p doping induced by the presence of boron, leading to the formation of an almost perfect semimetal band structure as shown.

Latea et al. (2010) examined the characteristics of field-effect transistors based on B-doped few-layer graphenes. Figure 6.9 shows their output characteristics and transfer characteristics. Both exhibited p-type behavior, that is, I_{ds} decreases (increases) as V_{gs} increases positively (negatively).

6.3.1.2 N-Doped Graphene

When an N atom is doped into graphene, it usually has three common bonding configurations within the carbon lattice: quaternary N (or graphitic N), pyridinic N, and pyrrolic N (Figure 6.10). Specifically, pyridinic N bonds with two C atoms at the edges or defects of graphene and

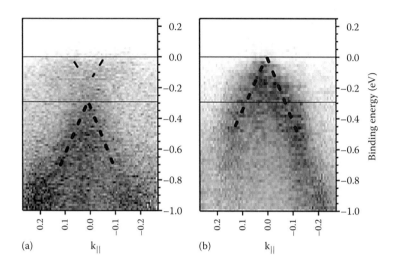

Figure 6.8 Dispersion of the valence band spectra of pure and B-doped graphene layers on polycrystalline Cu: (a, b) dispersion around the K-point, obtained as a snapshot of the analyzer 2D detector, for pure and B-doped graphene, respectively. (Reprinted with permission from Cattelan, M., Agnoli, S., Favaro, M. et al., Microscopic view on a chemical vapor deposition route to boron doped graphene nanostructures, *Chem. Mater.*, 25, 1490–1495, 2013. Copyright 2013, American Chemical Society.)

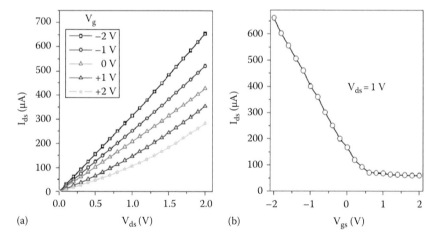

Figure 6.9   (a) Output characteristics (I_{ds} vs V_{ds}) and (b) transfer characteristics (I_{ds} vs V_{gs}) of the FET-based on B-doped graphene.

contributes one p electron to the π system. Pyrrolic N refers to N atoms that contribute two p electrons to the π system. Quaternary N refers to N atoms that substitute C atoms in the hexagonal ring. Among these nitrogen types, pyridinic N and quaternary N are sp^2 hybridized and pyrrolic N is sp^3 hybridized. Apart from these three common nitrogen types, N oxides of pyridinic N have been observed in both N-graphene and N-CNT studies. In this configuration, the nitrogen atom bonds with two carbon atoms and one oxygen atom.

Compared with boron-doped graphene, more developments have been achieved on N-doped graphene. Wu et al. (2010) calculated the electronic structure of substitutional doping of N atom

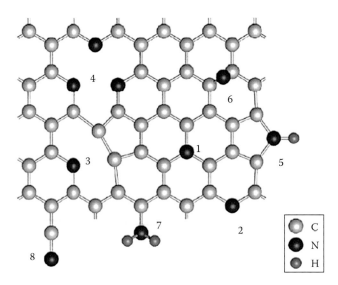

Figure 6.10   Possible configurations of nitrogen impurities in graphene: (1) substitutional or graphitic N, (2) pyridine-like N, (3) single N pyridinic vacancy, (4) triple N pyridinic vacancy, (5) pyrrole-like, (6) interstitial N or adatom, (7) amine, and (8) nitrile. (Reprinted with permission from Usachov, D., Vilkov, O., Grüneis, A. et al., Nitrogen-doped graphene: Efficient growth, structure, and electronic properties, *Nano Lett.*, 11, 5401–5407, 2011. Copyright 2011, American Chemical Society.)

into graphene. Since the N atom has one more electron than the carbon atom, the system exhibits electron-doping properties, that is, the Fermi level shifted approximately 0.7 eV above the Dirac point (Figure 6.11). Nevertheless, the nitrogen doping preserved the linear energy dispersion near the Dirac point like the boron doping. The system also shows a band gap of 0.14 eV around the Dirac point. The Bader analysis shows that 1.12 electrons transfer from the carbon atoms to the nitrogen atom, consistent with the electronegativity data of carbon and nitrogen atoms.

Large graphene planes containing various concentrations of N dopants randomly distributed in one sublattice are investigated using the TB model by Lherbier et al. (2013). The corresponding DOS are depicted in Figure 6.12. The band gap induced by sublattice asymmetry is robust with respect to such a random distribution of N dopants in one sublattice. The band gap dependence with the N concentration scales as $E_g(x_N) \propto x_N^{0.75}$ was found for various N concentrations $x_N = [0.5; 1; 4; 8]\%$ (inset of Figure 6.12), corresponding to $E_g = [45; 110; 340; 550]$ meV, respectively.

Experimentally, Usachov et al. (2011) studied N-doped graphene using angle-resolved photoemission spectroscopy (ARPES). Figure 6.13a shows a band structure of such N-graphene, which exhibits two striking features. First, the Dirac cone is shifted by ~0.3 eV toward higher binding energies. A part of the conduction band appears below EF, indicating n-type doping of graphene as a result of the charge transfer from N atoms. Second, a band gap of ~0.2 eV appears at the K-point, which is clearly visible in Figure 6.13b. The undoped gold-intercalated graphene is probably gapless, so the gap originates most likely from N impurities, which break the symmetry of the graphene lattice. From the XPS spectrum of the same sample at N 1s, the core level (Figure 6.13c) shows that most of the dopants are graphitic N. Graphitic N atoms contribute an average of ~50% of mobile electrons to the graphene conduction band, which is close to the value

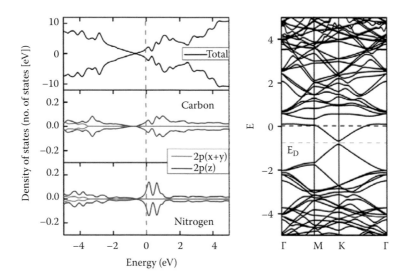

Figure 6.11 The DOS and PDOS of N-doped graphene, together with the electronic band structure. (Reproduced from Wu, M., Cao, C., and Jiang, J. 2010. Light non-metallic atom (B, N, O and F)-doped graphene: A first-principles study. *Nanotechnology* 21, 505202. With permission of IOP Publishing. All rights reserved.)

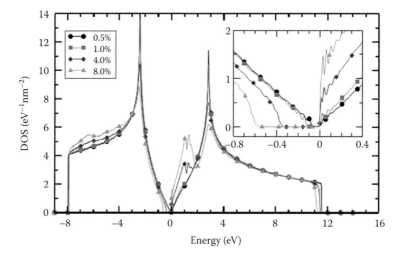

Figure 6.12 Density of states of graphene containing various concentrations of N dopants randomly distributed in one sublattice. Inset: Zoom of DOS in the band gap region. (Reprinted with permission from Lherbier, A., Botello-Méndez, A.R., Charlier, J.C., Electronic and transport properties of unbalanced sublattice N-doping in graphene, *Nano Lett.*, 13, 1446–1450, 2013. Copyright 2013, American Chemical Society.)

obtained with scanning tunneling techniques by Zhao et al. (2011). Zhang et al. (2011) showed that the band gap in N-graphene with 2.9 at.% nitrogen content was about 0.16 eV. According to theoretical calculation, generally speaking, the band gap of N-doped graphene will increase with the increment of nitrogen concentration. However, to date, only a few studies of N-doped graphene have reported a band gap in their work. Moreover, the relation between band gap and

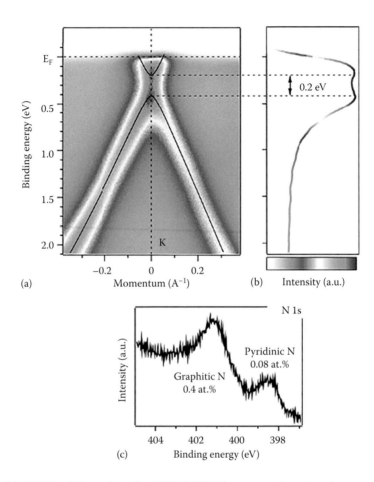

Figure 6.13 (a) ARPES of N-graphene/Au/Ni(111)/W(110), measured at the photon energy of 35 eV, through the K-point, few degrees off the direction, perpendicular to the ΓK. (b) PE spectrum at the K-point. (c) XPS spectrum of the same sample at N 1s core level. (Reprinted with permission from Usachov, D., Vilkov, O., Grüneis, A. et al., Nitrogen-doped graphene: Efficient growth, structure, and electronic properties, *Nano Lett.*, 11, 5401–5407, 2011. Copyright 2011, American Chemical Society.)

nitrogen concentration is not comparable among different works because the nitrogen bonding configurations are different due to different synthesis methods.

When applied into field-effect transistors, pure graphene exhibits p-type behavior because of the adsorption of oxygen or water in air (Figure 6.14d). By contrast, all the nitrogen-doped graphene exhibits typical n-type semiconductor behavior (Li et al. 2009b, Wang et al. 2009, Wei et al. 2009, Guo et al. 2010, Jin et al. 2011, Zhang et al. 2011), which is in agreement with theoretical calculations. Compared with the mobility of pure graphene (300–1200 cm^2 V^{-1} s^{-1}), the mobility of N-graphene decreases to 200–450 cm^2 V^{-1} s^{-1}, proving that the electron mobility of graphene decreases with the increment of band gap in most cases. Although several theoretical studies and the recent work about N-doped graphene nanoribbon by Wang et al. (2009) have shown that mobility of N-graphene may be maintained by controlling the nitrogen content at a low doping level and specific doping site, the scattering center induced by the defects will drastically

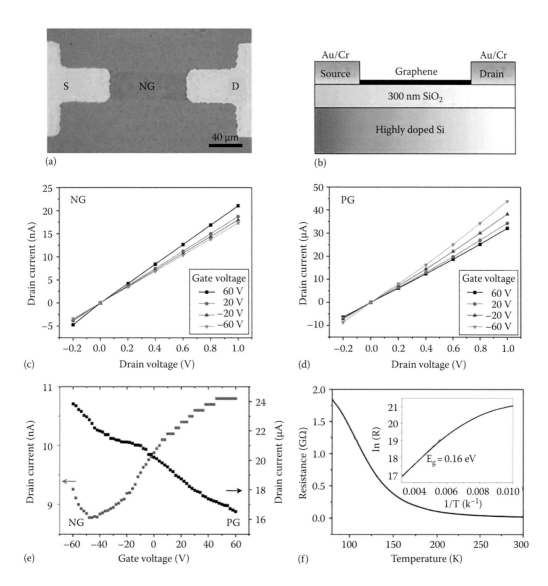

Figure 6.14 Electrical properties of nitrogen-doped graphene (NG) and pristine graphene (PG) in vacuum. (a) Optical microscope image of NG FET device. (b) Sketch of the device structure. (c, d) I_{ds}–V_{ds} output characteristics of NG3 (N/C = 1.6 at.%) and PG at variable back-gate voltages starting from –60 to 60 V in a step of 40 V, respectively. (e) Transfer characteristics (I_{ds}–V_g) of NG3 (red) and PG (black) devices at V_{ds} = 0.5 V. (f) Temperature dependence of the electrical resistance of NG2 (N/C = 2.9 at.%). The inset shows the change of ln(R) as a function of T^{-1} in the temperature range from 100 to 300 K.

decrease the mobility of N-graphene in practice. In addition, the grain boundaries in N-graphene may also contribute to a decrement of mobility (Reina et al. 2009).

6.3.1.3   Boron Nitrogen Co-Doped Graphene

BN is isoelectronic with carbon for similar structures, such as cubic-BN and diamond, as well as h-BN and graphite. BN sheets have wide band gaps, so these gaps may be opened effectively by alloying graphene with BN. In reference to the earlier sections, our overview will start from theoretical calculations. We would like to point out that theoretical simulations can only give qualitative results, and because of limited calculating capability, the atomic model is far from the real sample that we obtained in the experiment. Most of the time, we do not even know the exact atomic model of the samples that we get. When comparing the simulation result with the experimental result, extensive care should be taken.

B–N and C–C bonds with large binding energies are stronger than B–C and N–C bonds, so phase segregation occurs between carbon and BN during direct synthesis. However, the thermal effect in the growth process makes the uniform formation of the BN domain impossible. As a response to the possible clustering and violation of the uniformity of the BN domains, Fan et al. (2012) constructed two models that have 288 lattice sites (shown in Figure 6.15a, b) with BN domains separated randomly. The concentrations of BN are 6.25% and 4.17% for two models, respectively. Figure 6.15c and d demonstrates the band structures of both models. The energy gap can be opened effectively when BN domains merge, regardless of the disorderly arrangement of domains with different sizes. As a result of the isoelectronic properties of BN and C2, localized defect states are not found in the band structure, and the Fermi level is still around the Dirac

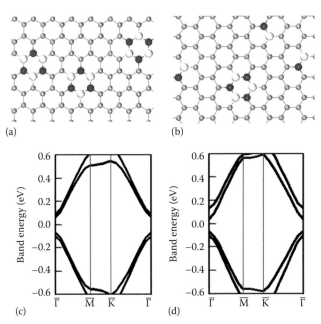

Figure 6.15 Schematic structures and electronic properties of graphene with the nonuniform BN domains including the model with three (BN)₃ domains in the lattice with 288 atoms (a, c) and that with one (BN)₃ domain and three BN domains in the lattice with 288 atoms (b, d). (Fan, X., Shen, Z., Liu, A.Q., and Kuo, J.L., Band gap opening of graphene by doping small boron nitride domains, *Nanoscale*, 4, 2157–2165, 2012. Reproduced by permission of The Royal Society of Chemistry.)

point without shift. The disordered effects just break the degeneracy of edges of the conduction band and valance band. The coupling of BN domains with different sizes can result in a relatively large splitting of degenerate states, especially the states of the conduction band edge.

The effect of a small domain with a low concentration to the band gap opening is very different from that of a large domain with a high concentration. In case of a small domain with a low concentration, the band gap opening in BN co-doped graphene is attributed to the breaking of the inherent equivalence of graphene sublattices. A large domain of BN with a high concentration induces the formation of a localized region for the graphene part. Therefore, the quantum confinement effect plays the key role for the opening of the band gap. The band gap induced by quantum confinement is comparably small. Using the tight-binding method, Ci et al. (2010) estimated that the gap of the BN domain with a size of ~42 nm is about 18 meV for the BCN films with 50 at.% C. Thus, BN doping can effectively open the π–π* band gap of graphene, but a steady increase in BN concentration leads to a decrease in band gap. This theoretical supposition has been verified by Chang et al. (2013) through synchrotron radiation-based x-ray absorption near-edge structure (XANES)–x-ray emission spectroscopy (XES) measurements on BN co-doped graphene films. The selected spectra are magnified at the π–π* region as shown in Figure 6.16b, for clarity. The two

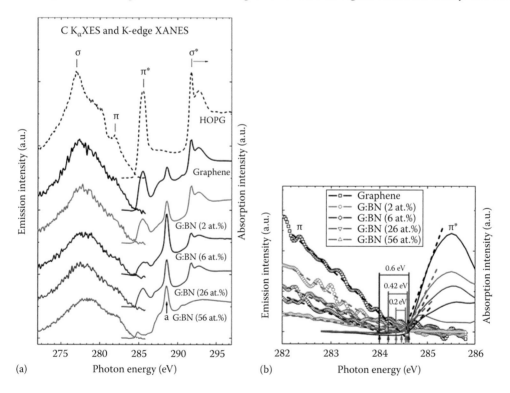

Figure 6.16 (a) Normalized C K-edge XANES and K_α XES spectra of the BNG films (left side, XES; right side, XANES) with various concentrations of B–N doping, pristine graphene, and HOPG. (b) Magnified selected spectra at the π–π* region. (Feature "a" present at 288.5 eV could be contributed by either C=N bonding with the B–N dopants or the C=O/C–OH bonding with the carbonyl group on the surface.) (Reprinted with permission from Chang, C., Kataria, S., Kuo, C. et al., Band gap engineering of chemical vapor deposited graphene by in situ BN doping, *ACS Nano*, 7, 1333–1341, 2013. Copyright 2013, American Chemical Society.)

extrapolated lines clearly intersect each other for pure graphene and films with BN concentration of 56%, indicating that these two samples have no band gap similar to that of the metallic highly ordered pyrolytic graphite (HOPG). However, in low-BN samples with increasing BN concentration, a monotonically increasing band gap up to 600 meV for the 6BNG sample was observed. However, the band gap decreased to 420 meV when BN content increased to 26%.

6.3.1.4 BCN Ternary Nanotubes

Soon after the discovery of CNTs, BCN ternary nanotubes or nanosheets were proposed for theoretical simulation. Researchers have found many possible atomic models of BCN ternary nanotubes, from stoichiometric BC_2N (Miyamoto et al. 1994), to random doping of BN in carbon (Lammert et al. 2001), to heterojunctions (Blase et al. 1997, Tomura et al. 2000). When unlimited bonding possibility is combined with the chirality of nanotubes, things become more complicated. Theoretically, giving a universal picture of the electronic structure of a BCN ternary nanotube is impossible. Ab initio calculation has mostly been applied to microscopically ordered, homogeneous nanotube models, which is far from the real experimental random-doped samples. On the other hand, most of the BN-doped nanotubes are multiwalled, and the morphologies are bamboo like and cone like. Such morphologies are quite different from that of the standard graphene roll-up model.

Only a few theoretical works concerning random doping models are available. Lammert et al. (2001) showed that under doping with boron and nitrogen, the microscopic doping inhomogeneity of CNTs is larger than that in normal semiconductors, leading to charge fluctuations that can be used to design nanoscale devices (diode behavior). In such structures, new acceptor and donor features are observed in the local density of states of the valence and conduction bands when compared with graphene.

Experimentally, the presented authors (Xu et al. 2008) fabricated field-effect transistor using BN co-doped single-walled CNT. Electronic transport measurements elucidate that 97.6% of their BCN-SWNTs are semiconducting, indicating that metallic C-SWNTs can be converted into semiconductors by BN co-doping (Figure 6.17). The presented authors also realized direct synthesis of

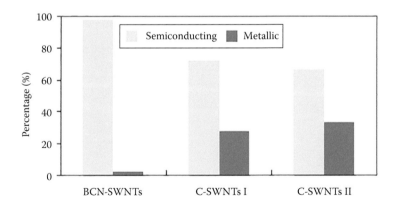

Figure 6.17 Histogram of the number of the semiconducting and metallic nanotubes in as-grown BCN-SWNT, C-SWNT I (using H_2 and CH_4 as reactant gases), and C-SWNT II (using B_2H_6 and CH_4 as reactant gases) samples, respectively. The percentages of the semiconducting nanotubes in the three types of samples are 97.6%, 72.2%, and 66.4%, respectively.

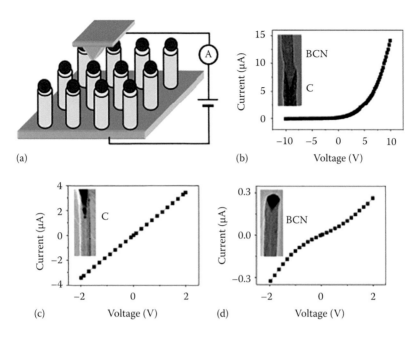

Figure 6.18 (a) Scheme of the experimental setup for the electrical transport measurement of individual nanotubes by a conductive atomic force microscope. (b–d) I–V curves for a single BCN/C nanotube junction, carbon nanotube, and BCN nanotube; the insets of panels (b–d) schematically showing the corresponding different nanotubes. (Reprinted with permission from Liao, L., Liu, K., Wang, W. et al., Multiwall boron carbonitride/carbon nanotube junction and its rectification behavior, *J. Am. Chem. Soc.*, 129, 9562–9563, 2007. Copyright 2007, American Chemical Society.)

massive BCN/C nanotube junctions via a bias-assisted hot-filament CVD method (Liao et al. 2007, Figure 6.18). The electrical transport measurements of individual nanotube junctions were carried out on a conductive atomic force microscopy. It is found that the BCN/C nanotube junctions show a typical rectifying diode behavior. The earlier two experiments prove that the BCN ternary nanotubes are semiconducting. But the prospective band gap tuning by BN doping is still unrealized.

6.3.2 Luminescence

Luminescence can provide a quick and reliable characterization of structure, morphology, and physical properties in general. But only a few references reporting the luminescence of BCN ternary nanotubes, nanofibers, or nanosheets are available. The band gap is predicted to be highly dependent on atomic arrangement and crystallinity. Experimental information on atomic arrangement is limited, although several theoretical treatments have been conducted (Liu et al. 1989, Miyamoto et al. 1995, Nozaki and Itoh 1996, Kar et al. 1998). For example, Liu et al. (1989) calculated several possible atomic arrangements in the BC_2N monolayer together with the associated band structures using pseudopotential local-orbital approach. The calculated band structures indicated that two of the structural models for BC_2N would be semiconductors with indirect band gaps of 1.6 or 0.5 eV, and a third possible structure that is a metal.

Watanabe et al. (1996) measured the photoluminescence (PL) properties of BC_2N films prepared by CVD using acetonitrile and boron trichloride. The PL peak energy was found to be 600 nm at room temperature and 580 nm at 4.2 K, corresponding to 2.07 and 2.14 eV, respectively. Chen et al. (1999) employed high resolution electron energy loss spectroscopy (HREELS) in ultra-high vacuum to investigate the band gap of BC_2N (prepared by CVD using boron trichloride and acetonitride) by measuring the electronic excitation spectrum from the valence band to the conduction band as a function of parallel momentum transfer. The excitation threshold (apparent gap) shows a negative dispersion, which indicates an indirect band structure (Figures 6.19 and 6.20).

One of the most frequently cited works in this field was done by the present authors, in which Bai et al. (2000b) reported that the color of the light emitted from BCN MWNTs could be controlled by tuning B_2H_6 concentration during CVD synthesis. The synthesis equipment is bias-assisted HFCVD, as introduced in Section 6.2.3. A mixture of high-purity N_2, H_2, CH_4, and B_2H_6 (diluted in N_2 at a concentration of 10 vol.%) was employed as a precursor. Figure 6.21 shows the room-temperature PL spectra from five different samples of BCN nanofibers, which correspond to inlet B_2H_6 concentrations of 2.5%, 2.0%, 1.5%, 1.0%, and 0.5%, respectively. With a change in the inlet B_2H_6 concentrations in the reactive gas mixture, the peak centers shift in the range of 470–390 nm. Bai stated the following in their paper: "During measurement, blue or violet light emission with a strong and stable brightness can also be observed clearly by the naked eye." However, in this pioneer work, the boron and nitrogen concentrations of these BCN nanotubes have not been studied. The present research is the first to achieve tunable properties of BCN ternary nanostructures.

Later, Yin et al. (2005) reported an ultraviolet emission (peaked at 319 nm) from BCN nanofibers with B/C/N atomic ratio of 1:1:1, suggesting a band gap energy of ~3.89 eV for this present porous BCN tubular material. Figure 6.22a shows the cathodoluminescence (CL) spectra taken from the BCN fiber (stretching across a large hole of a copper TEM grid) at 20 K. The CL spectra denoted as 1, 2, 3, and 4 were taken from different spots within or around the sample (Figure 6.22b). Sharp and intense ultraviolet emission centered at 319 nm at 20 K is visible

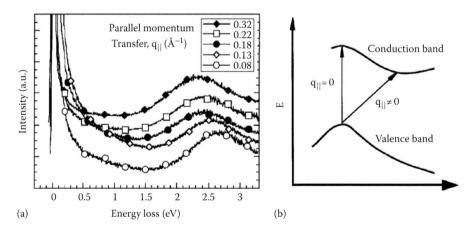

(a)   Energy loss (eV)   (b)

Figure 6.19   (a) HREELS spectra from a thin film of BC_2N for different parallel momentum transfers. The broad loss features observed above a threshold at 1–2 eV are due to interband transitions. (b) Schematic illustration of two interband transitions with $q_\parallel = 0$ (vertical) and $q_\parallel \neq 0$. (Reprinted with permission from Chen, Y., Barnard, J.C., Palmer, R.E. et al., Indirect band gap of light-emitting BC_2N, *Phys. Rev. Lett.*, 83, 2406–2408, 1999. Copyright 1999, American Physical Society.)

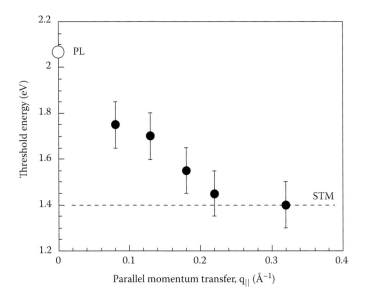

Figure 6.20 Threshold energies of the interband transitions for BC$_2$N thin films as a function of parallel momentum transfer. Also shown is the vertical band gap derived from PL measurements (open circle at q$_{||}$ = 0) and that indicated by STM measurements, where q$_{||}$ is not conserved (horizontal line). (Reprinted with permission from Chen, Y., Barnard, J.C., Palmer, R.E. et al., Indirect band gap of light-emitting BC$_2$N, *Phys. Rev. Lett.*, 83, 2406–2408, 1999. Copyright 1999, American Physical Society.)

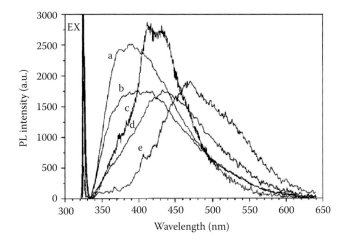

Figure 6.21 Room-temperature PL spectra from the BCN nanofibers, where the inlet B2H6 concentrations in the reactive gas mixture are (a) 2.5%, (b) 2.0%, (c) 1.5%, (d) 1.0%, and (e) 0.5%, respectively. EX indicates excitation light of 325 nm from the He–Cd laser. (Reprinted with permission from Bai, X.D., Wang, E.G., and Yu, J., Blue–violet photoluminescence from large-scale highly aligned boron carbonitride nanofibers, *Appl. Phys. Lett.*, 77, 67–69, 2000b. Copyright 2000, American Institute of Physics.)

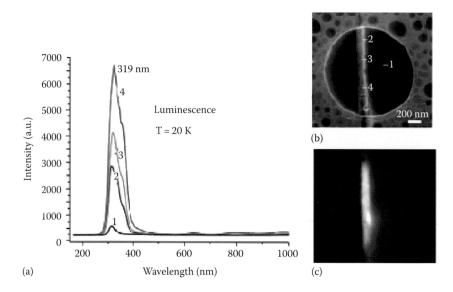

(a)

(b)

(c)

Wavelength (nm)

Figure 6.22 (a) CL spectra recorded at 20 K from various spots along an individual porous BCN tubular fiber in (b). The fiber reveals a band gap energy of 3.89 eV. (c) CL luminescence image from the fiber inside SEM. (Reprinted with permission from Yin, L.W., Bando, Y., and Golberg, D., Porous BCN nanotubular fibers: Growth and spatially resolved cathodoluminescence, *J. Am. Chem. Soc.*, 127, 16354–16355, 2005. Copyright 2005, American Chemical Society.)

(Figure 6.22c). Yang et al. (2006) synthesized BCN nanofibers with even higher BN concentrations. The B/C/N atomic ratio was 3:1:3, but the PL emissions peaked at about 370 nm (3.35 eV) and 700 nm (1.77 eV).

The band gap is predicted to be highly dependent on atomic arrangement and crystallinity, making the samples from different groups incomparable. During the booming research activities on graphene, rare band-edge luminescence reports on single and/or multilayer BCN nanosheets emerged. Calculations indicate that B- and/or N-doped graphenes have a direct band gap, but experiments show opposite results. This conflict indicates that the real sample we possess may be different from the ideal model. Furthermore, to date, the reported band gaps of BCN nanosheets, regardless of methods used, are all below 1 eV, which is far from the visible light range.

6.3.3 Magnetic Property

Magnetism in materials that contain only s and p electrons is of fundamental and technological interest, with potential applications in spintronic devices and molecular magnets for quantum computing (Ivanovskii 2007). The recent spate of research on graphene has generated several reports of intrinsic magnetism in graphene nanodots (Hod et al. 2008, Wang et al. 2008), zigzag nanoribbons (Son et al. 2006, Hod et al. 2007, Yazyev and Katsnelson 2008, Ramasubramaniam 2010), and defective sheets (Vozmediano et al. 2005, Lehtinen et al. 2004, Yazyev 2007). Calculations also predict that defects could introduce spin polarization in BN nanotubes and nanosheets. These defects include substitutional C dopant in both B and N sites (Wu et al. 2005, Guo et al. 2006, Moradian and Azadi 2008, Azevedo et al. 2009, Ramasubramaniam and Naveh 2011, Ukpong and Chetty 2012), vacancy (Ma et al. 2004b, Si and Xue 2007, Moradian and Azadi 2008,

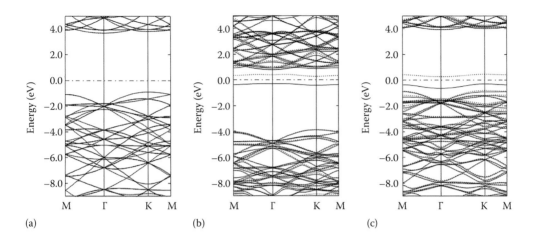

Figure 6.23 The calculated band structures of the pristine graphitic-BN (a), with a boron atom (b) and a nitrogen atom (c) substituted by a carbon atom. The solid lines represent the spin-up branches, while the dotted lines represent the spin-down branches. The Fermi level is set to zero. (Reproduced from Wu, R.Q., Peng, G.W., Liu, L., and Feng, Y.P. 2006. Possible graphitic-boron-nitride-based metal-free molecular magnets from first principles study. *J. Phys.: Condens. Matter* 18:569–575. With permission of IOP Publishing. All rights reserved.)

Azevedo et al. 2009), adatom (Yang et al. 2010), institutional Si or O dopant (Liu and Cheng 2007), and C/BN heterostructure (Choi et al. 2003, Pruneda 2010). In this section, we will focus on carbon (from single atom to graphene island) substitutional doped BN nanosheets, discussing the originality of spin polarization and the generation of magnetism in such sp^2 system.

C atoms can substitute in B and N sites. The formation energies of these two structures (in eV/at.) are 0.03 and 0.04, respectively, which are quite lower than the formation energy of B and N vacancy, as well as boron and nitrogen antisite defects (Azevedo et al. 2007). When one C atom substitutes B atom or N atom in h-BN lattice, one unpaired electron is left on the C atom, which will generate a local magnetic moment. Wu (2006) conducted first-principle calculations on this system and further expounded it. Figure 6.23 shows the spin-polarized band structure of pure h-BN, B atom, and N atom substituted by a C atom system. Figure 6.24 presents the corresponding DOS plots. For the pure graphitic-BN (Figure 6.23a), all states are in twofold degeneracy, indicating the absence of spin polarization. When a B atom is substituted by a C atom (Figure 6.23b), Fermi level of the system is lifted and a flat band near the Fermi level emerges, which is caused by the carbon atom acting as an n-type dopant. This flat band near the Fermi level is split into two branches. The spin-up branch is occupied, and the spin-down branch is left empty, leading to a spontaneous polarization in the doped system with a net magnetic moment of around 1.0 μB. The same effect is observed when a nitrogen atom is substituted by a carbon atom (Figure 6.23c); the only difference is that the Fermi level is pushed down because in this case, carbon acts as a p-type dopant. The same character was observed in the majority and minority DOS of the three systems (Figure 6.24). The majority DOS exceeds the minority DOS by the occupied peak just below the Fermi level. The minority DOS peak near the Fermi level, however, is unoccupied. Thus, both doped systems are spin polarized. The space distributions of the magnetic moment of these two doping situations are clearly shown in the magnetization density (difference in densities of majority and minority spins) mapping

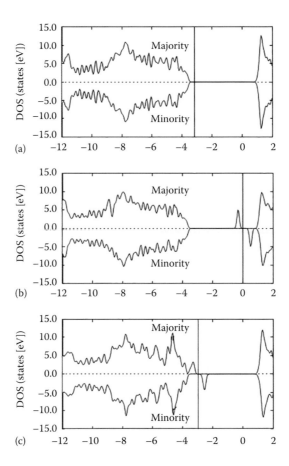

Figure 6.24 The calculated majority and minority DOS for (a) pristine, (b) a boron atom, and (c) a nitrogen atom substituted by carbon atom systems. The Fermi level is denoted by the vertical lines. (Reproduced from Wu, R.Q., Peng, G.W., Liu, L., and Feng, Y.P. 2006. Possible graphitic-boron-nitride-based metal-free molecular magnets from first principles study. *J. Phys.: Condens. Matter* 18:569–575. Permission of IOP Publishing. All rights reserved.)

(Figure 6.25). Clearly, the magnetization density originates from the carbon atom and localizes there. This behavior suggests that the spin polarization results from the unpaired electron of the carbon atom.

From the foregoing discussion, nonmagnetic C dopant can induce finite magnetic moments in a BN nanosheet. Next, we will consider if a correlation exists between two local magnetic moments. If coupling is present between the nearest-neighbor magnetic moments, long-range magnetic order will be established, which will result in ferromagnetism or antiferromagnetism. The interaction (J, i.e., the exchange energy) based on the Heisenberg type of spin coupling between the nearest-neighbor magnetic moments calculated by Liu and Cheng (2007) is summarized in Figure 6.26. For the systems that involved C substituting B (designated as C_B), when the C dopants are separated by 1 nm, the calculated 8 J are less than 5 meV (within the computational accuracy), and almost no interaction is observed between them. For systems that involved C substituting N (designated as C_N), very weak antiferromagnetic interaction (8 J ≈ −8 meV) is shown. Figure 6.25

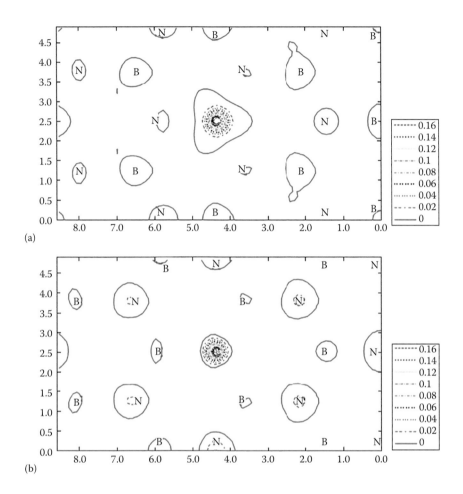

Figure 6.25 The magnetization density (difference in charge densities of majority and minority spins) distribution of the graphitic-BN with a boron atom (a) and a nitrogen atom (b) substituted by a carbon atom. Unit: spins $Å^{-3}$. (Reproduced from Wu, R.Q., Peng, G.W., Liu, L., and Feng, Y.P. 2006. Possible graphitic-boron-nitride-based metal-free molecular magnets from first principles study. *J. Phys.: Condens. Matter* 18:569–575. Permission of IOP Publishing. All rights reserved.)

shows that because of the large band gap of BN, no free electrons are present in this material; the magnetic moment on C atom is well localized. One nanometer is too large to have couplings between two C dopants. When distance decreases, interactions will become stronger. However, Liu did not present results when the distance between dopants is below 1 nm. Finally, in systems that involve O substituting N (designated as O_N), when defect spacing is about 1 nm, the exchange energy is as high as 76 meV, indicating strong ferromagnetic coupling with Curie temperature above 50 K.

For the C dopant composed of four carbon atoms (Figure 6.27), Ramasubramaniam and Naveh (2011) calculated the exchange energy (difference in ground-state energies between antiferromagnetic [AFM] and ferromagnetic [FM] coupled islands) as a function of island spacing. The result is similar to that of single-atom doping shown by Liu and Chang (2007). The two graphene sublattices are antiferromagnetically coupled to each other when spacing is small (N = 6, 9). When

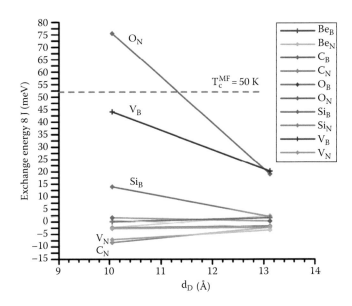

Figure 6.26   Exchange energy 8 J versus defect distances where $J \equiv 1/8(E_{AFM} - E_{FM})$. The energy range from −5 to 5 meV is the estimated region where the magnitude of 8 J is too small to be used to identify the sign of 8 J in our calculations. The dashed line indicates the value of exchange energy corresponding to a Curie temperature (T_c^{MF}) of 50 K under the mean-field approximation. (Reprinted with permission from Liu, R.F. and Cheng, C., Ab initio studies of possible magnetism in a BN sheet by nonmagnetic impurities and vacancies, *Phys. Rev. B*, 76, 014405–014412, 2007. Copyright 2007, American Physical Society.)

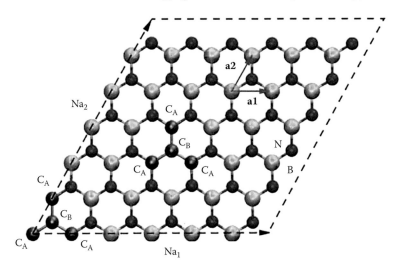

Figure 6.27   Hexagonal array of graphene islands in an h-BN lattice with N primitive cells on edge. The primitive vectors **a1** and **a2** of h-BN are of length $a_0 = 2.488$ Å. C atoms on the A and B sublattice are indicated by C_A and C_B. The center-to-center distance between islands is $r_{nn} = Na_0/\sqrt{3}$. Dotted lines enclose the periodic simulation cell. (Reprinted with permission from Ramasubramaniam, A. and Naveh, D., Carrier-induced antiferromagnet of graphene islands embedded in hexagonal boron nitride, *Phys. Rev. B*, 84, 075405–075411, 2011. Copyright 2011, American Physical Society.)

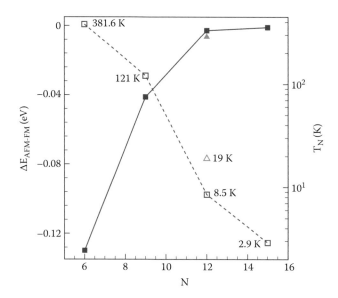

Figure 6.28 Energy difference between AFM- and FM-coupled C_4 islands as a function of separation N (solid squares) and Neel temperatures for AFM-coupled C_4 islands (hollow squares). AFM coupling is distinctly favored at small island spacings with correspondingly higher Neel temperatures. The lines are a guide to the eye. Triangles are for a C_9 island in an N = 12 simulation cell, indicating that increasing the island size while keeping the island spacing constant can serve to tune the Neel temperature of the array. (Reprinted with permission from Ramasubramaniam, A. and Naveh, D., Carrier-induced antiferromagnet of graphene islands embedded in hexagonal boron nitride, *Phys. Rev. B*, 84, 075405–075411, 2011. Copyright 2011, American Physical Society.)

spacing is larger than 1.7 nm (corresponding to N = 12 in Figure 6.28), the interaction becomes very weak. For all AFM cases, the magnetic moments per island is approximately ± 2 μB for weakly interacting islands (N = 12, 15). For the N = 6 cell, an average moment of ± 1.19 μB per island has been determined. The departure from the expected value of ± 2 μB is indicative of strong island–island interactions. The charge differences between AFM- and FM-coupled C4 islands and the relative spin density for each case for N = 9 cell and N = 15 cell are shown in Figure 6.29. In case of N = 15, the charge difference between AFM and FM is almost unnoticeable.

The exchange interaction of two C atoms and two graphene islands shows the same trend. The magnetics induced by C atoms in a h-BN lattice are almost fully revealed by calculation. Both individual C atoms and C clusters (graphene islands) could generate spontaneous spin polarization, and because of the ionic nature of BN bond, such magnetic moment is well localized. Thus, we could draw a rough physical image of the magnetic property of BCN nanosheets. When the dopants are far from each other (corresponding to a very low doping level in the experiment), no exchange interaction occurs between them. Thus, the materials are paramagnetic or superparamagnetic. When the dopants' domains become closer (the doping level becomes higher), antiferromagnetic coupling effect emerges, and the materials will present antiferromagnetic behavior. However, the calculations do not include all possibilities. It is reasonable to believe that in some cases, the carbon-doped BN may exhibit ferromagnetic characteristics. The magnetics of BCN nanosheets will be well tuned by the doping level. Producing molecular magnets by engineering C-doping BN nanosheets thus may be possible.

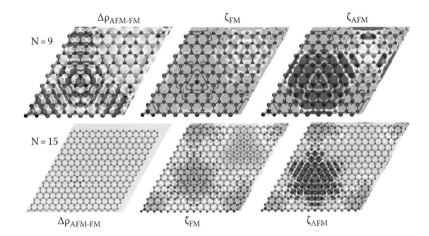

Figure 6.29   Charge difference $\Delta\rho_{AFM-FM}$ between AFM- and FM-coupled C$_4$ islands and relative spin density $\zeta = (\rho\uparrow - \rho\downarrow)/(\rho\uparrow + \rho\downarrow)$ for each case projected on the atomic plane for an N = 9 cell (upper row) and N = 15 cell (lower row). Light and dark regions correspond to values of $\Delta\rho_{AFM-FM} \geq 10^{-4}$ e/Å^2 and $\leq -10^{-4}$ e/Å^2, respectively, as well as $\zeta \geq 10^{-4}$ and $\leq -10^{-4}$, respectively. As seen from the ζ plots, each C$_4$ island spin-polarizes its BN neighbors due to its two unpaired electrons thereby inducing exchange interactions between islands. The $\Delta\rho$ plots show alternating belts of charge depletion and accumulation that are indicative of varying degrees of charge transfer between the islands and the h-BN matrix depending upon the magnetic coupling between islands (FM or AFM). The charge transfer is mediated entirely via the p$_z$ orbitals of C, B, and N. The differences are more obvious for the N = 9 case; for N = 15, the differences are very small (near-degenerate AFM and FM configurations) and are localized to the immediate vicinity of the islands. (Reprinted with permission from Ramasubramaniam, A. and Naveh, D., Carrier-induced antiferromagnet of graphene islands embedded in hexagonal boron nitride, *Phys. Rev. B*, 84, 075405–075411, 2011. Copyright 2011, American Physical Society.)

6.4   CONCLUDING REMARKS

In this chapter, we have briefly reviewed the progress of BCN nanotubes and nanosheets before, during, and after synthesis, property prediction, and property measurement. BCN nanostructures are theoretically expected to show a rich variety of physical properties and to enable numerous possible technological applications in the fields of nano-electronics, optical devices, field emission, catalysis, lubrication, and gas storage. Especially, the BCN nanosheet magnetics is an emerging new area in which a limited number of experiments have been done. The primary obstacle in current studies is the lack of well-controlled materials. The atomic structure of BCN nanosheets grown by different methods is different from each other, and the experimental result derived from these different groups is not comparable. Even in the same batch of growth, doping is not uniform from nanotube to nanotube. The synthesis method offers macroscale control, whereas the characterization method provides average result from microscale. However, what are currently expected are properties from atomic-scale structures. For example, graphene electronics or spintronics devices with exact edge structures have been designed. The magnetism of BCN nanosheets requires sub-1 nm doping position control. The BCN ternary nanotubes and nanosheets provide more possibilities and prospects than pure CNT and graphene. An infinitely fascinating nanoworld will open when people could really manipulate atoms.

REFERENCES

Azevedo, S.; Kaschny, J. R.; Castilho, C. M. C.; Mota, F. B. 2007. A theoretical investigation of defects in a boron nitride monolayer. *Nanotechnology* 18: 495707.

Azevedo, S.; Kaschny, J. R.; Castilho, C. M. C.; Mota, F. B. 2009. Electronic structure of defects in a boron nitride monolayer. *Eur. Phys. J. B* 67: 507–512.

Bai, X. D.; Guo, J. D.; Yu, J.; Wang, E. G.; Yuan, J.; and Zhou, W. 2000a. Synthesis and field-emission behavior of highly oriented boron carbonitride nanofibers. *Appl. Phys. Lett.* 76: 2624–2626.

Bai, X. D.; Wang, E. G.; Yu, J. 2000b. Blue–violet photoluminescence from large-scale highly aligned boron carbonitride nanofibers. *Appl. Phys. Lett.* 77: 67–69.

Blackburn, J. L.; Yan, Y.; Engtrakul, C.; Parilla, P. A.; Jones, K.; Gennett, T.; Dillon A. C.; Heben, M. J. 2006. Synthesis and characterization of boron-doped single-wall carbon nanotubes produced by the laser vaporization technique. *Chem. Mater.* 18: 2558–2566.

Blase, X.; Charlier, J. C.; De-Vita, A.; Car, R. 1997. Theory of composite $B_xC_yN_z$ nanotube heterojunctions. *Appl. Phys. Lett.* 70: 197–199.

Cattelan, M.; Agnoli, S.; Favaro, M. et al. 2013. Microscopic view on a chemical vapor deposition route to boron doped graphene nanostructures. *Chem. Mater.* 25: 1490–1495.

Chang, C.; Kataria, S.; Kuo, C. et al. 2013. Band gap engineering of chemical vapor deposited graphene by in situ BN doping. *ACS Nano* 7: 1333–1341.

Chen, Y.; Barnard, J. C.; Palmer, R. E. et al. 1999. Indirect band gap of light-emitting BC_2N. *Phys. Rev. Lett.* 83: 2406–2408.

Choi, J.; Kim, Y. H.; Chang, K. J.; Tomanek, D. 2003. Itinerant ferromagnetism in heterostructured C/BN nanotubes. *Phys. Rev. B* 67: 125421–125425.

Ci, L.; Song, L.; Jin, C. et al. 2010. Atomic layers of hybridized boron nitride and graphene domains. *Nat. Mater.* 9: 430–435.

Gai, P. L.; Stephan, O.; McGuire, K.; Rao, A. M.; Dresselhaus, M. S.; Dresselhaus G.; Colliex, C. 2004. Structural systematics in boron-doped single wall carbon nanotubes. *J. Mater. Chem.* 14: 669–675.

Droppa, Jr. R.; Hammer, P.; Carvalho, A. C. M.; Santos, M. C. dos; Alvarez, F. J. 2002. Incorporation of nitrogen in carbon nanotubes. *Non-Cryst. Solids* 299: 874–879.

Ewels, C. P.; Glerup, M. 2005. Nitrogen doping in carbon nanotubes. *J. Nanosci. Nanotechnol.* 5: 1345–1363.

Fan, X.; Shen, Z.; Liu, A. Q.; Kuo, J. L. 2012. Band gap opening of graphene by doping small boron nitride domains. *Nanoscale* 4: 2157–2165.

Golberg, D.; Bando, Y.; Bourgeois, L.; Sato, T. 2000. Large-scale synthesis and HRTEM analysis of single-walled B- and N-doped carbon nanotube bundles. *Carbon* 38: 2017–2027.

Golberg, D.; Bando, Y.; Han, W.; Kurashima, K.; Sato, T. 1999. Single-walled B-doped carbon, B/N-doped carbon and BN nanotubes synthesized from single-walled carbon nanotubes through a substitution reaction. *Chem. Phys. Lett.* 308: 337–342.

Guo, C. S.; Fan, W. J.; Zhang, R. Q. 2006. Spin polarization of the injected carriers in C-doped BN nanotubes. *Solid State Commun.* 137: 246–248.

Guo, J. D.; Zhi, C. Y.; Bai, X. D.; Wang, E. G. 2002. Boron carbonitride nanojunctions. *Appl. Phys. Lett.* 80: 124–126.

Han, W.; Bando, Y.; Kurashima, K.; Sato, T. 1998. Synthesis of boron nitride nanotubes from carbon nanotubes by a substitution reaction. *Appl. Phys. Lett.* 73: 3085.

Han, W.; Redich, P.; Ernst, F.; Ruhle, M. 1999. Formation of $(BN)_xC_y$ and BN nanotubes filled with boron carbide nanowires. *Chem. Mater.* 11: 3620–3623.

Hod, O.; Barone, V.; Peralta, J. E.; Scuseria, G. E. 2007. Enhanced half-metallicity in edge-oxidized zigzag graphene nanoribbons. *Nano Lett.* 7: 2295–2299.

Hod, O.; Barone, V.; Scuseria, G. E. 2008. Half-metallic graphene nanodots: A comprehensive first-principles theoretical study. *Phys. Rev. B* 77: 035411–035416.

Ivanovskii, A. L. 2007. Magnetic effects induced by sp impurities and defects in nonmagnetic sp materials. *Physics-Uspekhi* 50: 1031–1052.

Jin, Z.; Yao, J.; Kittrell, C.; Tour, J. M. 2011. Large-scale growth and characterizations of nitrogen-doped monolayer graphene sheets. *ACS Nano* 5: 4112–4117.

Kar, T.; Cuma, M.; Scheiner, S. 1998. Structure, stability, and bonding of BC2N: An ab initio study. *J. Phys. Chem.* 102: 10134–10141.

Kaner, R. B.; Kouvetakis, J.; Warble, C. E. et al. 1987. Boron-carbon-nitrogen materials of graphite-like structure. *Mater. Res. Bull.* 22: 399–404.

Keskar, G.; Rao, R.; Luo, J.; Hudson, J.; Chen, J.; Rao, A. M. 2005. Growth, nitrogen doping and characterization of isolated single-wall carbon nanotubes using liquid precursors. *Chem. Phys. Lett.* 412: 269–273.

Kim, K. S.; Zhao, Y.; Jang, H. et al. 2009. Large-scale pattern growth of graphene films for stretchable transparent electrodes. *Nature* 457: 706–710.

Latea, D. J.; Ghoshab, A.; Subrahmanyama, K. S. et al. 2010. Characteristics of field-effect transistors based on undoped and B- and N-doped few-layer graphenes. *Solid State Commun.* 150: 734–738.

Lherbier, A.; Botello-Méndez, A. R.; Charlier, J. C. 2013. Electronic and transport properties of unbalanced sublattice N-doping in graphene. *Nano Lett.* 13: 1446–1450.

Li, X.; Cai, W.; An, J. et al. 2009a. Large-area synthesis of high-quality and uniform graphene films on copper foils. *Science* 324: 1312–1314.

Li, X.; Wang, H.; Robinson, J. T.; Sanchez, H.; Diankov, G.; Dai, H. 2009b. Simultaneous nitrogen doping and reduction of graphene oxide. *J. Am. Chem. Soc.* 131: 15939–15944.

Liao, L.; Liu, K.; Wang, W. et al. 2007. Multiwall boron carbonitride/carbon nanotube junction and its rectification behavior. *J. Am. Chem. Soc.* 129: 9562–9563.

Liu, A. Y.; Wentzcovitch, R. M.; Cohen, M. L. 1989. Atomic arrangement and electronic structure of BC_2N. *Phys. Rev. B* 39: 1760–1765.

Liu, J.; Carroll, D. L.; Cech, J.; Roth, S. 2006. Single-walled carbon nanotubes synthesized by the pyrolysis of pyridine over catalysts. *J. Mater. Res.* 21: 2835–2840.

Liu, R. F.; Cheng, C. 2007. Ab initio studies of possible magnetism in a BN sheet by nonmagnetic impurities and vacancies. *Phys. Rev. B* 76: 014405–014412.

Liu, Z.; Ma, L.; Shi, G. et al. 2013. In-plane heterostructures of graphene and hexagonal boron nitride with controlled domain sizes. *Nat. Nanotechnol.* 8: 119–124.

Lammert, P. E.; Crespi, V. H.; Rubio, A. 2001. Stochastic heterostructures and diodium in B/N-doped carbon nanotubes. *Phys. Rev. Lett.* 87: 136402.

Lehtinen, P. O.; Foster, A. S.; Ma, Y.; Krasheninnikov, A. V.; Nieminen, R. M. 2004. Irradiation-induced magnetism in graphite: A density functional study. *Phys. Rev. Lett.* 93: 187202.

Levendorf, M. P.; Kim, C. J.; Brown, L. et al. 2012. Graphene and boron nitride lateral heterostructures for atomically thin circuitry. *Nature* 488: 627–632.

Ma, R.; Golberg, D.; Bando Y.; Sasaki, T. 2004a. Syntheses and properties of B–C–N and BN nanostructures. *Philos. Trans. R. Soc. Lond. A* 362: 2161–2186.

Ma, Y.; Lehtinen, P. O.; Foster, A. S.; Nieminen, R. M. 2004b. Magnetic properties of vacancies in graphene and single-walled carbon nanotubes. *New J. Phys.* 6: 68.

Miyamoto, Y.; Cohen, M. L.; Louie, S. G. 1995. Ab initio calculation of phonon spectra for graphite, BN, and BC_2N sheets. *Phys. Rev. B* 52: 14971–14975.

Miyamoto, Y.; Rubio, A.; Marbin, L. et al. 1994. Chiral tubules of hexagonal BC_2N. *Phys. Rev. B* 50: 4976–4979.

Moradian, R.; Azadi, S. 2008. Magnetism in defected single-walled boron nitride nanotubes. *Europhys. Lett.* 83: 17007.

Nozaki, H.; Itoh, S. 1996. Lattice dynamics of BC_2N. *Phys. Rev. B* 53: 14161–14170.

Oshima, C.; Nagashima, A. 1997. Ultra-thin epitaxial films of graphite and hexagonal boron nitride on solid surfaces. *J. Phys. Condens. Matter* 9: 1–20.

Panchakarla, L. S.; Subrahmanyam, K. S.; Saha, S. K. 2009. Synthesis, structure, and properties of boron- and nitrogen-doped graphene. *Adv. Mater.* 21: 4726–4730.

Pruneda, J. M. 2010. Origin of half-semimetallicity induced at interfaces of C-BN heterostructures. *Phys. Rev. B* 81: 161409–161412.

Ramasubramaniam, A. 2010. Electronic structure of oxygen-terminated zigzag graphene nanoribbons: A hybrid density functional theory study. *Phys. Rev. B* 81: 245413.

Ramasubramaniam, A.; Naveh, D. 2011. Carrier-induced antiferromagnet of graphene islands embedded in hexagonal boron nitride. *Phys. Rev. B* 84: 075405–075411.

Reina, A.; Jia, X. T.; Ho, J.; Nezich, D. et al. 2009. Large area, few-layer graphene films on arbitrary substrates by chemical vapor deposition. *Nano Lett.* 9: 30–35.

Rosei, R.; Crescenzi, M.; Sette, F. et al. 1983. Structure of graphitic carbon on Ni(111): A surface extended-energy-loss fine-structure study. *Phys. Rev. B* 28: 1161–1164.

Rubio, A.; Corkill, J. L.; Cohen, M. L. 1994. Theory of graphitic boron nitride nanotubes. *Phys. Rev. B* 49: 5081–5084.

Si, M. S.; Xue, D. S. 2007. Magnetic properties of vacancies in a graphitic boron nitride sheet by first-principles pseudopotential calculations. *Phys. Rev. B* 75: 193409–193412.

Son, Y. W.; Cohen, M.; Louie, S. G. 2006. Half-metallic graphene nanoribbons. *Nature* 444: 347–349.

Stephan, O.; Ajayan, P. M.; Colliex, C. et al. 1994. Doping graphitic and carbon nanotube structures with boron and nitrogen. *Science* 266: 1683–1685.

Terrones, M.; Grobert, N.; Terrones, N. 2002. Synthetic routes to nanoscale $B_xC_yN_z$ architectures. *Carbon* 40: 1665–1684.

Tomura, N.; Takahashi, S.; Kato, K. et al. 2000. Properties of composites $B_xC_yN_z$ nanotubes and related hetero-junctions. *Comput. Mater. Sci.* 17: 107–114.

Ukpong, A. M.; Chetty, N. 2012. Half-metallic ferromagnetism in substitutionally doped boronitrene. *Phys. Rev. B* 86: 195409–195423.

Usachov, D.; Vilkov, O.; Grüneis, A. et al. 2011. Nitrogen-doped graphene: Efficient growth, structure, and electronic properties. *Nano Lett.* 11: 5401–5407.

Vozmediano, M. A. H.; Lopez-Sancho, M. P.; Stauber, T.; Guinea, F. 2005. Local defects and ferromagnetism in graphene layers. *Phys. Rev. B* 72: 155121–155125.

Wang, S.; Zhang, L.; Xia, Z. et al. 2012. BCN graphene as efficient metal-free electrocatalyst for the oxygen reduction reaction. *Angew. Chem. Int. Ed.* 51: 4209–4212.

Wang, W. L.; Meng, S.; Kaxiras, E. 2008. Graphene nano flakes with large spin. *Nano Lett.* 8: 241–245.

Wang, W. L.; Bai, X. D.; Wang, E. G. 2007. Towards the single-walled B- and/or N-doped carbon nanotubes. *Int. J. Nanosci.* 6: 431–442.

Wang, W. L.; Bai, X. D.; Liu, K. H. et al. 2006. Direct synthesis of B–C–N single-walled nanotubes by bias-assisted hot filament chemical vapor deposition. *J. Am. Chem. Soc.* 128: 6530–6531.

Wang, X. R.; Li, X. L.; Zhang, L. et al. 2009. N-doping of graphene through electrothermal reactions with ammonia. *Science* 324: 768–771.

Watanabe, M. O.; Itoh, S.; Sasaki, T. 1996. Visible-light-emitting layered BC_2N semiconductor. *Phys. Rev. Lett.* 77: 187–189.

Wei, D.; Liu, Y.; Wang, Y. et al. 2009. Synthesis of N-doped graphene by chemical vapor deposition and its electrical properties. *Nano Lett.* 9: 1752–1758.

Wengsieh, Z.; Cherrey, K.; Chopra, N. G. et al. 1995. Synthesis of $B_xC_yN_z$ nanotubules. *Phys. Rev. B* 51: 11229.

Wu, M.; Cao, C.; Jiang, J. 2010. Light non-metallic atom (B, N, O and F)-doped graphene: A first-principles study. *Nanotechnology* 21: 505202.

Wu, R. Q.; Liu, L.; Peng, G. W.; Feng, Y. P. 2005. Magnetism in BN nanotubes induced by carbon doping. *Appl. Phys. Lett.* 86: 122510.

Wu, R. Q.; Peng, G. W.; Liu, L.; Feng, Y. P. 2006. Possible graphitic-boron-nitride-based metal-free molecular magnets from first principles study. *J. Phys. Condens. Matter* 18: 569–575.

Xu, Z.; Lu, W.; Wang, W. et al. 2008. Converting metallic single-walled carbon nanotubes into semiconductors by boron/nitrogen co-doping. *Adv. Mater.* 20: 3615–3619.

Yang, J.; Qiu, T.; Shen, C. Y. 2006. New BCN fibres for strong ultraviolet and visible light luminescence. *Chin. Phys. Lett.* 23: 2573–2575.

Yang, J. H.; Kim, D.; Hong, J.; Qian, X. 2010. Magnetism in boron nitride monolayer: Adatom and vacancy defect. *Surf. Sci.* 604: 1603–1607.

Yazyev, O. V.; Katsnelson, M. I. 2008. Magnetism in disordered graphene and irradiated graphite. *Phys. Rev. Lett.* 100: 047209.

Yazyev, O. V.; Helm, L. 2007. Defect-induced magnetism in graphene. *Phys. Rev. B* 75: 125408.

Yin, L. W.; Bando, Y.; Golberg, D. 2005. Porous BCN nanotubular fibers: Growth and spatially resolved cath-odoluminescence. *J. Am. Chem. Soc.* 127: 16354–16355.

Zhang, C.; Fu, L.; Liu, N. et al. 2011. Synthesis of nitrogen-doped graphene using embedded carbon and nitro-gen sources. *Adv. Mater.* 23: 1020–1024.

Zhao, L.; He, R.; Rim, K. T. et al. 2011. Visualizing individual nitrogen dopants in monolayer graphene. *Science* 333: 999–1003.

Zheng, Y.; Jiao, Y.; Ge, L. et al. 2013. Two-step boron and nitrogen doping in graphene for enhanced synergistic catalysis. *Angew. Chem. Int. Ed.* 52: 3110–3116.

Zhi, C. Y.; Bai, X. D.; Wang, E. G. 2004. Boron carbonitride nanotubes. *J. Nanosci. Nanotechnol.* 4: 35–51.

Zhi, C. Y.; Guo, J. D.; Bai, X. D.; Wang, E. G. 2002. Adjustable boron carbonitride nanotubes. *J. Appl. Phys.* 91: 5325–5334.

Chapter 7

Fabrication, Structure, and Properties of Graphene Platelets

Jun Ma, Qingshi Meng, Jiabin Dai, and Nasser Saber

CONTENTS

7.1 INTRODUCTION

This chapter reviews the past and current states of the emerging research on the fabrication, structure, properties, and major applications of graphene platelets (GnPs). While graphene is hailed as the brightest rising star in materials science and engineering, graphene oxide and its reduced derivatives are actually the form of graphene widely used. We, in this chapter, start with graphene and graphene oxide, elaborate on GnPs, and compare them with graphene and graphene oxide. We also focus on the surface modification of the platelets to build up a strong interface for polymer composites.

Graphene comprises a single layer of sp^2-hybridized carbon atoms creating a 2D hexagonal lattice; as a unique monolayer crystal sheet, graphene displays a room-temperature quantum Hall effect, and it is a fundamental building block of graphite, nanotubes, and fullerenes. Although the first experimental effort on graphene started as early as 1962 (Boehm et al. 1962), individual graphene had not been isolated until 2004 by Geim and Novoselov (Novoselov et al. 2004, 2005, 2007; Zhang et al. 2005) using a *scotch tape* method. Graphene has attracted enormous interest due to its exceptional properties—Young's modulus 1 TPa, fracture strength 130 GPa, and higher electrical/thermal conductivities than copper (Geim and Novoselov 2007; Balandin et al. 2008; Kuilla et al. 2010). Compounding graphene with other materials is a major means to harvest its striking properties, such as fabrication of polymer/graphene composites, because most polymers lack electrical and thermal functionalities and sufficient mechanical strength. Nevertheless, pristine graphene is not well mixable with polymers due to its inert surface. Hence, reduced graphene oxide has been widely used, but it is limited by the low structural integrity implying unsatisfactory electrical and thermal conductivities. As a result, GnPs (also called graphene nanoplatelets) consisting of a few-layer graphene have been developed, which feature cost-effectiveness, high structural integrity, and modified surface for compounding with other materials (Zaman et al. 2012a,b).

A GnP refers to a nanosheet consisting of a few stacked graphene layers, with a thickness below 10 nm. GnPs, a very recently developed graphene derivative mainly for polymer composites, are different from graphite nanosheets or nanoplatelets in that the latter is thicker than 10 nm. The thickness determines whether these platelets can take maximum advantages of the unique structure and striking functional and mechanical performance of graphene.

It is of great importance to keep the GnP thickness as low as possible, because (1) the total number of GnPs and their surface area in a given volume of a composite abruptly increase with reduction in the thickness and (2) low thickness reduces the negative effect of the poor through-plane functional and mechanical properties of graphene. To gauge the importance of the thickness, a model was developed that computes the total number and surface area of GnPs in a composite by assuming the following: (1) a total volume of 10 μm^3, (2) GnPs at 1 vol%, (3) each platelet to be treated as a rectangular cuboid whose lateral dimensions (length and depth) are 1×1 μm, and (4) each graphene layer to be 1 nm in thickness with layer spacing inclusive.

The total number of GnPs in the composite equals

$$\frac{10,000,000,000 \times 0.01}{1,000 \times 1,000 \times \text{platelet thickness}}$$

The total surface area of GnPs is

$$\left(1000 \times 1000 \times 2 + 4 \times 1000 \times \text{platelet thickness}\right) \times \text{total number}$$

Figure 7.1 shows the effect of GnP thickness on the total number and surface area of GnPs in the composite. Both the number and total surface area reduce dramatically when the GnP thickness increases from 1 to 10 nm; the reduction becomes far less obvious with the thickness over 10 nm. Therefore, we propose 10 nm as the upper limit of the GnP thickness.

Keblinski demonstrated that, with increasing graphene layers from one to five, the through-plane thermal conductivity dropped from 110 to 85 MW/mK, only 23% reduction (Hu et al. 2011); by contrast, polymers are thermally insulative. The electrical conductivity of GnPs was measured as 1460 S/cm (Meng et al. 2014), close to that of chemical vapor deposition (CVD)-grown graphene (2000 S/cm) (Wu et al. 2009). Therefore, these GnPs should be promising for improvement of the electrical and thermal conductivities of polymers.

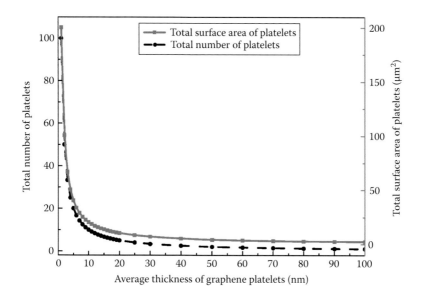

Figure 7.1 Effect of platelet thickness on the number and total surface area of platelets in a composite of given volume and fraction.

7.2 FABRICATION OF GRAPHENE PLATELETS

Three well-known methods developed for generating *pristine* graphene include micromechanical exfoliation of graphite, CVD, and epitaxial growth on SiC substrate. These methods do not suit polymer composites because of their inability to produce (1) graphene of suitable lateral dimensions and (2) an appropriate interface for compounding with polymers. As a result, graphene oxide has been developed for polymer composites, because (1) their basic plane structure is somewhat similar to graphene, (2) their lateral sizes suit the composites, and (3) they contain functional groups that can be used for the interface modification of the composites, although too many functional groups damage the structural integrity leading to unsatisfactory electrical and thermal conductivity and mechanical performance.

Heavily oxidizing graphite followed by thermal treatment is well known for the fabrication of graphene and GnPs of relatively low structural integrity (Potts et al. 2011). Nevertheless, this chapter focuses on the cost-effective fabrication of GnPs starting from a commercial graphite intercalation compound (GIC), also called expandable graphite; the platelets fabricated feature high structural integrity. Figure 7.2 contains a schematic of GIC, where chemicals intercalate between graphene layers; the layer number between the intercalated chemicals is called a stage.

Sonication is effective in splitting graphite into thinner sheets, but couples of sonication hours produced many defects to the product, leading to a reduction in the lateral size of GnPs and an I_D/I_G ratio of ~0.3 (Khan et al. 2011; O'Neill et al. 2011). Hence, the sonication time should be tuned to split graphite effectively, yet not too long causing many defects. It was found that sonication needs to be operated for 30 min below 30°C to produce an efficient exfoliation of GIC; that is, the sonication bath should connect to a freezer, or ice cubes should be added to counter the internal heat rise caused by sonication (Zaman et al. 2011). By randomly selecting 10 platelets,

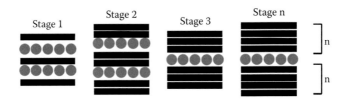

Figure 7.2   Schematic of staging phenomenon in GIC. (Ma, J., Meng, Q., Michelmore, A. et al., Covalently bonded interfaces for polymer/graphene composites, *J. Mater. Chem. A*, 1(13), 4255–4264, 2013. Reproduced with permission from the Royal Society of Chemistry.)

the thickness was measured as 11.4 ± 2.2 nm by atomic force microscopy (AFM). When these platelets were mixed with epoxy at 2.5 vol%, the resulting composite unfortunately showed no reduction in electrical volume resistivity, although these platelets can get rolled to produce nanotubes in the matrix (Zaman et al. 2011).

Combining sonication with thermal shock was proved effective in reducing the thickness of platelets and thus would reduce the electrical resistance of polymer/graphene composites (Zaman et al. 2012a). In a typical procedure, 0.1 g GIC was carefully transferred into a crucible that had been preheated in a furnace at 700°C. After 1 min of heat treatment, the crucible was moved out to sit on a ceramic for 30 s. Then the expanded product was transferred into a container. The operator in this process must wear a respirator, safety glasses, heat-resistant gloves, and closed shoes; the furnace should be placed in a fume cupboard. The raw material GIC was manufactured by treating highly crystalline natural flake graphite with a mixture of sulfuric acid and certain other oxidizing agents that aid in the *catalysis* of the sulfate intercalation, and the resultant product is a highly intumescent form of graphite (http://asbury.com/materials/graphite/). Upon a sudden thermal shock at 700°C, these intercalates generated an immense volume of gases (CO_2, CO_2, H_2O, etc.) that exceeded the van der Waals forces holding the GIC sheets together, leading to a high volume expansion of 200–300 times. The key point for a maximum volume expansion is to transfer GIC directly into a preheated crucible.

The expanded product was dispersed in *N*-methyl-2-pyrrolidone (NMP) at 1.0 wt% in a metal container by a mechanical stirrer, followed by sonication of 60 min below 30°C. When the suspension was diluted to 0.0004 wt%, AFM measurements of randomly selected 20 platelets indicated a thickness of 2.51 ± 0.39 nm. It is worth noting that the GnP thickness depends on the solvent chosen. Once NMP was replaced by tetrahydrofuran (THF), the thickness increased to 3.57 ± 0.50 nm. Representative measurements of the two values are shown in Figure 7.3. Since previous research has shown that corrugation of graphene can increase its thickness to ~1 nm (Bourlinos et al. 2009), each GnP may comprise three to four layers of graphene when dispersed in THF. It is worth to point out that the layered structure in each GnP should be retained through the modification and subsequent compounding with polymers. But GnPs should be able to expand or even exfoliate themselves with a proper surface modification and compounding process.

The evolution from a GIC to GnPs was investigated by x-ray diffraction (XRD). Figure 7.4 contains XRD patterns of GIC (Asbury 3494), washed GIC, the expanded product, and GnPs. A GIC is formed by intercalating atomic or molecular layers of different chemical species, such as alkali metals and bisulfate, between graphene layers. The number of graphene layers between the adjacent intercalated chemicals is known as a stage number ranging from 1 to 5 (Weller et al. 2005; Shih et al. 2011). Washing GIC leads to an increase in the diffraction intensity (from diffraction a to diffraction b in Figure 7.4) because most or part of these intercalated chemicals were removed by washing, leading to an increase in the stacking coherence; this is

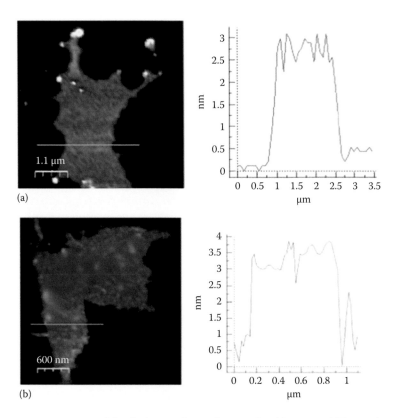

(a)

(b)

Figure 7.3 AFM measurement of the thickness of GnPs dispersed in (a) NMP and (b) THF. (Zaman, I., Kuan, H.-C., Ma, J. et al., From carbon nanotubes and silicate layers to graphene platelets for polymer nanocomposites, *Nanoscale*, 4(15), 4578–4586, 2012a. Reproduced with permission from the Royal Society of Chemistry.)

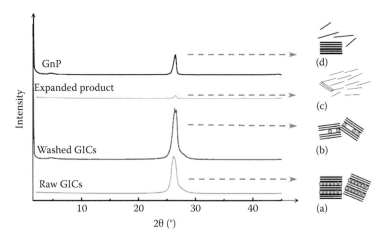

Figure 7.4 XRD patterns of a raw GIC, washed GIC, expanded product, and GnPs. (Araby, S., Zaman, I., Meng, Q., Kawashima, N., Michelmore, A., Kuan, H.C., Majewski, P., Ma, J., and Zhang, L., Melt compounding with graphene to develop functional, high-performance elastomers, *Nanotechnology*, 24, 165601, 2013. Copyright IOP Publishing. With permission.)

shown in the illustration next to Figure 7.4. Figure 7.4c illustrates that upon expansion, all stages may separate each other or even exfoliate, but the layered graphene structure should be retained in each stage, and this explains why only a small diffraction (c in Figure 7.4) is seen after the thermal treatment. The diffraction pattern of GnPs shows no shift in 2θ because the stages were expanded and even disorderly exfoliated in the fabrication while the layered structure in each stage should be retained; GnPs show an intense diffraction, in which the platelets may stack themselves through processes of sonication, washing, and drying. This conclusion is in agreement with two previous studies (Malesevic et al. 2008; Kun and Wéber 2011). The diffraction intensity and width (d in Figure 7.4) increases obviously through the sonication because GnPs stacked themselves through the densification process.

This process is further elucidated using Fourier transform infrared spectroscopy (FTIR). In Figure 7.5a, the washed compound was composed of hydroxyls (broad peak at 3050–3800 cm^{-1}),

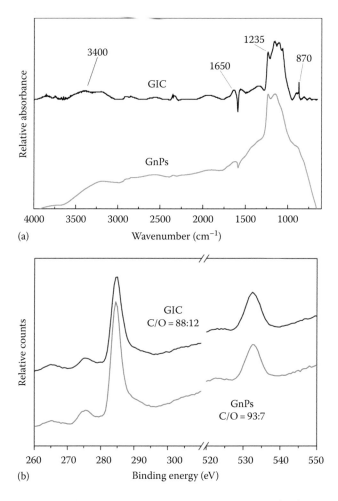

(a)

(b)

Figure 7.5 Characterization of the graphite intercarlation compound (GIC) and GnPs: (a) FTIR and (b) XPS. (Zaman, I., Kuan, H.-C., Ma, J. et al.: A facile approach to chemically modified graphene and its polymer nanocomposites. *Adv. Funct. Mater.* 2012b. 22. 2735–2743. Copyright Wiley-VCH Verlag GmbH & Co. KGaA. Reproduced with permission.)

carbonyls (1750–1850 cm⁻¹), carboxyls (1650–1750 cm⁻¹), C=C (1500–1600 cm⁻¹), and ethers and epoxides (1000–1280 cm⁻¹). After the thermal treatment, the spectrum indicates that most of the hydroxyl and carboxyl groups were removed. Importantly, the intensities of the bands assigned to epoxides and ethers were obviously strengthened, implying that more ether groups may be produced by the treatment. Hydroxyl and carboxyl groups on the compound require lower temperatures for desorption than epoxy and carbonyl groups, in consistence with a previous study where 63.5% epoxide groups were found in graphite oxide when heat treated at ~700°C (Niyogi et al. 2006).

The thermal shock and sonication raised the C:O ratio from 88:12 to 93:7 from X-ray photoelectron spectroscopy (XPS) analysis in Figure 7.5b; it is noteworthy that this ratio, achieved without a quartz tube and inert gas, is higher than that of reduced graphene oxide (Lomeda et al. 2008; Wang et al. 2008a, 2009b). Although combining thermal shock with ultrasonication had been reported to produce GnPs (Jiang et al. 2010), this fabrication approach combining thermal shock with sonication yielded the highest C:O ratio implying the maximum retainment of functionality and mechanical performance from its parent graphene.

7.3   STRUCTURE AND PROPERTIES OF GRAPHENE PLATELETS

Graphene features high electrical and thermal conductivities, while graphene oxide is not electrically conductive owing to its low structural integrity as usually measured by a Raman spectrometer. In Figure 7.6, both the GIC and GnPs show absorption at D, G, and 2D bands at around 1340, 1585, and 2690 cm⁻¹, respectively. The D band intensity corresponds to in-plane vibration of sp³-hybridized carbon atoms, while the G band intensity refers to that of sp²-hybridized

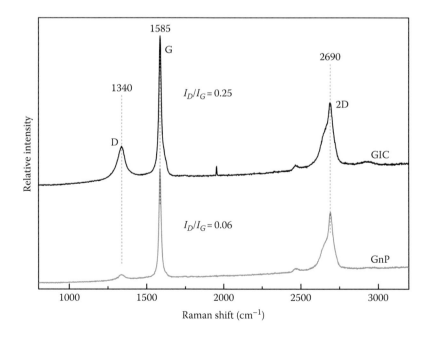

Figure 7.6   Characterization of GnPs using a Raman spectrometer.

TABLE 7.1 CARBON TO OXYGEN ATOMIC RATIO OF GRAPHITE INTERACTION COMPOUND (GIC) AND GRAPHENE PLATELETS (GnPs)

Materials	C:O Atomic Ratio	
	Elemental Analysis	XPS
GIC	84:16	88:12
GnPs	90:10	93:7

carbon atoms. The I_D/I_G of GIC is merely 0.25, much lower than the graphite oxide originated from the oxidation method (Wang et al. 2009a,b; Hsiao et al. 2010; Wang and Hu 2011), implying that the starting GIC possesses far lower oxidation degree and thus sound structural integrity for high functionality and mechanical performance. Through thermal treatment using a common furnace, the ratio reduces to 0.06, and this means an increase in the quantity of sp^2-hybridized carbon atoms. No virtual difference in the 2D band intensity between GIC and GnPs is seen, because both testings were conducted on powder samples. The high sp^2–sp^3-hybridized carbon atoms of GIC and the increased quantity of sp^2-hybridized carbon through thermal treatment are further confirmed by elemental analysis in Table 7.1, where the high C–O ratio observed for GIC is further enhanced through thermal treatment.

Transmission electron microscopy (Figure 7.7) was used to characterize the GnPs. In comparison with diffraction patterns of previous single graphene (Meyer et al. 2007; Li et al. 2008), the diffraction pattern suggests a well-crystallized two-layer graphene structure.

As discussed in Figure 7.1, the thickness of GnPs poses a significant effect on the number of platelets in a given volume fraction of composites. With the increasing number of graphene layers from 1 to 3, stiffness does not change while fracture strength reduces 23% from 130 to 101 GPa (Lee et al. 2009a). Given that the fracture strength of most polymers ranges from 1 to 80 MPa, these GnPs are indeed sufficiently robust to toughen or reinforce polymers.

The most promising feature of graphene lies in its electrical conductivity—6000 S/cm (Geim 2009). In practice, a single graphene layer made by CVD shows an electric conductivity of 2000 S/cm in Table 7.2. The conductivity of graphene oxide is as low as 0.05 S/cm, and it increases to 5 S/cm through chemical reduction and to 350 S/cm through thermal reduction (Marcano et al. 2010); all of these values are significantly lower than the CVD-grown graphene, which, however, cannot

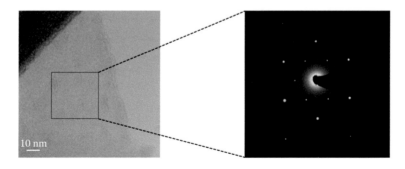

Figure 7.7 Transmission electron microscopy micrograph and its selected area diffraction of a GnP.

TABLE 7.2   ELECTRIC CONDUCTIVITIES OF NANOMATERIALS AND THEIR APPROXIMATE COST

Material	Electric Conductivity (S/cm)	Cost (US$/kg)
CVD-grown grapheme	2000 (Wu et al. 2009)	Unknown
As-prepared graphene oxide	0.05 (Marcano et al. 2010)	20–30 (Kim et al. 2010a)
Chemically reduced graphene oxide	5 (Marcano et al. 2010)	20–30
Thermally reduced graphene oxide	350 (Marcano et al. 2010)	20–30
Multiwalled carbon nanotubes (VGCF-X)	21.4	~300
Graphene platelets (GnPs)	1460	10–20

be processed in solution. The unsatisfactory electric conductivity of graphene oxide is caused by the high defect and disorder content, since its fabrication started with highly oxidized graphite. Nevertheless, reduced graphene oxide has been extensively used in energy storage devices mainly due to its improved conductivity through reduction and high surface area, as indicated by the publication numbers since 2008 in Figure 7.8. Highly conducting yet solution-processable graphene is highly desired.

Carbon nanotubes have been extensively studied over the past decades due to their promising electric conductivity and mechanical strength. In the measurement at room temperature using a standard four-point probe configuration (Keithley-4200), multiwalled carbon nanotubes (MWCNTs) show a conductivity of only 21.4 S/cm. By contrast, GnPs demonstrate a significant conductivity at 1460 S/cm, representing 6722% and 29100% increments over MWCNTs and chemically reduced graphene oxide, and more importantly they cost 10–20 US$/kg.

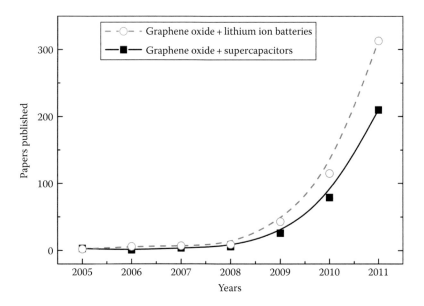

Figure 7.8   SciFinder search using graphene oxide for LiBs and supercapacitors.

7.4   COVALENT MODIFICATION OF GRAPHENE PLATELETS

Three factors determining the functional and mechanical properties of polymer composites include matrix, dispersion phase, and interface; of these, interface has attracted the most intensive interest in research, because interface (1) transfers stress from the matrix to dispersion phase, which thus shares a fraction of loading, and (2) restrains the deformation of matrix molecules in the vicinity of the dispersion phase. A strong interface can transfer stress and restrain molecular deformation more efficiently. Therefore, the graphene surface must be modified to build up a strong interface with polymer matrixes. Based on extensive research on polymer/nanolayer composites (Ma et al. 2003, 2004, 2005), we proposed two strategies for the creation of a strong interface for polymer/graphene composites: One is to utilize the molecular entanglement between matrix molecules and the long chains of the grafted surfactant on silicate layers, and another is to have the grafted surfactant react with matrix molecules to produce a covalent linkage between the matrix and dispersion phase. Both strategies involve the surface modification of nanosheets.

Surface modification is also indispensible in producing a stable graphene colloid. Most current graphene studies take advantage of the large fractions of hydroxyl, carbonyl, epoxy, and carboxylic groups to modify the surface of graphene oxide (GnO) by three major methods: (1) transforming the carboxyl groups on the GnO edges into chlorides using $SOCl_2$, followed by reaction with amine-terminated surfactants (Niyogi et al. 2006; Li et al. 2010; Zhuang et al. 2010); (2) reacting GnO with isocyanate (Lee et al. 2009b; Kim et al. 2010b; Zaman et al. 2011); and (3) bridging GnO with surfactants through the reaction of the GnO carboxylic groups with the surfactant amine groups (Geng et al. 2008). These methods have been partially successful in modifying the GnO surface: the first two methods, involving dangerous chemicals, are difficult for scaled-up production; the bonding produced by method (3) absorbs moisture in a humid environment. It is noteworthy that the reaction between the GnO epoxide group and the surfactant amine group has also been adopted to modify GnO (Wang et al. 2008b; Park et al. 2009; Yang et al. 2009a,b). However, all these methods started with highly oxidized graphite and short-chain surfactants, and this implies the necessity for reduction, which is inappropriate for mass production of conducting graphene.

Since the thermal treatment of graphite oxide at 727°C produced a maximum quantity of epoxide groups (Niyogi et al. 2006), a certain amount of epoxide groups may exist on the GnP surface. Epoxide groups of GnPs are actually found by XPS analysis in Figure 7.9. It is worth to mention that these epoxide groups are not sufficiently active to react with end-amine groups of organic molecules, and they also lack quantity. The epoxide groups of epoxy resins take the form of an equilateral triangle that causes a straining of the atoms. The straining causes the epoxides to exhibit a high reactivity for reaction with hardeners. Thus, a catalyst triisopropanolamine (TIPA) was adopted to provide an intense alkaline environment for this reaction.

An appropriate solvent must be found prior to surface modification of GnPs. The platelets were dispersed at 0.5 vol% in six solvents: acetone, THF, benzene, toluene, dichlorobenzene, *N,N'*-dimethylformamide, and NMP. While no clear difference was seen 30 min after sonication as shown in Figure 7.10, leaving these vials for 4 h proved that NMP is the best in suspending GnPs. Thus, NMP was adopted to modify GnPs in the following modification process.

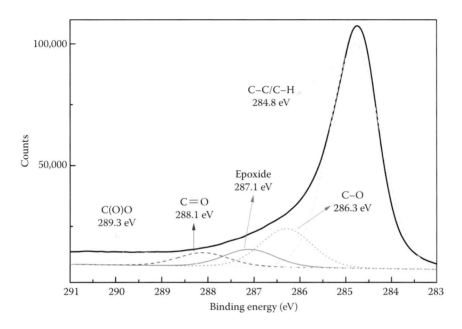

Figure 7.9 XPS analysis of GnPs. (Ma, J., Meng, Q., Michelmore, A. et al., Covalently bonded interfaces for polymer/graphene composites, *J. Mater. Chem. A*, 1(13), 4255–4264, 2013. Reproduced with permission from the Royal Society of Chemistry.)

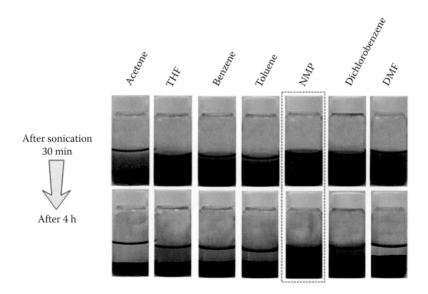

Figure 7.10 Dispersion of GnPs in seven organic solvents. (Zaman, I., Kuan, H.-C., Ma, J. et al.: A facile approach to chemically modified graphene and its polymer nanocomposites. *Adv. Funct. Mater.* 2012b. 22. 2735–2743. Copyright Wiley-VCH Verlag GmbH & Co. KGaA. Reproduced with permission.)

7.4.1   Modification of Graphene Platelets by a Long-Chain Surfactant

A commercial surfactant polyoxyalkyleneamine (B200, $M_w = 2000$) was selected to modify GnPs. Figure 7.11 shows the reaction mechanism.

   A series of experiments were designed to optimize the modification parameters. Of all the O atoms of GnPs, 5 atom% epoxide groups were assumed capable of reacting with the amine-terminated surfactant. TIPA was added as a catalyst to provide an alkaline environment. To promote the reaction, the mole numbers of surfactant and catalyst were doubled. Thus, Recipe 1 was produced where 0.1 g GnPs was mixed with 1.6 g surfactant and 0.15 g catalyst in 20 g NMP for reaction at 90°C for 4 h. While keeping the GnP fraction and surfactant quantity fixed, an orthogonal design was employed to create five more recipes for identifying the effect of catalyst quantity, temperature, and time on the modification of GnPs (Table 7.3). These modification products were investigated by FTIR (Figure 7.12). While no marked absorption was found for Recipe 1, all others show obvious absorption suggesting a new substance grafted with GnPs. While absorptions at 2980 and 2902 cm^{-1} represent the stretching vibration of CH$_2$ group, absorptions at 3677 and 1065 cm^{-1} correspond to the N–H and C–N groups of B200, respectively. This proves that B200 grafted with GnPs, in agreement with the change in the C:O ratio. It is worth to mention that the chemical bonding is stronger than the π–π physisorption of polymers on graphene. Temperature and the ratio of catalyst to GnPs are the two major factors determining the grafting efficiency. Since the Recipe 4 product has the least intensive absorptions at y_1, y_2, and y_3, it is used as a benchmark to demonstrate the absorption intensity of other products (Table 7.3). Recipe 6 indicates the most intensive absorptions and thus was adopted to modify GnPs.

   The modification markedly improved the suspension of GnPs in the solvent, since modified graphene platelets (m-GnPs) were able to suspend in THF at 0.1 wt% for 7 days, while GnPs remained suspended only for 2 days. When m-GnPs were dispersed in NMP, no precipitation was observed after at least 5 months of storage. This stable colloidal suspension of m-GnPs will provide a platform for a wide range of research and development of graphene-related products.

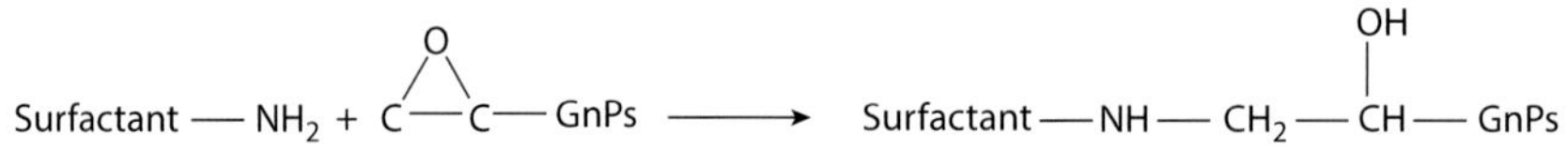

Figure 7.11   Mechanism of surface modification of GnPs.

TABLE 7.3   RECIPES FOR PRODUCING CHEMICALLY MODIFIED GRAPHENE PLATELETS

Recipe	Amount of Catalyst (g)	Temperature (°C)	Time (h)	Relative y_1 Intensity	Relative y_2 Intensity	Relative y_3 Intensity
1	0.15	90	4	N/A	N/A	N/A
2	0.15	150	12	1.5	1.4	1.7
3	0.46	90	12	1.3	1.4	1.3
4	0.16	150	1	1.0	1.0	1.0
5	1.38	90	1	1.5	1.5	1.6
6	1.38	150	4	2.6	2.4	2.6

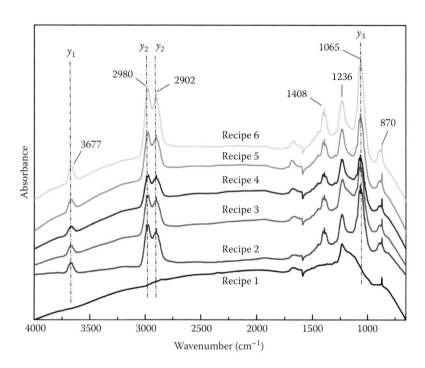

Figure 7.12 FTIR spectra of *m*-GnPs produced by six recipes. (Zaman, I., Kuan, H.-C., Ma, J. et al.: A facile approach to chemically modified graphene and its polymer nanocomposites. *Adv. Funct. Mater.* 2012b. 22. 2735–2743. Copyright Wiley-VCH Verlag GmbH & Co. KGaA. Reproduced with permission.)

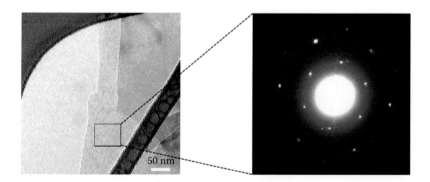

Figure 7.13 TEM micrograph of a typical *m*-GnP. (Zaman, I., Kuan, H.-C., Ma, J. et al.: A facile approach to chemically modified graphene and its polymer nanocomposites. *Adv. Funct. Mater.* 2012b. 22. 2735–2743. Copyright Wiley-VCH Verlag GmbH & Co. KGaA. Reproduced with permission.)

Transmission electron microscopy and electron diffraction (Figure 7.13) were used to characterize the *m*-GnPs. In comparison with those diffraction patterns of previous single graphene (Meyer et al. 2007; Li et al. 2008), the pattern suggests a well-crystallized, two- to three-layered graphene structure, in agreement with the thickness of unmodified GnPs.

Through covalent functionalization, the C:O ratio reduces to 90:10 (Figure 7.14a) due to the high C:O ratio 2.9 of the grafted modifier. Figure 7.14b shows Raman spectra of thoroughly

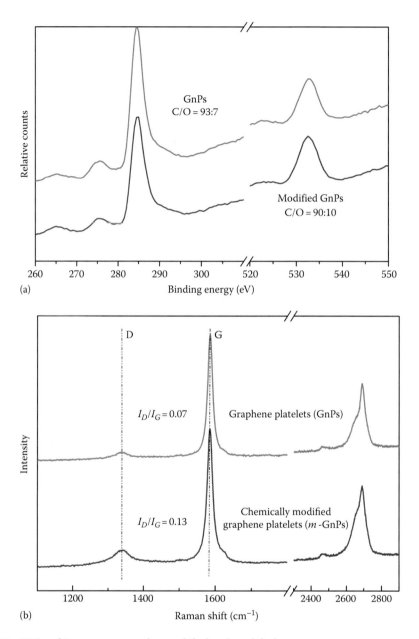

Figure 7.14 XPS and Raman spectra of unmodified and modified.

washed samples of GnPs and chemically modified GnPs. All samples show absorptions at 1340 and 1585 cm^{-1} corresponding to D band and G band, respectively. However, the broadness and intensity of these absorption patterns were different. The ratio increases markedly from 0.07 to 0.13 through the chemical modification, indicating the existence of B200 molecules on graphene surface produced by the grafting, corresponding to the XPS and FTIR analysis.

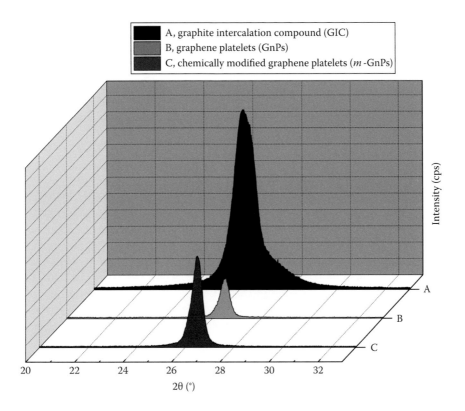

Figure 7.15 XRD patterns of washed GIC, GnPs, and *m*-GnPs. (Zaman, I., Kuan, H.-C., Ma, J. et al.: A facile approach to chemically modified graphene and its polymer nanocomposites. *Adv. Funct. Mater.* 2012b. 22. 2735–2743. Copyright Wiley-VCH Verlag GmbH & Co. KGaA. Reproduced with permission.)

The XRD spectra (Figure 7.15) show a strong diffraction peak at 26.6° fitting into the diffraction of the (002) plane with 0.34 nm intergraphene spacing. Although there is no obvious difference in the locations of diffraction peaks, their intensities were found to drastically reduce in an order: the intercalation compound > *m*-GnPs > GnPs, which is explained later. When thermally treated, the compound intercalants generated an immense volume of gases (CO_2, CO, H_2O, etc.), which led to a pressure as high as 100 MPa (McAllister et al. 2007). The high pressure produced highly wrinkled GnPs with a volume expansion of over 200 times. Wrinkling may start from aligned or adjacent functional groups such as hydroxide, carbonyl, and epoxide groups. This caused a significant reduction in the diffraction intensity. When GnPs were modified, B200 and a few cycles of washing and separation rearranged and stacked these wrinkled GnPs, leading to an increase in diffraction intensity.

C, H, N elemental microanalysis gave the empirical formula of *m*-GnPs to be $C_{2.0}O_{0.08}H_{0.07}$–$(B200)_{0.004}$. This indicates that the surface functionalization has taken place, with a ratio of one B200 chain for approximately every 160 hexagonal cells of the graphene basal plane. Such a low ratio may explain why *m*-GnPs show a similar XRD diffraction angle to GnPs (Figure 7.15).

7.4.2 Modification of Graphene Platelets by a Reactive Surfactant Containing Two End-Amine Groups

Interface for a composite refers to the boundary that transfers stress and refrains deformation between matrix and dispersion phase. Satisfactory mechanical properties are generally derived from a composite where interface is created by covalent bonding. A small molecule of two end-amine groups, 4,4′-diaminodiphenyl sulfone (DDS), was chosen to react with the epoxide groups of GnPs (Ma et al. 2013). The quantity of DDS tripled the stoichiometric requirement, in order to reduce the chance of DDS bridging adjacent layers to produce thicker platelets. The ideal effect would be only one end-amine group of every DDS molecule grafting to GnPs. These DDS-modified GnPs were further reacted theoretically with only one end-epoxide group of a diglycidyl ether of bisphenol A (DGEBA) molecule. Specifically, a calculated amount of GnPs was suspended in NMP (0.1 wt%) in a metal container, followed by sonication for 30 min at a temperature below 30°C. Calculated amounts of TIPA as a catalyst and DDS as a modifier were added to the mixture while stirring to dissolve fully, followed by another sonication at a temperature below 30°C for 1 h. The mixture was then transferred into a round-bottom flask with a condenser and kept reacted at 150°C for 32 h, to produce DDS-modified GnPs. The weight ratio of TIPA/DDS/GnPs was controlled at 8.28/5.00/0.10. The mixture was then washed by acetone at least three times to remove the excess DDS. The final process of graphene modification proceeded through the suspension of a calculated amount of DDS-modified GnPs in NMP (0.1 wt%) using a metal container followed by sonication for 30 min at a temperature below 30°C. DGEBA was then added to the mixture, followed by another sonication for 30 min below 30°C. After sonication, the mixture was transferred into a round-bottom flask with a condenser and kept reacted at 150°C for 4 h. Finally, the sample was washed again using acetone at least three times to remove excess DGEBA that was not used in the grafting stage. This produced further m-GnPs.

The reaction between the GnPs' epoxide groups and DDS's end-amine groups was investigated in terms of the effect of reaction times on the grafting efficiency. Upon reaction, new absorptions would be expected on the DDS-grafted GnPs. Figure 7.16 contains FTIR spectra of unmodified GnPs and these m-GnPs with different time slots, where four obvious new absorptions are seen for the m-GnPs: (1) Two medium-intensity absorptions at 1604 and 1670 cm^{-1} are due to complex molecular motions of the entire rings of DDS, (2) one weak absorption at 2913 cm^{-1} corresponds to C–H stretching, and (3) a broad and intense absorption in the 3100–3400 cm^{-1} range would be caused by the end-amine groups of DDS. These provide a solid evidence for the grafting between the GnPs' epoxide groups and DDS's end-amine groups. To determine which reaction time is ideal in producing the highest grafting density, software Origin was employed to calculate the area under the absorption at 1604 cm^{-1} in Figure 7.16. In Table 7.4, the absorption at 32 h witnessed the highest intensity, and hence, a mixing time of 32 h was chosen to produce DDS-modified GnPs.

After a thorough washing process, DDS-modified GnPs were mixed and reacted with a superfluous amount of DGEBA (monomer of epoxy), which was thoroughly washed again to remove nonreacted DGEBA molecules. In this process, the end-amine groups of DDS-modified GnPs would react with the epoxide groups of DGEBA to produce m-GnPs. Figure 7.17 contains spectra of unmodified GnPs and m-GnPs. The following absorption evolution is observed for m-GnPs: (1) two absorptions at 2920 and 2848 cm^{-1} correspond to C–H and CH$_2$–O bonds; (2) one at 1660 cm^{-1} disappeared, implying the reaction of the grafted end-amine groups with DGEBA; and (3) one at 1509 cm^{-1} may be caused by C–H$_3$ asymmetric stretching or C–H$_2$ stretching

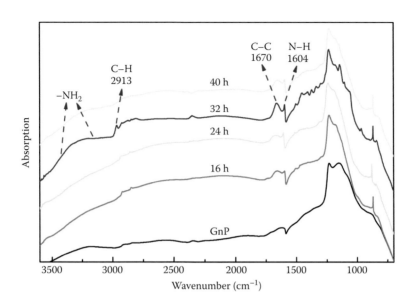

Figure 7.16   FTIR spectra of GnPs modified by DDS. (Ma, J., Meng, Q., Michelmore, A. et al., Covalently bonded interfaces for polymer/graphene composites, *J. Mater. Chem. A*, 1(13), 4255–4264, 2013. Reproduced with permission from the Royal Society of Chemistry.)

TABLE 7.4 EFFECT OF REACTION TIME ON THE GRAFTING EFFICIENCY OF DDS				
Time (h)	16	24	32	40
Relative intensity at 1604 cm^{-1}	1.00	1.11	2.22	1.44

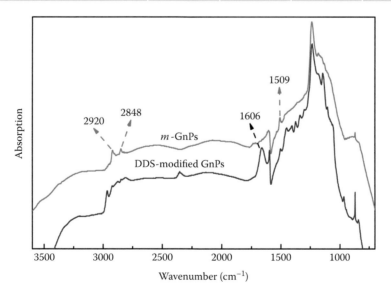

Figure 7.17   FTIR spectra of DDS-modified GnPs and *m*-GnPs. (Ma, J., Meng, Q., Michelmore, A. et al., Covalently bonded interfaces for polymer/graphene composites, *J. Mater. Chem. A*, 1(13), 4255–4264, 2013. Reproduced with permission from the Royal Society of Chemistry.)

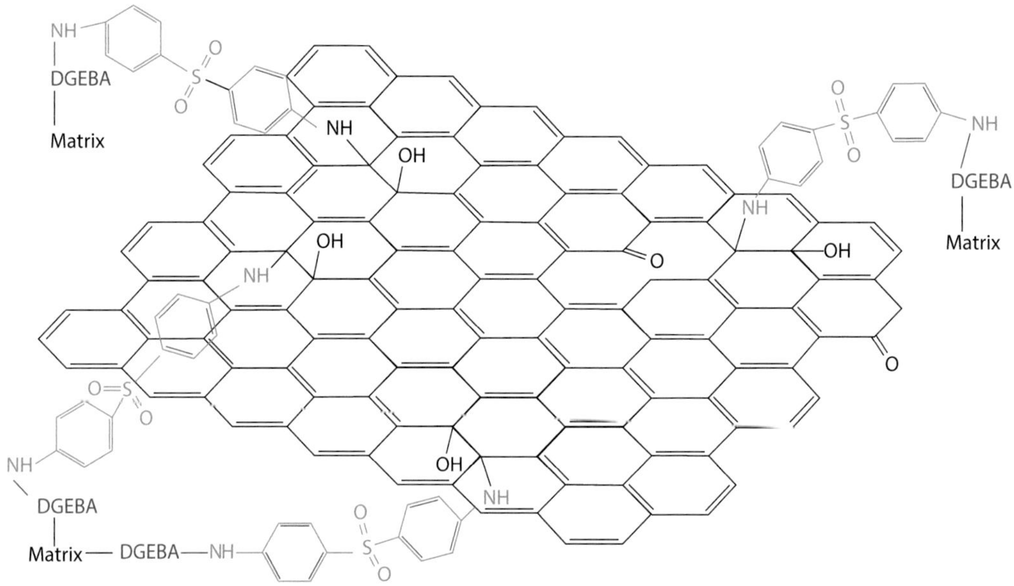

Figure 7.18 Atomic structures of covalently bonded interface between GnPs and matrix. (Ma, J., Meng, Q., Michelmore, A. et al., Covalently bonded interfaces for polymer/graphene composites, *J. Mater. Chem. A*, 1(13), 4255–4264, 2013. Reproduced with permission from the Royal Society of Chemistry.)

(Pretsch et al. 2009). All these point toward the reaction of DGEBA with DDS-modified GnPs. This two-step modification can create a covalently bonded interface for epoxy/GnP composites, as schematically shown in Figure 7.18.

Graphene features an exceptionally high in-plane electrical conductivity in comparison with orders of lower through-plane conductivity; while graphene is the stiffest and strongest material ever measured, graphite consisting of graphene layers is a well-known lubricator. These phenomena demonstrate the importance of keeping GnPs as thin as possible. A low thickness produces the following benefits: (1) retaining the high mechanical strength of graphene; (2) reducing the negative effect of through-plane conductivity; and (3) yielding high specific surface area, which implies more platelets in a given volume and fraction of composite. Since the bifunctional DDS and DGEBA used for GnP surface modification may covalently link adjacent GnPs, an increase in the GnP thickness was expected. Thus, the GnP thickness was examined by AFM.

Figure 7.19 contains a representative AFM micrograph and its height profile. Three types of GnPs are observed: (1) thick platelets, as shown by an exceptionally large white platelet in the left figure (its thickness was measured as ~20 nm; these platelets are rare); (2) platelets of ~10 nm in thickness, as shown by a gray arrow; and (3) platelets of ~2.5 nm in thickness, as shown by a white arrow. Since type (2) and (3) platelets are far more popular, over 40 platelets of these types were randomly selected, and their thickness was measured to be 6.4 ± 2.6 nm. Because surface modification was started with GnPs of 2–4 nm in thickness, the increase in platelet thickness after modification must be caused by the organic molecules used to modify GnPs. Both DDS and DGEBA have bifunctional end groups, thus capable of grafting two adjacent layers. These grafted adjacent layers cannot be separated any more in the following processes, thus producing

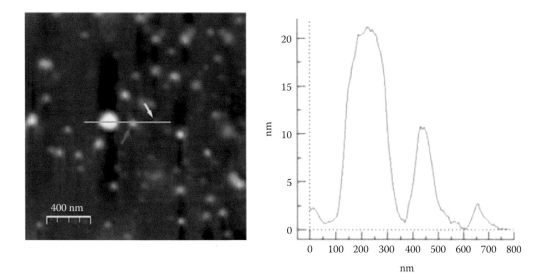

Figure 7.19   Characterization of GnPs using AFM. (Ma, J., Meng, Q., Michelmore, A. et al., Covalently bonded interfaces for polymer/graphene composites, *J. Mater. Chem. A*, 1(13), 4255–4264, 2013. Reproduced with permission from the Royal Society of Chemistry.)

thicker GnPs. This implies that higher fractions of GnPs may be needed for the following fabrication of electrically conductive polymer composites. Since the composite electric conductivity is mainly determined by its fillers' conductivity, the structural integrity of GnPs was measured by a Raman spectrometer—a high structural integrity of graphene always leads to an excellent electrical conductivity.

Both GnPs and *m*-GnPs show two absorptions at around 1350 and 1575 cm^{-1} in Figure 7.20. D band refers to the absorption at 1350 cm^{-1}, and its intensity indicates the quantity of disordered structure such as voids caused by oxidation and reduction; the G band intensity at 1575 cm^{-1} corresponds to an ordered structure of sp^2-hybridized carbon. Hence, the I_D/I_G ratio indicates a disorder degree. Practically, this ratio can be obtained by measuring the height or area of these two bands using Excel or other software; it is worth to note both height and area dependent on how to choose a base line. Since the height ratio is more sensitive, it was adopted in this study. Both thermally and chemically reduced graphene oxide produced an I_D/I_G ratio of ~1.0; by contrast, GnPs demonstrate a significantly lower I_D/I_G ratio of 0.07 in Figure 7.20. The ratio decreases to 0.03 after the modification, because GnPs were heated in NMP at 150°C for 36 h, and this reduced GnPs. The comparison demonstrates the much higher structural integrity of *m*-GnPs. In previous graphene studies, dozens of sonication hours produced a lot of defects to the product, leading to a reduction in the lateral size of GnPs and an increase in the I_D/I_G ratio (Khan et al. 2011; O'Neill et al. 2011). In comparison, the sonication time in this study is no longer than 2 h in total, and thus, its effect on the lateral dimension is trivial. It is worth to note that GnPs show D and G bands at 1359.2 and 1584.3 cm^{-1}, while these two bands shift to 1351.3 and 1581.7 cm^{-1} for *m*-GnPs, respectively. These red shifts must be caused by the surface modification of GnPs. No obvious change is seen on the 2D absorption.

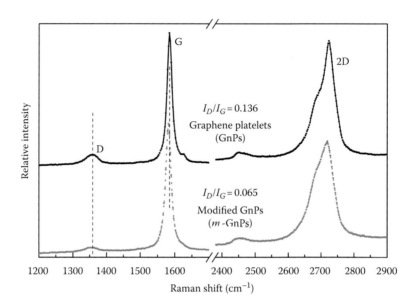

Figure 7.20 Raman spectra of GnPs and *m*-GnPs. (Ma, J., Meng, Q., Michelmore, A. et al., Covalently bonded interfaces for polymer/graphene composites, *J. Mater. Chem. A*, 1(13), 4255–4264, 2013. Reproduced with permission from the Royal Society of Chemistry.)

The two-step surface modification grafted organic molecules on GnPs, leading to an increase in the GnP thickness and their structural integrity. This may also produce some difference in the GnP layer spacing, which needs to be investigated by XRD.

Figure 7.21 shows XRD patterns of the GIC, GnPs, and *m*-GnPs, all of which were thoroughly washed before characterization. GIC retaining graphene-stacked structure exhibits a typical sharp and high-intensity diffraction at 26.18°, indicating the presence of a large amount of a crystalline phase in the specimen with a distance of 0.34 nm interlayer spacing; this implies a high structural integrity of graphene, opposite to graphene oxide. GnPs demonstrate a diffraction pattern at the same angle, implying the retainment of a graphene layer spacing of 0.34 nm. However, the intensity is largely reduced and this is explained in the light of the GnP fabrication process. After thermal expansion and ultrasonication, GIC converted into GnPs. The thermal expansion and ultrasonication caused corrugation, voids, reduction in lateral dimension, and increase in layer spacing between stages or even exfoliation, all of which contribute to the diffraction intensity reduction. *m*-GnPs show a slightly increased intensity, since the washing and modification procedures instigated a realigning and stacking of the wrinkled GnPs.

Since the diffraction angle does not change, only a small fraction of organic molecules was expected to graft with GnPs, which was identified by the following thermo-gravimetric analysis (TGA) analysis. Both GnPs and *m*-GnPs show an obvious mass loss from 100°C to 200°C, attributing to the deintercalation of H_2O. While GnPs demonstrate a little further loss until 600°C, an obvious loss is observed for *m*-GnPs, which must be caused by the grafted DDS and DGEBA. From room temperature to 600°C, the loss values of GnPs and *m*-GnPs are 3.3 and 9.5 wt%, respectively (Figure 7.22). The difference in these two values yields 4.5 wt%—the least weight fraction of the grafted molecules.

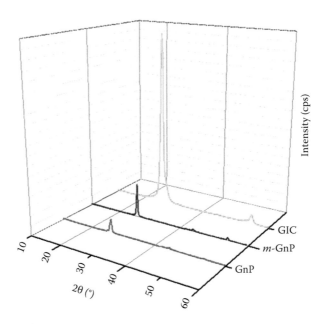

Figure 7.21   XRD patterns of the GIC, GnPs, and *m*-GnP. (Ma, J., Meng, Q., Michelmore, A. et al., Covalently bonded interfaces for polymer/graphene composites, *J. Mater. Chem. A*, 1(13), 4255–4264, 2013. Reproduced with permission from the Royal Society of Chemistry.)

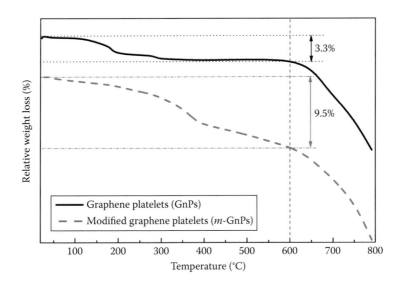

Figure 7.22   Weight loss of GnPs and *m*-GnPs in air. (Ma, J., Meng, Q., Michelmore, A. et al., Covalently bonded interfaces for polymer/graphene composites, *J. Mater. Chem. A*, 1(13), 4255–4264, 2013. Reproduced with permission from the Royal Society of Chemistry.)

7.5 POLYMER/GRAPHENE PLATELET COMPOSITES

Of engineering materials, polymers have seen the largest increase in applications due to their low manufacturing cost and high specific strength. Nevertheless, most polymers are limited by absolute strength or ductility and lack of functional properties such as electrical and thermal conductivities. Polymer composites have thus been developed to address these limitations. Appropriate interface modification is indispensable for fabricating polymer/graphene composites of high mechanical performance and conductivity, because a modified interface can efficiently transfer load, electron, and phonon. From the fabrication, structure, and property perspective, we hereby review epoxy/graphene composites of three types of interface strength, including unmodified composites, the composites modified by a long-chain surfactant, and the reactively modified composites.

7.5.1 Fabrication of Epoxy/Graphene Composites

7.5.1.1 Unmodified Epoxy/GnP Composites

A quantity of GnPs was suspended in THF using a metal container, and then DGEBA (182–196 g/equiv.) was added to the suspension and stirred until DGEBA dissolved completely, followed by sonication under 30°C for 1 h. After evaporating THF, hardener Jeffamine D-230 (J230) was added. The mixture was degassed at room temperature and poured into preheated and greased rubber molds, followed by curing (Zaman et al. 2012a).

7.5.1.2 Interface-Modified Epoxy/GnP Composites by a Long-Chain Surfactant

A long-chain surfactant, polyoxyalkyleneamine (B200, M_w = 2000 g/mol), was used to modify GnPs as elaborated in Figure 7.23a. For each B200 molecule (Figure 7.23a), the end-amine group reacted with the epoxide groups of GnPs, while the long chain can molecularly entangle with polymer matrix molecules, and this would promote the exfoliation and dispersion of GnPs. The m-GnPs were used to fabricate epoxy composites using a procedure similar to the unmodified composites (Zaman et al. 2012b).

GnPs were reactively modified by DDS, as detailed in Figure 7.23b, and the m-GnPs were similarly mixed with DGEBA and its hardener J230 (Ma et al. 2013).

7.5.2 Comparison of These Composites Regarding Functional and Mechanical Properties

To illuminate the effect of this covalently bonded interface on the composites' structure, the diffraction pattern of the 0.244 vol% composite was compared in Figure 7.24 with the same fraction composite fabricated with unmodified and B200-modified GnPs as described in Section 7.5.1. In the epoxy/unmodified GnP composite, a sharp diffraction at 26° is seen, which is due to the stacked GnPs. However, this diffraction is nearly invisible in both interface-modified composites, implying a lower degree of GnP stacking since the layered structure in each GnP should be retained all the time. The GnP surface modification builds up either a molecularly entangled or a

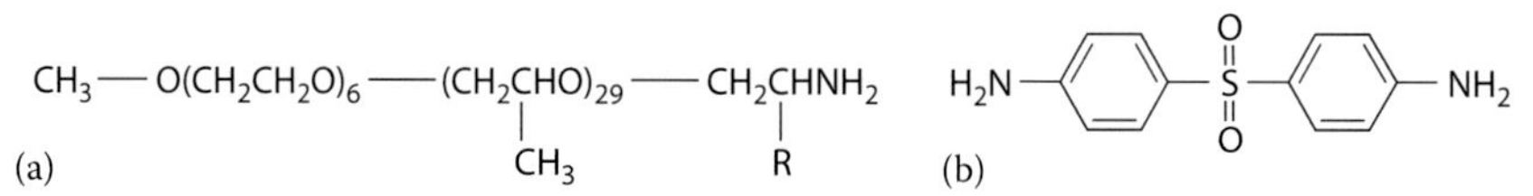

Figure 7.23   Molecular formula of (a) polyoxyalkyleneamine (B200) and (b) 4,4′-DDS.

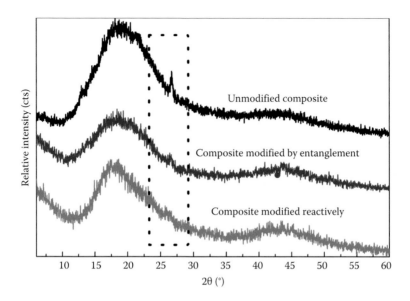

Figure 7.24   XRD patterns of epoxy/unmodified GnP nanocomposite with the modified ones at 0.244 vol%.

covalently bonded interface, promoting the dispersion and delamination of GnPs in matrix, and this explains the absence of the sharp 26° diffraction.

7.5.2.1   Unmodified Epoxy/GnP Composites

In Table 7.5, the unmodified GnPs markedly improve the fracture toughness of epoxy, although only marginal increases in the stiffness and electrical conductivity are seen. This implies that GnPs can disperse in the matrix relatively well in spite of lack of surface modification, since they are hydrophobic, carbon based, and thus somewhat compatible with epoxy. However, the dispersion needs a lot of improvement as no electrically conductive network is formed.

7.5.2.2   Epoxy/GnP Composites Modified by a Long-Chain Surfactant

In comparison with the unmodified system, the interface modification taking advantage of a long-chain surfactant B200 created a significant effect on the functional and mechanical properties of the composites—an order of 7–8 reduction in the electrical volume resistivity and far more

TABLE 7.5   FUNCTIONAL AND MECHANICAL PROPERTIES OF EPOXY/GnP COMPOSITES

Materials	GnP Fractions (vol%)	Electrical Volume Resistivity (Ω cm)	Increment in Young's Modulus (%)	Increment in Fracture Energy Release Rate (%)
Unmodified composites	0.24	2.95×10^{17}	3.2	109
	0.49	3.52×10^{16}	4.2	160
Interface-modified composites by molecular entanglement	0.24	2.05×10^{9}	17.5	212
	0.49	1.13×10^{7}	26.8	295
Interface-modified composites by reactive bonding	0.24	5.55×10^{12}	44.9	110
	0.4	1.58×10^{10}	57.4	196

increments in stiffness and fracture toughness. This is explained in light of interface modification that grafted B200 molecules onto GnPs. Each molecule, 15 nm in length and consisting of propylene oxide and ethylene oxide, is compatible and can entangle with the monomer DGEBA and harder J230. Thus, the *m*-GnPs can exfoliate and disperse better in the matrix than the unmodified ones, leading to more platelets in a given volume. These platelets contact each other to produce a conductive network and also to transfer stress and restain the movement of matrix molecules around their vicinity. Therefore, the B200-modified composites demonstrate far higher improvements in electrical conductivity, stiffness, and fracture toughness.

7.5.2.3 Epoxy/GnP Composites Modified Reactively

Since each DDS molecule contains two end-amine groups that can bridge adjacent graphene layers producing 6 nm thick platelets, there must be far less quantity of platelets in a given volume of the composites. This explains the moderate increments in electrical conductivity and fracture toughness in Table 7.5. Surprisingly, this system shows the highest stiffness improvement. Since the modifier DDS can be attached to GnPs by π–π interaction, there would be more DDS molecules in the system than the grafted ones. Each DDS molecule contains benzene groups and sulfone, which implies high stiffness imparted to the composites.

The covalent bonding modification of GnPs causes two conflicting effects to the increment of electrical conductivity. On the one hand, the grafted organic molecules act as a conductivity barrier. On the other hand, the covalently bonded interface between the GnPs and matrix promotes the exfoliation and dispersion of GnPs, preventing GnPs stacking, and this helps to form a conductivity network at a low GnP fraction. In Table 7.5, this modification markedly improves all the properties, implying a highly positive effect.

DDS-modified composites should exhibit more toughening, since a covalently bonded interface should be more effective in transferring load from the matrix to dispersion phase than the B200-modified composites. The lower actual toughness improvement was caused by the reduction of the platelet number in a given volume fraction of composites, associated with the increase in GnPs' thickness from ~3 to ~6 nm.

7.6 CONCLUSIONS AND CHALLENGES

7.6.1 What Was Covered and What Was Missed: Further Reading

This chapter has covered the fabrication, structure, properties, and surface modification of GnPs (where each platelet is thinner than 10 nm) as well as their applications in polymer composites. As a cost-effective derivative (10–20 US$/kg) of graphene, GnPs feature high structural integrity, while their surface can be covalently modified. These platelets may have a wide range of applications not only in composites and energy storage but in sensors, catalysis, drug/gene delivery, biological sensing and imaging, antibacterial materials, and biocompatible scaffold for cell culture as well.

Although graphene oxide has been extensively studied as electrodes for supercapacitors and lithium batteries, GnPs have not yet been reported for these applications likely because they are relatively new. For the platelets thicker than 10 nm, refer to comprehensive review papers (Jang and Zhamu 2008; Sengupta et al. 2011).

7.6.2 Future Challenges for Graphene Platelets

A number of key challenges confronting GnPs are as follows:

1. More surface functional groups are desired for GnPs. After surface functionalization by a long-chain surfactant (M_w = 2000), the grafting ratio is only at around 6 wt%, and this ratio needs further improvement to promote the suspension of GnPs in solvent and also their complete exfoliation and uniform dispersion in polymer matrixes. A higher grafting ratio also means that when used as an electrode, GnPs have more surface area to facilitate ionic and electric transportation. More surface functional groups should be produced with no sacrifice of high structural integrity.
2. GnPs need to be thinner. As discussed in Figure 7.1, the specific surface area of platelets in a given fraction and volume of a composite reduces significantly with an increase in thickness for platelets thinner than 10 nm. The low thickness also obviously alleviates the negative effect of poor through-plane conductivities of graphene.

REFERENCES

Araby, S., Zaman, I., Meng, Q., Kawashima, N., Michelmore, A., Kuan, H.C., Majewski, P., Ma, J., and Zhang, L. 2013. Melt compounding with graphene to develop functional, high-performance elastomers. *Nanotechnology* 24(16):165601.
Balandin, A.A., Ghosh, S., Bao, W. et al. 2008. Superior thermal conductivity of single-layer graphene. *Nano Letters* 8 (3):902–907.
Boehm, H., Clauss, A., Fischer, G., and Hofmann, U. 1962. Surface properties of extremely thin graphite lamellae. Paper read at *Proceedings of the Fifth Conference on Carbon.* Pergamon, New York.
Bourlinos, A.B., Georgakilas, V., Zboril, R., Steriotis, T.A., and Stubos, A.K. 2009. Liquid-phase exfoliation of graphite towards solubilized graphenes. *Small* 5 (16):1841–1845.
Geim, A.K. 2009. Graphene: Status and prospects. *Science* 324 (5934):1530–1534.
Geim, A.K. and Novoselov, K.S. 2007. The rise of graphene. *Nature Materials* 6 (3):183–191.
Geng, Y., Li, J., Wang, S.J., and Kim, J.K. 2008. Amino functionalization of graphite nanoplatelet. *Journal of Nanoscience and Nanotechnology* 8 (12):6238–6246.
Hsiao, M.-C., Liao, S.-H., Yen, M.-Y. et al. 2010. Preparation of covalently functionalized graphene using residual oxygen-containing functional groups. *ACS Applied Materials & Interfaces* 2 (11):3092–3099. http://asbury.com/materials/graphite.
Hu, L., Desai, T., and Keblinski, P. 2011. Thermal transport in graphene-based nanocomposite. *Journal of Applied Physics* 110 (3):033517-1-033517-5.
Jang, B.Z. and Zhamu, A. 2008. Processing of nanographene platelets (NGPs) and NGP nanocomposites: A review. *Journal of Materials Science* 43 (15):5092–5101.
Jiang, B., Tian, C., Wang, L. et al. 2010. Facile fabrication of high quality graphene from expandable graphite: Simultaneous exfoliation and reduction. *Chemical Communications* 46 (27):4920–4922.
Khan, U., Porwal, H., O'Neill, A. et al. 2011. Solvent-exfoliated graphene at extremely high concentration. *Langmuir* 27 (15):9077–9082.
Kim, H., Abdala, A.A., and Macosko, C.W. 2010a. Graphene/polymer nanocomposites. *Macromolecules* 43 (16):6515–6530.
Kim, H., Miura, Y., and Macosko, C.W. 2010b. Graphene/polyurethane nanocomposites for improved gas barrier and electrical conductivity. *Chemistry of Materials* 22 (11):3441–3450.
Kuilla, T., Bhadra, S., Yao, D. et al. 2010. Recent advances in graphene based polymer composites. *Progress in Polymer Science* 35 (11):1350–1375.

Kun, P. and Wéber, F. 2011. Cs. Balázsi: Preparation and examination of multilayer graphene nanosheets by exfoliation of graphite in high efficient attritor mill. *Central European Journal of Chemistry* 9 (1):47–51.

Lee, C., Wei, X., Li, Q. et al. 2009a. Elastic and frictional properties of graphene. *Physica Status Solidi (B)* 246 (11–12):2562–2567.

Lee, Y.R., Raghu, A.V., Jeong, H.M., and Kim, B.K. 2009b. Properties of waterborne polyurethane/functionalized graphene sheet nanocomposites prepared by an in situ method. *Macromolecular Chemistry and Physics* 210 (15):1247–1254.

Li, G.L., Liu, G., Li, M. et al. 2010. Organo-and water-dispersible graphene oxide—Polymer nanosheets for organic electronic memory and gold nanocomposites. *The Journal of Physical Chemistry C* 114 (29):12742–12748.

Li, X., Zhang, G., Bai, X. et al. 2008. Highly conducting graphene sheets and Langmuir–Blodgett films. *Nature Nanotechnology* 3 (9):538–542.

Lomeda, J.R., Doyle, C.D., Kosynkin, D.V., Hwang, W.-F., and Tour, J.M. 2008. Diazonium functionalization of surfactant-wrapped chemically converted graphene sheets. *Journal of the American Chemical Society* 130 (48):16201–16206.

Ma, J., Meng, Q., Michelmore, A. et al. 2013. Covalently bonded interfaces for polymer/graphene composites. *Journal of Materials Chemistry A* 1 (13):4255–4264.

Ma, J., Wang, R., Jin, J. et al. (xxxx). Highly conducting yet processable graphene.

Ma, J., Xiang, P., Mai, Y.W., and Zhang, L.Q. 2004. A novel approach to high performance elastomer by using clay. *Macromolecular Rapid Communications* 25 (19):1692–1696.

Ma, J., Xu, J., Ren, J.-H., Yu, Z.-Z., and Mai, Y.-W. 2003. A new approach to polymer/montmorillonite nanocomposites. *Polymer* 44 (16):4619–4624.

Ma, J., Yu, Z.Z., Kuan, H.C., Dasari, A., and Mai, Y.W. 2005. A new strategy to exfoliate silicone rubber/clay nanocomposites. *Macromolecular Rapid Communications* 26 (10):830–833.

Malesevic, A., Vitchev, R., Schouteden, K. et al. 2008. Synthesis of few-layer graphene via microwave plasma-enhanced chemical vapour deposition. *Nanotechnology* 19 (30):305604.

Marcano, D.C., Kosynkin, D.V., Berlin, J.M. et al. 2010. Improved synthesis of graphene oxide. *ACS Nano* 4 (8):4806–4814.

McAllister, M.J., Li, J.-L., Adamson, D.H. et al. 2007. Single sheet functionalized graphene by oxidation and thermal expansion of graphite. *Chemistry of Materials* 19 (18):4396–4404.

Meng, Q., Jin, J., Wang, R. et al. 2014. Processable 3-nm thick graphene platelets of high electrical conductivity and their epoxy composites. *Nanotechnology* 25 (12). doi: 10.1088/0957-4484/25/12/125707.

Meyer, J.C., Geim, A., Katsnelson, M. et al. 2007. The structure of suspended graphene sheets. *Nature* 446 (7131):60–63.

Niyogi, S., Bekyarova, E., Itkis, M.E. et al. 2006. Solution properties of graphite and graphene. *Journal of the American Chemical Society* 128 (24):7720–7721.

Novoselov, K.S., Geim, A.K., Morozov, S.V. et al. 2004. Electric field effect in atomically thin carbon films. *Science* 306 (5696):666–669.

Novoselov, K.S., Geim, A.K., Morozov, S. et al. 2005. Two-dimensional gas of massless Dirac fermions in graphene. *Nature* 438 (7065):197–200.

Novoselov, K.S., Jiang, Z., Zhang, Y. et al. 2007. Room-temperature quantum Hall effect in graphene. *Science* 315 (5817):1379–1379.

O'Neill, A., Khan, U., Nirmalraj, P.N., Boland, J., and Coleman, J.N. 2011. Graphene dispersion and exfoliation in low boiling point solvents. *The Journal of Physical Chemistry C* 115 (13):5422–5428.

Park, S., Dikin, D.A., Nguyen, S.T., and Ruoff, R.S. 2009. Graphene oxide sheets chemically cross-linked by polyallylamine. *The Journal of Physical Chemistry C* 113 (36):15801–15804.

Potts, J.R., Dreyer, D.R., Bielawski, C.W., and Ruoff, R.S. 2011. Graphene-based polymer nanocomposites. *Polymer* 52 (1):5–25.

Pretsch, E., Bèuhlmann, P., Buhlmann, P., and Badertscher, M. 2009. *Structure Determination of Organic Compounds: Tables of Spectral Data.* Berlin, Germany: Springer-Verlag.

Sengupta, R., Bhattacharya, M., Bandyopadhyay, S., and Bhowmick, A.K. 2011. A review on the mechanical and electrical properties of graphite and modified graphite reinforced polymer composites. *Progress in Polymer Science* 36 (5):638–670.

Shih, C.-J., Vijayaraghavan, A., Krishnan, R. et al. 2011. Bi-and trilayer graphene solutions. *Nature Nanotechnology* 6 (7):439–445.

Wang, G., Shen, X., Wang, B., Yao, J., and Park, J. 2009a. Synthesis and characterisation of hydrophilic and organophilic graphene nanosheets. *Carbon* 47 (5):1359–1364.

Wang, G., Yang, J., Park, J. et al. 2008a. Facile synthesis and characterization of graphene nanosheets. *The Journal of Physical Chemistry C* 112 (22):8192–8195.

Wang, H. and Hu, Y.H. 2011. Effect of oxygen content on structures of graphite oxides. *Industrial & Engineering Chemistry Research* 50 (10):6132–6137.

Wang, H., Robinson, J.T., Li, X., and Dai, H. 2009b. Solvothermal reduction of chemically exfoliated graphene sheets. *Journal of the American Chemical Society* 131 (29):9910–9911.

Wang, S., Chia, P.J., Chua, L.L. et al. 2008b. Band-like transport in surface-functionalized highly solution-processable graphene nanosheets. *Advanced Materials* 20 (18):3440–3446.

Weller, T.E., Ellerby, M., Saxena, S.S., Smith, R.P., and Skipper, N.T. 2005. Superconductivity in the intercalated graphite compounds C6Yb and C6Ca. *Nature Physics* 1 (1):39–41.

Wu, Z.-S., Ren, W., Gao, L. et al. 2009. Synthesis of graphene sheets with high electrical conductivity and good thermal stability by hydrogen arc discharge exfoliation. *ACS Nano* 3 (2):411–417.

Yang, H., Li, F., Shan, C. et al. 2009a. Covalent functionalization of chemically converted graphene sheets via silane and its reinforcement. *Journal of Materials Chemistry* 19 (26):4632–4638.

Yang, H., Shan, C., Li, F. et al. 2009b. Covalent functionalization of polydisperse chemically-converted graphene sheets with amine-terminated ionic liquid. *Chemical Communications* (26):3880–3882.

Zaman, I., Kuan, H.-C., Ma, J. et al. 2012a. From carbon nanotubes and silicate layers to graphene platelets for polymer nanocomposites. *Nanoscale* 4 (15):4578–4586.

Zaman, I., Kuan, H.-C., Ma, J. et al. 2012b. A facile approach to chemically modified graphene and its polymer nanocomposites. *Advanced Functional Materials* 22 (13):2735–2743.

Zaman, I., Phan, T.T., Ma, J. et al. 2011. Epoxy/graphene platelets nanocomposites with two levels of interface strength. *Polymer* 52 (7):1603–1611.

Zhang, Y., Tan, Y.-W., Stormer, H.L., and Kim, P. 2005. Experimental observation of the quantum Hall effect and Berry's phase in graphene. *Nature* 438 (7065):201–204.

Zhuang, X.D., Chen, Y., Liu, G. et al. 2010. Conjugated-polymer-functionalized graphene oxide: Synthesis and nonvolatile rewritable memory effect. *Advanced Materials* 22 (15):1731–1735.

Chapter 8

Helium Ion Microscopy for Graphene Characterization and Modification

Daniel Fox, Yangbo Zhou, and Hongzhou Zhang

CONTENTS

8.1 INTRODUCTION

The helium ion microscope (HIM) is a recently developed complementary tool to electron microscopes (Morgan et al., 2006, Ward et al., 2006, 2007, Notte et al., 2007, Postek et al., 2007b, Ananth et al., 2008, Postek and Vladar, 2008, Scipioni, 2008, Scipioni et al., 2008) and has great potential to play a crucial role in advanced material characterization and modification (Joy et al., 2007, Postek et al., 2007a, Postek and Vladar, 2008, Jepson et al., 2009a,b). The advantage and benefit of using the HIM as a high-resolution imaging tool have been demonstrated, which is expected to particularly address challenges encountered in nanoscale metrology. The invasiveness of the observation also arouses scientists' concern, especially for fairly sensitive samples like graphene. Nevertheless, a destructive method for a confined scale characterization, on the other hand, may uncover potential for precise modification: the HIM has a demonstrated capacity to directly and accurately tailor nanostructures.

In this chapter, we discuss controllable HIM modification and sub-nanometer metrology, with graphene used as an example. Challenges in graphene characterization and modification will be first introduced in the following section. To understand the HIM's potential to address these challenges, the basic principles of the tool will be explained. The next two sections will be focused on the applications of HIM in high-quality imaging and ultrafine modification. Our discussion will not be limited within the graphene research. To broaden the scope, we will introduce applications of the HIM in other relevant fields wherever appropriate.

8.2   CHALLENGES IN GRAPHENE CHARACTERIZATION AND MODIFICATION

Graphene, the world's first one-atom-thick material, has received a huge amount of interest since its first isolation (Novoselov et al., 2004) because of its extraordinary properties and remarkable potential applications. As a new form of carbon allotropes, graphene consists of a single-layer hexagonal network of sp^2-hybridized carbon atoms and hence represents a real 2D crystal (Elias et al., 2009). Due to its 2D nature, it exhibits unique electronic properties, which include large carrier mobility (Bolotin et al., 2008) ($\sim$200,000 cm^2 V^{-1} s^{-1}), high charge concentration ($\sim$10^{13} cm^{-2}), linear band dispersion at Dirac point (Peres, 2009), zero bandgap (Geim and Novoselov, 2007), ballistic transport over a large distance (Du et al., 2008) (>1 μm at 300 K), and chiral behavior. Graphene also has the highest recorded thermal conductivity (Balandin et al., 2008, Ghosh et al., 2008) ($\sim$5000 W m^{-1} K^{-1}) and the record-breaking mechanical strength (Lee et al., 2008) (42 N m^{-1}). It allows several daunting quantum effects such as the Klein paradox to be studied using a desktop instrument at room temperature and beyond (Katsnelson and Novoselov, 2007, Tan et al., 2007).

The combination of graphene's mechanical strength, thermal dissipation potential, and outstanding electronic properties offers industrial opportunities in a wide range of fields, including mechanical enhancement of materials (Wakabayashi et al., 2008), transparent electrodes for flexible display screens (Verma et al., 2010), solar cells (Wang et al., 2007), fuel cells (Shang et al., 2010), hydrogen storage (Srinivas et al., 2010), gas purification (Li et al., 2010), nanoelectromechanical systems (Bunch et al., 2007), gas sensors (Schedin et al., 2007), and biosensors (Shao et al., 2010, Chowdhury et al., 2011). In addition, novel graphene device concepts such as veselago lens (Cheianov et al., 2007), spintronics (Son et al., 2006), and pseudospintronics (San-Jose et al., 2009) devices have been proposed as well.

In terms of next-generation electronic devices in the semiconductor industry, graphene is also a promising material to sustain scaling of complementary metal–oxide–semiconductor (CMOS) technology (Liao et al., 2010, Lin et al., 2010, Schwierz, 2010), the fundamental building unit of integrated digital logic circuits, so that Moore's law of technology integration will be kept alive to continue fueling the world economy and improving the lives of billions. According to the International Technology Roadmap for Semiconductors (ITRS, 2009), the projected performance metrics of transistors in 2020 is beyond the capability of conventional bulk silicon CMOS technology, since both the silicon channel and the copper interconnect will be affected severely by size effects as device dimensions shrink to and below 16 nm. Graphene has the potential to be an alternate channel and interconnect material enabling the projected scaling of CMOS. Indeed, variants of graphene field effect transistors including bottom gating (Novoselov et al., 2004), top gating (Gu et al., 2007, Lemme et al., 2007, Kedzierski et al., 2008), and dual gating (Kim et al., 2009) have been recently demonstrated.

Despite the rapid progress in this field, working on graphene devices is still at an early stage, and there are obstacles to the realization of graphene's potential in information technology. In this section, we review two important challenges encountered in graphene research in general and its applications in the semiconductor sector in particular: *a reliable high-throughput method for nondestructive high-resolution characterization and a technique to engineer the geometry of graphene at the nanometer scale.*

8.2.1   Characterization of Graphene

The key demands for graphene characterization include determining *the number of layers* across the sample; *critical dimension* (CD) with atomic precision, especially the atomic configuration of graphene nanoribbon (GNR) edges; *sample cleanliness*; the *quality of the substrate* that the graphene sits on; and the *stacking configuration* of few-layered graphene (FLG). A variety of microscopy and metrology techniques have been a key enabler for the determination of graphene properties by addressing these demands, while for industry applications, characterization methodologies must be high throughput and compatible with the current technologies.

Optical microscopy, for example, Rayleigh imaging (Blake et al., 2007, Casiraghi et al., 2007, Jung et al., 2007), imaging ellipsometry (Gaskell et al., 2010, Wurstbauer et al., 2010), and Raman spectroscopy and imaging (Graf et al., 2007, Tang et al., 2010, Lui et al., 2011, Nolen et al., 2011), is capable of high-throughput large-area identification of graphene and counting of graphene layers. Graphene's electronic structure can be uniquely captured in its Raman spectrum and thus used as fingerprints for distinguishing single-layer graphene (SLG) and FLG flakes (less than five layers) (Ferrari et al., 2006). Raman characterization is also sensitive to crystalline quality, doping, and interaction between graphene sheets and the substrate. SLG is, however, visible only under very specific conditions in an ordinary optical microscope. The underneath substrate and illumination conditions strongly affect the optical contrast and the visibility. In addition, the lateral resolution of these optical characterization techniques is limited by the micrometer-sized optical probe.

In contrast, *scanning probe microscopy* (*SPM*), both atomic force microscopy (AFM) (Pandey et al., 2008) and scanning tunneling microscopy (STM) (Sutter et al., 2009, Neubeck et al., 2010), enables atomic resolution and crystallographic investigation. Local electrical properties and packing orders of FLG have been investigated by STM (Sutter et al., 2009, Nirmalraj et al., 2011). Performing characterization at the atomic scale using SPMs is, however, time consuming and reduces throughput, which is not compatible with industry applications.

Transmission electron microscopy (TEM), especially recently developed spherical aberration corrected (*Cs*-corrected) TEMs, extracts sub-nanometer information such as the local atomic arrangement of grain boundaries in polycrystalline graphene (Warner et al., 2009, Huang et al., 2011, Kim et al., 2011), the evolution and interaction of point defects in an SLG (Gass et al., 2008), quantitative atomic packing of bilayer graphene (Jinschek et al., 2011), and in situ investigation of graphene sublimation and multilayer edge reconstructions (Huang et al.). Atom-by-atom elemental identification and electronic state analysis are also demonstrated by using electron energy loss spectroscopy (EELS) in the TEM (Suenaga and Koshino, 2010). However, extensive image simulation is required for the interpretation of high-resolution TEM images. It is also impossible to get the aforementioned information acquired in the TEM from graphene layers that are incorporated into devices, because TEM can analyze only electron-transparent thin samples mounted on special holders.

8.2.2 Modification of Graphene

For graphene's digital-circuit applications, the primary challenge at present is to open a bandgap in graphene in a controlled and practical way, since the graphene's zero bandgap will result in a leakage current in the OFF state and hence a very small I_{on}/I_{off} ratio (Schwierz, 2010)—the absence of clearly defined logic states is *unacceptable* for digital logic operations. To open up a bandgap, one approach is to reduce the lateral dimension of the graphene sheet (i.e., its width) and fabricate GNRs. The bandgap of a GNR is determined by its width, edge geometry (Han et al., 2007), and edge stabilization. The electronic properties of graphene can therefore be controlled through proper patterning and functioning of GNRs.

To obtain adequately large bandgaps in GNRs, sub 5 nm widths are required (Schwierz, 2010, Terrones et al., 2010). However, the edge roughness and edge defects for such narrow widths can cause large variations of electronic characteristics, since electrons interact with the irregular edges frequently when the GNR width becomes comparable to the intrinsic mean free path of non-patterned 2D graphene. The existence of a gap in the band structure of all measured GNRs has been mostly attributed to edge roughness, and a small variation in features' size (e.g., one-tenth of the nominal dimension) often causes significant changes in device properties. This naturally asserts strict criteria for evaluating the performance of GNR fabrication methods.

To fabricate GNRs, a variety of physical and chemical modification techniques have been proposed (Huang et al., 2009, Bai and Huang, 2010, Terrones et al., 2010, Jia et al., 2011) such as a focused electron beam in a TEM (Fischbein and Drndic, 2008) instrument, a tip of the AFM (Nemes-Incze et al., 2010) or STM (Dobrik et al., 2010), a focused helium ion beam in the HIM (Lemme et al., 2009), using thermally activated nanoparticles (Datta et al., 2008a,b), and oxidative etching (Liu et al., 2008, Wang and Dai, 2010). The strengths and weaknesses of the state-of-the-art approaches of these GNR fabrication methods are outlined in Table 8.1.

Table 8.1 shows that high-quality GNRs have been fabricated via a range of methods, while there is still a gap to a high-throughput contamination-free industry-compatible fabrication method of GNRs with desirable dimensions. In addition to tailoring graphene geometry, other approaches to locally open up the graphene's bandgap have been proposed, for example, a tunable bandgap can be introduced by varying the electric field applied to a bilayer graphene (Castro et al., 2007). The electronic properties of graphene can also be modified by introducing defects and/or dopants and engineering its surrounding materials, since graphene is not immune to structural imperfection and/or foreign-atom incorporation including boundaries and interfaces. The adsorption of molecules on graphene, such as NO_2 and NH_3, results in p-type and n-type doping, respectively (Leenaerts et al., 2008). Defects in acid-treated graphene have also been found to affect the electronic structure of graphene (Coleman et al., 2008). Oxidative etching of graphene was attempted to modify a graphene surface at an elevated temperature in an oxygen-rich environment (Liu et al., 2008). Control of graphene doping has been demonstrated by engineering its contacting substrate, for example, F4-TCNQ for p-type (Pinto et al., 2009)

TABLE 8.1   STRENGTHS AND WEAKNESSES OF GNR FABRICATION TECHNOLOGIES

Method	Strengths	Weaknesses
CVD (Li et al., 2008b, Terrones et al., 2010)	• Large quantity production • Open edge	• Multistep processing • Large GNRs >20 nm
AFM (Nemes-Incze et al., 2010), STM (Biro and Lambin, 2010, Dobrik et al., 2010)	• Precise patterning along different crystallographic orientations • Small GNRs: 2.5–15 nm • Promising electrical results	• Specialized equipment • Low throughput
Unzipping CNTs (Jiao et al., 2009, Kosynkin et al., 2009)	• Smooth edges • Widths ~10–20 nm • High yields of GNRs	• Breaking of the GNRs • Multistep processing • Contamination due to chemicals used • Lack control of edge morphology • Significant chemical modifications and poor electrical conductivity
TEM (Egerton et al., 2004, Fischbein and Drndic, 2008, Schneider et al., 2010)	• In situ observation of the formation of GNRs • No long-range distortions • Stable GNRs	• Specialized equipment • Low throughput • Difficult to use the GNRs in a device
EBL (Chen et al., 2007, Han et al., 2007, Grigorescu and Hagen, 2009)	• Well-established technique in semiconductor industry • Patterning of relatively fine features	• Contamination due to chemicals used • Large GNRs >20 nm • Rough edges • Uncontrolled under-etching and unavoidable structural damage
Gas-phase etching (Liu et al., 2008, Wang and Dai, 2010)	• Narrow ribbon <10 nm • A high ON/OFF ratio ~10^4 at RT • Controlled etching	• Thickness-dependent etching • Oxygen-related hole doping
Metallic NP etching (Datta et al., 2008b)	• Patterning along different crystallographic orientations • Small GNRs ~25 nm	• Use of metal NPs • Low throughput

Notes: CVD, chemical vapor deposition; CNTs, carbon nanotubes; EBL, electron beam lithography; RT, room temperature; NP, nanoparticles.

and SiO_2 for *n*-type (Romero et al., 2008). These methods suffer from a common disadvantage that it is not selective in the area that is modified, and the whole sample is subjected to the modification.

With its sub-nanometer spatial resolution and milling capability, the HIM will be a key enabling technology for nanoscience and nanotechnology (Scipioni et al., 2008, Postek and Vladar, 2008, Behan et al., 2010, Bazou et al., 2011). It also facilitates graphene research and has great potential to address the two important challenges. In the following section, we will discuss the fundamentals of the HIM, and its recent applications in graphene research will be introduced in the next two sections.

8.3   FUNDAMENTALS OF HELIUM ION MICROSCOPY

The HIM was the most recent development in the field of ion microscopes. The key technology that eventually enabled its commercialization in 2006 was the invention of a reliable in situ process that can repeatedly perform atomic scale tip sharpening to give the gas field ionization source (GFIS) for ion emission. With the ongoing development of the platform and the source technique, the HIM has evolved three generations in the past few years, that is, Zeiss Orion, Orion Plus, and Orion Nanofab, and the image resolution has improved from 1 to 0.4 nm with enhanced capability of nanofabrication in the newest version. Details of the tool development can be found in the review article (Economou et al., 2012). In this section, we will discuss the basic structure of the microscope and its fundamental principles with a focus on the mechanisms of image formation and ion milling.

8.3.1   Principle of HIM

In the HIM, a beam of helium ions (called the primary beam) is generated, accelerated, and focused into a He$^+$ probe, which may be scanned in a raster over a region of a sample surface. The interaction between the ions and the atoms of the sample surface produces abundant signals such as Rutherford backscattered ions (RBIs), secondary electrons (SEs), and Auger electrons. The signals carry the information of the sample and can be recorded by appropriate detectors. The design of the HIM is similar to that of the conventional scanning electron microscope (SEM) or the focused ion beam (FIB), but the HIM finds itself somewhere between an SEM and an FIB with advantages over both. For example, the resolution of the tool surpasses what is currently achievable in bulk sample imaging with an SEM. Material removal is also possible, but with much higher precision than an FIB. We will first describe the anatomy of the tool with a brief introduction to the beam–sample interaction that is responsible for the image formation in the HIM and then highlight the specific features of the tool that are relevant to the graphene research.

8.3.1.1   Structure and Operation of the HIM

As shown in Figure 8.1, the HIM consists of several subsystems such as ion source, ion optics, column, vacuum, power supply, control electronics, and computers. The source consists of the helium ion emitter, a cryogenically cooled sharpened tungsten tip whose apex is a pyramid of atoms with a terminating layer of just three atoms (the so-called trimer). During operation, the emitter is held at a positive bias of 25–35 kV, and helium gas flows into the source region. The electric field is enhanced greatly at the apex due to its sharp geometry. The electric field adjacent to the trimer atoms is so large that helium atoms are preferentially ionized and become helium ions at the trimer region. The helium ions are then accelerated by the electric field through a hole in the extractor plate and continue down the column of the microscope. The emission current can be regulated from 1 fA to 100 pA by adjusting the He gas pressure in the ion source. The source brightness is beyond $\sim 4 \times 10^9$ A cm^{-2} sr^{-1}. The energy spread of the primary beam is in the range of 0.25–0.5 eV. High brightness, narrow energy spread, and enough beam current are the prerequisites to the formation of a fine ion probe, which in turn is crucial to the image

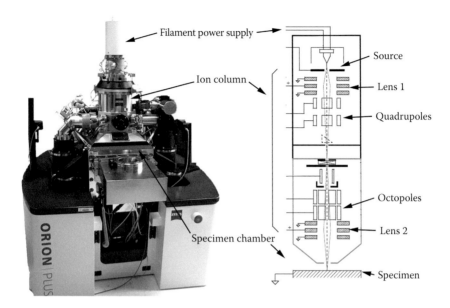

Figure 8.1   A photograph of the HIM (left) and a diagram of its ion optics (right): the ion source emitting helium ions that were accelerated and focused by the column toward the sample. The highly focused He⁺ probe scans across the sample surface to produce signals that can be detected to form images revealing the morphology of the surface.

TABLE 8.2   CHARACTERISTICS OF SEVERAL PRINCIPAL CHARGED BEAM SOURCES

	Electron Sources (100 kV)			Ion Sources	
Beam Sources	**Tungsten**	**LaB$_6$**	**FE**	**GFIS He⁺ (30 keV)**	**LMIS Ga⁺ (21 keV)**
Brightness (A cm^{-2} sr^{-1})	~10^5	~5×10^6	~10^9	~4×10^9	~10^6
Energy spread (eV)	3	1.5	0.3	0.2–0.5	50–100

Notes: LMIS, liquid metal ion source; GFIS, gas field ionization source; FE, field emission source.

resolution and milling precision. Table 8.2 summarizes the characteristics of several principal charged beam sources showing the advantages of the He⁺ source.

A series of electrostatic lenses, quadrupoles, and octopoles are used to focus the helium ion beam into a sub-nanometer probe. Unlike electron microscopes, the HIM utilizes electrostatic lenses rather than electromagnetic lenses to focus the beam. Since a helium ion is about 8000 times heavier than an electron, an electron acquires much larger velocity than a He⁺ does when they are accelerated through the same electrical potential. It means electromagnetic lenses are less effective at focusing ions due to the small force they can apply on a slow-moving helium ion ($F = qv \times B$). The force experienced by a particle in electrostatic lenses is, however, independent of its velocity, $F = qE$. Electrostatic lenses are thus used in the HIM to focus the He⁺ beam. The divergence of the beam can be limited to a small value to mitigate the severe aberrations linked to electrostatic lenses, and the diffraction effect is negligible because of the larger mass and a shorter de Broglie wavelength of the helium ions. This approach results in a practical absence of significant aberrations in the system. As long as the octopoles (see Figure 8.1) are optimized by the user to minimize the astigmatism of the probe, then sub-nanometer imaging is easily achieved with 0.24 nm resolution having been demonstrated. However, this resolution is dependent not only on the size of the probe but also on the size of the region in the sample from which signal is generated. This will be discussed in detail in the following part.

8.3.1.2 He⁺–Sample Interaction

Once the ion enters the sample, it interacts with the potential of sample atoms and collisions occur through which energy is transferred and the trajectory of the ion is deflected. The collisions may originate from two interactions, elastic and inelastic scatterings. In the elastic scattering, primary ions are scattered by the nuclear potential of the sample atoms (it is also called nuclear scattering). The elastic scattering may cause large deflection of the primary ions and change their trajectory significantly. Indeed, elastic scattering is responsible for the backscattered ions. The energy transferred through the elastic scattering may be significant as well. When a He ion of mass M_1 collides with a nucleus of mass M_2, the energy, E_2, transferred to the recoiling carbon atom is given by

$$E_2 = E_0 \frac{4 M_1 M_2}{\left(M_1 + M_2 \right)^2} \sin^2 \frac{\theta_c}{2}$$

where
  E_0 is the energy of incident He ion
  θ_c is the ion scattering angle

For a 180° collision, the energy transferred from a 30 keV He⁺ to a carbon atom is about 14 keV, which exceeds the atomic displacement threshold energy, of 22 eV, for carbon and can cause damage to the lattice. The probability of the large-angle scatterings is, however, smaller than small-angle scatterings. The probability is described by the cross section, and the Rutherford cross section can be used to estimate the probability in most of the HIM experiments (Nastasi et al., 1996).

Inelastic scattering is the direct interaction between the primary ion and the atomic electrons, which would cause the excitation of the electrons, that is, the kinetic energy would be converted into other forms of energy. The excited electrons may escape from the sample, that is, the SEs, and be collected to form an image. The more the energy of the primary beam dissipates in this process, the more SEs it produces and the better the image collection.

The physical quantity describing the energy transferred from the beam to the sample is the stopping power that is defined as the energy loss of the particle per length along which it travels in the sample, that is, it simply refers to the rate of transfer of energy from the beam to the sample (NIST, 2011). Figure 8.2a depicts the stopping power of the electron beam and the helium ion beam as a function of beam energy while they travel in carbon. The typical beam energy in HIM and SEM operations falls in the range shown in Figure 8.2a. As the electron energy decreases from 40 to 5 keV, the stopping power increases by a factor of 10. The electron beam interacts more and more strongly with the sample as it loses energy. The opposite is true for helium ions within the same energy range. Helium ions deposit energy to the sample at the greatest rate when they first enter the sample. In addition, the stopping power of the He⁺ beam is about one order larger than that of the electron beam with the same initial energy (35 keV). In Figure 8.2b, the stopping powers for both elastic and inelastic interactions are plotted separately as a function of ion energy for Ga⁺ and He⁺ traveling in carbon (Ziegler, 2004). The Ga⁺ interacts with the atom nuclei more strongly—20 times as large as the He⁺ beam, and it is also larger than its electronic contribution. For the He⁺

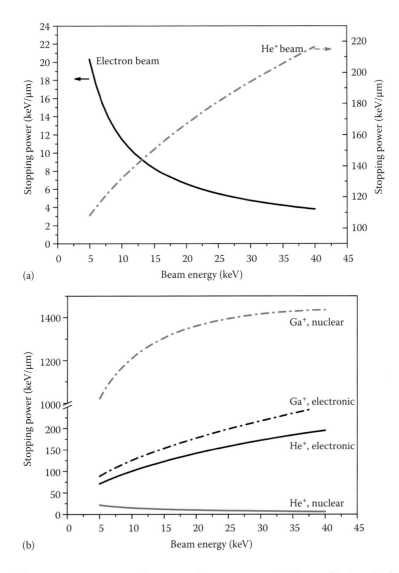

Figure 8.2 (a) Stopping power as a function of beam energy: He⁺ beam (dashed line) and electron beam (solid line). They follow opposite trends as the beam energy varies; (b) the contributions of electronic and nuclear scatterings to the stopping power versus beam energy: Ga⁺ beam (dashed line) and He⁺ beam (solid line).

beam in carbon, the electronic contribution is indeed three times as large as that from the nuclei. It means that the He⁺ can generate much more SEs for imaging through the electronic scatterings while creating less damage induced by these nuclear scattering, since the electronic and nucleate contributions are responsible for SE generation and sample lattice damage, respectively. It is the fundamental fact that a He⁺ beam could be used as an imaging tool.

For a thick sample, the primary charged particles (e.g., He⁺, Ga⁺, and electrons) will eventually stop inside. During the HIM operation, the trajectories of a large number of ions that are injected into the sample define the extent of the interaction (the interaction volume), while the recoils,

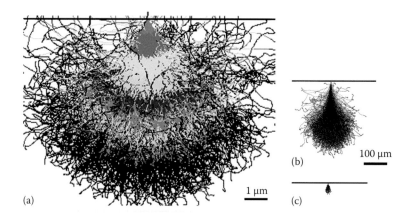

Figure 8.3 Simulated interaction volume of 30 keV charged particle beam with carbon: (a) the electron beam, (b) the helium ion beam, and (c) the gallium ion beam. Note the scale bars (b, c share the same)—the magnification for the He⁺ and Ga⁺ is 10 times larger than the electron.

the sample atoms participating in the collision, may acquire enough energy to diffuse inside the sample and cause secondary effects as well. The trajectory depends on the beam energy and the sample composition. We simulated the interaction volumes for 30 keV electrons, He⁺ and Ga⁺ ions in carbon (Figure 8.3). The trajectory of the helium ions in a sample is different from that of electrons. As well as penetrating deep into the sample, the beam electrons in Figure 8.3a experience large-angle deflections and backscattering in the sample. In comparison, the interaction of a 30 keV helium ion beam (Figure 8.3b) is far more localized within the sample. The gallium ions seem even less spread (Figure 8.3c), while it also means that most of the energy is being deposited at the surface and causes significant damage. Another important feature revealed in the simulation is that the helium ion beam remains well collimated in the near-surface region.

8.3.1.3 Secondary Electron Imaging in the HIM

The HIM/SEM images are gray-level pictures. Each pixel in the picture corresponds to a point on the sample where the beam scanned, and the brightness of the pixel correlates to the intensity of the SEs collected from the point. As discussed in the previous section, a fine probe can be formed in the HIM, and the He⁺–sample interaction is in favor of SE generation. Probe size and resolution are two strongly related terms; however, the difference is important when comparing SEM and HIM. The resolution is dependent not only on the size of the probe but also on the volume of the sample from which the signal is generated. Since SEs, by definition, have low energies (<50 eV) and the energy will still be dissipated while they diffuse toward the surface, only these produced near the surface may have enough energy left to overcome the surface barrier and escape. Therefore, the lower the energy distribution of SEs, the shallower the SE imaging can detect. Petrov and Vyvenko demonstrated that the SEs generated by 30 keV helium ions indeed have a lower-energy distribution compared with those generated by 30 keV electrons (Petrov et al., 2010). The majority of SEs produced by the 30 keV helium ion beam have an energy of less than 2 eV. This results in the mean escape depth of the SEs produced by helium ions being limited to typically 2 nm or less resulting in greater surface sensitivity than SEM imaging.

In addition, in the HIM, the backscatter yield of ions from the sample is very low, typically 1% or less, especially for low-Z materials. Since backscattered particles have almost the same energy as the primary beam, they may generate SEs as well (called SE2). The backscattered particles may exit from a location several micrometers away from the probe, and the SE2 produced there can still be detected, and this process reduces the resolution in SEM. However, due to the low backscattered yield of helium ions, SE1, the SEs generated directly by the probe, is the dominant signal produced in the HIM, and SE1 is localized and produced from a small region defined by the size of the probe (~0.2 nm in the HIM). The combination of the beam remaining more collimated within the sample near the surface, with reduced SE escape depth, and with large SE1/SE2 ratio results in higher-resolution and more surface-sensitive images than those that can be achieved in the SEM.

8.3.2 Advantages of HIM in Graphene Applications

Using helium ions as the primary beam rather than electrons brings enormous advantages (Bell et al., 2008), which makes the tool particularly suitable for graphene characterization, modification, and device fabrication. Against the challenges in graphene research, the advantages of the HIM are highlighted as follows.

Sub-nanometer resolution for graphene metrology: The HIM has a bright source ($\sim 4 \times 10^9$ A cm^{-2} sr^{-1}), which is about the size of a single atom, the smallest possible source (Morgan et al., 2006, Hill et al., 2008). The emission current can be regulated from 1 fA to 100 pA. Compared with electrons, helium ions have a much larger mass, and they exhibit a much shorter wavelength when accelerated in the same electrical field. The diffraction effect of the He$^+$ beam in terms of illumination apertures is thus negligible at its normal operation acceleration voltage (~30 kV).

Due to the ultimate small source size, the small energy dispersion, the high gun brightness, and the small illumination aperture, a probe size of 0.25 nm (at a beam convergent angle of 0.5 mrad) can be obtained with a beam current of 1 pA (Morgan et al., 2006, Hill et al., 2008, Williams and Carter, 2009). It thus brings out a resolution of <0.4 nm and a large depth of focus, which is close to resolving graphene edge configurations and enabling to discern irregularities as well as measure the size of the graphene flakes with great accuracy.

Surface sensitivity for graphene identification: In the HIM, the image using SEs illustrates the morphology of sample surface, and it is very sensitive to the change in local surface environment, such as electrical fields, magnetic fields, and thin layer of contaminations. In terms of graphene characterization, it is expected that the SE imaging can provide thickness measurement for identifying the number of layers in a graphene flake, clarify the contamination that may be invisible in SEM observation, and reveal minor topographical variations (e.g., cracks and rugged surfaces). HIM SE images disclose the surface features of the sample much better than SEM SE images.

Soft milling for fabrication of GNRs: Similar to other FIB systems, the strength of He$^+$–material interaction can be controlled by adjusting the dose and the beam energy. The HIM can thus work in both nondamage imaging and sputtering modes. In addition to its superior imaging capability, the HIM is also a nanofabrication workstation, and patterning of graphene can be done directly without using resist-based techniques (Chen et al., 2007, Han et al., 2007, Grigorescu and Hagen, 2009). Because of its sub-nm probe and the controllable strength of He$^+$–material interaction, the HIM provides higher fabrication precision and allows patterns with feature sizes an order of magnitude smaller than those of electron beam lithography (EBL) or Ga$^+$ FIB machining to be achieved. The HIM nanofabrication also avoids implanting metallic contaminants, which is inevitable in the Ga$^+$ FIB.

In addition, in the HIM, the ion dose required to pattern is approximately two orders of magnitude greater than that required to image the sample (Pickard and Scipioni, 2009), and imaging can, therefore, be carried out by the same tool in situ without causing further damage. This direct patterning is, however, not possible in the SEM as no sputtering of material occurs (Scipioni, 2009). *Nanostructuring of graphene is most convincingly demonstrated by helium ion microscopy* (Zhou and Loh, 2010). We will review the state-of-the-art achievements in HIM graphene imaging and modification in the following two sections.

8.4   HIM IMAGING FOR SAMPLE CHARACTERIZATION

As described in the previous section, HIM images reveal surface morphology of a sample. With the remarkable resolution and enhanced surface sensitivity, the HIM is expected to play important roles in metrology, especially the CD measurement in semiconductor sector (ITRS, 2009): the HIM has a particular advantage over the conventional SEM-based CD measurement while dealing with *insulating* samples. Similarly, it can find applications in graphene metrology in terms of measuring the lateral dimension of a graphene flake and/or the edge profiles of a graphene nanostructure (Arey et al., 2010, Bell, 2011, Postek et al., 2011, Fox et al., 2013). In addition, the HIM images of graphene samples can provide detailed information of surface contamination, local defects (i.e., macroscopic cracks, ruptures, folds, and voids), smoothness, uniformity, and thickness, which may be invisible in other microscopy methods.

The HIM is very promising in biological and medical research as well, since biological samples are normally beam sensitive and insulating, and the HIM has unique features suitable for imaging these samples. The HIM has large range of field of view (FOV) and large depth of focus (DOF), which can particularly facilitate the investigation of interactions between large biological samples (cells) and nanomaterials. For inorganic crystal structures, the strong channeling effect in the HIM may bring additional benefits to the study of grain structures, and this may have real applications in catalysis research. Some effort has been spent on developing the analytical capability of the tool; however, it is under development, and a viable attachment to analyze the compositions of the sample in the HIM is still absent.

In this section, we first discuss the metrology aspect of HIM imaging with a focus on the charge compensation and graphene measurement, and then introduce the graphene thickness contrast observed in the HIM and its biological applications. Finally, we briefly describe the channeling effects in HIM imaging and its applications in inorganic samples and outline the development of its analytical capabilities.

8.4.1   Charge Compensation and Graphene Metrology

In the charged-beam microscopes (e.g., HIM, SEM, and FIB), charged particles are injected into a sample, and secondary particles (i.e., secondary ions, electrons, and recoiled atoms), which contain charged particles, escape from the sample surface. The electrical neutralization of the sample is normally achieved via a proper grounding channel through the sample stage. The neutrality is of importance to the imaging process, since the local electrical field of a charged sample may smear the primary probe, vary landing energy of the beam, distort the signal, and cause severe sample drifting, for example, the bright strips in Figure 8.4a, an SEM image of the pore structure in a polymer sample (Daly et al., 2012). Due to the interruption of the ground channel, it is indeed a real challenge to maintain the neutrality while an insulating sample is being imaged.

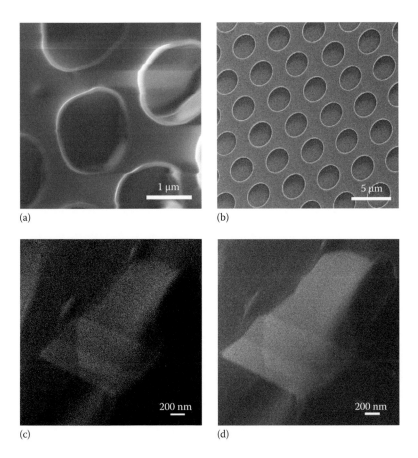

Figure 8.4   Charging effects and charging compensation in SEM and HIM. (a) An SEM image of micrometer-sized pores in a polymer sample shows artifacts caused by the charging and (b) a HIM image of the same pores, with charge compensation enabled, reveals the surface clearly; (c) and (d) are HIM images of graphene embedded in insulting polymers without and with the flood gun, that is, the charge compensator in the HIM.

In the SEM, this charging effect cannot be easily compensated, for it may be either positively or negatively charged depending on the local geometry and the beam energy—the charging is very much a dynamic process and varies constantly. In the HIM, the charging is, however, always positive since the incoming particles are positive He^+ and the exiting signals are electrons (negative charges). The local buildup of positive charge can be effectively neutralized by a broad beam electron flood gun. To observe an insulating sample in the HIM, the flood gun is pulsed on between scans of the ion beam in order to replenish the area of the sample with electrons. As shown in Figure 8.4b, the pore structures, which show drastic charging effect in the SEM image (Figure 8.4a), are resolved clearly in the HIM by using the flood gun. Figure 8.4c is an HIM image of a graphene flake embedded in an insulating polymer matrix with the flood gun switched off; Figure 8.4d is an image of the same area after optimizing the flood gun. It is clear that the effective charge compensation can greatly improve the image quality and reduce the charging artifacts.

For accurate topographical and dimensional analysis, the HIM provides very useful information (Behan et al., 2012, Fox et al., 2013). Figure 8.5a and b shows SEM and HIM images of Ni/Co oxide nanoparticles, respectively. The two images have a similar magnification. The edge profiles

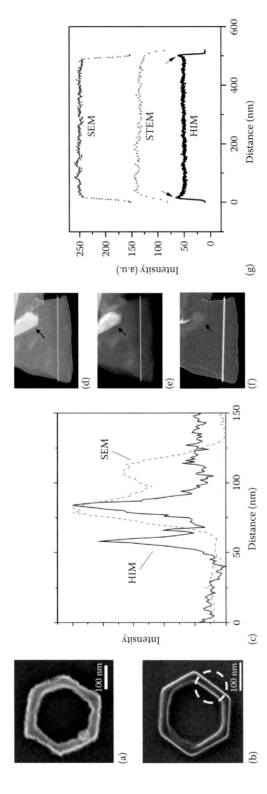

Figure 8.5 HIM metrology. (a) and (b) are SEM and HIM images of Ni/Co nanoparticles, respectively. (c) The signal intensity was drawn as a function of the position along a line perpendicular to its side wall. (d through f) SEM, STEM, and HIM images of a graphene flake. The intensity along the line indicated in the images was drawn in (g).

of the two particles are plotted in Figure 8.5c, and the peaks in the plot correspond to the edges in the image. We can see that the HIM provides the sharper edge profile and thus defines the size of the nanoparticle with less ambiguity. Figure 8.5d and e shows images of a single graphene flake acquired in three different microscopes: (d) the SEM in-lens image, (e) the STEM (scanning TEM)–HAADF (high-angle annular dark field) image, and (f) the HIM image. The corresponding intensity profiles are plotted in Figure 8.5f. A peak-like small intensity increase at the edge in the HIM profile is evident (Fox et al., 2013). It allows a straightforward evaluation of the lateral size of the flake by measuring the peak-to-peak distance.

Another interesting feature in Figure 8.5d and e is that the HIM imaging is very surface sensitive. In SEM and STEM images, bright regions appear around the top-right corner (Figure 8.5d and e, indicated by the arrows), while in the HIM image, it is hardly visible. The contrast in the SEM and STEM images results from another flake underneath the major one. The STEM image exploits the transmitted electrons and *sense* the whole sample volume along the path of the electron beam that penetrates the sample. It is, therefore, expected to see the underneath flake. In the SEM case, the SEs from the underneath flake can contribute to the image significantly, while the SE from the very surface makes the dominant contribution in the HIM. This shows that HIM SEs have a lower energy distribution, and the top layer attenuates the SEs from the underneath flake effectively. In the following section, we discuss this effect and explore the HIM contrast of graphene thickness.

8.4.2   Thickness Contrast of Graphene

The SEs in the HIM enable high lateral resolution and surface-sensitive images, and they are also the principal signal in the SEM imaging. The research of graphene SE thickness contrast dates from 2010 (Hiura et al., 2010). Mechanically exfoliated FLG flakes on thick insulating substrates (e.g., SiO_2, sapphire, and mica) were examined using the SEM with various beam energies, and a linear relationship between the SE intensity and graphene thickness was found. A practical method to determine the number of graphene layers was also proposed. Compared with the conventional optical observation, this method could characterize sub-micrometer graphene samples on various insulating substrates.

The mechanism of graphene thickness contrast has been controversial. Cazaux (2011) studied the SE emission of few-layered graphene on a SiC substrate and suggested that the thickness contrast is attributed to the graphene surface work function. This model is reasonable because the SEs must overcome the vacuum barrier at the surface to escape from the sample; the height of the barrier, that is, the work function, thus regulates the yield of the SE. Indeed, the work function Φ of graphene samples increases with the number of the layers from ~4.3 (one layer) to ~4.6 eV (four layers) and saturates for samples with more layers (Hibino et al., 2009, Ziegler et al., 2011). According to Cazaux's (2011) calculations, the SE emissions could decrease by >20% from monolayer to four-layer graphene due to the increase in the work function. Some experimental results seem to support the work function model (Park et al., 2012). However, Kochat et al. (2011) have found distinguishable contrast for more than four layers, which cannot be attributed to the work function variation because of its saturation over four layers. They have proposed a model based on the attenuation effect: the graphene flake acts as an absorbing layer to the electrons that have escaped from the substrate, and the thicker the graphene, the weaker the signal. The problems in this model are as follows: the mean free path of the SEs is significantly underestimated, and the attenuation, by itself, could not explain the layer contrast.

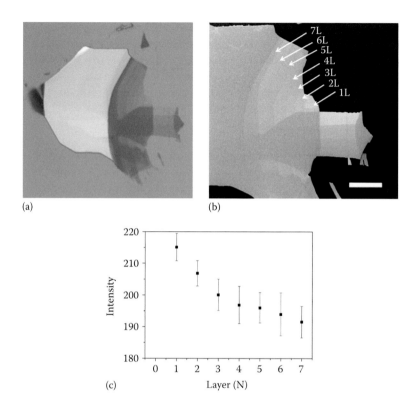

(a)

(b)

(c)

Figure 8.6 A large piece of mechanically exfoliated graphene flake with regions of different thicknesses on a SiO$_2$/Si substrate: (a) optical image showing the boundaries of the regions, (b) HIM image of the same flake (the scale bar in the image is 10 µm), and (c) the intensity as a function of the number of layers.

Since the energy spectra of HIM SEs are quite different from those obtained from the SEM, the HIM may offer a suitable platform to study the thickness contrast in graphene and help to settle the debates. Little work has been published on this topic, and it might be due to the limited access to the tool. Nevertheless, we have recently investigated the thickness contrast in the HIM. Some of the results are illustrated in Figure 8.6, where an optical image of a large graphene flake is presented side by side with its HIM image. In the HIM image, we can clearly discern graphene samples at least up to seven layers, though the difference in contrast reduces as the thickness increases (Figure 8.6c). There is no doubt that the HIM imaging is able to distinguish FLG samples by their different thicknesses. Combined with the sub-nanometer lateral resolution, the HIM may be a superior tool for graphene characterization. However, we must evaluate the beam damage for some applications where a nondestructive method is required. We will review HIM beam damage in the next section, while in the following parts of this section, we extend our discussion on HIM imaging for biological samples and bulk crystals since we expect that the HIM may become a key-enabling tool in these fields as well.

8.4.3 Biological Material Imaging

The HIM provides the unique ability to directly image the surface of biological samples. For biological material imaging, the advantages of helium ion microscopy over the other microscopic methodologies include (1) sub-nanometer resolution, (2) efficient charging control, (3) small beam damage, and (4) high depth of focus (Bazou et al., 2011, Santos-Martinez et al., 2012). In the SEM characterization, to circumvent the charging issue, a thin conducting layer may have to be applied to biological samples and nonconductive and/or sensitive materials in general. This approach may effectively minimize the sample charging. With a thin and uniform coating, the sample surface morphology can be preserved and revealed. However, the surface details are inevitably concealed by the coating layer. For example, Figure 8.7a is an SEM image of coated human Caco2 cells, which may indicate the splitting of a cell or simply two adjacent cells being connected by the coating layer. As discussed in previous sections, the HIM has a unique charge compensation mechanism, and the coating is no longer required. This allows detailed surface analysis to be done with the HIM. In Figure 8.7b, the two cells are clearly resolved with individual connections between them, which might be very well covered if a coating layer had been applied. It is clear that the application of a coating layer buries much of the surface information and fills voids on the surface. Any measurements of surface porosity or surface area would be greatly affected by the coating layer. The HIM may find potential applications in these areas by discarding the coating layer.

The depth of focus (DOF) is the largest difference in height between two objects that remain in focus in an image. Since very small apertures are used in the HIM compared with the SEM,

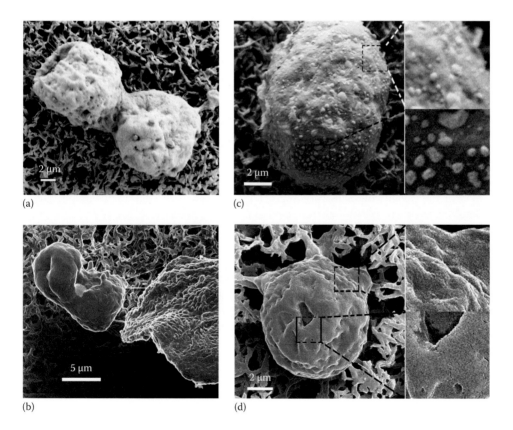

Figure 8.7   (a) SEM image of a coated Caco2 cells and (b) HIM image of uncoated Caco2 cells. The effect of the large depth of the focus in the HIM can be seen from (c) SEM image and (d) HIM image of Caco2 cells.

the helium ion beam can be made more collimated with a smaller convergence angle, and the beam will remain better focused over a larger range of sample height. Therefore, the HIM has a quite large DOF. Figure 8.7c and d shows SEM and HIM images of two large micrometer-sized cancer cells, respectively. The two images were both more or less focused on the top of the cells. To evaluate the DOF, we magnified two areas from each image that correspond to two regions at different heights. It is evident that both areas from the HIM image show clear and sharp features, indicating they are both well in focus. On the contrary, the bottom part from the SEM image is much blurred due to the relatively short DOF of the SEM. The large DOF in the HIM may be very useful for imaging large but sensitive samples where excessive focusing adjustment may cause significant sample damage.

8.4.4   Analytical Capability of the HIM

Nowadays, it is almost the standard configuration that electron microscopes are equipped with an attachment for chemical microanalysis, such as the spectrometers used for Energy Dispersive X-ray Spectroscopy (EDX), Wavelength Dispersive X-ray Spectroscopy (WDS), and Electron Energy Loss Spectroscopy (EELS). These instruments can provide qualitative and quantitative compositional analysis and greatly facilitate material characterization. EDX and WDS detect the characteristic x-rays, a secondary signal, which is generated by the inner-shell ionization of the material atoms and the succeeding x-ray photon emission due to the electron–hole recombination. The prerequisite for a viable x-ray analysis is that the primary electron must exhibit a reasonable large inner-shell ionization probability, that is, the ionization cross section, to give detectable signal strength. The velocity of the primary particle determines the cross section. In the typical energy range of the electron beam exploited in the electron microscopes (several keV to several hundreds of keV), the velocity of the primary electron is comparative to that of the inner-shell electrons of the material atom. The inner-shell ionization probability for the electron–sample interaction is, therefore, large enough to validate the x-ray analysis.

In the HIM, since the mass of a helium ion is about 7800 times larger than that of an electron, the energy required for the helium ion to acquire similar velocity to that of the electron in the electron microscope has to be as large as tens of MeV. For the typical beam energy in the HIM (tens of keV), the velocity of the helium ion is too low to generate a detectable amount of x-rays. It is, therefore, not a feasible method for chemical analysis in the HIM (Joy and Griffin, 2011).

Rutherford backscattering spectrometry (RBS) is an established ion beam analytical technique. In the RBS, MeV ions are scattered by the sample, and the energy spectra of the backscattered ions are analyzed to determine sample composition and its depth distribution. The energy loss in the elastic scattering that the backscattered ion experienced is a fingerprint of the colliding atom, which provides the compositional information. The ion energy also dissipates while traveling in the sample, and the energy loss correlates with the distance traveled, which is the underneath principle that reveals the depth profile of the sample through a quantitative model based on the stopping power. Naturally, despite the low beam energy exploited in the HIM, RBS has been attempted to enhance HIM analytical capability (Sijbrandij et al., 2008, Scipioni et al., 2009b, Behan et al., 2010, 2012). A typical Rutherford backscattered ion spectrum taken from a thin gold layer on top of a silicon substrate is shown in Figure 8.8a. A simulation program, SIMNRA (Scipioni et al., 2009b), is then used to fit the experimental data points to give the thickness of the film. Four Au films with different thicknesses are deposited, and the thicknesses of the films are then extracted by using the RBS measurement and depicted in Figure 8.8b. Thickness measurement gives good agreement with the nominal thickness. It indicates that the

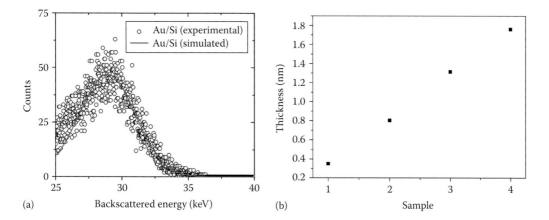

Figure 8.8   (a) A typical Rutherford backscattered ion spectra taken from a thin gold layer on top of a silicon substrate and (b) the thicknesses extracted from the RBI spectra are consistent with the nominal thicknesses that measured from the deposition.

RBS may be applicable to analyze thin films. However, due to the relatively low energy and the poor performance of the detector, the energy resolution of the spectrometer is inferior in terms of analyzing samples of unknown composition. In addition, it can hardly provide any useful information for insulating samples (Behan et al., 2010). To extend the analytical capability of the HIM, current research is focused on secondary ion mass spectroscopy, and it is expected to provide a practical method for chemical analysis in the new tri-beam (He$^+$, Ne$^+$, and Ga$^+$) tool (Winston et al., 2011, Rahman et al., 2012, Wirtz et al., 2012).

Although a quantitative chemical analysis is absent in the HIM, the HIM images show strong material contrast (Inai et al., 2007, Scipioni et al., 2009a), especially using the backscattered ions to form an RBI image. The RBI image can also provide texture data or crystallographic information (Temprano et al., 2011, Veligura et al., 2012). For thin samples, the HIM can be operated in a transmission mode, similar to the TEM, and crystallographic contrast has been acquired (Notte et al., 2010). In the past couple of years, the HIM has burgeoned into a new imaging methodology with superior resolution, novel contrast, and extensive potential applications, and it may become a key enabling tool in graphene characterization and biological and material research as well. In addition to these advantages in imaging, the HIM has shown great promise in nanoscale modification and fabrication, which is discussed in next section.

8.5   NANOSCALE MODIFICATION AND FABRICATION USING THE HIM

The HIM is an FIB microscope, much like the gallium ion FIB. The Ga$^+$ FIB is a conventional tool for nanoscale machinery, which has found applications in diverse fields, for example, ultrafine lamella preparation for TEM observation (Fox et al., 2012b), nano-tomography for 3D reconstruction (Ritter and Midgley, 2010, Villinger et al., 2012), and failure analysis in semiconductor industry (Yuan and Xiaowen, 2010). Although the beam–sample interactions in both microscopes are quite different, the HIM still has the ability to remove sample material by the process of sputtering. The sputtering yield of the He$^+$ beam in a material is, however, approximately much less than that of a Ga$^+$ beam (with the same beam energy). The low sputtering yield and the very well-focused probe of helium ions, coupled with the low backscatter yield, allow the HIM to define features as small as 4 nm, which makes the HIM ideally suited to the accurate removal of small volumes of material.

The mild and ultrafine milling capability of the HIM is particularly promising for graphene research. A precise and reliable method for tailoring the properties of graphene will greatly facilitate their applications. For example, opening of a bandgap is one feature that graphene requires for device applications (Zhang et al., 2007). Graphene modification has been achieved by many methods with varying degrees of success to date. Controlled defect introduction in graphene can be used to alter its electronic structure (Jafri et al., 2010) as well as its mechanical (Ansari et al., 2012), thermal (Hao et al., 2011), optical (Lucchese et al., 2010), and magnetic properties (Yazyev and Helm, 2007). GNRs are desirable for electronic (Li et al., 2008a, Schwierz, 2010) as well as potential spintronic (Kim and Kim, 2008) applications, while graphene nanopores have potential as DNA sequencers (Schneider et al., 2010), gas separation membranes (Jiang et al., 2009), and in hydrogen storage (Reunchan and Jhi, 2011). The HIM may provide a suitable solution to modify graphene properties with a high level of control.

In this section, we first review the modification of graphene in terms of defect creation and introduce some recent progress on the HIM graphene modification. The fabrication of GNRs will then be discussed followed by some additional examples on HIM milling and patterning of nanostructures. Finally, we provide a brief overview of helium beam chemistry, that is, beam-induced deposition and ion beam lithography.

8.5.1   Controllable Defect Creation in Graphene

The properties of graphene are strongly influenced by its defects, which may come from the synthesis process, the interaction with the substrate and environment, and intentional modification. Several researchers have tried to introduce defects using Ar^+ or proton bombardment, and the Fermi velocity of the irradiated graphene sample was substantially reduced (Tapaszto et al., 2008). The structural transition route of graphene under focused Ga^+ beam irradiation has been studied and correlated to its electrical transition (Zhou et al., 2010), and the electrical transport behavior of the irradiated graphene can thus be tailored from weak to strong localization regime. However, the gallium contamination may affect the measured conduction properties. The effect of low-energy electron beam irradiation on graphene has been investigated as well (Teweldebrhan and Balandin, 2009). An SEM was used to expose a graphene sample to electrons and measured the resulting effect with Raman spectroscopy. They found that even relatively low-energy irradiation of graphene (<20 keV) caused a significant shift in the material's optical response. This was attributed to bond breaking and sample heating, which, over a period of several hours, resulted in disordering and partial amorphization of the graphene lattice. Using low-energy electron beam to introduce defects is controversial due to the complication of contamination (Jones et al., 2009), while defects, even nanopores, can undoubtedly be created in graphene by the low-energy electron beam (as low as 1 keV) with the assistance of nitrogen gas (Fox et al., 2011).

Compared with these modification approaches, the HIM has the advantages of a small probe and controllable interaction strength. The relationship between the helium ion dose and the defect density has been recently experimentally established (Fox et al., 2013). The effects of the dose on both freestanding and supported graphene have been compared. As shown in Figure 8.9a, the defect density, indicated by the intensity of the D peak in the Raman spectra, increases with the dose. The defects are localized within the predefined beam-scanning areas (see Figure 8.9b).

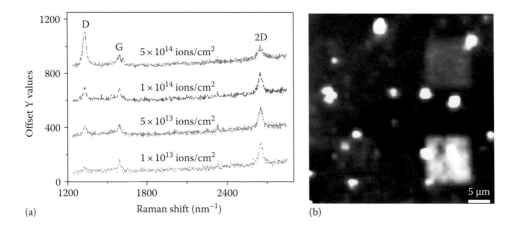

(a)

(b)

Figure 8.9 (a) Raman spectra of He⁺-irradiated graphene samples NS and (b) Raman mapping of the He⁺-irradiated areas: the brightness of the image represents the intensity of the D peak in (a).

The defect density is also correlated to the quality of high-resolution images (Boden et al., 2011, Fox et al., 2013). A high-magnification image with a good signal-to-noise ratio must introduce some defects into the graphene. A safe imaging dose on the order of 10^{13} He⁺/cm² is established, with both freestanding and supported graphene samples becoming highly defective at doses over 5×10^{14} He⁺/cm². With this information, a balance between the acceptable level of damage for applications and the information required from the characterization can then be achieved.

8.5.2 Tailoring the Morphology of Nanostructures Using HIM

As discussed in the previous section, one limitation of graphene that must be overcome is its lack of an electronic bandgap. Some of the first work on graphene patterning with the HIM was conducted in 2009 (Bell et al., 2009a,b). A 10 nm wide trench across a graphene flake was milled with in situ electrical measurements used to confirm the cut. It was also found that contamination deposited on the substrate led to residual conductivity, and the dose required to mill graphene on a substrate was lower than for suspended graphene. GNRs with the line widths as narrow as 3 nm were fabricated in 2010, and the lattice damage due to the milling recovers at a distance of about 1 nm from the patterned edge (Pickard et al., 2010). Our results show that the freestanding graphene can be patterned with the milling dose approximately 10^{18} He⁺/cm², and 4 nm wide dot and 7 nm wide ribbon arrays can be obtained (see Figure 8.10). The milling dose on supported graphene is, however, one order smaller. The helium ion patterned graphene structure is expected to have narrow and precise width and well-defined edge profile. Since a very large dose is used to mill graphene, the influence of induced defects must be further considered in the future. Preliminary results show a GNR array exhibits a strong Raman defect peak (D peak). There is no sign of amorphous carbon indicating that this peak is due to edge defects and a high crystal quality remains in the un-milled area.

The milling in the HIM is also suitable for changing the morphology and property of other nanomaterials and thin films. For example, Figure 8.11a is a Ni/Co nanoring (Behan et al., 2012) with a diameter of 300 nm and edge width of 23 nm, and the image was taken without significant He⁺ exposure. After the particle had been repeatedly scanned by the helium ion beam and received an excessive ion dose up to 10^{17} He⁺/cm², the edge width of the particle started to decrease

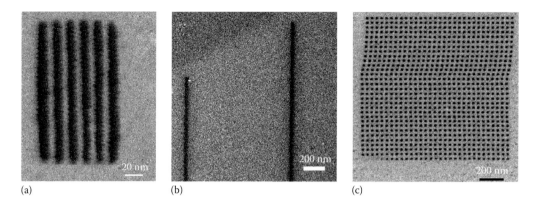

(a)   (b)   (c)

Figure 8.10   (a) A 10 nm wide suspending graphene nanoribbon array fabricated by the HIM, (b) 20 nm trenches milled in a supported graphene by the HIM, and (c) graphene anti-dot arrays milled by the HIM.

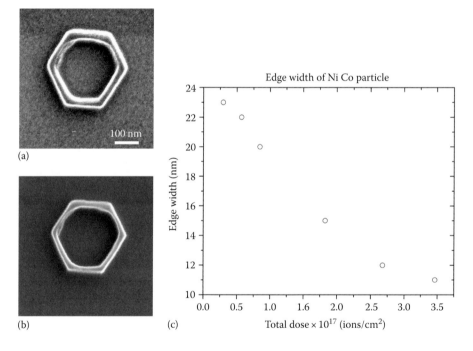

(a)

(b)   (c)

Figure 8.11   The HIM can modify the morphology of nanoparticles: (a) a Ni/Co nanoparticle with minimal helium ion exposure, (b) the same particle having been irradiated with excessive dose, and (c) the edge width decreases with the dose.

(Figure 8.11b). The decrease in the edge width seemed to correlate with the beam dose linearly as shown in Figure 8.11c. With a dose of 3.5×10^{17} He$^+$/cm^2, the edge width was shrunk to 11 nm, and it reduced by about 50%. The diameter of the particle kept unchanged. This result suggests that the HIM can be used as nano-scissor to modify the geometry of nanoparticles, while the beam damage, for example, amorphization, needs further investigation. An unusual logarithmic material removal rate was recorded while we were investigating the magnetic evolution of Co/Pt

(a)   (b)   (c)

Figure 8.12   Complex patterns milled by the HIM: (a) a nanosized Ireland map and (b) a more structured logo of Dublin GAA and (c) its nanoscale counterpart milled by the HIM.

multilayers as a function of ion dose using the HIM (Fowley et al., 2013). The unusual milling rate might be due to beam ablation rather than sputtering. Although the milling mechanism is still unclear, the HIM can feasibly be used to locally pattern magnetic or metallic layers. To extend this feasibility, the HIM can write complex patterns by uploading a binary image into the software program. An example of a pattern of the Ireland map and a Gaelic Athletic Association logo generated by the tool is shown in Figure 8.12. The integration of an external scan control system is required for the writing of patterns where more control over the beam is required. Parameters that have an effect on the milling that require an external scan control system include the direction of scanning or even the order in which features are milled.

One potential application of the HIM milling is to improve TEM lamellae fabrication (Rudneva et al., 2013). We have demonstrated the milling ability of the HIM to improve the quality of FIB prepared TEM lamellae (Fox et al., 2012a). The modified sample was shown to be thinner and smoother while having less gallium contamination and still retaining its high-quality crystal structure. The HIM has also been used to fabricate a wedge structure with a penetration in a TEM lamellae, demonstrating a high-level control of sample thickness. The extension of the lattice modification, from crystalline to amorphous, was directly observed to extend several hundred nanometers from the irradiated region under high dose conditions. Similarly, solid-state nanopores of 4 nm in diameter have also been fabricated by the HIM, which have been used to perform biomolecular analysis by measuring the passage of double-strand DNA (Yang et al., 2011).

8.5.3   Lithography and Beam-Assisted Deposition

Helium ion beam lithography has been investigated recently (Sidorkin et al., 2009, Winston et al., 2009, 2012, 2011, Alkemade et al., 2012). The helium ion beam remains well collimated in the resist layer leading to a reduction in forward scattering. There are also a greatly reduced number of backscatter events in the HIM. These factors result in smaller feature sizes, greater pattern densities, and no observable proximity effect. A series of lines have been written in a 5 nm thick HSQ resist layer, and these lines have a 6 nm line width and a 15 nm pitch. These patterns demonstrate feature sizes and densities that are beyond the limits of EBL technology.

The HIM can be fitted with an aftermarket gas injection system (GIS). The GIS is used to introduce a gas species into the region of the sample surface and decomposed by the ion beam. The gases most commonly used allow deposition of metals such as platinum or tungsten and insulators such as silicon oxide. This deposition process is known as ion beam–induced deposition. Decomposition of metals by the ion beam has been demonstrated to produce features within close proximity. This indicates that the precursor decomposition is quite well confined to the region of incidence of the ion beam (Alkemade et al., 2011). On the other hand, the injected gas can facilitate ion etching. For example, a fluorine compound such as XeF_2 can be decomposed by the ion beam and the fluorine gas then enhances the material removal rate when attempting to mill a sample, which helps to reduce the subsurface void formation and subsequent swelling effect (Alkemade et al., 2012).

8.6   SUMMARY

The interaction between a highly focused helium ion probe and sample surface not only generates abundant signals that disclose detailed sample morphology but also enables direct and controllable sample modification. HIM imaging offers sub-nanometer resolution and material sensitivity for surface investigation of a wide range of samples. Particularly, it is now possible to measure the thickness and geometry of graphene by HIM secondary imaging. HIM may also address metrology challenges encountered in the continuing semiconductor miniaturization and find applications in the characterization of nanostructures. Biological and medical research can benefit from HIM as well due to its large FOV and high efficient charge compensation mechanism. On the other hand, the helium ions can knock off atoms from sample surface directly, especially for low-Z materials. Because of the absence of beam strangling in 2D materials and the controllable milling strength, layered materials exemplified by graphene can be tailored to specific geometries with sub-nanometer precision and achieve desirable properties by HIM modification. It can be further extended to thin films structuring for the fabrication of various functional devices, such as plasmonic structures and DNA sequencing units. HIM has become a burgeoning field. The technology is rapidly evolving, in-depth knowledge concerning its imaging and modification mechanisms is being established, and more applications are being attempted and demonstrated. It is expected that HIM will bring great impact to nanoscience and nanotechnology.

ACKNOWLEDGMENTS

We would like to thank staff at the CRANN Advanced Microscopy Laboratory at Trinity College for their assistance and useful discussions. The work at the School of Physics and the Centre for Research on Adaptive Nanostructures and Nanodevices at Trinity College Dublin is supported by Science Foundation Ireland under Grants 11/PI/1105, 12/TIDA/I2433, and 07/SK/I1220a.

REFERENCES

Alkemade, P. F. A., E. M. Koster, E. van Veldhoven, and D. J. Maas. Imaging and nanofabrication with the helium ion microscope of the Van Leeuwenhoek laboratory in Delft. *Scanning* 34, (2012): 90–100.
Alkemade, P. F. A., H. Miro, E. van Veldhoven, D. J. Maas, D. A. Smith, and P. D. Rack. Pulsed helium ion beam induced deposition: A means to high growth rates. *Journal of Vacuum Science & Technology B* 29, (2011): 06FG05.
Ananth, M., L. Scipioni, and J. Notte. The helium ion microscope: The next stage in nanoscale imaging. *American Laboratory* 40, (2008): 42–46.

Ansari, R., S. Ajori, and B. Motevalli. Mechanical properties of defective single-layered graphene sheets via molecular dynamics simulation. *Superlattices and Microstructures* 51, (2012): 274–289.

Arey, B. W., V. Shutthanandan, and W. Jian. Helium ion microscopy versus scanning electron microscopy. *Microscopy & Microanalysis 2010* Portland, OR, (2010): http://www.microscopy.org/MandM/2010/arey.pdf.

Bai, J. W. and Y. Huang. Fabrication and electrical properties of graphene nanoribbons. *Materials Science & Engineering R-Reports* 70, (2010): 341–353.

Balandin, A. A., S. Ghosh, W. Bao, I. Calizo, D. Teweldebrhan, F. Miao, and C. N. Lau. Superior thermal conductivity of single-layer graphene. *Nano Letters* 8, (2008): 902–907.

Bazou, D., G. Behan, C. Reid, J. J. Boland, and H. Z. Zhang. Imaging of human colon cancer cells using He-ion scanning microscopy. *Journal of Microscopy* 242, (2011): 290–294.

Behan, G., J. F. Feng, H. Z. Zhang, P. N. Nirmalraj, and J. J. Boland. Effect of sample bias on backscattered ion spectroscopy in the helium ion microscope. *Journal of Vacuum Science & Technology A* 28, (2010): 1377–1380.

Behan, G., D. Zhou, M. Boese, R. M. Wang, and H. Z. Zhang. An investigation of nickel cobalt oxide nanorings using transmission electron, scanning electron and helium ion microscopy. *Journal of Nanoscience and Nanotechnology* 12, (2012): 1094–1098.

Bell, D. Contrast performance: Low voltage electrons vs. helium ions. *Microscopy & Microanalysis* 17, (2011): 660–661.

Bell, D. C., M. C. Lemme, L. A. Stern, and C. M. Marcus. Precision material modification and patterning with He ions. *Journal of Vacuum Science & Technology B* 27, (2009a): 2755–2758.

Bell, D. C., M. C. Lemme, L. A. Stern, J. R. Williams, and C. M. Marcus. Precision cutting and patterning of graphene with helium ions. *Nanotechnology* (2009b): 455301.

Bell, D. C., L. A. Stern, L. Farkas, and J. A. Notte. Helium ion microscope: Advanced contrast mechanisms for imaging and analysis of nanomaterials. In *EMC 2008 14th European Microscopy Congress*, September 1–5, 2008, Aachen, Germany. Luysberg, M., K. Tillmann, and T. Weirich (eds.), 2008, pp. 527–528. Springer, Berlin, Germany.

Berger, M. J., J. S. Coursey, M. A. Zucker, and J. Chang. Stopping-power and range tables for electrons, protons, and helium ions. NIST. 2011. http://physics.nist.gov/physrefdata/star/text/intro.html (accessed on October 2, 2014.)

Biro, L. P. and P. Lambin. Nanopatterning of graphene with crystallographic orientation control. *Carbon* 48, (2010): 2677–2689.

Blake, P., E. W. Hill, A. H. C. Neto, K. S. Novoselov, D. Jiang, R. Yang, T. J. Booth, and A. K. Geim. Making graphene visible. *Applied Physics Letters* 91, (2007): 063124.

Boden, S., Z. Moktadir, D. Bagnall, H. Rutt, and H. Mizuta. Beam-induced damage to graphene in the helium ion microscope. In *Graphene 2011 Conference*, Bilbao, Spain, April 11–14, 2011: http://eprints.soton.ac.uk/272304/.

Bolotin, K. I., K. J. Sikes, Z. Jiang, M. Klima, G. Fudenberg, J. Hone, P. Kim, and H. L. Stormer. Ultrahigh electron mobility in suspended graphene. *Solid State Communications* 146, (2008): 351–355.

Bunch, J. S., A. M. van der Zande, S. S. Verbridge, I. W. Frank, D. M. Tanenbaum, J. M. Parpia, H. G. Craighead, and P. L. McEuen. Electromechanical resonators from graphene sheets. *Science* 315, (2007): 490–493.

Casiraghi, C., A. Hartschuh, E. Lidorikis, H. Qian, H. Harutyunyan, T. Gokus, K. S. Novoselov, and A. C. Ferrari. Rayleigh imaging of graphene and graphene layers. *Nano Letters* 7, (2007): 2711–2717.

Castro, E. V., K. S. Novoselov, S. V. Morozov, N. M. R. Peres, J. M. B. L. dos Santos, J. Nilsson, F. Guinea, A. K. Geim, and A. H. C. Neto. Biased bilayer graphene: Semiconductor with a gap tunable by the electric field effect. *Physical Review Letters* 99, (2007): 216802.

Cazaux, J. Calculated dependence of few-layer graphene on secondary electron emissions from SiC. *Applied Physics Letters* 98, (2011): 013109.

Cheianov, V. V., V. Fal'ko, and B. L. Altshuler. The focusing of electron flow and a veselago lens in graphene P-N junctions. *Science* 315, (2007): 1252–1255.

Chen, Z., Y.-M. Lin, M. J. Rooks, and P. Avouris. Graphene nano-ribbon electronics. *Physica E: Low-Dimensional Systems and Nanostructures* 40, (2007): 228–232.

Chowdhury, R., S. Adhikari, P. Rees, S. P. Wilks, and F. Scarpa. Graphene-based biosensor using transport properties. *Physical Review B* 83, (2011): 045401.

Coleman, V., R. Knut, O. Karis, H. Grennberg, U. Jansson, R. Quinlan, B. Holloway, B. Sanyal, and O. Eriksson. Defect formation in graphene nanosheets by acid treatment: An X-ray absorption spectroscopy and density functional theory study. *Journal of Physics D: Applied Physics* 41, (2008): 062001.

Daly, R., J. E. Sader, and J. J. Boland. Existence of micrometer-scale water droplets at solvent/air interfaces. *Langmuir* 28, (2012): 13218–13223.

Datta, S., D. Strachan, S. Khamis, and A. Johnson. Crystallographic etching of few-layer graphene. *Nano Letters* 8, (2008a): 1912–1915.

Datta, S. S., D. R. Strachan, S. M. Khamis, and A. T. C. Johnson. Crystallographic etching of few-layer graphene. *Nano Letters* 8, (2008b): 1912–1915.

Dobrik, G., L. Tapasztó, P. Nemes-Incze, P. Lambin, and L. Biró. Crystallographically oriented high resolution lithography of graphene nanoribbons by stm lithography. *Physica Status Solidi (B)* 247, (2010): 896–902.

Du, X., I. Skachko, A. Barker, and E. Andrei. Approaching ballistic transport in suspended graphene. *Nature Nanotechnology* 3, (2008): 491–495.

Economou, N. P., J. A. Notte, and W. B. Thompson. The history and development of the helium ion microscope. *Scanning* 34, (2012): 83–89.

Egerton, R. F., P. Li, and M. Malac. Radiation damage in the TEM and SEM. *Micron* 35, (2004): 399–409.

Elias, D. C., R. R. Nair, T. M. G. Mohiuddin, S. V. Morozov, P. Blake, M. P. Halsall, A. C. Ferrari et al. Control of graphene's properties by reversible hydrogenation: Evidence for graphane. *Science* 323, (2009): 610–613.

Ferrari, A. C., J. C. Meyer, V. Scardaci, C. Casiraghi, M. Lazzeri, F. Mauri, S. Piscanec et al. Raman spectrum of graphene and graphene layers. *Physical Review Letters* 97, (2006): 187401.

Fischbein, M. D. and M. Drndic. Electron beam nanosculpting of suspended graphene sheets. *Applied Physics Letters* 93, (2008): 113107.

Fowley, C., Z. Diao, C. C. Faulkner, J. Kally, K. Ackland, G. Behan, H. Z. Zhang, A. M. Deac, and J. M. D. Coey. Local modification of magnetic anisotropy and ion milling of Co/Pt multilayers using a He$^+$ ion beam microscope. *Journal of Physics D: Applied Physics* 46, (2013): 195501.

Fox, D., Y. Chen, C. C. Faulkner, and H. Zhang. Nano-structuring, surface and bulk modification with a focused helium ion beam. *Beilstein Journal of Nanotechnology* 3, (2012a): 579–585.

Fox, D., A. O'Neill, D. Zhou, M. Boese, J. N. Coleman, and H. Z. Zhang. Nitrogen assisted etching of graphene layers in a scanning electron microscope. *Applied Physics Letters* 98, (2011): 243117.

Fox, D., R. Verre, B. J. O'Dowd, S. K. Arora, C. C. Faulkner, I. V. Shvets, and H. Zhang. Investigation of coupled cobalt–silver nanoparticle system by plan view TEM. *Progress in Natural Science: Materials International* 22, (2012b): 186–192.

Fox, D., Y. B. Zhou, A. O'Neill, S. Kumar, J. J. Wang, J. N. Coleman, G. S. Duesberg, J. F. Donegan, and H. Z. Zhang. Helium ion microscopy of graphene: Beam damage, image quality and edge contrast. *Nanotechnology* 24, (2013): 335702.

Gaskell, P. E., H. S. Skulason, W. Strupinski, and T. Szkopek. High spatial resolution ellipsometer for characterization of epitaxial graphene. *Optics Letters* 35, (2010): 3336–3338.

Gass, M. H., U. Bangert, A. L. Bleloch, P. Wang, R. R. Nair, and A. K. Geim. Free-standing graphene at atomic resolution. *Nature Nanotechnology* 3, (2008): 676–681.

Geim, A. K. and K. S. Novoselov. The rise of graphene. *Nature Materials* 6, (2007): 183–191.

Ghosh, S., I. Callzo, D. Teweldebrhan, E. P. Pokatilov, D. L. Nika, A. A. Balandin, W. Bao, F. Miao, and C. N. Lau. Extremely high thermal conductivity of graphene: Prospects for thermal management applications in nanoelectronic circuits. *Applied Physics Letters* 92, (2008): 151911.

Graf, D., F. Molitor, K. Ensslin, C. Stampfer, A. Jungen, C. Hierold, and L. Wirtz. Raman imaging of graphene. *Solid State Communications* 143, (2007): 44–46.

Grigorescu, A. E. and C. W. Hagen. Resists for sub-20-nm electron beam lithography with a focus on HSQ: State of the art. *Nanotechnology* 20, (2009): 292001.

Gu, G., S. Nie, R. M. Feenstra, R. P. Devaty, W. J. Choyke, W. K. Chan, and M. G. Kane. Field effect in epitaxial graphene on a silicon carbide substrate. *Applied Physics Letters* 90, (2007): 253507.

Han, M. Y., Ouml, B. Zyilmaz, Y. Zhang, and P. Kim. Energy band-gap engineering of graphene nanoribbons. *Physical Review Letters* 98, (2007): 206805.

Hao, F., D. Fang, and Z. Xu. Mechanical and thermal transport properties of graphene with defects. *Applied Physics Letters* 99, (2011): 041901.

Hibino, H., H. Kageshima, M. Kotsugi, F. Maeda, F. Z. Guo, and Y. Watanabe. Dependence of electronic properties of epitaxial few-layer graphene on the number of layers investigated by photoelectron emission microscopy. *Physical Review B* 79, (2009): 125437.

Hill, R., J. Notte, and B. Ward. The alis He ion source and its application to high resolution microscopy. *Physics Procedia* 1, (2008): 135–141.

Hiura, H., H. Miyazaki, and K. Tsukagoshi. Determination of the number of graphene layers: Discrete distribution of the secondary electron intensity stemming from individual graphene layers. *Applied Physics Express* 3, (2010): 095101.

Huang, B., Q. M. Yan, Z. Y. Li, and W. H. Duan. Towards graphene nanoribbon-based electronics. *Frontiers of Physics in China* 4, (2009): 269–279.

Huang, J. Y., F. Ding, B. I. Yakobson, P. Lu, L. Qi, and J. Li. In situ observation of graphene sublimation and multilayer edge reconstructions. *Proceedings of the National Academy of Sciences* 106(25), (2009): 10103–10108.

Huang, P. Y., C. S. Ruiz-Vargas, A. M. van der Zande, W. S. Whitney, M. P. Levendorf, J. W. Kevek, S. Garg et al. Grains and grain boundaries in single-layer graphene atomic patchwork quilts. *Nature* 469, (2011): 389–393.

Inai, K., K. Ohya, and T. Ishitani. Simulation study on image contrast and spatial resolution in helium ion microscope. *Journal of Electron Microscopy* 56, (2007): 163–169.

ITRS. The International Technology Roadmap for Semiconductors. http://www.itrs.net/.2009 (accessed on October 2, 2014.)

Jafri, S. H. M., K. Carva, E. Widenkvist, T. Blom, B. Sanyal, J. Fransson, O. Eriksson et al. Conductivity engineering of graphene by defect formation. *Journal of Physics D: Applied Physics* 43, (2010): 045404.

Jepson, M. A. E., B. J. Inkson, X. Liu, L. Scipioni, and C. Rodenburg. Quantitative dopant contrast in the helium ion microscope. *EPL* 86, (2009a): 26005.

Jepson, M. A. E., B. J. Inkson, C. Rodenburg, and D. C. Bell. Dopant contrast in the helium ion microscope. *EPL* 85, (2009b): 46001.

Jia, X. T., J. Campos-Delgado, M. Terrones, V. Meunier, and M. S. Dresselhaus. Graphene edges: A review of their fabrication and characterization. *Nanoscale* 3, (2011): 86–95.

Jiang, D.-E., V. R. Cooper, and S. Dai. Porous graphene as the ultimate membrane for gas separation. *Nano Letters* 9, (2009): 4019–4024.

Jiao, L., L. Zhang, X. Wang, G. Diankov, and H. Dai. Narrow graphene nanoribbons from carbon nanotubes. *Nature* 458, (2009): 877–880.

Jinschek, J. R., E. Yucelen, H. A. Calderon, and B. Freitag. Quantitative atomic 3-D imaging of single/double sheet graphene structure. *Carbon* 49, (2011): 556–562.

Jones, J. D., P. A. Ecton, Y. Mo, and J. M. Perez. Comment on modification of graphene properties due to electron-beam irradiation [*Appl. Phys. Lett.* 94, (2009): 013101]. *Applied Physics Letters* 95, (2009): 246101.

Joy, D. C. and B. J. Griffin. Is microanalysis possible in the helium ion microscope? *Microscopy & Microanalysis* 17, (2011): 643–649.

Joy, D. C., B. J. Griffin, J. Notte, L. Stern, S. McVey, B. Ward, and C. Fenner. Device metrology with high-performance scanning ion beams—Art. no. 65181i. In *Conference on Metrology, Inspection, and Process Control for Microlithography XXI*, San Jose, CA, 2007, pp. I5181–I5181.

Jung, I., M. Pelton, R. Piner, D. A. Dikin, S. Stankovich, S. Watcharotone, M. Hausner, and R. S. Ruoff. Simple approach for high-contrast optical imaging and characterization of graphene-based sheets. *Nano Letters* 7, (2007): 3569–3575.

Katsnelson, M. I. and K. S. Novoselov. Graphene: New bridge between condensed matter physics and quantum electrodynamics. *Solid State Communications* 143, (2007): 3–13.

Kedzierski, J., H. Pei-Lan, P. Healey, P. W. Wyatt, C. L. Keast, M. Sprinkle, C. Berger, and W. A. de Heer. Epitaxial graphene transistors on sic substrates. *Electron Devices, IEEE Transactions on* 55, (2008): 2078–2085.

Kim, K., Z. Lee, W. Regan, C. Kisielowski, M. F. Crommie, and A. Zettl. Grain boundary mapping in polycrystalline graphene. *ACS Nano* 5, (2011): 2142–2146.

Kim, S., J. Nah, I. Jo, D. Shahrjerdi, L. Colombo, Z. Yao, E. Tutuc, and S. K. Banerjee. Realization of a high mobility dual-gated graphene field-effect transistor with Al[sub 2]O[sub 3] dielectric. *Applied Physics Letters* 94, (2009): 062107.

Kim, W. Y. and K. S. Kim. Prediction of very large values of magnetoresistance in a graphene nanoribbon device. *Nature Nanotechnology* 3, (2008): 408–412.

Kochat, V., A. N. Pal, E. S. Sneha, A. Sampathkumar, A. Gairola, S. A. Shivashankar, S. Raghavan, and A. Ghosh. High contrast imaging and thickness determination of graphene with in-column secondary electron microscopy. *Journal of Applied Physics* 110, (2011): 014315.

Kosynkin, D. V., A. L. Higginbotham, A. Sinitskii, J. R. Lomeda, A. Dimiev, B. K. Price, and J. M. Tour. Longitudinal unzipping of carbon nanotubes to form graphene nanoribbons. *Nature* 458, (2009): U872–U875.

Lee, C., X. Wei, J. Kysar, and J. Hone. Measurement of the elastic properties and intrinsic strength of monolayer graphene. *Science* 321, (2008): 385–388.

Leenaerts, O., B. Partoens, and F. M. Peeters. Adsorption of H_2O, NH_3, CO, NO_2, and NO on graphene: A first-principles study. *Physical Review B* 77, (2008): 125416.

Lemme, M., D. Bell, J. Williams, L. Stern, B. Baugher, P. Jarillo-Herrero, and C. Marcus. Etching of graphene devices with a helium ion beam. *ACS Nano* 3, (2009): 2674–2676.

Lemme, M. C., T. J. Echtermeyer, M. Baus, and H. Kurz. A graphene field-effect device. *Electron Device Letters, IEEE* 28, (2007): 282–284.

Li, X., X. Wang, L. Zhang, S. Lee, and H. Dai. Chemically derived, ultrasmooth graphene nanoribbon semiconductors. *Science* 319, (2008a): 1229–1232.

Li, X. L., X. R. Wang, L. Zhang, S. W. Lee, and H. J. Dai. Chemically derived, ultrasmooth graphene nanoribbon semiconductors. *Science* 319, (2008b): 1229–1232.

Li, Y. F., Z. Zhou, P. W. Shen, and Z. F. Chen. Two-dimensional polyphenylene: Experimentally available porous graphene as a hydrogen purification membrane. *Chemical Communications* 46, (2010): 3672–3674.

Liao, L., Y.-C. Lin, M. Bao, R. Cheng, J. Bai, Y. Liu, Y. Qu, K. L. Wang, Y. Huang, and X. Duan. High-speed graphene transistors with a self-aligned nanowire gate. *Nature* 467, (2010): 305–308.

Lin, Y., C. Dimitrakopoulos, K. Jenkins, D. Farmer, H. Chiu, A. Grill, and P. Avouris. 100-Ghz transistors from wafer-scale epitaxial graphene. *Science* 327, (2010): 662.

Liu, L., S. M. Ryu, M. R. Tomasik, E. Stolyarova, N. Jung, M. S. Hybertsen, M. L. Steigerwald, L. E. Brus, and G. W. Flynn. Graphene oxidation: Thickness-dependent etching and strong chemical doping. *Nano Letters* 8, (2008): 1965–1970.

Lucchese, M. M., F. Stavale, E. H. M. Ferreira, C. Vilani, M. V. O. Moutinho, R. B. Capaz, C. A. Achete, and A. Jorio. Quantifying ion-induced defects and raman relaxation length in graphene. *Carbon* 48, (2010): 1592–1597.

Lui, C. H., Z. Q. Li, Z. Y. Chen, P. V. Klimov, L. E. Brus, and T. F. Heinz. Imaging stacking order in few-layer graphene. *Nano Letters* 11, (2011): 164–169.

Morgan, J., J. Notte, R. Hill, and B. Ward. An introduction to the helium ion microscope. *Microscopy Today* 14, (2006): 24–31.

Nastasi, M., J. W. Mayer, and J. K. Hirvonen. *Ion-Solid Interactions: Fundamentals and Applications.* Cambridge University Press, Cambridge, U.K., 1996.

Nemes-Incze, P., G. Magda, K. Kamarás, and L. Biró. Crystallographically selective nanopatterning of graphene on SiO_2. *Nano Research* 3, (2010): 110–116.

Neubeck, S., Y. M. You, Z. H. Ni, P. Blake, Z. X. Shen, A. K. Geim, and K. S. Novoselov. Direct determination of the crystallographic orientation of graphene edges by atomic resolution imaging. *Applied Physics Letters* 97, (2010): 053110.

Nirmalraj, P. N., T. Lutz, S. Kumar, G. S. Duesberg, and J. J. Boland. Nanoscale mapping of electrical resistivity and connectivity in graphene strips and networks. *Nano Letters* 11, (2011): 16–22.

Nolen, C. M., G. Denina, D. Teweldebrhan, B. Bhanu, and A. A. Balandin. High-throughput large-area automated identification and quality control of graphene and few-layer graphene films. *ACS Nano* 5, (2011): 914–922.

Notte, J., R. Hill, S. M. McVey, R. Ramachandra, B. Griffin, and D. Joy. Diffraction imaging in a He+ ion beam scanning transmission microscope. *Microscopy & Microanalysis* 16, (2010): 599–603.

Notte, J., B. Ward, N. Economou, R. Hill, R. Percival, L. Farkas, and S. McVey. An introduction to the helium ion microscope. In *International Conference on Frontiers of Characterization and Metrology for Nanoelectronics*, Gaithersburg, MD, 2007, pp. 489–496.

Novoselov, K. S., A. K. Geim, S. V. Morozov, D. Jiang, Y. Zhang, S. V. Dubonos, I. V. Grigorieva, and A. A. Firsov. Electric field effect in atomically thin carbon films. *Science* 306, (2004): 666–669.

Pandey, D., R. Reifenberger, and R. Piner. Scanning probe microscopy study of exfoliated oxidized graphene sheets. *Surface Science* 602, (2008): 1607–1613.

Park, M. H., T. H. Kim, and C. W. Yang. Thickness contrast of few-layered graphene in SEM. *Surface and Interface Analysis* 44, (2012): 1538–1541.

Peres, N., M.R. Graphene, new physics in two dimensions. *Europhysics News* 40, (2009): 17–20.

Petrov, Y. V., O. F. Vyvenko, and A. S. Bondarenko. Scanning helium ion microscope: Distribution of secondary electrons and ion channeling. *Journal of Surface Investigation-X-Ray Synchrotron and Neutron Techniques* 4, (2010): 792–795.

Pickard, D., B. Oezyilmaz, J. Thong, K. P. Loh, V. Viswanathan, A. hongkai, S. Mathew et al. Graphene nanoribbons fabricated by helium ion microscope. *APS March Meeting 2010* (2010): http://adsabs.harvard.edu/abs/2010APS.MARH21008P.

Pickard, D. and L. Scipioni. Graphene nano-ribbon patterning in the orion plus. *Carl Zeiss* (2009). Application Note.

Pinto, H. et al. P-type doping of graphene with F4-TCNQ. *Journal of Physics: Condensed Matter* 21, (2009): 402001.

Postek, M. T., A. E. Vladar, and J. Kramar. The helium ion microscope: A new tool for nanomanufacturing—Art. No. 664806. In *Conference on Instrumentation, Metrology and Standards for Nanomanufacturing*, San Diego, CA, 2007a, pp. 64806–64806.

Postek, M. T., A. E. Vladar, J. Kramar, L. A. Stern, J. Notte, and S. Mcvey. Helium ion microscopy: A new technique for semiconductor metrology and nanotechnology. *Frontiers of Characterization and Metrology for Nanoelectronics: 2007* 931, (2007b): 161–167.

Postek, M. T. and A. E. Vladar. Helium ion microscopy and its application to nanotechnology and nanometrology. *Scanning* 30, (2008): 457–462.

Postek, M. T., A. Vladár, C. Archie, and B. Ming. Review of current progress in nanometrology with the helium ion microscope. *Measurement Science and Technology* 22, (2011): 024004.

Rahman, F. H. M., S. McVey, L. Farkas, J. A. Notte, S. D. Tan, and R. H. Livengood. The prospects of a subnanometer focused neon ion beam. *Scanning* 34, (2012): 129–134.

Reunchan, P. and S.-H. Jhi. Metal-dispersed porous graphene for hydrogen storage. *Applied Physics Letters* 98, (2011): 093103.

Ritter, M. and P. A. Midgley. A practical approach to test the scope of FIB-SEM 3D reconstruction. *Journal of Physics: Conference Series* 241, (2010): 012081.

Romero, H. E., N. Shen, P. Joshi, H. R. Gutierrez, S. A. Tadigadapa, J. O. Sofo, and P. C. Eklund. N-type behavior of graphene supported on Si/SiO_2 substrates. *ACS Nano* 2, (2008): 2037–2044.

Rudneva, M., E. van Veldhoven, S. K. Malladi, D. Maas, and H. W. Zandbergen. Novel nanosample preparation with a helium ion microscope. *Journal of Materials Research* 28, (2013): 1013–1020.

San-Jose, P., E. Prada, E. McCann, and H. Schomerus. Pseudospin valve in bilayer graphene: Towards graphene-based pseudospintronics. *Physical Review Letters* 102, (2009): 247204.

Santos-Martinez, M. J., I. Inkielewicz-Stepniak, C. Medina, K. Rahme, D. M. D'Arcy, D. Fox, J. D. Holmes, H. Zhang, and M. W. Radomski. The use of quartz crystal microbalance with dissipation (QCM-D) for studying nanoparticle-induced platelet aggregation. *International Journal of Nanomedicine* 7, (2012): 243–255.

Schedin, F., A. K. Geim, S. V. Morozov, E. W. Hill, P. Blake, M. I. Katsnelson, and K. S. Novoselov. Detection of individual gas molecules adsorbed on graphene. *Nature Materials* 6, (2007): 652–655.

Schneider, G. G. F., S. W. Kowalczyk, V. E. Calado, G. g. Pandraud, H. W. Zandbergen, L. M. K. Vandersypen, and C. Dekker. DNA translocation through graphene nanopores. *Nano Letters* 10, (2010): 3163–3167.

Schwierz, F. Graphene transistors. *Nature Nanotechnology* 5, (2010): 487–496.

Scipioni, L. Recent applications development with the helium ion microscope. *Microscopy & Microanalysis* 14(S2), (2008): 1224–1225.

Scipioni, L. Nano-machining of graphene in the orion plus. *Carl Zeiss*, (2009). Application Note.

Scipioni, L., C. A. Sanford, J. Notte, B. Thompson, and S. McVey. Understanding imaging modes in the helium ion microscope. *Journal of Vacuum Science & Technology B* 27, (2009a): 3250–3255.

Scipioni, L., L. A. Stern, J. Notte, S. Sijbrandij, and B. Griffin. Helium ion microscope. *Advanced Materials & Processes* 166, (2008): 27–30.

Scipioni, L., W. Thompson, S. Sijbrandij, S. Ogawa, and IEEE. Material analysis with a helium ion microscope. *47th Annual International Reliability Physics Symposium*, Montreal, IEEE international. (2009b). DOI: 10.1109/IRPS.2009.5173271.

Shang, N., P. Papakonstantinou, P. Wang, and S. R. P. Silva. Platinum integrated graphene for methanol fuel cells. *The Journal of Physical Chemistry C* 114, (2010): 15837–15841.

Shao, Y. Y., J. Wang, H. Wu, J. Liu, I. A. Aksay, and Y. H. Lin. Graphene based electrochemical sensors and biosensors: A review. *Electroanalysis* 22, (2010): 1027–1036.

Sidorkin, V., E. van Veldhoven, E. van der Drift, P. Alkemade, H. Salemink, and D. Maas. Sub-10-nm nanolithography with a scanning helium beam. *Journal of Vacuum Science & Technology B* 27, (2009): L18–L20.

Sijbrandij, S., B. Thompson, J. Notte, B. W. Ward, and N. P. Economou. Elemental analysis with the helium ion microscope. In *52nd International Conference on Electron, Ion and Photon Beam Technology and Nanofabrication*, Portland, OR, 2008, pp. 2103–2106.

Son, Y.-W., M. L. Cohen, and S. G. Louie. Half-metallic graphene nanoribbons. *Nature* 444, (2006): 347–349.

Srinivas, G., Y. W. Zhu, R. Piner, N. Skipper, M. Ellerby, and R. Ruoff. Synthesis of graphene-like nanosheets and their hydrogen adsorption capacity. *Carbon* 48, (2010): 630–635.

Suenaga, K. and M. Koshino. Atom-by-atom spectroscopy at graphene edge. *Nature* 468, (2010): 1088–1090.

Sutter, E., D. P. Acharya, J. T. Sadowski, and P. Sutter. Scanning tunneling microscopy on epitaxial bilayer graphene on ruthenium (0001). *Applied Physics Letters* 94, (2009): 133101.

Tan, Y. W., Y. Zhang, K. Bolotin, Y. Zhao, S. Adam, E. H. Hwang, S. Das Sarma, H. L. Stormer, and P. Kim. Measurement of scattering rate and minimum conductivity in graphene. *Physical Review Letters* 99, (2007): 246803.

Tang, B., G. X. Hu, and H. Y. Gao. Raman spectroscopic characterization of graphene. *Applied Spectroscopy Reviews* 45, (2010): 369–407.

Tapaszto, L., G. Dobrik, P. Nemes-Incze, G. Vertesy, P. Lambin, and L. P. Biro. Tuning the electronic structure of graphene by ion irradiation. *Physical Review B* 78, (2008): 233407.

Temprano, I., G. Goubert, G. Behan, H. Zhang, and P. H. McBreen. Spectroscopic and structural characterization of the formation of olefin metathesis initiating sites on unsupported [Small Beta]-Mo2C. *Catalysis Science & Technology* 1, (2011): 1449–1455.

Terrones, M., A. R. Botello-Mendez, J. Campos-Delgado, F. Lopez-Urias, Y. I. Vega-Cantu, F. J. Rodriguez-Macias, A. L. Elias et al. Graphene and graphite nanoribbons: Morphology, properties, synthesis, defects and applications. *Nano Today* 5, (2010): 351–372.

Teweldebrhan, D. and A. A. Balandin. Modification of graphene properties due to electron-beam irradiation. *Applied Physics Letters* 94, (2009): 013101.

Veligura, V., G. Hlawacek, R. van Gastel, H. J. W. Zandvliet, and B. Poelsema. Channeling in helium ion microscopy: Mapping of crystal orientation. *Beilstein Journal of Nanotechnology* 3, (2012): 501–506.

Verma, V. P., S. Das, I. Lahiri, and W. Choi. Large-area graphene on polymer film for flexible and transparent anode in field emission device. *Applied Physics Letters* 96, (2010): 203108.

Villinger, C., H. Gregorius, C. Kranz, K. Höhn, C. Münzberg, G. Wichert, B. Mizaikoff, G. Wanner, and P. Walther. FIB/SEM tomography with TEM-like resolution for 3D imaging of high-pressure frozen cells. *Histochemistry and Cell Biology* 138, (2012): 549–556.

Wakabayashi, K., C. Pierre, D. A. Dikin, R. S. Ruoff, T. Ramanathan, L. C. Brinson, and J. M. Torkelson. Polymer–graphite nanocomposites: Effective dispersion and major property enhancement via solid-state shear pulverization. *Macromolecules* 41, (2008): 1905–1908.

Wang, X. and H. Dai. Etching and narrowing of graphene from the edges. *Natural Chemistry* 2, (2010): 661–665.

Wang, X., L. Zhi, and K. Mullen. Transparent, conductive graphene electrodes for dye-sensitized solar cells. *Nano Letters* 8, (2007): 323–327.

Ward, B., J. Notte, and N. Economou. Helium-ion microscopy. In *Photonics Spectra*. 8, (2007). http://www.photonics.com/Article.aspx?PID=5&VID=21&IID=144&Tag=Features&AID=30461 (accessed October 2, 2014.).

Ward, B. W., J. A. Notte, and N. P. Economou. Helium ion microscope: A new tool for nanoscale microscopy and metrology. In *50th International Conference on Electron, Ion, and Photon Beam Technology and Nanofavrication* Baltimore, MD, 2006, pp. 2871–2874.

Warner, J. H., M. H. Rummeli, T. Gemming, B. Buchner, and G. A. D. Briggs. Direct imaging of rotational stacking faults in few layer graphene. *Nano Letters* 9, (2009): 102–106.

Williams, D. B. and C. B. Carter. Lenses, apertures and resolution. In *Transmission Electron Microscopy: A Textbook for Materials Science*, 2009, pp. 91–114. Springer, New York.

Winston, D., B. M. Cord, B. Ming, D. C. Bell, W. F. DiNatale, L. A. Stern, A. E. Vladar et al. Scanning-helium-ion-beam lithography with hydrogen silsesquioxane resist. *Journal of Vacuum Science & Technology B* 27, (2009): 2702–2706.

Winston, D., J. Ferrera, L. Battistella, A. E. Vladar, and K. K. Berggren. Modeling the point-spread function in helium-ion lithography. *Scanning* 34, (2012): 121–128.

Winston, D., V. R. Manfrinato, S. M. Nicaise, L. L. Cheong, H. Duan, D. Ferranti, J. Marshman et al. Neon ion beam lithography (NIBL). *Nano Letters* 11, (2011): 4343–4347.

Wirtz, T., N. Vanhove, L. Pillatsch, D. Dowsett, S. Sijbrandij, and J. Notte. Towards secondary ion mass spectrometry on the helium ion microscope: An experimental and simulation based feasibility study with He$^+$ and Ne$^+$ bombardment. *Applied Physics Letters* 101, (2012): 041601.

Wurstbauer, U., C. Roling, W. Wegscheider, M. Vaupel, P. H. Thiesen, and D. Weiss. Imaging ellipsometry of graphene. *Applied Physics Letters* 97, (2010): 231901.

Yang, J. J., D. C. Ferranti, L. A. Stern, C. A. Sanford, J. Huang, Z. Ren, L. C. Qin, and A. R. Hall. Rapid and precise scanning helium ion microscope milling of solid-state nanopores for biomolecule detection. *Nanotechnology* 22, (2011): 285310.

Yazyev, O. V. and L. Helm. Defect-induced magnetism in graphene. *Physical Review B* 75, (2007): 125408.

Yuan, C. and Z. Xiaowen. Focused ion beam technology and application in failure analysis. *Electronic Packaging Technology & High Density Packaging (ICEPT-HDP), 2010 11th International Conference on* 2010, pp. 957–960.

Zhang, H. Z., R. M. Wang, L. P. You, J. Yu, H. Chen, D. P. Yu, and Y. Chen. Boron carbide nanowires with uniform CNX coatings. *New Journal of Physics* 9, (2007): 1–9.

Zhou, Y. B., Z. M. Liao, Y. F. Wang, G. S. Duesberg, J. Xu, Q. Fu, X. S. Wu, and D. P. Yu. Ion irradiation induced structural and electrical transition in graphene. *Journal of Chemical Physics* 133, (2010): 234703.

Zhou, Y. and K. P. Loh. Making patterns on graphene. *Advanced Materials* 22, (2010): 3615–3620.

Ziegler, D., P. Gava, J. Guttinger, F. Molitor, L. Wirtz, M. Lazzeri, A. M. Saitta, A. Stemmer, F. Mauri, and C. Stampfer. Variations in the work function of doped single- and few-layer graphene assessed by kelvin probe force microscopy and density functional theory. *Physical Review B* 83, (2011): 235434.

Ziegler, J. F. SRIM-2003. *Nuclear Instruments and Methods in Physics Research Section B: Beam Interactions with Materials and Atoms* 219–220, (2004): 1027–1036.

Chapter 9

Two-Dimensional Metal Oxide Nanosheets

Novel Building Blocks for Advanced Functional Materials

Delai Ye, Hua Yu, and Lianzhou Wang

CONTENTS

9.1 INTRODUCTION

Two-dimensional (2D) nanosheets, because of their ultrasmall thickness in atomic or molecular range and infinite lateral dimensions, have been widely considered as the thinnest functional materials [1]. The original work of this 2D materials can be dated back to the 1950s, when some clay minerals dispersed well in water, and a colloidal suspension was yielded due to the spontaneous exfoliation [2,3]. Shortly after that, with the appropriate selection of interlayer cations and solvents, a variety of inorganic layered compounds have been delaminated successfully, including metal chalcogenides [4–6], metal phosphates and phosphonates [7,8], metal oxides [1,9–17], and hydroxides [18–24]. Different from the spontaneous exfoliation of the clay minerals, these layered compounds need an extra modification of the chemical compositions or interlayer species displacement to weaken the interlayer charge density before the delamination. Particularly, in the past decades, some layered perovskite-type oxides, titanium oxide, and manganese oxides have been exfoliated successfully into 2D nanosheets with new chemical, physical, and/or optical properties [1,12–17].

Induced by the quantum size effect and unique 2D shape, these nanosheets possess unique functionalities in physics, chemistry, and other areas that are rarely seen in the bulk materials [25–28].

More importantly, these nanosheets are charge bearing on the surface that facilitates them, as architectural building blocks, to be fabricated into a variety of new nanostructures, including self-assembled thin films, rolling-up tubes, and hollow nanocapsules [25,26]. Thus, sophisticated nanodevices can be rationally designed by combining certain nanosheets and heterogeneous species with precise control at the atomic scale, which broadens their potential application in a wide range.

In the family of nanosheets, metal oxide nanosheets are of particular interest due to their structural diversity and distinguished electronic properties, which make them promising candidates in many areas ranging from catalysis to electronics [29,30]. In this chapter, we will start with the exfoliation process and characterization of various metal oxide nanosheets, followed by their assembly and applications as building blocks. Some recent progress of their applications for energy harvesting and energy storage will be highlighted. Finally, their potential applications will be summarized.

9.2 PREPARATION OF METAL OXIDE NANOSHEETS

Naturally, some layered materials with low interlayer interaction can be delaminated into single sheets spontaneously in aqueous suspension [31]. And the peeling-off process of graphite layer during writing with pencils is another good demonstration of layered compound delamination. However, it is a much more complicated process to get most of the metal oxide nanosheets due to their much higher layer charge density [26].

To date, a variety of layered metal oxides, such as $Cs_{0.7}Ti_{1.825}O_4$ [1,13], $K_{0.45}MnO_2$ [32–34], $K_4Nb_6O_{17}$ [35], and $RbTaO_3$ [36], have been delaminated into single sheets, and all of these metal compounds are made up of negatively charged MO_6 (M = Ti, Mn, Nb, Ta) octahedral slabs and alkali metal ions (K^+, Rb^+, and Cs^+) between the transition metal layers. Based on the fact that the interlayer cations of these layered compounds are exchangeable, which facilitates the chemical modification of the interlayer space at room temperature, a multistep exfoliation process of these layered metal oxides has been well established.

Figure 9.1 shows a typical soft-chemistry exfoliation procedure to prepare metal oxide nanosheets [37]. Generally, the layered parent compounds are synthesized via a high-temperature (800°C–1300°C) solid-state calcination process. Subsequently, acidic solutions are applied, and most of the interlayer alkali metal cations are replaced with hydrion, obtaining the protonic form of these layered compounds, for example, $H_{0.7}Ti_{1.825}O_4 \cdot H_2O$, $H_{0.13}MnO_2 \cdot 0.7H_2O$, $K_{0.8}H_{3.2}Nb_6O_{17} \cdot nH_2O$, $Rb_{0.1}H_{0.9}TaO_3 \cdot 1.3H_2O$. After that, these protonic oxides are treated with aqueous solutions containing bulky organic ions, for example, the tetrabutylammonium (TBA^+) cations at suitable concentrations, and the interlayer gallery goes through a swelling behavior in a stepwise manner, that is, mono-, bi-, or trilayer solvents intercalated into the gallery [38]. As a result, the electrostatic interaction between the host layers and the interlayer cations is substantially weakened, and with the help of external forces like mechanical shaking, the negatively charged metal oxide nanosheets, $Ti_{0.91}O_2^{0.36-}$ [1,13], $MnO_2^{0.4-}$ [24,32,39], $Nb_6O_{17}^{4-}$ [35], and TaO_3^- [36], are finally derived.

An electrical double layer is formed on the surface of the nanosheets during the swelling and exfoliation process, and the negative charges covered by this double layer provide enough electrostatic repulsion force between the nanosheets, obtaining a colloidal suspension that can keep stable for a long period. Normally, the thickness of the nanosheets depends on the thickness of the

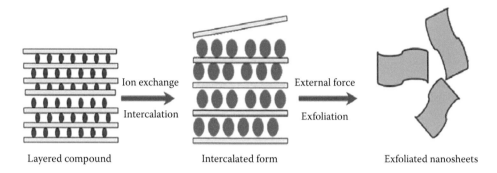

Layered compound   Intercalated form   Exfoliated nanosheets

Figure 9.1 Illustration of the general exfoliation procedures of layered compounds. (From Wang, L. et al., *Bio-Inspired Mater. Synth.*, 99, 2011, Chapter 4. With permission.)

parent host layers, while the lateral size is controllable and flexible. Since nanosheets are resulted from delamination of layered compounds, which consist of platy microcrystals varying from one to a few micrometers in the 2D dimension, their lateral size is always between hundreds of nanometers and a few micrometers when the effect of mechanical shaking is considered [1,33,40].

9.3 CHARACTERIZATION OF METAL OXIDE NANOSHEETS

To better understand the exfoliation process and characterize the resulting lamellar nanosheets, a series of modern techniques, including x-ray diffraction (XRD), transmission electron microscopy (TEM), atomic force microscopy (AFM), and x-ray absorption fine-structure spectroscopy have been employed.

Before the final exfoliation, the layered parent compounds generally experience a highly swollen phase with substantially enlarged interlayer space, which can be well identified using the XRD techniques [1,33]. A series of Bragg peaks could gradually shift to a lower angular region when the swelling condition is progressively intensified until any ordering in c-axis of the layered compounds disappears, indicating the ultimate separation of the nanosheets. In consistence with the clay minerals, the swelling degree of the metal oxide compounds is inversely related to the electrolyte concentration [1,33]. That's because the intercalation of big organic cations like tetrabutylammonium ions (TBA^+) involves a fast neutralization process of the alkaline TBA^+OH^- with the interlayer H^+, and electrical double layers on both sides of the host slabs are formed in this step to balance the interlayer charge. Therefore, lower TBA^+ concentrations mean more volume of solution has to diffuse into the interlayer space to keep the charge neutrality and subsequently results into osmotic hydration with larger interlayer gallery. Figure 9.2 is good evidence to this phenomenon using the XRD patterns of $H_{0.13}MnO_2 \cdot 0.7H_2O$ treated with different concentrations of TBA^+OH^-, strongly indicating the dominated role of these organic species in controlling the swelling/exfoliation process [13]. Once the swelling degree as well as the interlayer expansion is high enough, the exfoliation can be achieved easily with the presence of certain external mechanical forces. After the exfoliation, the nanosheets dispersed in the colloidal suspension can be well deposited onto microscopic substrates, which allow direct observation of their unilamellar feature by TEM and AFM. Figure 9.3 is a TEM image as well as the electron diffraction (ED) of titanium oxide nanosheets [41].

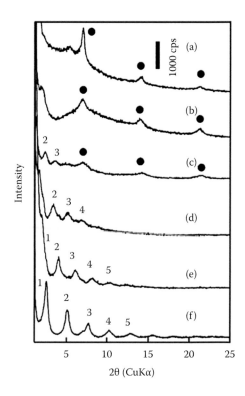

Figure 9.2 XRD patterns for a colloidal aggregate centrifuged from an aqueous mixture of $H_{0.13}MnO_2 \cdot 0.7H_2O$ and TBAOH. The molar ratios of TBA/H are (a) 5, (b) 10, (c) 25, (d) 50, (e) 70, and (f) 100. Numerals next to the peaks represent the order of basal reflections attributable to osmotic swelling. Circles denote the basal reflection from the TBA-intercalated phase (1.25 nm). (Reprinted with permission from Omomo, Y., Sasaki, T., Wang, L., and Watanabe, M., Redoxable nanosheet crystallites of MnO_2 derived via delamination of a layered manganese oxide, *J. Am. Chem. Soc.*, 125, 2003, 3568–3575. Copyright 2003 American Chemical Society.)

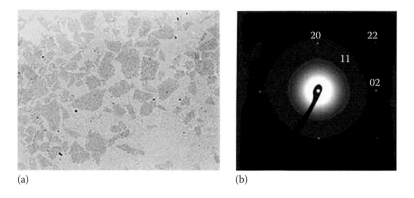

Figure 9.3 Transmission electron micrograph of (a) $Ti_{1-x}O_2^{4\delta-}$ crystallites and (b) diffraction patterns from a single nanosheet. (Reprinted with permission from Sasaki, T., Ebina, Y., Kitami, Y., Watanabe, M., and Oikawa, T., Two-dimensional diffraction of molecular nanosheet crystallites of titanium oxide, *J. Phys. Chem. B*, 105, 2001, 6116–6121. Copyright 2001 American Chemical Society.)

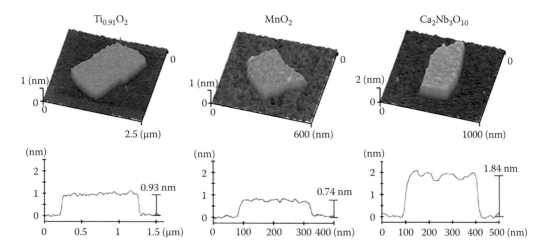

Figure 9.4 AFM images of $Ti_{0.91}O_2$, MnO_2, and $Ca_2Nb_3O_{10}$ nanosheets. Height profiles are shown in the bottom panels. (Osada, M. and Sasaki, T., Exfoliated oxide nanosheets: New solution to nanoelectronics, *J. Mater. Chem.*, 19, 2009, 2503–2511. Copyright 2009 Reproduced by permission of The Royal Society of Chemistry.)

The nanosheets with random shapes lying on the substrate have a very faint but homogeneous contrast, implying their ultrathin thickness and uniform distribution in the aqueous suspension. The ED data taken from a single nanosheet often display a symmetric spot pattern (Figure 9.3b), indicating the good retaining of the 2D atomic arrangement inherited from the parent layered compounds [41,42]. However, TEM can only qualitatively indicate the ultrathin nature of the nanosheets, while AFM is more powerful in identifying the unilamellar nature by providing a quantitative estimate of the nanosheet thickness. Figure 9.4 shows the typical AFM images of $Ti_{0.91}O_2^{0.36-}$, $MnO_2^{0.4-}$, and $Ca_2Nb_3O_{10}^{-}$ with the thicknesses of 1.12 ± 0.07 nm, 0.77 ± 0.05 nm, and 2.30 ± 0.09 nm, respectively, all of which are comparable to the host layers of the parent compounds if the surface adsorption of solvents or guest species on nanosheets is considered [26].

9.4 ASSEMBLY OF METAL OXIDE NANOSHEETS

Metal oxide nanosheets bear a negative charge on both surfaces and are always dispersed in an aqueous colloidal suspension. Besides, they are extremely high in 2D anisotropy with approximately 1 nm in thickness but submicrometers to 100 μm in lateral size. All these features enable them to be assembled and tailored into a variety of new functional nanomaterials by solution-based techniques.

Some macroscale assembling methods of the nanosheets have been successfully applied in the earlier stage. For example, freeze-drying has been used to restack a limited number of titanium oxide nanosheets and get a very thin lamella [43]. Besides, some conventional wet-state film fabrication methods including spin coating, dip coating, and electrophoretic deposition have also been employed to prepare highly oriented films parallel to the substrate surface with these nanosheets [44–46]. To further precisely control the formation and properties of the nanosheet-based composites, some novel methods have been well developed recently as illustrated in Figure 9.5 [47].

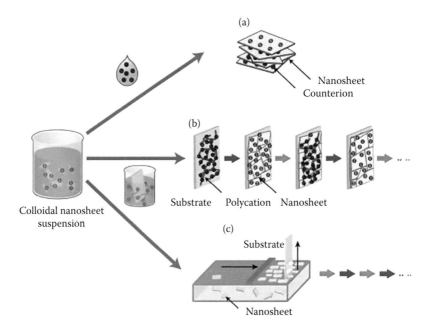

Figure 9.5 Schematic assembling processes of (a) flocculation, (b) electrostatic sequential adsorption, and (c) Langmuir–Blodgett for nanosheets. (From Ma, R. and Sasaki, T.: Nanosheets of oxides and hydroxides: Ultimate 2D charge-bearing functional crystallites. *Adv. Mater.* 22. 5082–5104. 2010. Copyright Wiley-VCH Verlag GmbH & Co. KGaA, Weinheim, Germany. Reproduced with permission.)

Generally, the flocculation and layer-by-layer (LBL) assembly are very effective in organizing oxide nanosheets into various new nanostructures with a wide range of interlayer counterions, such as organic polyelectrolytes, metal complexes, clusters, and sometimes metal hydroxide nanosheets with positively charged surfaces [39,48–54]. Flocculation is commonly applied for mesoporous lamellar solid preparation, while the LBL assembly (electrostatic sequential adsorption and Langmuir–Blodgett deposition) is preferable to build up multilayer nanofilms on a substrate or core–shell nanostructures, both of which will be discussed in the following part.

9.4.1   Flocculation

With the presence of negative charges on both sides of the surface, flocculation of the metal oxide nanosheets can be easily realized by mixing them with some foreign cations, generally polycations [55]. Induced by the interaction of different charges, the suspended nanosheets immediately become unstable and randomly restack into 3D crystal structures with the addition of electrolytes.

Through this method, a number of functional species can be easily incorporated into the nanosheet galleries to obtain new nanocomposites that are difficult to be prepared through traditional methods, for instance, ion exchange. These new composites are characteristic of some unique properties including disordered lamellar structure, homogeneously dispersed interlayer species, high surface area, and porosity, all of which offer them great opportunities in many areas, such as catalysis, photocatalysis, and electrochemistry [55–58].

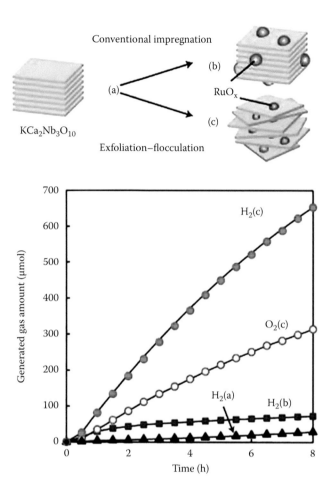

Figure 9.6   Photocatalytic activity of $KCa_2Nb_3O_{10}$ with RuO_x. (Reprinted with permission from Ebina, Y., Sakai, N., and Sasaki, T., Photocatalyst of lamellar aggregates of RuO_x-loaded perovskite nanosheets for overall water splitting, *J. Phys. Chem. B*, 109, 2005, 17212–17216. Copyright 2005 American Chemical Society.)

The formation of flocculated $KCa_2Nb_3O_{10} \cdot nH_2O$ using $Ca_2Nb_3O_{10}^{-}$ nanosheets and a mixed aqueous solution of KCl with trace amount of $RuCl_3$ is a good example to represent the advantages of designing new nanocomposites via this method as depicted in Figure 9.6 [59–61]. Being heat treated at 500°C, the Ru ions homogeneously dispersed between the nanosheets are transferred into RuO_x and enable the restacked $Ca_2Nb_3O_{10}^{-}$ nanosheets to stoichiometrically decompose pure water under UV light, which can't be observed in micrometer $KCa_2Nb_3O_{10}$ particles that are incorporated with RuO_x only on the surface by conventional impregnation.

In some cases, more than one layer of foreign species can be incorporated between the nanosheets if some larger polyoxocations like Al_{13} Keggin ion ($[AlO_4Al_{12}(OH)_{24}(H_2O)_{12}]^{7+}$) are applied with high concentration. For example, TiO_2 [62,63] and MnO_2 [55] nanosheets restacked with Al_{13} Keggin ions can attract a large amount of such ions on both sides; as a result, double layers of Al_{13} Keggin ions are intercalated between the nanosheets, and a novel pillared structure with enhanced mesoporosity, higher surface area, and approximately double interlayer space of the normal monolayer pillar structure is generated during the quick flocculation.

Besides the common polycations, another type of flocculating agent, the metal hydroxide nanosheets, that bears a positive charge on the surface is also popularly employed. Two types of nanosheets with oppositely charged surfaces are restacked on each other due to the neutralization of the different surface charges and yield a lamellar architecture with regular basal spacing, for instance, 1.2 and 2.0 nm for the flocculated solids of $[Mg_{2/3}Al_{1/3}(OH)_2]^{1/3+}$ and oxide nanosheets ($Ti_{0.91}O_2^{0.36-}$ and $Ca_2Nb_3O_{10}^-$), respectively [54]. Both of the basal spacing values are roughly consistent with the sum of the thickness of positively charged $[Mg_{2/3}Al_{1/3}(OH)_2]^{1/3+}$ and negatively charged $Ti_{0.91}O_2^{0.36-}$ or $Ca_2Nb_3O_{10}^-$, separately, indicating their alternately restacking manner.

9.4.2   Layer-by-Layer Fabrication

In many cases, metal oxide nanosheets are fabricated into multilayer thin films, which widely broaden their applications in both scientific research and practical approaches. As mentioned previously, some traditional solution-based processes, for example, the spin-coating technique, are efficient in preparing thin films with these nanosheets; however, to precisely control the film thickness in around 1 nm scale and the layer number of the nanosheets, no one can do the job as well as the LBL assembly, which generally refers to the electrostatic sequential deposition and the Langmuir–Blodgett (LB) deposition technique. More importantly, various types of organic and inorganic intersheet species with unique functionalities can be intentionally selected to interact with the nanosheets during the LBL process, thus offering the nanosheet-based materials more interesting properties and opportunities in both scientific study and practical applications.

The electrostatic sequential deposition firstly introduced by Decher has been demonstrated to be a very powerful tool for fabricating metal oxide nanosheets into multilayer thin films due to their charged surface, ultrathin thickness, and colloidal nature [64]. Simply, a pre-treated substrate like quartz glass, conducting glass, or silicon wafer with negatively charged surface is dipped into a cationic polyelectrolyte solution, such as polyethyleneimine (PEI) or poly (diallyldimethylammonium chloride) (PDDA) as a start. The polycations are adsorbed onto the negatively charged substrate surface immediately due to the electrostatic attraction, resulting into reverse charges on the surface of the substrate. Generally, only one layer of polycations can be loaded in this step and further adsorption over the saturated limit will be electrostatically repelled. A thorough washing with water is often followed to ensure the monolayer adsorption. After that, the substrate surface is deposited with a layer of nanosheets when being dipped in the nanosheet colloidal suspension, and meanwhile, the surface charges reverse again. As these two steps alternatively continue, a multilayer thin film is gradually built up consisting of this periodic bilayer as the basic building blocks with fixed thickness. Some characterization techniques, such as UV–Vis absorption and XRD, are very useful to confirm the stepwise growth of the thin film. For example, as can be seen in Figure 9.7, the electrostatic sequential deposition of $Ti_{0.91}O_2^{0.36-}$ and $MnO_2^{0.4-}$ nanosheets with PDDA into multilayer thin film is well monitored by the linear enhancement of the absorption band as the deposition cycle increases [39,65]. Also the periodic deposition of metal oxide nanosheets can be reflected in the XRD patterns from the gradually intensified Bragg peaks.

Currently, almost all the metal oxide nanosheets have been applied to fabricate multilayer thin films, and in most of the cases, PDDA plays the role of nanosheet glue to prepare the PDDA/nanosheet nanocomposites, such as $PDDA/Ti_{0.91}O_2^{0.36-}$ and $PDDA/MnO_2^{0.4-}$ [39,65]. In addition,

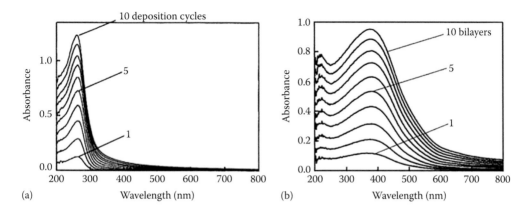

Figure 9.7 (a) UV–Vis absorption spectra in the multilayer buildup process of the titania nanosheets and a polydiallyldimethylammonium (PDDA) solution. UV–Vis absorption spectra monitoring sequential deposition of nanosheets. (Reprinted with permission from Ebina, Y., Tanaka, T., Harada, M., Watanabe, M., and Decher, G., Layer-by-layer assembly of titania nanosheet/polycation composite fims, *Chem. Mater.*, 13, 2001, 4661–4667. Copyright 2001 American Chemical Society.) (b) UV-vis absorption spectra of multilayer films of PEI/MnO$_2$/(PDDA/MnO$_2$)$_{n-1}$ prepared on quartz glass substrate. (Reprinted with permission from Wang, L., Omomo, Y., Sakai, N., Fukuda, K., Nakai, I., Ebina, Y., Takada, K., Watanabe, K., Sasaki, T., Fabrication and characterization of multilayer ultrathin fims of exfoliated MnO$_2$ nanosheets and polycations, *Chem. Mater.*, 15, 2003, 2873–2878. Copyright 2003 American Chemical Society.)

some inorganic species like Al$_{13}$ Keggin ions instead of organic PDDA can also deposit nanosheets into multilayer thin films, thus obtaining pure inorganic thin films.

Recently, another significant progress in nanosheet LBL assembly is the preparation of hollow sphere structures using these oxide nanosheets as nanocoatings on a removable core. These inorganic hollow sphere structures are of particular interest because of their low density, high surface area, and stability. The typical fabrication process starts with the nanosheets coating on a sacrificial template, for example, polymers, in an electrostatic sequential deposition manner [66]. Note that the nanosheets are extremely thin and flexible, which significantly facilitates the wrapping and replication of the spherical template. After a number of desired depositions, the core–shell structures have been achieved, which evolve into highly crystallized hollow spheres with precisely controlled shell thickness in a subsequent calcination when the polymer cores are thermally decomposed. To date, some metal oxide nanosheets including Ti$_{0.91}$O$_2$$^{0.36-}$ [55], MnO$_2$$^{0.4-}$ [67], and TaO$_3$$^-$ [68] have been fabricated into ultrathin hollow shells. Applications have been found in electronics, catalysis, drug delivery, etc. [69,70], and more intriguing potentials are expected.

Electrostatic sequential deposition has been demonstrated to be efficient in building thin films with nanosheets; however, the products are not so satisfactory because of the high density of overlaps and gaps between the nanosheets inevitably formed during the random nanosheet deposition. The Langmuir–Blodgett method is another novel technique for multilayer thin-film deposition with nanosheets, involving the loading of a floating monolayer on a liquid surface in a Langmuir trough and a following compression of the air–water interface, which significantly enhances the dense lateral packing while the overlaps and gaps are greatly prevented, thus making high-quality nanofilms.

Generally, to disperse the metal oxide nanosheets homogeneously on the water surface, some cationic surfactants like dioctadecyldimethylammonium are used in the nanosheets' colloidal

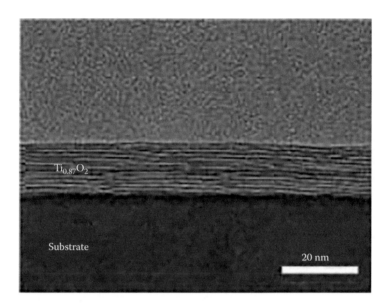

Figure 9.8   Cross-sectional TEM image of the 10-layer film. A lamellar fringe corresponding to stacked nanosheets is clearly resolved. (Reprinted with permission from Akatsuka, K., Haga, M.-A., Ebina, Y., Osada, M., Fukuda, K., Sasaki, T., Construction of highly ordered Lamellar nanostructures through Langmuir–Blodgett deposition of molecularly thin titania nanosheets tens of micrometers wide and their excellent dielectric properties, *ACS Nano*, 3, 1097–1106, Copyright 2009 American Chemical Society.)

suspension prior to the assembly and transfer of nanosheets at a high density onto a dipping substrate [48,71]. For example, the titanium oxide nanosheets have been proved to be able to adhere to amphiphilic ammonium cations at the air–water interface due to electrostatic attraction, thus floating on the interface to enable the normal LB procedure for nanosheet film fabrication [71]. The TEM image in Figure 9.8 shows an ultrathin film of oversized TiO_2 nanosheets neatly and regularly deposited on a substrate via the LB fashion [72]. The lamellar fringe of the nanosheet film can be clearly observed, indicating a highly ordered structure.

9.5   PROPERTIES AND APPLICATIONS OF METAL OXIDE NANOSHEETS

As discussed previously, a range of new nanocomposites based on the fabrication of nanosheets and interlayer counterparts have emerged with well-controlled composition and structure. Meanwhile, the family of nanosheets keeps growing steadily. For example, a recent superstar, graphene, which is an atomic layer of carbon nanosheet that firstly delaminated from graphite [73], has been incorporated with some metal oxide nanosheets, getting new nanomaterials with promising electronic and optoelectronic properties. Because of such a huge diversity, the nanosheet-based nanocomposites with new physical and chemical properties have shown great potential in broad areas. Especially, renewable and clean energy utilization has become one of the most important aspects in the new century [74]. It requires not only high-efficiency energy conversion devices but also advanced-energy storage alliances, such as photocatalytic devices, solar cells, lithium ion battery, and supercapacitors [75,76]. Nanosheets represent a diverse and largely untapped source of 2D systems with exotic electronic properties, high 2D anisotropy, and specific surface areas that are important for energy conversion and energy storage applications [26]. Herein, we offer a brief review of the applications of these metal oxide nanosheet–based composites mainly in the areas of solar and electrochemical energy conversion and storage.

9.5.1 Solar Energy Applications

Solar energy is the most promising renewable energy source that can fulfill the rising global demand for sustainable energy. Photocatalysis is one of the most effective approaches to harvest solar energy by photoinduced production of H_2 and O_2 gases or photodegradation of organic pollutants employing semiconductor photocatalysts. With the recent advancement in nanotechnology, research interests on the semiconductor photocatalysts have been extended to 2D nanostructured metal oxide materials due to its low cost and efficient photocatalytic activity. Many metal oxide nanosheets with unique structure are designed and explored as photocatalysts for water splitting or organic pollutant degradation, such as TiO_2 [77], $K_{0.8}H_{3.2}Nb_6O_{17}$ [35], TaO_3 [36], $Ca_2Ta_3O_{9.7}N_{0.2}$ [78], and $HTiNbO_5$ [79].

Currently, TiO_2-based photocatalysts are generally considered as the most promising candidate for solar energy utilization, and many advances have been made ever since the first observation of water photolysis on TiO_2 electrodes [80]. However, compared with the bulk TiO_2 crystals, titania nanosheets are less responsible to the visible light mainly because of the quantum size effect that increases the band gap to 3.8 eV, corresponding to the UV-light region with the wavelength less than 300 nm [81]. With the LBL deposition method, Harada et al. fabricated multilayer ultrathin films with Fe- or Ni-doped titania nanosheets as photocatalysts [82]. In this system, the UV absorption band was extended to the visible light region and stronger visible light response was detected than the pure titania nanosheets, in spite of the weakened photocatalytic activity from the dopants that aggravated the charge–carrier recombination. To enhance the visible light harvest, a new strategy of nitrogen doping has been developed recently by Liu et al. [29]. Using gaseous ammonia in the solid-state reaction, nitrogen was doped into the parent layered titanate particles and finally retained in the titania nanosheets (Figure 9.9). The optical absorption of the nitrogen-doped titania nanosheets was shifted to the visible light region, resulting in effective enhancement of photocurrent in the thin-film photoanodes based on the doped titania nanosheets. Calcium tantalum oxynitride $[Ca_2Ta_3O_{9.7}N_{0.2}]^-$ nanosheets were also prepared by exfoliating a layered perovskite oxynitride ($CsCa_2Ta_3O_{9.7}N_{0.2}$) via proton exchange and two-step intercalation of ethylamine and TBA ions [78]. The $[Ca_2Ta_3O_{9.7}N_{0.2}]^-$ nanosheets exhibited visible light photocatalytic activity for H_2 evolution in contrast with $CsCa_2Ta_3O_{9.7}N_{0.2}$ that exhibited very low photocatalytic activity for H_2 evolution under the visible light irradiation, even when methanol was added to water as a sacrificial agent [78]. The improved photocatalytic activity originates from the characteristics of nanosheets such as their molecular thickness and large surface area.

Zhai et al. [79] synthesized $HTiNbO_5$ nanosheets by exfoliating layered $HTiNbO_5$ in TBA hydroxide (TBAOH) to obtain $HTiNbO_5$ nanosheets [79]. The resulting nanosheets were then reassembled with TiO_2 colloids and finally heated with urea in air at 450°C. The mechanism and TEM image are shown in Figure 9.10 [79]. It was found that the TiO_2 nanoparticles existed in the anatase phase and the titanoniobate nanosheets were still maintained after nitrogen doping. The obtained nitrogen-doped nanohybrids showed a greatly expanded surface area with a mesoporous structure, and the doped nitrogen atoms were located in the interstitial sites of TiO_2,

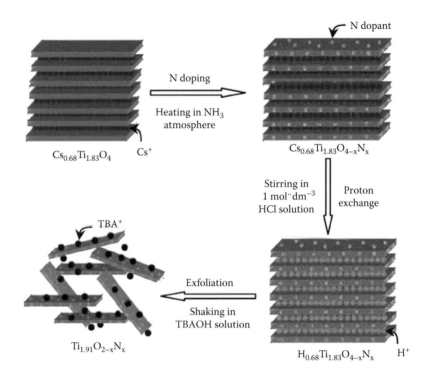

Figure 9.9 Schematic of procedures for preparing nitrogen-doped $Ti_{0.91}O_2$ nanosheets. TBA_+: tetrabutylammonium ion. (Liu, G. et al., Nitrogen-doped titania nanosheets toward visible light response, *Chem. Commun.*, 11, 1383–1385, 2009. Copyright 2009 reproduced by permission of The Royal Society of Chemistry.)

giving rise to the visible light response. The obtained N-doped nanohybrid had a higher activity than N-doped $HTiNbO_5$ nanosheets and N-doped TiO_2, indicating the synergetic effect of TiO_2 nanoparticles and $HTiNbO_5$ nanosheets [79].

Kim et al. prepared hybrid nanosheet photocatalysts for H_2 generation using subnanometer-thick layered titanate nanosheets incorporated with CdS quantum dots (QDs) (particle size ≈ 2.5 nm) [83]. The CdS QDs are strongly coupled with that of the layered titanate nanosheets, leading to an efficient electron transfer between them. As a consequence of the promoted electron transfer, the photoluminescence of CdS QDs is nearly quenched after hybridization, indicating the almost suppression of electron–hole recombination. These CdS-layered titanate nanohybrids show much higher photocatalytic activity for H_2 production than the precursor CdS QDs and layered titanate due to the increased lifetime of the electrons and holes and the expansion of the surface area upon hybridization [83]. This finding highlights the validity of 2D semiconductor nanosheets as effective building blocks for exploring highly visible-light-active photocatalysts for H_2 production.

Interestingly, with specified species incorporated into the nanosheets, new nanostructures can be rationally designed with new physical and chemical properties. For example, Seger et al. prepared thin polyaniline (PANI)–$Ti_{0.91}O_2$ film photoelectrodes with PANI and $Ti_{0.91}O_2$ nanosheets alternatively deposited in a controlled LBL manner [84]. Due to the switchable PANI from n-type to p-type at different potentials, the PANI–$Ti_{0.91}O_2$ thin film can perform as either n-type or p-type photoelectrodes based on the types of electrolyte applied.

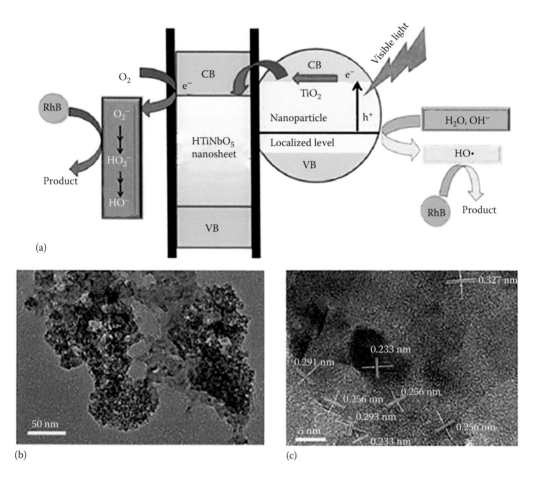

Figure 9.10 The mechanism of photocatalytic degradation of RhB over the HTiNbO$_5$ under (a) visible light irradiation; (b) TEM and (c) High-resolution transmission electron microscopy (HRTEM) images of HTiNbO$_5$ nanosheets. (Zhai, Z. et al., Nitrogen-doped mesoporous nanohybrids of TiO$_2$ nanoparticles and HTiNbO$_5$ nanosheets with a high visible-light photocatalytic activity and a good biocompatibility, *J. Mater. Chem.*, 22, 19122–19131, 2012. Copyright 2013 reproduced by permission of The Royal Society of Chemistry.)

9.5.2 Electrochemical Energy Applications

As the energy crisis has become a global challenge, many efforts have been put in the renewable energy sources. Electrochemical energy conversion and storage materials are one of the critical issues to solve this problem and have been extensively studied and developed during the past decades [85]. Since the metal oxide nanosheets often consist of transition metals with multiple valence and redox capability, some nanomaterials fabricated from these nanosheets have been demonstrated electrochemically active and reversible, while the unique properties of these nanosheets, for example, high surface area, are well maintained, which enable their application in the area of battery, supercapacitor, etc.

Flocculated MnO$_2$ nanosheets with Li ions were firstly applied in the Li-ion battery as the cathode material by Wang et al. [30]. After being dried thoroughly and then fabricated into coin cell, this new material is electrochemically tested and the charge–discharge curves of it

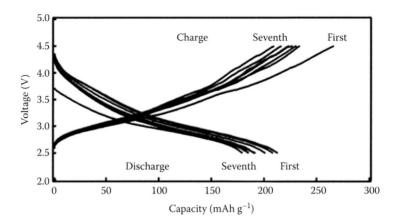

Figure 9.11   Charge–discharge curves of flocculated $Li_{0.36}MnO_2$. (Reprinted with permission from Wang, L. et al., Synthesis of a Li-Mn-oxide with disordered layer stacking through flocculation of exfoliated MnO_2 nanosheets, and its electrochemical properties, *Chem. Mater.*, 15, 4508–4514. Copyright 2003 American Chemical Society.)

in the first a few cycles are shown in Figure 9.11. An initial discharge capacity of 193 $mAh \cdot g^{-1}$ is generated with a continuous slope while no apparent plateau appears. Thereafter, the capacity gradually decreases, however in a rather limited rate upon cycling. It is worth noticing that this restacked material performs quite differently from the conventional layered $NaMnO_2$-derived $LiMnO_2$, which suffers from huge capacity loss due to the phase transformation into spinel during cycling. It is expected that the disordered lamellar structure results into another oxygen atom packing, which may increase the energy barrier required for the phase changing, thus inhibiting the phase evolution into spinel on cycling. Moreover, a later study on this material shows that it's also promising as a candidate of high-rate battery materials, which is especially important for the development of electric vehicles [86]. A fairly large discharge capacity of 79 $mAh \cdot g^{-1}$ at the current density of 2 $A \cdot g^{-1}$ is observed, which is 52% of 151 $mAh \cdot g^{-1}$ at the density of 50 $mA \cdot g^{-1}$. Another study on the octatitanate ($H_2Ti_8O_{17} \cdot H_2O$) synthesized via the exfoliation–flocculation process showed a similar lithium intercalation phenomenon [87]. The reassembled sample had a reversible capacity of 175 $mAh \cdot g^{-1}$ and energy efficiency of 91%, which is higher than the conventional octatitanate. Besides, a smaller overvoltage and better cycling stability were also observed, further confirming the advantages of this restacked structure.

Apart from the application in batteries, the restacked nanosheets have also shown great potential in the area of supercapacitors. Bulk ruthenium oxide crystal is a good candidate for supercapacitor electrodes due to its high electrical conductivity and electrochemical stability; however, their proton conductivity is still a big concern when used in an electrochemical cell.

To solve this problem, ruthenic oxide nanosheets were studied by Sugimoto et al. as the basic building blocks for thin-film supercapacitor electrode [88]. Earlier in 2003, layered ruthenic oxide crystal was exfoliated into monolayer nanosheets with the help of some alkyl ammonium molecules and then reloaded with TBA to form a thin-film supercapacitor electrode for the first time [88]. Thanks to the extra active sites from the interlayer surface and good proton conductivity from the interlayer water, a surprisingly high specific capacitance of 658 $F \cdot g^{-1}$ at 2 $mV \cdot s^{-1}$ was achieved, which is about one order lager than that of the bulk RuO_2 and close to that of the amorphous ruthenium oxide hydrate. Soon later, these ruthenic acid nanosheets were reassembled

Figure 9.12 HRTEM image of colloidal nanosheets from the mixed colloidal suspension with an RGO/ MnO_2 ratio of 0.3. (From Lee, Y.R., Kim, I.Y., Kim, T.W., Lee, J.M., and Hwang, S.J.: Mixed colloidal suspensions of reduced graphene oxide and layered metal oxide nanosheets: Useful precursors for the porous nanocomposites and hybrid films of graphene/metal oxide. *Chem.-A Euro. J.* 2012. 18. 2263–2271. Copyright 2012 Wiley-VCH Verlag GmbH & Co. KGaA. Reproduced with permission.)

into flexible and transparent supercapacitor electrodes on different substrates by the same group using the electrophoretic deposition method [89]. The one on Au substrate showed both high mass-specific capacitance of $620 \text{ F} \cdot \text{g}^{-1}$ and geometric specific capacitance of $0.82 \text{ F} \cdot \text{cm}^{-2}$ at room temperature, indicating the great potential of the electrophoretic deposition method to fabricate these nanosheets effectively. Then in a recent report, the capacitance of this reassembled RuO_2 nanosheets was further enhanced to around $700 \text{ F} \cdot \text{g}^{-1}$ in an acidic electrolyte, making it one of the high-capacitance materials and very attractive as the next-generation supercapacitor electrodes. Another recent study conducted by Lee et al. successfully deposited MnO_2 nanosheets onto the surface of graphene by adding Li^+ into mixed colloidal suspensions of MnO_2 nanosheets and graphene, yielding porous nanocomposites of Li–graphene–MnO_2 [90]. As shown in Figure 9.12, homogeneous mixing of the MnO_2 nanosheets and graphene sheets was realized in the colloidal suspension due to the highly negatively charged surface of both nanosheets. The flocculated Li–graphene–MnO_2 sample has a high capacitance of $210 \text{ F} \cdot \text{g}^{-1}$, and more than 95% capacitance retained after 1000 cycles, which is superior to the other samples without MnO_2 nanosheets or graphene separately, and the extremely expanded surface area as high as $70–100 \text{ m}^2 \cdot \text{g}^{-1}$ and the incorporation of highly conductive graphene are expected to account for the high electrochemical activity. Meanwhile, Zhang et al. also prepared a nanocomposite material via electrostatic co-precipitation of MnO_2 nanosheets and functionalized graphene [91]. And the resulting material exhibited an enhanced capacitive performance over Na-typed birnessite (Na/MnO_2) as well as a good cycling stability over 1000 times, which may mainly be attributed to the good electrical conductivity of graphene and pseudocapacitance of ultrathin MnO_2 nanosheets.

9.5.3 Applications in Other Areas

Most of the metal oxide nanosheets are insulating or semiconducting due to the combined electronic structure of the empty d orbitals of metal atoms (like Ti^{4+}, Nb^{5+}, Ta^{5+}, W^{6+}) and the full p orbitals of oxygen atoms [92]. Also, they are highly stable in composition and resistant to temperature variation [81,93]. All these properties offer great potential as dielectric alternatives in the design and application of new nanoelectronics.

Titania nanosheets ($Ti_{0.87}O_2$) with only TiO_6 octahedra in 2D plane have been intensively studied as ideal building blocks for high-κ dielectronics [94]. On one hand, they inherit the high dielectric constant of bulk TiO_2 materials; on the other hand, the Ti vacancies in $Ti_{0.87}O_2$ nanosheets keep them away from oxygen vacancies, which always lead to serious carrier trapping and current leakage in bulk TiO_2 [92,95]. In a recent study conducted by Osada et al., titania nanosheet–based nanofilms with thicknesses down to 10 nm were fabricated on flat $SrRuO_3$ substrates via the LBL deposition method [94]. Unlike traditional high-κ materials sensitively degrading upon size reduction, a surprisingly high κ value of ~125 was achieved with such extremely small thickness, strongly demonstrating the feasibility of using titania nanosheet–based high-κ nanoblocks in new nanodevices. Similar phenomenon was also revealed in multilayer films of perovskite nanosheets ($LaNb_2O_7$, $Ca_2Nb_3O_{10}$, and $Sr_2Nb_3O_{10}$) [96,97]. Although the film thickness was less than 5 nm, an even higher dielectric constant of more than 200 was detected as well as low leakage current densities ($<10^{-7}\,A\cdot cm^{-2}$).

Metal oxides, for example, TiO_2, are possible alternatives of SiO_2 as high-κ gate insulator in future small devices. Unfortunately, they generally suffer from large leakage currents due to the narrow band gaps and instabilities at high temperature on the Si substrates [98–101]. Compared with their bulk counterparts, metal oxide nanosheets exhibit more possibilities as the high-κ gate insulators due to their better thermal stability and broader band gaps [42,93]. For instance, a low leakage current ($<10^{-1}\,A\cdot cm^{-2}$) and small equivalent oxide thickness (EOT) values (~0.3 nm) were observed recently in multilayer nanofilms of $Ti_{0.87}O_2$ nanosheets directly growing on Si substrates, which provides experimental evidence for the availability of building high-κ gate insulating films with these metal oxide nanosheets [101].

Besides, with slight doping of other transition metal ions at the first stage of parent bulk materials synthesis, the metal oxide nanosheets have been modified in composition and found potential in nanoscale ferromagnets [102–105]. In 2006, multilayer films of $Ti_{0.8}Co_{0.2}O_2$ nanosheets were reported with room-temperature ferromagnetic properties for the first time [102]. A maximum magnetic moment of 1.4 μ_B/Co was detected due to the anisotropy induced by the 2D spin orbit polarization. This is a great enhancement compared with the theoretical spin moment of 1 μ_B/Co for low-spin Co^{2+} or 0.3 μ_B/Co and 1.1 μ_B/Co for Co-doped anatase semiconductors and insulators, respectively. And similarly improved ferromagnetic performance has also been found later in a series of transition metal–doped (Fe, Co, and Mn) titania nanosheets [103–105]. With respect to the ultrathin thickness of ferromagnetic nanosheets, they have an intrinsic advantage in the application of next-generation magneto-optical (MO) and magnetoelectronic nanodevices. For example, a remarkably MO response of ~104° cm^{-1} at a short operating wavelength of approximately 280 nm was exclusively found in the thin films of $Ti_{0.8}Co_{0.2}O_2$ and $Ti_{0.6}Fe_{0.4}O_2$ nanosheets [103,104]. And an even stronger MO absorption (~3*$10^{5°}\cdot cm^{-1}$) at 400–550 nm was reported in an alternately deposited $(Ti_{0.8}Co_{0.2}O_2/Ti_{0.6}Fe_{0.4}O_2)_5$ thin film due to the d–d transition interaction (Co^{2+}–Fe^{3+}) between adjacent layers [106]. With such superior MO properties, the metal oxide nanosheets are very promising for future practical applications in magnetic data storage and optical communication.

9.6 SUMMARY

In this chapter, the current research status of metal oxide nanosheets was addressed, reviewing a broad spectra from exfoliation of parent layered materials to reassembly of 2D nanosheets, from properties of single nanosheets to the application of nanosheet-based nanocomposites, especially in the area of solar energy and electrochemistry–energy conversion and storage.

Metal oxide nanosheets are ultimate 2D nanomaterials that exhibit many intriguing phys-iochemical properties and exciting functionalities that cannot be observed in bulk materials. Besides, the composition and crystal structure of nanosheets can be well controlled and tailored with well-developed methods in a 2D atomic level with good in-plane crystallinity. As a result, they provide an excellent platform to study the phenomena in 2D systems and the mechanisms behind them. To date, a series of metal oxide nanosheets have been prepared and shown great potentials in a wide range of areas. With the boom of nanotechnology, new types of metal oxide nanosheets are being exploited and their family grows steadily. More importantly, with the well-dispersed colloidal state at room temperature and highly charged surface, these nanosheets can be deposited or aggregated by a moderate aqueous reaction with charge-bearing interlayer counterparts like polyions, metal ions, or even metal hydroxide nanosheets that are hard to be intercalated in bulk materials. By intentionally selecting the interlayer species, new 3D nano-structures with combined functionalities can be rationally designed and applied in many more areas in the future.

At this stage, encouraging properties and performances of metal oxide nanosheets and their derivatives have been discovered and studied; however, considering the high cost and multi-step exfoliation process, it is still very difficult to precisely control the lateral size distribution and morphology variation in a large-scale production; in the meantime, how to arrange the nanosheets in a desirable way and position on large scale is also a great challenge. Therefore, priority for the future development on metal oxide nanosheets should be given to novel technolo-gies that can narrow the gap between fundamental research and scaled practical utilization. Another prospect could be a deeper understanding on the exfoliation mechanisms and chemis-try with the help of more reliable characterization tools and techniques.

Benefit from the recent intensive studies on the 2D graphene, metal oxide nanosheets also belonging to the ultimate 2D family have attracted increasing attention. With the growing under-standing of many interesting phenomena and extraordinary performance in the 2D structures, more sophisticated nanodevices and nanostructures will be designed and more unique proper-ties can be expected. Yet the technology is still at its early stage; the promising future of these metal oxide nanosheets can be foreseen as the basic building blocks for next-generation high-end nanomaterials in a broad spectrum of applications such as energy and environment.

REFERENCES

1. Sasaki, T., Watanabe, M., Hashizume, H., Yamada, H., and Nakazawa, H. Macromolecule-like aspects for a colloidal suspension of an exfoliated titanate. Pairwise association of nanosheets and dynamic reassembling process initiated from it. 1996. *Journal of the American Chemical Society* 118: 8329–8335.
2. Walker, G. Macroscopic swelling of vermiculite crystals in water. 1960. *Nature* 187: 312–313.
3. Jacobson, A. J., Colloidal dispersions of compounds with layer and chain structures. 1994. *Trans Tech Publ* 152: 1–12.
4. Lerf, A. and Schöllhorn, R. Solvation reactions of layered ternary sulfides A_xTiS_2, A_xNbS_2, and A_xTaS_2. 1977. *Inorganic Chemistry* 16: 2950–2956.
5. Joensen, P., Frindt, R., and Morrison, S. R. Single-layer MoS_2. 1986. *Materials Research Bulletin* 21: 457–461.
6. Yang, D., Sandoval, S. J., Divigalpitiya, W., Irwin, J., and Frindt, R. Structure of single-molecular-layer MoS_2. 1991. *Physical Review B* 43: 12053.
7. Alberti, G., Casciola, M., and Costantino, U. Inorganic ion-exchange pellicles obtained by delamination of α-zirconium phosphate crystals. 1985. *Journal of Colloid and Interface Science* 107: 256–263.

8. Yamamoto, N., Okuhara, T., and Nakato, T. Intercalation compound of $VOPO_4 \cdot 2H_2O$ with acrylamide: Preparation and exfoliation. 2001. *Journal of Materials Chemistry* 11: 1858–1863.

9. Rebbah, H., Borel, M., and Raveau, B. Intercalation of alkylammonium ions and oxide layers| $TiNbO_t$ |⁻. 1980. *Materials Research Bulletin* 15: 317–321.

10. Rebbah, H., Pannetier, J., and Raveau, B. Localization of hydrogen in the layer oxide $HTiNbO_5$. 1982. *Journal of Solid State Chemistry* 41: 57–62.

11. Nazar, L., Liblong, S., and Yin, X. T. Aluminum and gallium oxide-pillared molybdenum oxide MoO_3. 1991. *Journal of the American Chemical Society* 113: 5889–5890.

12. Treacy, M., Rice, S., Jacobson, A. J., and Lewandowski, J. Electron microscopy study of delamination in dispersions of the perovskite-related layered phases $K[Ca_2Na_{n-3}Nb_nO3_{n+1}]$: Evidence for single-layer formation. 1990. *Chemistry of Materials* 2: 279–286.

13. Sasaki, T. and Watanabe, M. Osmotic swelling to exfoliation. Exceptionally high degrees of hydration of a layered titanate. 1998. *Journal of the American Chemical Society* 120: 4682–4689.

14. Fang, M., Kim, C. H., and Mallouk, T. E. Dielectric properties of the lamellar niobates and titanoniobates $AM_2Nb_3O_{10}$ and $ATiNbO_5$ (A = H, K, M = Ca, Pb), and their condensation products $Ca_4Nb_6O_{19}$ and $Ti_2Nb_2O_9$. 1999. *Chemistry of Materials* 11: 1519–1525.

15. Schaak, R. E. and Mallouk, T. E. Perovskites by design: A toolbox of solid-state reactions. 2002. *Chemistry of Materials* 14: 1455–1471.

16. Schaak, R. E. and Mallouk, T. E. Prying apart ruddlesden-popper phases: Exfoliation into sheets and nanotubes for assembly of perovskite thin films. 2000. *Chemistry of Materials* 12: 3427–3434.

17. Kim, J. Y., Chung, I., Choy, J. H., and Park, G. S. Macromolecular nanoplatelet of Aurivillius-type layered perovskite oxide, $Bi_4Ti_3O_{12}$. 2001. *Chemistry of Materials* 13: 2759–2761.

18. Adachi-Pagano, M., Forano, C., and Besse, J. P. Delamination of layered double hydroxides by use of surfactants. 2000. *Chemical Communications* 1: 91–92.

19. Leroux, F. et al. Delamination and restacking of layered double hydroxides. 2001. *Journal of Materials Chemistry* 11: 105–112.

20. Hibino, T. and Jones, W. New approach to the delamination of layered double hydroxides. 2001. *Journal of Materials Chemistry* 11: 1321–1323.

21. Hibino, T. Delamination of layered double hydroxides containing amino acids. 2004. *Chemistry of Materials* 16: 5482–5488.

22. Li, L., Ma, R., Ebina, Y., Iyi, N., and Sasaki, T. Positively charged nanosheets derived via total delamination of layered double hydroxides. 2005. *Chemistry of Materials* 17: 4386–4391.

23. Liu, Z. et al. Synthesis, anion exchange, and delamination of Co-Al layered double hydroxide: Assembly of the exfoliated nanosheet/polyanion composite films and magneto-optical studies. 2006. *Journal of the American Chemical Society* 128: 4872–4880.

24. Liu, Z. et al. General synthesis and delamination of highly crystalline transition-metal-bearing layered double hydroxides. 2007. *Langmuir* 23: 861–867.

25. Sasaki, T. Fabrication of nanostructured functional materials using exfoliated nanosheets as a building block. 2007. *Journal of the Ceramic Society of Japan* 115: 9–16.

26. Osada, M. and Sasaki, T. Exfoliated oxide nanosheets: New solution to nanoelectronics. 2009. *Journal of Materials Chemistry* 19: 2503–2511.

27. Bizeto, M. A., Shiguihara, A. L., and Constantino, V. R. L. Layered niobate nanosheets: Building blocks for advanced materials assembly. 2009. *Journal of Materials Chemistry* 19: 2512–2525.

28. Ma, R., Liu, Z., Li, L., Iyi, N., and Sasaki, T. Exfoliating layered double hydroxides in formamide: A method to obtain positively charged nanosheets. 2006. *Journal of Materials Chemistry* 16: 3809–3813.

29. Liu, G. et al. Nitrogen-doped titania nanosheets towards visible light response. 2009. *Chemical Communications* 11: 1383–1385.

30. Wang, L. et al. Synthesis of a Li-Mn-oxide with disordered layer stacking through flocculation of exfoliated MnO_2 nanosheets, and its electrochemical properties. 2003. *Chemistry of Materials* 15: 4508–4514.

31. Bergaya, F., Theng, B. K. G., and Lagaly, G. *Handbook of Clay Science*. Vol. 1. 2006. London, U.K.: Elsevier Science.

32. Liu, Z., Ooi, K., Kanoh, H., Tang, W., and Tomida, T. Swelling and delamination behaviors of birnessite-type manganese oxide by intercalation of tetraalkylammonium ions. 2000. *Langmuir* 16: 4154–4164.

33. Omomo, Y., Sasaki, T., Wang, L., and Watanabe, M. Redoxable nanosheet crystallites of MnO$_2$ derived via delamination of a layered manganese oxide. 2003. *Journal of the American Chemical Society* 125: 3568–3575.

34. Liu, Z., Ma, R., Ebina, Y., Takada, K., and Sasaki, T. Synthesis and delamination of layered manganese oxide nanobelts. 2007. *Chemistry of Materials* 19: 6504–6512.

35. Saupe, G. B. et al. Nanoscale tubules formed by exfoliation of potassium hexaniobate. 2000. *Chemistry of Materials* 12: 1556–1562.

36. Fukuda, K., Nakai, I., Ebina, Y., Ma, R., and Sasaki, T. Colloidal unilamellar layers of tantalum oxide with open channels. 2007. *Inorganic Chemistry* 46: 4787–4789.

37. Wang, L., Zhu, Y., Tang, F., and Lu, M. G. Exfoliated two-dimensional nanosheets for self-assembly of new nanostructures. In Yangeng Gao (Ed.), *Bio-Inspired Materials Synthesis* (pp. 99–122) Kerala, India: Research Signpost, 2010.

38. Pinnavaia, T. J. and Beall, G. W. Eds., Polymer-clay nanocomposites. Chapters 11 and 13. 2000. Wiley: New York.

39. Wang, L. et al. Fabrication and characterization of multilayer ultrathin films of exfoliated MnO$_2$ nanosheets and polycations. 2003. *Chemistry of Materials* 15: 2873–2878.

40. Kim, T. W. et al. Soft-chemical exfoliation route to layered cobalt oxide monolayers and its application for film deposition and nanoparticle synthesis. 2009. *Chemistry-A European Journal* 15: 10752–10761.

41. Sasaki, T., Ebina, Y., Kitami, Y., Watanabe, M., and Oikawa, T. Two-dimensional diffraction of molecular nanosheet crystallites of titanium oxide. 2001. *The Journal of Physical Chemistry B* 105: 6116–6121.

42. Sasaki, T. and Watanabe, M. Semiconductor nanosheet crystallites of quasi-TiO$_2$ and their optical properties. 1997. *The Journal of Physical Chemistry B* 101: 10159–10161.

43. Sasaki, T., Nakano, S., Yamauchi, S., and Watanabe, M. Fabrication of titanium dioxide thin flakes and their porous aggregate. 1997. *Chemistry of Materials* 9: 602–608.

44. Abe, R. et al. Preparation of porous niobium oxides by soft-chemical process and their photocatalytic activity. 1997. *Chemistry of Materials* 9: 2179–2184.

45. Yui, T. et al. Synthesis of photofunctional titania nanosheets by electrophoretic deposition. 2005. *Chemistry of Materials* 17: 206–211.

46. Sugimoto, W., Terabayashi, O., Murakami, Y., and Takasu, Y. Electrophoretic deposition of negatively charged tetratitanate nanosheets and transformation into preferentially oriented TiO$_2$ (B) film. 2002. *Journal of Materials Chemistry* 12: 3814–3818.

47. Ma, R. and Sasaki, T. Nanosheets of oxides and hydroxides: Ultimate 2D charge-bearing functional crystallites. 2010. *Advanced Materials* 22: 5082–5104.

48. Umemura, Y., Shinohara, E., Koura, A., Nishioka, T., and Sasaki, T. Photocatalytic decomposition of an alkylammonium cation in a Langmuir–Blodgett film of a titania nanosheet. 2006. *Langmuir* 22: 3870–3877.

49. Wang, L., Ebina, Y., Takada, K., and Sasaki, T. Ultrathin films and hollow shells with pillared architectures fabricated via layer-by-layer self-assembly of titania nanosheets and aluminum keggin ions. 2004. *The Journal of Physical Chemistry B* 108: 4283–4288.

50. Wang, L., Sakai, N., Ebina, Y., Takada, K., and Sasaki, T. Inorganic multilayer films of manganese oxide nanosheets and aluminum polyoxocations: Fabrication, structure, and electrochemical behavior. 2005. *Chemistry of Materials* 17: 1352–1357.

51. Wang, Z. S., Ebina, Y., Takada, K., Watanabe, M., and Sasaki, T. Inorganic multilayer assembly of titania semiconductor nanosheets and Ru complexes. 2003. *Langmuir* 19: 9534–9537.

52. Yui, T. et al. Photoinduced one-electron reduction of MV^{2+} in titania nanosheets using porphyrin in mesoporous silica thin films. 2005. *Langmuir* 21: 2644–2646.

53. Zhou, Y., Ma, R., Ebina, Y., Takada, K., and Sasaki, T. Multilayer hybrid films of titania semiconductor nanosheet and silver metal fabricated via layer-by-layer self-assembly and subsequent UV irradiation. 2006. *Chemistry of Materials* 18: 1235–1239.

54. Li, L. et al. Layer-by-layer assembly and spontaneous flocculation of oppositely charged oxide and hydroxide nanosheets into inorganic sandwich layered materials. 2007. *Journal of the American Chemical Society* 129: 8000–8007.

55. Wang, L., Ebina, Y., Takada, K., Kurashima, K., and Sasaki, T. A new mesoporous manganese oxide pillared with double layers of alumina. 2004. *Advanced Materials* 16: 1412–1416.

56. Letaïef, S., Martín-Luengo, M. A., Aranda, P., and Ruiz-Hitzky, E. A Colloidal route for delamination of layered solids: Novel porous-clay nanocomposites. 2006. *Advanced Functional Materials* 16: 401–409.

57. Paek, S. M. et al. Exfoliation and reassembling route to mesoporous titania nanohybrids. 2006. *Chemistry of Materials* 18: 1134–1140.

58. Park, J. H. and Jana, S. C. Mechanism of exfoliation of nanoclay particles in epoxy-clay nanocomposites. 2003. *Macromolecules* 36: 2758–2768.

59. Ebina, Y., Sasaki, T., Harada, M., and Watanabe, M. Restacked perovskite nanosheets and their Pt-loaded materials as photocatalysts. 2002. *Chemistry of Materials* 14: 4390–4395.

60. Ebina, Y., Sakai, N., and Sasaki, T. Photocatalyst of lamellar aggregates of RuO_x-loaded perovskite nanosheets for overall water splitting. 2005. *The Journal of Physical Chemistry B* 109: 17212–17216.

61. Hata, H., Kobayashi, Y., Bojan, V., Youngblood, W. J., and Mallouk, T. E. Direct deposition of trivalent rhodium hydroxide nanoparticles onto a semiconducting layered calcium niobate for photocatalytic hydrogen evolution. 2008. *Nano Letters* 8: 794–799.

62. Kooli, F., Sasaki, T., and Watanabe, M. A new pillared structure with double-layers of alumina. 1999. *Chemical Communications* 2: 211–212.

63. Kooli, F., Sasaki, T., Rives, V., and Watanabe, M. Synthesis and characterization of a new mesoporous alumina-pillared titanate with a double-layer arrangement structure. 2000. *Journal of Materials Chemistry* 10: 497–501.

64. Decher, G. Fuzzy nanoassemblies: Toward layered polymeric multicomposites. 1997. *Science* 277: 1232–1237.

65. Sasaki, T. et al. Layer-by-layer assembly of titania nanosheet/polycation composite films. 2001. *Chemistry of Materials* 13: 4661–4667.

66. Caruso, F., Caruso, R. A., and Möhwald, H. Nanoengineering of inorganic and hybrid hollow spheres by colloidal templating. 1998. *Science* 282: 1111–1114.

67. Wang, L., Ebina, Y., Takada, K., and Sasaki, T. Ultrathin hollow nanoshells of manganese oxide. 2004. *Chemical Communications* 9: 1074–1075.

68. Huang, J. et al. Layer-by-Layer assembly of TaO_3 nanosheet/polycation composite nanostructures: Multilayer film, hollow sphere, and its photocatalytic activity for hydrogen evolution. 2010. *Chemistry of Materials* 22: 2582–2587.

69. Caruso, F. Hollow inorganic capsules via colloid-templated layer-by-layer electrostatic assembly. 2003. *Colloid Chemistry II* 227: 145–168.

70. Caruso, R. A., Susha, A., and Caruso, F. Multilayered titania, silica, and laponite nanoparticle coatings on polystyrene colloidal templates and resulting inorganic hollow spheres. 2001. *Chemistry of Materials* 13: 400–409.

71. Yamaki, T. and Asai, K. Alternate multilayer deposition from ammonium amphiphiles and titanium dioxide crystalline nanosheets using the Langmuir–Blodgett technique. 2001. *Langmuir* 17: 2564–2567.

72. Akatsuka, K. et al. Construction of highly ordered Lamellar nanostructures through Langmuir–Blodgett deposition of molecularly thin titania nanosheets tens of micrometers wide and their excellent dielectric properties. 2009. *ACS Nano* 3: 1097–1106.

73. Novoselov, K. et al. Electric field effect in atomically thin carbon films. 2004. *Science* 306: 666–669.

74. Panwar, N. L., Kaushik, S. C., and Kothari, S. Role of renewable energy sources in environmental protection: A review. 2011. *Renewable and Sustainable Energy Reviews* 15: 1513–1524.

75. Oelhafen, P. and Schüler, A. Nanostructured materials for solar energy conversion. 2005. *Solar Energy* 79: 110–121.

76. Arico, A. S., Bruce, P., Scrosati, B., Tarascon, J.-M., and van Schalkwijk, W. Nanostructured materials for advanced energy conversion and storage devices. 2005. *Nature Materials* 4: 366–377.

77. Tanaka, T., Ebina, Y., Takada, K., Kurashima, K., and Sasaki, T. Oversized titania nanosheet crystallites derived from flux-grown layered titanate single crystals. 2003. *Chemistry of Materials* 15: 3564–3568.

78. Ida, S., Okamoto, Y., Matsuka, M., Hagiwara, H., and Ishihara, T. Preparation of tantalum-based oxynitride nanosheets by exfoliation of a layered oxynitride, $CsCa_2Ta_3O_{10-x}N_y$, and their photocatalytic activity. 2012. *Journal of the American Chemical Society* 134: 15773–15782.

79. Zhai, Z. et al. Nitrogen-doped mesoporous nanohybrids of TiO$_2$ nanoparticles and HTiNbO$_5$ nanosheets with a high visible-light photocatalytic activity and a good biocompatibility. 2012. *Journal of Materials Chemistry* 22: 19122–19131.

80. Fujishima, A. and Honda, K. Photolysis-decomposition of water at the surface of an irradiated semiconductor. 1972. *Nature* 238: 37–38.

81. Sakai, N., Ebina, Y., Takada, K., and Sasaki, T. Electronic band structure of titania semiconductor nanosheets revealed by electrochemical and photoelectrochemical studies. 2004. *Journal of the American Chemical Society* 126: 5851–5858.

82. Harada, M., Sasaki, T., Ebina, Y., and Watanabe, M. Preparation and characterizations of Fe- or Ni-substituted titania nanosheets as photocatalysts. 2002. *Journal of Photochemistry and Photobiology A: Chemistry* 148: 273–276.

83. Kim, H. N., Kim, T. W., Kim, I. Y., and Hwang, S.-J. Cocatalyst-free photocatalysts for efficient visible-light-induced H$_2$ production: Porous assemblies of CdS quantum dots and layered titanate nanosheets. 2011. *Advanced Functional Materials* 21: 3111–3118.

84. Seger, B. et al. An n-type to p-type switchable photoelectrode assembled from alternating exfoliated titania nanosheets and polyaniline layers. 2013. *Angewandte Chemie International Edition* 52: 6400–6403.

85. Winter, M. and Brodd, R. J. What are batteries, fuel cells, and supercapacitors? 2004. *Chemical Reviews* 104: 4245–4269.

86. Sato, K., Suzuki, S., and Miyayama, M. Electrochemical properties of lithium titanate synthesized by reassembly of nanosheets. 2007. *Key Engineering Materials* 350: 139–142.

87. Suzuki, S. and Miyayama, M. Lithium intercalation properties of octatitanate synthesized through exfoliation/reassembly. 2006. *The Journal of Physical Chemistry B* 110: 4731–4734.

88. Sugimoto, W., Iwata, H., Yasunaga, Y., Murakami, Y., and Takasu, Y. Preparation of ruthenic acid nanosheets and utilization of its interlayer surface for electrochemical energy storage. 2003. *Angewandte Chemie International Edition* 42: 4092–4096.

89. Sugimoto, W., Yokoshima, K., Ohuchi, K., Murakami, Y., and Takasu, Y. Fabrication of thin-film, flexible, and transparent electrodes composed of ruthenic acid nanosheets by electrophoretic deposition and application to electrochemical capacitors. 2006. *Journal of the Electrochemical Society* 153: A255–A260.

90. Lee, Y. R., Kim, I. Y., Kim, T. W., Lee, J. M., and Hwang, S. J. Mixed colloidal suspensions of reduced graphene oxide and layered metal oxide nanosheets: Useful precursors for the porous nanocomposites and hybrid films of graphene/metal oxide. 2012. *Chemistry – A European Journal* 18: 2263–2271.

91. Zhang, J., Jiang, J., and Zhao, X. Synthesis and capacitive properties of manganese oxide nanosheets dispersed on functionalized graphene sheets. 2011. *The Journal of Physical Chemistry C* 115: 6448–6454.

92. Sato, H., Ono, K., Sasaki, T., and Yamagishi, A. First-principles study of two-dimensional titanium dioxides. 2003. *The Journal of Physical Chemistry B* 107: 9824–9828.

93. Fukuda, K. et al. Unusual crystallization behaviors of anatase nanocrystallites from a molecularly thin titania nanosheet and its stacked forms: Increase in nucleation temperature and oriented growth. 2007. *Journal of the American Chemical Society* 129: 202–209.

94. Osada, M. et al. High-κ dielectric nanofilms fabricated from titania nanosheets. 2006. *Advanced Materials* 18: 1023–1027.

95. Wang, Y. et al. Lattice distortion oriented angular self-assembly of monolayer titania sheets. 2010. *Journal of the American Chemical Society* 133: 695–697.

96. Li, B.-W. et al. A-site-modified perovskite nanosheets and their integration into high-kappa dielectric thin films with a clean interface. 2010. *Japanese Journal of Applied Physics* 49: 09MA01.

97. Li, B.-W. et al. Solution-based fabrication of perovskite nanosheet films and their dielectric properties. 2009. *Japanese Journal of Applied Physics* 48: 09KA15-1-5.

98. Kingon, A. I., Maria, J.-P., and Streiffer, S. Alternative dielectrics to silicon dioxide for memory and logic devices. 2000. *Nature* 406: 1032–1038.

99. Wilk, G. D., Wallace, R. M., and Anthony, J. High-κ gate dielectrics: Current status and materials properties considerations. 2001. *Journal of Applied Physics* 89: 5243–5275.

100. Robertson, J. High dielectric constant gate oxides for metal oxide Si transistors. 2006. *Reports on Progress in Physics* 69: 327.
101. Osada, M. and Sasaki, T. Two-dimensional dielectric nanosheets: Novel nanoelectronics from nanocrystal building blocks. 2012. *Advanced Materials* 24: 210–228.
102. Osada, M. et al. Ferromagnetism in two-dimensional $Ti_{0.8}Co_{0.2}O_2$ nanosheets. 2006. *Physical Review B* 73: 153301.
103. Osada, M., Ebina, Y., Takada, K., and Sasaki, T. Gigantic magneto–optical effects in multilayer assemblies of two-dimensional titania nanosheets. 2006. *Advanced Materials* 18: 295–299.
104. Osada, M. et al. Gigantic magneto-optical effects induced by (Fe/Co)-cosubstitution in titania nanosheets. 2008. *Applied Physics Letters* 92: 253110.
105. Dong, X. et al. Synthesis of Mn-substituted titania nanosheets and ferromagnetic thin films with controlled doping. 2009. *Chemistry of Materials* 21: 4366–4373.
106. Osada, M. et al. Orbital reconstruction and interface ferromagnetism in self-assembled nanosheet superlattices. 2011. *ACS Nano* 5: 6871–6879.

Computation and Modeling

Chapter 10

First-Principle Engineering Magnetism in Low-Dimensional Boron Nitride Materials

Aijun Du

CONTENTS

10.1 INTRODUCTION

The epoch-making discoveries of fullerene (Kroto et al. 1985), carbon nanotube (CNT) (Iijima 1991), and graphene (Novoselov et al. 2004) have stimulated intense research efforts regarding low-dimensional nanostructures constructed from graphitic honeycomb-like networks (Tang and Zhou 2013, Xu et al. 2013). Boron nitride (BN) materials consisting equal numbers of boron and nitrogen atoms are among the most promising inorganic nanoscale system that has been explored so far (Golberg et al. 2010, Pakdel et al. 2012). Following the initial report of CNTs in 1991 (Iijima 1991), BN nanotubes were theoretically predicted and experimentally identified in 1994 and 1995, respectively (Rubio et al. 1994, Chopra et al. 1995). Other one-dimensional (1D) BN nanoarchitectures including nanowires, nanoribbons, nanofibers, and nanorods have been produced by using a wide range of experimental synthesis methods (Golberg et al. 1996, Loiseau et al. 1996, Han et al. 1998, Chen et al. 1999, Golberg et al. 1999, Novoselov et al. 2005, Kim et al. 2008, Zhi et al. 2009, Nag et al. 2010, Li et al. 2011). Zero-dimensional (0D) octahedral BN fullerenes were experimentally realized in 1998 (Golberg et al. 1998). Two-dimensional (2D) BN nanosheet on metallic substrate was initially reported in 2004 (Corso et al. 2004), and freestanding 2D BN flakes were peeled off from a BN crystal 1c year later (Novoselov et al. 2005). Figure 10.1a through c presented low-dimensional BN nanostructures, that is, 2D BN nanosheet, 1D nanotube, and 0D fullerene, respectively.

Spintronics seek to exploit electron spins in addition to the electrical charges for logic and memory devices, and it ignites a revolution in information processing (Wolf et al. 2001). Recently,

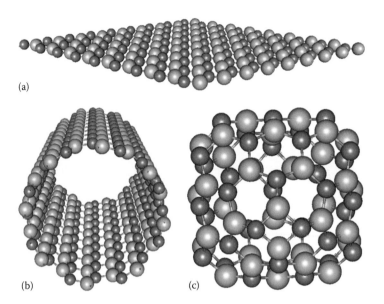

Figure 10.1   Low-dimensional BN nanostructures: (a) 2D nanosheet, (b) 1D nanotube, and (c) 0D $B_{36}N_{36}$ cage. Gray and dark gray balls represent boron (B) and nitrogen (N) atoms, respectively.

magnetism is found to be induced to CNTs (graphene) when they are unzipped (cut) into 1D graphene nanoribbon (GNNR; Son et al. 2006, Ugeda et al. 2010). One interesting application of the magnetism in zigzag GNNRs is the novel half-metallicity under an electric field, which has opened an exciting pathway for the development of next-generation metal-free spintronics (Son et al. 2006). Structurally, BN is a very close analogue of the carbon system, and magnetism could also be created along edge sites in low-dimensional BN nanostructures, which has not been explored much compared to the carbon system.

The goal of this chapter is to briefly survey recent work in low-dimensional BN-based nano-materials with a focus on discussing magnetic properties from a theoretical perspective. The inspiration is the active research development of magnetism in BN materials in recent years for potential magnetic device and spintronic applications (Li et al. 2008, 2009, Du et al. 2009). Our particular attention will be mainly focused on BN nanotubes, nanoribbons, and nanosheets while placing an emphasis on computational studies utilizing the ground-state density functional the-ory (Hohenberg and Kohn 1964, Kohn and Sham 1965). Some important topics such as the optical, mechanical, and thermal properties of low-dimensional BN nanomaterials (Golberg et al. 2010, Pakdel et al. 2012) will be excluded for conciseness. In light of the tremendous amount of informa-tion accumulated in the past decades, this chapter is intended to serve as a starting point for the interested readers and does not convey a thorough account of the status of BN-based research. Therefore, we apologize for having to neglect many important contributions to this exciting field.

10.2   INTRINSIC EDGE MAGNETISM IN BN NANOMATERIALS

Defect-free low-dimensional BN materials are expected to display nonmagnetic ground state as in the case of carbon materials. However, when a 2D BN nanosheet or 1D BN nanotube (BNNT) is cut into nanoribbons, strong magnetism can be generated due to the unpaired electron at bare B or N edges. By using first-principle calculations, Hao et al. (2006) discovered tremendous spin splitting (over 1.0 eV) in BN nanotube with open B and N ends. This is the first report on the spin polarization in BN system without a d electron. Furthermore, such a deep-gap state has been found to be highly stable under the external perturbation interaction with ferromagnetic samples. The magnetic character and spin-polarized electronic structure of the BN nanotube with an open end strongly depend on the tube chirality. As shown in Figure 10.2a, the local magnetic moment is attributed to the unpaired electrons of B and N atoms at the open end. High spin polarizations combined with high stable spin configuration highlight a promising candidate for developing future spintronic devices, for example, spin-polarized electron field emitters and scanning tunneling microscopy tips.

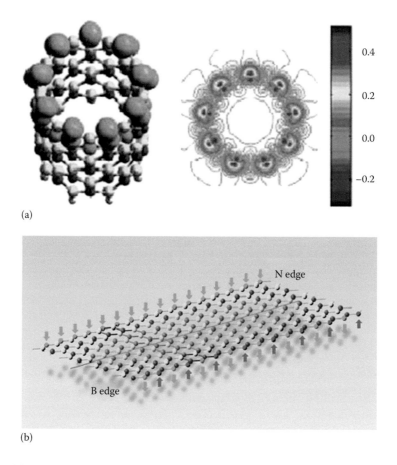

(a)

(b)

Figure 10.2   (a) Three- and two-dimensional plots of spin density [in units of e/(au)^3] at the open mouth of N-rich-ended (9,0) BNNT; (b) the high spin state, with a magnetic moment of 1 μB at each edge atom in BNNR. Gray and dark gray balls represent B and N atoms, respectively. (Reproduced with permission from Hao, S., Zhou, G., Duan, W.H., Wu, J., and Gu, B.L., Tremendous spin-splitting effects in open boron nitride nanotubes: Application to nanoscale spintronic devices, *J. Am. Chem. Soc.*, 128, 8453, 2006; Barone, V. and Peralta, J., Magnetic boron nitride nanoribbons with tunable electronic properties, *Nano Letts.*, 8, 2008, 2210. Copyright 2008 American Chemical Society.)

Barone and Peralta (2008) reported BN nanoribbons (BNNRs) with tunable electronic and magnetic properties as illustrated in Figure 10.2b. BNNRs with bare zigzag edges are found to be magnetic semiconductors with an energy gap that decreases slightly with the ribbon width. The spin configurations in zigzag BNNRs could be thermally accessible at room temperature as in the case of GNNRs. However, the target state for practical applications in the case of zigzag BNNRs is the high spin state (4 μ_B), whereas in graphene, it is the antiferromagnetic coupling state. The high spin state in zigzag BNNRs could be further stabilized by applying an external magnetic field or transition metal doping. The transverse electric field produces electron reorganization toward the B edge or N edge, depending on its direction, enabling an external control of the band gap of zigzag BNNRs to produce metallic ↔ semiconducting ↔ half-metallic transitions.

10.3 CARBON-DOPING-INDUCED MAGNETISM IN BN MATERIALS

Carbon is a typical dopant in low-dimensional BN-based nanomaterials. The substitution of B or N atoms with carbon atoms will inject electron and hole, respectively, offering possibility to tune charge carrier density and magnetism in BN materials. Experimentally, carbon doping in BN-based nanomaterials has been demonstrated to be easily controlled as evidenced by the electron microscopy image in Figure 10.3a (Krivanek et al. 2010). Theoretically, Wu et al. (2005) have carried out systematic first-principle calculations on the carbon-doped armchair and zigzag BN nanotubes. It was found that carbon substitution at either single boron site or a single nitrogen site in the BN nanotubes can induce spontaneous magnetization. The electronic band-structure calculation reveals a spin-polarized and flat band near the Fermi energy and the magnetization can be attributed to the carbon 2p electrons. Du et al. (2007) have investigated structural and electronic properties in BNNRs with both zigzag and armchair-shaped edges using spin-polarized density functional calculations. Figure 10.3b through d presents density of state plots in C-doped 1D zigzag and armchair BNNRs, respectively. As in the case of C-doped BN nanotube, C-doping in BNNRs could also show spontaneous magnetization. The earlier results collectively suggest that carbon atom doping on BNNTs and BNNRs may have potential applications in the development of BN-based magnetic nanodevices.

10.4 MAGNETISM INDUCED FROM CHEMICAL FUNCTIONALIZATION

Chemical functionalization such as hydrogenation can induce the structural transformation from sp^2 to sp^3 hybridization in graphene, leading to ferromagnetic order in semihydrogenated graphene (Zhou et al. 2009). It could be also an effective approach to produce magnetism in low-dimensional BN nanomaterials. Li et al. (2008) have demonstrated that the chemisorption of fluorine (F) atoms on the B atoms of BNNTs could generate spontaneous magnetization, whereas no magnetism is observed when the B and N atoms are equally fluorinated. This approach can be used to tune the magnetism in BNNTs as well as provide a synthetic route toward the realization of metal-free magnetic materials. Recently, tunable ferromagnetic spin ordering in BNNTs through topological fluorine adsorption has been reported by Zhang and

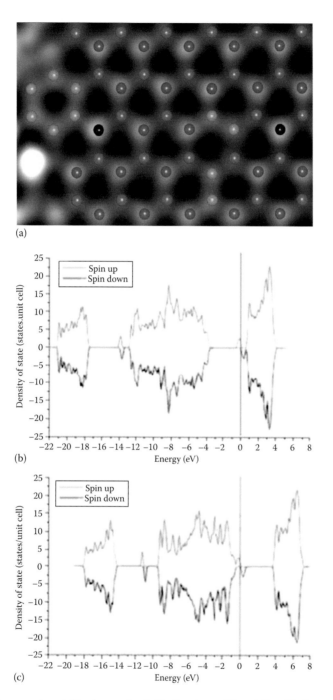

Figure 10.3   (a) The experimentally observed substitution impurities at B or N sites overlaid on the corresponding part of the experimental image. The atomic structure for a single BN layer is determined by DFT calculation. White, light gray, gray and dark gray atoms represented B, C, N, and O atoms, respectively. (From Krivanek, O.L. et al., *Nature*, 464, 571, 2010.) (b and c) are density of states for C-doped armchair BN nanoribbons at B site and N site, respectively. (From Du, A.J. et al., *J. Am. Chem. Soc.*, 131, 17354, 2007.)   *(Continued)*

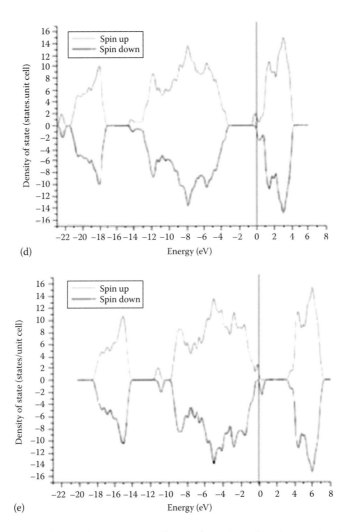

(d)

(e)

Figure 10.3 (Continued)   (d and e) are similar to (b and c) but for C-doped zigzag BNNRs *at B and N sites, respectively* [Du 2007]. (a) is reprinted with permission from *Nature* publishing group. (From Krivanek, O.L. et al., *Nature*, 464, 571, 2010 and (b–e) are reproduced from Du, A.J. et al., *Chemical Physics Letters* 447, 181, 2007 with permission from Elsevier Ltd. 2012.)

Guo (2009) (Figure 10.4). F atoms adsorbed on BNNTs in a well-defined order are found to induce long-range ferromagnetic spin ordering, offering strong spin polarization around the Fermi level. Interestingly, the spin polarization and magnetic moment are increased significantly and even give rise to half-metal as tube radius decreases. However, no magnetic moment is predicted in a flat BN sheet with the same topological fluorine arrangement as that in BNNTs. The radius-dependent behavior in F-BNNTs can be developed into a local curvature modulation procedure to efficiently enhance or quench the ferromagnetic ordering. This finding suggests a new route to facilitate the design of tunable spin devices based on low-dimensional BN materials via fluorine functionalization.

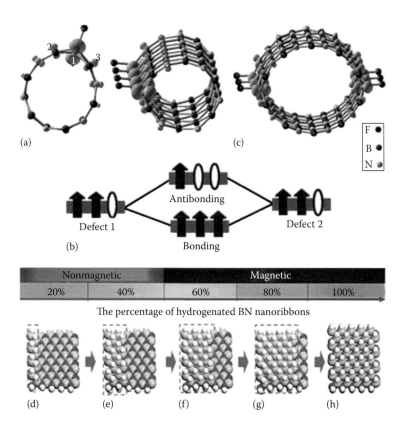

Figure 10.4   (a) Top and side views of spin density plots of the F-BN nanotubes with isosurface value of 0.15 e/Å³ for a (6,0) F-BNNT. (b) Illustration of the origin of exchange coupling mechanism for ferromagnetism in the F-BNNT. The arrow and ellipse represent the electron in a spin direction and the hole, respectively. (c) Spin density plot for a (10,0) F-BNNT with two opposed F chains. (d–h) presented structural and electronic properties of hydrogenated zigzag BNNRs at different H-coverages. The ratio of numbers of hydrogenated blocks to numbers of pristine BNNR blocks is increased from left to right. (Reproduced with permission from Zhang, Z.H. and Guo, W.L., Tunable ferromagnetic spin ordering in boron nitride nanotubes with topological fluorine adsorption, *J. Am. Chem. Soc.*, 131, 2009, 6874–6879. Copyright 2009 American Chemical Society; Reproduced with permission from Chen, W., Li, Y.F., Yu, G.T., Li, C., Zhang, S.B., Zhou, Z., and Chen, Z.F., Hydrogenation: A simple approach to realize semiconductor-half-metal-metal transition in boron nitride nanoribbons, *J. Am. Chem. Soc.*, 132, 2010, 1699–1705, Copyright 2010 American Chemical Society.)

Chen et al. (2010) have investigated electronic and magnetic properties of H-BNNRs at different coverages by using the first-principle computation. Fully hydrogenated armchair BNNRs are found to be all nonmagnetic semiconductors, while the zigzag counterparts exhibit diverse electronic and magnetic properties after hydrogenation. They are nonmagnetic when the percentage of hydrogenation in BNNRs is lower. However, a semiconductor → half-metal → metal transition occurs with increasing H-coverage, accompanied by a nonmagnetic to magnetic transfer *as shown in* Figure 10.4d through h. The half-metallic property has been predicted when the hydrogenation ratio is very large, but this behavior is sustained for partially hydrogenated zigzag BNNRs with a smaller degree of hydrogenation. So controlling the hydrogenation ratio can precisely modulate the electronic and magnetic properties of zigzag BNNRs, endowing low-dimensional BN nanomaterials many potential applications in the novel integrated functional nanodevices.

10.5   VACANCY MAGNETISM

Recently, significant progress has been made on the experimental fabrications of a single layer of h-BN sheets. By using electron irradiation and layer-by-layer sputtering, individual boron and nitrogen atoms in an h-BN nanosheet can be easily knocked off and then N or B vacancy will be created (Jin et al. 2009). Remarkably, the single boron vacancies are dominant, and the large vacancies with nitrogen atom–terminated zigzag edge can also be identified with atomic resolution. By controlling energetic irradiation and beam intensity, it allows to fabricate functional devices with the well-defined defect structures in h-BN nanosheets as shown in Figure 10.5. The experimentally produced B or N vacancies are all triangular shaped (see Figure 10.5c). To provide an in-depth understanding, Du et al. (2009) reported triangular-shaped nanohole, nanodot, and lattice antidot structures in h-BN monolayer sheets (see Figure 10.5d and e) by utilizing density functional theory (DFT) calculations based on the local spin density approximation. Such structures are found to exhibit very large magnetic moments and associated spin splitting as illustrated by the density of state plots as shown in Figure 10.5f and g, respectively. N-terminated nanodots and antidots show strong spin anisotropy around the Fermi level, that is, half-metallicity. While B-terminated nanodots are shown to lack magnetism due to edge reconstruction, B-terminated nanoholes can retain magnetic character due to the enhanced structural stability of the surrounding 2D matrix. In spite of significant lattice contraction due to the presence of multiple holes, antidot super lattices are predicted to be stable, exhibiting amplified magnetism as well as greatly enhanced half-metallicity. These results indicate new opportunities for designing h-BN-based nanoscale devices with potential applications in the areas of spintronics.

10.6   ADATOM-INDUCED MAGNETISM

Controlling charge and magnetic states is crucial to the development of magnetic nanostructures into quantum information devices. Adatom on low-dimensional BN materials may display significant magnetism due to the change in local structure around the adsorption site. Duan's group at Tsinghua University, by using spin-polarized density functional calculations, reported that nonmagnetic C adatoms could induce significant magnetism on BN nanotubes and hexagonal sheets (Li et al. 2009). As can be seen clearly from Figure 10.6a through f, the C adatom impurity states are strongly localized, leading to strong spin polarization in the hybrid C/BN system. The local magnetic moment is calculated to be as high as 2.0 μB per C atom regardless of the diameter of BN tubes. There are two valence electrons participating in the bonding, and the remaining two electrons of the C adatom are confined at the adsorption site and contribute to the large magnetic moment. Most recently, Huang et al. (2012) demonstrated an effective way to control the charge and magnetic states for transition metal atoms adsorbed on single-layer BN nanosheets through the internal defect engineering and external electric fields. Figure 10.6g presents the optimized geometry for transition metal atoms adsorbed onto B vacancy, N vacancy, and double BN vacancy sites in a single-layer BN nanosheet, and the corresponding magnetic moments are described in Figure 10.6h. Most interestingly, the application of an external electric

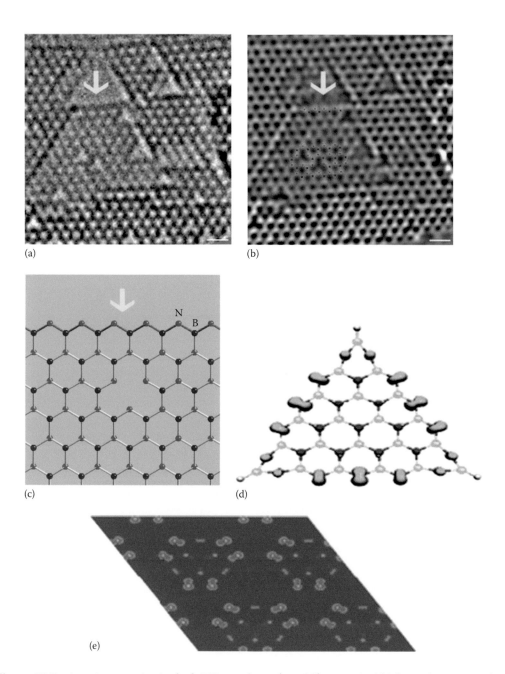

Figure 10.5   Atomic vacancies in the h-BN monolayer. (a and b) are typical high-resolution tunneling electron microscopy images showing the lattice defects in h-BN (single B or N vacancies) and even larger triangular-shaped vacancies with the same orientation. (c) Proposed models for B and N vacancies in the h-BN nanosheet. (d and e) present 3D spin magnetic charge density for triangular BN nanodot and antidot, respectively.   (*Continued*)

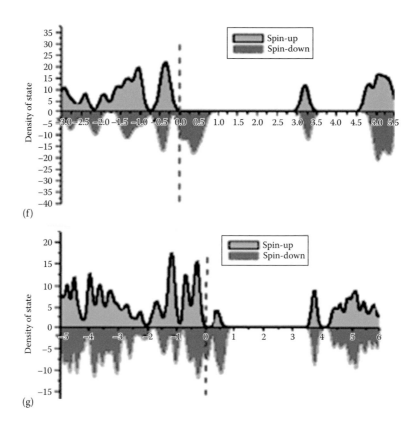

Figure 10.5 (Continued)   (f and g) plots correspond to spin-resolved density of states for (d and e), respectively. Gray and dark gray balls represent B and N atoms, respectively. (Reproduced with permission from Jin, C.H., Fang, S., Kazu, S., and Iijima, S., *Phys. Rev. Letts.*, 102, 195505–195509, 2009. Copyright 2009 by the American Physics Society; Reproduced with permission from Du, A.J., Chen, Y., Zhu, Z.H., Amal, R., Lu, G.Q., and Smith, S., Dots versus antidots: Computational exploration of structure, magnetism, and half-metallicity in boron-nitride nanostructures, *J. Am. Chem. Soc.*, 131, 17354–17359. Copyright 2009 American Chemical Society.)

field can easily modify the size of the crystal-field splitting of d orbitals of transition metal (see Figure 10.6i), leading to the spin crossover for transition metal on h-BN sheet. These findings suggest potential application of transition metal–doped BN nanosheet as field-driven nonvolatile memory devices.

10.7   CONCLUSION AND PERSPECTIVES

In this chapter, we have given a short overview of recent progress on first-principle engineering magnetism in low-dimensional BN-based nanomaterials. Theoretically, extensive studies have been carried out to predict magnetism in BN nanotubes and nanosheets as a result of vacancy, zigzag edge, chemical functionalization, adatom, and atomic doping. However, magnetism in low-dimensional BN materials has never been confirmed experimentally. One possible reason could be that all the proposed theoretical strategies for realizing magnetism in BN materials

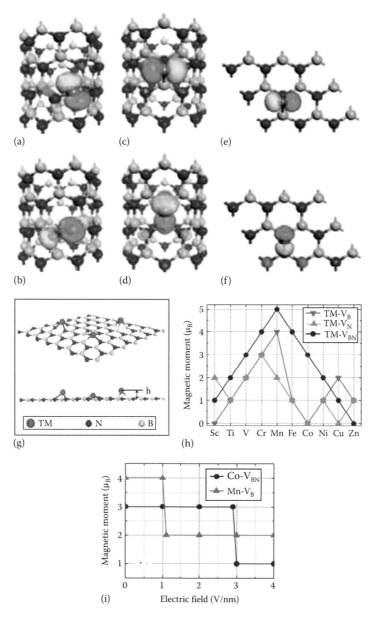

Figure 10.6   Band decomposed state plots for highest occupied molecular orbital (HOMO, a and c) and HOMO-1 (b and d) for C adatom adsorbed on the perpendicular-bridge and parallel-bridge sites of an (8,0) BN nanotube, respectively; (e) HOMO and (f) HOMO-1 band states for the C adsorbed on an h-BN sheet. (g and h) presented the schematic structure and the magnetic moments for transition metal atoms adsorb on B vacancy, N vacancy, and double BN vacancy in a single-layer BN nanosheet, respectively. (i) The magnetic moments of Co and Mn atoms deposited on BN vacancy and B vacancy as a function of external electric field. (Reproduced with permission from Li, J., Zhou, G., Chen, Y., Gu, B.L., and Duan, W.H., Magnetism of C adatoms on BN nanostructures: Implications for functional nanodevices, *J. Am. Chem. Soc.*, 131, 2009, 1796–1801. Copyright 2009 American Chemical Society; Reproduced with permission from Huang, B., Xiang, H.J., Yu, J., and Wei, S.H., *Phys. Rev. Letts.*, 108, 206802, 2012. Copyright 2012 American Physical Society.)

strongly depend on carefully doping or arranging vacancy or adatom in a well-defined order; thus, significant experimental challenges remain. Long-range magnetic exchange coupling between vacancy and adatoms is fully unexplored, and this might explain the absence of magnetism experimentally at ambient condition. Additionally, BNs are highly polarized materials compared to carbon materials, and the electronic properties can be significantly affected by strain, potentially leading to the change in magnetic moment around vacancy and adatom sites.

It is important to note that experimental methods usually tend to become more expensive with time, while computational methods will become cheaper as computers become faster. In combination with new developments in electronic structure theory and computational methods, computational approaches for the discovery and development of novel magnetic devices based on low-dimensional BN nanomaterials will hold tremendous promise for utilization in close cooperation with experiments. Looking forward, great research efforts in experiment are still urgently needed to verify existing theoretical prediction. State-of-the-art ab initio approaches will continue to furnish a clear interpretive picture of magnetism in low-dimensional BN nanomaterials and suggest new avenues for future development of BN-based magnetic and spintronic nanodevices.

ACKNOWLEDGMENTS

The author acknowledges generous grants of high-performance computer time from cluster computing facility at the Australian Institute for Bioengineering and Nanotechnology in the University of Queensland, Queensland Cyber Infrastructure Foundation (QCIF), and the Australian Partnership for Advanced Computing National Facility, which is supported by the Australian Commonwealth Government. The financial support of the Australian Research Council (ARC) under the Discovery Project (DP110101239) and ARC Queen Elizabeth Fellowship is also greatly appreciated.

REFERENCES

Barone, V. and Peralta, J. (2008), Magnetic boron nitride nanoribbons with tunable electronic properties. *Nano Letters* 8: 2210–2214.

Chen, W., Li, Y. F., Yu, G. T., Li, C., Zhang, S. B., Zhou, Z., and Chen, Z. F. (2010), Hydrogenation: A simple approach to realize semiconductor-half-metal-metal transition in boron nitride nanoribbons. *Journal of the American Chemical Society* 132: 1699–1705.

Chen, Y., Chadderton, L., Gerald, J. F., and Williams, J. S. (1999), A solid-state process for formation of boron nitride nanotubes. *Applied Physics Letters* 74: 2960–2963.

Chopra, N., Luyken, R. J., Cherrey, K., Crespi, V., Cohen, M., Louie, S. G., and Zettl, A. (1995), Boron nitride nanotubes. *Science* 269: 966–967.

Corso, M., Auwarter, W., Muntwiler, M., Tamai, A., Greber, T., and Osterwalder, J. (2004), Boron nitride nanomesh. *Science* 303: 217–220.

Du, A. J., Chen, Y., Zhu, Z. H., Amal, R., Lu, G. Q., and Smith, S. (2009), Dots versus antidots: Computational exploration of structure, magnetism, and half-metallicity in boron-nitride nanostructures. *Journal of the American Chemical Society* 131: 17354–17359.

Du, A. J., Smith, S., and Lu, G. Q. (2007), First-principle studies of electronic structure and C-doping effect in boron nitride nanoribbon. *Chemical Physics Letters* 447: 181–186.

Golberg, D., Bando, Y., Eremets, M., Takemura, K., Kurashima, K., and Yusa, H. (1996), Nanotubes in boron nitride laser heated at high pressure. *Applied Physics Letters* 69: 2045–2047.

Golberg, D., Bando, Y., Han, W., Kurashima, K., and Sato, T. (1999), Single-walled B-doped carbon, B/N-doped carbon and BN nanotubes synthesized from single-walled carbon nanotubes through a substitution reaction. *Chemical Physics Letters* 308: 337–342.

Golberg, D., Bando, Y., Huang, Y., Terao, T., Mitome, M., Tang, C., and Zhi, C. (2010), Boron nitride nanotubes and nanosheets. *ACS Nano* 4: 2979–2993.

Golberg, D., Bando, Y., Stephan, O., and Kurashima, K. (1998), Octahedral boron nitride fullerenes formed by electron beam irradiation. *Applied Physics Letters* 73: 2441–2443.

Han, W., Bando, Y., Kurashima, K., and Sato, T. (1998), Synthesis of boron nitride nanotubes from carbon nanotubes by a substitution reaction. *Applied Physics Letters* 73: 3085–3087.

Hao, S., Zhou, G., Duan, W. H., Wu, J., and Gu, B. L. (2006), Tremendous spin-splitting effects in open boron nitride nanotubes: Application to nanoscale spintronic devices. *Journal of the American Chemical Society* 128: 8453–8458.

Hohenberg, P. and Kohn, W. (1964), Inhomogeneous electron gas. *Physical Review B* 136: 864–871.

Huang, B., Xiang, H. J., Yu, J., and Wei, S. H. (2012), Effective control of the charge and magnetic states of transition-metal atoms on single-layer boron nitride. *Physical Review Letters* 108: 206802.

Iijima, S. (1991), Helical microtubules of graphitic carbon. *Nature* 354: 56–58.

Jin, C. H., Fang, S., Kazu, S., and Iijima, S. (2009), Fabrication of a freestanding boron nitride single layer and its defect assignments. *Physical Review Letters* 102: 195505–195509.

Kim, M., Chatterjee, S., Kim, S. M., Stach, E. A., Bradley, M. G., Pender, M. J., Sneddon, L. G., and Maruyama, B. (2008), Double-walled boron nitride nanotubes grown by floating catalyst chemical vapor deposition. *Nano Letters* 8: 3298–3302.

Kohn, W. and Sham, L. J. (1965), Self-consistent equations including exchange and correlation effects. *Physical Review A* 140: 1133–1138.

Krivanek, O. L., Chisholm, M. F., and Nicolosi, V. et al. (2010), Atom-by-atom structural and chemical analysis by annular dark-field electron microscopy. *Nature* 464: 571–574.

Kroto, H., Heath, J. R., O'Brien, S. C., Curl, R. F., and Smalley, R. E. (1985), C60: Buckminsterfullerene. *Nature* 318: 162–163.

Li, F., Zhu, Z. H., Yao, X. D., Lu, G. Q., Zhao, M. W., Xia, Y. Y., and Chen, Y. (2008), Fluorination-induced magnetism in boron nitride nanotubes from ab initio calculations. *Applied Physics Letters* 92: 102515.

Li, J., Zhou, G., Chen, Y., Gu, B. L., and Duan, W. H. (2009), Magnetism of C adatoms on BN nanostructures: Implications for functional nanodevices. *Journal of the American Chemical Society* 131: 1796–1801.

Li, L. H., Chen, Y., Behan, G., Zhang, H. Z., Petraic, M., and Glushenkov, A. M. (2011), Large-scale mechanical peeling of boron nitride nanosheets by low-energy ball milling. *Journal of Materials Chemistry* 21: 11862–11866.

Loiseau, A., Willaime, F., Demoncy, N., Hug, G., and Pascard, H. (1996), Boron nitride nanotubes with reduced numbers of layers synthesized by arc discharge. *Physical Review Letters* 76: 4737–4740.

Nag, A., Raidongia, K., Hembram, K., Datta, R., Waghmare, U. V., and Rao, C. N. R. (2010), Graphene analogues of BN: Novel synthesis and properties. *ACS Nano* 4: 1539–1544.

Novoselov, K. S., Geim, A. K., Morozov, S. V., Jiang, D., Zhang, Y., Dubonos, S. V., Grigorieva, I. V., and Firsob, A. A. (2004), Electric field effect in atomically thin carbon films. *Science* 306: 666–669.

Novoselov, K. S., Jiang, D, Schedin, F., Booth, T. J., Khotkevich, V. V., Morozov, S. V., and Geim, A. K. (2005), Two-dimensional atomic crystals. *Proceedings of the National Academy of Sciences of the United States of America* 102: 10451–10453.

Pakdel, A., Zhi, C. Y., Bando, Y., and Golberg, D. (2012), Low-dimensional boron nitride nanomaterials. *Materials Today* 15: 256–264.

Rubio, A., Corkill, J., and Cohen, M. (1994), Theory of graphitic boron nitride nanotube. *Physical Review B* 49: 5081–5084.

Son, Y. W., Cohen, M., and Louie, S. G. (2006), Half-metallic graphene nanoribbon. *Nature* 444: 347–349.

Tang, Q. and Zhou, Z. (2013), Graphene analogous low-dimensional materials. *Progress in Material Science* 58: 1244–1315.

Ugeda, M. M., Brihuega, I., Guinea, F., and Gómez-Rodríguez, J. M. (2010), Missing atom as a source of carbon magnetism. *Physical Review Letters* 104: 096804.

Wolf, S. A., Awschalom, D. D., Buhrman, R. A., Daughton, J. M., von Molnar, S., Roukes, M. L., Chtchelkanova, A. Y., and Treger, D. M. (2001), Spintronics: A spin-based electronics vision for the future. *Science* 294: 1488–1495.

Wu, R. Q., Liu, L., Peng, G. W., and Feng, Y. P. (2005), Magnetism in BN nanotubes induced by carbon doping. *Applied Physics Letters* 86: 122510–122513.

Xu, M. S., Liang, T., Shi, M., and Chen, H. (2013), Graphene-like two-dimensional materials. *Chemical Review* 113: 3766–3798.

Zhang, Z. H. and Guo, W. L. (2009), Tunable ferromagnetic spin ordering in boron nitride nanotubes with topological fluorine adsorption. *Journal of the American Chemical Society* 131: 6874–6879.

Zhi, C., Bando, Y., Tang, C., Kuwahara, H., and Golberg, D. (2009), Large-scale fabrication of boron nitride nanosheets and their utilization in polymeric composites with improved thermal and mechanical properties. *Advanced Materials* 21: 2889–2893.

Zhou, J., Wang, Q., Sun, Q., Chen, X. S., Kawazoe, Y., and Jena, P. (2009), Ferromagnetism in semi-hydrogenated graphene sheet. *Nano Letters* 9: 3867–3870.

Chapter 11

Ultrafast Transport in Nanotubes and Nanosheets

Aaron W. Thornton, Afsana Ahmed, Majumder Mainak,
Ho Bum Park, and Anita J. Hill

CONTENTS

11.1 INTRODUCTION

One of the most fascinating properties of nanotubes (NTs) and nanosheets is their smooth surfaces that offer essentially resistance-free transport of small molecules in the form of gases, vapors, and liquids. For example, it has been shown on multiple occasions that the transport through a NT is faster than if there were no material there at all. This super transport can be visualized by holding a garden hose but delivering as much water as a fire hose 10 times its size. This offers innumerable opportunities in applications such as efficient desalination, carbon capture, drug delivery, and energy delivery that require fast transport of small molecules (see Figure 11.1). The research journey so far has included alternating steps in theoretical predictions and

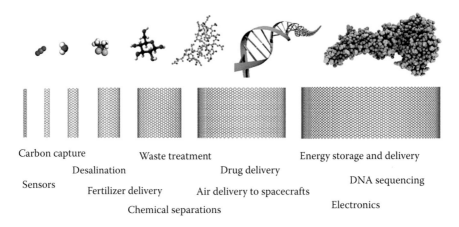

Carbon capture
Desalination
Sensors
Fertilizer delivery
Chemical separations
Waste treatment
Drug delivery
Air delivery to spacecrafts
Energy storage and delivery
DNA sequencing
Electronics

Figure 11.1   Ultrafast transport opportunities for gas, water, hydrocarbons, drugs, proteins, sugars, nutrients, contaminants, and DNA.

experimental verification but has yet translated to industrial-scale applications. It is of interest in this chapter to answer the following questions:

- What are the theoretical speed limits of transport?
- What are the actual speeds observed in NTs to date?
- What further developments are required to increase speeds toward the theoretical limits?
- What implications do these results have on our fundamental understanding of transport and the promising applications in the real world?

This chapter outlines the origins of ultrafast transport that began with predictions from molecular simulation techniques followed by experimental efforts to achieve such predictions. The discoveries are remarkable, making it to high-impact journals such as *Science* and *Nature*. There has been considerable media coverage for applications such as desalination, where NTs could facilitate pure water from seawater at remarkably efficient rates. In addition, NTs are conductive, and therefore, there is an opportunity to desalinate water while simultaneously capturing electrostatic energy for energy recovery systems, a phenomenon experimentally demonstrated with graphene. Ultrafast transport is also a necessary property for balance within the human body. For example, a single aquaporin embedded within a cell facilitates water transport at a rate of roughly three billion water molecules per second while gating undesired components. NTs also demonstrate this ability to separate mixtures at enormous rates as well as capture and deliver drugs to targeted destinations. Companies such as IBM are interested in using NTs for the ultrafast sequencing of DNA.

The concept of ultrafast separation can be likened to the philosophical riddle of Maxwell's demon whereby a demon separates molecules without performing work. To achieve such a phenomenon would impact numerous applications in terms of cost and efficiency. Section 11.2 summarizes the available platforms for experimentally measuring transport rates. Section 11.3 provides an overview of gas transport studies with molecular simulations tools, continuum models, and the comparison with experimental results. Section 11.4 also follows the same outline as Section 11.3 but for water transport. Finally, Section 11.5 presents progress in the research of nanosheets in the form of pristine graphene and graphene oxide (GO) layers.

11.2 EXPERIMENTAL PLATFORMS FOR MEASURING TRANSPORT IN CARBON NANOTUBE COMPOSITE MEMBRANES

Membranes composed of vertically aligned carbon nanotube (CNT) arrays made using microfabrication techniques were the experimental platforms that enabled measurements of transport; however, production of nanocomposite membranes composed of CNTs in a scalable and reproducible manner is still a persisting challenge. There is a broad body of literature that has appeared on the fabrication of nanocomposite membranes composed of CNTs. Here, comparison of these studies is presented by compiling the results from three different routes of CNT membrane fabrication with respect to the construction method (Figure 11.2).

Type 1 CNT membranes are composed of as-grown aligned CNTs filled with Si_3N_4 conformally to seal the gaps between NTs that have to be argon ion milled and plasma etched to open the CNT tips in subsequent steps (Fornasiero et al. 2008). This category of membrane has very-well-aligned fluidic transport channels, and mass transport through the open-ended CNT is the dominant transport path. Type 2 CNT membranes are synthesized by dispersing CNTs in a liquid with a surfactant, in this case zwitterionic molecules, which is vacuum filtered on a polyester support to partially align the CNTs to provide a supported membrane structure

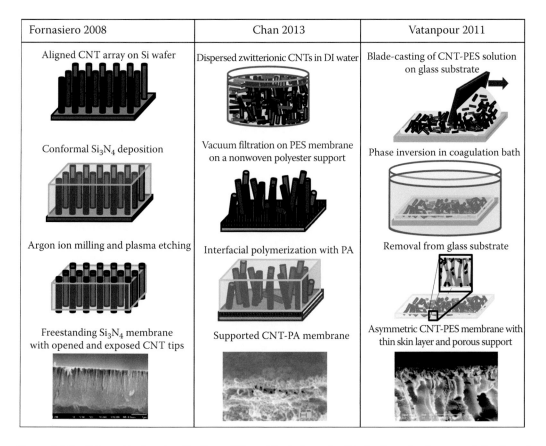

Figure 11.2 Schematic showing the three different techniques for fabrication of carbon nanotube–based nanocomposite membranes.

(Chan et al. 2013). The nanocomposite membrane is created by the polymerization of polyamide around the partially aligned CNTs. The procedure for the fabrication of Type 3 CNT membranes is based on the well-established technique of casting and phase inversion. Instead of the polymeric solution as the precursor, a dispersion of CNTs in polymer is utilized (Vatanpour et al. 2011). The nanocomposite membrane is asymmetric, and it has been reported that the dense skin layer has a higher concentration of CNTs as the amphiphilic nature of the NTs causes them to migrate to the solvent (acetone)/nonsolvent (water) interface. Transport rates and separation ratios through these platforms and their comparison with theoretical predictions will be explored further in the chapter.

11.3   ULTRAFAST GAS TRANSPORT IN NANOTUBES

NTs can be visualized as sheets of carbon rolled up into tubes. Without defects, the inner surface of an NT is more atomically smooth than any known surface. This means if a molecule was to collide with the surface, then the reflected trajectory would be mirrored hence losing no momentum. Similarly, if a molecule was to glide across the surface of the NT, then the molecule would not experience any *bumps* or *pot holes* allowing a frictionless slide like a puck in ice hockey. These superfast mechanisms are described in Section 11.3.2.

Since there is control over the tube diameter, NTs offer a natural size sieving of molecules therefore offering ultrafast transport only to the desired component of the mixture. Other opportunities for separation are in gating the entrance of the tube with switching or bulky functional groups described in Section 11.3.3. In addition, one could modify the surface of the tube to attract or repel one component over another, or simply make the surface more *bumpy* for one molecule over another.

The focus of this section will be to explore the ultrafast gas transport properties of NTs. Most of the research is with nitrogen, but there are also studies on carbon dioxide, hydrogen, methane, noble gases, and other light gases. There are works that address the effects of tube size, chirality, and smoothness as well as the effects of operating conditions such as temperature and pressure. Theoretical studies are split into molecular simulation techniques and continuum modeling, though there are some overlaps. There are only a few groups who have successfully synthesized aligned NT membranes due to the difficulty in growing pristine tubes, opening the tubes, distributing the tubes in resin or polymer, and etching the surface to expose the tubes. However, all groups have observed ultrafast transport properties above any porous system available. A final discussion linking conclusions will be given and the questions proposed in Section 11.1 will be addressed.

11.3.1   Molecular Simulation of Ultrafast Gas Transport

Molecular simulations are computational experiments that aim to mimic the kinetics and thermodynamics of a chosen molecular system, usually to predict a macro-scale property. The first simulations of transport in NTs were performed by Tuzun et al. (1996) followed by Mao and Sinnott (2000) that explored important effects such as gas–tube interactions, gas density, tube diameter, and tube flexibility. Skoulidas et al. (2002) were the first to apply this technique for the prediction of macro-scale transport properties (quantified as flux, permeation, permeance, or

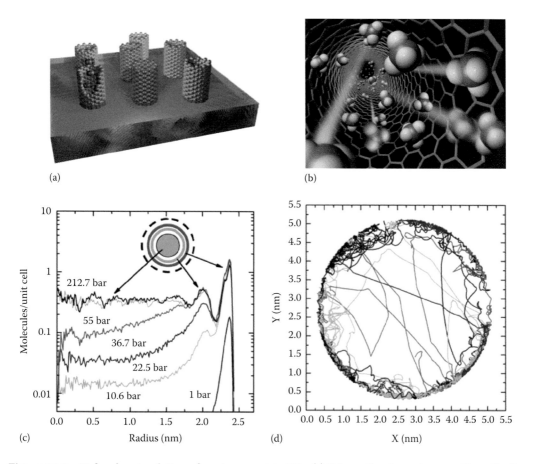

Figure 11.3 Molecular simulation of gas transport in NTs. (a) Schematic of membrane with vertically aligned NTs embedded in nonporous medium. (From Sholl, D.S. and Johnson, J.K., *Science*, 312(5776), 1003, 2006.) (b) Animation of methane molecule transporting through CNT (acknowledgment Scott Dougherty, LLNL). (c) Equilibrium gas concentrations throughout the radius of the tube and (d) dynamics of gas molecules diffusing inside the tube. (From Skoulidas, A.I. et al., *J. Chem. Phys.*, 124(5), 054708, 2006.)

permeability) of a membrane with vertically aligned NTs embedded within a nonporous matrix (see Figure 11.3a and b). The macroscopic flux (J) can be predicted using Fick's law:

$$J = -D_t \nabla c, \tag{11.1}$$

where
D_t is the transport diffusivity
∇c is the gas concentration gradient between upstream and downstream

Both transport diffusivity and concentration gradient can be derived from molecular simulation results. At first, a section of NT is constructed with periodic boundary conditions. The concentration gradient is then calculated using Grand Canonical Monte Carlo procedures that determine the concentration within the tube at thermodynamic equilibrium at the specified temperature and pressure (see Figure 11.3c). Since upstream pressure is usually much larger than the downstream pressure, it is sufficient to simply equate the concentration gradient with the equilibrium concentration at upstream pressure. Computed concentration can be compared with experimental isotherms to verify the interaction parameters between the gas and tube. The resulting cell packed with gas is then subject to molecular dynamics (MD) based on Newton's laws of motion. The net displacement of a particle is related to its self-diffusivity, D_s, according to the Einstein relation,

$$D_s = \lim_{t \to \infty} \frac{1}{6Nt} \left\langle \sum_{i=1}^{N} |\vec{r}_i(t) - \vec{r}_i(0)|^2 \right\rangle. \tag{11.2}$$

The self-diffusivity is equivalent to the transport diffusivity at low concentrations. At higher concentrations, the transport diffusivity is defined as the product of a corrected diffusivity and a thermodynamic correction factor developed by Theodorou et al. (1996). The corrected diffusivity D_0 is related to the mean-squared displacement of the center of mass of all the gas molecules as a group, as follows:

$$D_0 = \lim_{t \to \infty} \frac{1}{6Nt} \left\langle \left| \sum_{i=1}^{N} [\vec{r}_i(t) - \vec{r}_i(0)] \right|^2 \right\rangle. \tag{11.3}$$

Finally, the transport diffusivity is defined as

$$D_t = D_0 \left(\frac{\partial \ln f}{\partial \ln c} \right)_T. \tag{11.4}$$

For methane in a 10 µm thick membrane with (10, 10) CNTs, Skoulidas et al. predicted a remarkable flux of around 10,000 mol/s m² at an upstream pressure of 1 bar. This is equivalent to 224,000 L of methane passing through a square meter of membrane per second. Further studies confirmed that other gases also demonstrated fluxes at an order of magnitude higher than in any other nanoporous material (Chen and Sholl 2004). In fact, the diffusion is faster than in the bare gas phase without any material present. Comparisons were also made with silicate where the transport in NTs was one to three orders of magnitude faster than in silicate (Ackerman et al. 2003). Even the transport of gas mixtures was found to be three orders of magnitude higher than the zeolite ZSM-12 (Chen and Sholl 2004). Air separation of nitrogen and oxygen components was shown to be maximized by adjusting upstream and downstream pressures without a significant loss in flux (Arora and Sandler 2006).

Some of the early simulation studies were implemented with the tube atoms fixed in rigid positions, and hence, the effect of tube flexibility was not included. Tube flexibility occurs as a result of the thermal fluctuations across the atomic structure. At low pressure, the inclusion of NT flexibility was found to reduce transport diffusion by around an order of magnitude, while at higher pressure above 1 bar, the transport diffusivities were similar for both flexible and rigid NTs. Overall, the transport simulated in flexible NTs was still orders of magnitude above other nanoporous materials.

Further studies examined the frictionless nature of tube walls (Sokhan 2004). The low friction was found to be a result of the uniform structure and high density of carbon atoms closely weaved across the surface. Friction was characterized during simulation using the Maxwell coefficient of thermalization, which is the Stokesian friction coefficient divided by the collision frequency. It was found that friction was independent of gas density and temperature meaning that it is primarily a property of the surface. The shape of the gas molecule, however, does affect the friction, for example, nitrogen was found to have a lower friction coefficient than methane. In light of further studies, the interaction between the gas and tube also affects friction, which is the likely difference between nitrogen and methane.

Other studies compared the diffusive and viscous contributions to the transport coefficient (Bhatia et al. 2005). It was found that the viscous contribution is small even for a tube as large as 8.1 nm in diameter. This was because of the nearly specular nature of the diffusion contribution where the slip flow mechanism dominates. Hydrogen is found to be less specular than methane, which is due to the greater sensitivity to atomic detail of the surface that hydrogen experiences as a linear molecule compared with methane, which is more spherical in shape. Specular reflection is also reduced with increased density because the potential energy landscape is roughened by additional gas–gas interactions.

Temperature and tube size effects were explored by Jakobtorweihen et al. (2006). In small tubes (diameter <1 nm), the gas diffusion decreased with temperature, while in larger tubes (diameter >1 nm), the gas diffusion increased with temperature. Answers to this riddle were found in analysis of the radial position of the gas within the tube at various temperatures. Diffusivity decreases with temperature in small tubes because the gas moves closer to the wall with temperature and therefore encounters more collisions that consequently increase the resistance to diffusion. Diffusivity increases with temperature in larger tubes because the gas moves away from the wall and therefore encounters fewer collisions hence less friction.

As demonstrated in the study of Jakobtorweihen et al., the position of the gas molecules and the radial gas density profile proved to be one of the most insightful observations in molecular simulations. Another question that arose during simulation and experimental studies was the transport regime responsible for ultrafast transport properties. The next section will outline the transport regimes, but a critical result from simulation that is needed to confirm the dominant transport regime came from simulation, namely, the gas density profile. The gas density profile informs the researcher where the majority of gas sits during transport. Skoulidas et al. (2006) showed that the density of nitrogen at room temperature is 25 times greater at the wall compared with the center of the wall (see Figure 11.3c and d). The result enforced that most gas molecules that transport through the NT are in the surface diffusion regime. Results such as these also enlightened the gas storage community who were interested in storing gases at high densities at close to ambient conditions.

Most of the simulation studies used equilibrium molecular dynamics (EMD) to calculate transport diffusivity. Transport diffusivity can also be obtained using nonequilibrium molecular dynamics (NEMD) either by using the dual-control volume method (Heffelfinger and van Swol 1994) or by applying a gravity-like force (Cracknell et al. 1995). It has been shown that there is an excellent agreement in the Maxwell coefficient calculated using EMD and NEMD techniques (Sokhan 2004). The DCV method employs two control volumes for handling the particle creation/deletion at the upstream and downstream ends of the NT. In this way, the chemical potential gradient is controlled, hence mimicking a driving force in the flow direction. Particles flow down the chemical potential gradient and the flux is measured by counting particles flowing between the regions. This is a more direct method for measuring flux as opposed to the combination of transport diffusivity and concentration gradient described earlier. The main conclusions of NEMD simulations agree with EMD including the fact that the majority of gas molecules sit along the walls of the tube. In addition, NEMD allows the direct observation of entrance and exit effects at the tube ends. It was found that a traffic jam of gas molecules build up at the exit end of tube as gas molecules prefer the environment within the tube where the potential energy is strong rather than the external downstream environment (Lee and Sinnott 2004). Over longer simulation times, this mechanism continues until the whole tube fills with gas to saturation levels and linear viscous flow is observed.

Overall, molecular simulation offers a closer examination of transport properties and the effects of pressure, temperature, tube size, gas shape, and gas–tube interactions. The challenge then remains to develop or discover fundamental transport models that encompass molecular effects observed in simulation. Progress in continuum models and the remaining questions of theoretical speed limits of ultrafast transport are discussed in Section 11.3.2.

11.3.2 Transport Regimes for Ultrafast Gas Transport

Ultrafast transport in NTs was first discovered using molecular simulation techniques as described earlier. Further insightful discoveries such as the effects of temperature, pressure, and tube size were also observed using molecular simulation. However, this technique is incapable of answering questions such as "What is the theoretical limit of transport?" or offering any predictive power to explore all effects on transport. Fundamental transport models aim to satisfy this need. Transport models are often constructed for a particular regime. The main task is to determine the appropriate or dominant transport regime that usually depends on many factors such as tube size, temperature, pressure, tube wall roughness, gas shape, and gas–tube interactions.

Tube size or pore size has been known to be the most critical factor. According to IUPAC (Everett 1972) terminology, pore regimes are separated into micropores (pore diameter <2 nm), mesopores (2 nm < pore diameter < 50 nm), and macropores (pore diameter >50 nm). The most general transport regimes that correspond with these pore regimes are constrained activated diffusion, Knudsen diffusion, and viscous flow (bulk diffusion), described in a review by Verweij et al. (2007). Since NTs demonstrate unique properties compared with all other nanoporous materials, other transport regimes arise. Here, we summarize the key transport regimes appropriate for describing gas diffusion in NTs, namely, size sieving, surface diffusion, Knudsen diffusion, specular transport, and viscous flow, shown in Figure 11.4. It is noteworthy that there is always a combination of mechanisms during actual transport, especially when a pore size is close to the regime cutoff, and therefore, the final model could be simply a single model using the dominant transport mechanism or a dual model combining two mechanisms or a multimodel combining many mechanisms.

For ultrasmall pores, the size sieving mechanism is observed but is not relevant for describing ultrafast transport; however, the mechanism indeed is useful for separation applications that rely on the size and shape of the probe molecules. This mechanism can also be correlated with the micropore constrained transport described by Verweij et al. where molecules fit tightly rather than being completely excluded. The model proposed is based on a hopping model across Langmuir lattice sites that transition into a surface diffusion mechanism. The form of the equation follows a classic activation dependence:

$$J = -\alpha RT \nabla c, \tag{11.5}$$

where
  α is a proportionality parameter related to the energy barrier for a hopping event and the occupied/vacant fraction for each site
  R is the universal gas constant
  T is the temperature

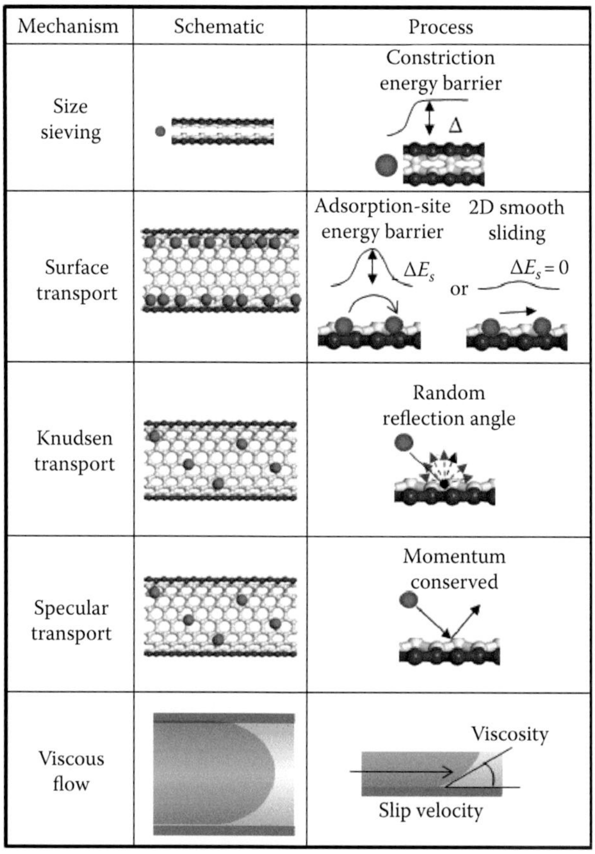

Mechanism	Schematic	Process
Size sieving		Constriction energy barrier Δ
Surface transport		Adsorption-site energy barrier ΔE_s or 2D smooth sliding $\Delta E_s = 0$
Knudsen transport		Random reflection angle
Specular transport		Momentum conserved
Viscous flow		Viscosity / Slip velocity

Figure 11.4 Transport regimes for gas diffusion in NTs.

Surface diffusion dominates when the majority of gas molecules are adsorbed upon the tube walls, which more often occurs with smaller tubes, stronger gas–tube interactions, and low temperatures. As flux can be separated into the two components of gas concentration and gas diffusivity according to Equation 11.1, surface diffusion will depend on the adsorbed phase density and the adsorbed phase diffusivity. Equations for density range from Henry's law (Everett and Powl 1976) or Boltzmann's law (Do et al. 2008) for low concentration to Langmuir model (Langmuir 1916) for single-layer adsorption to BET (Brunauer et al. 1938) for multilayer adsorption. If the walls are rough with bumps and troughs, then the diffusivity is also an activated diffusion process with the energy barrier associated with adsorption from one site to another site. Here, we offer the model summarized by Majumder et al. (2011):

$$J = -\rho_{app} D_s \mu_s \frac{dq}{dl}, \tag{11.6}$$

where

ρ_{app} is the apparent surface density
D_s is the surface diffusion coefficient dependent on energy barrier and temperature
μ_s is the reciprocal of tortuosity
dq/dl is the surface concentration gradient related to the heat of adsorption

If the walls are smooth, then the gas remains adsorbed but is free to glide upon the 2D surface only encountering hindrance when colliding with another gas molecule. Here, a simple 2D free gas model proposed by Verweij et al. describes the diffusivity dependence as follows:

$$D = \left(\frac{k_b T}{M}\right)^{1/2} (C_{ad}\sigma)^{-1}, \tag{11.7}$$

where
  k_B is the Boltzmann constant
  M is the gas molecular mass
  C_{ad} is the concentration of the adsorbed phase
  σ is the kinetic diameter of the gas molecule

In this model, a collision incident is incorporated using a mean free path concept related to the gas density and the gas kinetic diameter. Verweij et al. likened this incident to the scenario of motorbikes riding inside a cage, confined to the walls until a collision with another motorbike that sends the rider flying across the channel. An important fact about free surface diffusion is that the flux scales with the square root of molecular mass. This dependence has often been connected with Knudsen diffusion, and therefore, sometimes the incorrect mechanism is proposed. This complication will become evident by examining the experimental work in Section 11.3.3.

Knudsen transport is a well-known regime that typically occurs when the mean free path of a gas is much less than the pore diameter, that is, gas–wall collisions are more frequent than gas–gas collisions. The occurrence also depends on gas–tube interactions and temperature outlined by Thornton et al. (2009). As depicted in Figure 11.4, the incidence angle is randomly determined from a cosine rule, which means that momentum is lost and resistance is experienced upon every collision. The Knudsen model can expressed as

$$J = \frac{2\varepsilon_p \mu_{Kn} vr}{RTL}, \quad v = \left(\frac{8RT}{\pi M}\right)^{1/2}, \tag{11.8}$$

where
  ε_p is the porosity
  μ_{Kn} is the shape factor equal to the inverse of tortuosity
  v is the average molecule velocity
  r is the pore radius
  L is the membrane thickness

Overall, the Knudsen mechanism requires that the gas is in the gas phase and that the walls are rough such that the gas–wall collisions result in a random reflection angle.

Specular transport is the unique mechanism that only occurs for atomically smooth surfaces where a gas–wall collision results in a reflection angle identical to the incidence angle. Until the advent of NTs, no such nanostructure existed to test this mechanism. In practical terms, this mechanism means that the rate of gas molecules entering the tubes is identical to the rate of gas molecules exiting the tubes, that is, no longer rate limiting, no resistance. Therefore, the flux can be estimated from the external gas phase:

$$J = \frac{\Delta p}{\sqrt{2\pi RTM}}, \tag{11.9}$$

where Δp is the pressure gradient between upstream and downstream. This mechanism encompasses the result that transport is faster through the tube than if there was no material at all. This is because gas–gas collisions will occur in the bulk gas phase but not in the specular transport regime. As a comparison, Skoulidas et al. simulated methane flux of around 10,000 mol/s m² within the (10, 10) CNT in a 10 μm thick membrane, while the specular transport model predicts a flux of 10,628 mol/s m². It is important to note that although the model and simulation result comparisons are in good agreement, the mechanism may not be correct. As it was shown in other simulations, the majority of gas molecules are in adsorbed phase rather than the gas phase required by the specular transport model. To truly determine the correct mechanism, one needs to look at the trends with parameters such as temperature, molecular mass, and kinetic diameter.

Finally, the viscous or Poiseuille flow describes transport in larger tubes where gas–gas interactions are more dominant than gas–wall interactions and gas clusters form a bulk gas. The resulting expression is as follows:

$$J = \frac{\varepsilon_p r^2}{8RT\mu L\tau} P_m,$$ (11.10)

where
   μ is the gas viscosity
   τ is the tortuosity
   P_m is the mean pressure

The flow is driven by a pressure difference and the velocity at the pore wall is assumed to be zero, that is, no slip. The difference in flow rates between different gas molecules is only a result of differences in viscosity. Therefore, it is quite easy to determine whether this mechanism dominates transport by testing a variety of gases. It is worth noting that only Knudsen, specular, and viscous flow models are purely predictive with no adjustable parameters. Unfortunately, surface diffusion relies on a few parameters that are either unknown or difficult to measure and define.

Apart from the transport regimes described earlier, other fundamental theories include the description of the tangential momentum accommodation coefficient or a measure of specular versus Knudsen-type collisions. There is a Knudsen–Smoluchowski model where the tangential momentum accommodation coefficient (σ_v) is defined as the ratio of Knudsen diffusive collisions such that

$$D = D_{Kn}\left(\frac{2-\sigma_v}{\sigma_v}\right),$$ (11.11)

where
   D is the total diffusivity
   D_{Kn} is the predicted Knudsen diffusivity

This model has been extended to consider individual scattering events where the tangential momentum accommodation coefficient is defined as

$$D = D_{Kn}\left[1 - \frac{2\pi}{\ln\varepsilon}\frac{(1-\sigma_v)}{\sigma_v}\right],$$ (11.12)

where ε is the cutoff parameter that deals with incident angles that result in infinite collision times. These models have been utilized for 200 nm alumina pore filled with amorphous carbon,

10 nm polycarbonate pores, and simulated rough slit pores and recently applied to NTs. Other measures of friction or roughness include the Maxwell concept of thermalization where a fraction of molecules that transport through the tubes are trapped and lost upon the wall due to thermalization.

Overall, there are a team of fundamental transport models available to describe the ultrafast transport properties observed through NTs. Most likely, the total transport is a combination of surface diffusion, Knudsen diffusion, and specular transport mechanisms. There yet remains a unified model that predicts the correct ratio of mechanisms with no adjustable parameters. Therefore, there remain opportunities for future work in the transport model development.

11.3.3 Experimental Verification of Ultrafast Gas Transport

There are five groups who have experimentally demonstrated ultrafast gas transport in NTs, summarized in Figure 11.5 by Sears et al. (2010). Hinds et al. (2004) were the first to develop an array of aligned CNTs across a polymer film. The transport properties were originally described as close to the Knudsen prediction, although it was later shown that the open porosity was overestimated due to tube blocking with remaining Fe catalyst particles (Majumder et al. 2005). Therefore, the actual flux is likely to be around 40 times above that predicted by Knudsen theory. This means that there is only a discrepancy factor of 2 between experimental flux and that predicted by Chen et al. (2006) using the molecular simulation technique.

Holt et al. (2006) demonstrated fast mass transport through sub-2-nanometer CNTs achieving air flow rates from 16 to 120 times above the Knudsen prediction. The flow rates were also several orders of magnitude higher than those of commercial polycarbonate membranes, despite

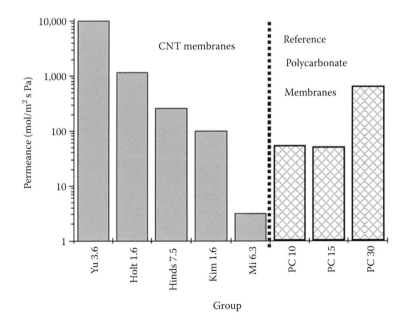

Figure 11.5 Nitrogen permeance in CNT membranes developed by Yu et al. (2010), Holt et al. (2006), Hinds et al. (2004), Kim et al. (2007), and Mi et al. (2007). Reference materials are included, polycarbonate with 10, 15, and 30 nm pores. (From Sears, K., *Materials*, 3(1), 127, 2010.)

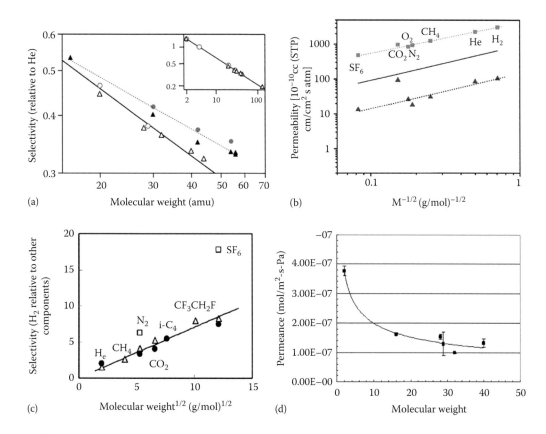

Figure 11.6 Correlation between gas permeability and molecular mass for (a) Holt et al. (2006), (b) Kim et al. (2007), (c) Yu et al. (2010), and (d) Majumder et al. (unpublished).

having pore sizes an order of magnitude smaller (see Figure 11.5). A correlation between flux and the square root of molecular mass was also observed (see Figure 11.6a). This further confirms that mass transport is through the sub-2-nanometer tubes rather than cracks or defects where no such correlation would be observed.

Mi et al. (2007) developed a vertically aligned CNT membrane on alumina supports. Enhanced factors of around 4 were achieved above Knudsen predictions. Once again the flux was found to be inversely proportional to the squared root of the molecular weight. The correlation suggests that the Knudsen diffusion mechanism is present, however, as described earlier and in the work of Verweij et al. (2007), the correlation does not necessarily mean Knudsen diffusion as specular and free surface diffusion mechanisms also scale with square root of molecular mass.

Kim et al. (2007) developed a scalable fabrication process for CNT/polymer nanocomposite membranes. Ultrafast mass transport was achieved for carbon dioxide, nitrogen, oxygen, methane, helium, hydrogen, and sulfur trifluoride. Correlation between permeance and inverse square root of molecular mass was also observed. Gas mixtures exhibited different properties than those observed using single-gas experiments, further confirming that non-Knudsen transport occurs.

Yu et al. (2009) produced pure CNT membranes achieving the highest density of NTs to date. This process was made possible using long (~750 μm) NTs where the capillary forces after solvent evaporation were strong enough to pull the NTs together forming a crack-free film. This means that there were interstitial pores between the CNTs that were not sealed as in the case of other membrane fabrication methods. Gas transport through the CNTs as well as the interstitial pores meant that these membranes achieved fluxes 1–4 orders of magnitude higher than all vertically aligned CNT membranes in the literature. In addition, permeabilities were approximately 450 times greater than those predicted by Knudsen, and their correlation with the square root of the molecular mass remained.

Finally, Majumder et al. fabricated CNT membranes and demonstrated ultrafast gas flow rates 15- to 30-fold above Knudsen predictions (Majumder 2011). Once again, a strong correlation between flux and molecular mass was observed at a slope close to that expected by Knudsen theory. A series of molecules with varying size were used to estimate the amount of open tubes that proved more accurate than the previous method of counting tubes from scanning electron micrograph images.

Overall, the total permeance mainly correlates with the NT density or the amount of porosity (see Figure 11.7). Yu et al. therefore achieved the highest flux. Other effects such as tube size and tube quality will also play a role, though there are not enough experimental data to confidently confirm the effects of tube size. However, tube quality can be estimated using the tangential momentum accommodation coefficient described in Section 11.3.2 (see Equations 11.11 and 11.12). Here, we calculate the coefficients for the available experimental data (see Figure 11.8). Holt et al., Yu et al., and Majumder et al. demonstrate the highest momentum coefficients that correspond with the fraction of specular permeation. This in turn can be interpreted as the highest-quality or smoothness NTs assuming that porosity is correctly estimated and that other effects such as entrance effects are ignored. Therefore, these results must be treated carefully since studies such as of Lee and Sinnott (2004) hinted that there are entrance effects that should not be ignored.

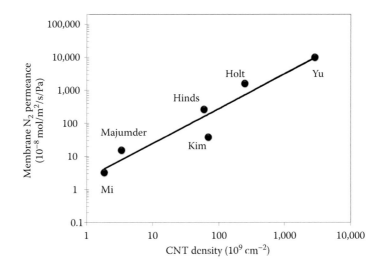

Figure 11.7   Correlation between nitrogen permeance and CNT density for Mi et al. (2007), Majumder (2011), Kim et al. (2007), Hinds et al. (2004), Holt et al. (2006), and Yu et al. (2010).

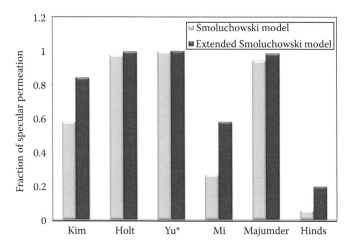

Figure 11.8 Tangential momentum accommodation coefficients for experimental nitrogen permeance calculated from Equations 11.11 and 11.12.

In conclusion, the questions proposed in the introduction section are revisited:

- What are the theoretical speed limits of transport? It is shown with theory and experimentation that transport through tubes is almost resistance free. This means that the major limiting factor is the velocity of gas molecules, which is, for example, around 500 m/s for nitrogen. Another limiting factor is the density of the gas as it flows through the tubes. Since there is fairly strong adsorption between the gas and tube at small tube diameters, the density is expected to be much higher than in other nanoporous materials. For a perfect tube with pure specular flow, the theoretical limit only depends on the porosity or density of tubes.
- What are the actual speeds observed in NTs to date? NT membranes have achieved a maximum nitrogen permeance of 10,000 mol/m² s Pa, which is several orders of magnitude above other nanoporous materials.
- What further developments are required to increase speeds toward the theoretical limits? Tube smoothness or quality and tube density are the key factors for further pushing experimental flow rates toward the theoretical limits.
- What implications do these results have on our fundamental understanding of transport and the promising applications in the real world? Unlike originally thought, the fundamental transport mechanism through NTs is not Knudsen but rather free surface diffusion. This mechanism utilizes the gas–tube interactions to maximize gas density and the smooth tube surfaces to maximize gas *sliding* across the tube interior. To inspire promising applications, one can imagine if a garden hose packed with continuous NTs were connected from a swimming pool of gaseous fuel to a spacecraft orbiting the planet, then it would take only 15 min to transport all the fuel to the spacecraft.

11.4 ULTRAFAST WATER TRANSPORT IN NANOTUBES

Several studies that include experiments and simulations have shown that water, driven by a pressure gradient, transports very rapidly through the interior of CNTs. Incorporating aligned CNTs within a membrane film has powerful implications of promising transport and separation properties. There is a growing interest in these membrane structures because it is envisioned that these membranes may lead to low-energy desalination technologies because the NTs can transport water at significantly faster rates, while the confinement and electrostatic interaction will lead to enhanced separation of salt molecules. In addition to desalination, there are interest in nanofluidic devices (Ulissi et al. 2011) for sensing, storing, and essentially controlling the flow of small molecules. Carbon surfaces such as those found in CNTs also have the advantage of biocompatibility, ready chemical modification to allow functionalization, and low friction properties.

This section outlines the research to date in molecular simulation, continuum modeling, and experimental verification. Simulation studies look at radial concentration profiles, friction coefficients, water–tube interactions, and water orientation. As with the gas simulation studies, there have been EMD and NEMD. Continuum models aim to explain ultrafast water transport using slip length and velocities. Good agreement is found between experiment, simulation, and continuum models that offer a complete encompassing framework for explaining and maximizing transport.

11.4.1 Molecular Simulation of Ultrafast Water Transport

Hummer et al. (2001) conducted several molecular simulation studies that revealed insightful features of water transport including tube hydrophobicity, water–tube interactions, and external forces. In an intuitive manner, they constructed thin films of NTs surrounded by water in a periodic cell. In the first study, phase transition in NTs such as the drying transition was shown to be drastically different from that found in bulk systems. In this situation, a vapor layer is formed that separates the bulk phase from the surface. This was the result of strong hydrogen bonding between water molecules, which causes the liquid to recede from the nonpolar surface of the CNT. In addition, pulse-like transmission of water molecules from the bulk into the tube was observed as a result of tight hydrogen bonding networks from the single-file water formation inside the tube (see Figure 11.9).

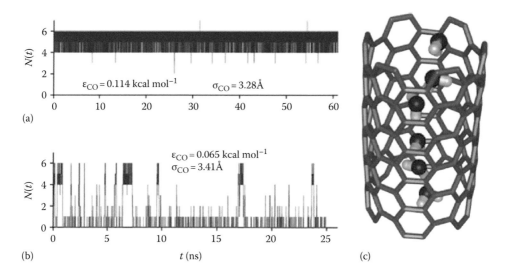

Figure 11.9   Water occupancy with (a) strong water–tube interactions and (b) weak water–tube interactions. (c) Single-file water transport is observed due to strong water–water hydrogen bonding. (From Hummer, G., *Nature*, 414, 188, 2001.)

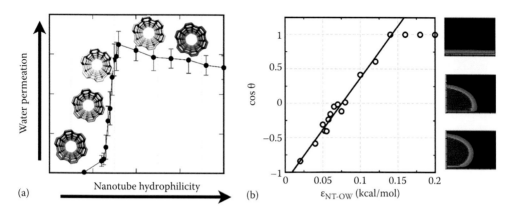

Figure 11.10   Effect of water–tube interaction strength (or NT hydrophilicity) on (a) water permeation in NTs and (b) contact angle of nanodroplet on graphene. (From Melillo, M. et al., *J. Phys. Chem. Lett.*, 2(23), 2978, 2011.)

With a minute increase in water–tube interactions, the tubes were completely filled with water as water preferred the tube environment over the bulk. Transport was later confirmed as stochastic in nature (Kalra et al. 2003). Other studies further explored the thermodynamics of water entering the tubes from a saline solution characterized by the excess chemical potential (Rasaiah et al. 2008). The chemical potential inside the tube depends on surface polarity, water–tube interactions, internal electric fields, and shape fluctuations, while the chemical potential outside the tube depends on pressure, osmolyte concentration, or external electric fields. Finding the right balance of chemical potential inside and outside the tube is the key to achieving high selectivity and high flow rates.

Melillo et al. (2011) used EMD to explore the effect of water–tube interaction strength on transport. As shown in Figure 11.10a, no water–tube interaction means that water–water interaction is more favorable so that water remains outside of the tube in the bulk water phase. As water–tube interactions increase, the environment within the tube is more favorable for water, and therefore, the water freely permeates into the tube. With even stronger water–tube interactions, the water finds the tube too favorable such that it sticks along the surface of the tube and diffuses with difficulty. This effect is also demonstrated by calculating the contact angle of a nanodroplet upon a graphene surface where the surface becomes completely hydrophilic at water–tube interaction strength of around 0.15 kcal/mol (see Figure 11.10b). This corresponds well with the work by Hummer et al. (2001) that showed slow pulse-like diffusion for water–tube interaction strength of 0.056 kcal/mol and ultrafast free flow using a water–tube interaction strength of 0.114 kcal/mol.

Joseph and Aluru (2008) addressed the question "Why are carbon nanotubes fast transporters of water?" By utilizing NEMD, they were able to closely examine the concentration profiles and the individual water orientations within carbon, boron nitride, and silicon NTs. It was clearly shown that carbon tubes offer less hydrogen bonds than silicon and boron nitride tubes; therefore, CNTs offer surfaces with lower friction. Furthermore, in the silicon NT, the oxygen atoms preferentially adsorbed directly over the six-membered rings across the surface forming a strongly coordinated layer of water, greatly restrictive for lateral motion. CNTs on the other hand offered no specific adsorption sites, and therefore, the surface appeared smooth to water molecules. Once again, the weak water–tube interactions in CNTs created a depletion region with low water concentrations close to the surface unlike silicon and boron nitride tubes with strong hydrogen binding and van der Waals interactions.

Another study by Corry (2008) systematically investigated water transport in four NTs of different diameters specifically for desalination. Sodium ions encountered energy barriers upon entry due mainly to the required shedding of the dehydration shell surrounding the ions. Therefore, ions could not transport through tubes smaller than about 0.5 nm, while water encountered no such impediment due to the formation of stable and favorable hydrogen bonds. Water transport reached around 2000 times greater than existing membrane technology FILMTEC SW30HR-380.

Self-diffusion of water, methanol, and ethanol in aluminosilicate NTs was investigated by Zhang et al. using MD simulations. Diffusivity decreases for all three liquids with increasing pore loading. The inner surface of aluminosilicate NTs contains hydroxyl groups that form strong hydrogen bonds with water creating a hopping mechanism with a *hop* energy barrier of around 17.4 kJ/mol. The surface hopping mechanism is prevalent at lower temperatures where the water molecules specifically orientate into the lowest-energy minima. At higher temperatures, the hopping energy barriers are overcome as water molecules lose preferential alignment, allowing them to move more freely across the tube axis. At low loadings, water spends more than half of the total diffusion time hydrogen bonded to at least one other water molecule. This water clustering effect upon the surface is partly responsible for decreasing diffusivity with increased loading. However, at high loadings, the steep decrease in diffusivity is not observed unlike with methanol and ethanol. In fact, the diffusivity tends toward that of bulk water with high loading as water clusters away from the surface of the tube and behaves closer to the bulk phase. Overall, the fast water diffusivity in aluminosilicate NTs is about three orders of magnitude smaller than in CNTs but considerably larger than in most zeolites.

Boron nitride NTs exhibit slightly faster water flow rates than CNTs according to Suk et al. (2008). This was due to a stronger energy barrier at the entrance and along the interior of the (6, 6) CNT compared with the boron nitride tube. The different energy barrier at the entrance was a result of the stronger van der Waals interactions that the boron nitride tube offers, consequently attracting the water inside more so than for CNTs. The different energy barrier within the tube interior is likely due to the correlation between water positions inside the pore. Similarly, Won and Aluru (2007) demonstrated that water prefers to adsorb within a (5, 5) boron nitride rather than a (5, 5) CNT. It is worth noting that these studies used static partial charges for the boron and nitrogen atoms. It is likely that the atoms are further polarized with water loading, and this effect is shown to drastically decrease the permeation according to Hilder et al. (2009).

All the studies mention the effects of surface roughness and water–tube interactions. Falk et al. (2010) carefully studied the friction coefficient and its dependence on tube curvature (directly related to tube diameter). The potential energy landscape depicted in Figure 11.11 reveals that the roughness depends on curvature by comparing a section of graphene with the inside section of a (10, 10) NT. The roughness effect is likened to that of an egg carton. When it is rolled on the inside, the structure is compressed and the wells of the landscape are smoothed out. As a result, smaller tubes should be smoother than larger tubes. This effect agrees well with the work by Thomas and McGaughey where smoothness was characterized with slip length, which is an extrapolation of the extra tube radius required to give zero velocity at a hypothetical wall (Thomas and McGaughey 2008, Thomas et al. 2010). Slip length increases considerably with shrinking tube size along with a decrease in viscosity, therefore confirming less friction in smaller tubes. This strong relationship between slip length and flow enhancement could be demonstrated using EMD and NEMD (Kannam et al. 2013; see Figure 11.12). A closer analysis of the water configuration across the internal surface of small

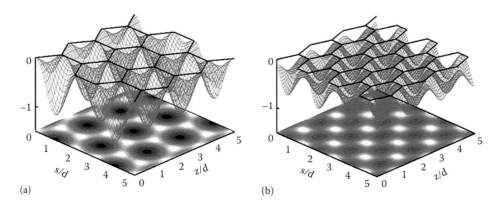

Figure 11.11 The potential energy landscape upon (a) a flat graphene surface and (b) the inside of a (10, 10) NT. (From Falk, K. et al., *Nano Lett.* 10(10), 4067, 2010.)

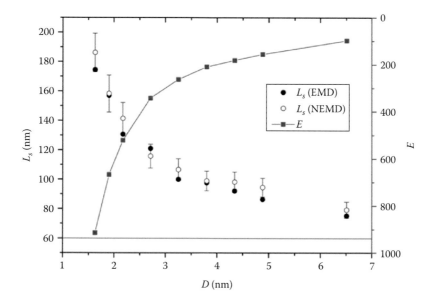

Figure 11.12 Slip length (L_s) predicted with EMD and NEMD and flow enhancement (E) with NT diameter (D). Black line at 60 (±6) is the slip length of water on a planar graphene surface. (From Kannam, S.K. et al., *J. Chem. Phys.*, 136(2), 024705, 2012; Kannam, S.K. et al., *J. Chem. Phys.*, 138(9), 094701, 2013.)

tubes revealed that water molecules do not align into adjacent adsorption sites, that is, the water positions are uncorrelated with the surface (Thomas et al. 2010).

Overall molecular simulations have covered every aspect of ultrafast water transport in NTs. The water–tube interaction strength determines the energy barrier for entrance, while the surface potential landscape determines the energy barrier for surface hopping. NTs offer the correct water–tube interaction strength and smooth potential landscape for ultrafast water transport orders of magnitude greater than conventional materials.

11.4.2 Continuum Models for Ultrafast Water Transport

The conventional understanding of liquid transport begins with the works of Laplace, Poisson, and Young, described in a summary by Whitby and Quirke (2007). The fluid velocity as a function of radial position, r, in the direction of flow, z, can be expressed as

$$U_z(r) = \frac{dP}{dz}\frac{1}{2\mu}\left[(L_s + R)^2 - (R - r)^2\right],$$

where
   μ is the viscosity
   L_s is the slip length
   R is the tube radius
   dP/dz is the pressure gradient along the flow direction

This results in a cross-sectional parabolic velocity profile across the radius of the tube. Integration of the fluid velocity over the cross-sectional area gives the Hagen–Poiseuille expression for flux (Q_{HP}),

$$Q_{HP} = \frac{\pi\left(R^4 + 4R^3 L_s\right)}{8\mu}\frac{\Delta P}{L}, \tag{11.13}$$

where ΔP is the pressure difference across the tube length (L). The critical question here is how well this classical equation can describe the transport at the nanoscale. The slip length L_s is conventionally assumed to be zero; however, molecular simulation has proved this assumption incorrect. In addition, the parabolic velocity profile is shown to deviate for tubes smaller than several nanometers (Travis et al. 1997, Sokhan 2004, Joseph and Aluru 2008), especially in sub-nanometer tubes where the single-file transport has no correlation with the continuum expression. Note that the Hagen–Poiseuille model assumes no variation in viscosity or density across the radius of the tube, which also has proved false from simulation studies.

The issue of the radial independent viscosity was addressed by Myers (2011) by incorporating a reduced viscosity model. Myers argued that perhaps the conventional term *slip length*, defined as the distance the velocity profile at the wall must be extrapolated to reach zero, is misleading. Instead, a length scale associated with the reduced viscosity makes more sense physically. The mathematical model takes the form of a biviscosity model in which the fluid in the cross section is split into two regions: a bulk region close to the center of the tube with viscosity set to that of bulk phase μ_b and a second region close to the tube wall with a reduced viscosity μ_r. The radial position (α) where the transition between phases occurs was set to a distance of $\delta = 0.7$ nm from the tube wall, taken from the simulation studies by Joseph and Aluru (2008) and Thomas et al. (2008). The viscosity of the second phase close to the wall was also chosen as an average viscosity of that observed close to the wall from the same simulation studies. As a result, the following modified expression for flux was constructed:

$$Q = Q_{HP}\frac{\alpha^4}{R^4}\left[1 + \frac{\mu_b}{\mu_r}\left(\frac{R^4}{\alpha^4} - 1\right)\right]. \tag{11.14}$$

In this case, the slip length may be defined in terms of the thickness of the depletion layer δ and the viscosity ratio μ_b/μ_r:

$$L_s = \delta\left(\frac{\mu_b}{\mu_r} - 1\right)\left[1 - \frac{3}{2}\frac{\delta}{R} + \left(\frac{\delta}{R}\right)^2 - \frac{1}{4}\left(\frac{\delta}{R}\right)^3\right]. \tag{11.15}$$

The elegance of this expression for flux and the new physical definition of slip length are evident. However, the predictive ability of the model depends on the thickness and viscosity of the depletion layer, parameters yet to be determined from first principles.

From a radically different approach, Chan and Hill (2011) constructed a mathematical model for the single-file transport of water based on phonon theory. It has been stated from multiple simulation authors that single-file diffusion is stochastic in nature (Kalra et al. 2003) though there are some deterministic forces such as chemical potential or entrance and surface hopping energy barriers. Since water transport is correlated with collective motion from strong water–water hydrogen bonds, it is justified that principles from classical phonon theory may be adopted. The model still takes advantage of the van der Waals and electrostatic interactions between water–water and water–tube as well as the entrance effects involving bulk water and water–tube considerations. In addition, the phonon theory offers the inclusion of spring constants throughout a chain of water molecules and the wave-like propagation effects from water–water interactions. The governing equation for motion of the jth water molecule is as follows:

$$\frac{d^2 u_j}{dt^2} = u_{j+1} - 2u_j + u_{j-1}, \tag{11.16}$$

where u_j is the vector position of the jth water molecule inside the tube. Equation 11.16 can be solved numerically or analytically by assuming the wave form for $u_{j\pm1}$, which results in an expression for the average group velocity:

$$v_g = a_s \left(\frac{c}{2m}\right)^{1/2}, \tag{11.17}$$

where
    a_s is the molecular spacing of water molecules
    c is the spring constant
    m is the mass of water

Excellent agreement is found with Hummer's hopping rate (Berezhkovskii and Hummer 2002), which is determined by Chan and Hill using nanomechanic principles. Numerical solutions of the model are order of magnitude computationally efficient over molecular simulation.

Hilder and Hill (2009) developed a model based on nanomechanic principles to investigate the entrance effects of water entering NTs. The model is based on discrete treatment of the water molecule and a continuum treatment of the NT. By considering potential orientations at the entrance, the energy barrier or advantage for entering the NT was determined for varying the NT diameter. A concept emerging from the model is the *suction energy*, where there are, in fact, scenarios where the potential energy environment within the tube is considerably more attractive than in the bulk water phase, thus creating a suction effect. This suction effect has also been considered by Hilder for gas transport in a variety of nanopores ranging in shape, size, and composition (Thornton et al. 2009). The final conclusions include the fact that the NT diameter necessary to accept a water molecule is 0.69 nm and the diameter that provides the maximum *suction energy* is 0.8 nm, which equates to a maximum entrance velocity of 2458 m/s. Considering that the average velocity of water in the bulk phase is roughly around 400 m/s, and assuming that the internal surface is truly frictionless, this incredible enhancement in velocity would in effect last throughout the length of the tube until reaching the end where an equivalent and opposite suction force is met. Fortunately, the additional kinetic energy already existing before entrance will overcome any exiting energy barrier, therefore the suction benefits are completely utilized.

In summary, the Hagen–Poiseuille continuum model has mainly been used to calculate an enhancement factor that NTs achieve above this model prediction. The model has been extended to consider the effects of radial-dependent viscosity, however lacking predictive capability. New theories incorporating classical phonon and nanomechanic principles have been developed and show promising predictivity and physical insights.

11.4.3 Experimental Verification of Ultrafast Water Transport

Fortunately, experimental verification of ultrafast water transport followed theoretical predictions; otherwise, this chapter may have never been written. Several groups have successfully synthesized aligned NT materials and measured enormous flux properties. However, there is still a large gap between the flow rates observed and those that are simulated. Here, we undertake to compare experiment, simulated, and classical theory results for nanoscale fluid flow. Nonetheless in retrospect, it is safe to state that water transport through NTs is faster than through any known tube.

Starting with the three fabrication methods for preparing CNT composite membranes outlined in Section 11.2, water flux and salt rejection results are compared in Table 11.1. To be noted that most of the fundamental studies on transport in CNTs have been based on the *rather appealing* Type 1 membrane architecture, however, they have several potential drawbacks: (1) the concentration of CNTs in the membrane even with a large areal density of $\sim 2 \times 10^{11}$ tubes/cm^2, the wt% of CNTs is quite small $\sim 0.6\%$, meaning that the overall flux through the membrane does not benefit significantly from the enhanced water transport through CNTs; (2) reproducible and scalable fabrication of the membrane is dependent on the ability to make large areas of aligned CNTs with narrow diameter distribution, which is a significant technical challenge. These studies although still show that with $\sim 45\%$ salt rejection from 75 ppm salt concentration, the flow rates through each CNT are still large. The salt rejection mechanism in this membrane is Donnan-type exclusion since the salt rejection decreases with salt concentration of the feed water.

The Type 2 CNT membranes appear to have a significantly larger wt% of NTs; however, the water flux is significantly smaller than Type 1 and in the range of typical reverse osmosis membranes, indicating that water transport is likely to occur by a solution–diffusion mechanism (see Table 11.1). This is further supported by the relatively large salt rejection at a large salt concentration of ~ 2500 ppm.

The nonaligned CNT membranes (Type 3) synthesized by the blade casting and phase inversion of a multiwalled nanotube/polyethersulfone (PES) dispersion have water fluxes comparable to Type 2; however, salt rejection is also significantly smaller, which is likely to arise from the large diameter of the NTs (~ 8 nm) as well as lack of targeted functionalization as in Type 2.

TABLE 11.1 COMPARISON OF THE PERFORMANCE OF DIFFERENT TYPES OF CNT NANOCOMPOSITE MEMBRANES

	Type 1—Grown CNT Forest (Fornasiero et al. 2008)	Type 2—Filtration Alignment (Chan et al. 2013)	Type 3—Nonaligned CNTs in Asymmetric Matrix (Vatanpour et al. 2011)
Flux in L/(m^2 h bar)	53.62[a]	1.33[b]	~2.25[c] for 1.1 wt% CNT membrane, ~2 for 2.2 wt%
No. of CNTs/cm^2	2.5 × 10^{11} (Holt et al. 2006)	—	—
CNT content by weight in polymer	0.6 wt%[d]	20 wt%	0.22, 1.1, and 2.2 wt%[e]
Salt rejection	45% Cl$^-$, 37% K$^+$ in 1.0 mM (~75 ppm)[f] KCl solution, ~80% cation and ~90% anion rejection for 1.0 mM (~330 ppm)[g] K$_3$Fe(CN)$_6$ solution at 0.69 bar[h]	98.6% for 43.5 mM (~2500 ppm) NaCl solution at 36.5 bar	200 ppm Na$_2$SO$_4$: 80%; 200 ppm NaCl: ~15% at 4 bar for 0.22 wt% CNT membrane
Membrane matrix material	Silicon nitride (Si$_3$N$_4$)	Polyamide (PA)	Polyethersulfone (PES), solvent: DMAc, nonsolvent: water
CNTs	Double-walled nanotubes (DWNTs) with sub-2 nm diameters	Zwitterion-functionalized CNTs with outer diameter of 1.5 nm	Multiwalled CNTs with OD ~10–30 nm

[a] Calculated from 37 μL/mm^2 h at 0.69 bar.
[b] Calculated from 28.7 GFD (gallons per square foot and day) at a pressure drop of 3.65 MPa.
[c] Calculated from ~9 kg/m^2 h and ~8 kg/m^2 h, respectively, in Figure 6 of Vatanpour et al. (2011).
[d] Calculated, with density for 2 nm DWCNTs = 2.5 g/cm^3 (Laurent 2010).
[e] Calculated from CNT concentrations of 0.04–0.4 wt% in casting solution containing 18 wt% PES.
[f] Calculated from 1 mM with M(KCl) = 74.55 g/mol.
[g] Calculated from 1 mM with M(K$_3$FE(CN)$_6$) = 329.24 g/mol.
[h] From supporting material for Fornasiero et al. (2008).

Majumder et al. (2005) grew NTs with inner diameters of about 7 nm and embedded them within a nonporous polystyrene matrix at a density of 5 × 10^{10} tubes/cm^2. With an applied pressure of about 1 atm, remarkable water permeabilities of around 1 cm^3/cm^2 min bar (or flow velocity ~10 cm/s) were observed, which is about five orders of magnitude above that expected by conventional flow theory and about half of that predicted by molecular simulation. The enhancement above the conventional flow theory greatly depends on the accuracy of measuring the tube density. In addition, extralong slip lengths of 39–54 nm were calculated, which are well above the 3.5 nm radius of the tubes, suggesting almost perfect frictionless interiors. Further work by Majumder et al. looked adding hydrophilic functional groups at the entrance of the tubes that significantly decreased the flow rates (Majumder et al. 2005, Majumder 2011).

Holt et al. (2006) followed this research by incorporating even smaller tubes (sub-2-nanometer) in a silicon nitride matrix via chemical vapor deposition. Water flow rates once again proved to be well above that predicted from conventional theory with slip lengths ranging from 140 to 1400 nm.

Qin et al. (2011) measured the water flow rate through an individual millimeter-long CNT using field effect transistors. A droplet of water is placed at the entrance of the NT, and the transistors measure the presence of water at intervals along the tube axis. The work demonstrates rate enhancements from 51 to 883 for NTs with diameters of 1.59–0.81 nm. Furthermore, the observed velocities fall into the range of 46–928 μm/s.

Du et al. (2011) developed superlong 4 mm vertically aligned CNTs within an epoxy resin and demonstrated flow of water, hexane, and dodecane. An alternating voltage was used to switch the water flow on and off. An enormous slip length of 485,000 nm was calculated with a flow rate of 0.27 mL/min and flow velocity of up to 3.36×10^{-2} m/s.

Kannam et al. compared slip lengths across experimental, simulated, and theoretical results (see Figure 11.13). For experimental studies, there is a remarkable variation in slip length varying over five orders of magnitude for a tube diameter range of 0.8–10 nm. For simulation studies, there is still a large variation in slip length between 1 and 1100 nm for the same range of tube diameters. The variation in experimental slip lengths is likely due to the large uncertainty in defining the actual accessible tube volume, while the variation in simulation results is accounted for by the differences in the molecular models used to simulate the system and the simulation details. Some experimental results here were not strictly NTs but can be called nanopipes made from amorphous carbon such as by Whitby et al. (2008) <10 nm. Nevertheless, Whitby et al. carbon nanopipes with diameters <10 nm still demonstrated pressure-driven flow 45 times above theoretical predictions.

Other related experimental work includes the study of contact angles of water droplets on electrochemical controlled carbon surfaces (Wang et al. 2007), confirming the relationship between hydrophilicity and water–tube interactions. DNA transport was demonstrated

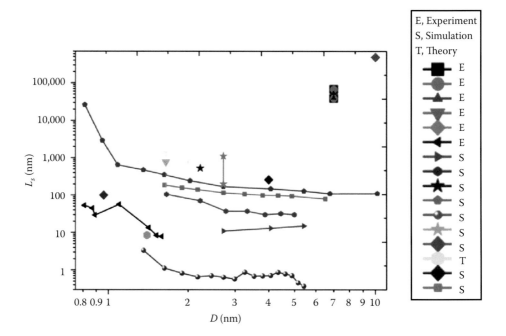

Figure 11.13 Slip lengths from the literature of water in CNTs. (From Kannam, S.K. et al., *J. Chem. Phys.*, 138(9), 094701, 2013.)

through small carbon tubes with varying diameters to help unravel the DNA for aligned transport (Cao et al. 2002). Finally, neutron scattering were performed to investigate soft dynamics of confined water molecules, leading to the discovery of icelike NTs with CNTs (Kolesnikov et al. 2004).

In conclusion, the questions proposed in the introduction can be answered as follows:

- What are the theoretical speed limits of transport? An individual water molecule is likely to exceed 2000 m/s due to the favorable suction forces at the entrance of the NT. For clusters or bulk water, an infinite slip boundary condition would allow transport at ultrafast rates that depend only on the pressure gradient.
- What are the actual speeds observed in NTs to date? In comparison with conventional flow with no-slip boundary conditions, actual water flow rates have exceeded conventional flow rates by a factor of 883.
- What further developments are required to increase speeds toward the theoretical limits? From the range of simulation and experimental work, it is clear that tailoring the chemical composition at different positions within the tube would enhance flow rates even further. For example, the entrance needs to be configured to favorably attract water. The interior of the tube needs to be configured to disrupt water–water and water–tube bonding in a way that allows free flow to occur. Finally, the exit of the tube needs to be configured so that the water is happy to exit the tube.
- What implications do these results have on our fundamental understanding of transport and the promising applications in the real world? These results confirm many other studies that describe water as the most unique fluid in the universe. Water's behavior depends so sensitively on the environment including the structure and interactions. With flow rates almost 1000 times faster than conventional flow, there are exciting possibilities for efficient environmental and energy applications such as desalination.

11.5 ULTRAFAST GAS AND WATER TRANSPORT IN NANOSHEETS

Graphene and GO have potential in applications such as energy storage, water filtration, sensing, barrier film technology, fuel cells, and flexible electronics. While graphene is hydrophobic, GO is hydrophilic, and therefore, there is potential for efficient water/oil separations. Graphene and GO platelets can be stacked layer by layer to form thin films with varying layer–layer distance. Alternatively, holes can be created using an electron beam to form selective pores for separating species of different sizes. Here, we review the simulation, theoretical, and experimental work to date in the area of gas and water transport.

Medhekar et al. (2010) employed MD simulations with reactive force fields to investigate the structural and mechanical properties of stacked GO platelets, shown in Figure 11.14. They found that strong hydrogen bonding networks form within the interlayer cavities. In the presence of water, the interlayer spacing increases by almost double as water interacts with the functional groups on the individual platelets as shown in Figure 11.14b. The flow rates of water were not calculated in this study and it is difficult to predict as was shown in the NT studies, superhydrophilicity can have a detrimental effect of flow rates. Although if water is capable of pushing layers apart and creating its own pathways, then it is possible to achieve ultrafast water flow.

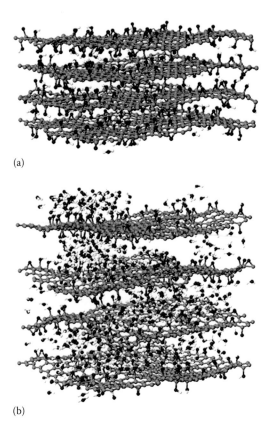

(a)

(b)

Figure 11.14   Atomic structure of hydrated graphene oxide containing (a) 0.9 and (b) 25.8 wt% of water. Layer–layer distance increases from (a) 5.1 to (b) 9.0 Å. Carbon, oxygen, and hydrogen atoms are represented by gray, black, and white spheres, respectively. (From Medhekar, N.V. et al., *ACS Nano*, 4(4), 2300, 2010.)

Pristine graphene layers were stacked using a virtual porous carbon algorithm by Biggs and Buts (2006), shown in Figure 11.15. By varying the stacking angle, the cavity size distribution is modified. With a fixed rotation angle of zero, platelets stack horizontally flat and the cavities arise from the gaps between platelets that are uniform is size. A random orientation angle of ±15° creates larger porosity and wider range of cavity sizes. The methane diffusivity, however, is faster in the uniform layers, which is possibly due to the smoother and straighter channels similar to that found in NTs.

Gas adsorption simulations within oxygen-functionalized graphene showed that the potential energy profile between layers significantly strengthened and roughened with functionalization (Gotzias et al. 2012). This can result in an increase of over 200% in gas uptake at cryogenic conditions. By varying the interlayer distance and the amount of functionalization, one can tune the adsorption properties of different gases for separation applications. Without diffusivity calculations, it is difficult to determine the gas flux properties of these materials, though high adsorption can play an important role in the flux.

Flux depends on film thickness, and therefore, a single layer of graphene with a thickness of one atomic diameter is an ideal membrane configuration. Suk and Aluru studied the water

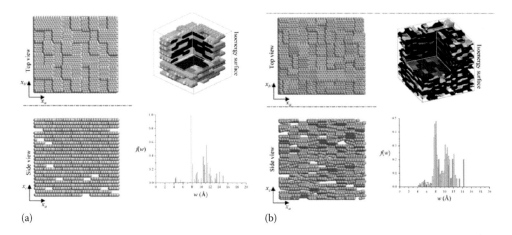

Figure 11.15 Platelets of graphene stacked (a) evenly and (b) unevenly. Isoenergy surfaces and pore size distributions are presented for each scenario. (From Biggs, M.J. and Buts, A., *Mol. Simul.*, 32(7), 579, 2006.)

transport through a single graphene layer with an aperture diameter of 0.75 and 2.75 nm, shown in Figure 11.16. In comparison with CNTs of similar diameters 0.78 and 2.71 nm, it was found that the flux through the monolayer graphene membrane only surpassed that of CNTs for the larger aperture ~2.7 nm, while the water flux in smaller apertures ~0.75 nm is greater in CNTs. In light of the discussion in Section 11.3, it is not surprising that transport is faster in smaller tubes where the water configuration is most favorable. Nevertheless, the transport through graphene is ultrafast and has promising properties for membrane-based separations. Another study by Du et al. (2011) explored a range of aperture sizes and shapes to separate nitrogen from hydrogen, achieving a selectivity of up to 6 at an incredibly fast rate of 3500 mol/m² s. Further work by Jiang et al. (2009) confirmed huge hydrogen over methane selectivity of 10²³ with an ultrafast hydrogen flux of 10 mol/cm² s.

Continuum models for transport in graphene sheets take the same form as the regimes outlined in Sections 11.3.2 and 11.4.2, but with slip geometry. Mechanisms such as size sieving, surface diffusion, Knudsen, specular, and viscous flow still apply. In addition, Su and Lua (2009) adopted a combination of Knudsen, configurational, and surface diffusion that incorporate effects such as gas concentration and energy barriers. Permeance (P) is expressed as

$$P = D_{ads} \frac{\mathrm{d}\ln p}{\mathrm{d}\ln C_{ads}} \frac{\theta}{\tau} \frac{\mathrm{d}C_{ads}}{\mathrm{d}p} + \frac{\theta}{\tau} \frac{D_{gas}}{RT}, \qquad (11.18)$$

where
   D_{ads} and D_{gas} are the diffusivities
   C_{ads} and C_{gas} are the concentrations for the adsorbed phase and gas phase, respectively
   θ is the porosity
   τ is the tortuosity
   p is the pressure

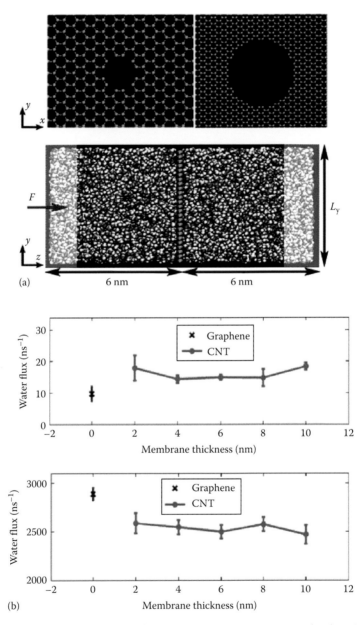

Figure 11.16   Graphene membrane with (a) nanopore of diameter 0.75 nm (left) and 2.75 nm (right) and (b) water flux through graphene nanopore of 0.75 nm compared with CNT of diameter 0.78 nm (top) and throughout graphene nanopore of 2.75 nm compared with CNT of diameter 2.71 nm (bottom). (From Suk, M.E. and Aluru, N.R., *J. Phys. Chem. Lett.*, 1(10), 1590, 2010.)

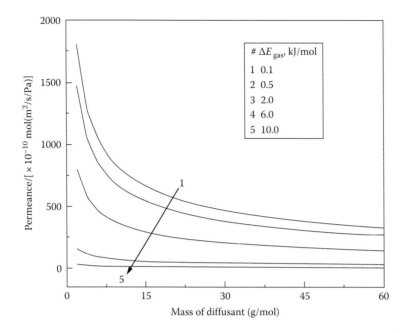

Figure 11.17 Theoretical permeances of gases with different masses and energy barriers. (From Su, J. and Lua, A.C., *Sep. Purif. Technol.*, 69(2), 161, 2009.)

The diffusivities and consequently the permeance are greatly dependent on the energy barriers as shown in Figure 11.17. As elucidated in the molecular simulation studies, the energy barriers can occur at the entrance of the channel, along the surface of the channel, or at the transitions from layer to layer in the case of graphene stacks. By controlling these energy barriers, one can drastically enhance the selectivity and flow rate of gas or liquid species.

Recently, it was reported that GO films in the dry state were not permeable to small gases (e.g., helium and argon), but water molecules permeated through hydrated GO films faster than expected (Nair et al. 2012), as shown in Figure 11.18a. The potential energy barrier for gas entrance, the tortuosity determined by the GO sheet size and thickness, the layer structure, residual water content, and possible deformation of thin-layered structures during dewetting are all important considerations when studying molecular transport in GO films. As such, the effective gas diffusional path is affected by the GO sheet size, stacking number, and stacking manner, assuming that the interlayer distances between GO sheets are sufficient to permit diffusion of small gases. Moreover, it is important to note that GO is not completely 2D, in contrast to graphene sheets, which suggests that GO should stack nonuniformly. For the earlier reasons, thin GO membranes should be permeable to small gases if sufficient transmembrane pressure is applied to overcome the energy barriers to pore entry and diffusion within pores or channels. The plausible explanation also simulated by Nair et al. and Medhekar et al. is the swelling or expansion of the interlayer distance with increased humidity. Reduced GO films did not exhibit the swelling effect, further confirming the necessary water–surface interactions to push the layers apart.

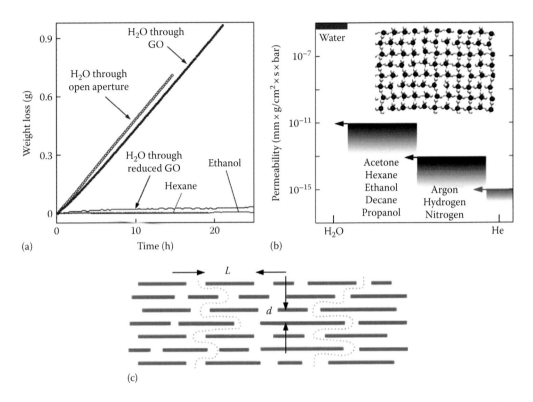

Figure 11.18 Permeation through graphene oxide. (a) Weight loss for a container sealed with a graphene oxide film. No loss was detected for ethanol and hexane, but water evaporated from the container as freely as through an open aperture. (b) Permeability of graphene oxide paper with respect to water and various small molecules. (c) Schematic view for possible permeation through the platelets of graphene oxide. (From Nair, R.R. et al., *Science*, 335(6067), 442, 2012.)

11.6   CONCLUSIONS

Overall, transport in nanosheets has been covered less than NTs, but there is great potential and the research area is growing. The ability to have adaptive and dynamic layer–layer distances is a unique attribute of stacked nanosheets. With careful control of the chemical functionalization of nanosheets, separation factors and ultrafast flow rates are possible. There are still many unanswered questions, for example, it has been demonstrated that nanosheets can be stacked to form barriers to gas while allowing free flow of water, which is difficult to explain. Nonetheless, simulation studies are continuing to shed light on the structure of nanosheets and their pathways for transport. Future research in nanosheets is likely to have great impact in the fields of nanofluidics, barrier technologies, and molecular separations.

REFERENCES

Ackerman, D. M., A. I. Skoulidas, D. S. Sholl, and J. K. Johnson. 2003. Diffusivities of Ar and Ne in carbon nanotubes. *Molecular Simulation* 29 (10–11):677–684. doi: 10.1080/0892702031000103239.
Arora, G. and S. I. Sandler. 2006. Air separation by single wall carbon nanotubes: Mass transport and kinetic selectivity. *The Journal of Chemical Physics* 124 (8):084702–084702-11.

Berezhkovskii, A. and G. Hummer. 2002. Single-file transport of water molecules through a carbon nanotube. *Physical Review Letters* 89 (6):064503.

Bhatia, S. K., H. B. Chen, and D. S. Sholl. 2005. Comparisons of diffusive and viscous contributions to transport coefficients of light gases in single-walled carbon nanotubes. *Molecular Simulation* 31 (9):643–649. doi: 10.1080/00268970500108403.

Biggs, M. J. and A. Buts. 2006. Virtual porous carbons: What they are and what they can be used for. *Molecular Simulation* 32 (7):579–593. doi: 10.1080/08927020600836242.

Brunauer, S., P. H. Emmett, and E. Teller. 1938. Adsorption of gases in multimolecular layers. *Journal of the American Chemical Society* 60 (2):309–319. doi: 10.1021/ja01269a023.

Cao, H., J. O. Tegenfeldt, R. H. Austin, and S. Y. Chou. 2002. Gradient nanostructures for interfacing microfluidics and nanofluidics. *Applied Physics Letters* 81 (16):3058–3060.

Chan, W.-F., H.-Y. Chen, A. Surapathi, M. G. Taylor, X. Shao, E. Marand, and J. K. Johnson. 2013. Zwitterion functionalized carbon nanotube/polyamide nanocomposite membranes for water desalination. *ACS Nano* 7 (6):5308–5319. doi: 10.1021/nn4011494.

Chan, Y. and J. M. Hill. 2011. A mechanical model for single-file transport of water through carbon nanotube membranes. *Journal of Membrane Science* 372 (1–2):57–65.

Chen, H., J. K. Johnson, and D. S. Sholl. 2006. Transport diffusion of gases is rapid in flexible carbon nanotubes. *The Journal of Physical Chemistry B* 110 (5):1971–1975. doi: 10.1021/jp056911i.

Chen, H. and D. S. Sholl. 2004. Rapid diffusion of CH_4/H_2 mixtures in single-walled carbon nanotubes. *Journal of the American Chemical Society* 126 (25):7778–7779. doi: 10.1021/ja039462d.

Corry, B. 2008. Designing carbon nanotube membranes for efficient water desalination. *The Journal of Physical Chemistry B* 112 (5):1427–1434. doi: 10.1021/jp709845u.

Cracknell, R. F., D. Nicholson, and N. Quirke. 1995. Direct molecular dynamics simulation of flow down a chemical potential gradient in a slit-shaped micropore. *Physical Review Letters* 74 (13):2463–2466.

Do, D. D., D. Nicholson, and H. D. Do. 2008. On the Henry constant and isosteric heat at zero loading in gas phase adsorption. *Journal of Colloid and Interface Science* 324 (1–2):15–24. doi: 10.1016/j.jcis.2008.05.028.

Du, F., L. Qu, Z. Xia, L. Feng, and L. Dai. 2011a. Membranes of vertically aligned superlong carbon nanotubes. *Langmuir* 27 (13):8437–8443. doi: 10.1021/la200995r.

Du, H., J. Li, J. Zhang, G. Su, X. Li, and Y. Zhao. 2011b. Separation of hydrogen and nitrogen gases with porous graphene membrane. *The Journal of Physical Chemistry C* 115 (47):23261–23266. doi: 10.1021/jp206258u.

Everett, D. H. 1972. Manual of symbols and terminology of physicochemical quantities and units, Appendix II: Definitions, terminology and symbols in colloid and surface chemistry. *Pure and Applied Chemistry* 31 (4):577–638. doi: 10.1351/pac197231040577.

Everett, D. H. and J. C. Powl. 1976. Adsorption in slit-like and cylindrical micropores in the Henry's law region. A model for the microporosity of carbons. *Journal of the Chemical Society, Faraday Transactions 1: Physical Chemistry in Condensed Phases* 72:619–636.

Falk, K., F. Sedlmeier, L. Joly, R. R. Netz, and L. Bocquet. 2010. Molecular origin of fast water transport in carbon nanotube membranes: Superlubricity versus curvature dependent friction. *Nano Letters* 10 (10):4067–4073. doi: 10.1021/nl1021046.

Fornasiero, F., H. G. Park, J. K. Holt, M. Stadermann, C. P. Grigoropoulos, A. Noy, and O. Bakajin. 2008. Ion exclusion by sub-2-nm carbon nanotube pores. *Proceedings of the National Academy of Sciences of the United States of America* 105 (45):17250–17255. doi: 10.1073/pnas.0710437105.

Gotzias, A., E. Tylianakis, G. Froudakis, and T. Steriotis. 2012. Theoretical study of hydrogen adsorption in oxygen functionalized carbon slit pores. *Microporous and Mesoporous Materials* 154:38–44. doi: 10.1016/j.micromeso.2011.10.011.

Heffelfinger, G. S. and F. van Swol. 1994. Diffusion in Lennard–Jones fluids using dual control volume grand canonical molecular dynamics simulation (DCV-GCMD). *The Journal of Chemical Physics* 100 (10):7548–7552.

Hilder, T. A., D. Gordon, and S.-H. Chung. 2009. Salt rejection and water transport through boron nitride nanotubes. *Small* 5 (19):2183–2190. doi: 10.1002/smll.200900349.

Hilder, T. A. and J. M. Hill. 2009. Maximum velocity for a single water molecule entering a carbon nanotube. *Journal of Nanoscience and Nanotechnology* 9:1403–1407.

Hinds, B. J., N. Chopra, T. Rantell, R. Andrews, V. Gavalas, and L. G. Bachas. 2004. Aligned multiwalled carbon nanotube membranes. *Science* 303:62–65.

Holt, J. K., H. G. Park, Y. Wang, M. Stadermann, A. B. Artyukhin, C. P. Grigoropoulos, A. Noy, and O. Bakajin. 2006. Fast mass transport through sub-2-nanometer carbon nanotubes. *Science* 312:1034–1037.

Hummer, G., J. C. Rasaiah, and J. P. Noworyta. 2001. Water conduction through the hydrophobic channel of a carbon nanotube. *Nature* 414:188.

Jakobtorweihen, S., F. J. Keil, and B. Smit. 2006. Temperature and size effects on diffusion in carbon nanotubes. *The Journal of Physical Chemistry B* 110 (33):16332–16336. doi: 10.1021/jp063424+.

Jiang, D., V. R. Cooper, and S. Dai. 2009. Porous graphene as the ultimate membrane for gas separation. *Nano Letters* 9 (12):4019–4024. doi: 10.1021/nl9021946.

Joseph, S. and N. R. Aluru. 2008. Why are carbon nanotubes fast transporters of water? *Nano Letters* 8 (2):452–458. doi: 10.1021/nl072385q.

Kalra, A., S. Garde, and G. Hummer. 2003. Osmotic water transport through carbon nanotube membranes. *Proceedings of the National Academy of Sciences of the United States of America* 100 (18):10175–10180. doi: 10.1073/pnas.1633354100.

Kannam, S. K., B. D. Todd, J. S. Hansen, and P. J. Daivis. 2012. Slip length of water on graphene: Limitations of non-equilibrium molecular dynamics simulations. *The Journal of Chemical Physics* 136 (2):024705–024705-9.

Kannam, S. K., B. D. Todd, J. S. Hansen, and P. J. Daivis. 2013. How fast does water flow in carbon nanotubes? *The Journal of Chemical Physics* 138 (9):094701–094701-9.

Kim, S., J. R. Jinschek, H. Chen, D. S. Sholl, and E. Marand. 2007. Scalable fabrication of carbon nanotube/polymer nanocomposite membranes for high flux gas transport. *Nano Letters* 7 (9):2806–2811. doi: 10.1021/nl071414u.

Kolesnikov, A. I., J.-M. Zanotti, C.-K. Loong, P. Thiyagarajan, A. P. Moravsky, R. O. Loutfy, and C. J. Burnham. 2004. Anomalously soft dynamics of water in a nanotube: A revelation of nanoscale confinement. *Physical Review Letters* 93 (3):035503.

Langmuir, I. 1916. The constitution and fundamental properties of solids and liquids. Part 1. Solids. *Journal of the American Chemical Society* 38 (11):2221–2295. doi: 10.1021/ja02268a002.

Lee, K.-H. and S. B. Sinnott. 2004. Computational studies of non-equilibrium molecular transport through carbon nanotubes. *The Journal of Physical Chemistry B* 108 (28):9861–9870. doi: 10.1021/jp036791j.

Majumder, M., N. Chopra, R. Andrews, and B. J. Hinds. 2005a. Nanoscale hydrodynamics: Enhanced flow in carbon nanotubes. *Nature* 438 (7064):44.

Majumder, M., N. Chopra, and B. J. Hinds. 2011. Mass transport through carbon nanotube membranes in three different regimes: Ionic diffusion, gas, and liquid flow. *ACS Nano* 5 (5):3867.

Majumder, M., N. Chopra, and B. J. Hinds. 2005b. Effect of tip functionalization on transport through vertically oriented carbon nanotube membranes. *Journal of the American Chemical Society* 127 (25):9062–9070. doi: 10.1021/ja043013b.

Majumder, M. and B. Corry. 2011. Anomalous decline of water transport in covalently modified carbon nanotube membranes. *Chemical Communications* 47 (27):7683–7685. doi: 10.1039/C1CC11134E.

Mao, Z. and S. B. Sinnott. 2000. A computational study of molecular diffusion and dynamic flow through carbon nanotubes. *The Journal of Physical Chemistry B* 104 (19):4618–4624. doi: 10.1021/jp9944280.

Medhekar, N. V., A. Ramasubramaniam, R. S. Ruoff, and V. B. Shenoy. 2010. Hydrogen bond networks in graphene oxide composite paper: Structure and mechanical properties. *ACS Nano* 4 (4):2300–2306. doi: 10.1021/nn901934u.

Melillo, M., F. Zhu, M. A. Snyder, and J. Mittal. 2011. Water transport through nanotubes with varying interaction strength between tube wall and water. *The Journal of Physical Chemistry Letters* 2 (23):2978–2983. doi: 10.1021/jz2012319.

Mi, W., Y. S. Lin, and Y. Li. 2007. Vertically aligned carbon nanotube membranes on macroporous alumina supports. *Journal of Membrane Science* 304 (1–2):1–7.

Myers, T. G. 2011. Why are slip lengths so large in carbon nanotubes? *Microfluidics and Nanofluidics* 10 (5):1141–1145. doi: 10.1007/s10404-010-0752-7.

Nair, R. R., H. A. Wu, P. N. Jayaram, I. V. Grigorieva, and A. K. Geim. 2012. Unimpeded permeation of water through helium-leak-tight graphene-based membranes. *Science* 335 (6067):442–444. doi: 10.1126/science.1211694.

Qin, X., Q. Yuan, Y. Zhao, S. Xie, and Z. Liu. 2011. Measurement of the rate of water translocation through carbon nanotubes. *Nano Letters* 11 (5):2173–2177. doi: 10.1021/nl200843g.

Rasaiah, J. C., S. Garde, and G. Hummer. 2008. Water in nonpolar confinement: From nanotubes to proteins and beyond. *Annual Review of Physical Chemistry* 59 (1):713–740. doi: 10.1146/annurev.physchem.59.032607.093815.

Sears, K., L. Dumee, J. Schutz, M. She, C. Huynh, S. Hawkins, M. Duke, and S. Gray. 2010. Recent developments in carbon nanotube membranes for water purification and gas separation. *Materials* 3 (1):127–149.

Sholl, D. S. and J. K. Johnson. 2006. Materials science: Making high-flux membranes with carbon nanotubes. *Science* 312 (5776):1003–1004. doi: 10.1126/science.1127261.

Skoulidas, A. I., D. M. Ackerman, J. K. Johnson, and D. S. Sholl. 2002. Rapid transport of gases in carbon nanotubes. *Physical Review Letters* 89 (18):185901.

Skoulidas, A. I., D. S. Sholl, and J. K. Johnson. 2006. Adsorption and diffusion of carbon dioxide and nitrogen through single-walled carbon nanotube membranes. *Journal of Chemical Physics* 124 (5):054708. doi: 10.1063/1.2151173.

Sokhan, V. P., D. Nicholson, and N. Quirke. 2004. Transport properties of nitrogen in single walled carbon nanotubes. *The Journal of Chemical Physics* 120 (8):3855–3863.

Sokhan, V. P. and N. Quirke. 2004. Interfacial friction and collective diffusion in nanopores. *Molecular Simulation* 30 (4):217–224. doi: 10.1080/08927020310001659106.

Su, J. and A. C. Lua. 2009. Experimental and theoretical studies on gas permeation through carbon molecular sieve membranes. *Separation and Purification Technology* 69 (2):161–167. doi: 10.1016/j.seppur.2009.07.014.

Suk, M. E. and N. R. Aluru. 2010. Water transport through ultrathin graphene. *The Journal of Physical Chemistry Letters* 1 (10):1590–1594. doi: 10.1021/jz100240r.

Suk, M. E., A. V. Raghunathan, and N. R. Aluru. 2008. Fast reverse osmosis using boron nitride and carbon nanotubes. *Applied Physics Letters* 92 (13):133120–133120-3.

Theodorou, D. N., R. Q. Snurr, and A. T. Bell. 1996. Molecular dynamics and diffusion in microporous materials. In *Comprehensive Supramolecular Chemistry*, G. Alberti and T. Bein (eds.), pp. 507–548. New York: Pergamon Press.

Thomas, J. A. and A. J. H. McGaughey. 2008. Reassessing fast water transport through carbon nanotubes. *Nano Letters* 8 (9):2788–2793. doi: 10.1021/nl8013617.

Thomas, J. A., A. J. H. McGaughey, and O. Kuter-Arnebeck. 2010. Pressure-driven water flow through carbon nanotubes: Insights from molecular dynamics simulation. *International Journal of Thermal Sciences* 49 (2):281–289. doi: 10.1016/j.ijthermalsci.2009.07.008.

Thornton, A. W., T. Hilder, A. J. Hill, and J. M. Hill. 2009. Predicting gas diffusion regime within pores of different size, shape and composition. *Journal of Membrane Science* 336 (1–2):101–108.

Travis, K. P., B. D. Todd, and D. J. Evans. 1997. Departure from Navier–Stokes hydrodynamics in confined liquids. *Physical Review E* 55 (4):4288–4295.

Tuzun, R. E., D. W. Noid, B. G. Sumpter, and R. C. Merkle. 1996. Dynamics of fluid flow inside carbon nanotubes. *Nanotechnology* 7 (3):241.

Ulissi, Z. W., S. Shimizu, C. Y. Lee, and M. S. Strano. 2011. Carbon nanotubes as molecular conduits: Advances and challenges for transport through isolated sub-2 nm pores. *The Journal of Physical Chemistry Letters* 2 (22):2892–2896. doi: 10.1021/jz201136c.

Vatanpour, V., S. S. Madaeni, R. Moradian, S. Zinadini, and B. Astinchap. 2011. Fabrication and characterization of novel antifouling nanofiltration membrane prepared from oxidized multiwalled carbon nanotube/polyethersulfone nanocomposite. *Journal of Membrane Science* 375 (1–2):284–294. doi: 10.1016/j.memsci.2011.03.055.

Verweij, H., M. C. Schillo, and J. Li. 2007. Fast mass transport through carbon nanotube membranes. *Small* 3 (12):1996–2004.

Wang, Z., L. Ci, L. Chen, S. Nayak, P. M. Ajayan, and N. Koratkar. 2007. Polarity-dependent electrochemically controlled transport of water through carbon nanotube membranes. *Nano Letters* 7 (3):697–702. doi: 10.1021/nl062853g.

Whitby, M., L. Cagnon, M. Thanou, and N. Quirke. 2008. Enhanced fluid flow through nanoscale carbon pipes. *Nano Letters* 8 (9):2632–2637. doi: 10.1021/nl080705f.

Whitby, M. and N. Quirke. 2007. Fluid flow in carbon nanotubes and nanopipes. *Nature Nanotechnology* 2 (2):87–94.

Won, C. Y. and N. R. Aluru. 2007. Water permeation through a subnanometer boron nitride nanotube. *Journal of the American Chemical Society* 129 (10):2748–2749. doi: 10.1021/ja0687318.

Yu, M., H. H. Funke, J. L. Falconer, and R. D. Noble. 2009. High density, vertically-aligned carbon nanotube membranes. *Nano Letters* 9 (1):225–229. doi: 10.1021/nl802816h.

Yu, M., H. H. Funke, J. L. Falconer, and R. D. Noble. 2010. Gated ion transport through dense carbon nanotube membranes. *Journal of the American Chemical Society* 132 (24):8285–8290. doi: 10.1021/ja9091769.

Chapter 12

Modeling Selective Transport and Desalination in Nanotubes

Michael Thomas, Ben Corry, Shin-Ho Chung, and Tamsyn A. Hilder

CONTENTS

12.1 MOTIVATION FOR MODELING NANOTUBES

At first, it would seem that nanotubes would make unlikely candidates for improving water permeation through membranes. The hydrophobic interior of a carbon nanotube seems a forbidding place for water molecules. Yet, it is this hydrophobic surface that imparts remarkable transport properties onto nanotubes.

The groundbreaking computer modeling work of Hummer et al. predicted that carbon nanotubes would spontaneously fill when immersed in water (Hummer et al. 2001). Using a technique called molecular dynamics (MD), the (6,6) CNT investigated in the study was found to fill with a single chain of hydrogen-bonded water molecules (Figure 12.1). These hydrogen bonds are, on average, longer lasting and more highly oriented than in bulk water, with each water molecule free to rotate about its hydrogen bond axes. This phase of water within the (6,6) nanotube has been described as vapor like and is stabilized through an increase in entropy (compared to bulk water) (Pascal et al. 2011). Ice-like and bulk liquid-like phases are observed in larger-diameter nanotubes.

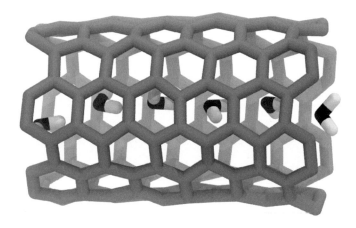

Figure 12.1   Single chain of hydrogen-bonded water molecules in a (6,6) carbon nanotube.

The computer modeling study conducted by Hummer et al. was the first in which the extraordinary rapid transport properties of nanotubes were alluded to. Water was seen to conduct through the nanotube at a similar rate to a biological counterpart, the transmembrane aquaporin-1 channel. Aquaporin allows the passage of water, but not ions and other species, in and out of a cell. Water conduction in the nanotubes occurs in small bursts that are said to flow with very little resistance. Only very small interactions are present between the hydrophobic surface of the nanotubes and the water molecules, resulting in the near-frictionless flow of water.

Landmark experimental studies followed this early modeling work. Aligned nanotube membranes constructed by Hinds et al. (2004) were examined for the gas and ion permeation properties. Although, it was not until Majumder et al. constructed membranes composed of arrays of aligned multiwalled carbon nanotubes with pore diameters of about 7 nm that water permeation was investigated (Majumder et al. 2005). Water flow rates through this membrane were found to be similar to the aquaporin-1 and many orders of magnitude larger than suggested by the Hagen–Poiseuille equation, a conventional continuum flow model. The slip length of these nanotubes was found to be incredibly large, three to four orders of magnitude larger than the pore diameters. These results indicate that the interior of the nanotube offered a near-frictionless surface for the water molecule to flow across, as predicted by computer modeling studies.

Experiments by Holt et al. (2006) demonstrated that these amazing rapid transport properties still occur in narrower nanotubes. Membranes composed of 1.3–2.0 nm diameter carbon nanotubes exhibited water flow rates three orders of magnitude larger than predicted by a continuum model and were consistent with the rates predicted from MD simulations by Hummer et al. Slip lengths were again found to be three to four orders of magnitude larger than the pore diameter, demonstrating that the near-frictionless flow of water also occurs in narrow nanotubes.

Nanotubes also have the ability to reject ions from permeating through the pore, while allowing water to flow. The dipole of the water molecules align to the charge of the ion solvated in bulk water. A monolayer of water molecules becomes bound to the ion, called the solvation shell. In wider nanotubes, ions are able to pass with the full complement of water molecules in their solvation shell. There is insufficient space for this to occur in narrow nanotubes. Some water molecules

C
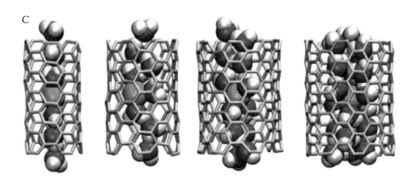

Figure 12.2 In narrow nanotubes (left), very few water molecules coordinate with a permeating ion (sodium). In wider nanotubes (right), water coordination is more bulk like, resulting in a smaller energy barrier for ion conduction. Oxygen is represented in black, hydrogen in white, and carbon in gray. (Reprinted with permission from Corry, B.A., Designing carbon nanotube membranes for efficient water desalination, *J. Phys. Chem. B*, 112(5), 2008, 1427–1434. Copyright 2008 American Chemical Society.)

must be removed from the solvation shell for the ion to fit inside the nanotube, and there is an energetic cost to do this. This makes it far less likely for ions to permeate narrow tubes. Modeling investigations predicted that the rejection of small ionic species, such as Na^+ and Cl^-, could be achieved by hydrophobic pores (Beckstein and Samsom 2004, Dzubiella and Hansen 2005, Corry 2006) and by pristine carbon nanotubes specifically (Peter and Hummer 2005). Ion rejection rates become larger as the nanotube becomes narrower, due to the need to partially desolvate an ion passing through narrow pores (Figure 12.2; Corry 2008).

Ion rejection by nanotubes (shown in Figure 12.3) was demonstrated experimentally by Fornasiero et al. (2008). Nanotubes with pore diameters 1–2 nm were shown to partially reject a number of salts, including Na^+ and Cl^-, with rejection reaching 98% under certain conditions. Nanotubes in this study were made by conformal deposition, followed by chemical etching to remove the caps at the end of the tube. The latter process functionalizes the ends of the nanotubes with carboxylic, carbonyl, and hydroxyl groups. Ion rejection was found to be dependent on pH, indicating that the electrostatic interactions between the ions and the functional groups dominate over steric effects; the desolvation of ions plays little role in salt rejection for nanotubes with pore diameters 1–2 nm. Additionally, the rejection rate was found to decrease as the concentration of the salt increased.

The rapid water transport and ion rejection properties make nanotubes ideal for a number of applications, including environmental sensors, antimicrobial agents, and renewable energy technologies. This chapter will focus on the use of nanotubes in water desalination, an application that pushes the boundaries of these two properties. Water desalination is the process in which seawater or brackish water is taken and progressively treated, removing smaller and smaller particles, until the water is potable. Various methods are employed to remove larger particles at each step. Traditionally, coagulants have been the most popular way of removing larger species by combining small particles into larger particles (Valavala et al. 2011). Other methods are also used, including adsorbents, oxidants, and granular media, depending on the quality of the intake water. More recently, ultrafiltration and microfiltration techniques have become more popular. These methods force water through membranes containing microscopic pores, while larger particles are excluded based on their size.

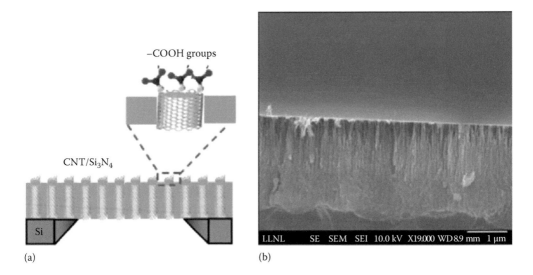

Figure 12.3   (a) A schematic of carbon nanotubes embedded in a silicon carbide matrix. Functionalization of the nanotube ends with carboxylate groups is shown in the inset. (b) A scanning electron image of the cross section of carbon nanotube/silicon nitride matrix. (Reprinted with permission from Fornasiero, F., Park, H.G., Holt, J.K. et al., Ion exclusion by sub-2-nm carbon nanotube pores, *Proc. Natl. Acad. Sci. USA*, 105, 2008, 17250–17255 and Copyright 2008 National Academy of Sciences, U.S.A.)

By far, the most energy-intensive step in the desalination process is the removal of small ionic species, particularly Na^+, K^+, and Cl^-, from water. Currently, the most popular method of removing these species is reverse osmosis filtration where the intake salt solution is placed under a large pressure and forced through a semipermeable membrane. This membrane allows water to permeate, but not ions, producing potable water. More than 95% of the ions must be rejected to produce potable water from typical seawater, although this number will vary depending on the type of input water. A large pressure must be used to overcome the osmotic pressure of the system: the desire for potable water to flow backward through the semipermeable membrane to equalize the salt concentrations. Additional pressure above the osmotic pressure must be applied to achieve usable potable water fluxes.

One measure of the membrane efficiency is the *permeability* of the reverse osmosis membrane, the amount of water flux per unit of force passing through the membrane. The best current reverse osmosis membranes are polyamide-based thin-film composites that can achieve a permeability of 3.5×10^{-12} m^3 m^{-2} Pa^{-1} s^{-1} (m^3 of water per m^2 of membrane per Pascal per second) and sufficient salt rejection. It is thought that nanotube-based desalination membranes may offer a way to increase the membrane permeability, and thus the efficiency of the reverse osmosis process, while maintaining the necessary ion rejection.

Computer modeling is at the forefront of nanotube-based device design, guiding the experimental processes needed to produce benefits to the wider community. Researchers are using a range of modeling techniques to unravel the fundamental concepts underlying transport through nanotubes. Using these results, new-generation technology, such as nanotube-based desalination membranes, can be constructed and implemented. The remainder of this chapter discusses how researchers model nanotubes computationally, as well as the contribution of computational modeling toward our understanding of the water permeation and salt rejection properties of nanotubes.

12.2 MODELING APPROACHES

12.2.1 Molecular Dynamics

Many computational methods have been used to model nanotubes and investigate their various properties: from highly detailed, but computationally intensive, quantum mechanical calculations to the fast, but less detailed, continuum models. MD is a computational method that provides a good balance between detail and speed that is routinely employed to model the dynamics and energetics of systems comprising up to a million atoms.

MD is a method of numerically solving classical equations of motion on a system composed of a number of particles. Atoms are modeled as charged spheres, incorporating the nucleus and electrons into a single particle. In quantum mechanical methods, these are modeled separately, while in continuum models, larger numbers of atoms are modeled as blocks of dielectric material.

A force field is a set of empirically and quantum mechanically derived values used to determine the strength of the interactions between atoms in MD simulations. The force field describes how atoms interact with one another. The ability of MD to produce accurate results depends on the ability of the force field to describe these interactions accurately. Much time and effort is spent developing force fields, many of which have been constructed and optimized for particular types of systems, such as proteins, liquids, and solids. Force fields can be polarizable or nonpolarizable; to include polarization is to account for an atom's change in response to a change in the surrounding electric field. Nonpolarizable force fields, such as CHARMM (MacKerell et al. 1998), Gromos (Scott et al. 1999), and Amber (Salomon-Ferrer et al. 2013), have been the most popular to date, as polarizable force fields require a lot of computing power. However, as computing power increases, the usefulness and popularity of polarizable force fields will also increase.

A force field is composed of parameters that describe a variety of interactions, such as

1. The stretching of bonds
2. The bending of angles
3. The torsion of dihedral angles
4. van der Waals interactions
5. Electrostatic interactions

The stretching, bending, and torsions are known as *bonded interactions*; they define how atoms covalently bonded to one another behave. The van der Waals and electrostatic interactions are known as "nonbonded" interactions as they describe how atoms interact through space. The parameters describing each of these interactions for each combination of atoms must be determined. For example, a bond strength, k, and an equilibrium separation distance, x_0, are required to determine the force, $F(x) = \frac{1}{2}k(x - x_0)^2$, of a bond stretch, which models the covalent bonding between atoms. Another example is the charge on each atom to calculate the force due to electrostatic interaction. These parameters are usually determined by high-level quantum calculations and are validated by their ability to replicate bulk properties of the particular molecule being described, such as heat of vaporization, dipole moment, or a variety of spectroscopic properties. These parameters describe only the average properties of a large number of a particular type of molecule, and so in conditions or environments that stray significantly from this, inaccurate results are likely to be produced. For instance, nonpolarizable force fields are unable to handle polarization explicitly; it is accounted for only in an average sense by being incorporated into other parameters.

Before the commencement of an MD simulation, a set of initial coordinates and velocities are required. Coordinates are usually obtained from experimental structure determination techniques, such as X-ray crystallography, or built using one of a number of various molecular visualization programs. Initial velocities are usually randomly assigned from a Maxwell–Boltzmann distribution centered on the desired temperature. Once the coordinates and velocities are assigned, the force acting on each atom in the initial coordinates is calculated using parameters from the force field. Using this force, we can describe how each atom will move over a very small amount of time, called a time step, which is usually 1–2 fs. Once the atoms have moved, the force experienced by each atom is recalculated. This process iterates for a defined number of time steps, creating a time evolution of the system described by the initial coordinates.

MD is an important tool for researchers investigating nanotubes, especially in the context of desalination. The atomic resolution of MD simulations allows for the investigation of many important features over timescales long enough to collect adequate statistics. It allows us to capture many permeation events of single ions and water molecules through nanotubes. Within MD, there are a variety of techniques used to describe particular aspects of nanotube function that other methods cannot offer, either due to insufficient detail or prohibitively long computing times. For example, these techniques can be used to calculate the energetics of ion and water permeation. This allows us to probe nanotubes to understand how their remarkable permeation properties arise and how we may optimize these properties.

MD does not have the ability to model bond formation or destruction as electrons are not modeled explicitly, as in quantum mechanical simulations. This means that MD is unable to capture any chemistry that may take place. This may be of particular importance when considering possible bond formation during chemical fouling of desalination membranes or the role that functional groups attached to the end of nanotubes may play. However, the exclusion of chemistry from MD allows us to conduct simulations over far greater timescales than can be achieved using quantum mechanical calculations.

12.2.2 Modeling Membranes

The pore size and chirality of nanotubes are defined by their chiral vector, $C = na_1 + ma_2 = (n,m)$, resulting in three general forms: armchair (n,n), chiral (n,m), and zigzag $(n,0)$, as shown in Figure 12.4. The majority of computational studies have focused on the armchair-type nanotube. It is generally assumed that the chirality of the nanotube does not affect transport properties (Alexiadis and Kassinos 2008), although there is some suggestions that it may (Won et al. 2006). Nanotubes are usually constructed using automated nanotube builders, such as the Carbon Nanostructure Builder available in the molecular visualization program VMD (Humphrey et al. 1996), providing the initial coordinates in order to commence an MD simulation.

It must be decided if the nanotube will be embedded in a membrane matrix or not. The matrix is an impermeable material that composes the majority of the membrane, through which nanotubes create pores. Whether or not a matrix should be included will depend on what aspect is to be investigated. Initial investigations by Hummer et al. modeled carbon nanotubes in bulk water in order to investigate their water filling and transport properties (Hummer et al. 2001). Most subsequent investigations have modeled nanotubes embedded in a matrix of some sort. In experiments, nanotubes are often embedded in silicon nitride (Holt et al. 2006) or polystyrene film (Hinds et al. 2004, Majumder et al. 2005) matrices. In MD simulations, nanotubes have been embedded in a variety of matrices including silicon nitride (Hilder et al. 2009, 2009a), lipid bilayers (Hilder 2012), and graphene bilayers (Gong 2010, Su and Guo 2012), as shown in Figure 12.5.

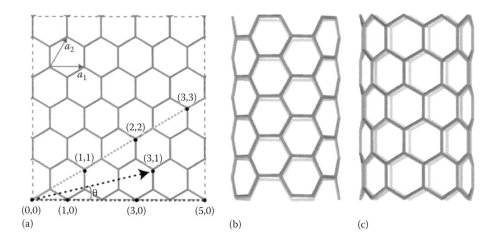

Figure 12.4   (a) Examples of chiral vectors on a graphene sheet. The dashed gray line follows the circumference of an armchair-type tube, the dashed black arrow follows a chiral-type tube, and the dashed horizontal line follows a zigzag-type tube. (b) A (5,5) nanotube, an example of an armchair-type tube. (c) A (9,0) nanotube, an example of a zigzag tube. (Reproduced from Maiti, A., *Nat. Mater.*, 2, 440, 2003.)

This creates a barrier, forcing water and ions to permeate through the nanotube if they are to get from one side to the other. This also means that additional forces can be added to the atoms in the system to model an applied pressure or an electric field in order to determine the water permeation and/or salt rejection properties. In this instance, the atoms composing the nanotube and matrix must be held in place by additional constraints to avoid their translation due to the applied pressure or electric field.

An alternative approach, developed by Zhu and Schulten (2003), uses a membrane constructed solely of nanotubes, rather than embedding the nanotube in a matrix, as shown in Figure 12.5d. By packing the nanotubes close together (e.g., hexagonal packing), there is not enough space for molecules to pass between the nanotubes, and so flow is confined to the interior of the tubes. The atoms of the nanotubes are held in place by artificial constraints, so as to allow a pressure difference to be added to the system without degradation of this membrane. This type of setup is popular as it increases the sampling rates of species permeating the nanotubes as it allows fluxes to be measured across many nanotubes instead of just one. This allows a reduction in the time needed for simulation to achieve the same accuracy in the flux.

Finally, water and salt ions are added to the nanotube/matrix system. Water molecules are usually added to either side of the membrane in a box shape, to which ions can be added. This shape allows for the system to be treated in a periodic fashion; water molecules and ions are able to wrap around from one edge to the other in the x, y, and z directions so that boundary effects do not affect the simulation.

12.2.3   Permeation Properties

It is the water flux and salt rejection properties of nanotubes that we wish to understand for their application as desalination membranes. The water flux of a membrane is the number of water molecules crossing through a given area of membrane in a particular period of time and has units of $m^3\,m^{-2}\,s^{-1}$. Water must be driven across the membrane by some force; this is often a hydrostatic pressure

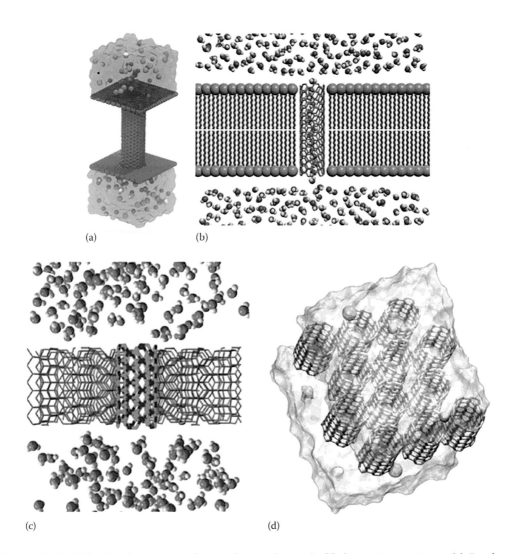

Figure 12.5 Molecular dynamic simulation of nanotubes embedded in various matrices. (a) Graphene bilayers. (b) Lipid bilayers. (Adapted with permission from Hilder, T.A., Yang, R., Gordon, D., Rendell, A.P., and Chung, S.-H., Silicon carbide nanotube as a chloride-selective channel, *J. Phys. Chem. C*, 116(7), 2012, 4465–4470. Copyright 2012 American Chemical Society.) (c) Silicon nitride membrane. (Adapted from Hilder, T.A. et al., *Small*, 5(19), 2183, 2009, image has been rotated and cropped.) (d) Array of nanotubes. (Adapted with permission from Corry, B.A., Designing carbon nanotube membranes for efficient water desalination, *J. Phys. Chem. B*, 112(5), 2008, 1427–1434. Copyright 2008 American Chemical Society.)

(as implemented by Zhu et al. [2002, 2004]) or sometimes an osmotic pressure (Kalra et al. 2003). The hydrostatic or osmotic pressure is quoted along with the water flux values so that comparisons can be made between different modeling investigations. A more explanative, albeit more time consuming, way to represent water permeation is the water permeability of a membrane. This measures how the water flux across the membrane responds to different pressures and is measured in units of $m^3\,m^{-2}\,s^{-1}\,Pa^{-1}$ (water flux per unit pressure). For MD investigations, the units of flux and permeability are often given as molecules $tube^{-1}\,ns^{-1}$ and molecules $tube^{-1}\,ns^{-1}\,MPa^{-1}$, respectively.

Salt rejection measures how good a membrane is at stopping the passage of salt ions. It is expressed as a ratio of the relative amount of permeant ions to the relative amount of permeant water molecules. For instance, the salt rejection is often described by

$$\text{Ion Rejection} = 1 - \frac{n_{\text{permeant ions}} / n_{\text{total ions}}}{n_{\text{permeant water}} / n_{\text{total water}}},$$

where

$n_{\text{permeant ions}}$ is the number of permeant salt ions

$n_{\text{total ions}}$ is the total number of salt ions in the simulation

$n_{\text{permeant water}}$ is the number of permeant water molecules

$n_{\text{total water}}$ is the total number of water molecules

It is usually expressed as a percentage. This definition can also be used for individual ion types to determine if the membrane is better at rejecting some ion types over others. This can be helpful for aiding the design of desalination membranes, as well as for other applications, such as ultra-sensitive biosensors (Hilder et al. 2011).

In order to measure the permeability of a nanotube-based membrane, a water flow through nanotubes must be induced. There are a number of ways this can be achieved, but as desalination facilities use pressure to achieve permeation, this is what is commonly used in MD. To do this, we need to add a pressure difference across the membrane. A hydrostatic pressure can be introduced by adding an additional force to a subset of water molecules during the force calculation step in the MD algorithm. This force acts to drive water molecules across the membrane, but as water flow across the membrane is limited, a high density of water will build up on this side of the membrane. On the other side of the membrane, water will be removed from near the membrane, but not replaced quickly enough, creating a region of low density. These two regions of differing density define the pressure difference. In practice, the force is applied to molecules only far from the membrane such that there is a region of constant density on each side.

Free energy profiles (like those displayed in Figure 12.6) allow us to observe the energetics of water and ion permeation through nanotubes. A useful MD technique for determining these free energy profiles is the potential of mean force (PMF) calculation (Torrie and Valleau 1974). It can be used to explain why water fluxes and salt rejection are different for different nanotubes. A PMF describes how the free energy of a simulation system changes as an atom, or groups of atoms, is moved along a particular reaction coordinate. This can be used, for example, to determine the energetics of an ion or water molecule permeating through a nanotube, as shown in Figure 12.6. Umbrella sampling is a commonly used technique that is used to determine a PMF. A number of simulations are conducted wherein atoms are biased to a particular position along the reaction coordinate in each simulation (in nanotubes, this is usually the radial and axial coordinates through the nanotube), and the value of the reaction coordinate is recorded. The combined set of results are then unbiased, usually using the weighted histogram analysis method (Kumar et al. 1992). This is done to achieve adequate sampling of unlikely configurations along the reaction coordinate to increase the accuracy of the free energy results.

Using these tools, we can now investigate various properties of nanotubes as follows:

- What nanotube dimensions offer the best water flux/permeability while maintaining greater than 95% salt rejection?
- Is water flow in the tube really *near-frictionless*?
- How do functional groups at the nanotube openings affect their function?
- How does the material that the nanotube is composed of (carbon, silicon carbide, boron nitride) affect transport properties?

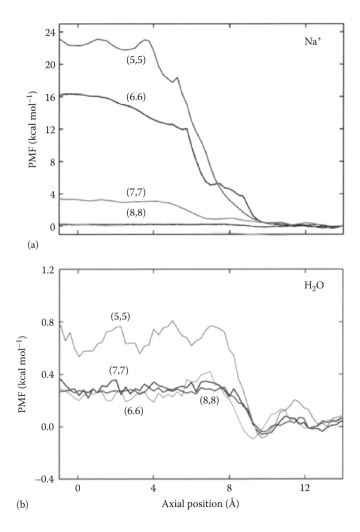

Figure 12.6   An example of a PMF calculation. A PMF of (a) a sodium ion and (b) a water molecule passing through a carbon nanotube. The (5,5) CNT has the smallest pore diameter, and the (8,8) has the largest. (Reprinted with permission from Corry, B.A., Designing carbon nanotube membranes for efficient water desalination, *J. Phys. Chem. B*, 112(5), 2008, 1427–1434. Copyright 2008 American Chemical Society.)

Using MD techniques, researchers have been able to provide answers to these questions. The subsequent section will explore what researchers have learned about nanotubes and their desalination properties through simulation with MD.

12.3   WATER AND ION PERMEATION PROPERTIES OF NANOTUBES

12.3.1   Spontaneous Water Filling

MD simulations of a (6,6), 0.81 nm diameter carbon nanotube immersed in water demonstrated that water molecules will spontaneously enter and remain in the pore (Hummer 2001). In contrast, a (5,5) carbon nanotube (diameter of 0.69 nm) will only partially fill with water molecules (Won and Aluru 2007) or alternate between empty and filled states (Corry 2008). All nanotubes with larger pore diameters will spontaneously fill. Typically, carbon nanotubes are modeled with a zero partial charge on each carbon atom. In contrast, boron nitride nanotubes and silicon nitride nanotubes have differing partial charges on each boron and nitrogen, and silicon and carbon atoms, respectively. These charges, combined with the van der Waals parameters, allow water molecules to hydrogen-bond with the nitrogen atoms (Won and Aluru 2008). As such, boron nitride tubes spontaneously fill at smaller pore diameters than carbon nanotubes, with the 0.69 nm (5,5) spontaneously filling, owing to the increased van der Waals and electrostatic interactions between the nanotube atoms and the water molecules filling the pore (Won and Aluru 2007, 2008, Hilder et al. 2009). Therefore, it is likely that the filling of silicon carbide nanotubes is similar to the filling of boron nitride nanotubes; water spontaneously enters smaller pore diameters than carbon nanotubes (Yang et al. 2011, Taghavi et al. 2013).

The confinement of water molecules within a nanotube is thermodynamically favorable, that is to say that the free energy is lower in a filled state than an unfilled state (Pascal et al. 2011). However, the driving force differs with the pore diameter:

- Filling is entropy driven in the smaller-diameter (5,5) and (6,6) carbon nanotubes (0.81–1.0 nm) due to the increased translations and rotations of water compared to water molecules in bulk water.
- Hydrogen bonds between water molecules in the larger (8,8) and (9,9) carbon nanotubes (1.1–1.2 nm) impart favorable enthalpic contributions due to a rigid hydrogen bonding network.
- Large translational entropy of water molecules in the 1.4 nm (10,10) nanotube and larger compared to bulk water induces spontaneous filling.

The splitting of the filling driving force into three domains of pore diameter also represents the structure of water in the nanotube in each case (Pascal et al. 2011). For the smaller pore sizes, the water is restricted to a single file along the pore axis, resulting in a gas-like phase of water molecules. The next largest domain, incorporating the (8,8) and (9,9) carbon nanotubes, is described as an ice-like phase where a slightly increased number of hydrogen bonds per water molecule, as compared to bulk water, restrict the water distribution to a torus perpendicular to the pore axis, as shown in Figure 12.7. As the pore diameter increases further, the structure of water in the nanotube becomes more and more liquid-like. Some layering of the water is still present at the interface with the inner nanotube wall, but this dissipates toward the center of the tube. Although a similar analysis on the driving force of filling in boron nitride and silicon carbide nanotubes has not been conducted, it is assumed that a similar process is at play, although the domains of entropy and enthalpic domination may differ.

12.3.2   Water Fluxes

The first suggestion of the impressive water permeation properties of nanotubes was the seminal MD work of Hummer et al. (2001). It was discovered that water permeated a (6,6) carbon nanotube in burst-like motions, during which water molecules move with very little resistance. This occurred solely from diffusion of water through the carbon nanotube; a force was not used to drive water through the nanotube. The average water flux over the course of the simulation was 17 molecules tube^{-1} ns^{-1}, comparable to the flux of a biological counterpart, the aquaporin-1 channel (Zeidel et al. 1992). Many MD studies exploring the water

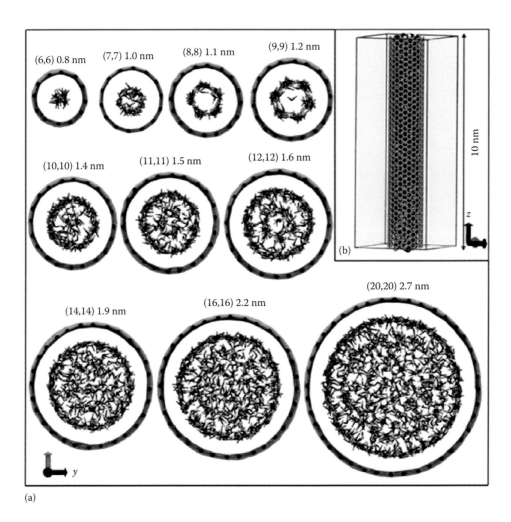

Figure 12.7 The structure of water inside carbon nanotubes of varying diameters. (a) The axial pore distribution looking along the pore axis. (b) A view of the carbon nanotube perpendicular to the pore axis. (Figure is reproduced from Pascal, T.A. et al., *Proc. Natl. Acad. Sci. USA*, 108(29), 11794, 2011.)

permeation properties of various forms of nanotubes followed. A summary of these results are presented in Table 12.1.

It has been suggested that the constant flow observed in nanotubes indicates that there is little to no friction. This is illustrated in Figure 12.8 in which MD studies have shown that the measured axial water velocity profile for carbon nanotubes remains constant to the wall of the nanotube in contrast to the decrease toward the wall in macroscopic flows (Majumder and Corry 2011). In macroscopic models of fluids flowing through a pipe or pore, the velocity of water flow through the pipe is at a minimum at the walls of the pipe and at a maximum at the center of the pipe, similar to the red line in Figure 12.8. Unlike in nanotubes, the friction between water and the pipe causes this slowdown at the walls.

As mentioned, most MD studies investigating water flow through carbon nanotubes assign a zero partial charge to each carbon atom in the nanotube. As a result, there is no electrostatic

TABLE 12.1   WATER FLUXES OF A RANGE OF PRISTINE CARBON (GREEN), BORON NITRIDE (ORANGE), AND SILICON CARBIDE (BLUE) NANOTUBES AS DETERMINED FROM MD SIMULATIONS

Tube Material	Neutral/ Partial Charges	Tube Type	Tube Diameter (nm)	Tube Length (nm)	Driving Force	Magnitude of DF (MPa)	Water Flux (Molecules/ ns/tube)	Reference
Carbon	Neutral	(5,5)	0.66	1.34	HP	208	10.4 ± 0.4	Corry (2008)
Carbon	Neutral	(10,0)	0.78	1.14	Diffusion	—	5.31	Won et al. (2006)
Carbon	Partial charges	(10,0)	0.78	1.14	Diffusion	—	6.69	Won et al. (2006)
Carbon	Neutral	(6,6)	0.81	1.34	Diffusion	—	17	Hummer et al. (2001)
Carbon	Neutral	(6,6)	0.81	1.34	Diffusion	—	5.9 ± 0.8	Zhu and Schulten (2003)
Carbon	Neutral	(6,6)	0.81	1.23	Diffusion	—	4.94	Won et al. (2006)
Carbon	Partial charges	(6,6)	0.81	1.23	Diffusion	—	9.13	Won et al. (2006)
Carbon	Neutral	(6,6)	0.81	2.09	Diffusion	—	15.0	Suk et al. (2008)
Carbon	Neutral	(6,6)	0.81	1.34	OP	—	5.8	Kalra et al. (2003)
Carbon	Neutral	(6,6)	0.81	1.4	HP	208	23.3 ± 0.3	Corry (2008)
Carbon	Neutral	(6,6)	0.81	1.34	HP	15	~8–18	Gong et al. (2008)
Carbon	Neutral	(7,7)	0.93	1.34	HP	208	43.7 ± 0.5	Corry (2008)
Carbon	Neutral	(7,7)	0.95	1.4	HP	100	20.9	Song and Corry (2009)
Carbon	Neutral	(7,7)	0.95	1.4	HP	200	41.3	Song and Corry (2009)
Carbon	Neutral	(8,8)	1.09	1.34	HP	208	81.5 ± 1.2	Corry (2008)
Carbon	Neutral	(8,8)	1.08	1.4	HP	100	39.4	Song and Corry (2009)

(Continued)

TABLE 12.1 (*CONTINUED*)   WATER FLUXES OF A RANGE OF PRISTINE CARBON (GREEN), BORON NITRIDE (ORANGE), AND SILICON CARBIDE (BLUE) NANOTUBES AS DETERMINED FROM MD SIMULATIONS

Tube Material	Neutral/ Partial Charges	Tube Type	Tube Diameter (nm)	Tube Length (nm)	Driving Force	Magnitude of DF (MPa)	Water Flux (Molecules/ ns/tube)	Reference
Carbon	Neutral	(8,8)	1.08	1.4	HP	200	78.8	Song and Corry (2009)
Carbon	Neutral	(8,8)	1.09	1.3	HP	246	107.8 ± 0.6	Corry (2011)
Carbon	Neutral	(9,9)	1.22	1.4	HP	100	57.4	Song and Corry (2009)
Carbon	Neutral	(9,9)	1.22	1.4	HP	200	110	Song and Corry (2009)
Carbon	Neutral	(50,50)	7	6.1	HP	290	4660 ± 33	Majumder and Corry (2011)
Boron nitride	Partial charges	(5,5)	0.79	1.4	HP	60–612	1.6–10.7	Hilder et al. (2009)
Boron nitride	Neutral	(6,6)	0.83	2.13	Diffusion	—	16.0	Suk et al. (2008)
Boron nitride	Neutral	(6,6)	0.83	2.13	HP	100–500	~12–49	Suk et al. (2008)
Silicon carbide	Partial charges	(5,5)	0.86	3.6	HP	100	17.16	Hilder et al. (2012)
Silicon carbide	Partial charges	(6,6)	1.03	3.6	HP	100	32.33	Hilder et al. (2012)
Silicon carbide	Partial charges	(7,7)	1.2	3.6	HP	100	69.26	Hilder et al. (2012)

Note: HP is hydrostatic pressure and OP is osmotic pressure.

interaction between the nanotube and the water molecules, only van der Waals interactions. The interaction between neutral carbon and water molecules tends to be quite weak. This allows water molecules to adopt particular orientations and hydrogen bonding at the water/ nanotube interface, making the surface of the nanotube *slippery* to water molecules (Joseph and Aluru 2008). The near-frictionless flow of water through nanotube implies that the water flux is independent of the length of the tube, which has been confirmed by various MD simulations studies (Kalra et al. 2003, Corry 2008, Nicholls et al. 2012). However, if nonzero partial

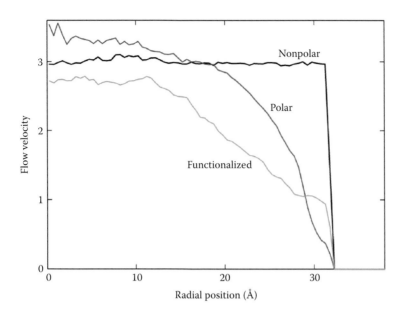

Figure 12.8   The axial velocity profile of water through various carbon nanotubes. Results are shown for a pristine uncharged carbon nanotube (nonpolar), a pristine partially charged carbon nanotube (polar), and a nanotube with bulky negatively charged functional groups attached along the length of the pore (functionalized). The center of the pore is at 0 Å, and the wall of the carbon nanotube is at 35 Å. (From Majumder, M. and Corry, B., Anomalous decline of water transport in covalently modified carbon nanotube membranes, *Chem. Commun.*, 47(27), 7683–7685, 2011. Adapted by permission of The Royal Society of Chemistry.)

charges are assigned to each atom in the carbon nanotube, stronger electrostatic interactions are possible. These interactions slow down the rate at which water can flow across the surface thus creating friction, as can be seen in Figure 12.8 (Majumder and Corry 2011).

Water molecules have been demonstrated to induce charges on carbon nanotube atoms (Lu et al. 2004). Won et al. (2006) determined the charges present in a (6,6) and a (10,0) carbon nanotube using quantum mechanical calculations and implemented these in MD simulations. The partial charges on the carbon atoms near the opening of the nanotubes were found to deviate significantly from zero, while the remainder were approximately zero. In both instances, the addition of these partial charges increased the water fluxes compared with uncharged tubes. Upon inspection of the PMFs in Figure 12.9, it can be seen that the addition of these charges lowers the free energy barrier that a water molecule must overcome to permeate through the pore between 0.32 and 0.56 $k_B T$. The free energy barrier for a water molecule to enter the pore is lowered due to the increased interactions with the partially charged pore opening. In the carbon nanotubes with zero partial charge on each atom, the transition of water from the bulk to the confined hydrophobic pore incurs a larger free energy penalty.

Due to the heterogeneous constituents of boron nitride and silicon carbide nanotubes, each atom in the nanotube will have a partial charge. Water fluxes in boron nitride tubes are potentially larger than those displayed by comparable diameter carbon nanotubes (Won and Aluru 2007, 2008). The partial charges lower the free energy barrier for water molecules to enter the pore even if there are slightly increased interactions along the length of the pore. If the partial charges of boron nitride and silicon carbide nanotubes are further increased, a decline in the water fluxes is seen (Won and Aluru 2008, Hilder et al. 2009). The electrostatic interactions

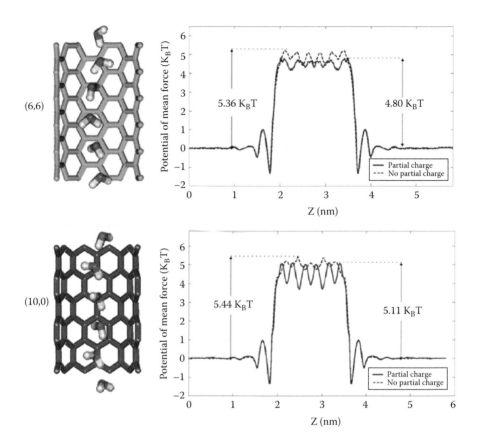

Figure 12.9   The free energy surface for a water molecule to permeate through partially charged (solid line) and uncharged (dashed line) (6,6) and (10,0) carbon nanotubes. (Adapted with permission from Won, C.Y., Joseph, S., and Aluru, N.R., Effect of quantum partial charges on the structure and dynamics of water in single-walled carbon nanotubes, *J. Chem. Phys.*, 125(1), 115701, 2006. Copyright 2006 American Institute of Physics.)

between the water molecules and the nanotube wall become so great with these increased partial charges that the water molecules interact with sections of the nanotube wall for long periods of time, introducing a large of amount of friction to the water flow.

12.3.3   Salt Rejection

Nanotubes have also been studied for their ability to reject salt ions. The experimental work by Fornasiero et al. (2008) demonstrated that 1–2 nm carbon nanotubes are capable of salt rejection. It was determined that salt rejection depends on a number of factors including solution pH and the valency of the ions. These factors indicate that rejection of the ions is occurring via electrostatic interactions between the carbon nanotube pore opening or membrane surface and the ions, rather than steric hindrance at the nanotube opening. Due to the sensitivity of salt rejection to pH, negatively charged carboxylate groups are thought to line the pore entrance of the nanotube. Rejection of 600 mM (roughly the concentration of salt in seawater) KCl solution was found to be 40%–50%, which decreased as this concentration increased. Other potassium salts were tested for salt rejection, demonstrating that salts containing higher valence anions

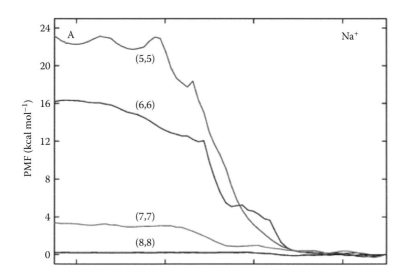

Figure 12.10   The free energy profile of Na$^+$ permeating through carbon nanotubes of various pore diameters. The left-hand side of the graph is the center of the nanotube and the right-hand side is bulk water. (Reprinted with permission from Corry, B.A., Designing carbon nanotube membranes for efficient water desalination, *J. Phys. Chem. B*, 112(5), 2008, 1427–1434. Copyright 2008 American Chemical Society.)

displayed better rejection for both anions and cations. However, in MD simulations, only NaCl and KCl salts have been modeled so far.

The potential of nanotubes to reject ions while maintaining large water fluxes was first noted in MD simulations by Kalra et al. (2003). Using an osmotic gradient, water molecules were shown to flow through a membrane composed of (6,6) carbon nanotubes from the freshwater reservoir to the saltwater reservoir, with Na$^+$ and Cl$^-$ ions never flowing in the opposite direction. Further investigations identified a large free energy barrier for Na$^+$ ions to permeate across (6,6), 0.81 nm diameter carbon nanotubes but a relatively small barrier for the larger 1.4 nm (10,10) nanotubes (Peter and Hummer 2005). This indicated that there may be a critical pore diameter below which the 95% salt rejection required for desalination could be achieved. A subsequent investigation quantitatively determined the salt rejection of (5,5), (6,6), (7,7), and (8,8) (0.66, 0.81, 0.93, 1.09 nm diameter, respectively) carbon nanotubes under a hydrostatic pressure of 208 MPa to be 100%, 100%, 95%, and 58% (Corry 2008). Displayed in Figure 12.10 is the PMF of a Na$^+$ ion passing through these tubes, (Corry 2008) showing the large energy barrier for the narrow (5,5) carbon nanotube and the very small barrier for the wider (8,8). These results indicate that fewer than 1 in 100 nanotubes in a reverse osmosis membrane could have a diameter larger than 0.93 nm tube to maintain an overall 95% salt rejection. This is an important result to guide the manufacturing of carbon nanotube membranes as their fabrication produces a wide distribution of pore diameters, rather than a single diameter.

The partial charges of water molecules are attracted to the charge of the ion resulting in the formation of a shell of water around the ion, called the solvation shell. The number of water molecules in this shell is the coordination number for that ion. For an ion to enter the narrower tubes, some of the water molecules in the solvation shell must be removed, so that the ion may fit. Removing these water molecules incurs an energetic penalty and so creates the energy barriers observed in Figure 12.10. For instance, as illustrated in Figure 12.11, a Na$^+$ ion has a coordination number of about 6 in bulk water, but this must reduce to 2 inside a (5,5), 0.66 nm carbon

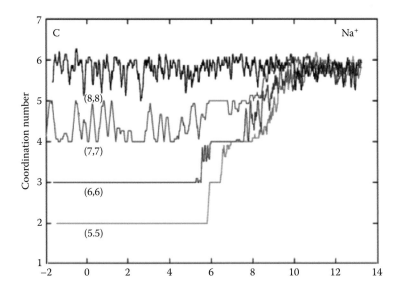

Figure 12.11 The coordination number of a Na⁺ ion as it moves from carbon nanotubes of various diameter pores (left-hand side of graph) to bulk water (right-hand side of graph). (Reprinted with permission from Corry, B.A., Designing carbon nanotube membranes for efficient water desalination, *J. Phys. Chem. B*, 112(5), 2008, 1427–1434. Copyright 2008 American Chemical Society.)

nanotube (Corry 2008). As the pore diameter increases, fewer water molecules are required to be removed from the solvation shell for the ion to move through the nanotube. At a diameter of 1.09 nm (a (8,8) carbon nanotube), the coordination number inside the tube is similar to that in bulk water.

Pristine boron nitride nanotubes also have the ability to reject salt. One MD study used hydrostatic pressure to force a NaCl solution through a single boron nitride nanotube embedded in a silicon nitride membrane (Hilder et al. 2009). The 0.69 nm diameter, (5,5) boron nitride nanotubes completely reject both Na⁺ and Cl⁻ ions. The large energy barrier for ion permeation through this nanotube (determined by calculating the PMF of each ion) was found to result from the removal of numerous water molecules from the coordination shell of each ion. This large dehydration barrier remains for Cl⁻ permeation through 0.83 nm diameter, (6,6) boron nitride nanotubes, but completely disappears for Na⁺. It is speculated that this results from the unique electrostatic interactions present in the boron nitride nanotubes due to the partial charges on the boron and nitrogen atoms in combination with the water structure inside the nanotube. A similar situation is found in the 0.97 nm diameter, (7,7) boron nitride nanotubes. In stark contrast, the salt rejection properties are reversed in the 1.1 nm (8,8) nanotubes; Na⁺ is completely rejected, while Cl⁻ is able to pass. The structure of the water molecules in this nanotube is highly ordered and has been described as ice-like, similar to the water structure in the (8,8) and (9,9) carbon nanotubes. The presence of Cl⁻ causes minimal deviation of water molecules from this structure, while Na⁺ causes a large reordering of water molecules, causing a large free energy barrier. This has also been recently observed in carbon nanotubes (He et al. 2013).

The ion rejection properties in (5,5), (6,6), and (7,7) silicon carbide nanotubes have also been determined through MD simulations (Hilder et al. 2012). The 0.86 nm diameter (5,5) nanotube completely rejected both Na⁺ and Cl⁻, much like its boron nitride counterpart. However, unlike boron nitride nanotubes, the rejection properties of Na^+ and Cl^- in the 1.0 nm (6,6) and 1.2 nm (7,7) silicon carbide nanotubes are reversed; Cl^- is able to permeate the pores, while Na^+ is rejected. This reversal is due predominantly to the large radial buckling present in the smaller-diameter silicon carbide nanotube (Alam and Ray 2007, 2008). The positive partially charged silicon atoms are closer to the center of the pore than the negatively partially charged carbon atoms. Due to simple electrostatic interactions, this makes the pore much more hospitable to the negatively charged Cl^- and inhospitable to the positively charged Na^+.

The mechanism of ion rejection in MD studies of pristine nanotubes differs from that in experimental investigations of nanotubes. The former focuses on narrow nanotubes, usually with pore diameters about 1 nm or less, whereas the latter investigates nanotubes with pores wider than 1 nm. Ions need to dehydrate to enter narrow tubes; a size exclusion mechanism is operating on hydrated and bare ions. Ion rejection in wider nanotubes is caused by electrostatic interactions between ions and functional groups at the nanotube pore opening.

12.3.4   Functionalization

During the manufacturing of membranes containing aligned nanotubes, the matrix in which the nanotubes are embedded must be etched to expose the tubes. Further etching removes the nanotube cap creating an open pore through the matrix. The etching process has been found to introduce chemical species that covalently bond to carbon atoms at the pore openings of nanotubes, known as functionalization (Majumder et al. 2005, Yang et al. 2005, Li et al. 2007). Typical functional groups include carboxylic (–COOH), hydroxyl (–OH), and carbonyl (C=O) groups. These functional groups can alter the water flux and ion rejection properties of nanotubes. The additional electrostatic interactions between these functional groups and water/ions are the main cause of these alterations, with steric and hydrodynamic effects playing only a small role (Fornasiero et al. 2008).

A detailed study by Corry determined how many different types of functionalizations can affect the properties of nanotubes (Corry 2011). In this investigation, 1.1 nm (8,8) carbon nanotubes were functionalized with varying numbers (either 2, 4, or 8) of COO^-, NH_3^+, OH (as shown in Figure 12.12), and $CONH_2$ functional groups, as well as a mixtures of NH_4^+ and COO^-, at the upstream pore opening. One or three of these functionalized carbon nanotubes were placed in an array with pristine carbon nanotubes (see Figure 12.5d as an example). Three configurations obtained 100% rejection of Na^+ and Cl^- ions, greater than the required 95%: 8 COO^-, $4NH_4^+$, and $3 \times 4NH_4^+$. The $3 \times 2NH_4^+$ obtained Na^+ and Cl^- rejection of 94% and 92%, just below the required 95%. Each of these improved on the rejection in pristine (8,8) nanotubes, which displayed Na^+ and Cl^- rejection of 28% and 86%, respectively. However, the most marked difference between the pristine and functionalized carbon nanotubes was the reduction in the water flux for all tested functionalized nanotubes, ranging from 67% to 13% of the pristine carbon nanotube flux. The three functionalized nanotubes obtaining sufficient Na^+ and Cl^- rejection displayed some of the lowest water fluxes with 13%, 24%, and 25% of the pristine fluxes for $8COO^-$, $4NH_4^+$, and $3 \times 4NH_4^+$, respectively. However, these water fluxes still equate to 3, 4.8, and 5.3 times larger than a common commercially used RO membrane SW30HR-380 assuming a packing density of carbon nanotubes in a matrix that has been achieved experimentally.

The reduction in water fluxes is due to the increased electrostatic interactions between water molecules and the functional groups, as well as steric blockages of the pore entrance. The strong electrostatic interactions between the water molecules and the functional groups increase the time that a water molecule spends at the pore entrance, slowing down the passage of following

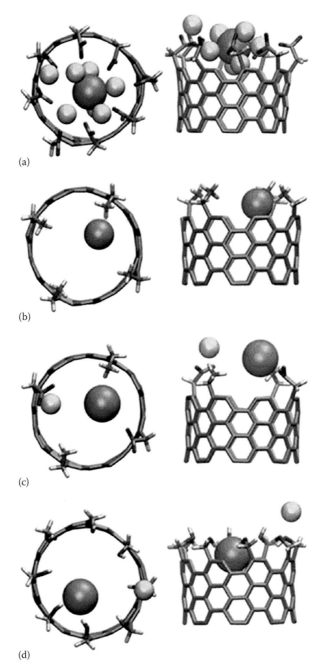

Figure 12.12   Top (left) and side (right) views of the position of ions near the pore opening of functionalized carbon nanotubes. Na^+ is represented by small spheres and Cl^- by large spheres. The carbon nanotubes depicted here are functionalized by (a) $8COO^-$, (b) $4NH_4^+$, (c) $2COO^-$, and $2NH_4^+$, and (d) $8OH$. (From Corry, B., Water and ion transport through functionalised carbon nanotubes: Implications for desalination technology, *Energy Environ. Sci.*, 4(3), 751–759, 2011. Reproduced by permission of The Royal Society of Chemistry.)

water molecules. Steric blockages can be caused by ions binding strongly to the functional groups or by the functional groups themselves; both these mechanisms reduce the area through which water can enter the nanotube pore.

The addition of functional groups at the pores' entrance can also affect the salt rejection of the nanotube. The situation is complicated by factors such as the pore diameter and the flexibility of the functional groups. For example, Cl^- rejection will increase if narrow nanotubes are functionalized with COO^- groups due to electrostatic repulsion. However, in wider nanotubes, Na^+ will aggregate around these COO^- groups, shielding the negative charge of the functional groups, allowing Cl^- to pass through the nanotube. This situation is illustrated in Figure 12.12a.

The mechanism of water retardation proposed by Corry is supported by further investigations by Hughes et al. (Hughes et al. 2012). In this study, arrays of aligned carbon nanotubes were functionalized with a range and combination of chemical moieties. It was found that electrostatic interactions between the water molecules and functional groups reduce water diffusion through the nanotube when compared to a pristine nanotube. The effect was greatest when the nanotube was functionalized with carboxylic acid (COOH) or carboxylate (COO^-) groups. Ion rejection is increased when the nanotubes are functionalized with these groups.

12.3.5   Ion Selectivity

Ion selectivity can be defined as a significant difference in the ability of a pore to pass/reject two ion types. This ability is very important in various biological processes; the selection between anions and cations is important for the regulation of blood pressure and organelle acidification, while the selection between Na^+ and K^+ is important for nerve conduction and maintaining electrochemical gradients across cells. Many proteins have developed the ability to discern between these ions, making these processes possible. One particular class of these proteins, called ion channels, are able to distinguish between ion types at near diffusion-limited rates. Nanotubes have the ability to mimic and replicate these properties for use in a range of applications, including ultrasensitive ion detection and antimicrobial agents.

The free energy barriers for ions to permeate a pristine nanotube can be different for different ion types due to the differences in dehydration energy for each ion type, making it more likely for some ions to pass than others. Ion selectivity has been investigated between Na^+, K^+, and Cl^- in pristine (5,5), (6,6), (7,7), (8,8), and (9,9) carbon nanotubes using MD simulations (Song and Corry 2009). For Na^+/K^+ selectivity, the PMF for each ion permeating each nanotube demonstrated that the free energy barrier was larger for K^+ in (5,5) and (6,6) carbon nanotubes, for Na^+ in (7,7) and (8,8) nanotubes, and roughly equal in (9,9) nanotubes. Na^+ has on average fewer (~6), but more tightly bound, water molecules in its coordination shell in bulk, whereas K^+ has more (~7), but less strongly bound water molecules. In the narrower (5,5) and (6,6) nanotubes, many water molecules are stripped from each ion, which is more energetically costly for Na^+ than K^+ as they are more tightly bound to Na^+. In the wider (7,7) and (8,8) carbon nanotubes, Na^+ is able to permeate with close to its full complement of water molecules, while some must still be stripped from K^+ for it to permeate. In the (9,9) carbon nanotubes, both ions are to pass with their full complement of water molecules, so there is little difference in their energy barriers. Similar principles have been shown for selection between anions in simplified narrow pores (Richards et al. 2012, 2012b).

Ion selectivity has been demonstrated in a range of other nanotube structures. For example, (9,9) carbon nanotubes functionalized with carbonyl groups at the pore openings are selective for Cl^- over Na^+ (Hilder et al. 2010). These nanotubes display similar conductance properties as

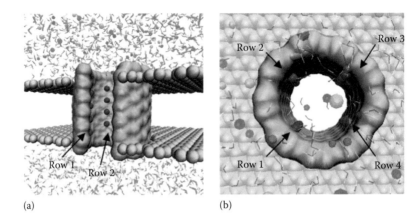

(a)   (b)

Figure 12.13   (a) Side and (b) top views of a (9,9) carbon nanotube functionalized with carbonyl (C=O) groups. The offset form is presented in this figure. (Reprinted with permission from Gong, X., Li, J., Xu, K., Wang, J., and Yang, H., A controllable molecular sieve for Na^+ and K^+ ions, *J. Am. Chem. Soc.*, 132(6), 2010, 1873–1877. Copyright 2010 American Chemical Society.)

biological Cl^--selective ion channels, such as ClC channels and GABA receptors. In addition, (6,6) and (7,7) boron nitride nanotubes are selective for Na^+ over Cl^- (Hilder et al. 2009a), while (10,10) boron nitride nanotubes are selective for Cl^- over K^+, the opposite of the selectivity found in (10,10) carbon nanotubes (Won and Aluru 2009). Differences in water structure, buckling, and ion dehydration energies produce different energy barriers for different ions to permeate the pore, resulting in selectivity. Pristine (6,6) and (7,7) silicon carbide nanotubes have an intrinsic selectivity for Cl^- over Na^+ (Hilder et al. 2012). The energy barrier for Na^+ permeation is much larger than Cl^- due to its water molecules being more strongly bound than Cl^-.

The interior of pores have been targeted as sites of functionalization, rather than the pore openings, in an attempt to mimic biological ion-selective structures (Gong et al. 2010). MD simulations of a (9,9) carbon nanotube functionalized with carbonyl oxygens, as shown in Figure 12.13, determined the selectivity between Na^+ and K^+. Three different configurations of carbonyl groups were studied, each producing marked differences in selectivity. The configuration mimicking the selectivity filter of a potassium channel (four sets of four carbonyl oxygens arranged in four rings) resulted in a Na^+-selective nanotube. When half of the carbonyl oxygen atoms were offset from the ring structure to a spiral structure (Figure 12.13a), the nanotube was nonselective. When two rows of carbonyl oxygen are removed from this structure, the nanotube exhibited K^+ selectivity. These results are surprising as the K^+ channel mimic selected for Na^+. It is difficult, if not impossible, to predict *a priori* the selectivity of functionalized nanotubes, given the number of factors involved in determining selectivity.

12.4 PRACTICAL FEASIBILITY AND CONCLUSIONS

The water and ion permeation properties of nanotubes make them ideal for use in many types of applications. Seawater desalination is a prime application for nanotube-based reverse osmosis membranes. MD simulations have demonstrated that some nanotubes have greatly increased water fluxes compared to commercially available membranes, while being able to maintain the necessary 95% salt rejection that is required for potable water. However, there are two issues that may limit the implementation of this technology.

A very narrow distribution of nanotube pore diameters is required in order to maintain the necessary salt rejection, a range of only 0.66–0.93 nm for unfunctionalized carbon nanotubes. Holt et al. (2006) and Fornasiero et al. (2008) have fabricated carbon nanotubes in membranes in the diameter range of 1–2 nm. Although smaller-diameter nanotubes have been fabricated—down to 0.4 nm (Guan 2008)—it is unclear how these nanotubes could be grown or placed in an aligned manner into a membrane matrix, while maintaining such a small range of pore sizes.

The second issue is the implementation of the nanotube-based reverse osmosis membranes into desalination facilities. Modern facilities typically operate at a recovery rate of 50%: half of the intake saltwater is desalinated, while the other half doubles in salt concentration. These facilities operate at a pressure a few percent above the osmotic pressure of this hypersaline solution. Most of the energy put into the reverse osmosis process is used to overcome this osmotic pressure, not to work against the friction encountered by water permeating through the membrane. Nanotube-based reverse osmosis membranes may save up to about 1% of energy used in this process, but some researchers argue that effort and money are better spent improving other processes in the desalination facility (Elimelech and Phillip 2011). It could be argued that nanotube-based membranes could save on capital costs by requiring less surface area than currently implemented membranes, to achieve the same potable water fluxes. However, for the foreseeable future, the cost of nanotube-based membranes will far outweigh the cost of purchasing and installing more traditional membrane modules. A possible advantage of having a high-flux nanotube membrane would lie in reduced size of the desalination plant, which may find specialized applications in places where size and weight are limited, for example, space missions.

Nanotube-based membranes may still prove to be more resistant to fouling, a common problem in currently used membranes. A number of organic and inorganic species are able to physically and chemically occlude pores, and therefore membranes must be regularly cleaned. Carbon nanotubes have been demonstrated to be able to be readily functionalized; perhaps the more chemically inert boron nitride and silicon nitride nanotubes will be more resistant to fouling. More research is required to determine the ability of these nanotubes to resist fouling. MD is not able to simulate chemical reactions taking place during chemical fouling of the membrane. There are other computational techniques that are available to study this such as quantum mechanical and hybrid quantum mechanical/molecular dynamic techniques. However, fouling in the form of physical blockage could potentially be modeled by MD for small fouling species.

The recycling of wastewater is becoming more popular as traditional water sources become scarcer. Filtration techniques employed by water treatment facilities allow some potentially harmful molecules such as endocrine-disrupting chemicals to pass through (Johnson and Sumpter 2001). In addition, it has been determined that traditional nanofiltration membranes allow the passage of hormones such as testosterone and progesterone (Ngheim et al. 2004). The selectivity properties of nanotube-based membranes make them ideal candidates as filtration membranes. The more rigid structure of nanotubes than polymers can allow for high levels of selectivity, which may be useful for filtering of these species.

MD is a computer simulation technique that provides a powerful tool to investigate transport in nanotubes. Nanotubes possess some very interesting properties, such as the high water throughputs and salt rejection. Water traverses through nanotube pores with almost frictionless flow, a unique property of nanotubes. Coupled with the salt rejection properties, this makes many types of nanotubes ideal for high-throughput saltwater desalination. There are many other applications of selective nanotube pores that are not discussed in this chapter, for example, biosensors and nanofluid devices, and we anticipate that they will find uses in a range of unexpected areas in the future.

REFERENCES

Alam, K. M. and Ray, A. K., A hybrid density functional study of Zigzag SiC nanotubes. *Nanotechnology* 18(49) (2007):495706.

Alam, K. M. and Ray, A. K., Hybrid density functional study of armchair SiC nanotubes. *Phys. Rev. B.*, 77(3) (2008):035436.

Alexiadis, A. and Kassinos, S., Molecular simulation of water in carbon nanotubes. *Chem. Rev.* 108(12) (2008):5014–5034.

Beckstein, O. and Samsom, M. S. P., The influence of geometry, surface character, and flexibility on the permeation of ions and water through biological pores. *Phys. Biol.* 1(1) (2004):42–52.

Corry, B., An energy-efficient gating mechanism in the acetylcholine receptor channel suggested by molecular and Brownian dynamics. *Biophys. J.* 90(3) (2006):799–810.

Corry, B., Water and ion transport through functionalised carbon nanotubes: Implications for desalination technology. *Energy Environ. Sci.* 4(3) (2011):751–759.

Corry, B. A., Designing carbon nanotube membranes for efficient water desalination. *J. Phys. Chem. B* 112(5) (2008):1427–1434.

Dzubiella, J. and Hansen, J. -P., Electric-field-controlled water permeation of a hydrophobic nanopore. *J. Chem. Phys.* 122(23) (2005):234706.

Elimelech, M. and Phillip W. A., The future of seawater desalination: Energy, technology, and the environment. *Science* 333(6043) (2011):712–717.

Fornasiero, F., Park, H. G., Holt, J. K. et al., Ion exclusion by sub-2-nm carbon nanotube pores. *Proc. Natl. Acad. Sci. U.S.A.* 105 (2008):17250–17255.

Gong, X., Li, J., Xu, K., Wang, J., and Yang, H., A controllable molecular sieve for Na^+ and K^+ ions. *J. Am. Chem. Soc.* 132(6) (2010):1873–1877.

Gong, X., Li, J., Zhang, H. et al., Enhancement of water permeation across a nanochannel by the structure outside the channel. *Phys. Rev. Lett.* 101(25) (2008):257801.

Guan, L., Suenaga, K., and Iijima, S., Smallest carbon nanotube assigned with atomic resolution accuracy. *Nano Lett.* 8(2)(2007):459–462.

He, Z., Zhou, J., Lu, X., and Corry, B., Ice-like water structure in carbon nanotube (8,8) induces cationic hydration enhancement. *J. Phys. Chem. C* 117(21) (2013):11412–11420.

Hilder, T. A., Gordon. D., and Chung, S. -H., Salt rejection and water transport through boron nitride nanotubes. *Small* 5(19) (2009):2183–2190.

Hilder, T. A., Gordon, D. and Chung S. -H., Boron nitride nanotubes selectively permeable to cations or anions. *Small* 5(24) (2009a):2870–2875.

Hilder, T. A., Gordon, D., and Chung, S. -H., Synthetic chloride-selective carbon nanotubes examined by using molecular and stochastic dynamics. *Biophys. J.* 99(6) (2010):1734–1742.

Hilder, T. A., Gordon, D., and Chung, S. -H., Computational modeling of transport in synthetic nanotubes. *Nanomedicine-UK* 7(6) (2011):702–709.

Hilder, T. A., Yang, R., Gordon, D., Rendell, A. P., and Chung, S. -H., Silicon carbide nanotube as a chloride-selective channel. *J. Phys. Chem. C* 116(7) (2012):4465–4470.

Hinds, B. J., Chopra, N., Rantell, T., Andrews, R., Gavalas, V., and Bachas, L. G., Aligned multiwalled carbon nanotube membranes. *Science* 303(5654) (2004):62–65.

Holt, J. K., Park, H. G., Wang, Y. et al., Fast mass transport through Sub-2-nanometer carbon nanotubes. *Science* 312(5776) (2006):1034–1037.

Hughes, Z. E., Shearer, C. J., Shapter, J., and Gale, J. D., Simulation of water transport through functionalized Single-Walled Carbon Nanotubes (SWCNTs). *J. Phys. Chem. C* 116(47) (2012):24943–24953.

Hummer, G., Rasaiah, J. C., and Noworyta, J. P., Water conduction through the hydrophobic channel of a carbon nanotube. *Nature* 414 (2001):188–190.

Humphrey, W., Dalke, A., and Schulten, K., VMD: Visual molecular dynamics. *J. Mol. Graphics*, 14(1) (1996):33–38.

Johnson, A. C. and Sumpter, J. P., Removal of endocrine-disrupting chemicals in activated sludge treatment works. *Environ. Sci. Technol.* 35(24) (2001):4697–4703.

Joseph, S. and Aluru, N. R., Why are carbon nanotubes fast transporters of water? *Nano Lett.* 8(2) (2008):452–458.

Kalra, A., Garde, S., and Hummer, G., Osmotic water transport through carbon nanotube membranes. *Proc. Natl. Acad. Sci. U.S.A.* 100(18) (2003):10175–10180.

Kumar, S., Bouzida, D., Swendsen, R. H., Kollman, P. A., and Rosenberg, J. M., The weighted histogram analysis method for free-energy calculations on biomolecules. I. The method. *J. Comput. Chem.* 13(8) (1992):1011–1021.

Li, P. H., Lim, X., Zhu, Y. et al., Tailoring wettability change on aligned and patterned carbon nanotube films for selective assembly. *J. Phys. Chem. B* 111(7) (2007):1672–1678.

Lu, D., Li, Y., Rotkin, S. V., Ravaioli, U., and Schulten, K., Finite-size effect and wall polarization in a carbon nanotube channel. *Nano Lett.* 4(12) (2004):2383–2387.

MacKerell, A. D. Jr., Bashford, D., Bellott, M. et al., All-atom empirical potential for molecular modeling and dynamics studies of proteins. *J. Phys. Chem. B* 102(18) (1998):3586–3616.

Maiti, A. Carbon nanotubes: Bandgap engineering with strain. *Nat. Mater.* 2 (2003):440–442.

Majumder, M., Chopra, N., Andrews, R., and Hinds, B. J., Enhanced flow in carbon nanotubes. *Nature* 438 (2005):44.

Majumder, M. and Corry, B., Anomalous decline of water transport in covalently modified carbon nanotube membranes. *Chem. Commun.* 47(27) (2011):7683–7685.

Nghiem, L. D., Schäfer, A. I., and Elimelech, M., Removal of natural hormones by nanofiltration membranes: Measurement, modeling, and mechanisms. *Environ. Sci. Technol.* 38(6) (2004):1888–1896.

Nicholls, W. D., Borg, M. K., Lockerby, D. A., and Reese, J. M., Water transport through (7,7) carbon nanotubes of different lengths using molecular dynamics. *Microfluid Nanofluid* 12(1–4) (2012):257–264.

Pascal, T. A., Goddard, W. A., and Jung, Y., Entropy and the driving force for the filling of carbon nanotubes with water. *Proc. Natl. Acad. Sci. U.S.A.* 108(29) (2011):11794–11798.

Peter, C. and Hummer, G., Ion transport through membrane-spanning nanopores studied by molecular dynamics simulations and continuum electrostatic calculations. *Biophys. J.* 89(4) (2005):2222–2234.

Richards, L. A., Schäfer, A. I., Richards, B. S., and Corry, B., The importance of dehydration in determining ion transport in narrow pores. *Small* 8(11) (2012):1701–1709.

Richards, L. A., Schäfer, A. I., Richards, B. S. and Corry, B., Quantifying barriers to monovalent anion transport in narrow non-polar pores. *Phys. Chem. Chem. Phys.* 14(33) (2012b):11633–11638.

Salomon-Ferrer, R., Case, D. A., and Walker, R. C. An overview of the amber biomolecular simulation package. *WIREs Comput. Mol. Sci.* 3(2) (2013):198–210.

Scott, W. R. P., Hünenberger, P. H., Tironi, I. G. et al., The GROMOS biomolecular simulation program package. *J. Phys. Chem. A* 103(19) (1999):3596–3607.

Song, C. and Corry, B., Intrinsic ion selectivity of narrow hydrophobic pores. *J. Phys. Chem. B* 113(21) (2009):7642–7649.

Su, J. and Guo, H., Effects of nanochannel dimension on the transport of water molecules. *J. Phys. Chem. B* 116(20) (2012):5925–5932.

Suk, M. E., Raghunathan, A. V., and Aluru, N. R., Fast reverse osmosis using boron nitride and carbon nanotubes. *Appl. Phys. Lett.* 92(13) (2008):133120.

Taghavi, F., Javadian, S., and Hashemianzadeh, S. M., Molecular dynamics simulations of single-walled silicon carbide nanotubes immersed in water. *J. Mol. Graph. Model.* 44 (2013):33–43.

Torrie, G. and Valleau, J., Monte Carlo free energy estimates using non-Boltzmann sampling: Application to the sub-critical lennard-jones fluid. *J. Chem. Phys. Lett.* 28(4) (1974):578–581.

Valavala, R., Sohn, J., Han, J., Her, N., and Yoon, Y., Pretreatment in reverse osmosis seawater desalination: A short review. *Environ. Eng. Res.* 16(4) (2011):205–212.

Won, C. Y., Joseph, S., and Aluru, N. R., Effect of quantum partial charges on the structure and dynamics of water in single-walled carbon nanotubes. *J. Chem. Phys.* 125(1) (2006):115701.

Won, C. Y. and Aluru, N. R., Water permeation through a subnanometer boron nitride nanotube. *J. Am. Chem. Soc.* 129(10) (2007):2748–2749.

Won, C. Y. and Aluru, N. R., Structure and dynamics of water confined in a boron nitride nanotube. *J. Phys. Chem. C* 112(6) (2008):1812–1818.

Won, C. Y. and Aluru, N. R., A chloride ion-selective boron nitride nanotube. *Chem. Phys. Lett.* 478(4–6) (2009):185–190.

Yang, D. -Q., Rochette, J. -F., and Sacher, E., Controlled chemical functionalization of multiwalled carbon nanotubes by kiloelectronvolt argon ion treatment and air exposure. *Langmuir* 21(18) (2005):8539–8545.

Yang, R., Hilder, T. A., Chung, S. -H., and Rendell, A., First-principles study of water confined in single-walled silicon carbide nanotubes. *J. Phys. Chem. C* 115(35) (2011):17255–17264.

Zeidel, M. L., Ambudker, S. V., Smith, B. L., and Agre, P., Reconstitution of functional water channels in liposomes containing purified red cell CHIP28 protein. *Biochemistry* 31(33) (1992):7436–7440.

Zhu, F. and Schulten, K., Water and proton conduction through carbon nanotubes as models for biological channels. *Biophys. J.* 85(1) (2003):236–244.

Zhu, F., Tajkhorshid, E., and Schulten, K., Pressure-induced water transport in membrane channels studied by molecular dynamics. *Biophys. J.* 83(1) (2002):154–160.

Zhu, F., Tajkhorshid, E., and Schulten, K. Theory and simulation of water permeation in aquaporin-1. *Biophys. J.* 86(1) (2004):50–57.

Functionalization

Chapter 13

Surface Chemistry of Two-Dimensional Layered Nanosheets

Motilal Mathesh, Hongbin Wang, Colin J. Barrow, and Wenrong Yang

CONTENTS

13.1   INTRODUCTION

Two-dimensional (2D) nanosheets have gained its interest since 2004, when Dr. Andre Geim and Dr. Konstantin Novoselov isolated free-standing single layer of graphene (Novoselov et al. 2005). This discovery challenged the theoretical views on 2D materials and showcased the research community about the stability and potential applications of the 2D nanosheets. The stability is governed by the fact that they might have infinitely small fluctuations at low temperatures owing to thermal fluctuations (Novoselov 2011). Another possibility is the stabilisation of nanosheets by ripples that extend 2D nanosheets into three-dimensional (3D) objects (Meyer et al. 2007). After isolating free-standing graphene, the work was extrapolated to other inorganic materials giving rise to MoS_2, BN, transitional metal dichalcogenides (TMDs), clay, and metal oxides and 2D nanosheets. Since then, there has been much interest in 2D nanosheets. These materials have a restricted dimension, which makes them disparate from its 3D form. One can discriminate this in terms of restriction in size or motion along length and breadth, which gives rise to a flat land (Novoselov 2011). Once graphene was isolated, its properties were significantly different from its counterpart, which could be used in fields like composite materials, electronics, and biomedicine. These plate-like materials could be tailored by stacking on top of each other to obtain layered 3D material, which has huge potential application. In this chapter, we discuss the synthesis, surface chemistry, and biological applications of 2D nanosheets.

13.1.1   Graphene and Graphene Oxide

Graphene and graphene oxide (GO) are the new frontiers in the field of nanomaterials, which have rich potential applications. Graphene has outstanding mechanical, electrical, and thermal properties, which are apt for its applications in the field of biosensors. On the other hand, GO is nonconductive; yet, it has gained interest because it is single sheet with rich functional groups with the possibility of being tuned at the molecular level, along with hydrophobic and hydrophilic groups that could be used to bind various biomolecules, thus paving way to a new era of nano-biotechnology. The detailed description of these carbon nanomaterials and their applications is discussed in further sections.

As shown in Figure 13.1 (Iwan and Chuchmala 2012), graphene has a 2D structure composed of single layer of carbon atoms connected by sp^2 covalent bonds (Castro Neto et al. 2009) forming a honeycomb-like structure. They are individual layers of graphite stacked on top of each other by van der Waals forces and $\pi-\pi$ interactions (Ratinac et al. 2011). Depending on the number of layers, these sheets can be divided into four groups, namely, single-layer graphene (Balandin et al. 2008), bilayer graphene (Ohta et al. 2006), few-layer graphene (Reina et al. 2008), and multilayer graphene (Johnson et al. 2010), each having its own unique properties. These sheets are flexible and not completely flat, which makes the free-standing papers of multilayer graphene to crumble (Castro Neto 2009). On the other hand, GO comprises a single-atom-thick carbon sheets (Figure 13.1) with carboxylate groups at the edges imparting negative charge, which is pH dependent (Mathesh et al. 2013). Its basal plane comprises hydroxyl and epoxide functional groups together with unmodified graphenic domains that are hydrophobic in nature and assists in $\pi-\pi$ interactions (Sanchez et al. 2011). These sheets thus result in giant amphiphilic sheet molecule that acts as a surfactant (Kim et al. 2010) and stabilizes hydrophobic molecules in solution (Guo et al. 2011).

With the isolation of single-layer graphene sheets, there was a notion of isolating other 2D nanosheets using micromechanical cleavage, which was later proved by Novoselov et al. (2005).

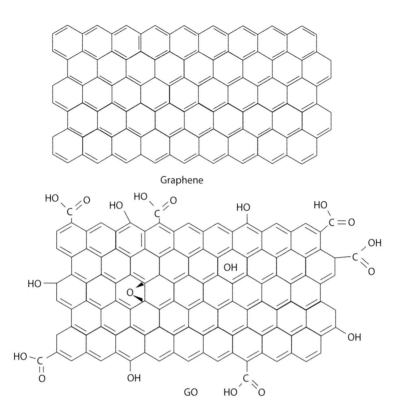

Figure 13.1 Chemical structure of graphene and graphene oxide (GO). (Adapted from *Prog. Poly. Sci.*, 37, Iwan, A. and Chuchmała, A., Perspectives of applied graphene: Polymer solar cells, 1805–1828, Copyright 2012, with permission from Elsevier.)

This gave rise to hexagonal boron nitride (*h*-BN), which possesses higher chemical and thermal stabilities (Yu et al. 2010), and MoS$_2$ (metal dichalcogenides) with tunable band gap. With a large variety of nanosheets available, it is possible to tailor their properties according to the needs and exploit them for practical applications.

13.1.2 Hexagonal Boron Nitride

h-BN also referred as *white graphene* comprises honeycomb-like lattice and is an isoelectric analogue of graphite (Lin et al. 2009). They do not absorb light in the visible region of the electromagnetic spectrum (Pakdel et al. 2013, Zeng et al. 2010). The individual layers are held together by van der Waals forces and have an interlayer distance of 0.33 nm (Pease 1952), which is similar to that of graphene. Herein, the hexagonal lattice comprising C–C bonds in graphene is replaced by polar, sp^2-hybridized B–N bonds that are covalent in nature (Song et al. 2010). They have a large band gap of 4–6 eV rendering them as an insulator (Zunger et al. 1976). Figure 13.2 shows the comparison between graphite plane and *h*-BN plane (Souche et al. 1998). The other important properties of BN are high-temperature resistance, chemical inertness, environmental safety, and poor wettability (Eichler and Lesniak 2008). They find applications in optoelectronic devices (Davis 1991, Shi et al. 2010) and dielectrics (Song et al. 2010). BN nanotubes have been chemically functionalized with molecular groups

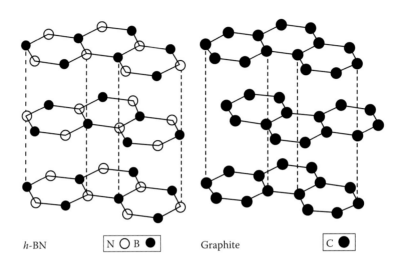

h-BN   N ◯ B ●   Graphite   C ●

Figure 13.2 Comparison between graphite and hexagonal boron nitride structures. (Adapted from *Micron*, 29, Souche, C., Jouffrey, B., Hug, G., and Nelhiebel, M., Orientation sensitive eels-analysis of boron nitride nanometric hollow spheres, 419–424, Copyright 1998, with permission from Elsevier.)

for the advent of technologically useful materials like biosensors (Sainsbury et al. 2007), which could be achieved in case of BN nanosheets as well. BN nanosheets have been shown to possess photoluminescence at 224 nm wavelength (Hua Li et al. 2012), which could be used for deep ultraviolet light-emitting diodes. On the other hand, they are also used as a template for graphene-based electronic device fabrication with reduced roughness and chemical reactivity (Dean et al. 2010).

13.1.3 MoS_2

TMDs have received great attention because they are semiconductors with fairly large band gap and are naturally available (Mak et al. 2010). MoS_2 belongs to this family, owning a honeycomb-like structure with covalently bonded S–Mo–S sheets, which in turn are bonded by van der Waals forces (Splendiani et al. 2010) with an interlayer spacing of 0.65 nm. Figure 13.3 shows the 3D structure of MoS_2 sheets (Radisavljevic et al. 2011). The band gap of bulk MoS_2 is 1.2 eV (Kam and Parkinson 1982), which changes to 1.8 eV (Mak et al. 2010) after exfoliation. This indirect-to-direct band gap transition from bulk- (3D) to single-layer sheets (2D) is due to quantum confinement effects, which is evidenced as photoluminescence (Chhowalla et al. 2013). The band gap could be chosen at will, as it is dependent on the number of layers. They could also be doped with transitional metal atoms giving rise to new properties such as net magnetic moment, which is not present in its 3D form. This makes them a good candidate for nanoelectronics and spintronic applications (Ataca and Ciraci 2011). Due to the presence of hexagonal structure in the basal planes, they could be used to bind aromatic and conjugated compounds with the help of van der Waals interaction (Moses et al. 2009). There have been reports suggesting intercalation of organic molecules between restacked single layers of MoS_2 with ferrocene (Divigalpitiya et al. 1989). This intercalation has been further extrapolated for photothermal therapy and quantitative measurements of oligonucleotides as discussed in Section 13.6.

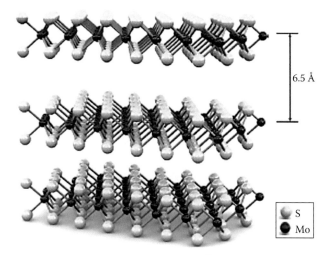

6.5 Å

S
Mo

Figure 13.3 3D representation of MoS_2 structure. (Adapted with permission from Macmillan Publishers Ltd. *Nat. Nanotechnol.*, Radisavljevic, B., Radenovic, A., Brivio, J., Giacometti, V., and Kis, A., Single-layer MoS_2 transistors, 2011, 6, 147–150, Copyright 2011.)

13.2 WHY 2D NANOSHEETS ARE IMPORTANT?

Once graphene could be isolated in 2D form, researchers synthesised to other 2D materials like MoS_2, h-BN, and $Bi_2Sr_2CaCu_2O_x$ (Novoselov et al. 2005). Major reasons that justify the importance of 2D nanosheets are peculiar properties of 2D materials, which are not observed in its counterpart 3D materials; open to manipulations; increased surface area; ease of functionalization; and fabrication of hybrid materials. These are explained in detail as follows:

1. The properties of 2D nanosheets are very different from its counterpart 3D materials from which they are exfoliated. This is due to the fact that their mobility is restricted to two dimensions rather than three dimensions. A well-known example for this is graphene, which is made from graphite. Graphene in its 2D form is light and transparent, yet dense enough to not allow helium atom to pass through it (Nair et al. 2012); more conductive than copper (Geim 2009) and stronger than steel (Savage 2012), which is not found in its 3D counterpart graphite.

2. According to Novoselov (2012a), the 3D materials have limitations related to modification and manipulation to achieve a desired function as the geometry could be easily affected by surrounding conditions such as radiation and the chemistry used for manipulation (for instance, doping) could be detrimental for other parameters like electron transfer (ET; Novoselov et al. 2012a). However, 2D nanosheets provide better control over the parameters given earlier, for example, the electronic properties could be easily controlled by strain, shear, and bending (Pereira et al. 2009), thus providing a superior authority on parameters for industrial applications.

3. One of the other reasons to use 2D nanosheets is increased surface area. Vermiculite that is used to purify water is an excellent example for this. The ion exchange ability of vermiculate increases 10^6 times when its surface area increases with expansion (Luckham and Rossi 1999).

4. In terms of the structure of 2D nanosheets, all the atoms are exposed on surface, which could be utilized for functionalization to synthesize hybrid materials for application in the field of drug delivery, biosensors, and diagnostic purposes.

5. It is easy to develop hybrid materials by stacking up 2D monolayers tuned for different properties by stacking up on each other by means of layer-by-layer (LBL) assembly or self-assembly. They would revolutionize the current technology with materials that has always been imagined. For instance, MoS_2 nanosheets have been grown on graphene to synthesize a hybrid material, with its application as excellent electrocatalytic activity in hydrogen evolution reaction (Li 2011b) and as cocatalyst for hydrogen production by water-splitting reaction (Xiang et al. 2012). On the other hand, bilayer graphene sandwiched between h-BN has been studied for the tunability of band gap for electronic applications (Ramasubramaniam et al. 2011) for biosensors. With respect to biological applications, diverse nanosheets could be functionalized with a variety of biomolecules to give a multifunctional 3D material that will eliminate the need for developing distinct materials for different applications. For example, GO and MoS_2 could be bound with various ssDNA and used for molecular probing of different targets of DNA at the same time.

13.3 NEED OF 2D NANOSHEETS FOR BIOLOGICAL APPLICATION

Graphene family sheets are known to have good thermal conductivity, stability, and mechanical strength, which could find its applications in the field of electronics (Gilje et al. 2007), photonics (Bonaccorso et al. 2010), sensors (Cheng et al. 2010), energy generation, and storage devices (Brownson et al. 2011). Much emphasis and research expertise has been put on the earlier-mentioned properties, but herein, we will focus on properties useful for biological applications. The most fascinating properties for biological and biomedical applications are the large surface area, control over size, high purity, and ease of functionalization (Novoselov et al. 2012b), which go well with other 2D nanosheets as well. Sanchez et al. (2011) have reported the properties of 2D materials that are useful for biological applications, which are discussed in the following text.

Derived from Brunauer-Emmett-Teller (BET) surface area measurements, a single graphene sheet has a surface area of 2600 $m^2\ g^{-1}$ (Stoller et al. 2008), which is much higher than other nanoparticles (NPs) that are studied in biological systems. Exposure of all its atoms on the surface is the most appreciating feature rather than few atoms, as in the case of single- or multi-walled carbon nanotubes, thus increasing surface coverage. This large surface area is useful for biological interactions. However, one of the major issues faced by graphene is restacking of its plate structure due to aggregation, which potentially reduces its surface area, which could be overcome with the help of pillared structures (Dimitrakakis et al. 2008) or with simple surface chemistry (Li et al. 2008). Researchers have shown the possibility of increasing the interlayer distance between the sheets with the assistance of carbon nanotubes (Zhang et al. 2010d) and, thus, helping in utilising of surface area in between two sheets.

Surface area available for utilization in a plate-like structure depends on the number of layers. As layer number increases, the available surface area decreases, due to the fact that stacked plates do not have accessibility to the bottom layers. In contrast, to increase the stiffness of graphene sheets, the number of layers should be more. Hence, there is a need for compensation with decrease in surface area.

The size of the NPs should be controllable for their appropriate use. In the case of graphene, the size could be easily controlled with the aid of ultrasonication, which breaks the larger flakes into smaller particles, which could then be characterized using atomic force microscopy or scanning electron microscopy. Dynamic light scattering is another tool that could be used for this purpose, but the drawback is the principles of DLS are based on spherical models rather than flat sheets (Sanchez et al. 2011). Particle size is responsible for in vivo distribution, toxicity, fate of NPs in biological systems, and ability to target the cells (Mohanraj and Chen 2007). Uptake rate of smaller particles is shown to be higher than bigger particles (Desai et al. 1997). In case of dichalcogenides like MoS_2, the control in shape and size can assist in controlling the surface chemistry (Chhowalla et al. 2013).

Biocompatibility of any given polymeric biomaterial depends on various factors that include interfacial free energy, type and density of surface charges, molecular weight, conformational flexibility, chemical structure and functional group, surface topography and roughness, and balance between hydrophobicity and hydrophilicity (Wang et al. 2004). Among them, the latter three plays a crucial role in case of nanosheets. The growth pattern of endothelial cells was studied with the help of gold alkanethiolates, which showed COOH groups to have the best growth rate (Tidwell et al. 1997), thus backing up GO as biomaterials. On the other hand, hydrophobic surfaces tend to adsorb more proteins as compared to hydrophilic surfaces (Elbert and Hubbel 1996), which causes blood clotting, because protein when in contact with foreign material causes platelet adhesion and activation, assisting in clot formation. Surface roughness, which is the other important parameter, has been shown to influence clot formation (Sheppard et al. 1994), suggesting materials should be flat to prevent thrombogenicity. Taken these points into consideration, there is much to be explored, as 2D nanosheets have not yet been fully studied for their toxic effects on cell lines. A significant group of researchers have taken efforts to study the biocompatibility of these NPs; yet, there is still a lack in understanding them for use in larger extent, which is a major drawback. Some of the research works that have made significant contribution to understand its biocompatibility are as follows:

- GO cells cause cytotoxicity, which is dose and time dependent. They have the ability to enter cytoplasm and have negative effects that include decreased cell adhesion (Wang et al. 2010a).
- Graphene papers have been shown to be biocompatible with mouse fibroblast cells (Chen et al. 2008a).
- Pegylated graphene has no possible toxicity at a dose 20 mg kg^{-1} to Balb c female mice (Yang et al. 2010a).
- Intravenous injection of GO showed accumulation in lungs for long period of times, inducing pulmonary toxicity, which is dose dependent (Zhang et al. 2011c).
- MoS_2 NPs have been synthesized by pulsed laser ablation in water and have been showed to have good solubility and biocompatibility using human embryonic epidermal fibroblast cells (Wu et al. 2011).

All these studies are at the preliminary level, and the mechanism of graphene interaction with cell lines is not yet clear which has to be investigated further.

13.4 SYNTHESIS OF 2D NANOSHEETS

2D nanosheets are synthesized by exfoliation from its 3D counterpart. There are many strategies adopted for the synthesis of 2D nanosheets. In this section, we will look in detail the various methods in application.

13.4.1 Mechanical Exfoliation

This technique involves the use of sheer force to pull out the top layer, by breaking weak bonds between the 3D stacked structure and separate individual sheets (Choi et al. 2010). Graphene was first isolated from graphite using scotch and tape method, which essentially involved breaking of π–π bond between the top layer and the one below. Figure 13.4 shows the procedure for mechanical exfoliation (Novoselov and Castro Neto 2012a). This has been extended to a variety of inorganic materials too. Single layers of BN, MoS_2, $NbSe_2$, and $Bi_2Sr_2CaCu_2O_x$ have been synthesized by rubbing its 3D crystal against the surface of one another (Novoselov et al. 2005). These leaves behind inhomogeneous layer of materials, but often single layers are found among them.

One of the important factors to be considered for mechanical exfoliation is that the interaction between the surfaces on which the material has to be deposited after cleavage should be stronger than the interaction between the layers of sheets that have been exfoliated from the crystal structure (Novoselov 2011). This process yields very high–purity materials but often results in very low yield, which is not suitable for large-scale production. To improve the yield, tailored ball milling has been adopted to produce BN nanosheets that could be applied to any layered materials (Li et al. 2011a). Even though the process is very easy, they cannot be put into action for industrial production due to poor yield. To circumvent these problems, researchers

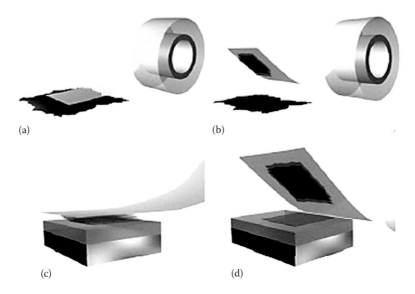

(a)   (b)

(c)   (d)

Figure 13.4   Schematic representation of micromechanical exfoliation of 2D crystals. (a) Adhesive tape is pressed against a 2D crystal and peeled; (b) the top few layers are attached to the tape. (c) The tape with crystals of layered material is pressed against a surface of choice. (d) Upon peeling off, the bottom layer is left on the substrate. (Adapted with permission Novoselov, K.S. and Castro Neto, A.H., Two-dimensional crystals-based heterostructures: Materials with tailored properties. *Physica Scripta*, 2012, Copyright Institute of Physics.)

considered to look for alternatives, which can produce large quantity of materials by compromising with the purity to some extent.

13.4.2 Chemical Vapor Deposition

This method involves the use of a surface that acts as a scaffold on which the 2D layer materials are grown and separated out. Often, the template is chosen based on the difference in properties between 2D nanosheets and the template that needs to be separated. One of the finest examples is surface-assisted growth of BN nanoribbons (Chen 2008b). The surface chosen for this purpose was ZnS nanoribbons, which could be easily synthesized in large quantity and also vaporized at 1100°C in the presence of hydrogen but BN do not vapourise at that condition, which helps in separation. Figure 13.5 represents the methodology followed for chemical vapor deposition (CVD; Novoselov and Castro Neto 2012a). Analogously, monolayer h-BN was also synthesized using ammonia-borane with copper foil as a substrate (Kim et al. 2011). Graphene has also been synthesized using this method, with the help of thermal decomposition of hydrocarbons on stainless steel surface, where stainless steel acts as substrate for decomposition, which could be etched away to obtain free-standing graphene. Although the nanosheet thickness could be tuned according to the needs, it is difficult to separate out the nanosheet from the template, which is a great disadvantage. Another disadvantage is, use of high temperature, which may cause trouble with the purity and the crystalline nature of the nanosheets. This evoked the need for other techniques.

13.4.3 Liquid Exfoliation

There has been various reports on the exfoliation of layered materials in liquids as it is a widely used method for large-scale production of nanosheets for industrial applications, hence

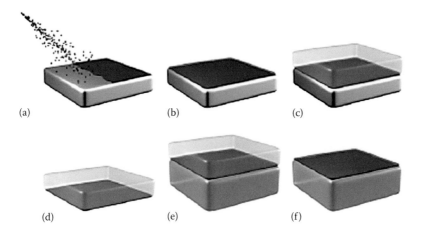

(a)   (b)   (c)

(d)   (e)   (f)

Figure 13.5 Schematic representation of transfer of chemical vapor deposition (CVD)-grown 2D crystals. (a and b) Two-dimensional crystals are grown by CVD on a surface of a metal. (c) A sacrificial layer is deposited on top of the 2D crystal. (d) The metal is etched away, leaving 2D crystal stuck on the sacrificial layer. (e) The sacrificial layer, together with the 2D crystal, is transferred onto the substrate of choice. (f) The sacrificial layer is removed. (Adapted with permission Novoselov, K.S. and Castro Neto, A.H., Two-dimensional crystals-based heterostructures: Materials with tailored properties. *Physica Scripta*, 2012, Copyright Institute of Physics.)

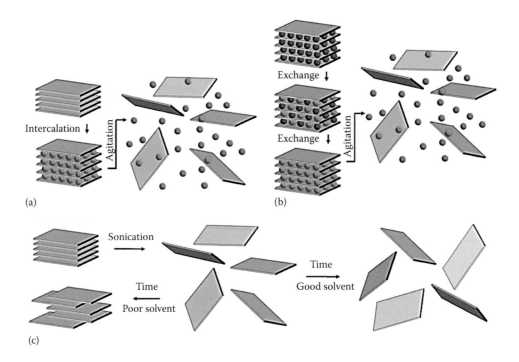

Figure 13.6 Schematic description of liquid exfoliation mechanisms. (a) Ion intercalation. Ions are intercalated between the layers in a liquid environment, swelling the crystal and weakening the inter-layer attraction. Then, agitation (such as shear, ultrasonication, or thermal) can completely separate the layers, resulting in an exfoliated dispersion. (b) Ion exchange. Some layered compounds contain ions between the layers so as to balance surface charge on the layers. These ions can be exchanged in a liquid environment for other often larger ions. Again, agitation results in an exfoliated dispersion. (c) Sonication-assisted exfoliation. The layered crystal is sonicated in a solvent, resulting in exfoliation and nanosheet formation. This mechanism also describes the dispersion of graphene oxide in polar solvents like water. (From Nicolosi, V., Chhowalla, M., Kanatzidis, M.G., Strano, M.S., and Coleman, J.N., Liquid exfoliation of layered materials, *Science*, 340, 1226419, 2013. Copyright, Adapted with permission of AAAS.)

overcoming the issues posed by mechanical cleavage. Other major advantage of this technique is the ease of separation of single layers in colloidal solution, with the aid of centrifugation. Liquid exfoliation could be divided into four categories (Nicolosi et al. 2013): oxidation followed by dispersion in solvents, intercalation, ion exchange, and ultrasonication. Figure 13.6 shows the schematic representation of liquid exfoliation techniques (Nicolosi et al. 2013).

13.4.3.1   Oxidation

Graphene is synthesized from GO using chemical oxidation of graphite flakes with the help of sulfuric acid and potassium permanganate (Su et al. 2009). This helps in the generation of oxygen functional groups on the edges and basal planes, which in turn increases the hydrophilicity, and helps in the dispersion of 2D flat nanosheet upon ultrasonication. This method was developed by Hummers initially and has been modified subsequently. These oxidized sheets could be further reduced using reducing agents like hydrazine hydrate (Li et al. 2008) or $NaBH_4$ (Shin et al. 2009), thereby synthesizing graphene sheets. The disadvantage of this method is the creation of structural defects in the due course of graphene synthesis, which is a compromising factor for large-scale production as mentioned earlier.

13.4.3.2 Intercalation

Materials with layered structure can take in molecules into the space between the two layers, which reduces the interlayer adhesion and boosts exfoliation process. Graphene and MoS_2 are synthesized using this method. Intercalants such as sodium dodecyl benzene sulfonate have been used to intercalate graphite oxide, followed by sonication yielding single- to few-layered graphene (Lotya et al. 2009). The advantage for this technique is the ease of controlling the shape of nanosheets (Ida et al. 2008), and the drawback is its sensitivity to conditions (Eda et al. 2011).

13.4.3.3 Ion Exchange

When it comes to metal oxide exfoliation such as manganese (Omomo et al. 2003), cobalt (Kim et al. 2009), and titanium oxide (Tanaka et al. 2003), electrostatic repulsions play an important role. These metal oxides can undergo protonation by exchange of counter ions, which destructs the stability of layered materials. This imparts electrostatic repulsion between the layers and causes them to swell. This swelling facilitates exfoliation through ultrasonication. Layered double hydroxides have been delaminated using dodecyl sulfate, which were exchanged for interlayer anions and refluxed with butanol (Ma et al. 2006).

13.4.3.4 Ultrasonication

Ultrasonication is widely used for the synthesis of 2D nanosheets in the presence of a solvent (Hernandez et al. 2008). In this case, the two governing factors are the ultrasonic waves and the solvent. The ultrasonic waves create cavitation bubbles, which disrupt the layered crystallites (Nicolosi et al. 2013). On the other hand, the solvent selection is governed by two criteria (Hernandez et al. 2008). Firstly, they should have the surface tension around 40–50 mJ m^{-2} (Hernandez et al. 2008), which is good for graphite exfoliation. Secondly, the interaction between the solvent and graphene should be stronger than the interaction between graphene layers. For example, N-methyl-pyrrolidone has been used as a solvent for liquid exfoliation of graphene (Coleman et al. 2011) and dimethylformamide for h-BN (Zhi et al. 2009). These solvents have the same surface energy as that of layers between crystals, which minimizes the energy gap for exfoliation. This method results in lower concentration of end product, which is its drawback.

All these methods have their own merits and demerits. For example, mechanical exfoliation produce high-quality graphene, but the concentration is really low; in liquid exfoliation by oxidation, the yield is high, but the sp^2 network is not fully restored (Geng et al. 2010b). For different application, different strategies could be used to exfoliate layered materials. If highly pure nanosheets are required, then mechanical exfoliation is used; on the other hand, if structural defects and intercalation of ions are required, then liquid exfoliation is used. Surface-assisted growth could be applied for multilayer requirement.

13.5 SURFACE CHEMISTRY AND FUNCTIONALIZATION

Surface of NPs plays a crucial role with respect to interaction with other molecules. Figure 13.7 shows various molecules that could be functionalized using graphene (Wang et al. 2011). The foundation for functionalization on GO is based on the presence of functional groups and its

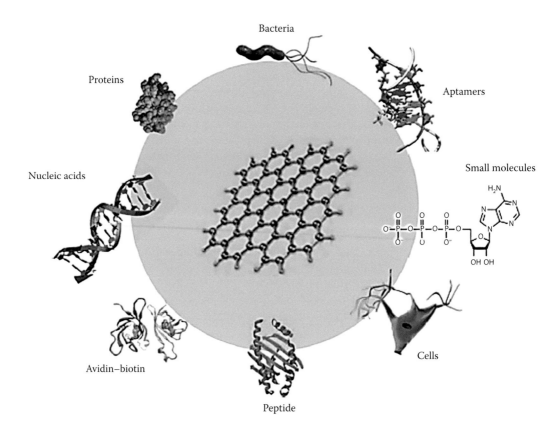

Figure 13.7 Graphene and its derivatives have been reported to be functionalized with biomolecules like avidin–biotin, peptides, NAs, proteins, aptamers, small molecules, bacteria, and cells through physical adsorption or chemical conjugation. The functionalized graphene biosystems with unique properties have been used for biological applications such as biological platforms, biosensors, and biodevices. (Adapted from *Trends Biotechnol.*, 29, Wang, Y., Li, Z., Wang, J., Li, J., and Lin, Y., Graphene and graphene oxide: Biofunctionalization and applications in biotechnology, 205–212, Copyright 2011, with permission from Elsevier.)

amphiphilic character, which could be exploited for the purpose of biological and biomedical applications. In case of MoS$_2$, thiol chemistry has been used to bind ligand, which acts as an artificial receptor for ß-galactosidase taking into account the ability of MoS$_2$ to absorb molecules at the internal and perimeter edges (Chou et al. 2013a). It has also been shown that single-layer MoS$_2$ could be functionalized through adatom absorption and by creating vacancy defects (Ataca and Ciraci 2011). The ease of surface functionalization is the main aspect that has bought the interest of many research groups. Based on the nature of interaction, the linkage could be classified into two, that is, covalent and noncovalent interactions, which are discussed in detail in the following sections.

13.5.1   Covalent Interaction

Graphene-related materials have rich amount of oxygen functional groups attached to them, which could help in molecular-level tuning and fabrication of hybrid materials. Covalent interaction basically takes place with the help of covalent bond formation. These bonds could be formed either on the basal planes or at the edges.

Carbon group forms the backbone of the graphene sheets, which are oxidized to synthesize GO. Most of the chemistry on this group is based on the diazonium salt reaction, which produces highly reactive free radicals on heating and attacks the sp^2-hybridized C=C atoms, thus forming a covalent bond (Georgakilas et al. 2012). This has already been in practice for binding aryl groups to glassy carbon (Delamar et al. 1992) and carbon nanotubes (Bahr et al. 2001). In case of graphene diazonium salt, addition reaction has been used to tune the electrical conductivity with the help of 4-nitrobenzene diazonium tetrafluoroborate (Sinitskii et al. 2010) and also to increase the solubility of graphene in polar aprotic solvents (Lomeda et al. 2008). This chemistry has also been used to graft styrene onto graphene sheet to improve the mechanical stability of polystyrene composite film. Hydroxylated aryl groups were used as initiators to synthesize polymer functionalized graphene sheets with the help of diazonium addition reaction and atom transfer radical polymerization (Fang et al. 2009). Similarly, aryl groups were covalently bound to basal plane by the reduction of diazonium salt to introduce band gap to fabricate graphene-based devices (Bekyarova et al. 2009). Another type of reaction employed to functionalize carbon center for applications in drug delivery systems is 1,3 dipolar cycloaddition (Kordatos et al. 2001). Extrapolating the work to graphene, this reaction was used to identify the active sites on graphene sheets (Quintana et al. 2010). Defect-free 2D graphene has been functionalized with azomethine ylide formed by 3,4-dihydroxybenzaldeyde and sarcosine condensation due to 1,3 dipolar cycloaddition to increase its dispersibility (Georgakilas et al. 2010). Highly functionalized graphene sheets with peptide have been demonstrated with the help of nitrene addition of azido-phenylalanine (Strom et al. 2010). The major application of these reactions is to increase the dispersibility of graphene and for the fabrication of polymer composites.

The carboxyl groups present at the edges of GO sheets impart hydrophilic character to GO. Even though these sheets are hydrophilic, they are not soluble in organic solvents and hence need to be functionalized, for which the COOH groups could be used as a potential site for covalent functionalization. One of the most common methods of edge functionalization is the acylation reaction (Loh et al. 2010). To carry on acylation, the COOH groups need to be activated, which could be achieved with the help of thionyl chloride ($SOCl_2$), 1-ethyl-3-(3-dimethylaminopropyl)-carbodiimide (EDC), and N, N'-dicyclohexylcarbodiimide (DCC). Once they are activated, they form intermediate products with COCl group or CONH group, which provides site for acylation by the formation of amide or ester groups. To increase the solubility of GO in solvents, octadecylamine was used to bind to the COCl groups, which were in turn activated by $SOCl_2$. Likewise, the thermal stability of GO was increased by binding with melamine-based compound, namely, 2-amino-4,6-didodecylamino-1,3,5-triazine (ADDT) via COCl group, activated as discussed earlier (Tang et al. 2011). EDC activated group has been used to impart solubility to nano GO (NGO) in biological solution by coupling it with polyethylene glycol (PEG), which could further be used for drug delivery purposes (Sun et al. 2008). The basic chemistry involved is the formation of carbodiimide bond between the GO and PEG. Lastly, DCC has been used to bind polyvinyl chloride to impart solubility to GO in organic solvents to ease solution-based characterization (Veca et al. 2009).

Epoxy and hydroxyl functional groups are present at the basal planes and arise due to oxidation reactions. These groups have not been explored much for functionalization with respect to other functional groups. The basic mechanism used for functionalization is nucleophilic attack on α-carbon by amine group (Dreyer et al. 2010). One of the finest example is the binding of polyallylamine, which has many amine groups attached to it. Amine groups attack the α-carbon resulting in cross-linking of GO sheets, thereby imparting mechanical strength (Park et al. 2009). The opening of the epoxy ring by amine group of 1-(3-aminopropyl)-3-methylimidazolium bromide, an ionic liquid, also helps in imparting solubility to GO in organic solvents due to the polarity of the final product (Yang et al. 2009). Phenyl isocyanate has been covalently bound to GO by the formation of ester bond with epoxy group, which has been extrapolated by research groups to improve the conductivity of composites (Stankovich et al. 2006a). In case of reduced GO (RGO), where most of the oxygen functional groups are removed by reduction, there exist residual epoxy groups that could be used for functionalization. Thermal stability of RGO has been shown to increase when epoxy group is used for grafting polymer rather than free radical grafting method (Hsiao et al. 2010).

The reactions at hydroxyl center are results of formation of carbamate esters. The functionalization involving carbamate esters has been shown with the help of binding of isocyanate compounds on GO by virtue of carbamate esters and amide bonds, by functionalization with hydroxyl and carboxyl groups on GO, respectively (Stankovich 2006b).

Functional groups present in GO can be divided into two categories: one on edges, the carboxyl groups, and the other on basal planes, which comprises epoxy, carbonyl, and hydroxyl groups. All these groups are the products of oxidation of the graphite flakes and are evident from FTIR studies (Mathesh et al. 2013), which helps in covalent interactions. Covalent interactions comprise of amide or ester bond formation between molecules comprising amine or hydroxyl groups with the carboxyl groups present on the edges. Apart from the edge functionalization, epoxy groups at the basal plane are also reactive and could be used for amide linkage. Functionalization with this group usually takes place via ring opening and amide linkage with molecules comprising amine group. The principle behind this is nucleophilic substitution, which takes place easily at room temperature and hence is a promising strategy (Kuila et al. 2012). Figure 13.8 shows different covalent functionalization chemistries for graphene nanosheets (Loh et al. 2010). This kind of interaction has been used to bind biocompatible molecules (poly-L-lysine) (Shan et al. 2009) to increase the solubility of graphene sheets and also acts as a linker between graphene sheets and the biomolecule.

13.5.2   Noncovalent Interaction

Covalent interactions result in change in the chemical nature of graphene or the molecule that is bound to it, which can change the properties of the individual. Suitable substitute for covalent interactions are the noncovalent interactions, which do not alter the chemical state or properties of the individual. In case of biomolecules, these kinds of interactions are prevalently applied. These could be divided into electrostatic interaction, π–π interaction, and hydrophobic interaction.

13.5.2.1   Electrostatic Interaction

GO has negatively charged carboxyl groups present on them, which provides site for functionalization using cationic molecules. This principle has been applied for the removal of cationic dyes (Ramesha et al. 2011) and also for the removal of copper ions (Yang et al. 2010d) from aqueous solutions. In case of graphene, charged groups could be introduced by reducing GO in the presence of polymers. Graphene synthesized from the reduction of GO in the presence of cationic poly(ethyleneimine) (PEI)-introduced cationic groups was used for self-assembly with acid-oxidized multiwalled carbon nanotubes (Yu and Dai 2009). In another

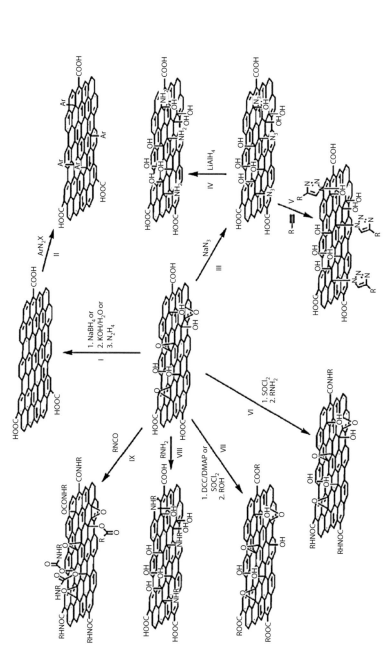

Figure 13.8 Schematic showing various covalent functionalization chemistry of graphene or GO. I: Reduction of GO into graphene by various approaches (1, NaBH$_4$; 2, KOH/H$_2$O; 3, N$_2$H$_4$). II: Covalent surface functionalization of reduced graphene via diazonium reaction (ArN$_2$X). III: Functionalization of GO by the reaction between GO and sodium azide. IV: Reduction of azide-functionalized GO (azide–GO) with LiAlH$_4$ resulting in the amino-functionalized GO. V: Functionalization of azide–GO through click chemistry (R–ChCH/CuSO$_4$). VI: Modification of GO with long alkyl chains (1, SOCl$_2$; 2, RNH$_2$) by the acylation reaction between the carboxyl acid groups of GO and alkylamine (after SOCl$_2$ activation of the COOH groups). VII: Esterification of GO by DCC chemistry or the acylation reaction between the carboxyl acid groups of GO and ROH alkylamine (after SOCl$_2$ activation of the COOH groups) (1, DCC/DMAP or SOCl$_2$; 2, ROH). VIII: Nucleophilic ring-opening reaction between the epoxy groups of GO and the amine groups of an amine-terminated organic molecular (RNH$_2$). IX: The treatment of GO with organic isocyanates leading to the derivatization of both the edge carboxyl and surface hydroxyl functional groups via formation of amides or carbamate esters (RNCO). (Loh, K.P., Bao, Q., Ang, P.K., and Yang, J., The chemistry of graphene, *J. Mater. Chem.*, 20, 2277–2289. Copyright 2010 Reproduced by permission of The Royal Society of Chemistry.)

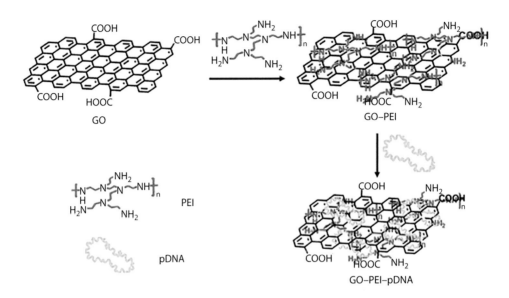

Figure 13.9   Schematic illustration showing the synthesis of GO–PEI–DNA complexes via an LBL assembly process. Graphene oxide (GO) was noncovalently functionalized by PEI polymers, forming positively charged GO–PEI complexes. GO–PEI complexes were used to load negatively charged pDNA by electrostatic interactions. (Feng, L., Zhang, S., and Liu, Z., Graphene based gene transfection, *Nanoscale*, 3, 1252–1257. Copyright 2011 Reproduced by permission of The Royal Society of Chemistry.)

instance, GO–PEI hybrid was used for high-efficiency loading of plasmid DNA (pDNA) and studied for gene transfection; pDNA self-assembled with cationic PEI resulted in (Figure 13.9) high transfection efficiency (Feng et al. 2011).

Furthermore, graphene was introduced with both positive and negatively charged groups by covalently binding poly(acrylamide) and poly(acrylic acid), which was then stacked by LBL assembly to form multilayered film (Shen et al. 2009).

13.5.2.2   π–π Interaction and Hydrophobic Interaction

The graphitic structure has delocalized π orbital, which can assist in π–π interaction. GO on their basal plane has small graphitic domains due to incomplete oxidation of graphite (Cote et al. 2010), which provides hydrophobic patches suitable for functionalization. The drug molecules with aromatic structures could be bound with GO sheets and used for drug delivery purposes. A nanocarrier hybrid system was fabricated with the assistance of π–π stacking between doxorubicin (DOX) and GO, which was further encapsulated with chitosan-conjugated folic acid (FA) and applied for delivery purposes (Depan et al. 2011). This hybrid material (Figure 13.10) is composed of chitosan, which provided biocompatibility, FA for targeted delivery, and DOX, an anticancer drug molecule.

The solubility of graphene sheets in aqueous solution is difficult due to strong π–π interaction between individual graphene sheets, which has to be overcome for dispersion in solvents. For industrial applications, it is necessary to prevent aggregation of graphene sheets as unique properties are available only in single sheets (Li et al. 2008). For this purpose, poly(N-vinyl-2-pyrrolidone) was used to functionalize graphene sheets via hydrophobic interactions, to prevent their aggregation in solution (Yoon and In 2011). Hydrophobic interactions have also been applied for binding triblock polymer (PEO-b–PPO-b–PEO) to graphene sheets, wherein hydrophobic PPO

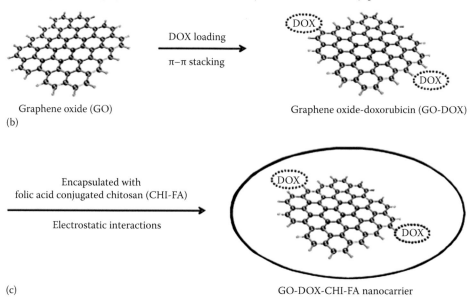

Preparation of folic acid-conjugated chitosan

Chitosan (CHI)

Folic acid (FA)

DCC, DMSO

pH 7.4

Chitosan-folic acid conjugate (CHI-FA)

(a)

Encapsulation of graphene oxide-doxorubicin nanohybrid with folic acid conjugated chitosan

DOX loading

π–π stacking

Graphene oxide (GO)

Graphene oxide-doxorubicin (GO-DOX)

(b)

Encapsulated with
folic acid conjugated chitosan (CHI-FA)

Electrostatic interactions

(c)   GO-DOX-CHI-FA nanocarrier

Figure 13.10 Scheme for the synthesis of nanocarrier: (a) conjugation of chitosan with folic acid, (b) loading of graphene oxide with doxorubicin, and (c) encapsulation of graphene oxide–loaded doxorubicin with folic acid–conjugated chitosan. (Adapted from *Mater. Sci. Eng. C*, 31, Depan, D., Shah, J., and Misra, R.D.K., Controlled release of drug from folate-decorated and graphene mediated drug delivery system: Synthesis, loading efficiency, and drug release response, 1305–1312, Copyright 2011, with permission from Elsevier.)

bound to graphene provided dispersion and PEO was used to form supramolecular hydrogel with α-cyclodextrin, by entering the cavities of cyclodextrin. This hybrid was proposed to be an ideal candidate for drug delivery and controlled drug release system (Zu and Han 2009).

The backbone of graphene family comprises aromatic rings that help π–π interaction with potential applications for drug delivery, as they can assist in the binding of water-insoluble aromatic drugs and act as carrier for the drugs. Small molecules such as pyrene derivates (Xu et al. 2008)

TABLE 13.1   GRAPHENE FUNCTIONALIZATION WITH VARIOUS BIOMOLECULES

Interaction	Force	Molecule	Application
Covalent	Amide bond	Glucose oxidase (Liu et al. 2010)	Biosensor
Covalent	In situ ring-opening polymerization of glycidol	Polyglycerol (Pham et al. 2010)	Anchor for magnetic particles
Covalent	Amide bond	Amine functionalised porphyrin (Xu et al. 2009)	Optoelectronic devices
Non covalent	π-π stacking	Porphyrin (Geng et al. 2010a)	Highly conductive materials
Non covalent	π-π stacking	SSDNA (Lv et al. 2010)	Biosensor
	π-π stacking	SN 38 (Liu et al. 2008)	Drug delivery
	π-π stacking and hydrophobic	Doxorubicin (Yang et al. 2008) (DOX) and camptothecin (CPT) (Zhang et al. 2010c)	Drug delivery
Non covalent	Hydrophobic	Horse radish peroxidase and oxalate oxidase (Zhang et al. 2012)	Enzyme immobilizaion
Non covalent	Electrostatic Interaction	Horse radish peroxidase (Zhang et al. 2010a)	Enzyme immobilizaion

and tetracyanoquinodimethane anion (Hao et al. 2008) could be bound to graphene owing to physisorption onto their basal planes via π–π interaction and van der Waals interactions (Dreyer et al. 2010). DOX hydrochloride has been bound by π–π interaction on GO sheets for controlled release and high-efficiency loading of drug, which was characterized by electrochemistry (Yang et al. 2008). The basal plane of GO has unoxidized graphenic domains, which give rise to hydrophobic domains, which can assist in hydrophobic interactions, and the edges of GO have negatively charged pH-dependent groups, which assist in electrostatic interaction; these interactions were used to bind horse radish peroxidase (Zhang et al. 2010b) and oxalate oxidase (Zhang et al. 2012b), respectively, to GO sheets, to demonstrate GO as a matrix for enzyme immobilization. With respect to other nanosheets, MoS_2 has been reported to be adhesive toward aromatic and conjugated molecules via physisorption (Moses et al. 2009). Table 13.1 gives few examples of hybrids produced with the assistance of these interactions. The basic difference between covalent and noncovalent interactions is the creation of defects on the graphene sheet in the former case.

13.6   RECENT DEVELOPMENTS IN 2D MATERIALS-BASED BIO-APPLICATIONS

13.6.1   Biosensor Applications

Detection of biological molecules is of high importance from biomedical and environmental standpoint (Pumera 2011), which draws attention toward biosensors. Biosensor is a device that can interact with an analyte and generate signals to be perceived by a transducer and quantify it. The main concern for this is to bind molecules to substrate and yield signals to be detected.

There are different types of biosensors available for biomolecules that are classified according to the principle used for detection. The major classification and the working principle with examples are explained in the following text (Pumera 2011).

Graphene, due to its good electrical conductivity, ease of functionalization, and good electrochemical properties, has been used as an effective biosensor for the detection of glucose oxidase (Liu et al. 2010b), NADH (Li et al. 2013), hemoglobin (Xu et al. 2010), cytochrome c (Wu et al. 2010), HRP (Zhou et al. 2010), and so on. The major reason behind using graphene is its high ET rate due to the presence of oxygen functional groups (Kuila et al. 2011), which makes them highly sensitive and increases the feasibility of detecting molecules at very low concentration. When molecules are bound to graphene, ET rate is increased thereby decreasing and increasing detection time and sensitivity, respectively. All this detection process generally works on the basis of binding a molecule to the surface of nanosheet, which decreases the ET distance between the molecules and the electrode surface (Shao et al. 2010). These sensors could be used for both quantitative and qualitative analyses. On the other hand, graphene, due to its 2D structure and the ability to recognize macromolecules, has been adopted for immunosensors, which can help in point care diagnosis (Zhang et al. 2013). The basic strategy utilizes antigen–antibody recognition by binding molecules via hydrophobic interactions, hydrogen bonding, or π–π stacking and subsequent detection by electrical or electrochemical properties of the sheets. For instance, an immunosensor for cancer biomarker (prostate specific antigen) was developed with the help of succinimidyl ester with a detection limit of 0.08 ng mL^{-1} (Yang and Gong 2010c).

Field-effect transistor (FET) biosensor relies on the recognition of biomolecules at the FET gate (Ahn et al. 2010). Graphene, being a zero-gap semiconductor together with the ease of surface modification and tunable band gap, makes it an ideal material (Wallace 1947) for this application. They have been used extensively for DNA sensing as they have charged group present on its backbone, which changes the electric charge distribution, thereby resulting in change in the conductivity of the channel (Mohanty and Berry 2008). Stine et al. (2010) were able to detect single-stranded (ss) DNA up to 10 nM. There has also been much interest in the use of graphene-based FET immunosensors as they have femtomolar-level protein detection limit (Kim et al. 2013). Micropatterned rGO FET displaying a p-type behavior was fabricated with the help of lithography and reduction, which was used for real-time detection of rotavirus. The sensitivity was found to be better than conventional ELISA technique (Liu et al. 2013). Similarly, flexible fluidic HIV-2 antibody immunosensors have also been developed for improved device performances and high sensitivity (Kwon et al. 2013).

The sensors that are based on sampling impedance of a system at the interface are known as impedimetric sensors (Katz and Willner 2003), which are generally regarded as a bio-recognition system. They have advantages such as sensitivity, require small potential for sensing, is a nondestructive method (Macdonald 1987), and does not require any electroactive moieties in the biomolecule (Bonanni et al. 2012). Experiments for the detection of DNA hybridization and polymorphism using electrochemical impedance spectroscopy were first developed by Bonanni and Pumera (Bonanni and Pumera 2011), wherein single-nucleotide polymorphism (SNP) responsible for Alzheimer's disease was detected at picomolar level. The principle involved partial release of hairpin DNA immobilized on graphene surface, due to hybridization with SNP, thereby resulting in decrease in charge transfer resistance (R$_{ct}$) value. Following this, different protocols were developed for impedimetric sensing for DNA probes (Dubuisson et al. 2011) and detection of DNA arrays (Gupta et al. 2013). Similarly, change in charge and conformation upon DNA immobilization and hybridization with graphene sheets were used for the detection of HIV-1 gene (Hu et al. 2011).

Furthermore, this type of sensor has also been utilized for the detection of trace amounts of pesticides, hormones, bacteria, and many other chemicals (Daniels and Pourmand 2007). Detection of sulfate-reducing bacteria was developed based on reduced graphene sheets/chitosan film, based on increase in charge transfer resistance (Qi et al. 2013, Wan et al. 2011). Recently, gold NP-modified graphene paper was used for the detection of *E. coli* O157:H7 based on impedimetric immunosensor with enhanced sensing performance, lower detection limit, and outstanding specificity (Wang et al. 2013). In this method, GO paper was reduced to prepare graphene film on which gold NPs were grown by electrodeposition technique. *E. coli* O157:H7 antibodies were immobilized via biotin–streptavidin system and used to detect bacteria. On a similar note, immunoglobulin G immunosensor was developed with chemically modified graphene. The detection was based on changes in impedance spectra of redox probe after the binding of IgG to its antibody (Loo et al. 2012).

Another kind of biosensor is FRET biosensor, which is associated with energy transfer between a donor and acceptor molecule, which can be utilized to probe biological molecules (Ha et al. 1996). Biosensors based on this phenomenon are based on either decrease (quench) or increase (enhance) in fluorescence emitted by molecules. Graphene has been observed to have excellent quenching properties toward various dyes (Ma et al. 2013). It is also known that graphene shows specificity toward ssDNA through noncovalent interaction (Husale et al. 2010); the above two properties were used as blueprint to design a FRET biosensor for selective detection of biomolecules (Lu et al. 2009). The principle involves binding of dye-labeled ssDNA to graphene, which quenches fluorescence from the dye by 97%, but when the labeled ssDNA binds to its complementary target molecule, dye-labeled DNA is released and hence fluorescence is mended back (Figure 13.11) (Wang et al. 2011). Similar FRET sensors have been fabricated for the detection of DNA with various detection limits (Liu et al. 2010a).

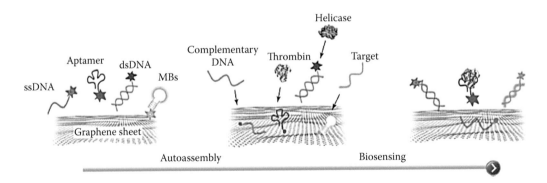

Figure 13.11 Principles of graphene-based FRET biosensors. ssDNA, aptamers, and MBs can be adsorbed onto the surfaces of graphene or graphene derivates (which also possess a planar surface and 2D structure). Fluorophore labels on the ends of probes are quenched rapidly when adsorbed onto the graphene surface. When analytes (e.g., cDNA, thrombin, and a designed complementary ssDNA or functional NAs like survivin mRNA) are introduced into the systems and bind their probes (ssDNA, aptamer and MB, respectively), the probe fluorescence is recovered, thus allowing detection. Conversely, dsDNA remains fluorescent before an enzyme (e.g., helicase) is introduced; ssDNA is then released, and fluorophore on the ssDNA is quenched by graphene-based nanomaterials. (Adapted from *Trend Biotechnol.*, 29, Wang, Y., Li, Z., Wang, J., Li, J., and Lin, Y., Graphene and graphene oxide: Biofunctionalization and applications in biotechnology, 205–212, Copyright 2011, with permission from Elsevier.)

FRET immunosensors have also been developed for pathogen detection, in which antibodies for rotavirus were immobilized on GO sheet via covalent interaction (carbodiimide-assisted amidation reaction), which were captured by specific antigens and detected by fluorescence quenching of GO (Jung et al. 2010).

MoS$_2$ has also been used for the binding of dye-labeled ssDNA via van der Waals forces between the basal planes of MoS$_2$ and nucleobases. Upon hybridizing with the complementary DNA, they form dsDNA, which interacts weakly with MoS$_2$; hence, fluorescence is observed, which assists in quantitative readout of target DNA (Zhu et al. 2013).

13.6.2   Bioimaging

From the diagnostic point of view, image-guided drug delivery system can be a boon for biomedical applications. To study the uptake of drug molecules, it is necessary to have a marker attached to the drug that could be utilized for imaging. GO due to its biocompatibility, cellular uptake, chemical modifications, and photoluminescence at near infra-red region can serve as a probe for bioimaging (Yang et al. 2013). PEG–GO has been shown to improve the solubility of GO in biological solution, which could then be bound with drug molecules via π–π interaction and observed, when excited by infrared light, to study the uptake of drug in vivo.

When the size of a semiconductor is reduced to nanometer scale, it results in quantum confinement, which could be applied for bioimaging applications (Yan et al. 2012), which goes well with graphene, being a semiconductor. Many research groups have fabricated graphene quantum dots (GQDs) (Pan et al. 2010, Tetsuka et al. 2012) and used for photoluminescence. Depending on the reaction temperature during synthesis by acid treatment, blue, green, or yellow GQDs could be synthesized (Peng et al. 2012). Researchers chose green luminescent GQDs to visualize the phase contrast image for human cancer cell lines (T47D) cells, by incubating them with T47D cells and nucleus stained with blue color. These images gave a proof for using GQDs for high-contrast imaging. Following the footsteps, Dong et al. (2012) incubated green GQDs with human breast cancer MCF-7 cells for 4 h and for the first time observed that GQDs could label cell nucleus (Dong et al. 2012). On the other hand, GQDs are biocompatible; research groups have shown that stem cells could also be labeled with GQDs with little cytotoxicity, but when labeled by semiconductor QDs like CdS NPs, the cells die, indicating low cytotoxicity and stable photostability (Zhang et al. 2012a) of GQDs with respect to semiconductor QDs.

Besides photoluminescent applications, GO has also been used for the purpose of magnetic resonance imaging and positron emission tomography (PET) techniques. For the former case, GO–Fe$_3$O$_4$ conjugate was internalized in Hela cells, as GO is bestowed with biocompatibility and physiological stability. Due to a decrease in local inhomogeneity, GO–Fe$_3$O$_4$ conjugate resulted in enhanced cellular imaging than isolated Fe$_3$O$_4$ NPs (Chen et al. 2011). For the latter case, GO was conjugated with TRC105 labeled with ^{64}Cu (Hong et al. 2012) and reported to open up new avenues for image-guided drug delivery and cancer therapy. TRC105 was covalently bound to GO, which targets CD105 (vascular marker for tumor angiogenesis), enabling imaging using radiolabeled GO, which accumulated in 4T1 murine breast cells, instantaneously. Molecular probing of ATP in mouse epithelial cells has also been demonstrated with the help of aptamer-FAM/GO nanosheets, showcasing GO as a carrier of DNA aptamers (Figure 13.12), protecting oligonucleotides from enzymatic cleavage and serving as a real-time sensing platform (Wang et al. 2010b).

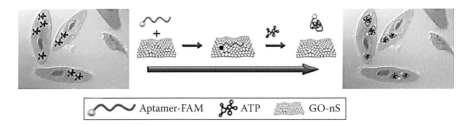

Aptamer-FAM   ATP   GO-nS

Figure 13.12   Schematic illustration of in situ molecular probing in living cells by using aptamer/GO-nS nanocomplex. (Adapted with permission from Wang, Y., Li, Z., Hu, D., Lin, C.T., Li, J., and Lin, Y. Aptamer/ graphene oxide nanocomplex for in situ molecular probing in living cells, *J. Am. Chem. Soc.*, 132, 2010b, 9274–9276. Copyright 2010 American Chemical Society.)

13.6.3   Oligonucleotide and Drug Delivery Vehicles

Graphene sheets due to high surface area and low toxicity have been considered and studied as delivery systems. They can act as a carrier of gene, siRNA, and drugs. Graphene is known to bind with ssDNA with a stronger interaction than dsDNA and also prevent enzymatic cleavage due to the presence of negatively charged groups that hinder the nuclease activity. These features could be exploited to deliver genes and develop a nonviral-based gene therapy. Researchers have shown effective loading of pDNA on graphene, which acts as a 2D nano-vector. Graphene bound with PEI polymers introduces positively charged groups, thus assisting the binding of pDNA (Feng et al. 2011). Due to the fact that graphene prevents enzymatic cleavage of DNA, it has been studied for the delivery of DNA to HeLa cells by molecular beacons (MBs) as shown in Figure 13.13 (Lu et al. 2010). This could be ideal for genetic engineering, as GO, apart from preventing enzymatic cleavage, has lower toxicity than SWCNTs and has high transfection efficiency. Other examples involve delivery vehicle for ribonucleic acid interface and antisense therapies, which are powerful tool for cancer therapy and acquired

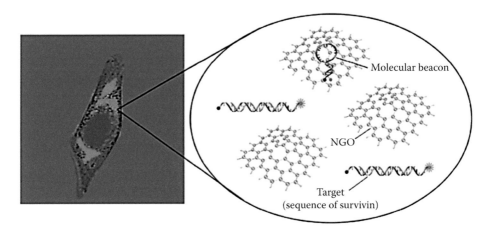

Molecular beacon

NGO

Target
(sequence of survivin)

Figure 13.13   Schematic representation of NGO delivery of MB into HeLa cells to detect survivin mRNA. (Adapted with permission from Lu, C.H., Zhu, C.L., Li, J., Liu, J.J., Chen, X., and Yang, H.H., Using graphene to protect DNA from cleavage during cellular delivery, *Chem. Commun.*, 46, 2010, 3116–3118. Copyright 2010 American Chemical Society.)

immunodeficiency syndrome (Zhang et al. 2013), as they need a platform for delivery without enzymatic cleavage, which is possible with the help of graphene.

For drug delivery systems, noncovalent interactions provide lots of advantages over covalent interaction (Doane and Burda 2012). Firstly, noncovalent interaction does not require any modification of the drug; thereby, it does not change any intrinsic property of drugs. Secondly, optimizing the binding for one drug could be extrapolated to other drugs with same properties. Thirdly, the release of drugs could be assisted with the change in the surrounding environment (pH, temperature, and hydrophobicity) rather than the use of external stimuli. The aromatic drug molecule camptothecin (CPT), which is not water soluble, needs a vehicle for its transport into the cells, which could be achieved by PEI-grafted GO. PEI–GO has also been used to deliver siRNA and DOX drug to cancer cells, in which PEI provides cationic groups and also increases solubility in biological solutions (Liu et al. 2008) and helps in the binding of siRNA through electrostatic interaction and drug is loaded via π–π stacking on GO. This facilitates a synergistic effect that enhances chemotherapy efficacy (Zhang et al. 2011a). In another instance, a comparative study for DOX delivery was carried out using NGO–PEG–Rituxan/DOX, free DOX, and NGO–PEG/DOX on Raji-B cells and was observed that the first bioconjugate worked better for DOX delivery than the latter two (Sun et al. 2008).

Targeted delivery of drugs has also been studied by functionalizing GO with FA molecules for recognition by folate receptors and binding with DOX and CPT via π–π stacking (Zhang 2010c). Similarly, thiolated rituxan was conjugated onto NGO–PEG through amine groups present on them and used for targeted eradication of cancer cells (Sun 2008). These studies show promising results for using graphene in the field of biomedicine.

13.6.4 Cancer Phototherapies

Cancer-related diseases are one of the growing causes of death in the present scenario. Currently available techniques for cancer treatment include chemotherapy and radiation therapy, which result in various side effects, as apart from affecting the cancerous cells, normal cells are affected too. To reduce side effects, it is required to have targeted therapy, which has less side effects. Graphene- and MoS_2-related particles are suitable candidates for this purpose. PEG–GO is known to have strong absorption at near-infrared region (Worle-Knirsch 2006) and can also bind drug molecules via π–π interaction. It also has the ability to be taken up by the cells and get deposited on tumors of mouse model. When they are irradiated at 808 nm wavelength, temperature increases and tumor ablation takes place (Yang 2010b). Studies conducted by other group shows that GO functionalized with FA/Ce6 can be used for photoablation of cancer cells (Penget al. 2011). On the other hand, MoS_2 when chemically exfoliated shows absorbance at the near-infrared region, which could be used for photothermal therapy of cancer. Together with these advantages, they also possess high drug-loading capacity equivalent to GO (Chou et al. 2013b).

Synergistic effects of photothermal therapy and drug delivery system have already been studied for improved cancer therapy (Zhang et al. 2011b). PEG–GO bound with DOX was used for this purpose, wherein DOX provided the chemotherapeutics and PEG–GO provided photothermal therapy when exposed to near-infrared region. In case of NGO, they have been shown to intercalate in between the dsDNA, in a mechanism similar to planar aromatic intercalation in DNA. In the presence of Cu^{2+} ions bound to the COOH groups of GO, together with the planar structure of GO, the catalysis of oxidative cleavage of DNA double helix is possible as shown in Figure 13.14, opening up a new class of chemotherapeutic agents (Ren et al. 2010).

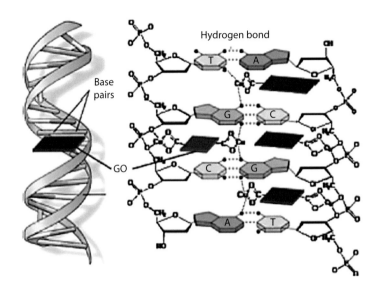

Figure 13.14   DNA cleavage system of nanosized graphene oxide sheets and copper ions. (Adapted with permission from Ren, H., Wang, C., Zhang, J. et al., DNA cleavage system of nanosized graphene oxide sheets and copper ions, *ACS Nano*, 4, 2010, 7169–7174. Copyright 2010 American Chemical Society.)

13.6.5   Cell Growth and Tissue Engineering

There has been an increasing demand to provide with cells or tissues, which can be used for replacing or restoring a tissue function (Langer and Vacanti 1993). But due to the limitation in scaffolds required for this purpose, progression in this field is slow (Williams et al. 1999). They basically require a scaffold that provides a medium for the growth of cells and prevent them from being rejected during transplantation. 3D scaffolds have problems associated with mechanical properties, electrical conductivity, adhesive forces, and lack of ability to adhere cells to 3D matrix (Dvir et al. 2011). These could be overcome with the help of graphene, which has been studied for its cell adhesion properties and could be bound with immunosuppressive drugs, which helps in preventing rejection by the body. Other than this equity, graphene has high elasticity, flexibility, and can adapt to irregular surfaces (Chen et al. 2008a, Lee et al. 2008, 2010), thereby fulfilling the requirements for scaffolds in tissue engineering. Chitosan, which is known to enhance bone formation (DiMartino et al. 2005), was reinforced with GO (Fan et al. 2010) to improve its mechanical properties with the help of carboxyl group forming an amide bond. These also helped in the maintenance of chitosan under harsh pH conditions, retaining shape and size and lower degradation rate. There has also been a keen interest in the study of graphene as a substrate for cell growth. Researchers have shown graphene to prevent the growth of prokaryotic cells due to the presence of free electron, nevertheless, affecting the growth of eukaryotic cells, hence increasing wound healing rate (Lu et al. 2012). Graphene paper has also been reported as an antibacterial agent (Hu et al. 2010), which shows high potential for clinical purposes. In one of the studies, GO and RGO nanowalls were compared for proliferation of *E. coli*, which revealed growth inhibition of bacteria due to the sharp edges present on them, breaking up the cell membrane, and showed RGO nanowalls to be more effective for this system due to better charge transfer ratio

(Akhavan and Ghaderi 2010). Apart from them, graphene has also been studied for stem cell differentiation. There is a search for stable platforms, which could be used for stem cell delivery, with capabilities for in vivo implant technology (Jung et al. 2001). Studies conducted to back up the notion for using graphene as platform for stem cell differentiation revealed graphene to serve as cell adhesion layer with electrical stimulation for cell differentiation (Park et al. 2011). The differentiation inductions differ on graphene and GO due to their surface chemistry, which could be controlled (Lee et al. 2011).

In addition to these properties, graphene could be functionalized with proteins with potential applications for tissue engineering (Goenka et al. 2014). Patterning of these structures has been carried out at microscale (Kodali et al. 2010) and nanoscale levels with the latter one termed as DNA origami structures (Yun et al. 2012).

13.7 PERSPECTIVE

With advancement in synthesizing 2D nanosheets, it is possible to tune them at the molecular level for the development of hybrid materials. Once the optical, thermal, electrical, mechanical properties of these nanosheets are well studied, it would be possible to functionalize them with molecules, which in turn will open up a new era of materials whose applications will be unmatchable. These 2D nanosheets could be hybridized with different biomolecules and stacked on top of each other to form heterostructures. On the other hand, the functionalization could also help in the development of novel materials that could be used for biomimetics and other emerging technologies. 2D nanosheets due to their excellent properties have become a rising star in the field of nanotechnology. The studies conducted show promising results for its use in the field of biology and biomedicine, as it could be easily functionalized via covalent and noncovalent interactions, thus opening up a great platform with huge potential applications. Most of the classical research on graphene is based on covalent interaction, but now there is a need to study noncovalent interactions, which are more useful in the case of delivery vehicles. On the contrary, these two interactions could be used simultaneously to develop a new class of chemotherapeutics, with ligands targeting receptors on tumors (covalent interaction via functional groups), controlled release of drugs (π–π interaction), and photothermal therapy (intrinsic property of GO), with applications in targeted drug delivery and photoablation. The intrinsic optical property of GO and MoS_2 could be used in bioimaging and diagnostic applications as they do not require any additional labeling for detection. Novel designs of biosensors can be developed with higher sensitivity for multiple molecule detection. The future of cancer therapy and tissue engineering scaffolds depends on certain factors that need to be investigated for clear idea. The demerits for the same, which need to be addressed, are characterization for their behavior under physiological conditions, cytotoxic effects, cellular uptake mechanism and interaction with cells, and fate of NPs after uptake, which will play a crucial role when used in living cells. One of the other areas of interest would be to control the size of the NPs as they are nonhomogeneous in size, and different sizes give rise to different properties. If all these problems are dealt precisely, 2D nanosheets can serve as a better candidate for use in biological and biomedical applications providing a better quality of life with new era of hybrid materials for cancer therapy and tissue engineering. Days are not far when drugs based on graphene will hit the market. The study of these 2D nanosheets is just a start, and still there is a long way to go for their properties to be thoroughly studied and applied for the betterment of mankind.

REFERENCES

Ahn, J.H., Choi, S.J., Han, J.W., Park, T.J., Lee, S.Y., and Choi, Y.K. (2010), Double-gate nanowire field effect transistor for a biosensor. *Nano Letters* 10: 2934–2938.

Akhavan, O. and Ghaderi, E. (2010), Toxicity of graphene and graphene oxide nanowalls against bacteria. *ACS Nano* 4: 5731–5736.

Ataca, C. and Ciraci, S. (2011), Functionalization of single-layer MoS_2 honeycomb structures. *The Journal of Physical Chemistry C* 115: 13303–13311.

Bahr, J.L., Yang, J., Kosynkin, D.V., Bronikowski, M.J., Smalley, R.E., and Tour, J.M. (2001), Functionalization of carbon nanotubes by electrochemical reduction of aryl diazonium salts: A bucky paper electrode. *Journal of the American Chemical Society* 123: 6536–6542.

Balandin, A.A., Ghosh, S., Bao, W. et al. (2008), Superior thermal conductivity of single-layer graphene. *Nano Letters* 8: 902–907.

Bekyarova, E., Itkis, M.E., Ramesh, P. et al. (2009), Chemical modification of epitaxial graphene: Spontaneous grafting of aryl groups. *Journal of the American Chemical Society* 131: 1336–1337.

Bonaccorso, F., Sun, Z., Hasan, T., and Ferrari, A. (2010), Graphene photonics and optoelectronics. *Nature Photonics* 4: 611–622.

Bonanni, A., Loo, A.H., and Pumera, M. (2012), Graphene for impedimetric biosensing. *Trends in Analytical Chemistry* 37: 12–21.

Bonanni, A. and Pumera, M. (2011), Graphene platform for hairpin-DNA-based impedimetric genosensing. *ACS Nano* 5: 2356–2361.

Brownson, D.A., Kampouris, D.K., and Banks, C.E. (2011), An overview of graphene in energy production and storage applications. *Journal of Power Sources* 196: 4873–4885.

Castro Neto, A.H., Guinea, F., Peres, N.M.R., Novoselov, K.S., and Geim, A.K. (2009), The electronic properties of graphene. *Reviews of Modern Physics* 81: 109–162.

Chen, H., Muller, M.B., Gilmore, K.J., Wallace, G.G., and Li, D. (2008a), Mechanically strong, electrically conductive, and biocompatible graphene paper. *Advanced Materials* 20: 3557–3561.

Chen, W., Yi, P., Zhang, Y., Zhang, L., Deng, Z., and Zhang, Z. (2011), Composites of aminodextran-coated Fe_3O_4 nanoparticles and graphene oxide for cellular magnetic resonance imaging. *ACS Applied Materials & Interfaces* 3: 4085–4091.

Chen, Z.-G., Zou, J., Liu, G. et al. (2008b), Novel boron nitride hollow nanoribbons. *ACS Nano* 2: 2183–2191.

Cheng, Z., Li, Q., Li, Z., Zhou, Q., and Fang, Y. (2010), Suspended graphene sensors with improved signal and reduced noise. *Nano Letters* 10: 1864–1868.

Chhowalla, M., Shin, H.S., Eda, G., Li, L.J., Loh, K.P., and Zhang, H. (2013), The chemistry of two-dimensional layered transition metal dichalcogenide nanosheets. *Nature Chemistry* 5: 263–275.

Choi, W., Lahiri, I., Seelaboyina, R., and Kang, Y.S. (2010), Synthesis of graphene and its applications: A review. *Critical Reviews in Solid State and Materials Sciences* 35: 52–71.

Chou, S.S., De, M., Kim, J. et al. (2013a), Ligand conjugation of chemically exfoliated MoS_2. *Journal of the American Chemical Society* 135: 4584–4587.

Chou, S.S., Kaehr, B., Kim, J. et al. (2013b), Chemically exfoliated MoS_2 as near-infrared photothermal agents. *Angewandte Chemie* 125: 4254–4258.

Coleman, J.N., Lotya, M., O'Neill, A. et al. (2011), Two-dimensional nanosheets produced by liquid exfoliation of layered materials. *Science* 331: 568–571.

Cote, L.J., Kim, J., Tung, V.C., Luo, J., Kim, F., and Huang, J. (2010), Graphene oxide as surfactant sheets. *Pure and Applied Chemistry* 83: 95–110.

Daniels, J.S. and Pourmand, N. (2007), Label-free impedance biosensors: Opportunities and challenges. *Electroanalysis* 19: 1239–1257.

Davis, R.F. (1991), III-V nitrides for electronic and optoelectronic applications. *Proceedings of the IEEE* 79: 702–712.

Dean, C.R., Young, A.F., Meric, I. et al. (2010), Boron nitride substrates for high-quality graphene electronics. *Nature Nanotechnology* 5: 722–726.

Delamar, M., Hitmi, R., Pinson, J., and Saveant, J.M. (1992), Covalent modification of carbon surfaces by grafting of functionalized aryl radicals produced from electrochemical reduction of diazonium salts. *Journal of the American Chemical Society* 114: 5883–5884.

Depan, D., Shah, J., and Misra, R.D.K. (2011), Controlled release of drug from folate-decorated and graphene mediated drug delivery system: Synthesis, loading efficiency, and drug release response. *Materials Science and Engineering: C* 31: 1305–1312.

Desai, M., Labhasetwar, V., Walter, E., Levy, R., and Amidon, G. (1997), The mechanism of uptake of biodegradable microparticles in Caco-2 cells is size dependent. *Pharmaceutical Research* 14: 1568–1573.

Di Martino, A., Sittinger, M., and Risbud, M.V. (2005), Chitosan: A versatile biopolymer for orthopaedic tissue-engineering. *Biomaterials* 26: 5983–5990.

Dimitrakakis, G.K., Tylianakis, E., and Froudakis, G.E. (2008), Pillared graphene: A new 3-D network nanostructure for enhanced hydrogen storage. *Nano Letters* 8: 3166–3170.

Divigalpitiya, W.M.R., Frindt, R.F., and Morrison, S.R. (1989), Inclusion systems of organic molecules in restacked single-layer molybdenum disulfide. *Science* 246: 369–371.

Doane, T.L. and Burda, C. (2012), The unique role of nanoparticles in nanomedicine: Imaging, drug delivery and therapy. *Chemical Society Reviews* 41: 2885–2911.

Dong, Y., Chen, C., Zheng, X. et al. (2012), One-step and high yield simultaneous preparation of single- and multi-layer graphene quantum dots from CX-72 carbon black. *Journal of Materials Chemistry* 22: 8764–8766.

Dreyer, D.R., Park, S., Bielawski, C.W., and Ruoff, R.S. (2010), The chemistry of graphene oxide. *Chemical Society Reviews* 39: 228–240.

Dubuisson, E., Yang, Z., and Loh, K.P. (2011), Optimizing label-free DNA electrical detection on graphene platform. *Analytical Chemistry* 83: 2452–2460.

Dvir, T., Timko, B.P., Kohane, D.S., and Langer, R. (2011), Nanotechnological strategies for engineering complex tissues. *Nature Nanotechnology* 6: 13–22.

Eda, G., Yamaguchi, H., Voiry, D., Fujita, T., Chen, M., and Chhowalla, M. (2011), Photoluminescence from chemically exfoliated MoS$_2$. *Nano Letters* 11: 5111–5116.

Eichler, J. and Lesniak, C. (2008), Boron Nitride (bn) and BN composites for high-temperature applications. *Journal of the European Ceramic Society* 28: 1105–1109.

Elbert, D.L. and Hubbell, J.A. (1996), Surface treatments of polymers for biocompatibility. *Annual Review of Materials Science* 26: 365–294.

Fan, H., Wang, L., Zhao, K. et al. (2010), Fabrication, mechanical properties, and biocompatibility of graphene-reinforced chitosan composites. *Biomacromolecules* 11: 2345–2351.

Fang, M., Wang, K., Lu, H., Yang, Y., and Nutt, S. (2009), Covalent polymer functionalization of graphene nanosheets and mechanical properties of composites. *Journal of Materials Chemistry* 19: 7098–7105.

Feng, L., Zhang, S., and Liu, Z. (2011), Graphene based gene transfection. *Nanoscale* 3: 1252–1257.

Geim, A.K. (2009), Graphene: Status and prospects. *Science* 324: 1530–1534.

Geng, J. and Jung, H.T. (2010a), Porphyrin functionalized graphene sheets in aqueous suspensions: From the preparation of graphene sheets to highly conductive graphene films. *The Journal of Physical Chemistry C* 114: 8227–8234.

Geng, J., Kong, B.S., Yang, S.B., and Jung, H.-T. (2010b), Preparation of graphene relying on porphyrin exfoliation of graphite. *Chemical Communications* 46: 5091–5093.

Georgakilas, V., Bourlinos, A.B., Zboril, R. et al. (2010), Organic functionalisation of graphenes. *Chemical Communications* 46: 1766–1768.

Georgakilas, V., Otyepka, M., Bourlinos, A.B. et al. (2012), Functionalization of graphene: Covalent and non-covalent approaches, derivatives and applications. *Chemical Reviews* 112: 6156–6214.

Gilje, S., Han, S., Wang, M., Wang, K.L., and Kaner, R.B. (2007), A chemical route to graphene for device applications. *Nano Letters* 7: 3394–3398.

Goenka, S., Sant, V., and Sant, S. (2014), Graphene-based nanomaterials for drug delivery and tissue engineering. *Journal of Controlled Release* 173: 75–88.

Guo, F., Kim, F., Han, T.H., Shenoy, V.B., Huang, J., and Hurt, R.H. (2011), Hydration-responsive folding and unfolding in graphene oxide liquid crystal phases. *ACS Nano* 5: 8019–8025.

Gupta, V.K., Yola, M.L., Qureshi, M.S., Solak, A.O., Atar, N., and Ustundag, Z. (2013), A novel impedimetric biosensor based on graphene oxide/gold nanoplatform for detection of DNA arrays. *Sensors and Actuators B-Chemical* 188: 1201–1211.

Ha, T., Enderle, T., Ogletree, D.F., Chemla, D.S., Selvin, P.R., and Weiss, S. (1996), Probing the interaction between two single molecules: Fluorescence resonance energy transfer between a single donor and a single acceptor. *Proceedings of the National Academy of Sciences of the United States of America* 93: 6264–6268.

Hao, R., Qian, W., Zhang, L., and Hou, Y. (2008), Aqueous dispersions of TCNQ-anion-stabilized graphene sheets. *Chemical Communications* 2008: 6576–6578.

Hernandez, Y., Nicolosi, V., Lotya, M. et al. (2008), High-yield production of graphene by liquid-phase exfoliation of graphite. *Nature Nanotechnology* 3: 563–568.

Hong, H., Yang, K., Zhang, Y. et al. (2012), in vivo targeting and imaging of tumor vasculature with radiolabeled, antibody-conjugated nanographene. *ACS Nano* 6: 2361–2370.

Hsiao, M.-C., Liao, S.H., Yen, M.Y. et al. (2010), Preparation of covalently functionalized graphene using residual oxygen-containing functional groups. *ACS Applied Materials & Interfaces* 2: 3092–3099.

Hu, W., Peng, C., Luo, W. et al. (2010), Graphene-based antibacterial paper. *ACS Nano* 4: 4317–4323.

Hu, Y., Li, F., Bai, X. et al. (2011), Label-free electrochemical impedance sensing of DNA hybridization based on functionalized graphene sheets. *Chemical Communications* 47: 1743–1745.

Hua Li, L., Chen, Y., Cheng, B.M., Lin, M.Y., Chou, S.L., and Peng, Y.C. (2012), Photoluminescence of boron nitride nanosheets exfoliated by ball milling. *Applied Physics Letters* 100: 261108.

Husale, B.S., Sahoo, S., Radenovic, A., Traversi, F., Annibale, P., and Kis, A. (2010), Ssdna binding reveals the atomic structure of graphene. *Langmuir* 26: 18078–18082.

Ida, S., Shiga, D., Koinuma, M., and Matsumoto, Y. (2008), Synthesis of hexagonal nickel hydroxide nanosheets by exfoliation of layered nickel hydroxide intercalated with dodecyl sulfate ions. *Journal of the American Chemical Society* 130: 14038–14039.

Iwan, A. and Chuchmała, A. (2012), Perspectives of applied graphene: Polymer solar cells. *Progress in Polymer Science* 37: 1805–1828.

Johnson, J.L., Behnam, A., Pearton, S.J., and Ural, A. (2010), Hydrogen sensing using pd-functionalized multilayer graphene nanoribbon networks. *Advanced Materials* 22: 4877–4880.

Jung, D.R., Kapur, R., Adams, T. et al. (2001), Topographical and physicochemical modification of material surface to enable patterning of living cells. *Critical Reviews in Biotechnology* 21: 111–154.

Jung, J.H., Cheon, D.S., Liu, F., Lee, K.B., and Seo, T.S. (2010), A graphene oxide based immuno-biosensor for pathogen detection. *Angewandte Chemie International Edition* 49: 5708–5711.

Kam, K.K. and Parkinson, B.A. (1982), Detailed photocurrent spectroscopy of the semiconducting group VIB transition metal dichalcogenides. *The Journal of Physical Chemistry* 86: 463–467.

Katz, E. and Willner, I. (2003), Probing biomolecular interactions at conductive and semiconductive surfaces by impedance spectroscopy: Routes to impedimetric immunosensors, DNA-sensors, and enzyme biosensors. *Electroanalysis* 15: 913–947.

Kim, D.-J., Sohn, I.Y., Jung, J.H., Yoon, O.J., Lee, N.E., and Park, J.S. (2013), Reduced graphene oxide field-effect transistor for label-free femtomolar protein detection. *Biosensors and Bioelectronics* 41: 621–626.

Kim, J., Cote, L.J., Kim, F., Yuan, W., Shull, K.R., and Huang, J. (2010), Graphene oxide sheets at interfaces. *Journal of the American Chemical Society* 132: 8180–8186.

Kim, K.K., Hsu, A., Jia, X. et al. (2011), Synthesis of monolayer hexagonal boron nitride on cu foil using chemical vapor deposition. *Nano Letters* 12: 161–166.

Kim, T.W., Oh, E.-J., Jee, A.-Y. et al. (2009), Soft-chemical exfoliation route to layered cobalt oxide monolayers and its application for film deposition and nanoparticle synthesis. *Chemistry—A European Journal* 15: 10752–10761.

Kodali, V.K., Scrimgeour, J., Kim, S. et al. (2010), Nonperturbative chemical modification of graphene for protein micropatterning. *Langmuir* 27: 863–865.

Kordatos, K., Da Ros, T., Bosi, S. et al. (2001), Novel versatile fullerene synthons. *The Journal of Organic Chemistry* 66: 4915–4920.

Kuila, T., Bose, S., Khanra, P., Mishra, A.K., Kim, N.H., and Lee, J.H. (2011), Recent advances in graphene-based biosensors. *Biosensors and Bioelectronics* 26: 4637–4648.

Kuila, T., Bose, S., Mishra, A.K., Khanra, P., Kim, N.H., and Lee, J.H. (2012), Chemical functionalization of graphene and its applications. *Progress in Materials Science* 57: 1061–1105.

Kwon, O.S., Lee, S.H., Park, S.J. et al. (2013), Large-scale graphene micropattern nano-biohybrids: High-performance transducers for FET-type flexible fluidic HIV immunoassays. *Advanced Materials* 25: 4177–4185.

Langer, R. and Vacanti, J.P. (1993), Tissue engineering. *Science* 260: 920–926.

Lee, C., Wei, X., Kysar, J.W., and Hone, J. (2008), Measurement of the elastic properties and intrinsic strength of monolayer graphene. *Science* 321: 385–388.

Lee, W.C., Lim, C.H.Y.X., Shi, H. et al. (2011), Origin of enhanced stem cell growth and differentiation on graphene and graphene oxide. *ACS Nano* 5: 7334–7341.

Lee, Y., Bae, S., Jang, H. et al. (2010), Wafer-scale synthesis and transfer of graphene films. *Nano Letters* 10: 490–493.

Li, D., Mueller, M.B., Gilje, S., Kaner, R.B., and Wallace, G.G. (2008), Processable aqueous dispersions of graphene nanosheets. *Nature Nanotechnology* 3: 101–105.

Li, L.H., Chen, Y., Behan, G., Zhang, H., Petravic, M., and Glushenkov, A.M. (2011a), Large-scale mechanical peeling of boron nitride nanosheets by low-energy ball milling. *Journal of Materials Chemistry* 21: 11862–11866.

Li, Y., Wang, H., Xie, L., Liang, Y., Hong, G., and Dai, H. (2011b), MoS_2 nanoparticles grown on graphene: An advanced catalyst for the hydrogen evolution reaction. *Journal of the American Chemical Society* 133: 7296–7299.

Li, Z., Huang, Y., Chen, L. et al. (2013), Amperometric biosensor for nadh and ethanol based on electroreduced graphene oxide–polythionine nanocomposite film. *Sensors and Actuators B: Chemical* 181: 280–287.

Lin, Y., Williams, T.V., and Connell, J.W. (2009), Soluble, exfoliated hexagonal boron nitride nanosheets. *The Journal of Physical Chemistry Letters* 1: 277–283.

Liu, F., Choi, J.Y., and Seo, T.S. (2010a), Graphene oxide arrays for detecting specific DNA hybridization by fluorescence resonance energy transfer. *Biosensors and Bioelectronics* 25: 2361–2365.

Liu, F., Kim, Y.H., Cheon, D.S., and Seo, T.S. (2013), Micropatterned reduced graphene oxide based field-effect transistor for real-time virus detection. *Sensors and Actuators B: Chemical* 186: 252–257.

Liu, Y., Yu, D., Zeng, C., Miao, Z., and Dai, L. (2010b), Biocompatible graphene oxide-based glucose biosensors. *Langmuir* 26: 6158–6160.

Liu, Z., Robinson, J.T., Sun, X., and Dai, H. (2008), Pegylated nanographene oxide for delivery of water-insoluble cancer drugs. *Journal of the American Chemical Society* 130: 10876–10877.

Loh, K.P., Bao, Q., Ang, P.K., and Yang, J. (2010), The chemistry of graphene. *Journal of Materials Chemistry* 20: 2277–2289.

Lomeda, J.R., Doyle, C.D., Kosynkin, D.V., Hwang, W.F., and Tour, J.M. (2008), Diazonium functionalization of surfactant-wrapped chemically converted graphene sheets. *Journal of the American Chemical Society* 130: 16201–16206.

Loo, A.H., Bonanni, A., Ambrosi, A., Poh, H.L., and Pumera, M. (2012), Impedimetric immunoglobulin G immunosensor based on chemically modified graphenes. *Nanoscale* 4: 921–925.

Lotya, M., Hernandez, Y., King, P.J. et al. (2009), Liquid phase production of graphene by exfoliation of graphite in surfactant/water solutions. *Journal of the American Chemical Society* 131: 3611–3620.

Lu, B., Li, T., Zhao, H. et al. (2012), Graphene-based composite materials beneficial to wound healing. *Nanoscale* 4: 2978–2982.

Lu, C.-H., Yang, H.-H., Zhu, C.-L., Chen, X., and Chen, G.-N. (2009), A graphene platform for sensing biomolecules. *Angewandte Chemie* 121: 4879–4881.

Lu, C.H., Zhu, C.L., Li, J., Liu, J.J., Chen, X., and Yang, H.H. (2010), Using graphene to protect DNA from cleavage during cellular delivery. *Chemical Communications* 46: 3116–3118.

Luckham, P.F. and Rossi, S. (1999), The colloidal and rheological properties of bentonite suspensions. *Advances in Colloid and Interface Science* 82: 43–92.

Lv, W., Guo, M., Liang, M.-H. et al. (2010), Graphene-DNA hybrids: Self-assembly and electrochemical detection performance. *Journal of Materials Chemistry* 20: 6668–6673.

Ma, H., Wu, D., Cui, Z. et al. (2013), Graphene-based optical and electrochemical biosensors: A review. *Analytical Letters* 46: 1–17.

Ma, R., Liu, Z., Li, L., Iyi, N., and Sasaki, T. (2006), Exfoliating layered double hydroxides in formamide: A method to obtain positively charged nanosheets. *Journal of Materials Chemistry* 16: 3809–3813.

Macdonald, J.R. 1987. *Impedance Spectroscopy: Emphasizing Solid Materials and Systems.* New York: Wiley.

Mak, K.F., Lee, C., Hone, J., Shan, J., and Heinz, T.F. (2010), Atomically thin MoS$_2$: A new direct-gap semiconductor. *Physical Review Letters* 105: 136805.

Mathesh, M., Liu, J., Nam, N.D. et al. (2013), Facile synthesis of graphene oxide hybrids bridged by copper ions for increased conductivity. *Journal of Materials Chemistry C* 1: 3084–3090.

Meyer, J.C., Geim, A.K., Katsnelson, M.I., Novoselov, K.S., Booth, T.J., and Roth, S. (2007), The structure of suspended graphene sheets. *Nature* 446: 60–63.

Mohanraj, V. and Chen, Y. (2007), Nanoparticles-a review. *Tropical Journal of Pharmaceutical Research* 5: 561–573.

Mohanty, N. and Berry, V. (2008), Graphene-based single-bacterium resolution biodevice and DNA transistor: Interfacing graphene derivatives with nanoscale and microscale biocomponents. *Nano Letters* 8: 4469–4476.

Moses, P.G., Mortensen, J.J., Lundqvist, B.I., and Norskov, J.K. (2009), Density functional study of the adsorption and van der waals binding of aromatic and conjugated compounds on the basal plane of MoS$_2$. *Journal of Chemical Physics* 130: 104709.

Nair, R.R., Wu, H.A., Jayaram, P.N., Grigorieva, I.V., and Geim, A.K. (2012), Unimpeded permeation of water through helium-leak–tight graphene-based membranes. *Science* 335: 442–444.

Nicolosi, V., Chhowalla, M., Kanatzidis, M.G., Strano, M.S., and Coleman, J.N. (2013), Liquid exfoliation of layered materials. *Science* 340: 1226419.

Novoselov, K.S. (2011), Nobel lecture: Graphene: Materials in the flatland. *Reviews of Modern Physics* 83: 837–849.

Novoselov, K.S. and Castro Neto, A.H. (2012a), Two-dimensional crystals-based heterostructures: Materials with tailored properties. *Physica Scripta* 2012: 014006.

Novoselov, K.S., Falko, V.I., Colombo, L., Gellert, P.R., Schwab, M.G., and Kim, K. (2012b), A roadmap for graphene. *Nature* 490: 192–200.

Novoselov, K.S., Jiang, D., Schedin, F. et al. (2005), Two-dimensional atomic crystals. *Proceedings of the National Academy of Sciences of the United States of America* 102: 10451–10453.

Ohta, T., Bostwick, A., Seyller, T., Horn, K., and Rotenberg, E. (2006), Controlling the electronic structure of bilayer graphene. *Science* 313: 951–954.

Omomo, Y., Sasaki, T., Wang, L., and Watanabe, M. (2003), Redoxable nanosheet crystallites of MnO$_2$ derived via delamination of a layered manganese oxide. *Journal of the American Chemical Society* 125: 3568–3575.

Pakdel, A., Wang, X., Bando, Y., and Golberg, D. (2013), Nonwetting "white graphene" films. *Acta Materialia* 61: 1266–1273.

Pan, D., Zhang, J., Li, Z., and Wu, M. (2010), Hydrothermal route for cutting graphene sheets into blue-luminescent graphene quantum dots. *Advanced Materials* 22: 734–738.

Park, S., Dikin, D.A., Nguyen, S.T., and Ruoff, R.S. (2009), Graphene oxide sheets chemically cross-linked by polyallylamine. *The Journal of Physical Chemistry C* 113: 15801–15804.

Park, S.Y., Park, J., Sim, S.H. et al. (2011), Enhanced differentiation of human neural stem cells into neurons on graphene. *Advanced Materials* 23: H263–H267.

Pease, R.S. (1952), An X-ray study of boron nitride. *Acta Crystallographica* 5: 356–361.

Peng H., Xu, C., Lin, J. et al. (2011), Folic acid-conjugated graphene oxide loaded with photosensitizers for targeting photodynamic therapy. *Theranostics* 1: 240–250.

Peng, J., Gao, W., Gupta, B.K. et al. (2012), Graphene quantum dots derived from carbon fibers. *Nano Letters* 12: 844–849.

Pereira, V.M., Castro Neto, A.H., and Peres, N.M.R. (2009), Tight-binding approach to uniaxial strain in graphene. *Physical Review B* 80: 045401.

Pham, T.A., Kumar, N.A., and Jeong, Y.T. (2010), Covalent functionalization of graphene oxide with polyglycerol and their use as templates for anchoring magnetic nanoparticles. *Synthetic Metals* 160: 2028–2036.

Pumera, M. (2011), Graphene in biosensing. *Materials Today* 14: 308–315.

Qi, P., Wan, Y., and Zhang, D. (2013), Impedimetric biosensor based on cell-mediated bioimprinted films for bacterial detection. *Biosensors and Bioelectronics* 39: 282–288.

Quintana, M., Spyrou, K., Grzelczak, M., Browne, W.R., Rudolf, P., and Prato, M. (2010), Functionalization of graphene via 1,3-dipolar cycloaddition. *ACS Nano* 4: 3527–3533.

Radisavljevic, B., Radenovic, A., Brivio, J., Giacometti, V., and Kis, A. (2011), Single-layer MoS$_2$ transistors. *Nature Nanotechnology* 6: 147–150.

Ramasubramaniam, A., Naveh, D., and Towe, E. (2011), Tunable band gaps in bilayer graphene–BN heterostructures. *Nano Letters* 11: 1070–1075.

Ramesha, G.K., Vijaya Kumara, A., Muralidhara, H.B., and Sampath, S. (2011), Graphene and graphene oxide as effective adsorbents toward anionic and cationic dyes. *Journal of Colloid and Interface Science* 361: 270–277.

Ratinac, K.R., Yang, W., Gooding, J.J., Thordarson, P., and Braet, F. (2011), Graphene and related materials in electrochemical sensing. *Electroanalysis* 23: 803–826.

Reina, A., Jia, X., Ho, J. et al. (2008), Large area, few-layer graphene films on arbitrary substrates by chemical vapor deposition. *Nano Letters* 9: 30–35.

Ren, H., Wang, C., Zhang, J. et al. (2010), DNA cleavage system of nanosized graphene oxide sheets and copper ions. *ACS Nano* 4: 7169–7174.

Sainsbury, T., Ikuno, T., Okawa, D., Pacile, D., Frechet, J.M.J., and Zettl, A. (2007), Self-assembly of gold nanoparticles at the surface of amine- and thiol-functionalized boron nitride nanotubes. *The Journal of Physical Chemistry C* 111: 12992–12999.

Sanchez, V.C., Jachak, A., Hurt, R.H., and Kane, A.B. (2011), Biological interactions of graphene-family nanomaterials: An interdisciplinary review. *Chemical Research in Toxicology* 25: 15–34.

Savage, N. (2012), Materials science: Super carbon. *Nature* 483: S30–S31.

Shan, C., Yang, H., Han, D., Zhang, Q., Ivaska, A., and Niu, L. (2009), Water-soluble graphene covalently functionalized by biocompatible poly-L-lysine. *Langmuir* 25: 12030–12033.

Shao, Y., Wang, J., Wu, H., Liu, J., Aksay, I.A., and Lin, Y. (2010), Graphene based electrochemical sensors and biosensors: A review. *Electroanalysis* 22: 1027–1036.

Shen, J., Hu, Y., Li, C., Qin, C., Shi, M., and Ye, M. (2009), Layer-by-layer self-assembly of graphene nanoplatelets. *Langmuir* 25: 6122–6128.

Sheppard, J.I., McClung, W.G., and Feuerstein, I.A. (1994), Adherent platelet morphology on adsorbed fibrinogen: Effects of protein incubation time and albumin addition. *Journal of Biomedical Materials Research* 28: 1175–1186.

Shi, Y., Hamsen, C., Jia, X. et al. (2010), Synthesis of few-layer hexagonal boron nitride thin film by chemical vapor deposition. *Nano Letters* 10: 4134–4139.

Shin, H.J., Kim, K.K., Benayad, A. et al. (2009), Efficient reduction of graphite oxide by sodium borohydride and its effect on electrical conductance. *Advanced Functional Materials* 19: 1987–1992.

Sinitskii, A., Dimiev, A., Corley, D.A., Fursina, A.A., Kosynkin, D.V., and Tour, J.M. (2010), Kinetics of diazonium functionalization of chemically converted graphene nanoribbons. *ACS Nano* 4: 1949–1954.

Song, L., Ci, L., Lu, H. et al. (2010), Large scale growth and characterization of atomic hexagonal boron nitride layers. *Nano Letters* 10: 3209–3215.

Souche, C., Jouffrey, B., Hug, G., and Nelhiebel, M. (1998), Orientation sensitive eels-analysis of boron nitride nanometric hollow spheres. *Micron* 29: 419–424.

Splendiani, A., Sun, L., Zhang, Y. et al. (2010), Emerging photoluminescence in monolayer MoS$_2$. *Nano Letters* 10: 1271–1275.

Stankovich, S., Dikin, D.A., Dommett, G.H.B. et al. (2006a), Graphene-based composite materials. *Nature* 442: 282–286.

Stankovich, S., Piner, R.D., Nguyen, S.T., and Ruoff, R.S. (2006b), Synthesis and exfoliation of isocyanate-treated graphene oxide nanoplatelets. *Carbon* 44: 3342–3347.

Stine, R., Robinson, J.T., Sheehan, P.E., and Tamanaha, C.R. (2010), Real-time DNA detection using reduced graphene oxide field effect transistors. *Advanced Materials* 22: 5297–5300.

Stoller, M.D., Park, S., Zhu, Y., An, J., and Ruoff, R.S. (2008), Graphene-based ultracapacitors. *Nano Letters* 8: 3498–3502.

Strom, T.A., Dillon, E.P., Hamilton, C.E., and Barron, A.R. (2010), Nitrene addition to exfoliated graphene: A one-step route to highly functionalized graphene. *Chemical Communications* 46: 4097–4099.

Su, C.-Y., Xu, Y., Zhang, W. et al. (2009), Electrical and spectroscopic characterizations of ultra-large reduced graphene oxide monolayers. *Chemistry of Materials* 21: 5674–5680.

Sun, X., Liu, Z., Welsher, K. et al. (2008), Nano-graphene oxide for cellular imaging and drug delivery. *Nano Research* 1: 203–212.

Tanaka, T., Ebina, Y., Takada, K., Kurashima, K., and Sasaki, T. (2003), Oversized titania nanosheet crystallites derived from flux-grown layered titanate single crystals. *Chemistry of Materials* 15: 3564–3568.

Tang, X.Z., Li, W., Yu, Z.Z. et al. (2011), Enhanced thermal stability in graphene oxide covalently functionalized with 2-amino-4,6-didodecylamino-1,3,5-triazine. *Carbon* 49: 1258–1265.

Tetsuka, H., Asahi, R., Nagoya, A. et al. (2012), Optically tunable amino-functionalized graphene quantum dots. *Advanced Materials* 24: 5333–5338.

Tidwell, C.D., Ertel, S.I., Ratner, B.D., Tarasevich, B.J., Atre, S., and Allara, D.L. (1997), Endothelial cell growth and protein adsorption on terminally functionalized, self-assembled monolayers of alkanethiolates on gold. *Langmuir* 13: 3404–3413.

Veca, L.M., Lu, F., Meziani, M.J. et al. (2009), Polymer functionalization and solubilization of carbon nanosheets. *Chemical Communications* 2565–2567.

Wallace, P.R. (1947), The band theory of graphite. *Physical Review* 71: 622–634.

Wan, Y., Lin, Z., Zhang, D., Wang, Y., and Hou, B. (2011), Impedimetric immunosensor doped with reduced graphene sheets fabricated by controllable electrodeposition for the non-labelled detection of bacteria. *Biosensors and Bioelectronics* 26: 1959–1964.

Wang, K., Ruan, J., Song, H. et al. (2010a), Biocompatibility of graphene oxide. *Nanoscale Research Letters* 6: 1–8.

Wang, Y., Li, Z., Hu, D., Lin, C.T., Li, J., and Lin, Y. (2010b), Aptamer/graphene oxide nanocomplex for in situ molecular probing in living cells. *Journal of the American Chemical Society* 132: 9274–9276.

Wang, Y., Li, Z., Wang, J., Li, J., and Lin, Y. (2011), Graphene and graphene oxide: Biofunctionalization and applications in biotechnology. *Trends in Biotechnology* 29: 205–212.

Wang, Y., Ping, J., Ye, Z., Wu, J., and Ying, Y. (2013), Impedimetric immunosensor based on gold nanoparticles modified graphene paper for label-free detection of *Escherichia coli* O157:H7. *Biosensors and Bioelectronics* 49: 492–498.

Wang, Y.X., Robertson, J., Spillman, W., Jr., and Claus, R. (2004), Effects of the chemical structure and the surface properties of polymeric biomaterials on their biocompatibility. *Pharmaceutical Research* 21: 1362–1373.

Williams, S.F., Martin, D.P., Horowitz, D.M., and Peoples, O.P. (1999), Pha applications: Addressing the price performance issue: I. Tissue engineering. *International Journal of Biological Macromolecules* 25: 111–121.

Worle-Knirsch, J.M., Pulskamp, K., and Krug, H.F. (2006), Oops they did it again! Carbon nanotubes hoax scientists in viability assays. *Nano Letters* 6: 1261–1268.

Wu, H., Yang, R., Song, B. et al. (2011), Biocompatible inorganic fullerene-like molybdenum disulfide nanoparticles produced by pulsed laser ablation in water. *ACS Nano* 5: 1276–1281.

Wu, J.F., Xu, M.Q., and Zhao, G.C. (2010), Graphene-based modified electrode for the direct electron transfer of cytochrome c and biosensing. *Electrochemistry Communications* 12: 175–177.

Xiang, Q., Yu, J., and Jaroniec, M. (2012), Synergetic effect of MoS$_2$ and graphene as cocatalysts for enhanced photocatalytic H$_2$ production activity of TiO$_2$ nanoparticles. *Journal of the American Chemical Society* 134: 6575–6578.

Xu, H., Dai, H., and Chen, G. (2010), Direct electrochemistry and electrocatalysis of hemoglobin protein entrapped in graphene and chitosan composite film. *Talanta* 81: 334–338.

Xu, Y., Bai, H., Lu, G., Li, C., and Shi, G. (2008), Flexible graphene films via the filtration of water-soluble noncovalent functionalized graphene sheets. *Journal of the American Chemical Society* 130: 5856–5857.

Yan, X., Li, B., and Li, L.S. (2012), Colloidal graphene quantum dots with well-defined structures. *Accounts of Chemical Research* 46: 2254–2262.

Yang, H., Shan, C., Li, F., Han, D., Zhang, Q., and Niu, L. (2009), Covalent functionalization of polydisperse chemically-converted graphene sheets with amine-terminated ionic liquid. *Chemical Communications* 2009: 3880–3882.

Yang, K., Wan, J., Zhang, S., Zhang, Y., Lee, S.-T., and Liu, Z. (2010a), In vivo pharmacokinetics, long-term biodistribution, and toxicology of pegylated graphene in mice. *ACS Nano* 5: 516–522.

Yang, K., Zhang, S., Zhang, G., Sun, X., Lee, S.-T., and Liu, Z. (2010b), Graphene in mice: Ultrahigh in vivo tumor uptake and efficient photothermal therapy. *Nano Letters* 10: 3318–3323.

Yang, M. and Gong, S. (2010c), Immunosensor for the detection of cancer biomarker based on percolated graphene thin film. *Chemical Communications* 46: 5796–5798.

Yang, S.T., Chang, Y., Wang, H. et al. (2010d), Folding/aggregation of graphene oxide and its application in Cu^{2+} removal. *Journal of Colloid and Interface Science* 351: 122–127.

Yang, X., Zhang, X., Liu, Z., Ma, Y., Huang, Y., and Chen, Y. (2008), High-efficiency loading and controlled release of doxorubicin hydrochloride on graphene oxide. *The Journal of Physical Chemistry C* 112: 17554–17558.

Yang, Y., Asiri, A.M., Tang, Z., Du, D., and Lin, Y. (2013), Graphene based materials for biomedical applications. *Materials Today* 16: 365–373.

Yoon, S. and In, I. (2011), Role of poly(n-vinyl-2-pyrrolidone) as stabilizer for dispersion of graphene via hydrophobic interaction. *Journal of Materials Science* 46: 1316–1321.

Yu, D. and Dai, L. (2009), Self-assembled graphene/carbon nanotube hybrid films for supercapacitors. *The Journal of Physical Chemistry Letters* 1: 467–470.

Yu, J., Qin, L., Hao, Y. et al. (2010), Vertically aligned boron nitride nanosheets: Chemical vapor synthesis, ultraviolet light emission, and superhydrophobicity. *ACS Nano* 4: 414–422.

Yun, J.M., Kim, K.N., Kim, J.Y. et al. (2012), DNA origami nanopatterning on chemically modified graphene. *Angewandte Chemie* 124: 936–939.

Zeng, H., Zhi, C., Zhang, Z. et al. (2010), "White graphenes": Boron nitride nanoribbons via boron nitride nanotube unwrapping. *Nano Letters* 10: 5049–5055.

Zhang, F., Zheng, B., Zhang, J. et al. (2010a), Horseradish peroxidase immobilized on graphene oxide: Physical properties and applications in phenolic compound removal. *The Journal of Physical Chemistry C* 114: 8469–8473.

Zhang, H., Gruner, G., and Zhao, Y. (2013), Recent advancements of graphene in biomedicine. *Journal of Materials Chemistry B* 1: 2542–2567.

Zhang, J., Zhang, F., Yang, H. et al. (2010b), Graphene oxide as a matrix for enzyme immobilization. *Langmuir* 26: 6083–6085.

Zhang, L., Lu, Z., Zhao, Q., Huang, J., Shen, H., and Zhang, Z. (2011a), Enhanced chemotherapy efficacy by sequential delivery of siRNA and anticancer drugs using pei-grafted graphene oxide. *Small* 7: 460–464.

Zhang, L., Xia, J., Zhao, Q., Liu, L., and Zhang, Z. (2010c), Functional graphene oxide as a nanocarrier for controlled loading and targeted delivery of mixed anticancer drugs. *Small* 6: 537–544.

Zhang, L.L., Xiong, Z., and Zhao, X.S. (2010d), Pillaring chemically exfoliated graphene oxide with carbon nanotubes for photocatalytic degradation of dyes under visible light irradiation. *ACS Nano* 4: 7030–7036.

Zhang, M., Bai, L., Shang, W. et al. (2012a), Facile synthesis of water-soluble, highly fluorescent graphene quantum dots as a robust biological label for stem cells. *Journal of Materials Chemistry* 22: 7461–7467.

Zhang, W., Guo, Z., Huang, D., Liu, Z., Guo, X., and Zhong, H. (2011b), Synergistic effect of chemo-photothermal therapy using pegylated graphene oxide. *Biomaterials* 32: 8555–8561.

Zhang, X., Yin, J., Peng, C. et al. (2011c), Distribution and biocompatibility studies of graphene oxide in mice after intravenous administration. *Carbon* 49: 986–995.

Zhang, Y., Zhang, J., Huang, X., Zhou, X., Wu, H., and Guo, S. (2012b), Assembly of graphene oxide–enzyme conjugates through hydrophobic interaction. *Small* 8: 154–159.

Zhi, C., Bando, Y., Tang, C., Kuwahara, H., and Golberg, D. (2009), Large-scale fabrication of boron nitride nanosheets and their utilization in polymeric composites with improved thermal and mechanical properties. *Advanced Materials* 21: 2889–2893.

Zhou, K., Zhu, Y., Yang, X., Luo, J., Li, C., and Luan, S. (2010), A novel hydrogen peroxide biosensor based on Au–graphene–HRP–chitosan biocomposites. *Electrochimica Acta* 55: 3055–3060.

Zhu, C., Zeng, Z., Li, H., Li, F., Fan, C., and Zhang, H. (2013), Single-layer MoS$_2$-based nanoprobes for homogeneous detection of biomolecules. *Journal of the American Chemical Society* 135: 5998–6001.

Zu, S.Z. and Han, B.-H. (2009), Aqueous dispersion of graphene sheets stabilized by pluronic copolymers: Formation of supramolecular hydrogel. *The Journal of Physical Chemistry C* 113: 13651–13657.

Zunger, A., Katzir, A., and Halperin, A. (1976), Optical properties of hexagonal boron nitride. *Physical Review B* 13: 5560–5573.

Chapter 14

Chemistry of Boron Nitride Nanosheets

Yunlong Liao, Zhongfang Chen, John W. Connell, and Yi Lin

CONTENTS

14.1 INTRODUCTION

Boron nitride nanosheets (BNNSs) are an exemplary member of the graphene-analogous 2D nanomaterial family (Golberg et al. 2010, Lin 2012a, Lu et al. 2012, Pakdel et al. 2012, Song et al. 2013). However, BNNSs had not captured much interest from scientists until graphene was exfoliated from graphite in 2004 by Novoselov et al. (2004). Shortly thereafter, BNNSs were exfoliated from hexagonal boron nitride (h-BN) by the same group of researchers (Novoselov et al. 2005). In both cases, the preparations were achieved via mechanical exfoliation by simply using a piece of adhesive tape, that is, the *Scotch tape method*. Sharing a nearly identical geometrical structure to graphene, BNNSs also have superb mechanical strength and high thermal conductivity. However, BNNSs have an intrinsically wide bandgap (~4–6 eV) in comparison to conductive graphene. These property differences make BNNSs potentially useful in applications where graphene may not be desirable. For example, BNNSs are attractive for use as thermally conductive but electrically insulating fillers for polymer or ceramic composites, thermal radiators, deep ultraviolet light emitter and lasers, and nanoelectronic devices. As h-BN is nicknamed *white graphite*, the monolayer BNNSs are often referred to as *white graphene* (Zeng et al. 2010, Liu et al. 2011).

Though graphite and bulk h-BN are isoelectronic and have similar layered structure, the individual layers stack in slightly different orientations. For h-BN, the crystal structure is hexagonal with the space group P6$_3$/mmc (no. 194) and the lattice constant being $a = b = 0.2504$ nm, $c = 0.6661$ nm, $\alpha = \beta = 90°$, $\gamma = 120°$, as shown in Figure 14.1 (Song et al. 2013). The B–N bonds are 1.45 Å in length with ionic characteristics, thus the AA′ stacking is slightly favored energetically (Pease 1950, Marom et al. 2010), in which the electron-deficient B atoms are located directly above or below the electron-rich N atoms in adjacent layers. In contrast, the C–C bonds in graphite

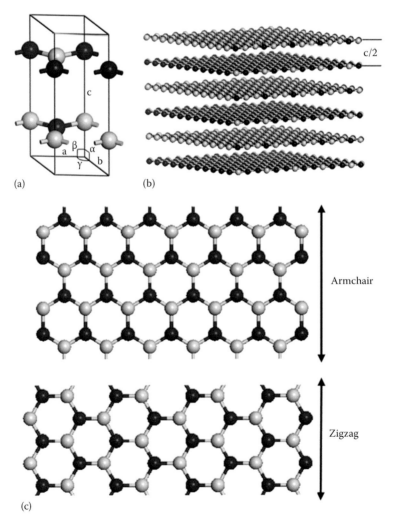

(a)   (b)

Armchair

Zigzag

(c)

Figure 14.1   (a) Unit cell of h-BN, (b) the structure of layered BNNSs in AA′ stacking pattern, and (c) geometric structure of armchair and zigzag boron nitride nanoribbons (BNNRs).

are completely symmetric with no ionic characteristics. Therefore, the preferred stacking for the graphene layers in graphite is offset, with the C atoms in one layer above or below the hexagonal centers of the adjacent layer.

BNNSs can be obtained via top-down (Han et al. 2008, Pacile et al. 2008, Zhi et al. 2009, Lin et al. 2010a, 2011b, Coleman et al. 2011) or bottom-up (Shi et al. 2010, Song et al. 2010, Kim et al. 2011b) approaches, which typically refer to the exfoliation of h-BN and the synthesis from B and N precursors, respectively. The available top-down methods include mechanical cleavage, direct solvent exfoliation, surfactant-assisted exfoliation, and chemical functionalization–induced exfoliation. The exfoliated BNNSs usually maintain the high crystallinity from their h-BN precursors. However, the lateral sizes for the commercially available h-BN particles are typically on the order of only 10 μm or less. This has resulted in limited lateral sizes of exfoliated BNNSs, which are often below 1 μm (Han et al. 2008, Pacile et al. 2008, Zhi et al. 2009, Lin et al. 2010a, Coleman et al. 2011). Recently, a few top-down methods have been reported on the unzipping of boron nitride nanotubes (BNNTs) into boron nitride nanoribbons (BNNRs) (Zeng et al. 2010, Erickson et al. 2011). BNNRs are a specific form of BNNSs that have definitive widths and thus edge structure (zigzag or armchair, shown in Figure 14.1 and discussed in Section 2). Alternatively, BNNSs could be synthesized from B and N precursors via bottom-up approaches such as chemical vapor deposition and thermal decomposition. The lateral dimensions of the synthesized BNNSs can reach a few centimeters (Shi et al. 2010, Song et al. 2010, Kim et al. 2011b).

While exfoliated BNNSs are useful for applications that require bulk quantities of materials such as polymeric nanocomposites, the bottom-up synthesized nanosheets as well as those from the Scotch tape mechanical exfoliation, typically supported on a metal or crystalline substrate, are more suitable for electronic or sensor applications requiring less quantities of material such as those with graphene as the active component (Kim et al. 2011a, Mayorov et al. 2011, Xue et al. 2011, Britnell et al. 2012).

BN polymorphs including BNNSs and BNNTs are excellent oxidation-resistant materials. They are stable up to ~800°C in air and inert against oxidative acid treatments. Note that the *chemical inertness* for BNNSs is relative. As given in the following detailed discussions, BNNSs do have rich noncovalent and ionic chemistry that can render them soluble in various solvents and processable for applications such as nanofillers in polymeric nanocomposites and supports for catalysts or sensors. However, it is relatively difficult to covalently modify the B–N bond network. Highly reactive species are required to achieve successful functionalization.

In this chapter, various aspects of BNNS chemistry from both the theoretical and experimental perspectives are reviewed. Most theoretical explorations have focused on the potential use of BNNSs and BNNRs in nanoelectronics or spintronics by means of density functional theory (DFT) computations. Therefore, the prediction of how to tune their electronic and magnetic properties is of the most interest. In contrast, the experimental chemistry studies tend to use BNNS materials into more diverse applications such as polymer composites, sensors, catalysis, and biological conjugations. Interestingly, the theoretical studies are far beyond the pace of experimental ones. It is expected that this seemingly unusual misalignment of research interests will become more balanced when more reliable synthesis of larger quantities of BNNSs and BNNRs with controllable dimensions and edge structures become more routine and the techniques to characterize the atomic structure with the fidelity needed to validate the theoretical predictions become readily available.

14.2 THEORETICAL INVESTIGATIONS OF BNNS CHEMISTRY

14.2.1 Electronic and Magnetic Properties of Pristine BNNSs and BNNRs

BNNSs retain the electrical insulating characteristics of h-BN and are wide-bandgap semiconductors. The computed bandgaps (4.3 eV for local density of approximation, 6.0 eV for GW correction) (Blasé et al. 1995) are generally smaller than the experimentally measured value (5.97 eV) (Watanabe et al. 2004). Inspired by the investigations of graphene nanoribbons (GNRs), numerous studies on both zigzag and armchair BNNRs were performed due to their distinctive electronic and magnetic properties closely associated with the effects of quantum confinement and symmetry, as well as the edge effects, which could not appear in nanosheets with indefinite planar dimensions.

The electronic and magnetic properties of BNNRs are significantly affected by the edge type as well as how the edges are passivated or modified. The armchair BNNRs with either bare or hydrogen-passivated edges are all nonmagnetic, wide-bandgap semiconductors (Du et al. 2007, Park and Louie 2008, Hu et al. 2012). In comparison, bare and passivated zigzag BNNRs exhibit different electronic and magnetic behaviors. For example, hydrogen-passivated zigzag BNNRs are nonmagnetic insulators. However, those with bare edges exhibit spin-polarized magnetism: for the lowest-energy configuration, a ferromagnetic spin state is located at N edge and an antiferromagnetic spin state is located at B edge (Figure 14.2) (Du et al. 2007, Barone and Peralta 2008, Topsakal et al. 2009). Note that bare zigzag BNNRs are higher in energy than bare armchair BNNRs and are also less stable than those with five- to seven-membered rings (Mukherjee and Bhowmick 2011, Wu et al. 2011).

The intrinsic wide bandgap seriously impedes the application of BNNSs in nanoelectronics and optical applications, which generally require the bandgap to be less than 3 eV. Thus, numerous methodologies, such as the introduction of defects and doping, as well as edge and surface modification, have been proposed to lower the bandgap of BNNSs as discussed in the next section.

14.2.2 Defects and Doping

The existence of impurity states could present a new pathway to tune the electronic properties making them more attractive for applications in electronic, magnetic, and optical devices. Structural defects in BNNSs and BNNRs, in the forms of intrinsic point defects and extrinsic substitution or doping, offer a very important opportunity to tune the electronic and magnetic properties.

Vacancy defects: Vacancy defects can be produced by electron beam or reactive ion etching in the monolayer BNNSs by mechanical or solvent exfoliation from few-layered BNNSs (Alem et al. 2009, Jin et al. 2009, Meyer et al. 2009, Warner et al. 2010, Suenaga et al. 2012). The vacancy defects observed by high-resolution transmission electron microscopy (HR-TEM) (Figure 14.3) always appear as triangles of various sizes. Although N vacancies are thermodynamically more stable (Jiménez et al. 1996, 1997, Azevedo et al. 2007), exhaustive electron microscopic analyses indicate that more defects are N-terminated (i.e., B vacancies) (Alem et al. 2009, Jin et al. 2009). Several possible reasons were proposed, including the lower knockon energy of B than N atoms (Zobelli et al. 2007) and selective stability of N-terminations under electron-rich conditions (Du et al. 2009, Okada 2009, Yin et al. 2010). A more plausible explanation might be a kinetic model, which suggests that B atoms have higher atomic displacement rate than N under high-energy electron irradiation conditions (Kotakoski et al. 2010).

Typically, the topological change, such as the earlier-mentioned vacancy defects in BNNSs and BNNRs, is always accompanied with modification of electronic and magnetic properties (Si and Xue 2007, Pan and Yang 2010, Tang and Cao 2010, Yang et al. 2010, Yin et al. 2010,

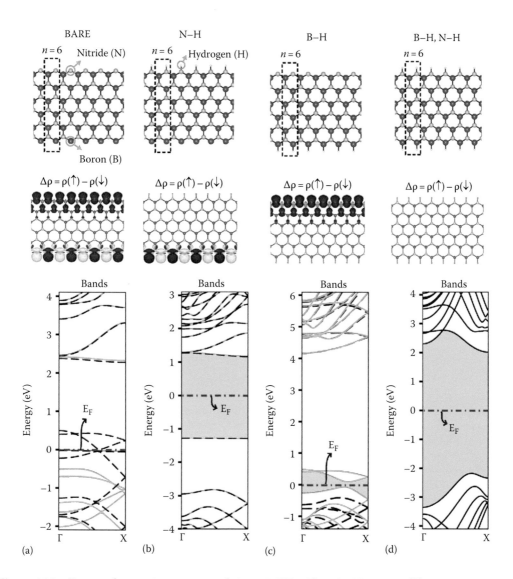

Figure 14.2   Top panels: atomic structures of zigzag BNNRs. The primitive unit cell has $n = 6$ B–N pairs delineated by dotted lines in the top panels. The unit cell is doubled due to the antiferromagnetic interaction between adjacent N atoms. Middle panels: isosurface plots of difference in charge density between up-spin and down-spin states, $\Delta\rho = \rho(\uparrow) - \rho(\downarrow)$. Bottom panels: energy-band structure with dotted (gray) and solid (white) lines showing spin-up and spin-down states, respectively. (a) Bare zigzag BNNR; (b) B-side free, but N-side is passivated by hydrogen atoms; (c) N-side free, but B-side is saturated by hydrogen atoms; (d) both sides are saturated by hydrogen atoms. The bands in (a) through (c) are calculated using double cell. (Reprinted with permission from Topsakal, M., Aktürk, E., and Ciraci, S., First-principles study of 1D and 2D honeycomb structures of boron nitride, *Phys. Rev. B*, 79, 115442, 2009. Copyright 2009, by the American Physical Society.)

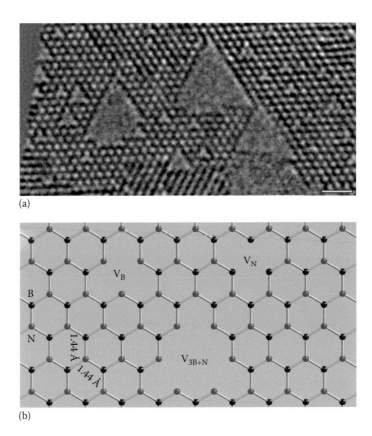

Figure 14.3   Atomic defects in monolayer of BNNSs. (a) A typical HR-TEM single frame showing the lattice defects in h-BN such as monovacancies (one lattice atom missing) and even larger vacancies, all of which are triangle shaped with the same orientation. (b) Models for the atomic defects in h-BN. V_B and V_N stand for boron and nitrogen monovacancy, respectively. Note that the V_B and V_N should have an opposite orientation, while the V_B and V_{3B+N} (missing three boron and one nitrogen atoms) are in the same orientation and surrounded by two coordinated nitrogen atoms. Scale bar = 1 nm. (Reprinted with permission from Jin, C., Lin, F., Suenaga, K., and Iijima, S., Fabrication of a freestanding boron nitride single layer and its defect assignments, *Phys. Rev. Lett.*, 102, 195505, 2009. Copyright 2009, by the American Physical Society.)

Zhang et al. 2011, Liu et al. 2012). For example, monolayer BNNS with a single B vacancy defect exhibits half-metallic behavior. When a monolayer of BNNS has many vacancy defects with different sizes and edge terminations, its magnetism becomes proportional to the number of unpaired electrons on the B/N atoms at the vacancy edges (Yin et al. 2010, Liu et al. 2012). For BNNRs, vacancy defects have slightly different effects due to a strengthened quantum effect. For example, they tend to appear in the vicinity of the B edge rather than at other sites. Variation of the defect location and concentration in zigzag BNNRs can significantly alter the band structure and lead to a spin-polarized or spin-gapless semiconductor, or even a half-metal (Pan and Yang 2010).

Additionally, defects may induce further property variations for BN nanomaterials. For instance, triangular vacancies in BNNRs may induce phonon scattering and reduce the thermal conductivity due to asymmetry of the triangular vacancies (Muralidharan et al. 2011, Yang et al. 2011).

Line defects: Line defects were experimentally observed on a BNNS film supported on a Ni(111) surface. The line defects were located between *fcc* and *hcp* triangular islands, which have opposite orientations. A possible model structure was proposed in which the N atoms terminate at two adjacent triangular islands and result in line defects (Auwärter et al. 2003). Additionally, Born–Oppenheimer molecular dynamic simulations indicated that pentagon–octagon–pentagon line defects could be created between two BN domains by inserting dimers (B_2, N_2, or C_2) and enable tuning of the electronic and magnetic properties of BNNSs via tailoring orientation arrangements (Figure 14.4) (Li et al. 2012).

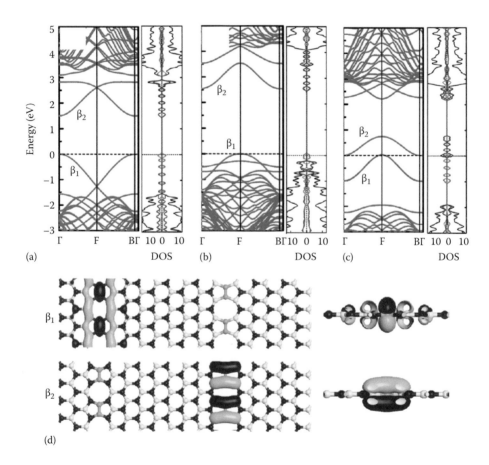

Figure 14.4 Computed band structures, density of states, and partial density of states of a line-defect-embedded BNNS with (a) B and N antisite line defects, (b) B_2 and N_2 5–8–5 line defects, and (c) C_2 5–8–5 line defects. (d) Top and side views of the profiles of β_1 and β_2 states at the Γ point for the line-defect-embedded BNNS with a pair of C_2 5–8–5 line defects. The Fermi level is set as zero. The total density of states and partial density of states on the two line defects are plotted in black, gray, and deep gray lines, respectively. The isosurface absolute value is 0.03 au. Symbols Γ, F, and B represent the (0, 0, 0), (0, 0.5, 0), and (0.5, 0, 0) k-points in the first Brillouin zone, respectively. The dark gray, light gray, and gray balls represent nitrogen, boron, and carbon atoms, respectively. (Reprinted with permission from Li, X., Wu, X., Zeng, X.C., and Yang, J., Band-gap engineering via tailored line defects in boron-nitride nanoribbons, sheets, and nanotubes, *ACS Nano*, 6, 4104–4112, 2012. Copyright 2012, American Chemical Society.)

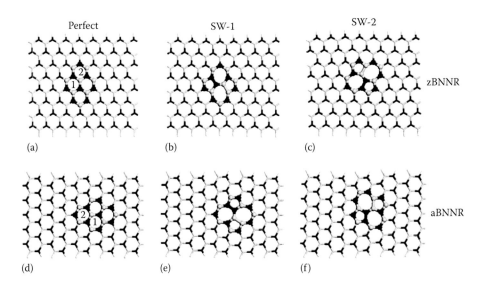

Figure 14.5   Optimized structures of BNNRs: (a, d) perfect and (b/e, c/f) with various SW defects. zBNNR and aBNNR represent zigzag and armchair BNNRs, respectively. (Reprinted with permission from Chen, W., Li, Y., Yu, G., Zhou, Z., and Chen, Z., Electronic structure and reactivity of boron nitride nanoribbons with stone-wales defects, *J. Chem. Theory Comput.*, 5, 3088–3095, 2009a. Copyright 2009, American Chemical Society.)

Stone–Wales (SW) defects: By rotating one bond of a traditional six-membered ring by 90°, an SW defect can be created that is comprised of two pairs of five-membered and seven-membered rings (5-7-7-5). Chen et al. (2009a) investigated the effect of the SW defect on BNNRs by means of DFT computations. Encouragingly, it was found that SW defects considerably reduce the bandgaps of BNNRs independent of the defect orientations (Figure 14.5). Thermodynamically, however, SW defects are not favorable due to unstable B–B and N–N bonds (Seifert et al. 1997, Pokropivny et al. 2000, Golberg et al. 2007, Rao and Govindaraj 2009, Wang 2010a).

Doping: Doping is defined as the substitution of atoms in the B–N bond network with other atoms. The most popular doping element of choice is carbon due to its similarity with B and N atoms. The resulting carbon-doped BNNSs or BNNRs may also be viewed as defective, not due to vacancies or reconstruction, but because of the presence of foreign carbon atoms that otherwise seamlessly fit into the honeycomb structure.

Carbon substitution with either B or N non-edge atoms in BNNRs could induce spontaneous magnetic and conducting properties (Du et al. 2007). DFT static and dynamic computations by Berseneva et al. (2011) indicate that the carbon substitution process (Figure 14.6) is governed by both the response of electron beam irradiation and the energetics of the atomic configurations. They also found that C–B substitution is energetically more favorable than C–N substitution, especially when the system is positively charged.

A recent achievement is the prediction of controllable and catalyst-free carbon doping of BNNSs with vacancies via CO molecules (Zhao et al. 2013). In this process, CO molecules have high selectivity for both B and N vacancies on BNNSs, and the activation energies of CO molecules on B and N vacancies could be as low as 0.30 and 0.37 eV, respectively.

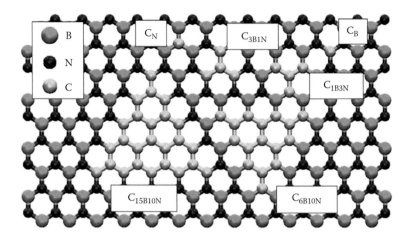

Figure 14.6   Examples of the structures (Zhao et al. 2013), including single–carbon atom impurities in BNNSs, four–carbon atom islands with boron ($4C_{1B3N}$ structure) and nitrogen ($4C_{3B1N}$) termination, as well as 16-atom ($16C_{6B10N}$ structure) and 25-atom ($25C_{15B10N}$ structure) islands. (Reprinted with permission from Berseneva, N., Krasheninnikov, A.V., and Nieminen, R.M., Mechanisms of postsynthesis doping of boron nitride nanostructures with carbon from first-principles simulations, *Phys. Rev. Lett.*, 107, 035501, 2011. Copyright 2011, by the American Physical Society.)

In addition to carbon, many other dopants were studied theoretically. These dopants could be main-group light elements (Be, B, C, N, O, Al, and Si) (Liu and Cheng 2007) or various transition (V, Cr, Fe, Co, Cu, and Mn) (Huang et al. 2012) or noble metals (Ag, Au, Pt, and Pd) (Zhou et al. 2011b). It was suggested that many of these dopants would evoke magnetism by introducing spin-polarized impurity states.

14.2.3   Edge Modification of BNNRs

Due to the quantum confinement effects, the bandgap values of zigzag and armchair BNNRs are quite different. For example, with hydrogen termination, the bandgap of zigzag BNNRs decreases monotonically with ribbon width and converges to a constant value (~3.5 eV), much smaller than BNNSs (~4.3 eV) due to the existence of strong edge states. In comparison, for the armchair BNNRs, the bandgap exhibits an oscillation following the general three-family structure behavior and converges to a same value (~4.5 eV) that is somewhat larger than BNNSs due to relatively weak edge states (Figure 14.7) (Du et al. 2007, Zhang and Guo 2008, Wang et al. 2011b).

Different passivation strategies are available for BNNRs, which can be divided into two types, sp^2 and sp^3, by considering the hybrid type of B or N atoms at BNNR edges (Figure 14.8). Passivation with sp^2 type has been widely applied in the investigation of BNNRs. Taking a hydrogenated edge as an example, if only the B edge is terminated with H atoms, zigzag BNNRs exhibit ferromagnetic half-metal behavior (Kan et al. 2011). On the contrary, hydrogen passivation only on the N edge would result in antiferromagnetic semiconductor properties with an indirect bandgap (Zheng et al. 2008, Lai et al. 2009). If passivated on both edges, as discussed in Section 2.1, zigzag BNNRs become nonmagnetic wide-gap semiconductors (Du et al. 2007, Barone and Peralta 2008, Topsakal et al. 2009). Similarly, the fully hydrogenated armchair and zigzag BNNRs are both

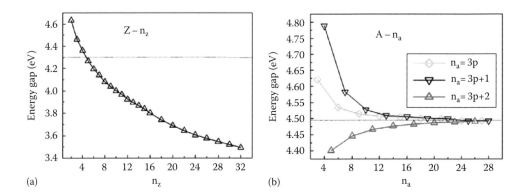

Figure 14.7 The variation of bandgaps of zigzag and armchair BNNRs is drawn as a function of widths in (a) and (b), respectively. Dotted lines denote the energy gap of an isolated BN sheet along the corresponding directions. (Reprinted with permission from Zhang, Z. and Guo, W., Energy-gap modulation of BN ribbons by transverse electric fields: First-principles calculations, *Phys. Rev. B*, 77, 075403, 2008. Copyright 2008, by the American Physical Society.)

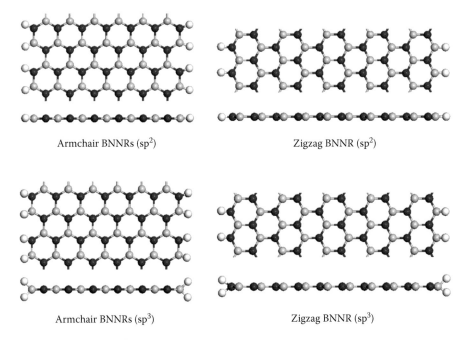

Figure 14.8 Top and side views of hydrogen passivated armchair and zigzag BNNRs (sp^2 and sp^3 hybridization).

nonmagnetic semiconductors, but with direct and indirect bandgaps, respectively. Interestingly, their respective bandgaps strongly depend on the external electric field (Park and Louie 2008).

For the BNNRs terminated by other atoms, O-terminated or S-terminated zigzag BNNRs are metallic (Lopez-Bezanilla et al. 2011), while O-passivated armchair BNNRs remain semiconducting (Lopez-Bezanilla et al. 2012). Furthermore, zigzag BNNRs have increased bandgap when passivated by F, but reduced bandgap when terminated by Cl, OH, and NO_2 (Wu et al. 2009). Computational studies from nonequilibrium Green's functions and the DFT indicated that fluorinated zigzag BNNRs could be of half-metal or semiconductor characteristic depending on the sites and ratio of fluorination (Zeng et al. 2012). Moreover, the negative differential resistance and varistor-type behaviors were observed, which may be useful in multifunctional molecular devices.

Various transition metals have also been explored to modify BNNR edges. Although the electronic and magnetic properties of armchair BNNRs are usually robust against modification, the edge termination by different transition metal atoms (Fe, Co, Cr, Ni, Cu, and Mn) can transform these ribbons from insulator into half-metal (Wang et al. 2011c).

The sp^3-hybrid-type passivation can also be realized. For example, under hydrogen-rich environment, B or N edge atoms can bond to two hydrogen atoms. Such sp^3-hybrid-type hydrogenation can transform zigzag BNNRs into ferromagnetic metals, but armchair BNNRs remain nonmagnetic semiconductors (Ding et al. 2009).

14.2.4 Surface Modifications

The surface of BNNSs could be chemically modified by covalent addition of many functional groups, such as H, F, O, OH, Cl, CH_3, and NH_2, and also by noncovalent adsorptions.

Double-sided covalent modifications: Hydrogenation is a representative for covalent additions. The fully hydrogenated BNNSs prefer the stirrup configuration rather than the boat or chair configuration (Figure 14.9) (Bhattacharya et al. 2010, Tang et al. 2013). All three hydrogenated BNNS conformers are direct wide-bandgap semiconductors, and the bandgaps are configuration dependent. At Perdew–Burke–Ernzerhof level of theory, the bandgaps are 3.27, 5.01, and 4.92 eV for the chair, boat, and stirrup configurations, respectively, while that of the pristine BNNSs is 4.51 eV at the same level (Tang et al. 2013).

For BNNRs, surface hydrogenation also plays a crucial role for tuning electronic and magnetic properties (Chen et al. 2010). Fully hydrogenated armchair BNNRs are nonmagnetic semiconductors, but fully hydrogenated zigzag BNNRs have ferromagnetic metallic behavior. Remarkably, partially hydrogenated zigzag BNNRs present a diverse array of properties. Specifically, the band structure transition (semiconductor–half-metal–metal) and magnetism transition (nonmagnetic–magnetic) occur with increasing hydrogenation (Figure 14.10). For the hydrogenated zigzag BNNRs, the bands around the Fermi level are dominated by edge B/N atoms with sp^3 hybridization and the edge-attached H atoms. As the hydrogenation ratio increases, both the highest energy valence and lowest energy conduction bands shift toward the Fermi level and eventually cross. High-fidelity control of the bandgaps and magnetic properties of hydrogenated zigzag BNNRs could enable a wide variety of applications especially in nanoelectronics.

Very recently, the interlayer B–H⋯H–N dihydrogen bonds were introduced in the hydrogenated bilayer BNNSs (Tang et al. 2013). The presence of such bonding can drastically reduce the intrinsic large bandgaps of chair-typed BN configuration (a meta-stable structure) due to interlayer charge transfer. While the insulating properties of the more stable stirrup-typed bilayer BN can be engineered by applying an electric field (Figure 14.9), the dihydrogen bonds can also exist in other hydrogenated 2D nanostructures, such as ZnO, AlN, and GaN. These fascinating results illuminate a new pathway for bandgap engineering of BNNSs and other 2D nanomaterials.

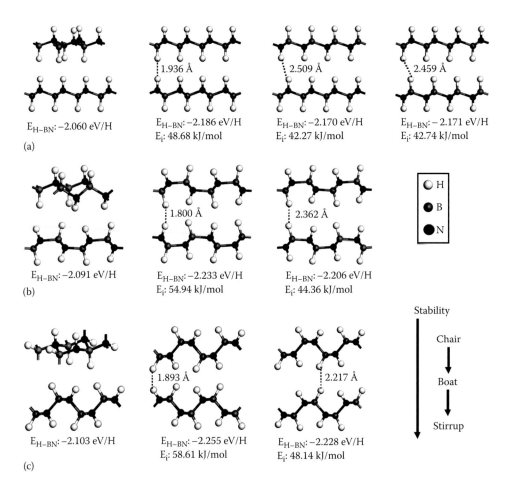

Figure 14.9 Geometries of fully hydrogenated monolayer and bilayer BN nanosheets in (a) chair-, (b) boat-, and (c) stirrup-type conformations. The first column shows the oblique (upper) and side (lower) views of fully hydrogenated monolayer. The second, third, and forth columns correspond to side views of fully hydrogenated bilayer with interlayer B–H···H–N, N–H···H–N, and B–H···H–B interactions, respectively. The hydrogenation energy (E_{H-BN}), interaction energy (E_i), and optimal interlayer H···H distance are indicated at the bottom of each conformation. (Tang, Q., Zhou, Z., Shen, P., and Chen, Z.: Band gap engineering of BN sheets by interlayer dihydrogen bonding and electric field control. *Chem. Phys. Phys. Chem.* 2013. 14. 1787–1792. Copyright Wiley-VCH Verlag GmbH & Co. KGaA. Reproduced with permission.)

Different from surface hydrogenation, surface functionalization of BNNSs by addition of groups with large atomic radius, such as F, Cl, CH_3, and NH_2, prefers the chair-like configuration (Bhattacharya et al. 2012). The bandgap of chemically functionalized BNNSs could vary from 0.3 to 3.1 eV depending on the functional group.

The surface oxidation of BNNSs produces BN oxide. The most favorable sites for chemisorption of oxygen atoms are over the N atom, which could stretch or even break the B–N bond.

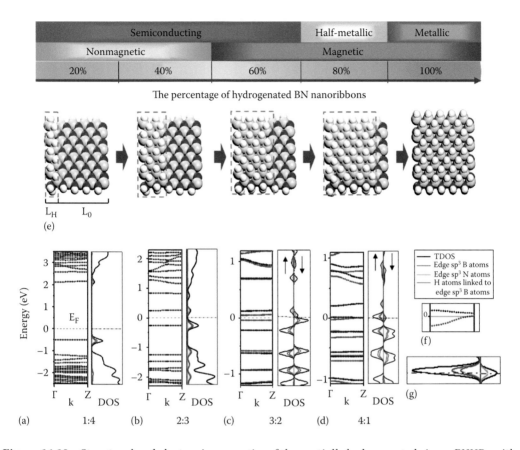

Figure 14.10   Structural and electronic properties of the partially hydrogenated zigzag BNNRs with the different ratios of L_H to L_0: (a) 1:4, (b) 2:3, (c) 3:2, and (d) 4:1. One unit corresponds to the part in the dash lines in (a), and the L_H and L_0 are respective unit numbers of hydrogenated and pristine BNNR blocks. The dark and gray dot lines denote the spin-up (↑) and spin-down (↓) channels, respectively, for the band structures of (c) and (d). (e) The structure of fully hydrogenated zigzag BNNR. (f) and (g) show the zooms on the region about the Fermi level of band structure and density of states of (d). (Reprinted with permission from Chen, W., Li, Y., Yu, G., Li, C.-Z., Zhang, S.B., Zhou, Z., and Chen, Z., Hydrogenation: A simple approach to realize semiconductor–half-metal–metal transition in boron nitride nanoribbons, *J. Am. Chem. Soc.*, 132, 1699–1705, 2010. Copyright 2010, American Chemical Society.)

Oxidation tends to form an O domain or O chain on the BNNSs, which may result in unzipping of BNNSs along zigzag direction (Zhao et al. 2012a).

Single-sided covalent addition: BNNSs with hydrogenation occurring on only one side have been investigated. For example, BNNSs semi-hydrogenated on only B atoms were predicted to be a ferromagnetic half-metal. They were more energetically favorable than those semi-hydrogenated on only N atoms because of the difference in electronegativity between B and N atoms (Wang 2010b). Recently, with a systematic computational examination, Li et al. (2013a) found that semi-hydrogenated BNNSs have a bandgap of 2.7 eV corresponding to blue light absorption.

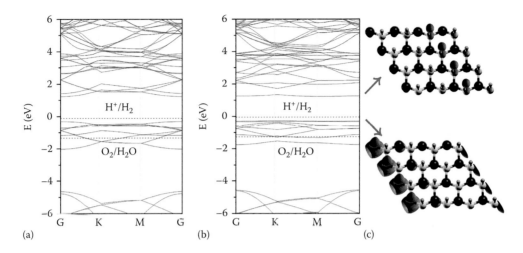

Figure 14.11 The electronic band structure for graphitic BNNSs semi-hydrogenated on nitrogen sub-lattice under (a) antiferromagnetic state and (b) strip-like antiferromagnetic state with HSE06 functional. The Fermi level is set to zero. The redox potentials of water splitting are shown by dashed lines. (c) The charge density of valence band (the lower part) and conduction band (the upper part) under strip-like antiferromagnetic state with isovalue 0.01 e/Å^3. (From Li, X. et al., *Sci. Rep.*, 3, 1858, 2013a. With permission.)

They thus proposed that semi-hydrogenated BNNSs are promising visible-light-driven metal-free photocatalysts for water splitting (Figure 14.11).

For half-fluorinated BNNSs, F energetically prefers binding to B atoms rather than N atoms. Under external strain, such half-fluorinated BNNSs exhibit interesting magnetic transitions between ferromagnetism and antiferromagnetism; especially, exerting 6% compression drives the half-fluorinated BNNSs into a half-metal. These phenomena were attributed to both through-bond and π–π direct interactions (Ma et al. 2011).

Noncovalent modifications: Few theoretical reports have been dedicated to noncovalent interactions between BNNSs and functional molecules with planar or aromatic moieties (Lin et al. 2011c, Tang et al. 2011, Bermúdez-Lugo et al. 2012, Zhao et al. 2012b).

Computations by Tang et al. demonstrated that both an electron acceptor (tetracyanoquinodimethane, TCNQ) and a donor (tetrathiafulvalene, TTF) can be noncovalently bonded onto BNNSs or BNNRs (Figure 14.12) (Tang et al. 2011). The optimized functional molecule–BNNS distance (3.3–3.7 Å) is close to the typical BN layer distance. The interfacial charge transfer between TCNQ or TTF and BNNSs or BNNRs significantly reduces the intrinsic wide bandgap of pristine BNNSs and consequently results in a *p*- or *n*-type semiconductor, respectively (Tang et al. 2011).

Duan and coworkers found that all five types of the deoxyribonucleic/ribonucleic acid (DNA/RNA) nucleobases energetically prefer to facially stack on BNNSs (Figure 14.13) (Lin et al. 2011c). The anions (N and O atoms) of nucleobases prefer to locate themselves above B atoms in BNNSs as much as possible with adsorption energies ranging from 0.5 to 0.69 eV. Unlike the interaction between nucleobases and graphene (Zheng et al. 2003, Tu and Zheng 2008, Ahmed et al. 2012) or CNT (Niyogi et al. 2002, Sun et al. 2002, Banerjee et al. 2005, Tasis et al. 2006) with π–π interaction, the interaction between DNA/RNA and BNNSs is governed by mutual polarization

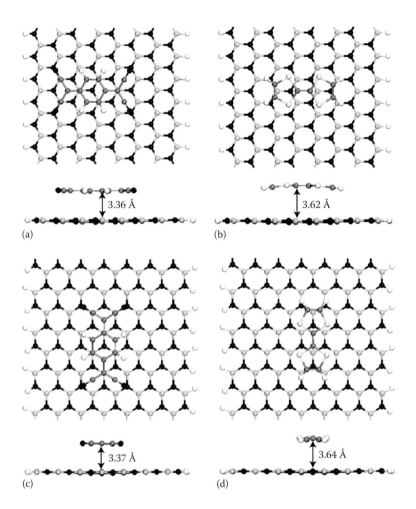

Figure 14.12 Optimized geometries of the TCNQ-doped zigzag (a) and armchair (c) BNNRs, as well as the TTF-doped zigzag (b) and armchair (d) BNNRs. (Reprinted with permission from Tang, Q., Zhou, Z., and Chen, Z., Molecular charge transfer: A simple and effective route to engineer the band structures of BN nanosheets and nanoribbons, *J. Phys. Chem. C*, 115, 18531–18537, 2011. Copyright 2011, American Chemical Society.)

due to the ionic nature of BN bonds. The fundamental properties of DNA/RNA on BNNSs are robust, thus BNNSs are promising template for biological applications that use such noncovalent chemistry.

CO$_2$ molecules can weakly adsorb on various BN nanostructures including BNNSs and BNNTs. Interestingly, the interaction between CO$_2$ molecules and BN nanostructures can be significantly enhanced when extra electrons are introduced to the BN surface (Figure 14.14) (Sun et al. 2013). Once the electrons are removed, CO$_2$ molecules can spontaneously desorb without any reaction barrier. Thus, CO$_2$ capture/release can be well controlled via charging/discharging the BN nanomaterials, demonstrating that BN nanomaterials are excellent substrates for reversible capture and release of CO$_2$.

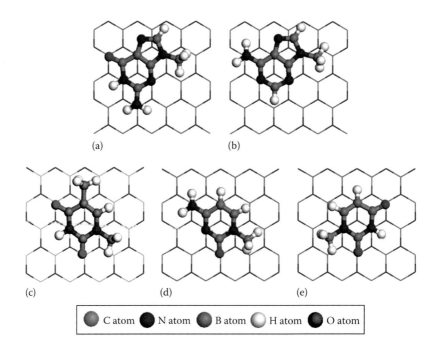

Figure 14.13   Optimized configurations of the five nucleobases adsorbed on a BNNS: (a) guanine (G), (b) adenine (A), (c) thymine (T), (d) cytosine (C), and (e) uracil (U). (Lin, Q., Zou, X., Zhou, G., Liu, R., Wu, J., Li, J., and Duan, W., Adsorption of DNA/RNA nucleobases on hexagonal boron nitride sheet: An ab initio study, *Phys. Chem. Chem. Phys.*, 13, 12225–12230, 2011c. Reproduced by permissions of The Royal Society of Chemistry.)

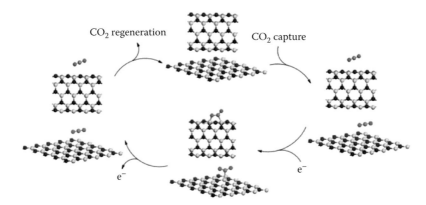

Figure 14.14   Charge-controlled CO_2 capture and release on BNNSs. (Reprinted with permissions from Sun, Q., Li, Z., Searles, D.J., Chen, Y., Lu, G., and Du, A., Charge-controlled switchable CO_2 capture on boron nitride nanomaterials, *J. Am. Chem. Soc.*, 135, 8246–8253, 2013. Copyright 2013, American Chemical Society.)

14.3 EXPERIMENTAL INVESTIGATIONS OF BNNS CHEMISTRY

Strictly speaking, several exploratory experimental studies on BNNS chemistry were carried out on h-BN, the parent material of BNNSs. These chemistry methods, including the direct solvent dispersion and conjugation with Lewis bases, were used as a means to exfoliate h-BN into nanosheets. The dispersed nanosheets were subsequently used for further modification such as conjugation with inorganic or biological molecules and covalent functionalization, which are the *true chemistry* of BNNSs.

14.3.1 Direct Liquid Dispersion: Solvents and Melts

Analogous to the *Scotch tape method* in a physicist's lab, the most direct means to obtain exfoliated BNNSs in a wet chemistry lab is to use a polar solvent and a water bath sonicator. Zhi et al. first reported the sonication-assisted exfoliation process of h-BN in N,N'-dimethylformamide (DMF) to obtain few-layered BNNSs (Zhi et al. 2009). DMF is a common polar aprotic organic solvent with high boiling point (153°C). It has been widely used in the wet fabrication of polymer composites and, importantly, the successful dispersion and processing of carbon allotropes including carbon nanotubes (Bergin et al. 2008) and graphene (Hernandez 2008). It has been proposed that the polarity of DMF molecules, accompanied by sonication, induces strong interactions with h-BN surface, which overcame the van der Waals forces between BN layers and leads to exfoliation. The solvent polarity effect has been investigated in more detail using both Fourier transform infrared spectroscopy (FT-IR) and theoretical calculations (Lian et al. 2011). It has been confirmed that the interactions of h-BN surface and DMF molecules are predominately the charge attraction of O atoms from the DMF to the B atoms on the h-BN surface in terms of the Lewis acid–base interactions.

Similar to DMF, many other polar organic solvents are also useful for the exfoliation of h-BN for BNNS dispersions. Among these, N-methyl-2-pyrrolidone (NMP), N,N'-dimethylacetamide (DMAc), 1,2-dichlorobenzene (DCB), and ethylene glycol (EthGly) were found to be quite comparable to or even more efficient than DMF in exfoliating h-BN into dispersed BNNSs (Figure 14.15) (Lin and Connell 2012a). Conversely, nonpolar solvents exhibited essentially no ability to disperse BNNSs. These results further support the qualitative solvent polarity effect theory in the h-BN exfoliation and dispersion.

Coleman et al. (2011) interpreted the same phenomenon using the more quantitative Hansen solubility parameter theory. According to this theory, the energy required to exfoliate any low-dimensional nanomaterials should match the solvent–solute interaction energy, which is the *surface energy* of the material (Hernandez 2008). The ideal surface energy for direct exfoliation of h-BN should be around 65 mJ/m^2; thus, isopropanol and NMP were considered to be two of the best solvents. Guided by this theory, the researchers successfully prepared exfoliated 2D nanosheets of various chemical compositions, including not only graphene (Hernandez 2008) and h-BN but also molybdenum disulfide (MoS$_2$) and tungsten disulfide (WS$_2$), among others (Coleman et al. 2011).

Compared to the use of individual solvents, Zhou et al. (2011a) proposed a mixed-solvent strategy for h-BN, MoS$_2$, and WS$_2$ using an ethanol/water mixture. The mixed solvent showed improved dispersion and exfoliation capability due to the near match of the surface energy and the required exfoliation energy.

Generally, water is not considered as an effective solvent to exfoliate BN. It has been suggested that BN nanomaterials are hydrophobic at room temperature due to their resistance against

Figure 14.15 (a) Pictures of direct organic dispersions of BNNSs from simple bath sonication right after preparation. The solvents from left to right are NMP, DMAc, DCB, DMF, EthGly, ethanol (EtOH), isopropanol (IPA), acetone, tetrahydrofuran (THF), chloroform (CHCl₃), ethyl acetate (EtAc), methanol (MeOH), dimethyl sulfoxide (DMSO), toluene, and hexanes. (b) The same dispersions were irradiated with a laser beam from the left side of each vial to visually indicate the Tyndall effect (more intense reflection suggesting higher concentration). (c) Pictures were taken after ~1 week (days 7–9). (Lin, Y. and Connell, J.W., Advances in 2D boron nitride nanostructures: Nanosheets, nanoribbons, nanomeshes, and hybrids with graphene, *Nanoscale*, 4, 6908–6939, 2012a. Reproduced by permission of The Royal Society of Chemistry.)

moisture (Dutta et al. 2011) and insolubility in water (Li et al. 2008, Lin et al. 2011b, Pakdel et al. 2011, Wang et al. 2011d). In addition, according to the Hansen–Coleman model discussed earlier, the much higher surface energy of water compared to polar organic solvents should have made it very unlikely to exfoliate and disperse low-dimensional nanomaterials in general.

This prediction however is not accurate. Our experimental studies showed that h-BN could be effectively exfoliated by water, resulting in soluble BNNSs with as-prepared concentrations up to 0.2 mg/mL by simple sonication (Lin et al. 2011b). A side effect was that some dispersed nanosheets become much smaller in lateral size compared to those obtained from organic polar solvents under very similar sonication conditions (Figure 14.16). It was proposed that the hydrolysis of the B–N network was the main cause for the nanosheet size reduction, yielding oxidized boron with ammonia evolution. Consistently, dissolved trace ammonia was detected in the BNNS dispersions. In addition, hydroxyl groups were found in the final BNNS samples but not in the pristine h-BN. These water-dispersed BNNSs were quite stable in both neutral and basic conditions, but quickly precipitated out from acidic solutions. Moreover, Wang et al. found that methylsulfonic acid (MSA), a strong protic acid also with high surface energy, was quite effective in exfoliating and dispersing BNNSs (Wang et al. 2011d). These seemingly contradictory observations warrant further theoretical and mechanistic investigations.

A related method to exfoliate h-BN is to use molten hydroxides, such as a mixture of NaOH and KOH (Li et al. 2013b). The exfoliation process was carried out at 180°C in an autoclave. The hydroxides, being solids at room temperature, melted into liquid phase under the experimental conditions, making the system somewhat similar to that using conventional solvents as discussed earlier. It was proposed that the cations are first adsorbed on the outermost h-BN layers of a thick h-BN crystal causing partial curling of the layers (Figure 14.17), then the hydroxyl anions adsorbed onto the opposite side of the same curled layer, further driving the curling and exfoliation. This mechanism repeatedly occurred on the newly exposed surfaces. As such, h-BN

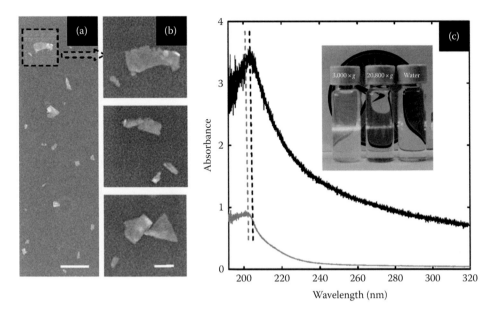

Figure 14.16   (a) and (b) Scanning electron microscopy (SEM) images (at lower and higher magnifications, respectively) for an aqueous BNNS dispersion obtained from further centrifugation at $20,800 \times g$. Scale bars are 200 and 50 nm, respectively. (c) Overlay of the optical absorption spectra of BNNS aqueous dispersions (from an 8 h sonication reaction) upon centrifugation at $3000 \times g$ for 15 min (black, acquired at 1/10 dilution and scaled back to avoid signal oversaturation) and further at $20800g$ for 30 min (gray, acquired directly). The bandgap peak positions are highlighted by the dashed lines of the respective color. Shown in the inset are BNNS dispersions (from a 24 h sonication reaction) upon centrifugation at $3,000 \times g$ for 15 min and further at $20,800 \times g$ for 30 min (middle). Pure water (right) is shown for comparison. Laser light shone from right, showing the Tyndall effects of the nanosheet dispersions. (Lin, Y., Williams, T.V., Xu, T.-B., Cao, W., Elsayed-Ali, H.E., and Connell, J.W., Aqueous dispersions of few-layered and monolayered hexagonal boron nitride nanosheets from sonication-assisted hydrolysis: Critical role of water, *J. Phys. Chem. C*, 115, 2679–2685. Copyright 2011, American Chemical Society.)

was peeled into BNNSs in a layer-by-layer fashion. Similar to the case using water, the molten hydroxide–induced exfoliation was also accompanied by the cutting of the nanosheets due to the reaction between h-BN and hydroxyl anions, yielding ammonia and boron oxide, even though no sonication was used.

The BNNSs from direct liquid exfoliation and dispersion are highly crystalline. Note that the surfaces of BNNSs can be considered impurity-free, allowing for further chemical modifications as discussed in some of the following sections.

14.3.2   Noncovalent Chemistry: Surfactants and Polymers

Using surfactants or polymers to induce dispersibility of various nanomaterials has been well studied, especially in the past decade. Most surfactants, or *surface-active agents*, are for aqueous dispersion of hydrophobic nanomaterials (Islam et al. 2003, Lotya et al. 2009). The surfactant molecules usually contain a hydrophilic *head* and a long lipophilic *tail*. During the dispersion process, which is often assisted by sonication, the long lipophilic tail adsorbs onto the hydrophobic surface of the nanomaterial, with the hydrophilic head pointing toward water environment, forming a half micelle-like

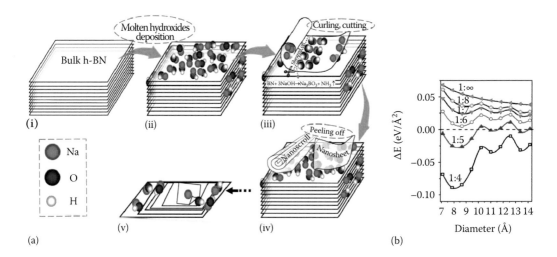

Figure 14.17 Illustration of the h-BN exfoliation mechanism using molten hydroxides. (a) Schematic representation of the exfoliating process following the sequence: (ii) deposition of hydroxides on h-BN, (iii) peripheral self-curling of sheets, (iv) insertion of hydroxides, (v) cut by the reaction of the reactants, and exfoliation of sheets. (b) The difference between the self-curling energy (ΔE) of Na^+ and OH^- adsorbed on a curved BN monolayer and the interlayer binding energy. The numbers above the lines indicate the coverage of Na^+ and OH^- on the BN monolayer represented by number of ions per five BN pairs on one surface. Negative ΔE values mean the self-curling process is energetically preferred. (Li, X., Hao, X., Zhao, M., Wu, Y., Yang, J., Tian, Y., and Qian, G.: Exfoliation of hexagonal boron nitride by molten hydroxides. *Adv. Mater.* 2013b. 25. 2200–2204. Copyright Wiley-VCH Verlag GmbH & Co. KGaA. Reproduced with permission.)

structure. The conventional use of surfactants to remove lipophilic waste using water is a physical process. For nanomaterial dispersion, however, the direct surfactant adsorption to the nanostructural surface is much more intimate in nature and may be categorized as *noncovalent interactions*.

In a comprehensive study dedicated mostly to MoS_2 exfoliation, the use of sodium cholate as an effective surfactant to obtain an aqueous dispersion of exfoliated BNNS was also briefly reported (Figure 14.18) (Smith et al. 2011). Sodium cholate was previously shown to be quite useful for the exfoliation of SWNTs (Moore et al. 2003) and graphite (Lotya et al. 2010). Synthetic surfactants are also viable options. For instance, by using pyrene-modified hyaluronan, Zhang et al. successfully exfoliated graphene, BNNSs, and MoS_2 (Zhang et al. 2013). Hyaluronan is a hydrophilic natural polysaccharide, while pyrene is a large aromatic molecule that can strongly interact with graphitic surfaces via π–π interactions (Chen et al. 2001). Therefore, the pyrene-modified hyaluronan behaved just like a surfactant molecule since pyrene moieties presumably attached onto BNNS surfaces via similar noncovalent interactions.

Interactions of the π–π type are different from hydrophobic interactions in that the former can be equally applied in nonaqueous systems. In fact, the first report on the exfoliation and dispersion of BNNSs used a conjugated polymer, namely, poly (*m*-phenylenevinylene-co-2, 5-dictoxy-*p*-phenylenevinylene), for their dispersion in dichloroethane (Han et al. 2008). This polymer belongs to a family well known as PPV and has been extensively employed to directly disperse SWNTs (Hirsch 2002, Tasis et al. 2006), graphene (Zheng et al. 2003, Zhu et al. 2010), and BNNTs (Zhi et al. 2005) via π–π interactions due to the highly conjugated structure of the polymers.

Linear polymers are also capable of exfoliating and dispersing BNNSs. The dispersion mechanism has been described as steric stabilization, in which the polymer chains partially adsorb

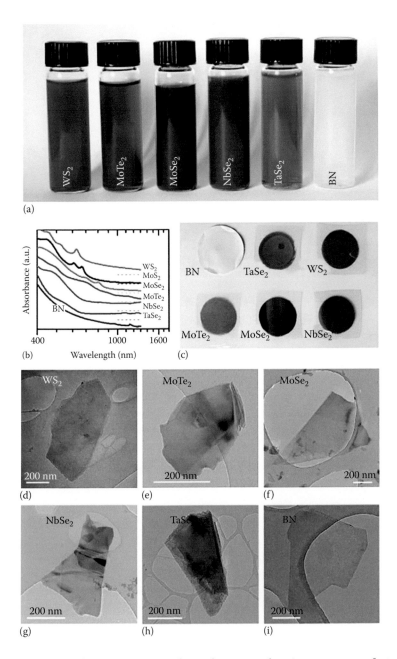

Figure 14.18 Dispersion of various inorganic layered compounds using aqueous surfactant solutions. (a) Photograph of dispersions of WS$_2$, molybdenum ditelluride (MoTe$_2$), molybdenum diselenide (MoSe$_2$), niobium diselenide (NbSe$_2$), tantalum diselenide (TaSe$_2$), and BN all stabilized in water by sodium cholate. (b) Absorption spectra of dispersions shown in (a). The spectra are vertically shifted for clarity. (c) Vacuum-filtered thin films of BN, TaSe$_2$, WS$_2$, MoTe$_2$, MoSe$_2$, and NbSe$_2$ (note that due to its brittleness, the BN film is shown supported by a porous cellulose membrane). (d through i) Transmission electron microscopy (TEM) images of flakes deposited on TEM grids from the dispersions in (a).   (*Continued*)

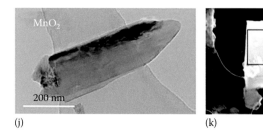

(j)   (k)

Figure 14.18 (Continued)   Dispersion of various inorganic layered compounds using aqueous surfactant solutions. (j) TEM image of a manganese dioxide (MnO_2) flake stabilized in water using sodium cholate. Such flakes were both exfoliated from an MnO_2 nanoparticulate powder where flakes were found as a minority phase. (k) An SEM image of an MnO_2 flake on a TEM grid. Energy dispersive x-ray spectral analysis, taken in the region marked by the box, confirmed the composition of this flake to be very close to MnO_2. (Smith, R.J., King, P.J., Lotya, M., Wirtz, C., Khan, U., De, S. et al.: Large-scale exfoliation of inorganic layered compounds in aqueous surfactant solutions. *Adv. Mater.* 2011. 23. 3944–3948. Copyright Wiley-VCH Verlag GmbH & Co. KGaA. Reproduced with permission.)

onto BNNSs while also partially extending into the solvent environment. May et al. studied the dispersions of several 2D nanomaterials including BNNSs with the assistance of a variety of polymers, such as poly(butadiene) (PBD), poly(styrene-co-butadiene) (PBS), polystyrene (PS), poly(vinyl chloride) (PVC), poly(vinyl acetate) (PVAc), polycarbonate (PC), poly(methyl methacrylate) (PMMA), poly(vinylidene chloride) (PVDC), and cellulose acetate (CA) (May et al. 2012). The solvent systems used were tetrahydrofuran (THF) and cyclohexanone (CXO). By comparing with the Hildebrand theory, they concluded that the matching of solubility parameters of the nanomaterial (i.e., BNNSs and others), polymer, and the solvent was critical to achieve good dispersion (Table 14.1). Similarly, using poly(vinyl alcohol), Coleman and coworkers obtained

TABLE 14.1   POLYMERS USED IN THE STUDY BY MAY ET AL. AND THEIR HILDEBRAND SOLUBILITY PARAMETERS[a]

Polymer	Solubility Parameter ($MPa^{1/2}$)	Graphene/THF $\langle C \rangle$ ($\mu g/mL$)	BN/THF $\langle C \rangle$ ($\mu g/mL$)	MoS_2/THF $\langle C \rangle$ ($\mu g/mL$)	Graphene/CXO $\langle C \rangle$ ($\mu g/mL$)
PBD	17.2	17	8	17	68
PBS	17.7	17	11	23	79
PS	18.1	16	17	29	110
PVAc	19.3	22	3	32	119
PVC	19.8	6	34	33	92
PC	21.9	—	—	—	129
PMMA	22.7	20	13	28	141
PVDC	25.0		2	20	
CA	26.7	1	10	—	126

Source: Solubility parameters were taken from the Polymer Handbook. Brandrup, J., Immergut, E.H., Grulke, E.A., Abe, A., Bloch, D.R., *Polymer Handbook*, 4th edn, John Wiley & Sons, Norvich, NY, 2005. Reproduced with permission Copyright 2012, American Chemical Society.

[a] Also shown are the mean concentrations for each dispersion.

exfoliated BNNSs dispersed in water (Khan et al. 2013). Polymeric composites containing these polymer-exfoliated BNNSs exhibited much enhanced mechanical properties in comparison to the neat polymers, similar to those containing graphene nanofillers (Kuilla et al. 2010).

14.3.3   Ionic Chemistry: Lewis Acid–Base Interactions

Different from the conjugated aromatic carbon graphitic structures, the B–N bond network for BN nanomaterials exhibits localized ionic characteristics with a denser electron cloud on the N atoms. Therefore, it is possible to use Lewis base molecules to attach to the electron-deficient B atoms or Lewis acid molecules to attach to electron-rich N atoms. The former has been studied in some detail for BNNTs (Xie et al. 2005, Wu et al. 2006, Pal et al. 2007, Ikuno et al. 2007, Maguer et al. 2009). Similarly, we functionalized h-BN with Lewis bases such as octadecylamine (ODA) and an amine-terminated oligomeric polyethylene glycol (PEG) (Lin et al. 2010b). The functionalization induced exfoliation of the layered structure and yielded BNNSs with mostly few layers (Figure 14.19). The successful exfoliation indicated that the intercalation of the functional molecules in between h-BN layers might have occurred during functionalization. The functionalized BNNSs were soluble in organic solvents and/or water, depending upon the functionalities used. Importantly, nuclear magnetic resonance (NMR) results revealed that the amine end of the functional groups was adjacent to the nanosheet surface. This strongly supported the notion that the mode of functionalization was Lewis acid (B atoms)—base (amino group) interactions. As seen from the atomic force microscopy images (Figure 14.19), the functional molecules nearly covered the entire BNNS surfaces, also consistent with the earlier mechanism.

We also intentionally introduced defects to the highly crystalline h-BN surface by mechanical ball-milling (Lin et al. 2010b). The presence of defects significantly enhanced the h-BN functionalization efficiency. The solubilization ratio of h-BN solid after 60 min ball-milling into THF after the functionalization reaction with ODA improved to 40% compared to the 10% uptake for nontreated h-BN (Figures 14.20 and 14.21). Note that the ball-milling conditions used were quite aggressive; the resulting BNNSs were crumpled. A few other reports suggested that, by using different ball-mill equipment, shear force might be more dominant than impact force, which could yield BNNSs with larger lateral sizes and less defects (Ghosh et al. 2008, Li et al. 2011).

Similar Lewis acid–base reactions were demonstrated using the bottom-up produced BNNSs reported by Nag et al. (2010). These high–surface area BNNSs were prepared in a high-temperature reaction between boric acid and urea. While the pristine BNNSs were found to be insoluble in toluene and other nonpolar solvents, sonication of the suspension mixtures with the presence of Lewis bases, such as trioctylamine or trioctylphosphine, resulted in clear dispersions.

14.3.4   Covalent Chemistry: Basal Plane Functionalization and Others

Direct modification of the B–N bond network requires highly reactive species. For example, Coleman and coworkers reported the use of nitrene radicals (Sainsbury et al. 2012a) and oxygen radicals (Sainsbury et al. 2012b) to directly modify the basal planes of BNNSs. In both cases, the electron-rich radicals attacked the electron-deficient B atoms. Different from the Lewis acid–base ionic reactions where the B–N bond network was intact, it was proposed that, because the radical species were highly reactive, the adjacent B–N bond might have been cleaved. The functionalized BNNSs were soluble in various solvents and could be further processed toward specific applications. For example, in the oxygen radical reaction (Figure 14.22), t-butoxy radical was used to attach to the B atoms. The resultant boronate ester was further hydrolyzed into hydroxyl groups.

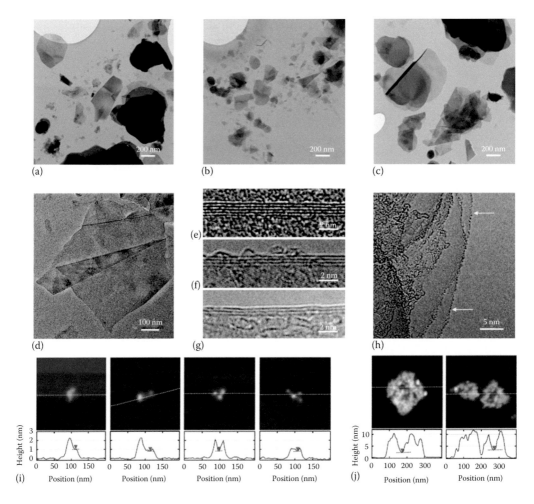

Figure 14.19 Typical low-magnification TEM images of (a) a PEG–BN sample, (b) the supernatant, and (c) the residue from the same sample subjected to further centrifugation at ~10,000g; (d) a TEM image of a BNNS in the same PEG–BN sample and (e) the high resolution-TEM (HR-TEM) image of a folded line of the nanosheet in (d); (f, g) HR-TEM images of some other BNNS folded edges; (h) a possible single layer (marked by arrows) extruded from a thicker structure; (i) atomic force microscopy (AFM) images (dimensions: 200 × 200 nm) and the corresponding height profiles of possibly monolayered functionalized BNNS species; and (j) AFM images (dimensions: 400 × 400 nm) and the corresponding height profiles of several few-layered functionalized BNNS species. (Lin, Y., Williams, T.V., Cao, W., Elsayed-Ali, H.E., and Connell, J.W., Defect functionalization of hexagonal boron nitride nanosheets, *J. Phys. Chem. C*, 114, 17434–17439. Copyright 2010b, American Chemical Society.)

These soluble covalently functionalized BNNSs were directly incorporated with various common polymer matrices, such as polycarbonate and poly(vinyl alcohol), for enhanced mechanical properties. The hydroxyl groups were further modified using a diisocyanate compound in order to improve compatibility of BNNSs with polyurethane. Subsequent testing of composites embedded with hydroxyl-modified BNNSs demonstrated the mechanical improvement in all aspects, including Young's modulus, ultimate strength, elongation at break, and toughness.

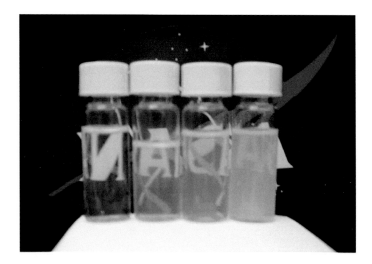

Figure 14.20 Photograph of as-extracted dispersions of ODA-functionalized BNNS samples started from (from left to right) pristine h-BN and h-BN ball-milled for 10, 30, and 60 min, respectively. (Lin, Y., Williams, T.V., Cao, W., Elsayed-Ali, H.E., and Connell, J.W., Defect functionalization of hexagonal boron nitride nanosheets. *J. Phys. Chem. C*, 114 17434–17439. Copyright 2010b, American Chemical Society.)

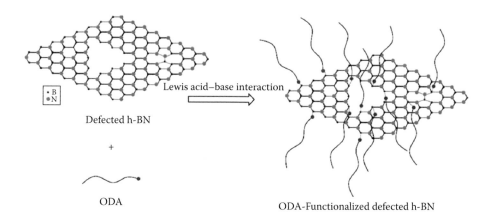

Figure 14.21 Functionalization of defected h-BN. (Lin, Y., Williams, T.V., Cao, W., Elsayed-Ali, H.E., and Connell, J.W., Defect functionalization of hexagonal boron nitride nanosheets, *J. Phys. Chem. C*, 114, 17434–17439. Copyright 2010b, American Chemical Society.)

As discussed in Section 3.1, the attachments of hydroxyl groups onto BNNSs could be achieved by direct sonication in water (Coleman et al. 2011). Alternatively, the use of various highly reactive chemicals also resulted in oxygen-containing functional groups on BNNSs. For example, Nazarov et al. (2012) reported the functionalization and exfoliation of h-BN using hydrazine and various strong oxidants such as H_2O_2, H_2SO_4/HNO_3 mixture, and oleum. The reactions were carried out in an autoclave at elevated temperature of 100°C–150°C. The functionalization induced exfoliation and solubilization of h-BN, resulting in BNNSs soluble in both water and DMF. Although h-BN

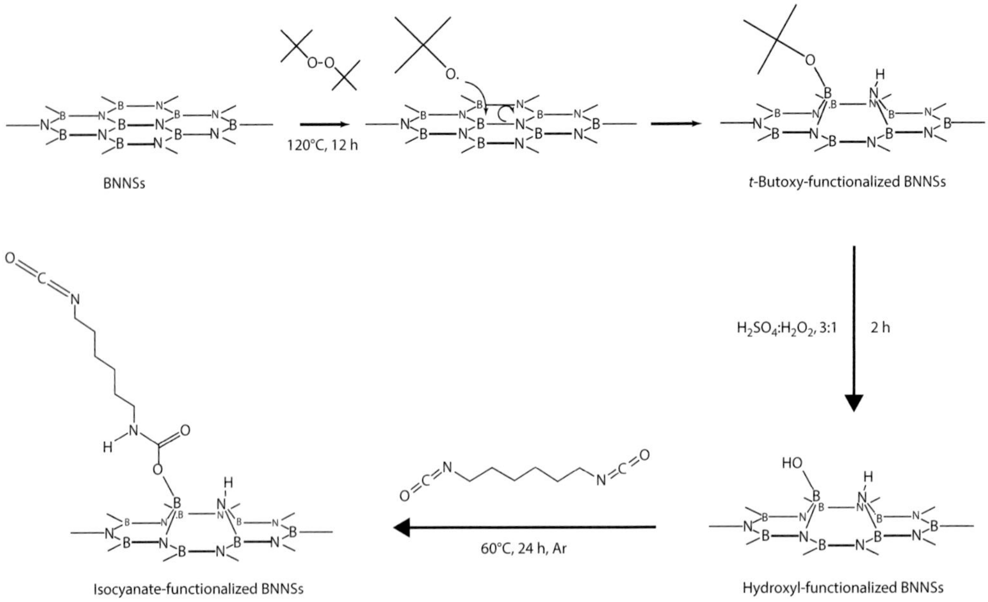

Figure 14.22 Covalent functionalization of BNNSs using oxygen radical and subsequent modifications. (Sainsbury, T., Satti, A., May, P., Wang, Z., McGovern, I., Gun'ko, Y.K., and Coleman, J., Oxygen radical functionalization of boron nitride nanosheets, *J. Am. Chem. Soc.*, 134, 18758–18771. Copyright 2012b, American Chemical Society.)

is known to be resistant to oxidation, the reported conditions were harsh enough to introduce ~6–7 wt% of oxygen into the exfoliated BNNS products. For the reaction of h-BN with hydrazine, it was proposed that the hydroxyl radicals from hydrazine decomposition under the autoclave conditions might be responsible for the functionalization. For the reaction of h-BN with H_2O_2, FT-IR of the exfoliated BNNS product showed the presence of hydroxyl groups as a result of oxidation. Interestingly, these hydroxyl-functionalized BNNSs exhibited very similar optical absorption and bandgap characteristics to the BNNSs obtained from direct water sonication. It was thus likely that the hydroxyl groups were preferentially attached to the nanosheet edges, with lateral surfaces remaining largely intact.

In another report, Yu et al. (2012) showed the covalent attachment of γ-aminopropyl-triethoxysilane to the hydroxyl groups that presumably already existed on h-BN. A hyperbranched aromatic polyamide was subsequently attached to the amine end of the short functional *bridge* (Figure 14.23). The functionalized exfoliated BNNSs were then used in polymeric nanocomposites, which showed improved nanofiller dispersion and, as a result, enhanced thermal stability and thermal conductivity in comparison to the neat epoxy matrix.

14.3.5 Decoration with Inorganic Nanoparticles

With high percentage of surface atoms, nanoparticles (NPs) present many unique properties. Compared to free-standing NPs, those supported on nanoscale substrates, such as nanotubes, nanowires, and nanosheets, often exhibit synergistically enhanced optical, electronic, electrical, thermal, and many other properties that are useful in a variety of applications (Wildgoose et al. 2006, Georgakilas et al. 2007, Kamat 2009, Bai and Shen 2012).

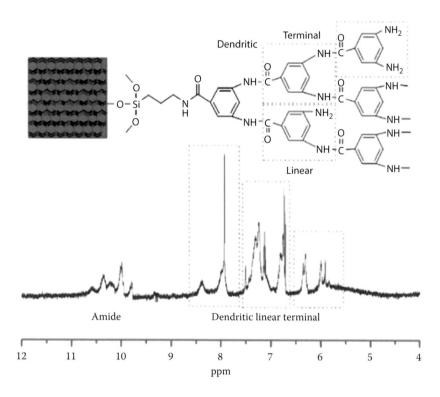

Figure 14.23 ¹H nuclear magnetic resonance (NMR) spectrum of hyperbranched polyamide-functionalized BNNS (structure also shown) in DMSO-d_6. (Reprinted from *Polymer*, 53, Yu, J., Huang, X., Wu, C., Wu, X., Wang, G., and Jiang, P., Interfacial modification of boron nitride nanoplatelets for epoxy composites with improved thermal properties, 471–480, 2012, Copyright 2011, with permission from Elsevier.)

Several groups have prepared metal–BNNS conjugates. Metallic NPs can be formed in situ by reduction or decomposition of precursors in the presence of BNNSs. For example, Wang et al. decorated gold (Au) and platinum (Pt) NPs on bottom-up synthesized BNNSs (Wang et al. 2011a). The decorations of Pt and Au on BNNSs were carried out using different synthetic processes. For the former, hydrogen was used to reduce H_2PtCl_6-impregnated BNNSs in solid state. For the latter, urea was used to reduce $HAuCl_4$ in the presence of BNNSs in an aqueous suspension. To enhance the Au–BNNS interaction, the nanosheets were pretreated with hydrogen peroxide to introduce functional groups. The Au NPs thus obtained were of smaller sizes (3.3 vs. 7.7 nm with nontreated BNNSs).

Lin et al. (2012b) recently anchored silver (Ag) NPs onto BNNSs in homogeneous aqueous dispersions. Ag NPs were formed by the reduction of silver acetate using hydrazine in the presence of water-exfoliated BNNSs. The adding sequence of the metal precursor and the reducing agent resulted in quite different morphologies of metal NPs, and thus different optical properties of the Ag-BNNS conjugates (Figure 14.24).

In a different approach, Singhal et al. (2013) prepared Au NPs separately by using a conventional reduction procedure, and then transferred the aqueously dispersed NPs into an organic phase with ODA as the stabilizer. The ODA-stabilized Au NPs were sonicated with a BNNS dispersion in DMF and allowed to incubate over time, resulting in Au–BNNS conjugates

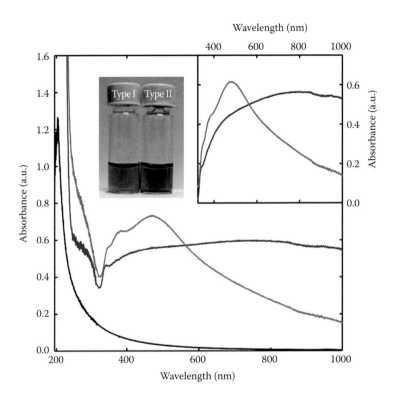

Figure 14.24 Optical absorption spectra and the corresponding photographs of type I (left; a dark-colored solution adding silver acetate solution to BNNS-hydrazine mixture) and type II (right; a dark gray-colored solution; adding hydrazine solution to BNNS-silver acetate mixture) Ag–BNNS nanohybrids in as-prepared aqueous dispersions. The spectrum of starting BNNS aqueous dispersion (diluted to the same equivalent BNNS concentration as in the Ag–BNNS dispersions) is also shown for comparison. Shown in the inset are the same spectra but subtracted with those of BNNS–hydrazine (Type I) and BNNS–silver acetate (Type II) starting solution mixtures, respectively, before the addition of the other reagent. (Lin, Y., Bunker, C.E., Fernando, K.A.S., and Connell, J.W., Aqueously dispersed silver nanoparticle-decorated boron nitride nanosheets for reusable, thermal oxidation-resistant surface enhanced Raman spectroscopy (SERS) devices, *ACS Appl. Mater. Interfaces*, 4, 1110–1117. Copyright 2012b, American Chemical Society.)

(Figure 14.25). Using more Au NPs resulted in a higher coverage on the BNNS surface. The conjugates exhibited enhanced electrical conductivity due to the percolation of the metallic NPs.

The decoration of metal or even metal oxide NPs onto BNNSs could also be achieved in solid-state reactions as demonstrated by Lin et al. (2009a,b, 2011a). Such a process is expected to be equally applicable to bottom-up synthesized or exfoliated BNNSs. Typically, an organic metal salt precursor was thermally decomposed in a solid mixture with h-BN without any reducing reagent, forming NPs dispersed on the sheet surface. The thermal energy for metal salt decomposition could be in the form of conventional oven heating (Lin et al. 2009b), ball-milling (Lin et al. 2009a), or microwave heating (Lin et al. 2011a). This versatile method was demonstrated more systematically using carbon nanosubstrates including carbon nanotubes and graphene.

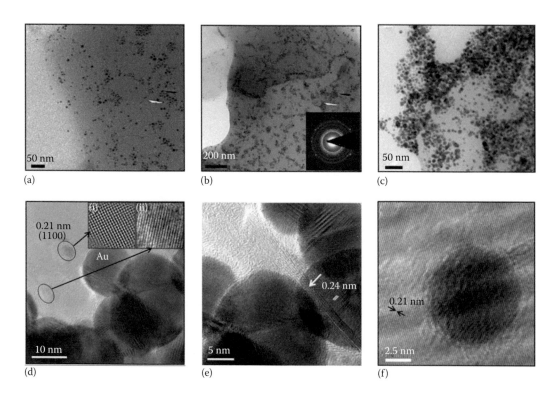

Figure 14.25 TEM images of Au-decorated BNNSs at different concentrations of BNNSs and Au: (a) 1:2, (b) 1:6, and (c) 1:10. (d) HR-TEM images showing the BN lattice (with outline); the inset shows the corresponding inverse fast Fourier transform (IFFT) image and the Au nanoparticles on and under the sheet. (e) Image showing the multilayer sheets and the Au nanoparticles with twin defects. (f) HR-TEM image elucidating the Au nanoparticle beneath the lattice-resolved single BNNS. (Singhal, S.K., Kumar, V., Stalin, K., Choudhary, A., Teotia, S., Reddy, G.B. et al.: Gold-nanoparticle-decorated boron nitride nanosheets: Structure and optical properties. *Particle Particle Syst. Charact.* 2013. 30. 445–452. Copyright Wiley-VCH Verlag GmbH & Co. KGaA. Reproduced with permission.)

Generally, while free-standing NPs often intend to aggregate, the presence of the nanosubstrate allows the individual dispersion of the NPs and makes them more stable against various functioning processes. Catalysis is one of the most popular applications that can benefit from the nanoscale dispersions of NPs. For example, a recent DFT study suggested that Au and Au_2 clusters attached on pristine and defective BNNSs have enhanced adsorption and catalytic activation ability of the molecular oxygen (Gao et al. 2012). The experimentally prepared Pt–BNNS and Au–BNNS conjugates discussed earlier were both found useful in catalyzing the oxidation of CO to CO_2 (Figure 14.26) (Wang 2011a).

Nanosubstrates with unique properties may also further benefit the performance of the NPs in specific applications. For instance, taking advantage of the unique oxidation resistance of BNNSs, Lin et al. demonstrated reusable surface-enhanced Raman spectroscopy (SERS) sensors in which Ag NPs were the active sensing component (Lin et al. 2012b). The Ag–BNNS conjugates from the in situ reduction method discussed earlier are water soluble; thus, it is easy to use solution processing techniques to form a thin film supported on quartz. The Ag–BNNS thin films were robust against repeated solvent washing and short heat pulses up to ~400°C.

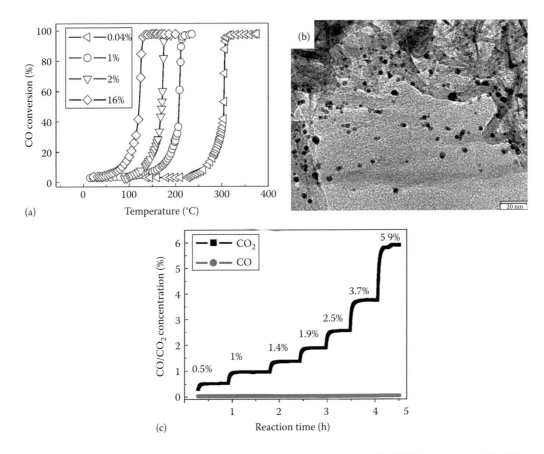

Figure 14.26 (a) The conversion of CO over temperatures by 40 mg Pt/BNNSs catalyst with different loadings (0.04%, 1%, 2%, and 16%) in the gas flow (1% CO, 10% O_2, and 89% N_2). (b) TEM image of 1% Pt/BNNSs. (c) The catalytic performance of 1% Pt/BNNSs with different CO concentrations from 0.5% to 5.9% at a flow rate of 66.7 mL/min at 250°C. The O_2 concentration was fixed at 10%. (Wang, L., Sun, C., Xu, L., and Qian, Y., Convenient synthesis and applications of gram scale boron nitride nanosheets, *Catal. Sci. Technol.*, 1, 1119–1123, 2011a. Reproduced by permission of The Royal Society of Chemistry.)

While no significant active material loss or NP aggregation was observed, these properties made Ag–BNNS films excellent reusable SERS devices, in which even trace amount of organic analytes are removed while maintaining the sensitivity after each measurement cycle.

14.3.6 Conjugates with Biomolecules

BN nanomaterials have been considered as benign and nontoxic (Chen et al. 2009b). In fact, h-BN has been an important component in many cosmetic products mainly as a white lubricant that is amenable for a variety of colorizations (Martin et al. 2007). One of the most desirable properties for a nanomaterial toward many bioapplications is to attain aqueous solubility, which is the first step toward *biocompatibility* with physiological conditions. For BNNSs, this was achieved via functionalization with hydrophilic moieties such as PEG (Lin et al. 2010a) or direct water dispersion (Lin et al. 2011b) discussed previously. While PEG-based (*PEGylated*) materials are widely

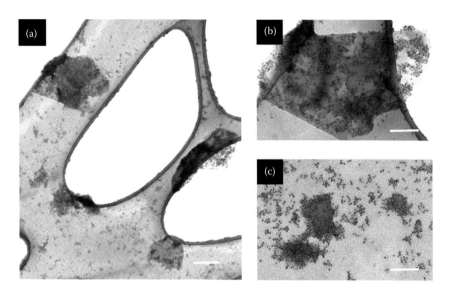

Figure 14.27   TEM images at (a) lower and (b, c) higher magnifications of a sample from incubation of a ferritin protein solution with the BNNS aqueous dispersion. Scale bars are 200, 100, and 100 nm, respectively. (Lin, Y., Williams, T.V., Xu, T.-B., Cao, W., Elsayed-Ali, H.E., and Connell, J.W., Aqueous dispersions of few-layered and monolayered hexagonal boron nitride nanosheets from sonication-assisted hydrolysis: critical role of water, *J. Phys. Chem. C*, 115, 2679–2685. Copyright 2011b, American Chemical Society.)

used in the *bio-nano* field, the direct water-dispersed BNNSs are impurity-free, with the whole surface of BNNSs available for further conjugation.

In a straightforward demonstration, Lin et al. (2011b) incubated an aqueous BNNS dispersion with a diluted ferritin solution. It was found that ferritin proteins were attached to the nanosheets and formed stable conjugates after prolonged standing (Figure 14.27). It is well known that most of the proteins, including ferritin, contain lysine moieties (i.e., pendant amino groups) on their surfaces. These lysine groups might thus be responsible for the stable ferritin–BNNS conjugates via Lewis acid–base (i.e., amino–B) interactions. Except for a theoretical study of noncovalent interactions of nucleic bases with the nanosheets (Tang et al. 2011), additional explorations on the conjugations of BNNSs with other biomolecules are yet to be reported.

14.3.7   Doping

Doping of BNNSs with foreign atoms may yield products with different electronic and electrical properties. As discussed in Section 2.2, carbon-doped BNNSs attracted significant attention because of the structural similarities between graphene and BNNS. Most of the current experimental studies have focused on the bottom-up synthesis of boron–carbon–nitrogen (BCN) hybrid nanosheets in which the C:BN stoichiometry and the electrical conductivity can be tunable (Ci et al. 2010). Importantly, theoretical studies suggested that the stable configuration for B, N, C atoms in a single nanosheet is to have maximal B–N and C–C bonds, which means that the h-BN and graphene island formation is more favorable than random atomic dispersion (da Rocha Martins and Chacham 2010). This was supported by various experimental data, including initial spectroscopic results (Ci et al. 2010) and, more recently, directly imaging via an advanced TEM technology called annular dark-field scanning TEM (ADF-STEM) (Figure 14.28) (Krivanek et al. 2010).

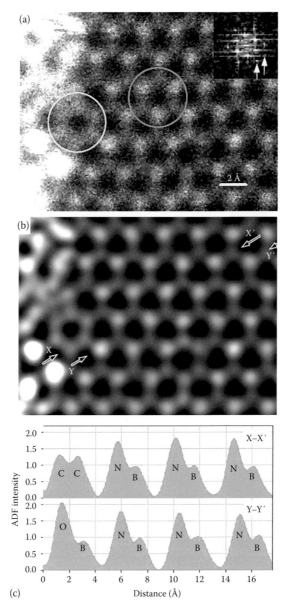

Figure 14.28 ADF-STEM image of a monolayer BNNS with a small amount of carbon impurity doping. (a) As recorded. (b) Corrected for distortion, smoothed, and deconvolved to remove probe tail contributions to nearest neighbors. (c) Line profiles showing the image intensity (normalized to equal one for a single boron atom) as a function of position in image b along X–X′ and Y–Y′. The elements giving rise to the peaks seen in the profiles are identified by their chemical symbols. Inset at top right in a shows the Fourier transform of an image area away from the thicker regions. Its two arrows point to $(11\bar{2}0)$ and $(20\bar{2}0)$ reflections of the hexagonal BN that correspond to recorded spacings of 1.26 and 1.09 Å. (Reprinted by permission from Macmillan Publishers Ltd. *Nature*, Krivanek, O.L., Chisholm, M.F., Nicolosi, V., Pennycook, T.J., Corbin, G.J., Dellby, N. et al., Atom-by-atom structural and chemical analysis by annular dark-field electron microscopy, 464, 571–574, Copyright 2010.)

BCN hybrid nanosheets were also prepared by replacing carbon atoms in graphene using B- and N-containing precursors. However, directly doping BNNSs with carbon is more related to the scope of this chapter. The only example available to date was to use a high-energy electron beam in a TEM microscope to irradiate BNNSs, BNNRs, or BNNTs in the presence of a carbon source such as paraffin wax (Figure 14.29) (Wei et al. 2011). It was proposed that the electron beam would knock out the B and N atoms, preferably at BN edges, with subsequent healing by C substitutions.

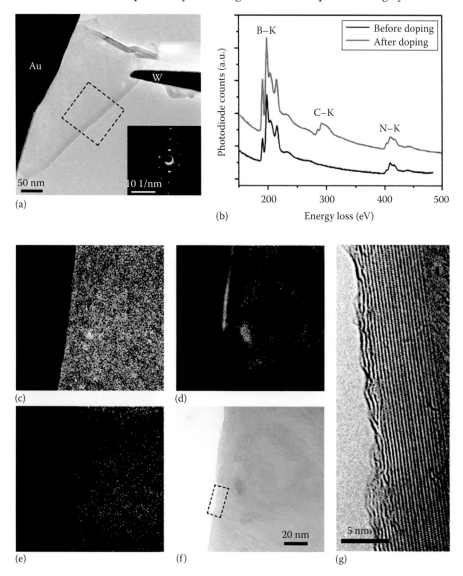

Figure 14.29 (a) TEM image showing a BNNS placed between an Au electrode and a tungsten (W) probe. The inset is a selected area electron diffraction pattern of the BNNS. (b) Electron energy loss spectra of the BNNS before and after doping. (c through f) Energy-filtered elemental maps (B (c), C (d), and N (e)) and corresponding zero-loss image (f) of the doped BNNS. (g) HR-TEM image of the framed area in (f).   (*Continued*)

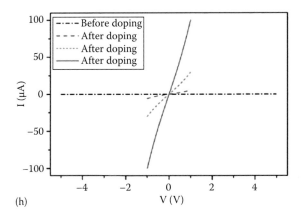

(h)

Figure 14.29 (Continued)   (h) Two-terminal current–voltage (I–V) curves of the BNNS before and after doping. The curves after doping correspond to different doping times. The longer the BNNS was doped, the higher conductance it showed. (Reprinted with permission from Wei, X., Wang, M.-S., Bando, Y., and Golberg, D., Electron-beam-induced substitutional carbon doping of boron nitride nanosheets, nanoribbons, and nanotubes, *ACS Nano*, 5, 2916–2922, 2011. Copyright 2011, American Chemical Society.)

Such substitutions could be carried out to a degree that all the atoms in the layer are replaced by C, making C:BN ratio also tunable similar to the bottom-up synthesized BCN hybrid nanosheets. With carbon doping, the BN nanostructures exhibited improved conductivity and transformed from insulators to semiconductors and, in some cases, conductors.

In addition to carbon, other atoms can also be used to dope the B–N bond network in the nanosheet structure. Very recently, Wu et al. prepared Ce-doped h-BN by heating a mixture of boric acid, melamine, and cerium (III) acetate to 1200°C in nitrogen (Wu et al. 2013). Spectroscopic data suggested that Ce might have been incorporated into the B–N bond network in the form of B–N–O–Ce. Importantly, the electronic bandgap of the resultant material lowered with increasing the concentration of incorporated Ce^{3+} ions and reached as low as 3.3 eV (in comparison to 4.2 eV of a control sample without Ce). Note that the as-prepared h-BN sheets in this work were of 30–50 nm thick, so not exactly within the *nanosheet* domain. In another publication, Xue et al. attempted to dope the solvent-exfoliated BNNSs with F (Xue et al. 2013). Although there was not much microscopic or spectroscopic evidence to confirm the F-doping, the resultant material did show improved electrical conductivity in comparison to the parent BNNSs.

14.4 SUMMARY

In this chapter, recent developments in the chemistry of BNNSs from both computational and experimental aspects are summarized. Computationally, much interest has been focused on tuning the large bandgap of BNNSs (especially BNNRs) as well as introduction of magnetic properties by means of defect introduction, doping, and edge or surface modifications. Experimentally, chemical modification strategies to exfoliate them into BNNSs with improved solubility in organic solvents and/or water have been demonstrated. However, production of BNNSs in bulk quantities (i.e., gram level and above) and high-fidelity chemical functionalization still remain a significant challenge. Nevertheless, examples presented demonstrate that exfoliated and dispersed BNNSs can be further modified by techniques such as covalent functionalization using appropriate reagents or conjugation with inorganic NPs or biomolecules.

Compared to the intense, detailed, and voluminous research of graphene, the work conducted on BNNSs thus far is just the tip of iceberg of 2D materials. It is envisioned that BNNSs should be used in applications that would take advantage of their unique properties including atomically smooth surface with no dangling bonds, high surface area, large bandgap, low color, thermal conductivity, electrically insulating, high mechanical strength, excellent thermal stability, and excellent oxidation-resistant properties. As discussed in this chapter, chemically modified BNNSs could be ideal candidates in applications such as graphene electronic devices, polymeric composites, and durable sensors. Future studies in both computational and experimental directions in BNNS chemistry will advance the fundamental understanding of chemical structure–property relationships and elucidate methodology of high-fidelity chemical manipulation to achieve the property combination required for their use in a variety of advanced electronic devices.

REFERENCES

Ahmed, T., Kilina, S., Das, T., Haraldsen, J. T., Rehr, J. J., and Balatsky, A. V. (2012), Electronic fingerprints of DNA bases on graphene. *Nano Letters*, 12: 927–931.

Alem, N., Erni, R., Kisielowski, C., Rossell, M. D., Gannett, W., and Zettl, A. (2009), Atomically thin hexagonal boron nitride probed by ultrahigh-resolution transmission electron microscopy. *Physical Review B*, 80: 155425.

Auwärter, W., Muntwiler, M., Osterwalder, J., and Greber, T. (2003), Defect lines and two-domain structure of hexagonal boron nitride films on Ni(111). *Surface Science*, 545: L735–L740.

Azevedo, S., Kaschny, J. R., Castilho, C. M. C. D., and Mota, F. D. B. (2007), A theoretical investigation of defects in a boron nitride monolayer. *Nanotechnology*, 18: 495707.

Bai, S. and Shen, X. (2012), Graphene-inorganic nanocomposites. *RSC Advances*, 2: 64–98.

Banerjee, S., Hemraj-Benny, T., and Wong, S. S. (2005), Covalent surface chemistry of single-walled carbon nanotubes. *Advanced Materials*, 17: 17–29.

Barone, V. and Peralta, J. E. (2008), Magnetic boron nitride nanoribbons with tunable electronic properties. *Nano Letters*, 8: 2210–2214.

Bergin, S. D., Nicolosi, V., Streich, P. V., Giordani, S., Sun, Z., Windle, A. H. et al. (2008), Towards solutions of single-walled carbon nanotubes in common solvents. *Advanced Materials*, 20: 1876–1881.

Bermúdez-Lugo, J., Perez-Gonzalez, O., Rosales-Hernández, M., Ilizaliturri-Flores, I., Trujillo-Ferrara, J., and Correa-Basurto, J. (2012), Exploration of the valproic acid binding site on histone deacetylase 8 using docking and molecular dynamic simulations. *Journal of Molecular Modeling*, 18: 2301–2310.

Berseneva, N., Krasheninnikov, A. V., and Nieminen, R. M. (2011), Mechanisms of postsynthesis doping of boron nitride nanostructures with carbon from first-principles simulations. *Physical Review Letters*, 107: 035501.

Bhattacharya, A., Bhattacharya, S., and Das, G. P. (2012), Band gap engineering by functionalization of BN sheet. *Physical Review B*, 85: 035415.

Bhattacharya, A., Bhattacharya, S., Majumder, C., and Das, G. P. (2010), First principles prediction of the third conformer of hydrogenated BN sheet. *Physica Status Solidi (RRL) – Rapid Research Letters*, 4: 368–370.

Blasé, X., Rubio, A., Louie, S. G., and Cohen, M. L. (1995), Quasiparticle band structure of bulk hexagonal boron nitride and related systems. *Physical Review B*, 51: 6868–6875.

Brandrup, J., Immergut, E. H., Grulke, E. A., Abe, A., Bloch, D. R., *Polymer Handbook*, 4th edn., 2005. Norvich, NY: John Wiley & Sons.

Britnell, L., Gorbachev, R. V., Jalil, R., Belle, B. D., Schedin, F., Mishchenko, A. et al. (2012), Field-effect tunneling transistor based on vertical graphene heterostructures. *Science*, 335: 947–950.

Chen, R. J., Zhang, Y., Wang, D., and Dai, H. (2001), Noncovalent sidewall functionalization of single-walled carbon nanotubes for protein immobilization. *Journal of the American Chemical Society*, 123: 3838–3839.

Chen, W., Li, Y., Yu, G., Li, C.-Z., Zhang, S. B., Zhou, Z., and Chen, Z. (2010), Hydrogenation: A simple approach to realize semiconductor–half-metal–metal transition in boron nitride nanoribbons. *Journal of the American Chemical Society*, 132: 1699–1705.

Chen, W., Li, Y., Yu, G., Zhou, Z., and Chen, Z. (2009a), Electronic structure and reactivity of boron nitride nanoribbons with stone-wales defects. *Journal of Chemical Theory and Computation*, 5: 3088–3095.

Chen, X., Wu, P., Rousseas, M., Okawa, D., Gartner, Z., Zettl, A., and Bertozzi, C. R. (2009b), Boron nitride nanotubes are noncytotoxic and can be functionalized for interaction with proteins and cells. *Journal of the American Chemical Society*, 131: 890–891.

Ci, L., Song, L., Jin, C., Jariwala, D., Wu, D., Li, Y. et al. (2010), Atomic layers of hybridized boron nitride and graphene domains. *Nature Materials*, 9: 430–435.

Coleman, J. N., Lotya, M., O'Neill, A., Bergin, S. D., King, P. J., Khan, U. et al. (2011), Two-dimensional nanosheets produced by liquid exfoliation of layered materials. *Science*, 331: 568–571.

da Rocha Martins, J. and Chacham, H. (2010), Disorder and segregation in B–C–N graphene-type layers and nanotubes: Tuning the band gap. *ACS Nano*, 5: 385–393.

Ding, Y., Wang, Y., and Ni, J. (2009), The stabilities of boron nitride nanoribbons with different hydrogen-terminated edges. *Applied Physics Letters*, 94: 233107.

Du, A., Chen, Y., Zhu, Z., Amal, R., Lu, G. Q., and Smith, S. C. (2009), Dots versus antidots: Computational exploration of structure, magnetism, and half-metallicity in boron–nitride nanostructures. *Journal of the American Chemical Society*, 131: 17354–17359.

Du, A. J., Smith, S. C., and Lu, G. Q. (2007), First-principle studies of electronic structure and C-doping effect in boron nitride nanoribbon. *Chemical Physics Letters*, 447: 181–186.

Dutta, R. C., Khan, S., and Singh, J. K. (2011), Wetting transition of water on graphite and boron-nitride surfaces: A molecular dynamics study. *Fluid Phase Equilibria*, 302: 310–315.

Erickson, K. J., Gibb, A. L., Sinitskii, A., Rousseas, M., Alem, N., Tour, J. M., and Zettl, A. K. (2011), Longitudinal splitting of boron nitride nanotubes for the facile synthesis of high quality boron nitride nanoribbons. *Nano Letters*, 11: 3221–3226.

Gao, M., Lyalin, A., and Taketsugu, T. (2012), Catalytic activity of Au and Au_2 on the h-BN surface: Adsorption and activation of O_2. *Journal of Physical Chemistry C*, 116: 9054–9062.

Georgakilas, V., Gournis, D., Tzitzios, V., Pasquato, L., Guldi, D. M., and Prato, M. (2007), Decorating carbon nanotubes with metal or semiconductor nanoparticles. *Journal of Materials Chemistry*, 17: 2679–2694.

Ghosh, J., Mazumdar, S., Das, M., Ghatak, S., and Basu, A. K. (2008), Microstructural characterization of amorphous and nanocrystalline boron nitride prepared by high-energy ball milling. *Materials Research Bulletin*, 43: 1023–1031.

Golberg, D., Bando, Y., Huang, Y., Terao, T., Mitome, M., Tang, C., and Zhi, C. (2010), Boron nitride nanotubes and nanosheets. *ACS Nano*, 4: 2979–2993.

Golberg, D., Bando, Y., Tang, C. C., and Zhi, C. Y. (2007), Boron nitride nanotubes. *Advanced Materials*, 19: 2413–2432.

Han, W.-Q., Wu, L., Zhu, Y., Watanabe, K., and Taniguchi, T. (2008), Structure of chemically derived mono- and few-atomic-layer boron nitride sheets. *Applied Physics Letters*, 93: 223103.

Hernandez, Y., Nicolosi, V., Lotya, M., Blighe, F. M., Sun, Z., De, S. et al. (2008), High-yield production of graphene by liquid-phase exfoliation of graphite. *Nature Nanotechnology*, 3: 563–568.

Hirsch, A. (2002), Functionalization of single-walled carbon nanotubes. *Angewandte Chemie International Edition*, 41: 1853–1859.

Hu, C., Ogura, R., Onoda, N., Konabe, S., and Watanabe, K. (2012), Quasiparticle band gaps of boron nitride nanoribbons. *Physical Review B*, 85: 245420.

Huang, B., Xiang, H., Yu, J., and Wei, S.-H. (2012), Effective control of the charge and magnetic states of transition-metal atoms on single-layer boron nitride. *Physical Review Letters*, 108: 206802.

Ikuno, T., Sainsbury, T., Okawa, D., Fréchet, J. M. J., and Zettl, A. (2007), Amine-functionalized boron nitride nanotubes. *Solid State Communications*, 142: 643–646.

Islam, M. F., Rojas, E., Bergey, D. M., Johnson, A. T., and Yodh, A. G. (2003), High weight fraction surfactant solubilization of single-wall carbon nanotubes in water. *Nano Letters*, 3: 269–273.

Jiménez, I., Jankowski, A., Terminello, L. J., Carlisle, J. A., Sutherland, D. G. J., Doll, G. L. et al. (1996), Near-edge x-ray absorption fine structure study of bonding modifications in BN thin films by ion implantation. *Applied Physics Letters*, 68: 2816–2818.

Jiménez, I., Jankowski, A. F., Terminello, L. J., Sutherland, D. G. J., Carlisle, J. A., Doll, G. L. et al. (1997), Core-level photoabsorption study of defects and metastable bonding configurations in boron nitride. *Physical Review B*, 55: 12025–12037.

Jin, C., Lin, F., Suenaga, K., and Iijima, S. (2009), Fabrication of a freestanding boron nitride single layer and its defect assignments. *Physical Review Letters*, 102: 195505.

Kamat, P. V. (2009), Graphene-based nanoarchitectures. Anchoring semiconductor and metal nanoparticles on a two-dimensional carbon support. *Journal of Physical Chemistry Letters*, 1: 520–527.

Kan, E., Wu, F., Xiang, H., Yang, J., and Whangbo, M.-H. (2011), Half-metallic dirac point in B-edge hydrogenated BN nanoribbons. *Journal of Physical Chemistry C*, 115: 17252–17254.

Khan, U., May, P., O'Neill, A., Bell, A. P., Boussac, E., Martin, A. et al. (2013), Polymer reinforcement using liquid-exfoliated boron nitride nanosheets. *Nanoscale*, 5: 581–587.

Kim, E., Yu, T., Sang Song, E., and Yu, B. (2011a), Chemical vapor deposition-assembled graphene field-effect transistor on hexagonal boron nitride. *Applied Physics Letters*, 98: 262103.

Kim, K. K., Hsu, A., Jia, X., Kim, S. M., Shi, Y., Hofmann, M. et al. (2011b), Synthesis of monolayer hexagonal boron nitride on Cu foil using chemical vapor deposition. *Nano Letters*, 12: 161–166.

Kotakoski, J., Jin, C. H., Lehtinen, O., Suenaga, K., and Krasheninnikov, A. V. (2010), Electron knock-on damage in hexagonal boron nitride monolayers. *Physical Review B*, 82: 113404.

Krivanek, O. L., Chisholm, M. F., Nicolosi, V., Pennycook, T. J., Corbin, G. J., Dellby, N. et al. (2010), Atom-by-atom structural and chemical analysis by annular dark-field electron microscopy. *Nature*, 464: 571–574.

Kuilla, T., Bhadra, S., Yao, D., Kim, N. H., Bose, S., and Lee, J. H. (2010), Recent advances in graphene based polymer composites. *Progress in Polymer Science*, 35: 1350–1375.

Lai, L., Lu, J., Wang, L., Luo, G., Zhou, J., Qin, R. et al. (2009), Magnetic properties of fully bare and half-bare boron nitride nanoribbons. *Journal of Physical Chemistry C*, 113: 2273–2276.

Li, G.-X., Liu, Y., Wang, B., Song, X.-M., Li, E., and Yan, H. (2008), Preparation of transparent BN films with superhydrophobic surface. *Applied Surface Science*, 254: 5299–5303.

Li, L. H., Chen, Y., Behan, G., Zhang, H., Petravic, M., and Glushenkov, A. M. (2011), Large-scale mechanical peeling of boron nitride nanosheets by low-energy ball milling. *Journal of Materials Chemistry*, 21: 11862–11866.

Li, X., Wu, X., Zeng, X. C., and Yang, J. (2012), Band-gap engineering via tailored line defects in boron-nitride nanoribbons, sheets, and nanotubes. *ACS Nano*, 6: 4104–4112.

Li, X., Zhao, J., and Yang, J. (2013a), Semihydrogenated BN sheet: A promising visible-light driven photocatalyst for water splitting. *Scientific Reports*, 3: 1858.

Li, X., Hao, X., Zhao, M., Wu, Y., Yang, J., Tian, Y., and Qian, G. (2013b), Exfoliation of hexagonal boron nitride by molten hydroxides. *Advanced Materials*, 25: 2200–2204.

Lian, G., Zhang, X., Tan, M., Zhang, S., Cui, D., and Wang, Q. (2011), Facile synthesis of 3D boron nitride nanoflowers composed of vertically aligned nanoflakes and fabrication of graphene-like BN by exfoliation. *Journal of Materials Chemistry*, 21: 9201–9207.

Lin, Q., Zou, X., Zhou, G., Liu, R., Wu, J., Li, J., and Duan, W. (2011c), Adsorption of DNA/RNA nucleobases on hexagonal boron nitride sheet: An ab initio study. *Physical Chemistry Chemical Physics*, 13: 12225–12230.

Lin, Y., Baggett, D. W., Kim, J.-W., Siochi, E. J., and Connell, J. W. (2011a), Instantaneous formation of metal and metal oxide nanoparticles on carbon nanotubes and graphene via solvent-free microwave heating. *ACS Applied Materials and Interfaces*, 3: 1652–1664.

Lin, Y., Bunker, C. E., Fernando, K. A. S., and Connell, J. W. (2012b), Aqueously dispersed silver nanoparticle-decorated boron nitride nanosheets for reusable, thermal oxidation-resistant surface enhanced Raman spectroscopy (SERS) devices. *ACS Applied Materials and Interfaces*, 4: 1110–1117.

Lin, Y. and Connell, J. W. (2012a), Advances in 2D boron nitride nanostructures: Nanosheets, nanoribbons, nanomeshes, and hybrids with graphene. *Nanoscale*, 4: 6908–6939.

Lin, Y., Watson, K. A., Fallbach, M. J., Ghose, S., Smith, J. G., Delozier, D. M. et al. (2009b), Rapid, solventless, bulk preparation of metal nanoparticle-decorated carbon nanotubes. *ACS Nano*, 3: 871–884.

Lin, Y., Watson, K. A., Ghose, S., Smith, J. G., Williams, T. V., Crooks, R. E. et al. (2009a), Direct mechano-chemical formation of metal nanoparticles on carbon nanotubes. *Journal of Physical Chemistry C*, 113: 14858–14862.

Lin, Y., Williams, T. V., Cao, W., Elsayed-Ali, H. E., and Connell, J. W. (2010b), Defect functionalization of hexagonal boron nitride nanosheets. *Journal of Physical Chemistry C*, 114: 17434–17439.

Lin, Y., Williams, T. V., and Connell, J. W. (2010a), Soluble, exfoliated hexagonal boron nitride nanosheets. *Journal of Physical Chemistry Letters*, 1: 277–283.

Lin, Y., Williams, T. V., Xu, T.-B., Cao, W., Elsayed-Ali, H. E., and Connell, J. W. (2011b), Aqueous dispersions of few-layered and monolayered hexagonal boron nitride nanosheets from sonication-assisted hydrolysis: Critical role of water. *Journal of Physical Chemistry C*, 115: 2679–2685.

Liu, R.-F. and Cheng, C. (2007), Ab Initio studies of possible magnetism in a BN sheet by nonmagnetic impurities and vacancies. *Physical Review B*, 76: 014405.

Liu, Y., Bhowmick, S., and Yakobson, B. I. (2011), BN white graphene with "colorful" edges: The energies and morphology. *Nano Letters*, 11: 3113–3116.

Liu, Y., Zou, X., and Yakobson, B. I. (2012), Dislocations and grain boundaries in two-dimensional boron nitride. *ACS Nano*, 6: 7053–7058.

Lopez-Bezanilla, A., Huang, J., Terrones, H., and Sumpter, B. G. (2011), Boron nitride nanoribbons become metallic. *Nano Letters*, 11: 3267–3273.

Lopez-Bezanilla, A., Huang, J., Terrones, H., and Sumpter, B. G. (2012), Structure and electronic properties of edge-functionalized armchair boron nitride nanoribbons. *Journal of Physical Chemistry C*, 116: 15675–15681.

Lotya, M., Hernandez, Y., King, P. J., Smith, R. J., Nicolosi, V., Karlsson, L. S. et al. (2009), Liquid phase production of graphene by exfoliation of graphite in surfactant/water solutions. *Journal of the American Chemical Society*, 131: 3611–3620.

Lotya, M., King, P. J., Khan, U., De, S., and Coleman, J. N. (2010), High-concentration, surfactant-stabilized graphene dispersions. *ACS Nano*, 4: 3155–3162.

Lu, F. S., Wang, F., Cao, L., Kong, C. Y., and Huang, X. C. (2012), Hexagonal boron nitride nanomaterials: Advances towards bioapplications. *Nanoscience and Nanotechnology Letters*, 4: 949–961.

Ma, Y., Dai, Y., Guo, M., Niu, C., Yu, L., and Huang, B. (2011), Strain-induced magnetic transitions in half-fluorinated single layers of BN, GaN and graphene. *Nanoscale*, 3: 2301–2306.

Maguer, A., Leroy, E., Bresson, L., Doris, E., Loiseau, A., and Mioskowski, C. (2009), A versatile strategy for the functionalization of boron nitride nanotubes. *Journal of Materials Chemistry*, 19: 1271–1275.

Marom, N., Bernstein, J., Garel, J., Tkatchenko, A., Joselevich, E., Kronik, L., and Hod, O. (2010), Stacking and registry effects in layered materials: The case of hexagonal boron nitride. *Physical Review Letters*, 105: 046801.

Martin, E., Christoph, L., Ralf, D., Bernd, R., and Jens, E. (2007). *Hexagonal Boron Nitride (hBN): Applications from Metallurgy to Cosmetics*. Baden-Baden, Allemagne: Göller, 84, pp. E49–E53.

May, P., Khan, U., Hughes, J. M., and Coleman, J. N. (2012), Role of solubility parameters in understanding the steric stabilization of exfoliated two-dimensional nanosheets by adsorbed polymers. *Journal of Physical Chemistry C*, 116: 11393–11400.

Mayorov, A. S., Gorbachev, R. V., Morozov, S. V., Britnell, L., Jalil, R., Ponomarenko, L. A. et al. (2011), Micrometer-scale ballistic transport in encapsulated graphene at room temperature. *Nano Letters*, 11: 2396–2399.

Meyer, J. C., Chuvilin, A., Algara-Siller, G., Biskupek, J., and Kaiser, U. (2009), Selective sputtering and atomic resolution imaging of atomically thin boron nitride membranes. *Nano Letters*, 9: 2683–2689.

Moore, V. C., Strano, M. S., Haroz, E. H., Hauge, R. H., Smalley, R. E., Schmidt, J., and Talmon, Y. (2003), Individually suspended single-walled carbon nanotubes in various surfactants. *Nano Letters*, 3: 1379–1382.

Mukherjee, R. and Bhowmick, S. (2011), Edge stabilities of hexagonal boron nitride nanoribbons: A first-principles study. *Journal of Chemical Theory and Computation*, 7: 720–724.

Muralidharan, K., Erdmann, R. G., Runge, K., and Deymier, P. A. (2011), Asymmetric energy transport in defected boron nitride nanoribbons: Implications for thermal rectification. *AIP Advances*, 1: 041703.

Nag, A., Raidongia, K., Hembram, K. P. S. S., Datta, R., Waghmare, U. V., and Rao, C. N. R. (2010), Graphene analogues of BN: Novel synthesis and properties. *ACS Nano*, 4: 1539–1544.

Nazarov, A. S., Demin, V. N., Grayfer, E. D., Bulavchenko, A. I., Arymbaeva, A. T., Shin, H.-J. et al. (2012), Functionalization and dispersion of hexagonal boron nitride (h-BN) nanosheets treated with inorganic reagents. *Chemistry – An Asian Journal*, 7: 554–560.

Niyogi, S., Hamon, M. A., Hu, H., Zhao, B., Bhowmik, P., Sen, R. et al. (2002), Chemistry of single-walled carbon nanotubes. *Accounts of Chemical Research*, 35: 1105–1113.

Novoselov, K. S., Geim, A. K., Morozov, S. V., Jiang, D., Zhang, Y., Dubonos, S. V. et al. (2004), Electric field effect in atomically thin carbon films. *Science*, 306: 666–669.

Novoselov, K. S., Jiang, D., Schedin, F., Booth, T. J., Khotkevich, V. V., Morozov, S. V., and Geim, A. K. (2005), Two-dimensional atomic crystals. *Proceedings of the National Academy of Sciences of the United States of America*, 102: 10451–10453.

Okada, S. (2009), Atomic configurations and energetics of vacancies in hexagonal boron nitride: First-principles total-energy calculations. *Physical Review B*, 80: 161404.

Orellana, W. and Chacham, H. (2001), Stability of native defects in hexagonal and cubic boron nitride. *Physical Review B*, 63: 125205.

Pacile, D., Meyer, J. C., Girit, C. O., and Zettl, A. (2008), The two-dimensional phase of boron nitride: Few-atomic-layer sheets and suspended membranes. *Applied Physics Letters*, 92: 133107.

Pakdel, A., Zhi, C., Bando, Y., Nakayama, T., and Golberg, D. (2011), Boron nitride nanosheet coatings with controllable water repellency. *ACS Nano*, 5: 6507–6515.

Pakdel, A., Zhi, C. Y., Bando, Y., and Golberg, D. (2012), Low-dimensional boron nitride nanomaterials. *Materials Today*, 15: 256–265.

Pal, S., Vivekchand, S. R. C., Govindaraj, A., and Rao, C. N. R. (2007), Functionalization and solubilization of BN nanotubes by interaction with Lewis bases. *Journal of Materials Chemistry*, 17: 450–452.

Pan, Y. and Yang, Z. (2010), Electronic structures and spin gapless semiconductors in BN nanoribbons with vacancies. *Physical Review B*, 82: 195308.

Park, C.-H. and Louie, S. G. (2008), Energy gaps and stark effect in boron nitride nanoribbons. *Nano Letters*, 8: 2200–2203.

Pease, R. S. (1950), Crystal structure of boron nitride. *Nature*, 165: 722–723.

Pokropivny, V. V., Skorokhod, V. V., Oleinik, G. S., Kurdyumov, A. V., Bartnitskaya, T. S., Pokropivny, A. V. et al. (2000), Boron nitride analogs of fullerenes (the fulborenes), nanotubes, and fullerites (the fulborenites). *Journal of Solid State Chemistry*, 154: 214–222.

Rao, C. N. R. and Govindaraj, A. (2009), Synthesis of inorganic nanotubes. *Advanced Materials*, 21: 4208–4233.

Sainsbury, T., Satti, A., May, P., O'Neill, A., Nicolosi, V., Gun'ko, Y. K., and Coleman, J. N. (2012a), Covalently functionalized hexagonal boron nitride nanosheets by nitrene addition. *Chemistry – A European Journal*, 18: 10808–10812.

Sainsbury, T., Satti, A., May, P., Wang, Z., McGovern, I., Gun'ko, Y. K., and Coleman, J. (2012b), Oxygen radical functionalization of boron nitride nanosheets. *Journal of the American Chemical Society*, 134: 18758–18771.

Seifert, G., Fowler, P. W., Mitchell, D., Porezag, D., and Frauenheim, T. (1997), Boron-nitrogen analogues of the fullerenes: Electronic and structural properties. *Chemical Physics Letters*, 268: 352–358.

Shi, Y., Hamsen, C., Jia, X., Kim, K. K., Reina, A., Hofmann, M. et al. (2010), Synthesis of few-layer hexagonal boron nitride thin film by chemical vapor deposition. *Nano Letters*, 10: 4134–4139.

Si, M. S. and Xue, D. S. (2007), Magnetic properties of vacancies in a graphitic boron nitride sheet by first-principles pseudopotential calculations. *Physical Review B*, 75: 193409.

Singhal, S. K., Kumar, V., Stalin, K., Choudhary, A., Teotia, S., Reddy, G. B. et al. (2013), Gold-nanoparticle-decorated boron nitride nanosheets: Structure and optical properties. *Particle and Particle Systems Characterization*, 30: 445–452.

Smith, R. J., King, P. J., Lotya, M., Wirtz, C., Khan, U., De, S. et al. (2011), Large-scale exfoliation of inorganic layered compounds in aqueous surfactant solutions. *Advanced Materials*, 23: 3944–3948.

Song, L., Ci, L., Lu, H., Sorokin, P. B., Jin, C., Ni, J. et al. (2010), Large scale growth and characterization of atomic hexagonal boron nitride layers. *Nano Letters*, 10: 3209–3215.

Song, X., Hu, J., and Zeng, H. (2013), Two-dimensional semiconductors: Recent progress and future perspectives. *Journal of Materials Chemistry C*, 1: 2952–2969.

Suenaga, K., Kobayashi, H., and Koshino, M. (2012), Core-level spectroscopy of point defects in single layer h-BN. *Physical Review Letters*, 108: 075501.

Sun, Q., Li, Z., Searles, D. J., Chen, Y., Lu, G., and Du, A. (2013), Charge-controlled switchable CO_2 capture on boron nitride nanomaterials. *Journal of the American Chemical Society*, 135: 8246–8253.

Sun, Y.-P., Fu, K., Lin, Y., and Huang, W. (2002), Functionalized carbon nanotubes: Properties and applications. *Accounts of Chemical Research*, 35: 1096–1104.

Tang, Q., Zhou, Z., and Chen, Z. (2011), Molecular charge transfer: A simple and effective route to engineer the band structures of BN nanosheets and nanoribbons. *Journal of Physical Chemistry C*, 115: 18531–18537.

Tang, Q., Zhou, Z., Shen, P., and Chen, Z. (2013), Band gap engineering of BN sheets by interlayer dihydrogen bonding and electric field control. *Chemical Physics and Physical Chemistry*, 14: 1787–1792.

Tang, S. and Cao, Z. (2010), Theoretical study of stabilities and electronic properties of the vacancy and carbon-doping defects in zigzag boron nitride nanoribbons. *Computational Materials Science*, 48: 648–654.

Tasis, D., Tagmatarchis, N., Bianco, A., and Prato, M. (2006), Chemistry of carbon nanotubes. *Chemical Reviews*, 106: 1105–1136.

Topsakal, M., Aktürk, E., and Ciraci, S. (2009), First-principles study of two- and one-dimensional honeycomb structures of boron nitride. *Physical Review B*, 79: 115442.

Tu, X. and Zheng, M. (2008), A DNA-based approach to the carbon nanotube sorting problem. *Nano Research*, 1: 185–194.

Wang, J., Lee, C. H., and Yap, Y. K. (2010a), Recent advancements in boron nitride nanotubes. *Nanoscale*, 2: 2028–2034.

Wang, Y. (2010b), Electronic properties of two-dimensional hydrogenated and semihydrogenated hexagonal boron nitride sheets. *Physica Status Solidi (RRL) – Rapid Research Letters*, 4: 34–36.

Wang, L., Sun, C., Xu, L., and Qian, Y. (2011a), Convenient synthesis and applications of gram scale boron nitride nanosheets. *Catalysis Science and Technology*, 1: 1119–1123.

Wang, S., Chen, Q., and Wang, J. (2011b), Optical properties of boron nitride nanoribbons: Excitonic effects. *Applied Physics Letters*, 99: 063114.

Wang, Y., Ding, Y., and Ni, J. (2011c), Electronic structures of Fe-terminated armchair boron nitride nanoribbons. *Applied Physics Letters*, 99: 053123.

Wang, Y., Shi, Z., and Yin, J. (2011d), Boron nitride nanosheets: Large-scale exfoliation in methanesulfonic acid and their composites with polybenzimidazole. *Journal of Materials Chemistry*, 21: 11371–11377.

Warner, J. H., Rümmeli, M. H., Bachmatiuk, A., and Büchner, B. (2010), Atomic resolution imaging and topography of boron nitride sheets produced by chemical exfoliation. *ACS Nano*, 4: 1299–1304.

Watanabe, K., Taniguchi, T., and Kanda, H. (2004), Direct-bandgap properties and evidence for ultraviolet lasing of hexagonal boron nitride single crystal. *Nature Materials*, 3: 404–409.

Wei, X., Wang, M.-S., Bando, Y., and Golberg, D. (2011), Electron-beam-induced substitutional carbon doping of boron nitride nanosheets, nanoribbons, and nanotubes. *ACS Nano*, 5: 2916–2922.

Wildgoose, G. G., Banks, C. E., and Compton, R. G. (2006), Metal nanoparticles and related materials supported on carbon nanotubes: Methods and applications. *Small*, 2: 182–193.

Wu, X., An, W., and Zeng, X. C. (2006), Chemical functionalization of boron–nitride nanotubes with NH_3 and amino functional groups. *Journal of the American Chemical Society*, 128: 12001–12006.

Wu, X.-J., Wu, M.-H., and Zeng, X. (2009), Chemically decorated boron-nitride nanoribbons. *Frontiers of Physics in China*, 4: 367–372.

Wu, M., Wu, X., Pei, Y., and Zeng, X. (2011), Inorganic nanoribbons with unpassivated zigzag edges: Half metallicity and edge reconstruction. *Nano Research*, 4: 233–239.

Wu, J., Yin, L., and Zhang, L. (2013), Tuning the electronic structure, bandgap energy and photoluminescence properties of hexagonal boron nitride nanosheets via a controllable Ce^{3+} ions doping. *RSC Advances*, 3: 7408–7418.

Xie, S.-Y., Wang, W., Fernando, K. A. S., Wang, X., Lin, Y., and Sun, Y.-P. (2005), Solubilization of boron nitride nanotubes. *Chemical Communications*, 3670–3672.

Xue, J., Sanchez-Yamagishi, J., Bulmash, D., Jacquod, P., Deshpande, A., Watanabe, K. et al. (2011), Scanning tunnelling microscopy and spectroscopy of ultra-flat graphene on hexagonal boron nitride. *Nature Materials*, 10: 282–285.

Xue, Y., Liu, Q., He, G., Xu, K., Jiang, L., Hu, X., and Hu, J. (2013), Excellent electrical conductivity of the exfoliated and fluorinated hexagonal boron nitride nanosheets. *Nanoscale Research Letters*, 8: 1–7.

Yang, J., Kim, D., Hong, J., and Qian, X. (2010), Magnetism in boron nitride monolayer: Adatom and vacancy defect. *Surface Science*, 604: 1603–1607.

Yang, K., Chen, Y., Xie, Y., Wei, X. L., Ouyang, T., and Zhong, J. (2011), Effect of triangle vacancy on thermal transport in boron nitride nanoribbons. *Solid State Communications*, 151: 460–464.

Yin, L.-C., Cheng, H.-M., and Saito, R. (2010), Triangle defect states of hexagonal boron nitride atomic layer: Density functional theory calculations. *Physical Review B*, 81: 153407.

Yu, J., Huang, X., Wu, C., Wu, X., Wang, G., and Jiang, P. (2012), Interfacial modification of boron nitride nanoplatelets for epoxy composites with improved thermal properties. *Polymer*, 53: 471–480.

Zeng, H. B., Zhi, C. Y., Zhang, Z. H., Wei, X. L., Wang, X. B., Guo, W. L. et al. (2010), "White graphenes": Boron nitride nanoribbons via boron nitride nanotube unwrapping. *Nano Letters*, 10: 5049–5055.

Zeng, J., Chen, K.-Q., and Sun, C. Q. (2012), Electronic structures and transport properties of fluorinated boron nitride nanoribbons. *Physical Chemistry Chemical Physics*, 14: 8032–8037.

Zhang, Z. and Guo, W. (2008), Energy-gap modulation of BN ribbons by transverse electric fields: First-principles calculations. *Physical Review B*, 77: 075403.

Zhang, A., Teoh, H. F., Dai, Z., Feng, Y. P., and Zhang, C. (2011), Band gap engineering in graphene and hexagonal BN antidot lattices: A first principles study. *Applied Physics Letters*, 98: 023105.

Zhang, F., Chen, X., Boulos, R. A., Md Yasin, F., Lu, H., Raston, C., and Zhang, H. (2013), Pyrene-conjugated hyaluronan facilitated exfoliation and stabilisation of low dimensional nanomaterials in water. *Chemical Communications*, 49: 4845–4847.

Zhao, Y., Wu, X., Yang, J., and Zeng, X. C. (2012a), Oxidation of a two-dimensional hexagonal boron nitride monolayer: A first-principles study. *Physical Chemistry Chemical Physics*, 14: 5545–5550.

Zhao, J.-X., Yu, Y.-Y., Bai, Y., Lu, B., and Wang, B.-X. (2012b), Chemical functionalization of BN graphene with the metal-arene group: A theoretical study. *Journal of Materials Chemistry*, 22: 9343–9350.

Zhao, J.-X., Wang, H.-X., Liu, Y.-J., Cai, Q.-H., and Wang, X.-Z. (2013), Catalyst-free achieving of controllable carbon doping of boron nitride nanosheets by CO molecules: A theoretical prediction. *RSC Advances*, 3: 4917–4926.

Zheng, F., Zhou, G., Liu, Z., Wu, J., Duan, W., Gu, B.-L., and Zhang, S. B. (2008), Half metallicity along the edge of zigzag boron nitride nanoribbons. *Physical Review B*, 78: 205415.

Zheng, M., Jagota, A., Semke, E. D., Diner, B. A., McLean, R. S., Lustig, S. R. et al. (2003), DNA-assisted dispersion and separation of carbon nanotubes. *Nature Materials*, 2: 338–342.

Zhi, C., Bando, Y., Tang, C., Kuwahara, H., and Golberg, D. (2009), Large-scale fabrication of boron nitride nanosheets and their utilization in polymeric composites with improved thermal and mechanical properties. *Advanced Materials*, 21: 2889–2893.

Zhi, C., Bando, Y., Tang, C., Xie, R., Sekiguchi, T., and Golberg, D. (2005), Perfectly dissolved boron nitride nanotubes due to polymer wrapping. *Journal of the American Chemical Society*, 127: 15996–15997.

Zhou, K.-G., Mao, N.-N., Wang, H.-X., Peng, Y., and Zhang, H.-L. (2011a), A Mixed-solvent strategy for efficient exfoliation of inorganic graphene analogues. *Angewandte Chemie International Edition*, 50: 10839–10842.

Zhou, Y. G., Yang, P., Sun, X., Wang, Z. G., Zu, X. T., and Gao, F. (2011b), First-principles study of the noble metal-doped BN layer. *Journal of Applied Physics*, 109: 084308.

Zhu, Y., Murali, S., Cai, W., Li, X., Suk, J. W., Potts, J. R., and Ruoff, R. S. (2010), Graphene and graphene oxide: Synthesis, properties, and applications. *Advanced Materials*, 22: 3906–3924.

Zobelli, A., Gloter, A., Ewels, C. P., Seifert, G., and Colliex, C. (2007), Electron knock-on cross section of carbon and boron nitride nanotubes. *Physical Review B*, 75: 245402.

Chapter 15

Functionalization of Carbon Nanotubes for Catalytic Applications

Li Wang, Lei Ge, Thomas Rufford, and Zhonghua Zhu

CONTENTS

15.1 INTRODUCTION

The unique rolled graphene structure of carbon nanotubes (CNTs) provides these materials with various properties that could be useful in their application as catalyst supports. For example, properties of CNTs that are of interest in catalyst supports include their good electrical conductivity, excellent mechanical strength, thermal and chemical stability, and special adsorption properties (Serp et al., 2003; Serp and Castillejos, 2010). CNT-supported catalysts have been reported for important liquid-phase (hydrogenation, hydroformylation) or gas-phase (Fischer–Tropsch process, ammonia decomposition, and preferential CO oxidation) reactions (Yin et al., 2004a; Vu et al., 2006b; Trepanier et al., 2009), for supported homogeneous catalysis (hydrogenation and cyanosilylation reactions) (Banerjee and Wong, 2002b; Baleizão et al., 2004), and for electrocatalysis (fuel cell electrodes) (Lee et al., 2006;

Wang et al., 2008b). However, one of the challenges in the development of CNT-supported catalysts is anchoring of active catalyst phases—most commonly metal or metal oxide nanoparticles—to CNTs because of the strength of the carbon–carbon bond, which makes CNTs very chemical stable, and the hydrophobicity of the graphene ring structure (Jiang and Gao, 2003; Jiang et al., 2003).

In this chapter, we provide a critical review of state-of-the-art methods for the functionalization of CNTs that may be used to facilitate the loading of active catalyst phases onto nanotube structures. In addition to the catalytic applications presented here, functionalized CNTs (f-CNTs) also have potential in biological applications such as drug delivery (Bianco et al., 2005). The incorporation of functional groups such as oxygen-containing carboxyl groups on the CNT surface can change the nanotube's surface chemistry to allow better wetting of the CNT surface by solutions containing catalyst precursors. Moreover, functional groups can provide stronger interactions with metal catalyst precursor molecules than the pristine graphene CNT surface and thus act as anchor sites for the active catalyst phase. If the distribution of functional groups across the CNT surface is well controlled, it is possible to obtain CNTs with well-dispersed catalyst particles, which is a desirable property for most catalysts. On the other hand, the functional groups on CNTs can change the electronic structure of metal catalysts deposited on them and the interaction between metals and CNTs, thereby altering the catalytic activity and selectivity.

15.2   PROPERTIES OF FUNCTIONALIZED CARBON NANOTUBES

Due to the inert and hydrophobic nature of the graphene structure of CNTs, surface functionalization is essential to enhance the interaction of metal precursors and metal catalyst nanoparticles with the sites on the CNTs. The attachment of functional groups to the tube ends and sidewalls of CNTs can enhance the hydrophilicity and improve the dispersion of the nanotubes in the aqueous solutions, or polar organic solutions, containing catalyst metal precursors. The dispersion of f-CNTs in solution facilitates more even sorption of metal precursors on the nanotube and ultimately can lead to a more uniform distribution of metal nanoparticles across the CNT surface.

In addition to the improved solubility of f-CNTs, functional groups can also serve as preferred anchor sites for metals that can assist in the control of the metal loading location (Rao et al., 2007; Wang et al., 2007a; Wepasnick et al., 2010). This preferential anchoring of metals to functional groups is due to the strong interactions permitted between the functional groups and metal ions. Some functional groups such as carboxyls may even participate in ion exchange with metal ions to form carboxylic acid groups. The interaction between the target metal ions and carboxyls on CNT surface is described in Equation 15.1, proposed by Ebbesen et al. (1996):

$$CNT-COOH + M^+X^- \rightarrow CNT-COO^-M^+ + HX \tag{15.1}$$

where a carboxylic group on the CNT surface (CNT–COOH) exchanges a proton with the metal ion (M^+). Subsequently, a reducing agent is added to reduce the surface metal ions into metal nanoparticles on the CNT surface. In this example, the functional groups serve as the preferred nucleation centers for the reduction of metal ions and can greatly improve the dispersion of metal catalysts on the CNT surfaces (Lee et al., 2006).

15.3 METHODS TO FUNCTIONALIZE CARBON NANOTUBES

Functional groups containing heteroatoms such as oxygen, nitrogen, and sulfur can be introduced to the graphene ring structure of CNTs by either covalent bonds or noncovalent bonds. Covalent functionalization involves the attachment of functional moieties to the CNT skeleton by a covalent bond created through nonreversible reactions at sidewall sites or defect sites, such as at the CNT end caps. Noncovalent functionalization of CNTs is based on the complexation of functional moieties to the CNT skeleton by attractive forces such as van der Waals, electrostatic, or π–π interactions (Saha and Kundu, 2010). These interactions are much weaker than covalent bonds, so noncovalent f-CNTs are not suitable for all applications. However, a key advantage of noncovalent f-CNTs is that typically the noncovalent attachment of moieties does not have a significant effect on the electronic structure of an single-walled carbon nanotube (SWCNT) skeleton.

15.3.1 Covalent Functionalization

There are many approaches reported for the covalent functionalization of CNTs including techniques that target defects in the as-synthesized CNT structure (Hirsch, 2002), techniques that create defects in a pristine CNT structure such as oxidation, and CNT synthesis techniques that aim to introduce heteroatoms such as nitrogen directly into the conjugate carbon ring structure of the nanotube. The most commonly applied method for creating defects in the CNTs is oxidation in a strong acid or oxidizing gas such as air.

Oxidation of CNTs can create edge defects in the carbon, but most commonly oxidation creates oxygen-containing functional groups such as carboxyls (–COOH), carbonyls (–C=O), and hydroxyls (–OH). The primary oxygen-containing functional groups found on f-CNTs are shown in Figure 15.1 (Kundu et al., 2008). Many of these functional groups are present in as-synthesized CNTs, and importantly, oxygen functional groups can be used as a reaction site to attach another moiety to the CNT skeleton. To increase the concentration of these groups on CNTs, various oxidants including nitric acid (Dujardin et al., 1998), sulfuric acid (Sumanasekera et al., 1999), hydrogen peroxide (Wang et al., 2007b), potassium permanganate (Zhang et al., 2003), air (Park et al., 2001), and steam (Tobias et al., 2006) have been employed at elevated temperatures. An added advantage of functionalization by oxidation is that impurities such as amorphous carbon and residual CNT synthesis metal catalysts can be removed from the CNT sample. An alternative approach to treatment in strong acids or oxidizing gases is the use of microwave irradiation to introduce carboxyls, carbonyls, hydroxyls, and allyls on CNT surfaces (Raghuveer et al., 2006).

Of these processes, treatment of CNTs in concentrated HNO_3 or a mixture of H_2SO_4 + HNO_3 is the commonly employed method to produce f-CNTs (Jiang and Gao, 2003; Solhy et al., 2008). The key parameters to be controlled during the oxidation treatment are the concentration of the acid, the temperature of the reaction, and the duration of the treatment process. One of the key challenges in the oxidation of CNTs is that the nanotubes tend to form aggregates or bundles, which may prevent access of the liquid-phase oxidizing agents to the surfaces of the CNTs. Therefore, it is essential to ensure that CNTs are properly dispersed within the oxidizing solution, and the most common method to disperse CNTs during acid treatment is sonication (Xing, 2004; Xing et al., 2005).

Oxidation of CNTs can also be used to open the nanotube caps and cut the tubes into shorter lengths, which provides opportunities to confine metal nanoparticles inside the CNT channels

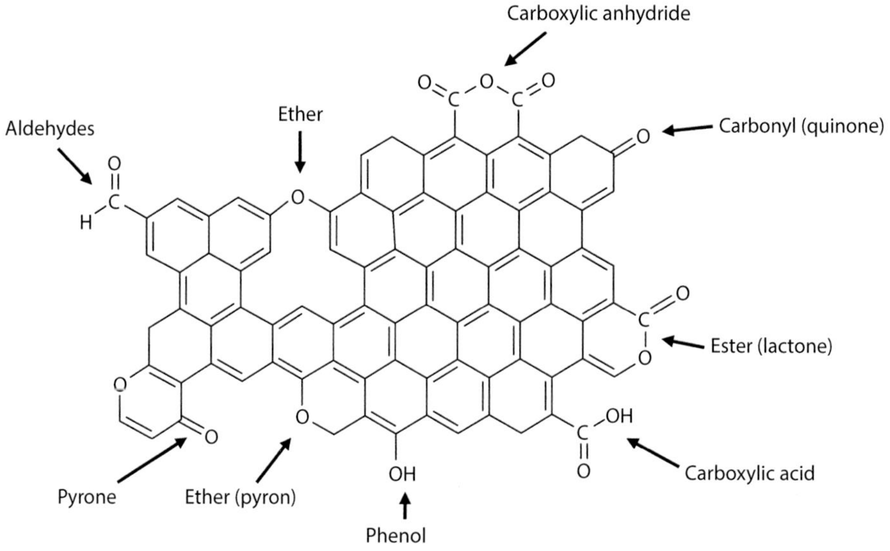

Figure 15.1   Schematic representation of oxygen functional groups present on CNT surfaces. (Reprinted with permission from Kundu, S., Wang, Y.M., Xia, W., and Muhler, M., Thermal stability and reducibility of oxygen-containing functional groups on multiwalled carbon nanotube surfaces: A quantitative high-resolution XPS and TPD/TPR study, *J. Phys. Chem. C*, 112, 2008, 16869–16878. Copyright 2008, American Chemical Society.)

to create nanoreactors. Although acid oxidation treatment to prepare CNTs for catalyst filling is a relatively inexpensive procedure that can be completed with commonly available glassware, the process must be controlled carefully to ensure the acid attack is targeted at the end caps and specific locations along the nanotube sidewalls. Otherwise, the integrity of the CNT may be lost, a wide distribution of tube lengths may be obtained, and low product yield could result. Recently, we reported the synthesis of Ru nanoparticle–filled CNTs (Wang et al., 2011) that had been cut using a Ag- or Fe-catalyzed oxidation method reported by others (Pan and Bao, 2008; Wang et al., 2008a). The weight loss of CNTs in this experiment was relatively small (~20 wt.%), and the length distribution of CNTs could be tuned by the amount of Ag or Fe loaded to the CNTs, the oxidation temperature, and the duration of treatment. This treatment procedure introduced phenol, ether, and carbonyl groups to the CNTs instead of carboxyl groups that are commonly generated by direct acid oxidation (Wang et al., 2011). Moreover, the defects in the catalytic oxidized f-CNTs were at the ends of the nanotubes with very few defects created along the sidewall sections.

Nitrogen-containing functional groups are another important type of modification to the surface structure and functionality of CNTs. Similar to oxygen functional groups, nitrogen groups can be used to enable the uniform assembly of metal nanoparticles on CNTs. Moreover, the nitrogen heteroatoms can modify the electron-donor properties of CNTs and can act as active basic sites in base-catalyzed chemical conversions (van Dommele et al., 2006). Nitrogen atoms can be incorporated directly into the nanotube structure during the CNT synthesis by the pyrolysis of mixtures of organometallic and nitrogen-containing organic compounds (Glerup et al., 2003; Kudashov et al., 2004; Maiyalagan et al., 2005), or by chemical vapor deposition of nitrogen-rich hydrocarbons like melamine over a heterogeneous iron, cobalt, or nickel catalyst (Tang et al., 2004; van Dommele et al., 2006). Alternatively, nitrogen-containing functional groups can be introduced by post-synthesis treatment such as annealing CNTs in ammonia atmosphere (Arrigo et al., 2008), microwave plasma treatment of CNTs in N_2 (Chen et al., 2010), and by electro-reduction to graft an aminobenzene monolayer onto the CNT surface (Guo and Li, 2005).

Functional groups containing sulfur, phosphorus (Cruz-Silva et al., 2008; Zhang et al., 2008), fluorine (Lee et al., 2003), chlorine (Barthos et al., 2005), and bromine (Unger et al., 2002) have also been covalently attached to CNTs to modify the CNT surface and electronic properties and to improve their performance in the catalytic application. For example, ball-milling nanotubes in a H_2S atmosphere has been reported as an effective method to create thiol groups on CNTs (Kónya et al., 2002), and the photolysis of cyclic disulfide with CNTs led to the formation of sulfur-containing functional groups and the modification of CNT sidewalls (Nakamura et al., 2006). Due to the strong affinity of thio and thiol groups to gold, sulfur f-CNTs are of particular interests in preparing catalysts with self-assembled Au monolayers.

Once a pristine CNT has been functionalized with an oxygen functional group, or other groups, this simple f-CNT can be further functionalized by forming covalent bonds with more complex moieties. The carboxyl groups on f-CNTs are the most commonly employed sites for further functionalization, since these groups enable the covalent coupling of molecules through the creation of amide and ester bonds. For example, (1) thiol groups were attached to carboxyl groups on CNTs by thiolation (Lim et al., 2003) and (2) dendrimers and nucleic acids have been attached to carboxylic groups on CNTs (Balasubramanian and Burghard, 2005). In another example, Liu and Adronov (2004) reacted the carboxylic groups on CNTs with thionyl chloride and then pentaerythritol to obtain an f-CNT with three primary alcohol functionalities generated for every reacted acid group (Figure 15.2). This approach allows the number of terminating functional groups on the CNT surface to be increased and thus may present an increased number of anchor sites for catalyst deposition.

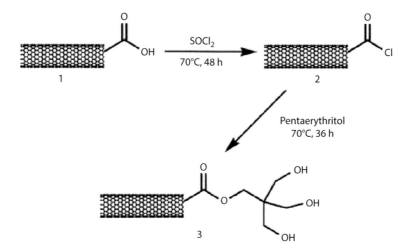

Figure 15.2 Schematic diagram of maximizing the functional group amount on CNTs by treating with pentaerythritol. (Reprinted with permission from Liu, Y. and Adronov, A., Preparation and utilization of catalyst-functionalized single-walled carbon nanotubes for ring-opening metathesis polymerization, *Macromolecules*, 37, 4755–4760, 2004. Copyright 2004, American Chemical Society.)

15.3.2 Noncovalent Functionalization

Since noncovalent functionalization methods offer the possibility of attaching functional moieties to CNTs without affecting the electronic network of the tubes, noncovalent functionalization has attracted extensive attention for the electrocatalysis applications of CNTs. In addition, the noncovalent functionalization of CNTs by wrapping with bioactive macromolecules such as protein and DNA opens the possibility of their application in biomedical fields such as drug delivery (Chen et al., 2001; Schnorr and Swager, 2010).

One of the most useful noncovalent methods to create f-CNTs for catalyst deposition is to utilize π–π stacking interactions between the graphitic sidewalls of CNTs and aromatic rings in functional moieties overlaying the sidewall. The terminal functional groups of the aromatic compounds can then act as anchor sites for metals or other heteroatoms. An example of the π–π stacking approach is the attraction of pyrene-containing compounds to CNT sidewalls, as shown in Figure 15.3. These pyrene compounds with a long aliphatic chain terminated by either a thiol group or a carboxylic group can function as linkers to deposit metal nanoparticles onto CNTs, as has been demonstrated by Georgakilas et al. (2005) and Liu et al. (2003). Similarly, bifunctional molecules such as 1-aminopyrene with amino as end functional groups to anchor metal nanoparticles have been employed to functionalize CNTs via π–π stacking (Wang et al., 2008b). Benzyl alcohol adsorbed on a CNT surface via π–π stacking interactions of benzyl ring with CNT surface can provide a hydroxyl that increases the hydrophilicity of the CNTs and then coordinate with metal precursors to facilitate better metal dispersion (Eder and Windle, 2008; Aksel and Eder, 2010).

Electrostatic interactions between low concentrations of carboxylic groups on a purified CNT and a cationic polyelectrolyte can be used to functionalize a CNT. For example, the cationic polyelectrolyte poly(diallyldimethylammonium) chloride (PDADMAC) was adsorbed to a CNT through the electrostatic interaction between the carboxyl groups and the polyelectrolyte

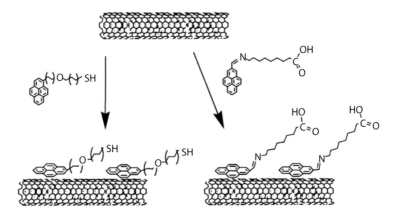

Figure 15.3 Noncovalent functionalization of CNTs by pyrene derivatives through π–π stacking interaction. (Reprinted from *Chem. Phys. Lett.*, 367, Liu, L., Wang, T., Li, J., Guo, Z.-X., Dai, L., Zhang, D., and Zhu, D., Self-assembly of gold nanoparticles to carbon nanotubes using a thiol-terminated pyrene as interlinker, 747–752, Copyright 2003, with permission from Elsevier; Reprinted with permission from Georgakilas, V., Tzitzios, V., Gournis, D., and Petridis, D., Attachment of magnetic nanoparticles on carbon nanotubes and their soluble derivatives, *Chem. Mater.*, 17, 2005, 1613–1617. Copyright 2005, American Chemical Society.)

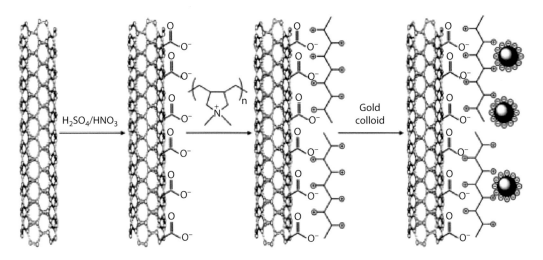

Figure 15.4 Noncovalent functionalization of CNTs by polyelectrolytes through electrostatic interaction for anchoring gold nanoparticles. (Reprinted with permission from Jiang, K., Eitan, A., Schadler, L.S., Ajayan, P.M., Siegel, R.W., Grobert, N., Mayne, M., Reyes-Reyes, M., Terrones, H., and Terrones, M., Selective attachment of gold nanoparticles to nitrogen-doped carbon nanotubes, *Nano Lett.*, 3, 275–277, 2003. Copyright 2005, American Chemical Society.)

chains, and then the polyelectrolyte served as anchor sites for negatively charged Au nanoparticles (Figure 15.4) (Jiang et al., 2003). Similar functionalization of CNTs by PDADMAC was employed to deposit Pt nanoparticles (Wang and Wang, 2008), and the amino-rich cationic polyethyleneimine was reported to effectively interact with CNTs and provide adsorption sites for the attachment of Au nanoparticles (Hu et al., 2005).

Electrostatic approaches also permit the layer-by-layer self-assembly of polyelectrolytes on CNTs. Cationic polyelectrolyte, PDADMAC, was firstly adsorbed onto CNT surfaces, followed by the adsorption of anionic polyelectrolyte, poly(sodium 4-styrenesulfonate) (PSS). The PDADMAC/PSS bilayer coating formed on CNT surfaces due to the oppositely charged surfaces could be used for anchoring positively charged metal nanoparticles (Kim and Sigmund, 2004). Here, the wrapping of CNTs by polyelectrolyte also contributes to the driving force for the functionalization of polyelectrolyte on CNT's sidewalls.

15.4 METAL CATALYST LOADING ONTO CARBON NANOTUBES

CNTs have been used directly as catalysts for a few reactions including the decomposition of methane to produce CO- and CO_2-free hydrogen (Muradov, 2001), the oxidative dehydrogenation of ethylbenzene (Pereira et al., 2004), and the decomposition of NO (Luo et al., 2000). f-CNTs, such as nitrogen-containing CNTs, have also been reported as active and stable solid-base catalysts for Knoevenagel condensation (a nucleophilic addition of an active hydrogen compound to a carbonyl group followed by a dehydration reaction) (van Dommele et al., 2006). Although CNTs and f-CNTs have some catalytically active properties, the more common application of CNTs proposed in the literature is that the nanotubes are used as supports for an active catalyst phase. Metal or metal oxide can be loaded onto CNTs using organometallic complexes, forming metal nanoparticles directly on the CNT surfaces by in situ deposition and reduction, or the nanoparticles can be preformed and deposited onto CNTs. In this section, we describe each of these three methods for attaching active catalyst phases to CNTs. Another form of CNT catalyst often used for biological reactions is enzymes immobilized on CNTs, which can be prepared by either noncovalent (Chen et al., 2001) or covalent functionalization (Wang et al., 2005). The immobilization of enzymes on CNTs is beyond the scope of the current chapter.

15.4.1 Grafting of Organometallic Complexes onto Carbon Nanotubes

Organometallic complexes have been reported as useful homogeneous catalysts for reactions such as ring-opening metathesis polymerization, which may be catalyzed by ruthenium alkylidene (Liu and Adronov, 2004). The recovery of the organometallic catalyst from the reaction solution can, however, be a difficult and costly process. Therefore, it is desirable to immobilize homogeneous catalysts on nanoparticles like CNTs that can be more readily separated from the reagents and products than a dissolved complex. To graft organometallic complexes onto CNTs requires first the functionalization of the CNT surface so that covalent bonds between the organometallic complex and the f-CNTs can be formed via surface reactions.

Figure 15.5 presents an example of grafting an amino-containing complex, octaamino-substituted erbium bisphthalocyanine (OAErPc$_2$), to the carboxylic groups on f-CNTs by an amide bond (Xu et al., 2005). Xu et al. (2005) observed strong intramolecular interactions and charge transfer from the phthalocyanine rings of OAErPc$_2$ to CNTs. Similar amide linkages were employed by Frehill et al. (2002) to graft the ruthenium complex [Ru(dcdpy)(bpy)$_2$] (PF$_6$)$_2$ to CNTs via the reaction of the carboxylic groups of ruthenium complex and amino groups of CNTs. More examples of grafting organometallic complexes onto f-CNTs are listed in Table 15.1.

The nature of the CNT functionalization will determine the pathway for the grafting of the organometallic complex to the nanotube. For example, both pristine and f-CNTs will react by different pathways with Vaska's compound (*trans*-IrCl(CO)(PPh$_3$)$_2$) to form covalent CNT–metal complexes: Vaska's compound is coordinated to pristine CNTs by η^2-coordination across the graphene double bonds, but with oxidized f-CNTs, this compound will react with oxygen functionalities to form a hexacoordinate structure around the compound's Ir atom (Figure 15.6; Banerjee and Wong, 2002a). In that study, Banerjee and Wong (2002a) also reported that the complexes of oxidized nanotubes with Vaska's compound were more soluble and stable in an organic solution. In another study, Banerjee and Wong (2002b) reported that a hexacoordinate structure could also be formed around the Rh atom in the coordinating Wilkinson's complex (RhCl(PPh$_3$)$_3$) grafted to oxidized CNTs through the Rh metal–oxygen bond.

Although most studies reported in the literature have coupled organometallic complexes to CNTs by covalent bonds, there are also examples of noncovalent bonding of organometallic complexes to CNTs. For example, Park et al. (2006) utilized π–π stacking interactions between the cyclopentadiene (Cp) rings of Cp$_2$ZrCl$_2$ and the graphitic sidewalls of CNTs to immobilize this metallocene-based catalyst on CNTs. The interactions between Cp rings and CNTs were found to be quite stable and the Cp$_2$ZrCl$_2$ was not removed by washing with toluene.

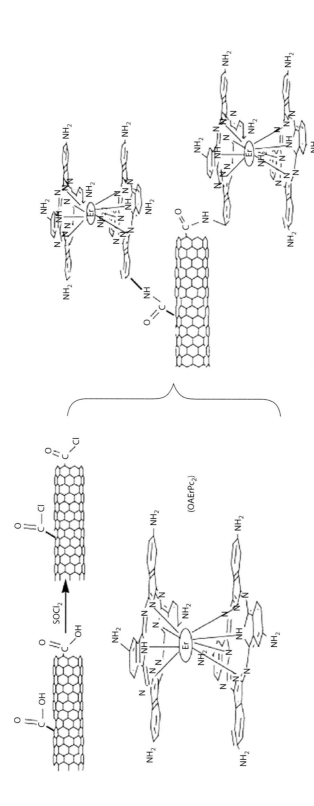

Figure 15.5 Synthetic route for OAErPc₂ covalently bonded to CNTs via the amide bond. (Reprinted from *Mater. Chem. Phys.*, 94, Xu, H.B., Chen, H.Z., Shi, M.M., Bai, R., and Wang, M., A novel donor–acceptor heterojunction from single-walled carbon nanotubes functionalized by erbium bisphthalocyanine, 342–346, Copyright 2005, with permission from Elsevier.)

TABLE 15.1   EXAMPLES OF GRAFTING ORGANOMETALLIC COMPLEXES ONTO f-CNTs

Functional Groups on CNTs	Functional Groups on Organometallic Complexes	Comments	Reference
Hydroxyl	Trialkoxysilane	Siloxane bonds were formed to immobilize the Rh complex onto CNTs.	Lemus-Yegres et al. (2006)
Thiol	Styryl	The grafting of the vanadyl Schiff base complex was achieved through the radical chain mechanism between thiol and vinyl groups.	Baleizão et al. (2004)
Nido–C_2B_9–carborane	—	CNT-supported nickel carborane complex was prepared by the in situ reaction of f-CNTs with dichlorobis (triphenylphosphine) nickel(II).	Zhu et al. (2006)

15.4.2 Formation of Metal Nanoparticles on Carbon Nanotube Surface

Metal and metal oxide nanoparticles have long been known to have exceptional catalytic properties for a wide range of important reactions. Porous alumina, zeolite, and activated carbon-supported metal catalysts have been used industrially for many decades. CNTs offer many advantages over the conventional porous support materials because CNTs provide excellent electrical conductivity, mechanical strength, thermal stability, and adsorption properties. CNT-supported metal nanoparticles can be synthesized via two different routes: (1) by the reduction of metal precursors to metal nanoparticles in the presence of CNTs or (2) by the inclusion of the preformed metal nanoparticles onto CNTs.

The metal precursors, usually from a solution of the metal salt, can be adsorbed onto CNTs and subsequently reduced to form metal nanoparticles on the surface of CNTs. A strong interaction between the precursor and the CNT surface is required to facilitate adsorption of the metal precursor on the CNTs, and as discussed in Section 15.2, functionalization on CNTs is required to achieve this strong interaction. Wet impregnation is the simplest and most versatile procedure to prepare supported metal catalysts, and this method has been widely used to prepare CNT-supported metal catalysts. In a typical impregnation procedure, f-CNTs are mixed with the metal precursor solution, and the mixture is stirred for sufficient time to disperse the metal precursor across the CNT surfaces. Sonication can be employed to disperse the CNTs and to assist the homogeneous distribution of precursor over nanotubes. The impregnated CNTs are dried and then can be calcined or reduced to convert the metal precursor to the desired active metal nanoparticle phase. As described in Section 15.3, the most common CNT functionalization method is oxidation in HNO_3 or H_2SO_4 + HNO_3 acid solutions. For example, Yu et al. (1998) reported that the mixed acid solution produced a higher density of surface functional groups on CNTs than just the HNO_3 acid treatment and led to a higher dispersion of Pt nanoparticles on CNT surfaces. In another example, Giordano et al. (2003) grafted the dimeric complex [$(Rh_2Cl_2(CO)_4$] on COONa–f-CNTs to prepare molecularly dispersed Rh carbonyl species, which were then reduced to Rh nanoparticles as illustrated by the scheme in Figure 15.7.

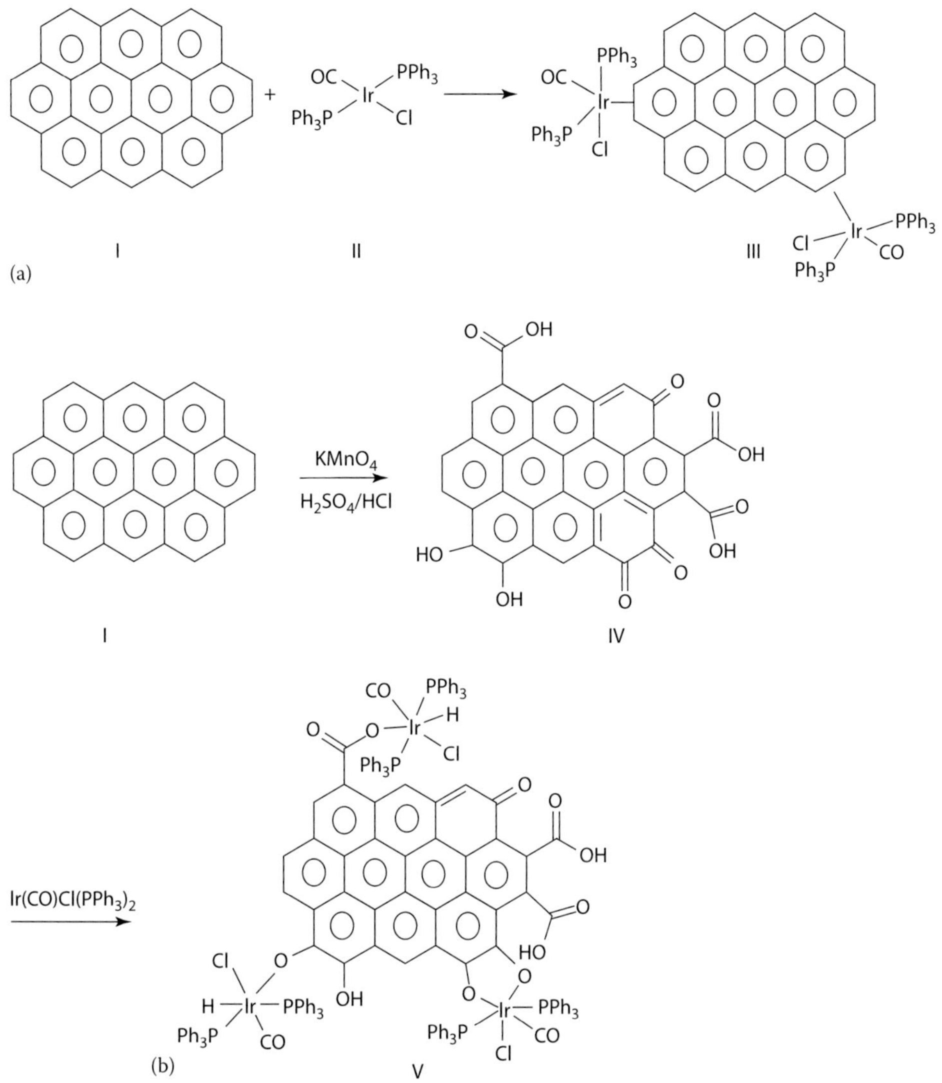

Figure 15.6   Proposed reaction scheme for the functionalization reactions of CNTs with *trans*-IrCl(CO) (PPh₃)₂. (a) Complexation of *trans*-IrCl(CO)(PPh₃)₂ with raw CNTs produced a theorized η²-coordination complex. (b) Addition of *trans*-IrCl(CO)(PPh₃)₂ to the oxidized CNTs yielded a nanotube–Vaska's compound adduct, with several possible modes of metal–nanotube coordination illustrated. (Reprinted with permission from Banerjee, S. and Wong, S.S., Functionalization of carbon nanotubes with a metal-containing molecular complex. *Nano Lett.*, 2, 49–53, 2002a. Copyright 2002, American Chemical Society.)

The impregnation conditions can be controlled to achieve selective deposition of metal nanoparticles either on the external CNT surface or inside the nanotubes. In order to deposit metal nanoparticles by impregnation inside the channels of CNTs, supercritical CO_2 and other liquids with low surface tension were employed as solvent to dissolve metal precursors (Pan and Bao, 2008). After sonication and stirring for an extended period, the metal precursor solutions

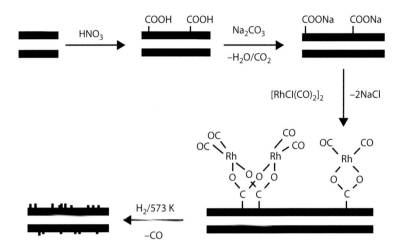

Figure 15.7 Schematic diagram of surface-mediated organometallic reaction between $[(Rh_2Cl_2(CO)_4]$ and COONa-modified CNTs to prepare highly dispersed Rh nanoparticles. (From Giordano, R., Serp, P., Kalck, P., Kihn, Y., Schreiber, J., Marhic, C., and Duvail, J.-L., Preparation of rhodium catalysts supported on carbon nanotubes by a surface mediated organometallic reaction, *Eur. J. Inorg. Chem.*, 2003, 610–617. Copyright Wiley-VCH Verlag GmbH & Co. KGaA. Reprinted with permission.)

filled the CNT channels by capillary forces (Pan and Bao, 2008). By careful selection of solvent for the metal precursor and the precise control of the volume of metal precursor solution, the selective deposition of nickel nanoparticles inside has also been reported (75% of nanoparticles in the internal cavity), and this process is illustrated in Figure 15.8.

Instead of loading a metal precursor on the CNT and then reducing in a second step, the metal precursor can be reduced in situ to deposit metal nanoparticles on CNT surfaces. Various metal nanoparticles, including Pt, Au, Pd, Ag, Rh, Ru, and Ni, have been supported on CNTs using one-step reduction and deposition procedures (Liu et al., 2002; Oh et al., 2005; Quinn et al., 2005; Li et al., 2006; Raghuveer et al., 2006). Nanotubes with high concentrations of oxygen-containing functional groups are preferred support CNT for single-step deposition–reduction procedures (Lordi et al., 2001; Liu et al., 2002). One of the key studies in this topic is the work by Rajalakshmi et al. (2005), who deposited Pt nanoparticles on CNTs by the reduction of H_2PtCl_6 with $NaBH_4$. Rajalakshmi and coworkers compared the dispersion of Pt nanoparticles deposited by this method on three types of CNTs: as-synthesized CNTs, sonicated CNTs, and acid-oxidized f-CNTs. The acid-oxidized f-CNTs were observed to have smaller Pt nanoparticles (3–5 nm) with a more uniform particle distribution than the nonoxidized CNTs.

Further supporting evidence of the advantage of f-CNTs for the deposition of metal nanoparticles is provided by Zanella et al. (2005). Zanella et al. investigated the functionalization of CNTs with a series of aliphatic bifunctional thiols through a solvent-free procedure. The f-CNTs were used as supports to deposit gold nanoparticles via the reaction of $HAuCl_4$ with citric acid. The particle size of Au nanoparticles deposited on the f-CNTs was dependent on the aliphatic bifunctional thiols employed: small gold particles, with a narrow particle size distribution around 1.7 nm, were obtained on 1,6-hexanedithiol-f-CNTs, but the average Au particle size was around 5.5 nm for CNTs functionalized with aminothiol.

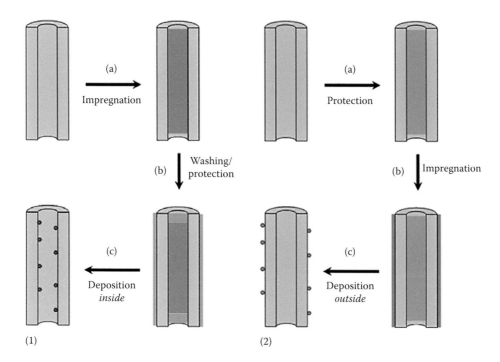

Figure 15.8   Schematic views of a longitudinal cross section of a CNT during different steps for the selective deposition of nanoparticles: (1) inside CNTs (including (a) impregnation with the ethanolic solution containing the metal precursor; (b) impregnation with distilled water to wash and protect the outer surface from metal deposition; and (c) after drying, calcination, and reduction, metal particles are decorating the inner surface of the CNT with a high selectivity are metal particles) and (2) outside CNTs (including (a) impregnation with the organic solvent to protect the inner tubule from metal deposition; (b) impregnation with the aqueous solution containing the metal precursor; (c) after drying, calcination, and reduction, metal particles are decorating the outer surface of the CNT with a high selectivity are metal particles). (Reprinted with permission from Tessonnier, J.P., Ersen, O., Weinberg, G., Pham-Huu, C., Su, D.S., and Schlogl, R., Selective deposition of metal nanoparticles inside or outside multiwalled carbon nanotubes, *ACS Nano*, 3, 2081–2089, 2009. Copyright 2009, American Chemical Society.)

Agents used to reduce metal precursors to metal nanoparticles on CNT include hydrogen, polyol, formaldehyde, and sodium borohydride. The selection of reducing agent can greatly influence the metal reduction kinetics and, subsequently, affect the growth and ultimate size of the metal nanoparticles. Li et al. (2003b) investigated two reducing agents—formaldehyde and ethylene glycol (EG)—to prepare Pt/CNTs from H_2PtCl_6. The Pt/CNTs reduced in EG exhibited a more homogeneous dispersion of spherical Pt nanoparticles with a narrower particle size distribution than the Pt/CNTs reduced in formaldehyde. Likewise, Rajalakshmi et al. reported sodium borohydride to be a more effective reducing agent to prepare Pt nanoparticles on CNTs than EG (Rajalakshmi et al., 2005).

Reduction of metal precursors on CNTs is usually performed at elevated temperatures to accelerate the reduction of metal precursors and the nucleation of metal particles. Microwave radiation, which can provide fast heating, has also been studied recently as a low-temperature method for the reduction of metal precursors in polyol. For example, Chen et al. (2005) and

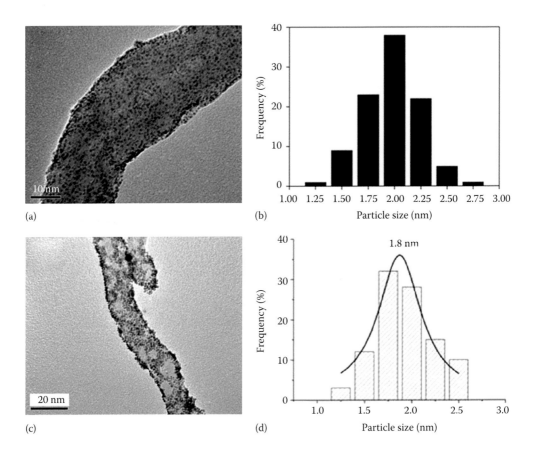

Figure 15.9 Transmission electron microscopy (TEM) images and particle size distribution histograms of Pt–Ru nanoparticles (a and b) and Pt nanoparticles (c and d) on polyelectrolyte–f-CNTs. The Pt–Ru and Pt loading were both 40 wt.%. (Reprinted with permission from Wang, S., Wang, X., and Jiang, S.P., Pt–Ru nanoparticles supported on 1-aminopyrene-functionalized multiwalled carbon nanotubes and their electrocatalytic activity for methanol oxidation, *Langmuir*, 24, 10505–10512. Copyright 2008 American Chemical Society; Reprinted with permission from Wang, S., Jiang, S.P., White, T.J., Guo, J., and Wang, X., Electrocatalytic activity and interconnectivity of Pt nanoparticles on multiwalled carbon nanotubes for fuel cells, *J. Phys. Chem. C*, 113, 18935–18945, 2009. Copyright 2009, American Chemical Society.)

Liu et al. (2005) reported that microwave heating can minimize the temperature and concentration gradients in the reactant solution and produce a more uniform environment for the nucleation and growth of metal nanoparticles. Wang et al. (2008b, 2009) reported CNTs functionalized with various polyelectrolytes through π–π stacking or electrostatic attractions that were used as supports for metal nanoparticles deposited using a microwave-assisted polyol process. Figure 15.9 shows Pt–Ru or Pt nanoparticles with small sizes and narrow particle size distributions that were uniformly distributed on CNTs in Wang et al.'s studies. In addition, the in situ reduction of metal ions can be conducted at room temperature under gamma irradiation as demonstrated with Ag, Pd, and Pt–Ru nanoparticles on CNTs by Oh et al. (2005).

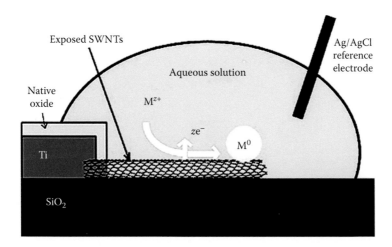

Figure 15.10   Schematic diagram of the CNT as an electrode used for electrodeposition of metal nanoparticles. (Reprinted with permission from Quinn, B.M., Dekker, C., and Lemay, S.G., Electrodeposition of noble metal nanoparticles on carbon nanotubes, *J. Am. Chem. Soc.*, 127, 2005, 6146–6147. Copyright 2005, American Chemical Society.)

Another strategy to form metal nanoparticles on CNT surfaces is electrodeposition (Wang et al., 2004a; Day et al., 2005; Quinn et al., 2005). For example, Au nanoparticles were deposited on CNTs by electrodeposition from a $HAuCl_4$ aqueous solution in a two-electrode cell as shown in Figure 15.10. The size and surface coverage of the metal nanoparticle can be tuned with the deposition potential, deposition time, and metal precursor concentration (Quinn et al., 2005).

15.4.3   Inclusion of Preformed Metal Nanoparticles onto Carbon Nanotubes

Direct deposition and reduction in a single-step process, as described in Section 15.4.2, is a simple and effective way to disperse metal nanoparticles on the CNT surfaces. However, in the single-step approach, it is difficult to control the nanoparticle size, shape, and composition, together with the absolute metal loading. The alternative approach that can potentially provide better control of particle properties is to attach preformed metal nanoparticles to the CNTs. Furthermore, solution-phase nanoparticle synthesis techniques that have been developed recently allow preparation of metal nanoparticles with precisely controlled size, shape, and composition (Xia et al., 2009). For example, colloidal metal nanoparticles synthesized by the reduction of metal precursors in solution, or by chemical reduction with alcohol (polyol in most cases), hydrogen, and borohydride, are commonly used. To stabilize the nanoparticles and avoid aggregation, stabilizers such as dendrimers, block copolymer micelles, and surfactants have been used as capping agents during the particle synthesis (Narayanan and El-Sayed, 2005).

Once preformed metal nanoparticles have been prepared, these particles may need to be modified with suitable functional groups that can be reacted, or interacted, with functional groups on the CNT surface. For example, an interlinker of a bifunctional molecule (17-(1-pyrenyl)-13-oxo-heptadecanethiol) terminated with pyrenyl unit at one end and thiol group at the other end has been reported to connect Au nanoparticles to CNTs through the π–π stacking interaction of a pyrenyl ring with the CNTs and the covalent bonding of thiol

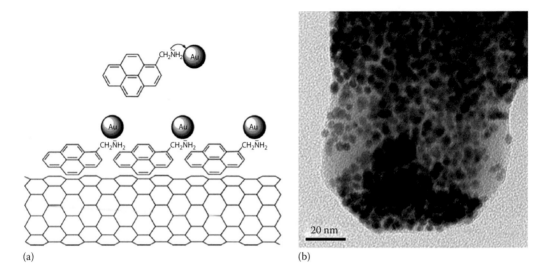

(a)   (b)

Figure 15.11   (a) Schematic diagram of gold nanoparticle assembly on CNTs through 1-pyrenemethyl-amine interlinker and (b) a TEM image of the derived Au nanoparticle–CNT composites. (Reprinted with permission from Ou, Y.-Y. and Huang, M.H., High-density assembly of gold nanoparticles on multiwalled carbon nanotubes using 1-pyrenemethylamine as interlinker, *J. Phys. Chem. B*, 110, 2006, 2031–2036. Copyright 2006, American Chemical Society.)

group with Au nanoparticles (Liu et al., 2003). Sulfur-containing functional groups have also been introduced on the sidewalls of CNTs via the photolysis of cyclic disulfides, and gold nanoparticles can then be attached onto the f-CNTs through the strong Au–S binding interaction (Nakamura et al., 2006). Similarly, Coleman et al. introduced thioether groups onto CNT surfaces through the Bingel reaction and grafted 5 nm gold colloids onto the CNTs (Coleman et al., 2003).

Metal nanoparticles can be modified with aromatic compounds, and through stacking onto the CNT surfaces by π–π interaction between the aromatic rings on nanoparticles and the CNT sidewalls, metal nanoparticles are connected to CNTs. Au nanoparticles were functionalized by 1-pyrenemethylamine through the alkylamine substituent and attached to CNTs via π–π stacking interaction (Figure 15.11a). Using this strategy, Au nanoparticles with diameters of 2–4 nm can be densely assembled on the sidewalls of CNTs (Figure 15.11b). Other molecules with similar structures to 1-pyrenemethylamine, such as N-(1-naphthyl)ethylene diamine and phenethyl-amine, can also be used as the interlink to form the high-density deposition of Au nanopar-ticles on the surface of CNTs (Ou and Huang, 2006). Pt nanoparticles with sizes of around 2 nm were also modified by triphenylphosphine (PPh₃) and deposited on the surface of CNTs by the π–π interaction between the phenyl rings of triphenylphosphine and the backbone of CNTs (Mu et al., 2005).

Hydrophobic interactions between the ligands' capping metal nanoparticles and CNT sur-faces have been used to immobilize metal nanoparticles onto CNTs. For example, octanethiol-capped Au nanoclusters of 1–3 nm diameter were anchored to the sidewalls of acetone-activated CNTs through the hydrophobic interactions between methyl groups in acetone (on CNTs) and alkyl chains in octanethiol (on Au nanoparticles) (Ellis et al., 2003). Han et al. used a molecu-larly mediated route to assemble alkanethiolate monolayer–capped Au nanoparticles on CNTs.

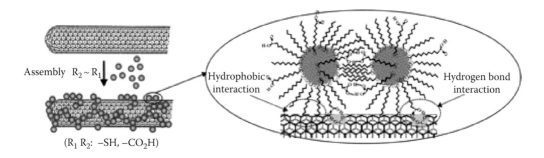

Figure 15.12 Schematic illustration of the assembly of monolayer-capped nanoparticles on CNTs through a combination of hydrophobic and hydrogen bond interactions. (Reprinted with permission from Han, L., Wu, W., Kirk, F.L., Luo, J., Maye, M.M., Kariuki, N.N., Lin, Y., Wang, C., and Zhong, C.-J., A direct route toward assembly of nanoparticle–carbon nanotube composite materials, *Langmuir*, 20, 2004, 6019–6025. Copyright 2004, American Chemical Society.)

Both the hydrophobic interaction between the alkyl chains of the capping molecules on nanoparticles and the hydrophobic sidewalls of CNT and the hydrogen bonds between the carboxylic groups of the nanoparticle capping molecules and those present on CNT surfaces contributed to the high stability of the resulted composite materials (Figure 15.12).

The anchoring of metal colloids to CNTs can also be achieved by electrostatic interactions between CNTs and metal nanoparticles. Figure 15.5 illustrates the attachment of negatively charged Au nanoparticles to positively charged CNT surfaces, which were modified by cationic polyelectrolytes (Jiang et al., 2003). Similarly, the positively charged metal nanoparticles were anchored on the negatively charged CNTs either from acid treatment or from acid treatment followed by layer-by-layer polyelectrolyte self-assembly. The density of nanoparticles binding to the former was lower than those to the latter, probably caused by the enlarged reactive sites (negatively charged sites) from polyelectrolyte coatings (Kim and Sigmund, 2004).

Despite the advantages of using preformed metal nanoparticles, there remains a significant challenge in using this procedure in catalyst preparation: usually the capping agents used to stabilize the nanoparticles in solution must be removed from the surface of nanoparticles. If not removed after immobilizing the nanoparticles on the CNT support, the capping agent could block the active catalytic sites. Thermal treatment is the most commonly used method to remove capping agents from the surface of nanoparticles. In a typical treatment process, Han et al. (2004) removed thiolate capping shells from Au nanoparticles at 300°C. The nanoparticles remained strongly adhered to the CNT surfaces, even when the CNTs were sonicated in a hydrophobic solvent the nanoparticles did not detach. Still after thermal treatment, some aggregation of the nanoparticles was observed, with the extent of particle size growth dependent on the initial loading of the nanoparticles. Thermal treatment was also used to remove PPh_3 capping layers surrounding Pt nanoparticles deposited on CNTs (Mu et al., 2005). The original average Pt nanoparticle size was 1.9 ± 0.5 nm; after thermal treatment, small particle size growth to 2.7 ± 1.0 nm occurred (Figure 15.13). Most Pt nanoparticles were smaller than 4 nm, and the size distribution remained to be narrow even after thermal treatment. Besides thermal treatment, UV–ozone oxidation has been reported as an effective method to remove the capping layers from the surface of Pt nanoparticles (Aliaga et al., 2009).

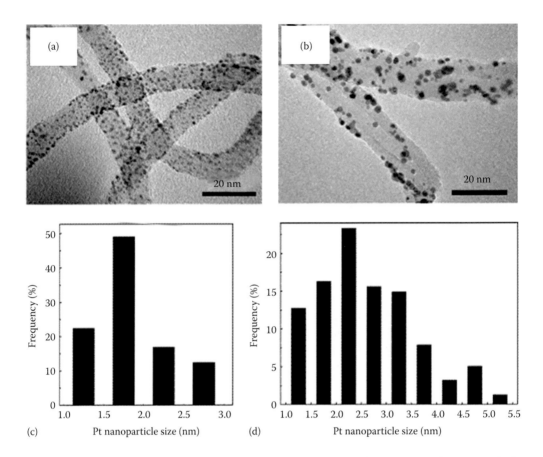

Figure 15.13 TEM images and particle size distribution histograms of Ru nanoparticles on CNTs before (a and c) and after (b and d) thermal treatment. The Pt loading was 24 wt.%. (Reprinted with permission from Mu, Y., Liang, H., Hu, J., Jiang, L., and Wan, L., Controllable Pt nanoparticle deposition on carbon nanotubes as an anode catalyst for direct methanol fuel cells, *J. Phys. Chem. B*, 109, 2005, 22212–22216. Copyright 2005, American Chemical Society.)

15.5 APPLICATIONS OF CARBON NANOTUBE–SUPPORTED METAL CATALYSTS

Since the first report on dispersing Ru nanoparticles on CNT surfaces for the catalytic reaction of cinnamaldehyde hydrogenation (Planeix et al., 1994), CNT-supported metal catalysts have been applied in various catalytic reactions, exhibiting higher activity and selectivity than those deposited on conventional catalyst supports such as activated carbon, zeolite, and Al_2O_3 (Serp et al., 2003; Georgakilas et al., 2007; Chu et al., 2010; Serp and Castillejos, 2010).

15.5.1 Supported Homogeneous Catalysis

Supported homogeneous catalysis using a metal complex immobilized on a solid support is regarded as a novel system that combines the advantages of both heterogeneous and homogeneous catalyses. Organometallic complexes grafted on CNT surfaces have been found active in many homogeneous reactions, including the hydrogenation of cyclohexene to cyclohexane (Banerjee and Wong, 2002b), ring-opening metathesis polymerization (Liu and Adronov, 2004), the cyanosilylation of aldehydes (Baleizão et al., 2004), and olefin polymerization (Park et al., 2006; Zhu et al., 2006).

In the hydrogenation of cyclohexene, Rh(NN)Si immobilized on CNTs exhibited a catalytic activity significantly higher than the free Rh(NN)Si molecules. The reasons could be the electronic modifications of the anchored complex through the interaction of metal atoms with CNT surfaces and also by the confinement inside the porosity of CNTs. The Rh complexes were not leached out after three catalytic runs, indicating the effective anchoring of the complex on the CNT surfaces. The catalytic performances of the hybrid catalysts were enhanced in the second and third runs compared to the initial run, which revealed the possibility of more active species resulted from the reaction conditions (Lemus-Yegres et al., 2006). The CNT-supported nickel(II) carborane complexes were also found to be more active than their homogeneous analogues in olefin polymerization (Zhu et al., 2006).

Many comparisons have been made between CNTs and other supports for homogeneous catalysis. In one study, Baleizão et al. (2004) reported that CNT-supported vanadyl complexes exhibited significantly higher activity for the catalytic cyanosilylation of aldehydes with trimethysilylcyanide than vanadyl complexes supported on activated carbon. The vanadyl complex/CNTs system was truly heterogeneous (no leaching) and reusable (no activity decrease) over five consecutive reaction runs. Similarly, Lemus-Yegres et al. (2006) reported higher catalytic activity for a hydrogenation reaction performed over Rh complexes supported on CNTs than over Rh complexes supported on activated carbon. The superior performance of CNTs as homogeneous catalyst supports was attributed to the unique structures and properties of the CNTs; however, detailed investigations on the inherent reasons for the higher activity of these CNT-supported catalysts have not been conducted.

15.5.2 Heterogeneous Catalysis

CNT-supported metal catalysts have been widely employed for various liquid-phase and gas-phase heterogeneous catalytic reactions, with some representative examples summarized in Table 15.2. It can be seen that CNT-supported metal catalysts present superior catalytic performance among various supports including activated carbon, graphite, SiO_2, CeO_2, MgO, TiO_2, and Al_2O_3. Their high catalytic activity can be attributed to several reasons, such as low diffusion resistances in the mesopores of CNTs, the high metal dispersion on CNTs, the high electronic conductivity of CNTs, and the confinement effect inside CNT channels.

The structure and properties of CNTs can be controlled and modified through different ways: (1) modulate CNT synthesis and pretreatment conditions to control the specific surface areas and/or internal diameters of the CNTs, (2) dope the CNTs with other elements such as nitrogen and boron, (3) functionalize the CNT surfaces through oxidative treatment or grafting, and (4) deposit metal particles either on the external CNT surfaces or in the internal CNT channels (Serp and Castillejos, 2010). These modified properties of CNTs can change the electronic structure of metal catalysts, metal dispersion, as well as metal–support interactions and in turn alter the activity and selectivity of CNT-supported catalysts.

Firstly, the diameter and graphitic degree of CNTs affect their electronic structure and adsorption behavior, thereby influencing the catalytic performance of CNT-supported metal catalysts. When applying different sized CNTs as metal support, different catalytic performances have been obtained. The activity of Fe nanoparticles deposited on narrow-pored CNTs was 2.5 times that of

TABLE 15.2   CATALYTIC PERFORMANCE OF CNT-SUPPORTED METAL CATALYSTS IN HETEROGENEOUS CATALYSIS

Reaction	Catalyst	Catalytic Performance	Reference
Fischer–Tropsch synthesis (FTS)	Fe	Higher catalytic activity and selectivity toward heavier hydrocarbons were obtained over Fe nanoparticles supported on CNTs with narrow pores due to higher metal dispersion and reduction degree.	Abbaslou et al. (2010)
	Fe	The confinement of Fe nanoparticle inside CNTs led to higher selectivity to long-chain hydrocarbons and retarded particle aggregation during reduction and reaction.	Abbaslou et al. (2009); Chen et al. (2008a)
	Co	For Co/CNT catalyst system, the addition of Ru increased the FTS rate while the K addition decreased the FTS rate; both Ru and K enhanced the selectivity of FTS toward heavier hydrocarbons.	Trepanier et al. (2009)
Ethanol production from syngas	Rh–Mn	The Rh–Mn nanoparticles confined inside CNTs presented an overall ethanol formation rate of 30 mol/mol Rh/h, an order of magnitude higher than those deposited outside CNTs, despite the higher accessibility of the latter.	Pan et al. (2007b)
NH_3 synthesis and decomposition	Ru, Rh, Pt, Pd, Ni, and Fe	CNT-supported Ru nanoparticles exhibited the highest NH_3 conversion rate among various active phases on different supports. Their performance could be further improved by modifying CNTs with KOH.	Yin et al. (2004b)
	Ru	The Ru nanoparticles deposited on exterior CNT surfaces showed higher activity than those confined inside nanotube channels in ammonia synthesis, although both had a similar Ru particle size.	Guo et al. (2010)
	Co, Fe	Fresh commercial CNTs containing residual Co or Fe nanoparticles were highly active for ammonia decomposition.	Zhang et al. (2007)
	Ru	Both the elimination of acidic groups from CNT surfaces and the nitrogen doping onto CNTs enhanced catalytic activity of Ru/CNTs in ammonia decomposition.	Garcia-Garcia et al. (2010)
Preferential CO oxidation (PROX)	Ru	The confinement of Ru nanoparticles inside CNTs improved the catalytic performance for the PROX reaction, owing to the reinforced reactivity sites and the reactant enrichment inside the nanochannels of CNTs.	Li et al. (2011)

(Continued)

TABLE 15.2 (*CONTINUED*)   CATALYTIC PERFORMANCE OF CNT-SUPPORTED METAL CATALYSTS IN HETEROGENEOUS CATALYSIS

Reaction	Catalyst	Catalytic Performance	Reference
	Pt	Pt supported on CNTs was extremely active for the PROX reaction at room temperature, compared to the Pt supported on active carbon and graphite.	Tanaka et al. (2008)
	Pt	The PROX reaction proceeded rapidly on Pt catalysts supported on CNTs, which was due to the promotion effect of residue Ni–MgO on CNTs from their synthesis.	Tanaka et al. (2010)
Hydrogenation	Ru	Ru/CNTs showed higher catalytic activity than Raney Ni/Al$_2$O$_3$ and Ru/SiO$_2$ in the glucose hydrogenation to sorbitol.	Pan et al. (2007a)
	Ru	CNTs were the most efficient support for Ru in the cellulose conversion (hydrolysis and hydrogenation) among SiO$_2$, CeO$_2$, MgO, Al$_2$O$_3$, and CNTs.	Deng et al.(2009)
	Pt–Ru	The confined Pt–Ru bimetallic nanoparticles of 2 nm size inside CNTs displayed excellent activity and selectivity in the selective hydrogenation of cinnamaldehyde.	Castillejos et al. (2009)
	Pt, Ru, Pt–Ru	In the selective hydrogenation of cinnamaldehyde, high activity was obtained when using CNTs as support compared to activated carbon; bimetallic nanoparticles were more selective than monometallic nanoparticles.	Vu et al. (2006a)
	Pt	CNT-encapsulated Pt nanoparticles showed high activity and enantioselectivity in the asymmetric hydrogenation of α-ketoesters, due to the ultrahigh enrichment of the chiral modifier and reactants inside the channels of CNTs.	Chen et al. (2011)
Dehydrogenation	Pt, Pd	The production of pure hydrogen was achieved by the dehydrogenation of cyclohexane or methylcyclohexane with Pt or Pd nanoparticles supported on CNTs, which was more efficient than the commercial Pt/Al$_2$O$_3$.	Wang et al. (2004b)
	CeO$_2$	CeO$_2$ particles filled inside CNTs showed superior catalytic performance of oxidative dehydrogenation of ethylbenzene to styrene, owing to the strengthened interaction between CeO$_2$ and inner CNT walls.	Rao et al. (2012)
C–C coupling	Pd	Pd/CNTs showed excellent catalytic activity and selectivity for Suzuki cross-coupling reactions, and the superior catalytic performance was maintained after several reaction cycles.	Chen et al. (2008b)

Fe nanoparticles on wide-pored CNTs in Fischer–Tropsch reaction. Both higher metal dispersion and higher reduction degree, resulted from the difference in the electronic properties of inner surfaces of the CNTs with different diameters, contributed to the more active and selective catalysts on narrow-pored CNTs (Abbaslou et al., 2010). Similarly, in CO oxidation, Cu nanoparticles supported on CNTs with smaller diameters presented better catalytic performance, which was attributed to their better Cu dispersion and higher thermal stability (Lu et al., 2009). Significant differences in the reaction selectivity have also been observed in cinnamaldehyde hydrogenation over Pt/CNTs with different CNT inner diameters (Ma et al., 2007).

The high degree of graphitization in CNTs has been discussed as a reason for the reactivities of CNT-supported metal catalysts, since graphitic carbon structure affects the electron conductivities of CNTs. A semiquantitative correlation between carbon graphitic degree and catalytic activity of carbon-supported catalysts has been obtained in ammonia decomposition reaction (Li et al., 2007). For the same catalytic reaction, Zheng et al. (2010) reported that CNT supports with higher graphitic order (i.e., CNTs with fewer defects) exhibited better catalytic performance than CNTs with many defects.

Secondly, the surface functionalization on CNTs is of vital importance to the performance of CNT-supported metal catalysts. Acid oxidation on CNTs has been found to be an effective way to decrease the particle size of metal nanoparticles deposited on CNTs through the generation of functional groups on CNT surfaces and therefore enhance the activity of many catalytic reactions, such as propylene decomposition (Li et al., 2003a) and Fischer–Tropsch reaction (Abbaslou et al., 2009). Different functional groups lead to varied interactions between CNTs and metal precursors, thereby affecting the metal dispersion and catalytic performance. Giordano et al. compared the Rh dispersion on the pristine, COOH-modified, and COONa-modified CNTs, and found the highest Rh dispersion on COONa-modified CNTs owing to the strongest interaction between the surface carboxylate groups and Rh precursor. The smallest Rh nanoparticles showed higher activity and selectivity toward C=C bond, compared with their large-sized counterparts on pristine and COOH-modified CNTs in the hydrogenation of *trans*-cinnamaldehyde and the hydroformylation of hex-1-ene (Giordano et al., 2003). The functional groups on CNTs can change the electronic structure of metal catalysts deposited on them through the interaction between metals and CNTs, and thus alter the catalytic activity and selectivity. In the selective hydrogenation of cinnamaldehyde, the rate of cinnamaldehyde conversion was significantly enhanced after the removal of the acidic oxygen groups from CNT surfaces, since the electron transfer from CNTs to metals was enhanced (Toebes et al., 2003; Vu et al., 2006b). The oxygen functional groups can withdraw electrons from metal centers and decrease the electron density of metals, leading to the deterioration of the rate-determining step of ammonia decomposition, associative desorption of nitrogen (Yin et al., 2004a,b; Chen et al., 2010). Other electron-withdrawing groups such as nitrogen groups also decreased the surface activity of Ru in ammonia decomposition (Chen et al., 2010). The effect of functional groups on the catalytic performance is dependent on the catalytic reactions. Wang et al. (2010) observed improved Pt dispersion after CNT functionalization. However, when these Pt/CNT catalysts were tested with different probe reactions, the functional groups showed varied effects: Some reactions were promoted, some reactions were inhibited, and some reactions were not significantly influenced by functionalization of the CNTs.

Thirdly, the well-defined tubular structure of CNTs provides an opportunity to confine metal nanoparticles inside the nanochannels of CNTs. An increasing number of studies have demonstrated that the confined metal nanoparticles present a different catalytic reactivity with respect to the same metals deposited on the CNT exterior surfaces. The confinement of metal nanoparticles inside CNTs has significantly improved the catalytic activity, compared with their counterparts dispersed outside CNTs, in Fischer–Tropsch synthesis (FTS; Chen et al., 2008a; Abbaslou et al., 2009), ethanol production from syngas (Pan et al., 2007b), preferential oxidation of CO in H_2-rich atmosphere (Li et al., 2011), and the selective hydrogenation of cinnamaldehyde (Castillejos et al., 2009). Not only catalytic activity, but also selectivity is altered by the encapsulation of catalyst nanoparticles inside CNTs. The confined Fe nanoparticles showed higher selectivity to long-chain hydrocarbons in FTS reaction than those nanoparticles dispersed on the outer surface of CNTs (Chen et al., 2008a; Abbaslou et al., 2009). In cinnamaldehyde hydrogenation, the bimetallic Pt–Ru nanoparticles confined inside CNTs also showed an improved selectivity toward cinnamyl alcohol formation (Castillejos et al., 2009). In preferential CO oxidation (PROX) reaction, the selectivity on CO oxidation was enhanced by confining Ru nanoparticles inside CNTs (Li et al., 2011). The confinement of metal catalyst inside CNTs does not always bring in improvement of catalytic performance. The Ru nanoparticles confined inside CNTs exhibited lower ammonia synthesis rate than those deposited on the exterior CNT surfaces (Guo et al., 2010).

Pan and Bao (2011) concluded the factors contributing to the confinement effect within CNTs on catalysis including the following: (1) The electronic interaction of the confined metals with CNTs can influence the electron transfer between reactants and catalysts, thereby influencing the catalytic activity; (2) the nanochannels of CNTs provide spatial restriction on metal particles and hamper their aggregation under reaction conditions; (3) the reactants can be enriched inside the CNT channels, creating a locally higher reactant concentration inside CNTs; and (4) the diffusion behavior of reactants inside the CNT channels can be altered. These four effects may influence the performance of metal-confined CNT catalysts to different extents, depending on the selected metals, the diameter of CNTs, and the specific reactions.

15.5.3 Electrocatalysis

Due to their high electronic conductivity, CNTs are promising support materials for electrocatalysts. To date, CNTs have been widely studied as support materials for Pt and Pt alloy catalysts in fuel cells such as polymer electrolyte membrane fuel cells (PEMFCs) and direct methanol fuel cells (DMFCs).

The CNT-supported metal catalysts have shown enhanced electrocatalytic activity and improved fuel cell performance in comparison to other carbon-supported catalysts. The oxygen reduction reaction (ORR) mass activity of Pt catalysts on CNTs was about six times as that of Pt on commercial Vulcan XC-72, and the current density of DMFC with the Pt/CNTs was 39% higher than that with the Pt/XC-72 (Li et al., 2003b). Pt/CNTs also excelled the commercial PEMFC catalyst (E-TEK) of the same Pt loading in terms of electrochemical activity (Xing, 2004). The bimetallic Pt–Ru nanoparticles supported on CNTs showed high activity for methanol oxidation reaction, and when using it as an anode catalyst in DMFC, the power density can be greatly enhanced and the metal loading can be reduced compared to Pt–Ru nanoparticles on carbon black (Li et al., 2006). In addition, CNTs were more corrosion resistant than carbon black under fuel cell operation condition; therefore, lower loss in active Pt area and ORR activity resulted with CNTs as support during fuel cell operation (Wang et al., 2006). The high electrocatalytic activity and durability of CNT-supported catalysts could be the result of their high electrical conductivity and electrochemical stability, smaller amount of organic impurities, better metal dispersion, and less reactant–product mass transport hindrance in their mesopores.

The surface functionalization on CNTs strongly affects the structure of CNT-supported metal nanoparticles and their electrocatalytic performance. It was found that Pt nanoparticles dispersed on acid-oxidized CNTs presented smaller size and more uniform particle distribution compared to those deposited on nonoxidized CNTs. A higher ORR activity was obtained with the oxidized CNT-supported Pt catalysts; when using them as the cathode catalyst, a cell voltage of 680 mV at 500 mA/cm² was obtained at fuel cell operating conditions (Rajalakshmi et al., 2005). The acid sonochemical technique was more effective to create functional groups on CNT surfaces for depositing small-sized and uniformly distributed Pt nanoparticles onto CNTs than acid reflux. High-loading Pt nanoparticles were uniformly dispersed on sonochemical-treated CNTs, which were highly active in the electrochemical hydrogen adsorption and desorption with cyclic voltammetry measurements (Xing, 2004).

During the rigid acid oxidation, the structural integrity of CNTs is damaged to some extent, which could exert negative effect on the electronic properties of CNTs. Noncovalent functionalization, on the other hand, can provide active sites for metal deposition, while maintaining the perfect graphitic structure of CNTs; therefore, it is advantageous in electrocatalysis applications. Compared to Pt–Ru nanoparticles on acid-treated CNTs, the Pt–Ru catalysts on 1-aminopyrene f-CNTs showed much higher electrochemically active surface area and electrocatalytic activity for the electro-oxidation of methanol in DMFC, as well as enhanced stability (Wang et al., 2008b). Similarly, superior electrocatalytic performance was obtained over Pt nanoparticles supported on PDADMAC–f-CNTs in comparison with the Pt nanoparticles on conventional acid-treated CNTs (Wang and Wang, 2008).

15.6   CONCLUSIONS

CNTs are promising catalyst support materials because of their unique structures and properties. However, to utilize CNTs as catalyst supports, the CNT should be appropriately functionalized before the deposition of active phases. Many approaches have been investigated to modify CNT surfaces with functional groups that can act as anchor sites for metals with advantages and disadvantages for each of the approach. The selection of the functionalization method and catalyst phase synthesis procedure depends on the specific metals to be deposited and the reaction to be catalyzed. Covalent functionalization is usually a simple approach that can produce a chemically stable modification of the CNT surface. However, covalent functionalization may also damage the integrity of CNT's graphitic structure, which may affect properties such as electronic conductivity. Noncovalent functionalization approaches modify the CNT surface without damaging the nanotube's integrity, but many of the noncovalent approaches require complex organic synthesis methods and the interaction between the functional moiety and CNTs is relatively weak.

Metals can be loaded on f-CNTs in the form of organometallic complexes or metal nanoparticles through many different procedures. Organometallic complexes can be grafted onto CNTs through the covalent or noncovalent interaction between functional groups of CNTs and ligands of organometallic complexes. Alternatively, metal nanoparticles can be formed directly on CNT surfaces by deposition and reduction, or the nanoparticles can be preformed and connected to CNTs. Careful control of the deposition and reduction conditions is required to achieve well-defined narrow particle sizes and shapes. The inclusion of preformed nanoparticles on CNTs can produce metal nanoparticles with well-defined particle sizes, but this approach is more complicated than wet impregnation methods, and the removal of capping or stabilization agents from the surface of nanoparticles remains an issue with this approach.

The catalytic performance of CNT-supported metal catalysts in supported homogeneous catalysis, heterogeneous catalysis, and electrocatalysis has been reviewed in this chapter. There are many reports of CNT-supported catalysts with superior catalytic performance to catalysts deposited on conventional supports such as activated carbon and porous alumina. The high catalytic performance of CNTs can be attributed to the unique structures and properties of CNTs, beneficial CNT features including low diffusion resistances in the CNT mesopore channels, high electronic conductivity of CNTs, potential to obtain well-dispersed metal particles on CNTs, modified redox properties of metal catalysts on CNTs, and the strong interaction between metals and CNTs.

The flexibility to control catalyst deposition sites and CNT properties such as length and diameter opens the opportunity to tailor the CNT-supported catalysts for various applications. The future research on CNT-supported metal system should focus on the thorough exploration of how the structure of CNTs affects the properties of catalysts and their catalytic performance, which can then be employed as guidance for tailoring CNT synthesis and pretreatment with desired structures and properties for the optimal catalytic performance.

REFERENCES

Abbaslou, R. M. M., Soltan, J., and Dalai, A. K. 2010. Effects of nanotubes pore size on the catalytic performances of iron catalysts supported on carbon nanotubes for Fischer–Tropsch synthesis. *Applied Catalysis A: General*, 379, 129–134.

Abbaslou, R. M. M., Tavassoli, A., Soltan, J., and Dalai, A. K. 2009. Iron catalysts supported on carbon nanotubes for Fischer–Tropsch synthesis: Effect of catalytic site position. *Applied Catalysis A: General*, 367, 47–52.

Aksel, S. and Eder, D. 2010. Catalytic effect of metal oxides on the oxidation resistance in carbon nanotube-inorganic hybrids. *Journal of Materials Chemistry*, 20, 9149–9154.

Aliaga, C., Park, J. Y., Yamada, Y., Lee, H. S., Tsung, C.-K., Yang, P., and Somorjai, G. A. 2009. Sum frequency generation and catalytic reaction studies of the removal of organic capping agents from Pt nanoparticles by UV–ozone treatment. *The Journal of Physical Chemistry C*, 113, 6150–6155.

Arrigo, R., Havecker, M., Schlogl, R., and Su, D. S. 2008. Dynamic surface rearrangement and thermal stability of nitrogen functional groups on carbon nanotubes. *Chemical Communications*, 40, 4891–4893.

Balasubramanian, K. and Burghard, M. 2005. Chemically functionalized carbon nanotubes. *Small*, 1, 180–192.

Baleizão, C., Gigante, B., Garcia, H., and Corma, A. 2004. Vanadyl salen complexes covalently anchored to single-wall carbon nanotubes as heterogeneous catalysts for the cyanosilylation of aldehydes. *Journal of Catalysis*, 221, 77–84.

Banerjee, S. and Wong, S. S. 2002a. Functionalization of carbon nanotubes with a metal-containing molecular complex. *Nano Letters*, 2, 49–53.

Banerjee, S. and Wong, S. S. 2002b. Structural characterization, optical properties, and improved solubility of carbon nanotubes functionalized with Wilkinson's catalyst. *Journal of the American Chemical Society*, 124, 8940–8948.

Barthos, R., Méhn, D., Demortier, A., Pierard, N., Morciaux, Y., Demortier, G., Fonseca, A., and Nagy, J. B. 2005. Functionalization of single-walled carbon nanotubes by using alkyl-halides. *Carbon*, 43, 321–325.

Bianco, A., Kostarelos, K., and Prato, M. 2005. Applications of carbon nanotubes in drug delivery. *Current Opinion in Chemical Biology*, 9, 674–679.

Castillejos, E., Debouttiere, P. J., Roiban, L., Solhy, A., Martinez, V., Kihn, Y., Ersen, O., Philippot, K., Chaudret, B., and Serp, P. 2009. An efficient strategy to drive nanoparticles into carbon nanotubes and the remarkable effect of confinement on their catalytic performance. *Angewandte Chemie International Edition*, 48, 2529–2533.

Chen, J., Zhu, Z. H., Wang, S., Ma, Q., Rudolph, V., and Lu, G. Q. 2010. Effects of nitrogen doping on the structure of carbon nanotubes (CNTs) and activity of Ru/CNTs in ammonia decomposition. *Chemical Engineering Journal*, 156, 404–410.

Chen, R. J., Zhang, Y., Wang, D., and Dai, H. 2001. Noncovalent sidewall functionalization of single-walled carbon nanotubes for protein immobilization. *Journal of the American Chemical Society*, 123, 3838–3839.

Chen, W., Fan, Z. L., Pan, X. L., and Bao, X. H. 2008a. Effect of confinement in carbon nanotubes on the activity of Fischer–Tropsch iron catalyst. *Journal of the American Chemical Society*, 130, 9414–9419.

Chen, W., Zhao, J., Lee, J. Y., and Liu, Z. 2005. Microwave heated polyol synthesis of carbon nanotubes supported Pt nanoparticles for methanol electrooxidation. *Materials Chemistry and Physics*, 91, 124–129.

Chen, X. C., Hou, Y. Q., Wang, H., Cao, Y., and He, J. H. 2008b. Facile deposition of Pd nanoparticles on carbon nanotube microparticles and their catalytic activity for Suzuki coupling reactions. *Journal of Physical Chemistry C*, 112, 8172–8176.

Chen, Z., Guan, Z., Li, M., Yang, Q., and Li, C. 2011. Enhancement of the performance of a platinum nanocatalyst confined within carbon nanotubes for asymmetric hydrogenation. *Angewandte Chemie International Edition*, 50(21): 4913–4917.

Chu, H. B., Wei, L., Cui, R. L., Wang, J. Y., and Li, Y. 2010. Carbon nanotubes combined with inorganic nanomaterials: Preparations and applications. *Coordination Chemistry Reviews*, 254, 1117–1134.

Coleman, K. S., Bailey, S. R., Fogden, S., and Green, M. L. H. 2003. Functionalization of single-walled carbon nanotubes via the Bingel reaction. *Journal of the American Chemical Society*, 125, 8722–8723.

Cruz-Silva, E., Cullen, D. A., Gu, L., Romo-Herrera, J. M., Muñoz-Sandoval, E., López-Urías, F., Sumpter, B. G. et al. 2008. Heterodoped nanotubes: Theory, synthesis, and characterization of phosphorus–nitrogen doped multiwalled carbon nanotubes. *ACS Nano*, 2, 441–448.

Day, T. M., Unwin, P. R., Wilson, N. R., and Macpherson, J. V. 2005. Electrochemical templating of metal nanoparticles and nanowires on single-walled carbon nanotube networks. *Journal of the American Chemical Society*, 127, 10639–10647.

Deng, W. P., Tan, X. S., Fang, W. H., Zhang, Q. H., and Wang, Y. 2009. Conversion of cellulose into sorbitol over carbon nanotube-supported ruthenium catalyst. *Catalysis Letters*, 133, 167–174.

Dujardin, E., Ebbesen, T. W., Krishnan, A., and Treacy, M. M. J. 1998. Purification of single-shell nanotubes. *Advanced Materials*, 10, 611–613.

Ebbesen, T. W., Hiura, H., Bisher, M. E., Treacy, M. M. J., Shreeve-Keyer, J. L., and Haushalter, R. C. 1996. Decoration of carbon nanotubes. *Advanced Materials*, 8, 155–157.

Eder, D. and Windle, A. H. 2008. Carbon–inorganic hybrid materials: The carbon-nanotube/TiO$_2$ interface. *Advanced Materials*, 20, 1787–1793.

Ellis, A. V., Vijayamohanan, K., Goswami, R., Chakrapani, N., Ramanathan, L. S., Ajayan, P. M., and Ramanath, G. 2003. Hydrophobic anchoring of monolayer-protected gold nanoclusters to carbon nanotubes. *Nano Letters*, 3, 279–282.

Frehill, F., Vos, J. G., Benrezzak, S., Koós, A. A., Kónya, Z., Rüther, M. G., Blau, W. J. et al. 2002. Interconnecting carbon nanotubes with an inorganic metal complex. *Journal of the American Chemical Society*, 124, 13694–13695.

Garcia-Garcia, F. R., Alvarez-Rodriguez, J., Rodriguez-Ramos, I., and Guerrero-Ruiz, A. 2010. The use of carbon nanotubes with and without nitrogen doping as support for ruthenium catalysts in the ammonia decomposition reaction. *Carbon*, 48, 267–276.

Georgakilas, V., Gournis, D., Tzitzios, V., Pasquato, L., Guldi, D. M., and Prato, M. 2007. Decorating carbon nanotubes with metal or semiconductor nanoparticles. *Journal of Materials Chemistry*, 17, 2679–2694.

Georgakilas, V., Tzitzios, V., Gournis, D., and Petridis, D. 2005. Attachment of magnetic nanoparticles on carbon nanotubes and their soluble derivatives. *Chemistry of Materials*, 17, 1613–1617.

Giordano, R., Serp, P., Kalck, P., Kihn, Y., Schreiber, J., Marhic, C., and Duvail, J.-L. 2003. Preparation of rhodium catalysts supported on carbon nanotubes by a surface mediated organometallic reaction. *European Journal of Inorganic Chemistry*, 2003, 610–617.

Glerup, M., Castignolles, M., Holzinger, M., Hug, G., Loiseau, A., and Bernier, P. 2003. Synthesis of highly nitrogen-doped multi-walled carbon nanotubes. *Chemical Communications*, 21, 2542–2543.

Guo, D.-J. and Li, H.-L. 2005. High dispersion and electrocatalytic properties of platinum on functional multi-walled carbon nanotubes. *Electroanalysis*, 17, 869–872.

Guo, S. J., Pan, X. L., Gao, H. L., Yang, Z. Q., Zhao, J. J., and Bao, X. H. 2010. Probing the electronic effect of carbon nanotubes in catalysis: NH_3 synthesis with Ru nanoparticles. *Chemistry: A European Journal*, 16, 5379–5384.

Han, L., Wu, W., Kirk, F. L., Luo, J., Maye, M. M., Kariuki, N. N., Lin, Y., Wang, C., and Zhong, C.-J. 2004. A direct route toward assembly of nanoparticle–carbon nanotube composite materials. *Langmuir*, 20, 6019–6025.

Hirsch, A. 2002. Functionalization of single-walled carbon nanotubes. *Angewandte Chemie International Edition*, 41, 1853–1859.

Hu, X., Wang, T., Qu, X., and Dong, S. 2005. in situ synthesis and characterization of multiwalled carbon nanotube/Au nanoparticle composite materials. *The Journal of Physical Chemistry B*, 110, 853–857.

Jiang, K., Eitan, A., Schadler, L. S., Ajayan, P. M., Siegel, R. W., Grobert, N., Mayne, M., Reyes-Reyes, M., Terrones, H., and Terrones, M. 2003. Selective attachment of gold nanoparticles to nitrogen-doped carbon nanotubes. *Nano Letters*, 3, 275–277.

Jiang, L. Q. and Gao, L. 2003. Modified carbon nanotubes: An effective way to selective attachment of gold nanoparticles. *Carbon*, 41, 2923–2929.

Kim, B. and Sigmund, W. M. 2004. Functionalized multiwall carbon nanotube/gold nanoparticle composites. *Langmuir*, 20, 8239–8242.

Kónya, Z., Vesselenyi, I., Niesz, K., Kukovecz, A., Demortier, A., Fonseca, A., Delhalle, J. et al. 2002. Large scale production of short functionalized carbon nanotubes. *Chemical Physics Letters*, 360, 429–435.

Kudashov, A. G., Okotrub, A. V., Bulusheva, L. G., Asanov, I. P., Shubin, Y. V., Yudanov, N. F., Yudanova, L. I., Danilovich, V. S., and Abrosimov, O. G. 2004. Influence of Ni-Co catalyst composition on nitrogen content in carbon nanotubes. *Journal of Physical Chemistry B*, 108, 9048–9053.

Kundu, S., Wang, Y. M., Xia, W., and Muhler, M. 2008. Thermal stability and reducibility of oxygen-containing functional groups on multiwalled carbon nanotube surfaces: A quantitative high-resolution XPS and TPD/TPR study. *Journal of Physical Chemistry C*, 112, 16869–16878.

Lee, K., Zhang, J., Wang, H., and Wilkinson, D. P. 2006. Progress in the synthesis of carbon nanotube- and nanofiber-supported Pt electrocatalysts for PEM fuel cell catalysis. *Journal of Applied Electrochemistry*, 36, 507–522.

Lee, Y. S., Cho, T. H., Lee, B. K., Rho, J. S., An, K. H., and Lee, Y. H. 2003. Surface properties of fluorinated single-walled carbon nanotubes. *Journal of Fluorine Chemistry*, 120, 99–104.

Lemus-Yegres, L., Such-Basáñez, I., De Lecea, C. S.-M., Serp, P., and Román-Martínez, M. C. 2006. Exploiting the surface –OH groups on activated carbons and carbon nanotubes for the immobilization of a Rh complex. *Carbon*, 44, 605–608.

Li, B. D., Wang, C., Yi, G. Q., Lin, H. Q., and Yuan, Y. Z. 2011. Enhanced performance of Ru nanoparticles confined in carbon nanotubes for CO preferential oxidation in a H_2-rich stream. *Catalysis Today*, 164, 74–79.

Li, C. H., Yao, K. F., and Liang, J. 2003a. Influence of acid treatments on the activity of carbon nanotube-supported catalysts. *Carbon*, 41, 858–860.

Li, L., Zhu, Z. H., Yan, Z. F., Lu, G. Q., and Rintoul, L. 2007. Catalytic ammonia decomposition over Ru/carbon catalysts: The importance of the structure of carbon support. *Applied Catalysis A: General*, 320, 166–172.

Li, W., Liang, C., Zhou, W., Qiu, J., Zhou, Sun, G., and Xin, Q. 2003b. Preparation and characterization of multiwalled carbon nanotube-supported platinum for cathode catalysts of direct methanol fuel cells. *The Journal of Physical Chemistry B*, 107, 6292–6299.

Li, W., Wang, X., Chen, Z., Waje, M., and Yan, Y. 2006. Pt–Ru supported on double-walled carbon nanotubes as high-performance anode catalysts for direct methanol fuel cells. *The Journal of Physical Chemistry B*, 110, 15353–15358.

Lim, J. K., Yun, W. S., Yoon, M.-H., Lee, S. K., Kim, C. H., Kim, K., and Kim, S. K. 2003. Selective thiolation of single-walled carbon nanotubes. *Synthetic Metals*, 139, 521–527.

Liu, L., Wang, T., Li, J., Guo, Z.-X., Dai, L., Zhang, D., and Zhu, D. 2003. Self-assembly of gold nanoparticles to carbon nanotubes using a thiol-terminated pyrene as interlinker. *Chemical Physics Letters*, 367, 747–752.

Liu, Y. and Adronov, A. 2004. Preparation and utilization of catalyst-functionalized single-walled carbon nanotubes for ring-opening metathesis polymerization. *Macromolecules*, 37, 4755–4760.

Liu, Z., Gan, L. M., Hong, L., Chen, W., and Lee, J. Y. 2005. Carbon-supported Pt nanoparticles as catalysts for proton exchange membrane fuel cells. *Journal of Power Sources*, 139, 73–78.

Liu, Z., Lin, X., Lee, J. Y., Zhang, W., Han, M., and Gan, L. M. 2002. Preparation and characterization of platinum-based electrocatalysts on multiwalled carbon nanotubes for proton exchange membrane fuel cells. *Langmuir*, 18, 4054–4060.

Lordi, V., Yao, N., and Wei, J. 2001. Method for supporting platinum on single-walled carbon nanotubes for a selective hydrogenation catalyst. *Chemistry of Materials*, 13, 733–737.

Lu, C.-Y., Tseng, H.-H., Wey, M.-Y., and Hsueh, T.-W. 2009. The comparison between the polyol process and the impregnation method for the preparation of CNT-supported nanoscale Cu catalyst. *Chemical Engineering Journal*, 145, 461–467.

Luo, J., Gao, L., Leung, Y., and Au, C. 2000. The decomposition of NO on CNTs and 1 wt% Rh/CNTs. *Catalysis Letters*, 66, 91–97.

Ma, H. X., Wang, L. C., Chen, L. Y., Dong, C., Yu, W. C., Huang, T., and Qian, Y. T. 2007. Pt nanoparticles deposited over carbon nanotubes for selective hydrogenation of cinnamaldehyde. *Catalysis Communications*, 8, 452–456.

Maiyalagan, T., Viswanathan, B., and Varadaraju, U. V. 2005. Nitrogen containing carbon nanotubes as supports for Pt—Alternate anodes for fuel cell applications. *Electrochemistry Communications*, 7, 905–912.

Malek Abbaslou, R. M., Tavasoli, A., and Dalai, A. K. 2009. Effect of pre-treatment on physico-chemical properties and stability of carbon nanotubes supported iron Fischer–Tropsch catalysts. *Applied Catalysis A: General*, 355, 33–41.

Mu, Y., Liang, H., Hu, J., Jiang, L., and Wan, L. 2005. Controllable Pt nanoparticle deposition on carbon nanotubes as an anode catalyst for direct methanol fuel cells. *The Journal of Physical Chemistry B*, 109, 22212–22216.

Muradov, N. 2001. Catalysis of methane decomposition over elemental carbon. *Catalysis Communications*, 2, 89–94.

Nakamura, T., Ohana, T., Ishihara, M., Tanaka, A., and Koga, Y. 2006. Sidewall modification of single-walled carbon nanotubes with sulfur-containing functionalities and gold nanoparticle attachment. *Chemistry Letters*, 35, 742–743.

Narayanan, R. and El-Sayed, M. A. 2005. Catalysis with transition metal nanoparticles in colloidal solution: Nanoparticle shape dependence and stability. *The Journal of Physical Chemistry B*, 109, 12663–12676.

Oh, S.-D., So, B.-K., Choi, S.-H., Gopalan, A., Lee, K.-P., Ro Yoon, K., and Choi, I. S. 2005. Dispersing of Ag, Pd, and Pt–Ru alloy nanoparticles on single-walled carbon nanotubes by γ-irradiation. *Materials Letters*, 59, 1121–1124.

Ou, Y.-Y. and Huang, M. H. 2006. High-density assembly of gold nanoparticles on multiwalled carbon nanotubes using 1-pyrenemethylamine as interlinker. *The Journal of Physical Chemistry B*, 110, 2031–2036.

Pan, J., Li, J., Wang, C., and Yang, Z. 2007a. Multi-wall carbon nanotubes supported ruthenium for glucose hydrogenation to sorbitol. *Reaction Kinetics and Catalysis Letters*, 90, 233–242.

Pan, X. and Bao, X. 2011. The effects of confinement inside carbon nanotubes on catalysis. *Accounts of Chemical Research*, 44, 553–562.

Pan, X. L. and Bao, X. H. 2008. Reactions over catalysts confined in carbon nanotubes. *Chemical Communications*, 47, 6271–6281.

Pan, X. L., Fan, Z. L., Chen, W., Ding, Y. J., Luo, H. Y., and Bao, X. H. 2007b. Enhanced ethanol production inside carbon-nanotube reactors containing catalytic particles. *Nature Materials*, 6, 507–511.

Park, S., Yoon, S. W., Lee, K.-B., Kim, D. J., Jung, Y. H., Do, Y., Paik, H.-J., and Choi, I. S. 2006. Carbon nanotubes as a ligand in Cp_2ZrCl_2-based ethylene polymerization. *Macromolecular Rapid Communications*, 27, 47–50.

Park, Y. S., Choi, Y. C., Kim, K. S., Chung, D. C., Bae, D. J., An, K. H., Lim, S. C., Zhu, X. Y., and Lee, Y. H. 2001. High yield purification of multiwalled carbon nanotubes by selective oxidation during thermal annealing. *Carbon*, 39, 655–661.

Pereira, M. F. R., Figueiredo, J. L., Órfão, J. J. M., Serp, P., Kalck, P., and Kihn, Y. 2004. Catalytic activity of carbon nanotubes in the oxidative dehydrogenation of ethylbenzene. *Carbon*, 42, 2807–2813.

Planeix, J. M., Coustel, N., Coq, B., Brotons, V., Kumbhar, P. S., Dutartre, R., Geneste, P., Bernier, P., and Ajayan, P. M. 1994. Application of carbon nanotubes as supports in heterogeneous catalysis. *Journal of the American Chemical Society*, 116, 7935–7936.

Quinn, B. M., Dekker, C., and Lemay, S. G. 2005. Electrodeposition of noble metal nanoparticles on carbon nanotubes. *Journal of the American Chemical Society*, 127, 6146–6147.

Raghuveer, M. S., Agrawal, S., Bishop, N., and Ramanath, G. 2006. Microwave-assisted single-step functionalization and in situ derivatization of carbon nanotubes with gold nanoparticles. *Chemistry of Materials*, 18, 1390–1393.

Rajalakshmi, N., Ryu, H., Shaijumon, M. M., and Ramaprabhu, S. 2005. Performance of polymer electrolyte membrane fuel cells with carbon nanotubes as oxygen reduction catalyst support material. *Journal of Power Sources*, 140, 250–257.

Rao, G. P., Lu, C., and Su, F. 2007. Sorption of divalent metal ions from aqueous solution by carbon nanotubes: A review. *Separation and Purification Technology*, 58, 224–231.

Rao, R., Zhang, Q., Liu, H., Yang, H., Ling, Q., Yang, M., Zhang, A., and Chen, W. 2012. Enhanced catalytic performance of CeO_2 confined inside carbon nanotubes for dehydrogenation of ethylbenzene in the presence of CO_2. *Journal of Molecular Catalysis A: Chemical*, 363–364, 283–290.

Saha, M. S. and Kundu, A. 2010. Functionalizing carbon nanotubes for proton exchange membrane fuel cells electrode. *Journal of Power Sources*, 195, 6255–6261.

Schnorr, J. M. and Swager, T. M. 2010. Emerging applications of carbon nanotubes. *Chemistry of Materials*, 23, 646–657.

Serp, P. and Castillejos, E. 2010. Catalysis in carbon nanotubes. *ChemCatChem*, 2, 41–47.

Serp, P., Corrias, M., and Kalck, P. 2003. Carbon nanotubes and nanofibers in catalysis. *Applied Catalysis A: General*, 253, 337–358.

Solhy, A., Machado, B. F., Beausoleil, J., Kihn, Y., Goncalves, F., Pereira, M. F. R., Orfao, J. J. M., Fiqueiredo, J. L., Faria, J. L., and Serp, P. 2008. MWCNT activation and its influence on the catalytic performance of Pt/MWCNT catalysts for selective hydrogenation. *Carbon*, 46, 1194–1207.

Sumanasekera, G. U., Allen, J. L., Fang, S. L., Loper, A. L., Rao, A. M., and Eklund, P. C. 1999. Electrochemical oxidation of single wall carbon nanotube bundles in sulfuric acid. *The Journal of Physical Chemistry B*, 103, 4292–4297.

Tanaka, K.-I., Shou, M., and Yuan, Y. 2010. Low temperature PROX reaction of CO catalyzed by dual functional catalysis of the Pt supported on CNT, CNF, graphite, and amorphous-C with Ni-MgO, Fe, and $Fe-Al_2O_3$: Oxidation of CO via HCOO intermediate. *The Journal of Physical Chemistry C*, 114, 16917–16923.

Tanaka, K.-I., Shou, M., Zhang, H., Yuan, Y., Hagiwara, T., Fukuoka, A., Nakamura, J., and Lu, D. 2008. An extremely active Pt/Carbon nano-tube catalyst for selective oxidation of CO in H_2 at room temperature. *Catalysis Letters*, 126, 89–95.

Tang, C., Bando, Y., Golberg, D., and Xu, F. 2004. Structure and nitrogen incorporation of carbon nanotubes synthesized by catalytic pyrolysis of dimethylformamide. *Carbon*, 42, 2625–2633.

Tessonnier, J. P., Ersen, O., Weinberg, G., Pham-Huu, C., Su, D. S., and Schlogl, R. 2009. Selective deposition of metal nanoparticles inside or outside multiwalled carbon nanotubes. *ACS Nano*, 3, 2081–2089.

Tobias, G., Shao, L., Salzmann, C. G., Huh, Y., and Green, M. L. H. 2006. Purification and opening of carbon nanotubes using steam. *Journal of Physical Chemistry B*, 110, 22318–22322.

Toebes, M. L., Prinsloo, F. F., Bitter, J. H., Van Dillen, A. J., and De Jong, K. P. 2003. Influence of oxygen-containing surface groups on the activity and selectivity of carbon nanofiber-supported ruthenium catalysts in the hydrogenation of cinnamaldehyde. *Journal of Catalysis*, 214, 78–87.

Trepanier, M., Tavasoli, A., Dalai, A. K., and Abatzoglou, N. 2009. Co, Ru and K loadings effects on the activity and selectivity of carbon nanotubes supported cobalt catalyst in Fischer–Tropsch synthesis. *Applied Catalysis A: General*, 353, 193–202.

Unger, E., Graham, A., Kreupl, F., Liebau, M., and Hoenlein, W. 2002. Electrochemical functionalization of multi-walled carbon nanotubes for solvation and purification. *Current Applied Physics*, 2, 107–111.

Van Dommele, S., De Jong, K. P., and Bitter, J. H. 2006. Nitrogen-containing carbon nanotubes as solid base catalysts. *Chemical Communications*, (46), 4859–4861.

Vu, H., Goncalves, F., Philippe, R., Lamouroux, E., Corrias, M., Kihn, Y., Plee, D., Kalck, P., and Serp, P. 2006a. Bimetallic catalysis on carbon nanotubes for the selective hydrogenation of cinnamaldehyde. *Journal of Catalysis*, 240, 18–22.

Vu, H., Gonçalves, F., Philippe, R., Lamouroux, E., Corrias, M., Kihn, Y., Plee, D., Kalck, P., and Serp, P. 2006b. Bimetallic catalysis on carbon nanotubes for the selective hydrogenation of cinnamaldehyde. *Journal of Catalysis*, 240, 18–22.

Wang, C., Waje, M., Wang, X., Tang, J. M., Haddon, R. C., and Yan 2004a. Proton exchange membrane fuel cells with carbon nanotube based electrodes. *Nano Letters*, 4, 345–348.

Wang, C. F., Guo, S. J., Pan, X. L., Chen, W., and Bao, X. H. 2008a. Tailored cutting of carbon nanotubes and controlled dispersion of metal nanoparticles inside their channels. *Journal of Materials Chemistry*, 18, 5782–5786.

Wang, H., Zhou, A., Peng, F., Yu, H., and Yang, J. 2007a. Mechanism study on adsorption of acidified multi-walled carbon nanotubes to Pb(II). *Journal of Colloid and Interface Science*, 316, 277–283.

Wang, L., Ge, L., Rufford, T. E., Chen, J., Zhou, W., Zhu, Z., and Rudolph, V. 2011. A comparison study of catalytic oxidation and acid oxidation to prepare carbon nanotubes for filling with Ru nanoparticles. *Carbon*, 49, 2022–2032.

Wang, S., Jiang, S. P., White, T. J., Guo, J., and Wang, X. 2009. Electrocatalytic activity and interconnectivity of Pt nanoparticles on multiwalled carbon nanotubes for fuel cells. *The Journal of Physical Chemistry C*, 113, 18935–18945.

Wang, S. and Wang, X. 2008. Polyelectrolyte functionalized carbon nanotubes as a support for noble metal electrocatalysts and their activity for methanol oxidation. *Nanotechnology*, 19, 265601.

Wang, S., Wang, X., and Jiang, S. P. 2008b. PtRu nanoparticles supported on 1-aminopyrene-functionalized multiwalled carbon nanotubes and their electrocatalytic activity for methanol oxidation. *Langmuir*, 24, 10505–10512.

Wang, X., Li, N., Webb, J. A., Pfefferle, L. D., and Haller, G. L. 2010. Effect of surface oxygen containing groups on the catalytic activity of multi-walled carbon nanotube supported Pt catalyst. *Applied Catalysis B: Environmental*, 101, 21–30.

Wang, X., Li, W., Chen, Z., Waje, M., and Yan, Y. 2006. Durability investigation of carbon nanotube as catalyst support for proton exchange membrane fuel cell. *Journal of Power Sources*, 158, 154–159.

Wang, Y., Iqbal, Z., and Malhotra, S. V. 2005. Functionalization of carbon nanotubes with amines and enzymes. *Chemical Physics Letters*, 402, 96–101.

Wang, Y., Shah, N., and Huffman, G. P. 2004b. Pure hydrogen production by partial dehydrogenation of cyclohexane and methylcyclohexane over nanotube-supported Pt and Pd catalysts. *Energy & Fuels*, 18, 1429–1433.

Wang, Y., Shan, H., Hauge, R. H., Pasquali, M., and Smalley, R. E. 2007b. A highly selective, one-pot purification method for single-walled carbon nanotubes. *Journal of Physical Chemistry B*, 111, 1249–1252.

Wepasnick, K. A., Smith, B. A., Bitter, J. L., and Fairbrother, D. H. 2010. Chemical and structural characterization of carbon nanotube surfaces. *Analytical and Bioanalytical Chemistry*, 396, 1003–1014.

Xia, Y., Xiong, Y., Lim, B., and Skrabalak, S. E. 2009. Shape controlled synthesis of metal nanocrystals: Simple chemistry meets complex physics? *Angewandte Chemie International Edition*, 48, 60–103.

Xing, Y. 2004. Synthesis and electrochemical characterization of uniformly-dispersed high loading Pt nanoparticles on sonochemically-treated carbon nanotubes. *The Journal of Physical Chemistry B*, 108, 19255–19259.

Xing, Y., Li, L., Chusuei, C. C., and Hull, R. V. 2005. Sonochemical oxidation of multiwalled carbon nanotubes. *Langmuir*, 21, 4185–4190.

Xu, H. B., Chen, H. Z., Shi, M. M., Bai, R., and Wang, M. 2005. A novel donor–acceptor heterojunction from single-walled carbon nanotubes functionalized by erbium bisphthalocyanine. *Materials Chemistry and Physics*, 94, 342–346.

Yin, S., Xu, B., Zhu, W., Ng, C., Zhou, X., and Au, C. 2004a. Carbon nanotubes-supported Ru catalyst for the generation of CO_x-free hydrogen from ammonia. *Catalysis Today*, 93, 27–38.

Yin, S. F., Zhang, Q. H., Xu, B. Q., Zhu, W. X., Ng, C. F., and Au, C. T. 2004b. Investigation on the catalysis of CO_x-free hydrogen generation from ammonia. *Journal of Catalysis*, 224, 384–396.

Yu, R., Chen, L., Liu, Q., Lin, J., Tan, K.-L., Ng, S. C., Chan, H. S. O., Xu, G.-Q., and Hor, T. S. A. 1998. Platinum deposition on carbon nanotubes via chemical modification. *Chemistry of Materials*, 10, 718–722.

Zanella, R., Basiuk, E. V., Santiago, P., Basiuk, V. A., Mireles, E., Puente-Lee, I., and Saniger, J. M. 2005. Deposition of gold nanoparticles onto thiol-functionalized multiwalled carbon nanotubes. *The Journal of Physical Chemistry B*, 109, 16290–16295.

Zhang, J., Comotti, M., Schuth, F., Schlogl, R., and Su, D. S. 2007. Commercial Fe- or Co-containing carbon nanotubes as catalysts for NH_3 decomposition. *Chemical Communications*, 2007, 1916–1918.

Zhang, J., Liu, X., Blume, R., Zhang, A., Schlögl, R., and Su, D. S. 2008. Surface-modified carbon nanotubes catalyze oxidative dehydrogenation of *n*-butane. *Science*, 322, 73–77.

Zhang, J., Zou, H., Qing, Q., Yang, Y., Li, Q., Liu, Z., Guo, X., and Du, Z. 2003. Effect of chemical oxidation on the structure of single-walled carbon nanotubes. *Journal of Physical Chemistry B*, 107, 3712–3718.

Zheng, W. Q., Zhang, J., Zhu, B., Blume, R., Zhang, Y. L., Schlichte, K., Schlogl, R., Schuth, F., and Su, D. S. 2010. Structure-function correlations for Ru/CNT in the catalytic decomposition of ammonia. *ChemSusChem*, 3, 226–230.

Zhu, Y., Sia, S. L. P., Carpenter, K., Kooli, F., and Kemp, R. A. 2006. Syntheses and catalytic activities of single-wall carbon nanotubes-supported nickel (II) metallacarboranes for olefin polymerization. *Journal of Physics and Chemistry of Solids*, 67, 1218–1222.

Chapter 16

Chemical Functionalization and Composites of Boron Nitride Nanotubes

Wenjun Meng and Chunyi Zhi

CONTENTS

16.1 INTRODUCTION

Boron nitride is an inorganic compound with a chemical formula BN, consisting of equal numbers of boron and nitrogen atoms. BN exists in various crystalline forms; among them, the hexagonal form is the most stable and softest. Boron nitride nanotubes (BNNTs) can be seen as a rolled graphite-like BN sheet, in a fashion shown in Figure 16.1 (Golberg et al., 2010). As a structural

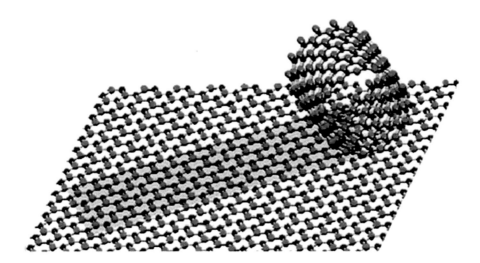

Figure 16.1   Atomic model of a single-layered BNNT made through wrapping of a planar monatomic BN nanosheet. (Reprinted with permission from Golberg, D., Bando, Y., Huang, Y., Terao, T., Mitome, M., Tang, C.C., and Zhi, C.Y., Boron nitride nanotubes and nanosheets, *ACS Nano*, 4, 2979–2993, 2010. Copyright 2010, American Chemical Society.)

analogue of carbon nanotubes (CNTs) (Iijima, 1991), the BNNTs were first predicted theoretically in 1994 (Blase et al., 1994; Rubio et al., 1994); since then, BNNTs have become one of the most intriguing non-CNTs.

In 1981, Ishii et al. reported the growth of h-BN whiskers by heating h-BN powder in nitrogen atmosphere. These whiskers possess a bamboo-like structure, which could be called bamboo-like BNNTs, but the authors did not use the term at that time. The nanoscaled BN with a perfect tubular structure was successfully synthesized 14 years later by arc discharge in 1995 (Chopra et al., 1995). In the following years, a variety of methods, such as laser ablation (Golberg et al., 1996), template synthesis (Han et al., 1998), autoclave (Dai et al., 2007), ball-mining (Chen et al., 1999), and chemical vapor deposition (CVD; Lourie et al., 2000), were invented and applied to synthesize BNNTs. The obtained BNNTs have different purities, structures, and diameters so that they can meet the requirements for detailed physical property investigations. Here, we briefly summarize the differences between BNNTs and CNTs in physical properties in Table 16.1.

The most obvious difference is their visible appearance. BNNTs are pure white (sometimes slightly yellowish due to N vacancies), while CNTs are totally black, as shown in Figure 16.2.

Although BNNT is structurally a very close analogue of the CNT, the B–N bond contains a significant ionic component because of the difference in the electronegativity of B and N atoms. This polarity can significantly alter both electronic and photophysical properties in the system (Liu and Marder, 2008). Primarily, in contrast to metallic or semiconducting CNTs (Mintmire et al., 1992), BNNTs are electrical insulators with a large band gap between 5.0 and 6.0 eV (Blase et al., 1994; Lauret et al., 2005), basically independent of tube chirality and morphology, at least for the realistic, experimentally verified diameters. The discrepancy in electronic structure results in different luminescence emission: BNNTs have violet or ultraviolet luminescence under excitation by electrons or photons (Wu et al., 2004; Lauret et al., 2005; Zhi et al., 2005d; Tang et al., 2007), while CNTs can emit infrared light whose wavelengths depend on their chiralities

TABLE 16.1   SUMMARY OF THE DIFFERENCES BETWEEN BNNTs AND CNTs IN PHYSICAL PROPERTIES

Materials	Bonding	Electronic Structure	Raman Active Mode	Luminescence	Young's Modulus (TPa)	Thermal Conductivity (W/mK) at r.t.	Thermal Stability
CNTs	Covalent bonds; bonding length: 1.400–1.463 Å	Metallic or semiconducting, dependent on chiralities	G band: 1580 cm⁻¹; D band: 1350 cm⁻¹; RBM model: (10, 10) CNT, 171.0 cm⁻¹	Infrared (wavelength: 800–1700 nm)	1.09–1.25 0.84–0.99 (theoretical) 0.27–0.95 (experimental)	~6000 (Theoretical, SWCNT); >3000 (experimental, MWCNT, D ~ 14 nm); ~1000 (experimental, MWCNT, D ~ 10 nm); ~300 (experimental, MWCNT, D ~ 35 nm)	Depends on sample, roughly between 500°C –700°C
BNNTs	Covalent bonds with ionic component; bonding length: 1.437–1.454 Å	5.0–6.0 eV band gap, independent of chiralities	A1 tangential mode: 1370 cm⁻¹; RBM model: (10, 10) BNNT, 153.0 cm⁻¹	Violet and/or ultraviolet (wavelength: 220–460 nm)	0.784–0.912 0.71–0.83 (theoretical) 0.5–0.7; 1.22 ± 0.24 (experimental)	>CNTs' (theoretical) ~180–300 (theoretical, SWBNNT); ~180–300 (experimental, MWBNNT);	High, up to 800°C –900°C in air

(a)   (b)

Figure 16.2   Images of (a) CNTs and (b) BNNTs. Their visible appearance is totally different: CNTs are black, while BNNTs are purely white. (Reprinted from *Mater. Sci. Eng. R*, Zhi et al., Boron nitride nanotubes, *Materials Science & Engineering R-Reports*, 70, 92–111, Copyright 2010, with permission from Elsevier.)

(O'Connell et al., 2002; Harutyunyan et al., 2009; Murakami et al., 2009). Moreover, Raman excitation in the UV (229 nm) allows the identification of the A1 tangential mode at 1370 cm^{-1} for BNNTs (Arenal et al., 2006); whereas CNTs possess a nominal G band at 1580 cm^{-1} and a defect-induced D band at 1350 cm^{-1} (Dresselhaus and Eklund, 2000). Both the two nanotubes have radial breathing modes with a slightly different frequency (Akdim et al., 2003).

Young's modulus of BNNTs is predicted to be 0.7–0.9 TPa by theoretical calculation, while the value of CNTs is a bit higher, around 1.0–1.2 TPa (Hernandez et al., 1998). The experimental Young moduli of both BNNTs and CNTs vary in different samples fabricated by different methods (Chopra and Zettl, 1998; Suryavanshi et al., 2004; Coleman et al., 2006; Golberg et al., 2007). However, all the data indicate that BNNTs exhibit excellent elastic properties and mechanical stiffness comparable to CNTs.

With respect to thermal properties, the electrically insulating BNNTs display very high thermal conductivity, which is attributed to the thermal transfer via the phonons (Chang et al., 2007). The experimental data of MWBNNTs is about 180–300 W/mK, which is comparable with that of CNTs at a similar diameter (Chang et al., 2006). Importantly, BNNTs possess higher resistance to oxidation and better thermal stability compared to CNTs (Golberg et al., 2001b; Chen et al., 2004; Xu et al., 2009). They are inert to most acids and alkalis, and stable up to at least 700°C in air.

The aforementioned exciting properties of BNNTs make the materials particularly useful for the fabrication of composites to obtain mechanical reinforcement, high thermal conductivity, and a low coefficient of thermal expansion in a matrix. In this chapter, we provide a state-of-the-art review in the development of BNNTs and thoroughly summarize valuable achievements with respect to BNNTs synthesis, chemical functionalization, and their prospective applications.

16.2 FABRICATION

The techniques known for the growth of CNTs, such as arc discharge (Chopra et al., 1995), laser ablation (Guo et al., 1995), and CVD (Endo et al., 1993), do not work well for the sister BNNT system. Therefore, the modified methods and alternative techniques have been developed to synthesize BNNTs. In this section, some of the most important methods developed for the growths of BNNTs are briefly reviewed.

16.2.1 Arc Discharge

The first successful fabrication of BNNTs was reported by Chopra et al. (1995) via an arc discharge between a BN-packed tungsten rod and a cooled copper electrode. Multiwalled BNNTs with inner diameters on the order of 1–3 nm and with lengths up to 200 nm were produced in a plasma arc-discharge apparatus, which is similar to that used for carbon fullerene production. The synthesized BNNTs revealed metallic nanoparticles encapsulated at the tube tip ends, which originated from the tungsten hollow electrode used for the starting of BN powder encapsulation. Since the insulating nature of bulk BN prevents its use as an electrode, conductive boron compounds, such as HfB_2 (Loiseau et al., 1996, 1998; Suenaga et al., 1997; Cumings and Zettl, 2000) and ZrB_2 (Saito and Maida, 1999), were used instead in most following-up improved experiments. High-purity N_2 was used as a nitrogen source as well as a protection gas. However, in all of these experiments, the yields of BNNTs were rather low, and the products contained various metal impurities originated from the electrode materials used.

16.2.2   Template Synthesis

Due to the analogous layered structures and close lattice constants between BNNTs and CNTs, it is supposed to get BNNTs by the chemical substitution of the C atoms in CNTs with B and N atoms. Han et al. (1998) initially synthesized BNNTs via the substitution reaction using multi-walled CNTs as natural consumable templates. During the synthesis, CNTs were modified by a highly reactive B_2O_3 vapor (B source) and a flowing N_2 or NH_3 gas (N source) at 1200°C–1700°C. The obtained BNNTs had diameters and lengths similar to those of the starting CNT templates. The major drawback of this method was the difficulty in removing the carbon residues in the tube lattice, even when a subsequent oxidation treatment was performed. Therefore, the resultant nanotubes were actually C-doped BNNTs. In the following-up improved experiment, MoO_3 was applied to improve the yield of pure BNNTs in the template synthesis routes (Golberg et al., 2000a). Moreover, the morphologies of synthesized BNNTs could be controlled by using corresponding CNT morphologies (Golberg et al., 1999, 2000b).

Another kind of the templates used for the growth of BNNTs was a porous alumina filter membrane. Bechelany et al. (2007) synthesized highly ordered arrays of BNNTs by combining a polymer thermolysis route and a template process using pores inside porous alumina. However, the crystallization of BNNTs was not satisfactory, and the yield was still limited.

16.2.3   Ball-Milling

Original ball-mining method was pioneered by Chen et al. (1999) to synthesize BNNTs. The nanotubes were produced by first ball-milling hexagonal BN powder to generate highly disordered or amorphous nanostructures, followed by annealing at temperatures up to 1300°C. The annealing leads to the nucleation and growth of hexagonal BNNTs of both cylindrical and bamboo-like morphology. No specific catalyst was used in the reaction, but the contaminant Fe particles from the stainless-steel milling container were supposed to catalyze the growth of the bamboo-like tubes. In the following years, many works relevant to this method were carried out to improve the quality and purity of ball-milled BNNTs by using a boron powder as the precursor, introducing NH_3 as a nitrogen source during ball-mining and optimizing annealing conditions (Chen et al., 2002; Yu et al., 2005; Singhal et al., 2008; Cao et al., 2009; Kim et al., 2012). The yield of BNNTs was improved, and their diameters were reduced to less than 10 nm. However, due to the residues of the bamboo-like structures and the B/B–N reactants (amorphous B particles and BN bulky flakes) contained in the products, the purity of the obtained BNNTs was still not too high.

16.2.4   Autoclave

Due to the high-crystallization temperature of BNNTs, an autoclave was thought to be a not suitable apparatus for the synthesis of BNNTs when safety problem was concerned. However, some special autoclaves used below 600°C were adopted (Xu et al., 2003; Rosas et al., 2005; Dai et al., 2007). In a typical experiment, multiwalled BNNTs were produced with a yield of about 50% by copyrolyzing NH_4BF_4, KBH_4, and NaN_3 in a temperature range from 450°C to 600°C (Xu et al., 2003). A mixture of Zn powder and reduced Fe powder was applied to the reaction system and supposed to be served as catalyst during the reaction. The diameter of obtained BNNTs was in the range of 60–350 nm and the length was 0.5–5 μm. The BNNTs fabricated by this method had special morphologies showing thin tube walls and large inner spaces.

16.2.5 Laser Ablation

Golberg et al. (1996) carried out the pioneering work on laser ablation of hexagonal or cubic BN targets at ultrahigh nitrogen pressure for the growth of BNNTs. Hexagonal or cubic BN single crystals were laser heated in a diamond anvil cell at high nitrogen pressure (5–15 GPa). Temperatures above 5000 K can be reached by focusing a stabilized CO_2 laser of power up to 240 W onto the edge of specimens. The circular or polygonal cross-sectional nanotubes having three to eight shells and a characteristic outer dimension cross section of 3–15 nm were found to have grown either in melted cubic BN or in hexagonal + amorphous BN. The drawback of the synthesis was a relatively small number of BNNTs compared to dominant standard cubic, hexagonal, or amorphous BN flakes. Several years later, Laude et al. (2000) synthesized BNNTs in a somewhat larger quantity under lower pressure through laser heating. The obtained BNNTs exhibited lengths of up to 40 μm. However, BN-coated B nanoparticles and BN flakes were also found in the product. Both multiwalled BNNTs (Yu et al., 1998) and single-walled BNNTs (Lee et al., 2001; Arenal et al., 2007) could be obtained by the similar experimental setup through laser ablation method.

An improved laser ablation method named pressurized vapor/condenser method, which might provide a very promising route for the large-scale production of BNNTs, has been developed in 2009 (Smith et al., 2009). The boron vapor is generated by heating target materials by a high-powered laser. The large density difference between the hot boron vapor (over 4000°C) and the surrounding high-pressure nitrogen (room temperature) generated a strong buoyancy force and a narrow vertical plume of boron vapor with a velocity profile. Then, a cooled metal wire traversed the boron plume and acted as a condenser and triggered homogeneous nucleation. No catalyst was ever employed, and hundreds of milligrams of few-walled BNNTs could be obtained by the synthetic route.

16.2.6 Chemical Vapor Deposition

CVD is a chemical process used to produce high-purity, high-performance solid materials. In a typical CVD process, the substrate is exposed to one or more volatile precursors, which react and/or decompose on the substrate surface to produce the desired deposit. For the synthesis of BNNTs, many works are involved in the CVD method (Lourie et al., 2000; Tang et al., 2002a, 2006; Terao et al., 2008; Cao et al., 2009; Kim et al., 2012).

Lourie et al. (2000) described the initial CVD synthesis of multiwalled BNNTs on nickel boride catalyst particles at 1000°C–1100°C. Borazine ($B_3N_3H_6$) generated in situ from a molten salt consisting of a mixture of $(NH_4)_2SO_4$, $NaBH_4$, and Co_3O_4 at 300°C–400°C was used as a precursor. The BNNTs exhibited lengths of up to 5 μm and often possessed bulbous, flag-like, and/or club-like tips.

Ma et al. (2001) prepared multiwalled BNNTs via a CVD route from a $B_4B_3O_2H$ precursor. The tips encapsulated boron oxynitride nanoclusters, which incorporate silicon, aluminum, and calcium and served as effective promoters for the growth of BNNTs. Wang et al. (2005) realized the CVD growth of BNNTs at a low temperature (600°C) by a plasma-enhanced pulsed-laser deposition technique. In another CVD experiment, BNNTs were fabricated by using a low pressure of a borazine precursor in conjunction with a floating nickelocene catalyst (Kim et al., 2008). The obtained BNNTs were double walled and were pure BNNTs with a high crystalline quality. This method is promising to be further optimized toward realization of a continuous growth of BNNTs.

To obtain large amount of highly pure BNNTs, Bando and his colleagues have also developed a so-called boron oxide CVD (BOCVD) method, which used a boron powder and a metal oxide as reactants (Tang et al., 2002a,c; Zhi et al., 2005c). In a typical experiment, a mixture

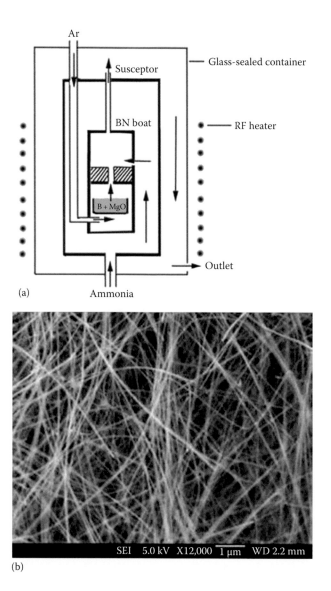

Ar

Susceptor

Glass-sealed container

BN boat

RF heater

B + MgO

Outlet

(a)

Ammonia

SEI   5.0 kV   X12,000   1 μm   WD 2.2 mm

(b)

Figure 16.3   Fabrication of BNNTs via a BOCVD method. (a) Illustration of the apparatus for the synthesis of BNNTs via a BOCVD method. (Tang, C., Bando, Y., Sato, T., and Kurashima, K., A novel precursor for synthesis of pure boron nitride nanotubes, *Chem. Commun.*, 1290–1291, 2002a. Reproduced by permission of The Royal Society of Chemistry.) (b) SEM image of BNNTs produced by BOVCD method. The tube length can be up to tens of micrometers. (Reprinted from *Mater. Sci. Eng. R*, Zhi et al., Boron nitride nanotubes, *Materials Science & Engineering R-Reports*, 70, 92–111, Copyright 2010, with permission from Elsevier.)

of B and MgO was loaded into a BN crucible at the bottom of the reaction chamber in a vertical induction furnace (Figure 16.3a). The mixture was used as the reactant and heated to approximately 1300°C to form highly reactive B_2O_3 and Mg vapors, which were subsequently argon-transported into a reaction chamber and reacted with supplying ammonia in the lower-temperature furnace zone. This method separates a boron precursor (boron powder plus metal oxide) from the as-grown BNNTs during the growth, which could protect BNNTs from contamination by the precursors and guarantee their ultimate purity, as displayed in an SEM image (Figure 16.3b). Various metal oxides were found to be effective for the BNNTs' growth. A mixture of MgO and FeO or MgO and SnO was found to be the best (Zhi et al., 2005c). This BOCVD method is currently successfully utilized for routine gram-quantity BNNT production.

16.2.7 Plasma-Jet Method

Shimizu et al. (1999) reported a plasma-jet method to synthesize multiwalled BNNTs. Nanotubes were prepared in a reaction chamber equipped with a water-cooled plasma torch, target holder, and a powder collector. A BN sintered disk was set on the target holder and was exposed to the high-temperature region of the plasma jet, which was generated at the power of 8 kW using Ar–H$_2$ gas for a plasma gas. Then, BNNTs could be collected from the surface of the sintered disk and from the powder collector. Although the crystallization of tubes was perfect, the purity of the product was low. Bengu and Marks (2001) synthesized a limited number of single-walled BNNTs using low-energy electron-cyclotron resonance plasma.

16.2.8 Purification

Only few works have been related to purification of BNNTs, which might be due to the stubborn nature of the specific impurities (BN particles, thick BN fibers, boron, etc.). Similar to CNTs, an acid was used to remove the metal catalysts remained in the BNNT product. BN nanoparticles were first converted to water-soluble B$_2$O$_3$ via a partial oxidation treatment, which could be dissolved in hot water (Tang et al., 2006). This technique utilizes the difference in the antioxidation ability of BN particles and BNNTs. Another method to remove the obstinate impurities is to adopt centrifugation separation based on the difference in geometry and weight between impurities and BNNTs (Zhi et al., 2006c; Kim et al., 2008).

16.3 MODIFICATION

By virtue of their superb thermal and oxidation stability and chemical inertness, BNNTs might be very useful in a variety of application fields. In this section, we will discuss the progress on the modification of BNNTs and emphasize tuning their electronic structure and expanding their applications by chemical functionalization. Due to the chemical inertness and poor wetting properties (Yum and Yu, 2006), works on experimental modifications of BNNTs are limited. Here, we briefly show the experimental strategies currently developed for the modification of BNNTs in Figure 16.4. The detailed methods and techniques will be included in the following descriptions.

16.3.1 Doping

In the area of semiconductor production, doping is used to intentionally introduce impurities into an extremely pure semiconductor to modulate its electrical properties. As far as BN doping is concerned, the ternary BN–C system has received remarkable research interests because of the marked similarities between layered BN and C materials (Zhi et al., 2004; An et al., 2007). However, the carbon doping process could not be well controlled. Two types of multiwalled nanotubes

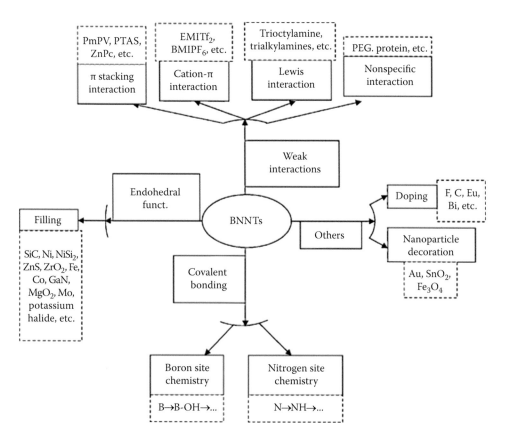

Figure 16.4 Schematic diagram of chemistry with BNNTs. (Reprinted from *Mater. Sci. Eng. R*, Zhi et al., Boron nitride nanotubes, *Materials Science & Engineering R-Reports*, 70, 92–111, Copyright 2010, with permission from Elsevier.)

having been considered during the formation of B–C–N compositions, that is, homogeneously structured B–C–N shells and sandwich-like heterogeneous BN–C structures (Golberg et al., 2004; Huang et al., 2007a). Figure 16.5 depicts a homogeneous B–C–N nanotube and a heterogeneous BN–C insulated B–C–N nanotube. Elemental profiles recorded across the homogeneous tube display perfect correction between the B and C peak intensities, while for the heterogeneous B–C–N tube, the brightest intensity strips on the elemental maps and the well-separated concentration profile peaks demonstrate that the tube is composed of a BN-rich external sheath and C-rich internal layers. Both tubes in Figure 16.5 were prepared using the conversion of C-based nanotube templates at high temperatures, although with slightly varying parameters. Thus, the formation of a particular B–C–N tube is a complicated function of the synthesis parameters, that is, precursors, temperature, flow rates, and vapor pressures of supply gases and gases generated in situ.

Tang et al. (2005) reported the growth of F-doped BNNTs using BF_3 and NH_3 as precursors by means of a typical CVD process. In the experiment, a lump of $MgCl_2$ was placed into the high-temperature area of the reactor, acting as a CVD substrate as well as a catalyst. Finally, fluorination of BN nanotubes results in highly curled tubular BN sheets and makes insulating BN nanotubes semiconducting. The phenomenon is assumed to be particularly important for the applications in the future nanoscale electronic devices with tunable properties.

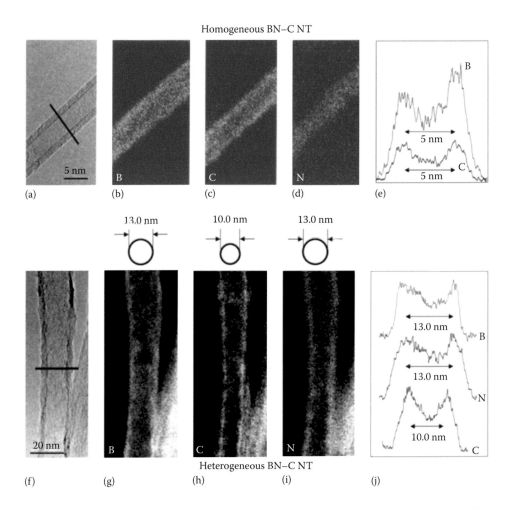

Figure 16.5 Two typical elemental distributions in BN–C nanotubes, as revealed by energy-filtering transmission electron microscopy: (b–e) homogeneous distribution of all constituent (B, C, and N) species; (g–j) phase-separated BN–C NT. Zero-loss TEM images (a, f), elemental maps (b–d; g–i), and cross-sectional elemental profiles (e, j) taken from individual tubes are shown. The tube cross sections of interests are marked with black bars on the zero-loss images (a, f). While the maps and concentration profiles correlate in the homogeneous tube (b–e), they display notable spatial separation in (g–j) with the BN-rich and C-rich domains enriching the outermost and the innermost NT fragments, respectively. (From Golberg, D. et al., *MRS Bull.*, 29, 38, 2004. Copyright 2004, Materials Research Society.)

Eu-doped BN bamboo-like fibers were produced by a ball-milling method using a mixture of B and Eu as precursors (Chen et al., 2007). A special in situ doping process was used to incorporate Eu^{2+} ions into nanotube walls during the growth. A dramatic change of luminescence properties was realized in the resultant Eu-doped BNNTs. The strong visible-light emission, generated from the entire doped nanotube, is expected to have many applications in nanosized lighting sources and nanospectroscopy. Doping has been proved to be a powerful tool to tune properties of semiconductors; but for BNNTS, challenges still exist. Until now, only the earlier-mentioned C, F, and Eu doping were successfully applied to BNNTs. Moreover, controlling the concentration of a dopant is extremely difficult for BNNTs; even the carbon doping of BNNTs could not be well controlled. Therefore, more techniques for the growth of doping of BNNTs should be developed.

16.3.2 Filling

Since BNNTs have very poor wetting properties(Yum and Yu, 2006), most of the well-established filling methods for CNTs using capillary or wet chemistry were not effective for BNNTs (Tang et al., 2002b). Therefore, most filled BNNTs were synthesized during a high-temperature growth. The filling processes were not under control, and in some cases, the filled BNNTs were only occasionally observed in the samples. However, in principle and theoretically, BNNTs filled with various materials, including metals and semiconductors, could be essentially useful (Mickelson et al., 2003; Li et al., 2005). In the following section, two important practices for filling of BNNTs, that is, cluster filling and wire filling, will be described.

16.3.2.1 Cluster Filling

Okada et al. (2001) reported first-principles total-energy electronic-structure calculations that provide energetics of encapsulation of C_{60} in BNNTs and electronic structures of resulting *BNC peapods*. They showed that the energy gain upon encapsulation is larger than that for carbon peapods and predicted that the BNNTs are promising candidates to accommodate the fullerenes inside to form a new ternary semiconductor with interesting structural hierarchy. High-yield C_{60} filling of BNNTs was realized by packing the spherical fullerene molecules into the interior spaces of BNNTs through uniformly heating a mixture of C_{60} and BNNTs in a sealed evacuated quartz ampoule (Mickelson et al., 2003). HRTEM images showed that a linear chain of nearly evenly spaced C_{60} molecules was filled inside the tubes (Figure 16.6). Moreover, the HRTEM image of the double-walled BNNTs, which contain amorphous BN *plugs* at the left and right portions of

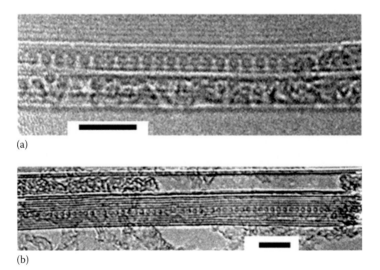

(a)

(b)

Figure 16.6   HRTEM image of C_{60}-treated BNNTs. (a) A double-walled nanotube is filled with a linear chain of C_{60}. (b) A five-walled nanotube is similarly filled. The C_{60}–C_{60} center to center distance in both cases is ~0.9 nm. Bar, 4 nm. (From Mickelson, W., Aloni, S., Han, W.Q., Cumings, J., and Zettl, A., Packing C-60 in boron nitride nanotubes, *Science*, 300, 467–469, 2003. Reprinted with permission of AAAS.)

the tube interior but a completely empty central portion, indicated that the C_{60} molecules enter BNNTs only through the ends of the tube and not through the side walls. The obtained C_{60}-filled BNNTs could work as a special nanoswitch owing to the insulating nature of BNNTs and an exciting possibility of shuffling/moving of C_{60} molecules inside the tube channel (Kwon et al., 1999; Hwang et al., 2005).

Moreover, metal clusters were also successfully encapsulated in BNNTs channels using metallic oxides as a filling medium. Golberg et al. (2001a) for the first time embedded pure isolated Mo clusters into BNNTs over their entire length. This was accomplished by means of two-step thermochemical treatment of chemically vapor-deposited CNT with B_2O_3, CuO, and MoO_3 oxides in a flowing N_2 atmosphere. The hybrid was demonstrated to be consisting of a conductive metal core and an insulating BN nanotubular shield. The clusters could levitate within the tubular channels because they do not wet the internal BN shells. Thus, the nanostructures may serve as natural *pipelines* for the delivery of tiny metallic clusters under applied thermal, magnetic, or electric fields.

16.3.2.2 Wire Filling

Han et al. (1999b) synthesized silicon carbide (SiC)-filled BNNTs in high yield by using CNTs as templates. They combined both CNT-substitution reaction and confined reaction to fill SiC nanowires into the entire length of BNNTs. They also embedded boron carbide nanowires into BNNTs by the same method (Han et al., 1999a). The diameters of the filled nanowires typically ranged from 7 to 20 nm, and the thickness of BN and C tubular layers was about 5 nm. The length of the filling could be up to the entire length of the nanotubes. Later, they developed a method to generate GaN nanorods with continuous BN coatings, by which the typical thickness of outer encapsulating BN shells was less than 5 nm (Han and Zettl, 2002). They also inserted one-dimensional crystals of potassium halides, including KI, KCl, and KBr, into BNNTs using a capillary technology (Han et al., 2004). They found that under continuous HRTEM electron irradiation (current density 500 nA/mm², exposure time typically 10 s), the KCl nanocrystal could be induced to cleave, resulting in two smaller independent KCl nanocrystals within the common BNNT housing (Han and Zettl, 2004).

Bando's group invented an original way of continuous BNNT filling with 3d-transition metals via capillarity through a two-step, high-temperature synthesis (Bando et al., 2001; Golberg et al., 2003b). The process consisted of preliminary filling of C-rich nanotubes with metal oxide melt or metal particles followed by C → BN conversion within tubular layers. By this process, the difficulties in wetting BNNTs, which is the biggest obstacle of filling, could be overcome by the using of CNTs alternatively as a template. Then, the CNT template synthesis, a well-established method for BNNTs growth, was utilized for the conversion. TEM images of resultant BNNTs clearly showed the filling of an Invar Fe–Ni alloy inside the BN tubular layers (Figure 16.7).

Golberg et al. (2003a) created MgO_2-based filling inside BNNTs by a BOCVD method. Interestingly, under moderate heating and/or room-temperature aging, the fillings easily decomposed, leaving stable Mg-containing phases inside the tubes, while the oxygen might be locally generated from the nanotubes. Because the investigated BNNTs possessed open tip ends, the oxygen generated would be released at a stable rate with long release term. Thus, the first nanotube-based oxygen generator has been realized.

16.3.3 Nanoparticle Decoration

It is possible to decorate BNNTs with various nanoparticles via intrinsic nonspecific interactions. Some nanoparticles, such as SnO_2, Au, Ag, and Fe_3O_4, might serve as new catalysts for

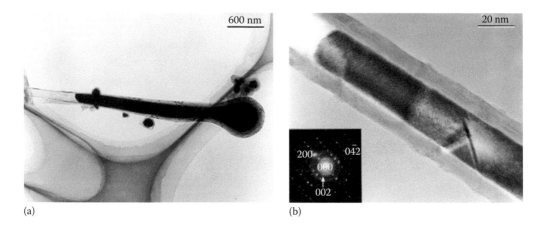

(a)   (b)

Figure 16.7   (a) TEM image of a BNNT filled with the Invar Fe–Ni alloy. (b) HRTEM image of a representative filled NTs. The inset shows a diffraction pattern taken from the filling. (Reprinted from *Chem. Phys. Lett.*, 347, Bando, Y., Ogawa, K., and Golberg, D., Insulating 'nanocables': Invar Fe-Ni alloy nanorods inside BN nanotubes, 349–354, 2001, Copyright 2010, with permission from Elsevier.)

the secondary growth of branched wires on BNNT surfaces producing interesting heterojunctions and nanoarchitectures (Han and Zettl, 2003; Zhi et al., 2006a; Sainsbury et al., 2007; Huang et al., 2009, 2010). Han and Zettl (2003) fully coated BNNTs with a thin nominally uniform layer of SnO_2 by a simple and efficient chemical reaction in solution. The resulting SnO_2 coatings on BNNTs were thin and uniform, with uninterrupted coverage (Figure 16.8). The characteristic lattice distances of both BNNTs and SnO_2 have been changed due to the strong interactions between them, which may imply the formation of Sn–N bonds or electrostatic tube/particle interactions (Zhi et al., 2006a).

By virtue of the functionalization of BNNTs with a high density of surface functional groups, such as amine- and thiol-functional group, gold nanoparticles could be immobilized at the surface of BNNTs via weak and strong covalent bonding, respectively (Sainsbury et al., 2007). Huang et al. (2009) prepared biotin-fluorescein-functionalized multiwalled BN nanotubes with

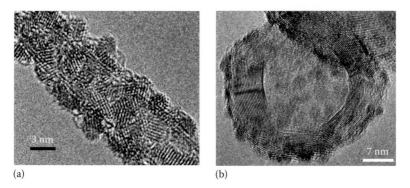

(a)   (b)

Figure 16.8   (a) HRTEM image of an individual double-walled BN nanotube fully coated with SnO_2. (b) HRTEM image of an SnO_2-coated BN fullerene-like nanoparticle. (Reprinted with permission from Han, W.Q. and Zettl, A., Functionalized boron nitride nanotubes with a stannic oxide coating: A novel chemical route to full coverage, *J. Am. Chem. Soc.*, 125, 2062–2063, 2003. Copyright 2003, American Chemical Society.)

anchored Ag nanoparticles. Ag nanoparticles were attached to BNNTs via the polydopamine coating by a secondary surface-mediated reaction. Intrinsic pH-dependent photoluminescence and Raman signals in attached fluorescein molecules enhanced by Ag nanoparticles allowed this novel nanohybrid to perform as a practical pH sensor.

Huang et al. (2010) developed an ethanol-thermal process for in situ formation of dense and uniformly distributed Fe_3O_4 nanoparticles on the surfaces of multiwalled BNNTs. Due to the superparamagnetic properties of the coated Fe_3O_4 nanoparticles, the functionalized BNNTs could be physically manipulated in a relatively low magnetic field. The novel BNNT-based magnetic nanocomposites may find a wide range of potential applications in magnetorheological devices, microelectromechanical systems, magnetic-targeted drug delivery, and boron neutron capture therapy. When BNNTs were decorated with various nanoparticles, their potential applications of BNNTs could be extended, owing to the combination of the diverse functional properties of nanoparticles with the large surface area of BNNTs. However, most works on nanoparticle decoration on BNNTs were related to multiwalled structures with only relatively low surface areas. Therefore, follow-up experiments on the nanoparticle decoration of single-walled BNNTs are awaited in the future.

16.3.4   Covalent Bonding

The covalent bonding modification of BNNTs is difficult, due to the high stability of B–N bonds. Thus, chemical functionalization via covalent bonds has solely relied on existing defect sites. The chemistry of BNNTs might go on either N sites and/or B sites, which could result in different properties of functionalized BNNTs (An et al., 2007; Cao et al., 2009).

N-site-functionalization is more plausible, because the amino group formed during the synthesis of BNNTs is an active functional group that may participate in many organic chemical reactions. The modification of BNNTs with long alkyl chains could be achieved by a chemical reaction between the amino groups of the synthesized BNNTs with the COCl group of stearoyl chloride (Zhi et al., 2005e). In contrast to the starting multiwalled BNNTs, which were insoluble in organic solvents, the resultant functionalized BNNTs were soluble in many organic solvents. The modification of long alkyl chains might also induce drastic changes in the band structure of the BNNTs, shown by the CL and UV/Vis absorption experiments. The BNNTs–polymer composites (BNNTs–PS, BNNTs–PMMA) were fabricated via covalent bonding using an atom transfer radical polymerization approach (Zhi et al., 2007a). The obtained composites were proved to have a perfect physical contact between the two phases (Figure 16.9). Moreover, carbonization of the BNNTs–polymer composites by simply heating would lead to carbon–BNNTs composites, which might exhibit better transport properties and electrochemical activities than pure BNNTs. Moreover, a long alkyl chain could also be connected to a BNNT surface at room temperature via an S_N2 nucleophilic substitution reaction catalyzed by a strong Lewis acid (Zhi et al., 2007c).

The earlier-mentioned modifications of BNNTs are based on amino groups that are formed simultaneously during the tube growth. Therefore, the reaction sites on BNNTs were not so sufficient. Ikuno et al. (2007) developed a method to increase the numbers of amine group on the BNNT surface by virtue of an aggressive nonequilibrium ammonia glow plasma treatment. The resultant functionalized BNNTs have the potential to be utilized as building blocks for nanoscale structures and composite materials since the amine functional groups present on the modified tubes can be used as binding sites for the immobilization of small molecules, polymers, and condensed phase materials.

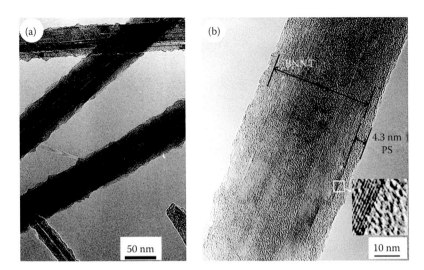

Figure 16.9 (a) Low-magnification and (b) high-resolution TEM images of BNNTs–PS. (Reprinted with permission from Zhi, C., Bando, Y., Tang, C., Kuwahara, H., and Golberg, D., Grafting boron nitride nanotubes: From polymers to amorphous and graphitic carbon, *J. Phys. Chem. C*, 111, 1230–1233, 2007a. Copyright 2007, American Chemical Society.)

On the other hand, the B-site covalent bonding modification is not that popular, because most of the organic chemical reactions are related to C and N species. The hydroxylated BNNTs were prepared by covalently connecting hydroxyl group (–OH) to the B sites using hydrogen peroxide treatment at high temperature and high pressure (Zhi et al., 2009b). The modification altered the hydrophobic property of pristine BNNTs since the functionalized BNNTs became water soluble. Moreover, the B-site modified BNNTs became chemically active and opened various routes for BNNTs' further modification, for example, an esterification reaction with carboxyl group could be performed.

16.3.5 Weak Interactions

Not only covalent bonds but also weak interactions can be utilized to functionalize BNNTs. Covalent functionalization can effectively modify BNNT properties, whereas functionalization via weak interactions is important when it is essential to preserve the intrinsic nanotube properties. Various materials could interact with BNNTs via π stacking interactions and cation–π interactions (Zhi et al., 2005b,f, 2007c, Huang et al., 2007c, Wang et al., 2008). Moreover, electrostatic interactions, interactions between Lewis acids and bases (Pal et al., 2007, Maguer et al., 2009), and some nonspecific interactions (Xie et al., 2005, Zhi et al., 2005a,b, Ikuno et al., 2007) were also used for the modification of BNNTs.

Since BNNTs possess a well-developed π electronic structure, π stacking interactions could be used to wrap polymers on BNNTs. The BNNTs could be solubilized via wrapping BNNTs with a conjugated polymer, poly [*m*-phenylenevinylene-*co*-(2,5-dioctoxy-*p*-phenylenevinylene)] (PmPV) (Zhi et al., 2005f). The resultant BNNT solution was highly transparent after dilution, in which no precipitation was observed during a long time keeping at ambient conditions (Figure 16.10). By virtue of soluble PmPV-wrapped BNNTs, a method was developed to purify BNNTs, which is effective to remove large BN particles from BNNTs (Zhi et al., 2006c).

Figure 16.10   A fully transparent solution of multiwalled BNNTs in chloroform. The light gray color of the solution is due to a PmPV polymer used for BNNT wrapping via π stacking interactions. (Reprinted with permission from Zhi, C.Y., Bando, Y., Tang, C.C., Xie, R.G., Sekiguchi, T., and Golberg, D., Perfectly dissolved boron nitride nanotubes due to polymer wrapping, *J. Am. Chem. Soc.*, 127, 15996–15997, 2005f. Copyright 2005, American Chemical Society.)

BNNTs were also found to induce a new phase formation in some ionic liquids, such as 1-ethyl-3-methylimidazolium trifluoromethanesulfonate (EMITf2) and 1-butyl-3-methylimidazolium hexafluorophosphate (BMIPF6), and form a gel with new molecule ordering through cation–π interactions (Zhi et al., 2007c). Other molecules, such as zinc phthalocyanine (Huang et al., 2007a), perylene-3,4,9,10-tetracarboxylic acid tetrapotassium salt (Wang et al., 2008), and poly-aniline (Zhi et al., 2005b), can strongly interact with BNNTs and induce dramatic variations in optical and chemical properties.

The Lewis acid nature of boron in BNNTs can be exploited to functionalize and solu-bilize their nanostructures through interaction with Lewis bases. Pal et al. (2007) dis-persed BNNTs in a hydrocarbon medium with the retention of the nanotube structure by interacting BNNTs with a trialkylamine or trialkylphosphine (Lewis bases). Moreover, based on the affinity of the nitrogen atoms of quinuclidine molecules (Lewis bases) for boron atoms incorporated in the nanotubes network (Lewis acids), single-walled and multiwalled BNNTs were functionalized (Maguer et al., 2009). Specific substitution of the quinuclidine skeleton permitted solubilization of the nanotubes in different media.

There have been some unspecific interactions used to modify BNNTs, the nature of which has not been specified so far. Xie et al. (2005) used a polymer (polyethylene glycol) to wrap BNNTs leading to a good solubility. It was suggested that the amino groups in the polymer can interact with the B sites of BNNTs. No detailed information was given about the nature of these interac-tions though. Some biomolecules, such as proteins and DNA, were also found to be immobilized on BNNTs based on some nonspecific interactions (Zhi et al., 2005a, 2007b).

16.4   BNNT USED AS NANOFILLERS FOR COMPOSITES

BNNTs have superb thermal and oxidation stability and chemical inertness, which make them attractive for applications in composite materials. However, due to the earlier-mentioned difficulties in the high-yield growth of pure BNNTs, the studies for BNNT composites were only initiated in recent years. The materials now fabricated include polymeric and ceramic composites, whose physical properties were improved by the formation of binary hybrid.

16.4.1   Polymeric BNNT Composites

The first polymeric BNNT composites were fabricated to enhance the mechanical properties of polystyrene (PS) (Zhi et al., 2006b). Tensile tests indicated that the elastic modulus of the composite films was increased by ~21% when a ~1 wt.% soluble BNNT fraction was in use. With the assistance of PmPV surfactant, the dispersion of BNNTs as well as the interfacial interactions between BNNTs and PS could be improved (Figure 16.11). Moreover, the composites have a better stability to oxidation and a slightly lowered glass transition temperature (T_g) than a blank polymer, which could imply that BNNTs could become an important reinforcement additive for polymers due to their excellent mechanical properties and thermal conductivity. In the following studies, the elastic modulus of polymethyl methacrylate (PMMA) (Zhi et al., 2007a, 2008), polycarbonate, and polyvinyl butyral (Zhi et al., 2007b) was shown to be increased by the addition of BNNTs. The better mechanical reinforcement of these composites was mainly attributed to the higher elastic modulus of the additive BNNTs.

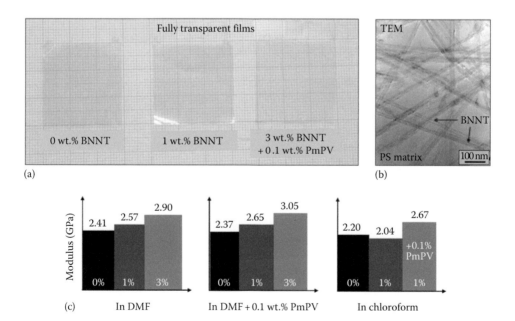

Figure 16.11   (a) Fully transparent polymeric films containing various fractions of BNNTs in PS matrices; (b) TEM image of BNNTs evenly dispersed in a polymer; (c) tensile tests of PS films with adding variable BNNT fractions dissolved in different solvents. (From Zhi, C. et al., *J. Mater. Res.*, 21, 2794, 2006b. Copyright Materials Research Society.)

The electrically insulating BNNTs also possess high thermal conductivity, so another important application of BNNTs as nanofillers is relied on their use for the fabrication of novel thermo-conductive insulating polymeric composites. The first example of improving thermal conductivity of polymer (PMMA) was achieved by the fabrication of polymeric BNNT composites (Zhi et al., 2008). The thermal conductivity of PMMA loaded with a 10 wt.% BNNT fraction was improved three times compared to blank PMMA. In the following-up study, four polymers (i.e., poly(methyl methacrylate), PS, poly(vinyl butyral), poly(ethylene vinyl alcohol)) were mixed with BNNTs to form the polymeric composites (Zhi et al., 2009a). More than 20-fold thermal conductivity improvement of the polymers was obtained by utilizing BNNTs as nanofillers.

16.4.2   Ceramic BNNT Composites

The robustness of BNNTs to oxidation is a solid advantage for the fabrication of ceramic BNNT composites. The first ceramic BNNT composites synthesized was barium calcium aluminosilicate glass composites reinforced with ~4 wt.% of BNNTs (Bansal et al., 2006). Compared with unreinforced glass, the composites with BNNTs showed 90% increase in the strength and 35% increase in fracture toughness. Later on, Huang et al. also fabricated Al_2O_3–BNNTs and Si_3N_4–BNNTs composites and investigated their mechanical properties. When a small amount of BNNTs (0.5 wt.%) was loaded, the Vickers hardness of Al_2O_3 was improved from 17.3 to 19.1 GPa, while it was decreased to 14.5 GPa when a 2.5 wt.% BNNT fraction was added (Huang et al., 2007b).

16.5   SUMMARY

BNNTs, a very close analogue of the CNTs, possess very good properties, including excellent electrically insulation, high thermal conductivity, good mechanical strength, and superb thermal and antioxidation stabilities. Nowadays various growth methods have been developed for the fabrication of this promising nanomaterial, though high-quality BNNTs are still difficult to fabricate in large quantities. Despite the main difficulties in the pure BNNT preparation, modification strategies, such as doping, filling, nanoparticles decoration, weak interactions, and covalent bonding, are developed to effectively tune the electronic structure and expand the applications of BNNTs. BNNTs are also utilized as nanofillers to form composite materials, including polymeric composites and ceramic composites, whose physical properties are improved.

REFERENCES

Akdim, B., Pachter, R., Duan, X. F., and Adams, W. W. 2003. Comparative theoretical study of single-wall carbon and boron-nitride nanotubes. *Physical Review B*, 67, 245404.

An, W., Wu, X., Yang, J. L., and Zeng, X. C. 2007. Adsorption and surface reactivity on single-walled boron nitride nanotubes containing stone–Wales defects. *The Journal of Physical Chemistry C*, 111, 14105–14112.

Arenal, R., Ferrari, A. C., Reich, S., Wirtz, L., Mevellec, J. Y., Lefrant, S., Rubio, A., and Loiseau, A. 2006. Raman spectroscopy of single-wall boron nitride nanotubes. *Nano Letters*, 6, 1812–1816.

Arenal, R., Stephan, O., Cochon, J. L., and Loiseau, A. 2007. Root-growth mechanism for single-walled boron nitride nanotubes in laser vaporization technique. *Journal of the American Chemical Society*, 129, 16183–16189.

Bando, Y., Ogawa, K., and Golberg, D. 2001. Insulating 'nanocables': Invar Fe-Ni alloy nanorods inside BN nanotubes. *Chemical Physics Letters*, 347, 349–354.

Bansal, N. P., Hurst, J. B., and Choi, S. R. 2006. Boron nitride nanotubes-reinforced glass composites. *Journal of the American Ceramic Society*, 89, 388–390.

Bechelany, M., Bernard, S., Brioude, A., Cornu, D., Stadelmann, P., Charcosset, C., Fiaty, K., and Miele, P. 2007. Synthesis of boron nitride nanotubes by a template-assisted polymer thermolysis process. *Journal of Physical Chemistry C*, 111, 13378–13384.

Bengu, E. and Marks, L. D. 2001. Single-walled BN nanostructures. *Physical Review Letters*, 86, 2385–2387.

Blase, X., Rubio, A., Louie, S. G., and Cohen, M. L. 1994. Stability and band-gap constancy of boron-nitride nanotubes. *Europhysics Letters*, 28, 335–340.

Cao, F. L., Ren, W., Ji, Y. M., and Zhao, C. Y. 2009. The structural and electronic properties of amine-functionalized boron nitride nanotubes via ammonia plasmas: A density functional theory study. *Nanotechnology*, 20, 145703.

Chang, C. W., Fennimore, A. M., Afanasiev, A., Okawa, D., Ikuno, T., Garcia, H., Li, D., Majumdar, A., and Zettl, A. 2006. Isotope effect on the thermal conductivity of boron nitride nanotubes. *Physical Review Letters*, 97, 085901.

Chang, C. W., Okawa, D., Garcia, H., Majumdar, A., and Zettl, A. 2007. Nanotube phonon waveguide. *Physical Review Letters*, 99, 045901.

Chen, H., Chen, Y., Li, C. P., Zhang, H., Williams, J. S., Liu, Y., Liu, Z., and Ringer, S. P. 2007. Eu-doped boron nitride nanotubes as nanosized visible light source. *Advanced Materials*, 19, 1845–1848.

Chen, Y., Chadderton, L. T., Fitzgerald, J., and Williams, J. S. 1999. A solid-state process for formation of boron nitride nanotubes. *Applied Physics Letters*, 74, 2960–2962.

Chen, Y., Conway, M., Williams, J. S., and Zou, J. 2002. Large-quantity production of high-yield boron nitride nanotubes. *Journal of Materials Research*, 17, 1896–1899.

Chen, Y., Zou, J., Campbell, S. J., and Le Caer, G. 2004. Boron nitride nanotubes: Pronounced resistance to oxidation. *Applied Physics Letters*, 84, 2430–2432.

Chopra, N. G., Luyken, R. J., Cherrey, K., Crespi, V. H., Cohen, M. L., Louie, S. G., and Zettl, A. 1995. Boron-nitride nanotubes. *Science*, 269, 966–967.

Chopra, N. G. and Zettl, A. 1998. Measurement of the elastic modulus of a multi-wall boron nitride nanotube. *Solid State Communications*, 105, 297–300.

Coleman, J. N., Khan, U., and Gun'ko, Y. K. 2006. Mechanical reinforcement of polymers using carbon nanotubes. *Advanced Materials*, 18, 689–706.

Cumings, J. and Zettl, A. 2000. Mass-production of boron nitride double-wall nanotubes and nanococoons. *Chemical Physics Letters*, 316, 211–216.

Dai, J., Xu, L. Q., Fang, Z., Sheng, D. P., Guo, Q. F., Ren, Z. Y., Wang, K., and Qian, Y. T. 2007. A convenient catalytic approach to synthesize straight boron nitride nanotubes using synergic nitrogen source. *Chemical Physics Letters*, 440, 253–258.

Dresselhaus, M. S. and Eklund, P. C. 2000. Phonons in carbon nanotubes. *Advances in Physics*, 49, 705–814.

Endo, M., Takeuchi, K., Igarashi, S., Kobori, K., Shiraishi, M., and Kroto, H. W. 1993. The production and structure of pyrolytic carbon nanotubes (PCNTS). *Journal of Physics and Chemistry of Solids*, 54, 1841–1848.

Golberg, D., Bando, Y., Dorozhkin, P., and Dong, Z. C. 2004. Synthesis, analysis, and electrical property measurements of compound nanotubes in the B-C-N ceramic system. *MRS Bulletin*, 29, 38–42.

Golberg, D., Bando, Y., Eremets, M., Takemura, K., Kurashima, K., and Yusa, H. 1996. Nanotubes in boron nitride laser heated at high pressure. *Applied Physics Letters*, 69, 2045–2047.

Golberg, D., Bando, Y., Fushimi, K., Mitome, M., Bourgeois, L., and Tang, C. C. 2003a. Nanoscale oxygen generators: MgO_2-based fillings of BN nanotubes. *Journal of Physical Chemistry B*, 107, 8726–8729.

Golberg, D., Bando, Y., Han, W., Kurashima, K., and Sato, T. 1999. Single-walled B-doped carbon, B/N-doped carbon and BN nanotubes synthesized from single-walled carbon nanotubes through a substitution reaction. *Chemical Physics Letters*, 308, 337–342.

Golberg, D., Bando, Y., Huang, Y., Terao, T., Mitome, M., Tang, C. C., and Zhi, C. Y. 2010. Boron nitride nanotubes and nanosheets. *ACS Nano*, 4, 2979–2993.

Golberg, D., Bando, Y., Kurashima, K., and Sato, T. 2000a. MoO_3-promoted synthesis of multi-walled BN nanotubes from C nanotube templates. *Chemical Physics Letters*, 323, 185–191.

Golberg, D., Bando, Y., Kurashima, K., and Sato, T. 2000b. Ropes of BN multi-walled nanotubes. *Solid State Communications*, 116, 1–6.

Golberg, D., Bando, Y., Kurashima, K., and Sato, T. 2001a. Nanotubes of boron nitride filled with molybdenum clusters. *Journal of Nanoscience and Nanotechnology*, 1, 49–54.

Golberg, D., Bando, Y., Kurashima, K., and Sato, T. 2001b. Synthesis and characterization of ropes made of BN multiwalled nanotubes. *Scripta Materialia*, 44, 1561–1565.

Golberg, D., Costa, P., Lourie, O., Mitome, M., Bai, X. D., Kurashima, K., Zhi, C. Y., Tang, C. C., and Bando, Y. 2007. Direct force measurements and kinking under elastic deformation of individual multiwalled boron nitride nanotubes. *Nano Letters*, 7, 2146–2151.

Golberg, D., Xu, F. F., and Bando, Y. 2003b. Filling boron nitride nanotubes with metals. *Applied Physics A: Materials Science and Processing*, 76, 479–485.

Guo, T., Nikolaev, P., Rinzler, A. G., Tomanek, D., Colbert, D. T., and Smalley, R. E. 1995. Self-assembly of tubular fullerenes. *Journal of Physical Chemistry*, 99, 10694–10697.

Han, W. Q., Bando, Y., Kurashima, K., and Sato, T. 1998. Synthesis of boron nitride nanotubes from carbon nanotubes by a substitution reaction. *Applied Physics Letters*, 73, 3085–3087.

Han, W. Q., Chang, C. W., and Zettl, A. 2004. Encapsulation of one-dimensional potassium halide crystals within BN nanotubes. *Nano Letters*, 4, 1355–1357.

Han, W. Q., Kohler-Redlich, P., Ernst, F., and Ruhle, M. 1999a. Formation of (BN)(x)C-y and BN nanotubes filled with boron carbide nanowires. *Chemistry of Materials*, 11, 3620–3623.

Han, W. Q., Redlich, P., Ernst, F., and Ruhle, M. 1999b. Synthesizing boron nitride nanotubes filled with SiC nanowires by using carbon nanotubes as templates. *Applied Physics Letters*, 75, 1875–1877.

Han, W. Q. and Zettl, A. 2002. GaN nanorods coated with pure BN. *Applied Physics Letters*, 81, 5051–5053.

Han, W. Q. and Zettl, A. 2003. Functionalized boron nitride nanotubes with a stannic oxide coating: A novel chemical route to full coverage. *Journal of the American Chemical Society*, 125, 2062–2063.

Han, W. Q. and Zettl, A. 2004. Nanocrystal cleaving. *Applied Physics Letters*, 84, 2644–2645.

Harutyunyan, H., Gokus, T., Green, A. A., Hersam, M. C., Allegrini, M., and Hartschuh, A. 2009. Defect-induced photoluminescence from dark excitonic states in individual single-walled carbon nanotubes. *Nano Letters*, 9, 2010–2014.

Hernandez, E., Goze, C., Bernier, P., and Rubio, A. 1998. Elastic properties of C and $B_xC_yN_z$ composite nanotubes. *Physical Review Letters*, 80, 4502–4505.

Huang, Q., Bando, Y., Sandanayaka, A., Tang, C. C., Wang, J. B., Sekiguchi, T., Zhi, C. Y. et al. 2007a. Photoinduced charge injection and bandgap-engineering of high-specific-surface-area BN nanotubes using a zinc phthalocyanine monolayer. *Small*, 3, 1330–1335.

Huang, Q., Bando, Y., Zhao, L., Zhi, C. Y., and Golberg, D. 2009. pH sensor based on boron nitride nanotubes. *Nanotechnology*, 20, 415501.

Huang, Q., Bando, Y. S., Xu, X., Nishimura, T., Zhi, C. Y., Tang, C. C., Xu, F. F., Gao, L., and Golberg, D. 2007b. Enhancing superplasticity of engineering ceramics by introducing BN nanotubes. *Nanotechnology*, 18, 1–7.

Huang, Q., Sandanayaka, A. S. D., Bando, Y., Zhi, C., Ma, R., Shen, G., Golberg, D. et al. 2007c. Donor-acceptor nanoensembles based on boron nitride nanotubes. *Advanced Materials*, 19, 934–938.

Huang, Y., Lin, J., Bando, Y., Tang, C. C., Zhi, C. Y., Shi, Y. G., Takayama-Muromachi, E., and Golberg, D. 2010. BN nanotubes coated with uniformly distributed Fe_3O_4 nanoparticles: Novel magneto-operable nanocomposites. *Journal of Materials Chemistry*, 20, 1007–1011.

Hwang, H. J., Choi, W. Y., and Kang, J. W. 2005. Molecular dynamics simulations of nanomemory element based on boron-nitride nanotube-to-peapod transition. *Computational Materials Science*, 33, 317–324.

Iijima, S. 1991. Helical microtubules of graphitic carbon. *Nature*, 354, 56–58.

Ikuno, T., Sainsbury, T., Okawa, D., Frehet, J. M. J., and Zettl, A. 2007. Amine-functionalized boron nitride nanotubes. *Solid State Communications*, 142, 643–646.

Kim, J., Lee, S., Seo, D., and Seo, Y. S. 2012. Synthesis of multiwall boron nitride nanotubes dependent on crystallographic structure of boron. *Materials Chemistry and Physics*, 137, 182–187.

Kim, M. J., Chatterjee, S., Kim, S. M., Stach, E. A., Bradley, M. G., Pender, M. J., Sneddon, L. G., and Maruyama, B. 2008. Double-walled boron nitride nanotubes grown by floating catalyst chemical vapor deposition. *Nano Letters*, 8, 3298–3302.

Kwon, Y. K., Tomanek, D., and Iijima, S. 1999. "Bucky shuttle" memory device: Synthetic approach and molecular dynamics simulations. *Physical Review Letters*, 82, 1470–1473.

Laude, T., Matsui, Y., Marraud, A., and Jouffrey, B. 2000. Long ropes of boron nitride nanotubes grown by a continuous laser heating. *Applied Physics Letters*, 76, 3239–3241.

Lauret, J. S., Arenal, R., Ducastelle, F., Loiseau, A., Cau, M., Attal-Tretout, B., Rosencher, E., and Goux-Capes, L. 2005. Optical transitions in single-wall boron nitride nanotubes. *Physical Review Letters*, 94, 037405.

Lee, R. S., Gavillet, J., De La Chapelle, M. L., Loiseau, A., Cochon, J. L., Pigache, D., Thibault, J., and Willaime, F. 2001. Catalyst-free synthesis of boron nitride single-wall nanotubes with a preferred zig-zag configuration. *Physical Review B*, 64, 121405.

Li, Y. B., Dorozhkin, P. S., Bando, Y., and Golberg, D. 2005. Controllable modification of SiC nanowires encapsulated in BN nanotubes. *Advanced Materials*, 17, 545–549.

Liu, Z. and Marder, T. B. 2008. B-N versus C-C: How similar are they? *Angewandte Chemie International Edition*, 47, 242–244.

Loiseau, A., Willaime, F., Demoncy, N., Hug, G., and Pascard, H. 1996. Boron nitride nanotubes with reduced numbers of layers synthesized by arc discharge. *Physical Review Letters*, 76, 4737–4740.

Loiseau, A., Willaime, F., Demoncy, N., Schramchenko, N., Hug, G., Colliex, C., and Pascard, H. 1998. Boron nitride nanotubes. *Carbon*, 36, 743–752.

Lourie, O. R., Jones, C. R., Bartlett, B. M., Gibbons, P. C., Ruoff, R. S., and Buhro, W. E. 2000. CVD growth of boron nitride nanotubes. *Chemistry of Materials*, 12, 1808–1810.

Ma, R. Z., Bando, Y., Sato, T., and Kurashima, K. 2001. Growth, morphology, and structure of boron nitride nanotubes. *Chemistry of Materials*, 13, 2965–2971.

Maguer, A., Leroy, E., Bresson, L., Doris, E., Loiseau, A., and Mioskowski, C. 2009. A versatile strategy for the functionalization of boron nitride nanotubes. *Journal of Materials Chemistry*, 19, 1271–1275.

Mickelson, W., Aloni, S., Han, W. Q., Cumings, J., and Zettl, A. 2003. Packing C-60 in boron nitride nanotubes. *Science*, 300, 467–469.

Mintmire, J. W., Dunlap, B. I., and White, C. T. 1992. Are fullerene tubules metallic. *Physical Review Letters*, 68, 631–634.

Murakami, Y., Lu, B., Kazaoui, S., Minami, N., Okubo, T., and Maruyama, S. 2009. Photoluminescence sidebands of carbon nanotubes below the bright singlet excitonic levels. *Physical Review B*, 79, 1–19.

O'Connell, M. J., Bachilo, S. M., Huffman, C. B., Moore, V. C., Strano, M. S., Haroz, E. H., Rialon, K. L. et al. 2002. Band gap fluorescence from individual single-walled carbon nanotubes. *Science*, 297, 593–596.

Okada, S., Saito, S., and Oshiyama, A. 2001. Semiconducting form of the first-row elements: C-60 chain encapsulated in BN nanotubes. *Physical Review B*, 64, 1303.

Pal, S., Vivekchand, S. R. C., Govindaraj, A., and Rao, C. N. R. 2007. Functionalization and solubilization of BN nanotubes by interaction with Lewis bases. *Journal of Materials Chemistry*, 17, 450–452.

Rosas, G., Sistos, J., Ascencio, J. A., Medina, A., and Perez, R. 2005. Multiple-walled BN nanotubes obtained with a mechanical alloying technique. *Applied Physics A: Materials Science and Processing*, 80, 377–380.

Rubio, A., Corkill, J. L., and Cohen, M. L. 1994. Theory of graphitic boron-nitride nanotubes. *Physical Review B*, 49, 5081–5084.

Sainsbury, T., Ikuno, T., Okawa, D., Pacile, D., Frechet, J. M. J., and Zettl, A. 2007. Self-assembly of gold nanoparticles at the surface of amine- and thiol-functionalized boron nitride nanotubes. *Journal of Physical Chemistry C*, 111, 12992–12999.

Saito, Y. and Maida, M. 1999. Square, pentagon, and heptagon rings at BN nanotube tips. *Journal of Physical Chemistry A*, 103, 1291–1293.

Shimizu, Y., Moriyoshi, Y., Tanaka, H., and Komatsu, S. 1999. Boron nitride nanotubes, webs, and coexisting amorphous phase formed by the plasma jet method. *Applied Physics Letters*, 75, 929–931.

Singhal, S. K., Srivastava, A. K., Pant, R. P., Halder, S. K., Singh, B. P., and Gupta, A. K. 2008. Synthesis of boron nitride nanotubes employing mechanothermal process and its characterization. *Journal of Materials Science*, 43, 5243–5250.

Smith, M. W., Jordan, K. C., Park, C., Kim, J.-W., Lillehei, P. T., Crooks, R., and Harrison, J. S. 2009. Very long single- and few-walled boron nitride nanotubes via the pressurized vapor/condenser method. *Nanotechnology*, 20, 505604.

Suenaga, K., Colliex, C., Demoncy, N., Loiseau, A., Pascard, H., and Willaime, F. 1997. Synthesis of nanoparticles and nanotubes with well-separated layers of boron nitride and carbon. *Science*, 278, 653–655.

Suryavanshi, A. P., Yu, M. F., Wen, J. G., Tang, C. C., and Bando, Y. 2004. Elastic modulus and resonance behavior of boron nitride nanotubes. *Applied Physics Letters*, 84, 2527–2529.

Tang, C., Bando, Y., Sato, T., and Kurashima, K. 2002a. A novel precursor for synthesis of pure boron nitride nanotubes. *Chemical Communications*, 12, 1290–1291.

Tang, C., Bando, Y., Zhi, C., and Golberg, D. 2007. Boron-oxygen luminescence centres in boron-nitrogen systems. *Chemical Communications*, 44, 4599–4601.

Tang, C. C., Bando, Y., Huang, Y., Yue, S. L., Gu, C. Z., Xu, F. F., and Golberg, D. 2005. Fluorination and electrical conductivity of BN nanotubes. *Journal of the American Chemical Society*, 127, 6552–6553.

Tang, C. C., Bando, Y., Sato, T., and Kurashima, K. 2002b. Uniform boron nitride coatings on silicon carbide nanowires. *Advanced Materials*, 14, 1046–1049.

Tang, C. C., Bando, Y., Shen, G. Z., Zhi, C. Y., and Golberg, D. 2006. Single-source precursor for chemical vapour deposition of collapsed boron nitride nanotubes. *Nanotechnology*, 17, 5882–5888.

Tang, C. C., Ding, X. X., Huang, X. T., Gan, Z. W., Qi, S. R., Liu, W., and Fan, S. S. 2002c. Effective growth of boron nitride nanotubes. *Chemical Physics Letters*, 356, 254–258.

Terao, T., Bando, Y., Mitome, M., Kurashima, K., Zhi, C. Y., Tang, C. C., and Golberg, D. 2008. Effective synthesis of surface-modified boron nitride nanotubes and related nanostructures and their hydrogen uptake. *Physica E-Low-Dimensional Systems, and Nanostructures*, 40, 2551–2555.

Wang, J. S., Kayastha, V. K., Yap, Y. K., Fan, Z. Y., Lu, J. G., Pan, Z. W., Ivanov, I. N., Puretzky, A. A., and Geohegan, D. B. 2005. Low temperature growth of boron nitride nanotubes on substrates. *Nano Letters*, 5, 2528–2532.

Wang, W., Bando, Y., Zhi, C., Fu, W., Wang, E., and Golberg, D. 2008. Aqueous noncovalent functionalization and controlled near-surface carbon doping of multiwalled boron nitride nanotubes. *Journal of the American Chemical Society*, 130, 8144–8145.

Wu, J., Han, W. Q., Walukiewicz, W., Ager, J. W., Shan, W., Haller, E. E., and Zettl, A. 2004. Raman spectroscopy and time-resolved photoluminescence of BN and $B_xC_yN_z$ nanotubes. *Nano Letters*, 4, 647–650.

Xie, S. Y., Wang, W., Fernando, K. A. S., Wang, X., Lin, Y., and Sun, Y. P. 2005. Solubilization of boron nitride nanotubes. *Chemical Communications*, 29, 3670–3672.

Xu, L. Q., Peng, Y. Y., Meng, Z. Y., Yu, W. C., Zhang, S. Y., Liu, X. M., and Qian, Y. T. 2003. A co-pyrolysis method to boron nitride nanotubes at relative low temperature. *Chemistry of Materials*, 15, 2675–2680.

Xu, Z., Golberg, D., and Bando, Y. 2009. In situ TEM-STM recorded kinetics of boron nitride nanotube failure under current flow. *Nano Letters*, 9, 2251–2254.

Yu, D. P., Sun, X. S., Lee, C. S., Bello, I., Lee, S. T., Gu, H. D., Leung, K. M., Zhou, G. W., Dong, Z. F., and Zhang, Z. 1998. Synthesis of boron nitride nanotubes by means of excimer laser ablation at high temperature. *Applied Physics Letters*, 72, 1966–1968.

Yu, J., Chen, Y., Wuhrer, R., Liu, Z. W., and Ringer, S. P. 2005. In situ formation of BN nanotubes during nitriding reactions. *Chemistry of Materials*, 17, 5172–5176.

Yum, K. and Yu, M. F. 2006. Measurement of wetting properties of individual boron nitride nanotubes with the Wilhelmy method using a nanotube-based force sensor. *Nano Letters*, 6, 329–333.

Zhi, C., Bando, Y., Tang, C., and Golberg, D. 2005a. Immobilization of proteins on boron nitride nanotubes. *Journal of the American Chemical Society*, 127, 17144–17145.

Zhi, C., Bando, Y., Tang, C., and Golberg, D. 2006a. SnO_2 nanoparticle-functionalized boron nitride nanotubes. *Journal of Physical Chemistry B*, 110, 8548–8550.

Zhi, C., Bando, Y., Tang, C., Honda, S., Kuwahara, H., and Golberg, D. 2006b. Boron nitride nanotubes/polystyrene composites. *Journal of Materials Research*, 21, 2794–2800.

Zhi, C., Bando, Y., Tang, C., Honda, S., Sato, K., Kuwahara, H., and Golberg, D. 2005b. Characteristics of boron nitride nanotube–polyaniline composites. *Angewandte Chemie International Edition*, 44, 7929–7932.

Zhi, C., Bando, Y., Tang, C., Kuwahara, H., and Golberg, D. 2007a. Grafting boron nitride nanotubes: From polymers to amorphous and graphitic carbon. *Journal of Physical Chemistry C*, 111, 1230–1233.

Zhi, C., Bando, Y., Terao, T., Tang, C., Kuwahara, H., and Golberg, D. 2009a. Towards thermoconductive, electrically insulating polymeric composites with boron nitride nanotubes as fillers. *Advanced Functional Materials*, 19, 1857–1862.

Zhi, C., Bando, Y., Wang, W., Tang, C., Kuwahara, H., and Golberg, D. 2007b. DNA-mediated assembly of boron nitride nanotubes. *Chemistry: An Asian Journal*, 2, 1581–1585.

Zhi, C., Bando, Y., Wang, W., Tang, C., Kuwahara, H., and Golberg, D. 2007c. Molecule ordering triggered by boron nitride nanotubes and "Green" chemical functionalization of boron nitride nanotubes. *Journal of Physical Chemistry C*, 111, 18545–18549.

Zhi, C. Y., Bai, X. D., and Wang, E. G. 2004. Boron carbonitride nanotubes. *Journal of Nanoscience and Nanotechnology*, 4, 35–51.

Zhi, C. Y., Bando, Y., Tan, C. C., and Golberg, D. 2005c. Effective precursor for high yield synthesis of pure BN nanotubes. *Solid State Communications*, 135, 67–70.

Zhi, C. Y., Bando, Y., Tang, C. C., Golberg, D., Xie, R. G., and Sekigushi, T. 2005d. Phonon characteristics and cathodolumininescence of boron nitride nanotubes. *Applied Physics Letters*, 87, 049902.

Zhi, C. Y., Bando, Y., Tang, C. C., Honda, S., Sato, K., Kuwahara, H., and Golberg, D. 2005e. Covalent functionalization: Towards soluble multiwalled boron nitride nanotubes. *Angewandte Chemie International Edition*, 44, 7932–7935.

Zhi, C. Y., Bando, Y., Tang, C. C., Honda, S., Sato, K., Kuwahara, H., and Golberg, D. 2006c. Purification of boron nitride nanotubes through polymer wrapping. *Journal of Physical Chemistry B*, 110, 1525–1528.

Zhi, C. Y., Bando, Y., Tang, C. C., Xie, R. G., Sekiguchi, T., and Golberg, D. 2005f. Perfectly dissolved boron nitride nanotubes due to polymer wrapping. *Journal of the American Chemical Society*, 127, 15996–15997.

Zhi, C. Y., Bando, Y., Terao, T., Tang, C. C., Kuwahara, H., and Golberg, D. 2009b. Chemically activated boron nitride nanotubes. *Chemistry: An Asian Journal*, 4, 1536–1540.

Zhi, C. Y., Bando, Y., Wang, W. L. L., Tang, C. C. C., Kuwahara, H., and Golberg, D. 2008. Mechanical and thermal properties of polymethyl methacrylate-BN nanotube composites. *Journal of Nanomaterials*, 18, 1–5.

Chapter 17

Plasma Functionalization of Nanotubes

Zhiqiang Chen, Lu Hua Li, Xiujuan J. Dai, Ying (Ian) Chen, and Xungai Wang

CONTENTS

17.1 INTRODUCTION

17.1.1 Plasma Technology

The term *plasma* was first used by Tonks and Langmuir in 1929 to describe a gas discharge (Perucca 2010). Plasma as the fourth state of matter is a fully or partially ionized gas. In a partially ionized gas, it is composed of free electrons, ions, neutral atoms and/or molecules, free radicals, and ultraviolet (UV) photons. Generally, the numbers of negative and positive charges are equal, and thus the overall charge of the plasma is neutral.

Plasma states can be divided into hot plasmas and cold plasmas (nonequilibrium plasmas), according to the degree of ionization of the plasma. Hot plasmas are almost fully ionized and have very high temperatures of electrons and heavy particles (atoms, molecules, or ions). Nonthermal plasmas, also known as cold plasmas, have a low degree of ionization and consist of low-temperature atoms, molecules or ions, and relatively high-temperature electrons (Meyyappan 2011). Nonthermal plasmas are a highly developed technology for surface functionalization. They have become a convenient and versatile tool for the surface modification of various materials used in a wide range of fields, including biomaterial and biomedicine (Cantini et al. 2012, Chen et al. 2007, Fridman et al. 2008, Gomathi et al. 2008, Kuzuya et al. 2008), nanotechnology (Ariga et al. 2012, Gupta et al. 2007, Meyyappan 2011), energy conversion (Tao et al. 2011), and environmental control (Gomez et al. 2009, Moreau et al. 2008, Sreenivasan and Gleason 2009, Vandenbroucke et al. 2011).

TABLE 17.1   SOME OF THE PROCESSES PRODUCING EXCITED SPECIES WHICH MAY OCCUR IN PLASMAS

Electron Impact	
Excitation	$A + e \rightarrow A^* + e$
Ionization	$A + e \rightarrow A^+ + 2e$
Dissociation	$A + e \rightarrow A\cdot + \cdot A + e$
De-excitation	$A^* \rightarrow A + h\nu$
Recombination	$A^+ + e \rightarrow A^* + h\nu$

Notes: e, $h\nu$ represents an electron and optical emission; A, atom or molecule; A^*, excited species; $A\cdot$, $\cdot A$, free radicals; A^+, ion.

Nonthermal plasmas can be divided into atmospheric-pressure plasmas and low-pressure plasmas. Compared to atmospheric-pressure plasmas, a low-pressure plasma allows more controllable and reproducible introduction of functional groups (Shishoo 2007). A low-pressure plasma apparatus typically consists of a power supply, a reaction chamber with electrodes, and gas feeding and pumping units including gas controllers and pressure gauge.

The positive, negative, and neutral species as well as UV photons in a cold plasma all play critical roles in surface modification. Nevertheless, electrons are most important. Electrons are the main contributors to the production of reactive species, since they are the lightest species in the plasma and therefore absorb the largest amount of energy from the electric field. The energy can be transferred via collisions with the gas atoms or molecules, resulting in excitation, ionization, and dissociation (Denis et al. 2010, Milella 2002), as shown in Table 17.1.

The reactive species in plasma can break bonds on the surface of a material, which is significant for the surface chemistry (Denes and Manolache 2004, Grace and Gerenser 2003). Free radicals can form cross-linked polymeric layers or graft onto the surface (Denes and Manolache 2004, Siow et al. 2006). UV photons emitted by the plasma also have some effects on the surface of substrates (Kang and Neoh 2002, Wertheimer et al. 1999).

There are many different categories of plasma reactions with the surface of inorganic and organic materials (Chan et al. 1996, Denes and Manolache 2004, Förch et al. 2005, Grace and Gerenser 2003, Kale and Desai 2011, Kang and Neoh 2002, Loh et al. 1987, Siow et al. 2006), but the plasmas that we discussed here can be classified into gas plasma treatment and plasma polymerization. Gas plasma treatment, which leads to surface etching, activation, and production of surface functional groups, is generally conducted using nonpolymerizing gases, while plasma polymerization uses polymerizing gases and monomers.

17.1.2   Nanotubes and Nanocomposites

The reinforcement of polymers with nanofillers provided a new class of advanced multifunctional materials with extraordinary mechanical properties, giving greatly commercial interests in many applications, such as aerospace components, automobiles, and sporting goods (Hussain et al. 2006, Keledi et al. 2012, Thostenson et al. 2005, Tjong 2006). Among nanofillers, nanotubes, such as carbon nanotubes (CNTs) and boron nitride nanotubes (BNNTs), are ideal filler materials for producing polymer nanocomposites with high mechanical, electrical, and thermal properties (Baughman et al. 2002, Byrne and Gun'ko 2010, Coleman et al. 2006) due to their excellent mechanical, electrical, and thermal properties. It has been reported that CNTs and BNNTs have similar Young's moduli in the range of 1.2 TPa (Zhi et al. 2010). However, in order to fully exploit their unique properties in composites, two main challenges need to be addressed. Firstly, the agglomeration of the nanotubes into bundles, owing to van der Waals interactions, makes them extremely difficult to be uniformly dispersed in a polymer matrix. As a result, the mechanical and electrical properties of the nanotube/polymer nanocomposites could be poor (Ma et al. 2010). Moreover, both CNTs and BNNTs have inert surfaces, which results in poor interfacial adhesions with the polymer matrix. This limits the efficiency of load transfer (González-Domínguez et al. 2011). Therefore, achieving uniform dispersion of the nanotubes and optimum adhesion with the matrix has been the focus of scientific research. To date, many methods have been reported to solve the problems, such as mechanical dispersion (Li et al. 1999, Ma et al. 2010) (ultrasonication, calendaring, and ball milling) and chemical functionalization (Byrne and Gun'ko 2010, Karousis et al. 2010). Although physical blending can greatly improve the dispersion of the nanotubes in a polymer, the poor interfacial bonding between them and the matrix impedes the ability of load transfer (Tseng et al. 2007). It has been reported that surface functionalization of CNTs can effectively improve their dispersion as well as interfacial adhesion to a polymer matrix (Ma et al. 2010, Rana et al. 2009).

17.2   PLASMA FUNCTIONALIZATION OF CNTs

Wet chemical methods are normally used to functionalize nanotubes. These approaches require multistep reactions to achieved the desired functionality and, in most cases, involve chemical functionalization conducted in solvents like strong acids, which could damage the nanotube structure and affect the unique advantages of nanotubes (Hu et al. 2010, Karousis et al. 2010). As an alternative, plasma treatment of nanotubes provides a one-step functionalization and hence avoids the complex procedures of filtration, washing, and drying. It is more environmentally benign, because it minimizes the use of solvents and acids. Moreover, plasma treatments are applied only to the surface and generally do not alter the bulk properties. In addition, the functional species and their quantities can be well controlled by choices of gases, monomer, and plasma parameters. For example, oxygen-containing gas plasmas, such as O_2 (Chirila et al. 2005, Felten et al. 2005, Scaffaro et al. 2012), Ar/O_2 (Chen et al. 2009, 2011), Ar/CO_2 (Bubert et al. 2003), or Ar/H_2O (Chen et al. 2010b), have been used to introduce oxygen-containing functional groups such as C–O, C=O, and O–C=O to improve the dispersion of CNTs and to achieve particular properties, such as in nanocomposites, biomaterials, and sensors. Plasma etching has been used to purify aligned CNTs, open their tips, and change their topography (Huang and Dai 2002, Liu et al. 2010).

Among the range of chemical functionalities that can be achieved on nanotubes, amine groups (especially primary amines—NH_2) are more important for biomaterials (Lee et al. 2011, Lu et al. 2009, Yoon et al. 2009, 2011, Luais et al. 2010) and polymer composites (Chen et al. 2008, Garg et al. 2010). It has been reported that the amine groups on the CNT surface not only improve the dispersion in the epoxy matrix but also allow direct covalent bonding to the epoxide groups (Ramanathan et al. 2005, Shen et al. 2007, Wang et al. 2006), therefore enhancing mechanical properties of nanocomposites (Geng et al. 2008, Gojny et al. 2004, Hosur et al. 2010, Lachman and Daniel Wagner 2010, Yang et al. 2009, Zhang et al. 2012). The focus of this section is the plasma functionalization of CNTs with amine groups, especially primary amines. There have been many plasma approaches to the functionalization of CNTs with amine groups (see Table 17.2). It can be seen that the approaches can be divided into three groups: nitrogen-containing gas plasma treatment, plasma-induced grafting polymerization, and plasma polymerization of amine-containing monomers, as summarized in Figure 17.1.

TABLE 17.2   PLASMA FUNCTIONALIZATION OF CNTs WITH AMINE GROUP

Plasma Type	Precursor	Sample Type	Sample Handling[a]	Functional Groups	NH$_2$/C[b]	Reference
RF	NH$_3$/Ar	MWCNTs	Deposited on the grounded electrode	Amine, amide, imine, and nitrile groups detected by XPS	—	Delpeux et al. (1999)
Atmos. pressure	N$_2$/NH$_3$ (35%)	Acid-treated MWCNTs	Deposited onto an Au-coated Si substrate	Amines and amides detected by near-edge x-ray absorption fine structure	—	Roy et al. (2006)
MW	NH$_3$	SWCNT film	—	Amine, amide, imine, and nitrile groups detected by XPS, FTIR, and Raman spectroscopy	—	Khare et al. (2004)
RF	NH$_3$	MWCNTs	Attached to a Scotch tape	Amine, amide, imine, nitrile, and imide groups detected by XPS	—	Felten et al. (2005)
RF	NH$_3$	MWCNT film	—	Primary amines were confirmed by a chemical derivatization with pentafluoro-benzaldehyde (PFB)	~1%	Yook et al. (2010)
MW	NH$_3$/Ar	MWCNTs	Placed in a glass beaker	Primary amines were confirmed by a chemical derivatization with 4-trifluoromethyl-benzaldehyde (TFB)	~1.7%	Chen et al. (2010)
MW	N$_2$	SWCNTs	Deposited on a CaF$_2$ disk	Amine, amide, and nitrile groups detected by XPS, FTIR, and Raman spectroscopy	—	Khare et al. (2005)
RF	N$_2$	MWCNT film	—	Amine, amide, and nitrile groups detected by XPS and Raman spectroscopy	—	Gohel et al. (2005)

			[a]		[b]	
Pulsed dc	N_2	Aligned MWCNT film	—	Amine and imine groups detected by XPS	—	Jones et al. (2008)
MW	N_2/Ar	MWCNTs	Not given	Amine, nitrile, amides, and oxime groups detected by XPS	—	Kalita et al. (2009)
MW	N_2/Ar	MWCNTs	Deposited on different supports (a conductive adhesive tape or a copper grid)	Amine, nitrile, amides, and oxime groups detected by XPS	—	Ruelle et al. (2007, 2009)
RF	N_2/H_2	SWCNTs	Placed on a silicon wafer	Amine, amide, and oxime groups detected by XPS	—	Yoon et al. (2009, 2011) and Lee et al. (2011)
MW	N_2/H_2	Aligned CNT film		Amine and amide groups detected by XPS	—	Luais et al. (2010)
MW	Ar plasma-assisted UV grafting of 1-vinylimidazole	SWNT film		The imidazole ring detected by XPS, and N/C was about 7.8%	—	Yan et al. (2005)
RF	Pretreatment Ar, graft polyacrylicnitrile (PAN)	MWCNTs	Deposited on the grounded electrode	Amine groups detected by XPS	—	Chen et al. (2010)
MW	N_2 plasma-induced chitosan (CS)	MWCNTs	Put in a stirring reactor	Amine and amide groups detected by FTIR	—	Shao et al. (2010)
RF plasma	Allylamine	MWCNTs	Put in a rotating reactor	Amine, amide, imine, and nitrile groups detected by FTIR, XPS	—	Abou Rich et al. (2012)

[a] Indicates how nanotube powder was handled in the plasma treatment.
[b] Indicates the percentage of primary amines if the quantification was performed.

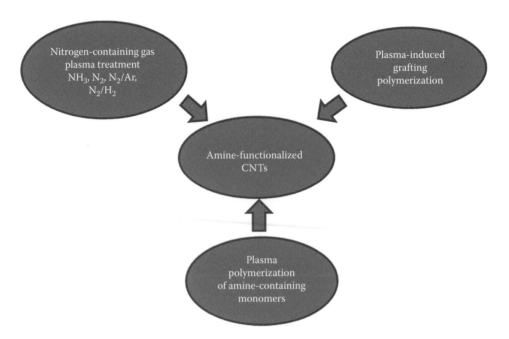

Figure 17.1   Plasma approaches used to create amine groups on CNT surfaces.

CNTs with amine-containing surfaces have been prepared using ammonia (NH_3) gas. Delpeux et al. (1999) reported that the residual oxygen in the reactor could impede the reactivity of MWCNTs with NH_3 plasma. They suggested that this was due to a competition between nitrogen and oxygen reactive species. They also found that NH_3 plasma treatment was less efficient when pristine MWCNTs were annealed to remove oxygen contaminations. The reason for this was not given. However, recently, Roy et al. (2006) reported using a dielectric barrier discharge plasma in an NH_3 atmosphere to modify acid-treated CNTs. They found that oxygen groups introduced on the CNT surface by acid treatment could act as activation sites attracting N atoms. The CNTs treated by NH_3 plasma attracted higher N concentration (about 5.5%) in acid treatment than nonactivated CNTs. Khare et al. (2004) confirmed the existence of amine groups on the SWCNTs that were functionalized by a microwave discharge of NH_3 through Fourier transform infrared (FTIR) spectroscopy with band assignment aided by computational modeling. This was further supported by XPS and Raman spectroscopy. Felten and his coworkers (2005) discovered that NH_3 plasmas could introduce various nitrogen-containing groups (amine, amide, and nitrile) on the MWCNT surface, and the concentration and type of functional groups were chiefly related to the plasma conditions, such as input power, treatment time, pressure, and position of the sample inside the chamber. Although those approaches have shown that the amine groups can be introduced onto the surface of the CNTs using NH_3 plasmas, the effect of the plasma treatment on the structure of the nanotubes was not reported.

N_2 and N_2 with H_2 or Ar have also been experimented to modify the surface of CNTs for the incorporation of amine groups. Khare et al. (2005) studied the functionalization of SWCNTs using a microwave-generated N_2 plasma. The authors found the functionalization strongly relied on the distance between the discharge source and the sample. At a relatively large distance, no functionalization was observed. They proposed this was due to the increase in the recombination of atomic nitrogen with the increase in distance. At a minimum distance, cyanides were observed. They suggested that the formation of cyanides was due to highly reactive species that either directly or indirectly form the cyanide. In the indirect route, a defect may be created first and then reacts to form the cyanide products. NH and HNH radicals have also been observed in N_2 plasma. It is found that UV photons can defunctionalize the nanotubes.

Gohel et al. (2005) examined field emission properties of N_2 plasma-treated MWCNT films. They found that the field emission properties of the modified MWCNT films were significantly improved. They suggested that this was attributable not only to the shortened nanotubes and reduced nanotube density but also to the nitrogen-containing groups on the MWCNT surface introduced by the plasma treatment. Jones et al. (2008) studied the functionalization of vertically aligned MWCNT films using a pulsed dc N_2/Ar plasma treatment. Pulsed dc power was used because a stable glow discharge could be observed only under a pulsed mode. SEM images of the treated MWCNT films showed ball-shaped structures at the ends of the bundled tubes. XPS results showed nitrogen-containing functional groups on the surface of the MWCNTs. Ruelle et al. (2007, 2009) modified MWCNTs in the postdischarge region of an N_2/Ar microwave plasma. XPS analysis showed the presence of amine, nitrile, amides, and oxime grafted on the MWCNT surface. Kalita et al. (2009) functionalized MWCNTs with N_2/Ar plasma for the fabrication of photovoltaic devices. XPS results showed the surface modification of MWCNTs with imine, amine, nitrile, and amide groups incorporated on the side walls, which resulted in a homogeneous distribution of MWCNTs in solvent. The performance of the devices was improved using the functionalized MWCNTs. Transmission electron microscopy (TEM) results showed that the inside wall of the functionalized MWCNTs was maintained, but some defects were found on the outermost walls.

Recently, Yoon et al. (2009) functionalized SWCNTs with primary amine groups with N_2/H_2 plasma in order to prepare poly [(D,L-lactic)-co-(glycolic acid)] (PLGA)-based nanofiber composites by electro-spinning. The N_2/H_2 plasma-introduced primary amine groups on the surface of SWCNT improved the dispersion and adhesion with PLGA. The influence of the plasma treatment on the structure of the SWCNTs was not discussed.

An alternative way to create amines on the surface of CNTs is plasma-induced grafting or plasma polymerization of amine-based polymers or monomers. Yan et al. (2005) studied Ar plasma-assisted UV grafting of 1-vinylimidazole. UV radical grafting of 1-vinylimidazole took place on the defect sites of the nanotubes created by the Ar plasma. Chen and colleagues (Chen et al. 2010c, Tseng et al. 2007) have studied primary amine functionalization of MWCNTs by plasma-induced grafting polymerization for the preparation of covalently integrated epoxy composites. First, the surface of MWCNTs was activated by the Ar plasma treatment. Subsequently, two polymers were used for grafting onto the MWCNTs: maleic anhydride (MA) and polyacrylic nitrile (PAN). After grafting MA onto the MWCNTs (MWCNTs-g-MA), they were mixed with the curing agent (diamine) in order to obtain primary amine groups that could covalently bond with the epoxy resin. For the PAN-grafted MWCNTs (MWCNTs-g-AN), a further reaction of the MWCNTs with hydroxylamine was required to produce the aminated MWCNTs (MWCNTs-g-mAN). Both methods were effective in improving the dispersion of functionalized MWCNTs as well as the mechanical and electrical properties of functionalized MWCNT/epoxy nanocomposites. Most recently, Abou Rich et al. (2012) have reported using CW plasma polymerization of allylamine to attach amine groups to the surface of MWCNTs. They evaluated the effect of the plasma conditions (radio frequency (RF) power and time) on the nitrogen content. Amine, amide, imine, and nitrile groups were detected on the MWCNTs by FTIR and XPS. The surface morphology of the functionalized MWCNTs and the thickness of plasma polymer were not reported.

As can be seen from the earlier review, many researchers have already performed functionalization of CNTs with amine groups by plasma approaches and confirmed nitrogen incorporation. However, in most cases, the density of primary amines was not quantified (Table 17.2), except two recent studies by Chen et al. (2010a) and Yook et al. (2010). Both studies used NH_3 plasma to functionalize the CNT surface and detected only moderate levels of primary amines ($NH_2/C < 2\%$) using chemical derivatization and XPS method.

In order to achieve a higher level of primary amines firmly attached to a CNT surface without causing serious damage to nanotube structure, we used a continuous wave plus pulsed plasma (CW + P) method using a mixture of N_2 and H_2 (15%) (Chen et al. 2012). This CW + P method was first proposed and used with Si and Ti from our previous work (Li et al. 2009). It has been shown that the plasma polymer obtained by this approach has a high retention of desired functional groups through the pulsed plasma polymerization and good adhesion to inert substrates through the CW plasma polymerization. This was based on previous work on achieving a controllable higher level of primary amines in a pulsed plasma polymerization of heptylamine (HA) by altering pulse duty cycles and average power (Dai et al. 2009). The amount of primary amines on the CNT surface after plasma treatment was quantified using chemical derivatization with 4-trifluoromethylbenzaldehyde (TFBA) followed by XPS analyses. Since the TFBA can react only with primary amines, the amount of primary amines can be determined from the amount of fluorine detected by XPS (Choukourov et al. 2004). The reaction is shown in Figure 17.2.

In our work, three different plasma modes—CW, pulsed (P), and combined (CW + P) were compared under the same total energy input (power × time). It was found that the combined mode achieved the highest recorded density of NH_2, which was about 2.3% (Figure 17.3a). Selectivity for primary amines (NH_2/N) for $N_2 + H_2$ CW + P plasma (about 30%) is also higher than any previous treatments (Figure 17.3b). Although the N/C ratios for both CW and CW + P-treated samples were very similar, the CW plasma introduced the lowest density of NH_2 (about 0.6) on the CNT surface. Moreover, the pulsed plasma modified sample has the highest level of nitrogen (nearly 12%), but its NH_2 density (1.4%) is relatively low compared to that for CW + P.

It was suggested that this is because the combined plasma captures the advantages of both CW and pulsed plasmas. In the first step, CW plasma etches away the amorphous carbon and creates active sites on the MWCNT surface. In the following pulsed plasma treatment, NH radicals formed in the pulsed plasma can be easily grafted to the cleaned and activated MWCNT surface to form primary amine groups. Moreover, •H radicals created in the pulsed plasma can reduce other nitrogen-containing groups to amine groups (Favia et al. 1996). The structure of treated CNTs was not damaged. Although the surface of the $N_2 + H_2$ plasma-treated MWCNTs appeared rougher than the untreated, the electron diffraction patterns of MWCNT clusters before and after $N_2 + H_2$ plasma treatment (Figure 17.4) were similar. This indicated that the integrity of the MWCNT structure was maintained.

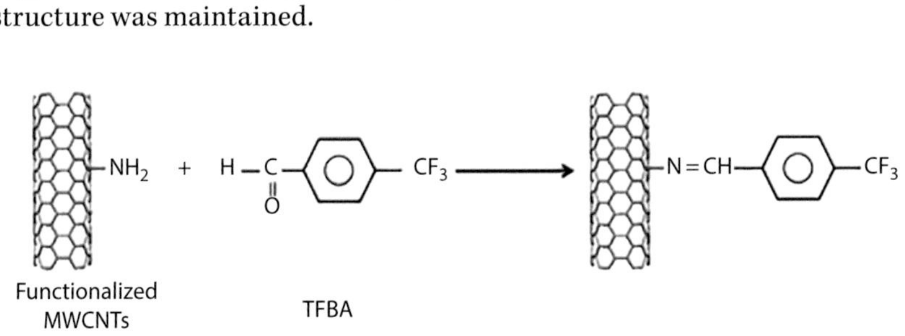

Figure 17.2 Schematic diagram of chemical derivatization by TFBA of primary anime groups. (From Chen, Z.Q. et al., *Plasma Processes and Polymers* 9(7), 733, 2012.)

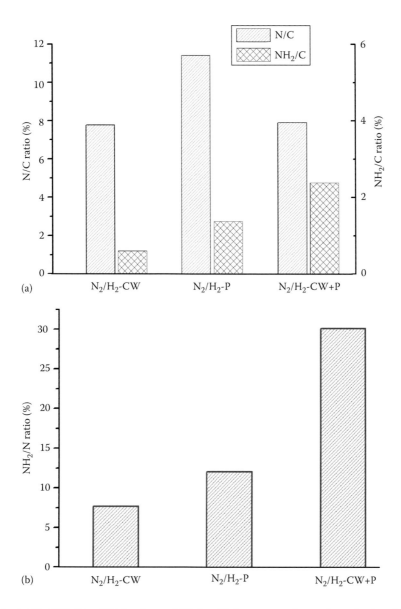

Figure 17.3   (a) The NH_2/C and N/C ratios and (b) the NH_2/N ratios at the surface of N_2/H_2 plasma-treated MWCNTs using three plasma modes.

Nanomaterials, such as CNTs, are hard to handle in plasma systems due to their aggregation and large surface area (Inagaki et al. 1992, 1993). Generally, CNTs are deposited on a substrate for treatment in the plasma system (Table 17.2). This can achieve a uniform functionalization but difficult on large-scale treatment. Suitable plasma reaction equipment is essential for processing nanotube functionalization. Recently, Shao et al. (2010) and Abou Rich et al. (2012) used stirring or rotating plasma reactors, in order to achieve a uniform treatment on CNTs, but the effects on uniformity of treatment were not clearly demonstrated.

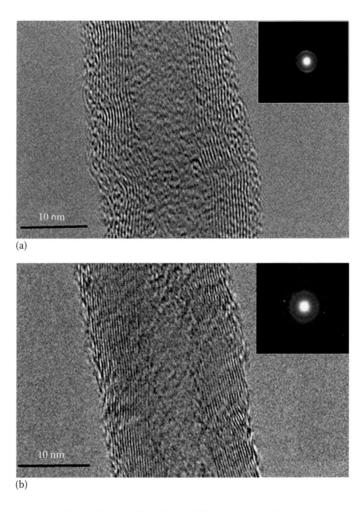

(a)

(b)

Figure 17.4   TEM images of CNTs before (a) and after (b) $N_2 + H_2$ CW plasma treatment. The insets show electron diffraction patterns from MWCNT clusters in a nearby region. (From Chen, Z.Q. et al., *Plasma Processes and Polymers* 9(7), 733, 2012.)

Most recently, our group used a stirring plasma to achieve a uniform coating on CNTs using a combination of CW + P plasma polymerization of HA (Chen et al. 2014). The schematic diagram of the stirring plasma system is shown in Figure 17.5, where CNTs are placed at the bottom of a round flask that is located on a magnetic stirrer. A magnetic bar is used to stir the powders during plasma polymerization. This system gave the uniform coating on the CNT surface, where CNTs were uniformly deposited with the HA plasma polymer and their diameter was obviously increased compared with that of untreated CNTs (Figure 17.6).

The combined-mode plasma polymerization captures the advantages of both CW and pulsed plasma polymerization, where CW plasma polymerization first produces a strong cross-linked plasma polymer as a good foundation for better bonding of a pulsed plasma polymer, in the immediately following pulse plasma polymerization, which retains the chemical

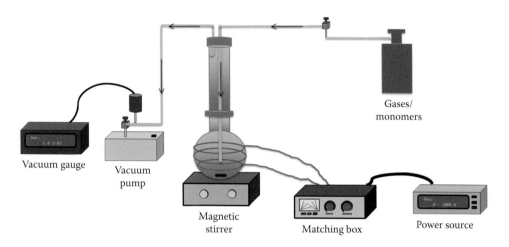

Figure 17.5 Schematic diagram of the stirring plasma system. (Chen, Z. et al., *Composites Part A: Applied Science and Manufacturing* 56, 172, 2014.)

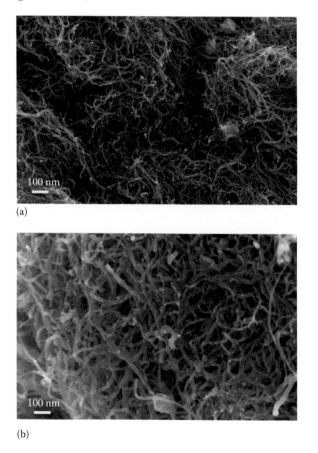

Figure 17.6 SEM image of CNTs: (a) untreated CNTs, (b) CW + P plasma–polymerized CNTs. (Chen, Z. et al., *Composites Part A: Applied Science and Manufacturing* 56, 172, 2014.)

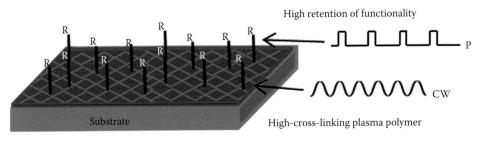

R: functional groups

Figure 17.7   The CW + P plasma polymerization.

functional groups from the monomer (Figure 17.7). It has also been shown that the CW + P plasma polymerization can give a higher density of NH_2 (about 3.6%) than using CW or P alone (Chen et al. 2012).

17.3 PLASMA AMINE-FUNCTIONALIZED CNT/EPOXY NANOCOMPOSITES

The presence of amine groups, especially primary amines, on the CNT surface should greatly improve their dispersion in, and interfacial bonding with, an epoxy resin, thus increasing the strength of CNT/epoxy nanocomposites. The first experimental work on the MWCNT/ epoxy interfacial strength was conducted by Cooper et al. (2002). They used a special pull-out test for individual MWCNTs from the epoxy matrix to quantify the interfacial shear strength. It ranged from 35 to 376 MPa. They suggested that covalent bonding between MWCNTs and epoxy could enhance the interfacial shear strength. Frankland et al. (2002) used molecular simulations to predict the shear strength of the SWCNT–polymer interface. They found that the interfacial interactions could be increased by the introduction of functional groups onto CNTs.

Many researchers have shown that CNTs that have been primary amine functionalized by chemical treatments can improve the properties of epoxy resin. Gojny et al. (2003) refluxed oxidized MWCNTs directly in the triethylenetetramine, resulting in primary amine–grafted MWCNTs. The TEM images of the primary amine–functionalized MWCNT/epoxy nano-composites showed an improved interfacial interaction with the matrix. The investigation of the thermo-mechanical properties of the nanocomposites showed that primary amine–functionalized MWCNTs had a bigger effect on T_g than untreated MWCNTs (Gojny and Schulte 2004). The same group also used commercial primary amine–functionalized double-walled carbon nanotubes (DWCNTs) to increase Young's modulus of epoxy by 6% by adding 0.1 wt% of functionalized DWCNTs (Gojny et al. 2004) and a 40% increase in the toughness of epoxy with 0.5% functionalized DWCNTs (Gojny et al. 2005). Commercially aminated MWCNTs have also been employed by Lachman and Wagner (2010) to reinforce an epoxy. Their results showed an increase in Young's modulus of 23.1% with 0.34 wt% addition of aminated MWCNTs. Yang et al. (2009) also showed that adding 0.6 wt% of MWCNTs, which were amine functionalized by a chemical method, increased the flexural modulus by 22%, compared with the neat epoxy. Recently, Chen et al. (2010c) reported using an argon plasma as a pretreatment to help graft

PAN onto the CNT surface in order to introduce primary amine functionality. The electrical properties of the resulting CNT/epoxy nanocomposites were found to be enhanced. Most recently, Zhang et al. (2012) introduced primary amine groups on the MWCNT surface using gamma ray irradiation and found that 1 wt% loading of the modified MWCNTs into the epoxy could lead to a 20% increase in tensile modulus.

Although a lot of work on primary amine–functionalized CNT/epoxy nanocomposites has been done, the application of plasma directly functionalized CNTs attached with primary amine groups in nanocomposite has not been reported so far. It was not till recently that we reported that amine-functionalized CNTs by a combined plasma mode (CW + P) in a nitrogen plus hydrogen gas mixture have improved the nano- and macromechanical properties of an epoxy resin (Chen et al. 2013). It was shown that the plasma amine functionalization of CNTs was effective in improving the dispersion and interfacial adhesion in an epoxy matrix. This can be seen from the SEM images of the fractured surfaces of the samples (Figure 17.8).

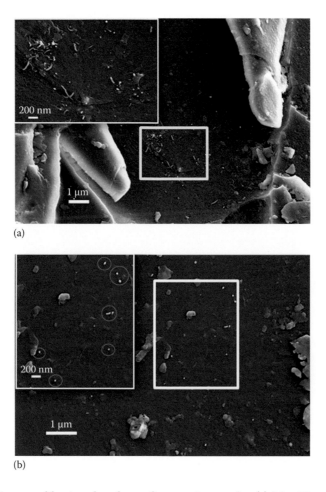

(a)

(b)

Figure 17.8   SEM images of fractured surfaces of composite samples: (a) 0.1 wt% untreated CNT/epoxy and (b) 0.1 wt% functionalized CNT/epoxy; insets are close-ups of the boxed region. (Chen, Z.Q. et al., *Composites Part A: Applied Science and Manufacturing* 45, 145, 2013.)

Some agglomerations of the untreated CNTs in the resin can be seen in Figure 17.8a. Many CNTs were visibly pulled out of the epoxy matrix suggesting poor adhesion. The dispersion of the functionalized CNTs (Figure 17.8b) appears to be improved as most of the visible nanotubes were well separated, where the functionalized CNTs were not pulled out but broken during the three-point bending test. This suggests good interfacial bonding between the modified CNTs and epoxy resin. As a result, the incorporation of only 0.1 wt% of functionalized MWCNTs greatly improved both the nano- and macromechanical properties of epoxy. The nanomechanical properties were determined by nanoindentation tests, which have been used as an accurate method for evaluating the deformation behavior, hardness, and elastic modulus of CNT-reinforced polymer composites (Li et al. 2004).

According to the Oliver–Pharr method (Oliver and Pharr 1992), which measures the load at a constant rate of loading up to a set peak load as a function of depth to calculate the mechanical

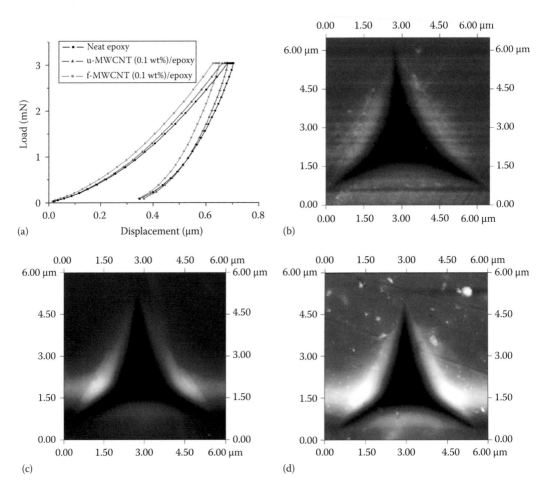

Figure 17.9 (a) Representative load–displacement curves of nanoindentations. AFM images of the indentation impressions for the epoxy and CNT-reinforced samples: (b) neat epoxy; (c) 0.1 wt% unmodified CNTs; (d) 0.1 wt% plasma-functionalized CNTs. (Chen, Z.Q. et al., *Composites Part A: Applied Science and Manufacturing* 45, 145, 2013.)

properties, the area of the indentation is inversely proportional to the hardness of the composite, and the slope of the unloading curve is directly proportional to the elastic modulus. From Figure 17.9, the depth and area of the indentations in the plasma-treated MWCNT-reinforced sample are smaller than those in the neat epoxy and untreated MWCNT/epoxy samples, while its unloading slope is the highest of all the samples. This indicates that the MWCNT/epoxy nanocomposites have better nanomechanical properties. In fact, the hardness and the elastic modulus of epoxy containing the plasma-functionalized MWNCTs have been increased by about 40% and 19%, respectively. According to thermomechanical and flexural analyses, the macromechanical properties of the nanocomposites were also enhanced greatly.

It has also been showed that CNTs that were uniformly coated by a thin plasma polymer with high levels of amine groups by a combination of CW and pulsed plasma polymerization of HA can effectively improve the mechanical properties of CNT/epoxy nanocomposites (Chen et al. 2014). The plasma-polymerized CNTs have improved the dynamic thermomechanical and flexural mechanical properties of the epoxy resin. The flexural modulus, flexural strength, and toughness of the epoxy resin with the addition of only 0.5 wt% plasma-polymerized MWCNTs increased by about 22%, 17%, and 70%, respectively.

17.4   PLASMA FUNCTIONALIZATION OF BORON NITRIDE NANOTUBES

BNNTs are insoluble in most organic solvents and aqueous media, greatly limiting many of their prospective applications, such as drug delivery, biology, and nanocomposite. Although it is possible to realize the solubilization of BNNTs via solution-based chemical methods (Xie et al. 2005, Zhi et al. 2005), the high chemical and thermal stability of BNNTs inhibits many traditional solution-based functionalization reactions. In addition, solution-based functionalization methods normally involve heating, stirring, and sonication and therefore are not suitable for the treatment of BNNT thin films without damaging the macroscopic structure of the film. So the functionalization of BNNTs via a solid-state route is of great importance, and plasma treatment is the only reported solid-state functionalization method till now.

The first plasma functionalization of BNNTs was realized using an aggressive nonequilibrium ammonia glow plasma (Ikuno et al. 2007). The BNNTs used in this study were produced by thermal reaction between B and MgO powders in ammonia gas at 1200°C. The pristine BNNTs showed a high crystallinity as revealed by the high-resolution TEM image in Figure 17.10a. The plasma treatment was conducted in a system equipped with a 1.5 kW microwave generator and a DC power supply for the sample holder. During the treatment, ammonia plasma of 200 W was applied to −100 V biased BNNTs for 10 min at room temperature with an estimated dose of 1.3×10^{15} ions/cm^2. After the plasma treatment, the six to eight surface shells of the BNNTs turned to wavy structures, indicating lattice bending and existence of a high level of defects caused by the plasma (Figure 17.10b). The damage to the BNNTs was attributed mainly to the bombardment of ammonia glow plasma–introduced ions and radicals, especially N_2^+ ions because of the negative bias applied. The N_2^+ ions had a high kinetic energy of about 100 eV around the BNNTs, which can easily break the B–N bond of 7–8 eV. The newly created defects were believed to greatly promote the functionalization of the BNNTs by the NH_3 plasma–generated radicals, such as •NH_2 at relatively low temperatures. The BNNTs functionalized by amine on the sidewall showed dramatically better dispersibility in chloroform, a commonly used organic solvent, and could form homogeneous suspension for several hours to several days (Figure 17.11). More interestingly,

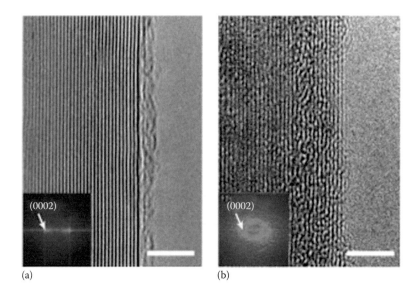

(a)   (b)

Figure 17.10   High-resolution TEM images and Fourier transform images of the sidewall of (a) pristine BNNT and (b) ammonia plasma–treated BNNT. Both scale bars are 4 nm. (Copyright 2007, with permission from Elsevier.)

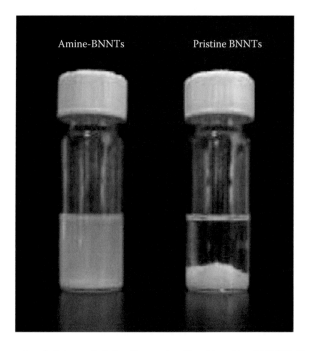

Figure 17.11   Photographs of vials containing the amine-functionalized BNNTs (left-hand side) and the pristine BNNTs (right-hand side) in chloroform after standing for 5 h. (Ikuno, T. et al., 2007. *Solid State Communications* 142(11), 643, 2007.)

the amine-functionalized BNNTs are much more amenable to further chemical treatment or functionalization than the pristine BNNTs. For example, 3-bromopropinoyl chloride (BPC) covalently attached BNNTs could be produced simply by sonication of the NH_3 plasma–treated BNNTs in BPC. In contrast, the sonication of the pristine BNNTs under the same condition did not result in any BPC attachment.

After the experimental study of the functionalization of BNNTs by ammonia plasma, density functional theory calculations provided detailed mechanism of the functionalization in an atomic scale (Cao et al. 2009). It is found that in both zigzag and armchair pristine BNNTs, •NH_2 radical can chemically bond to the B atoms but not the N atoms on the sidewall of the BNNTs. The chemical bond to •NH_2 changes the hybridization of the B atom from sp^2 to sp^3 and pulls the B atoms out of the nanotube wall. With the increase in the diameter, BNNTs become less reactive to •NH_2 radicals due to the decreased curvature. As NH_3 plasma treatment can create a large amount of defects on the surface of BNNTs, it is essential to understand the interaction between •NH_2 radical and defective BNNT structures. Among the four calculated defects, namely, N antisite, Stone–Wales, B vacancy, and nitrogen vacancy, N vacancy shows an extraordinarily large reaction energy to NH_2^* due to the preference of •NH_2 to bond to unsaturated dangling bond of B atoms. On the other hand, divacancies (two B vacancies, two N vacancies, or adjacent B and N vacancy) are much more favorable for •NH_2 functionalization than single-vacancy defects. The electronic properties of the •NH_2-grafted BNNTs could be very different from those of the pristine BNNTs. For example, the calculations of the band structure and density of states (DOS) of pristine and •NH_2-functionalized (8, 0) BNNTs disclose that the bonded •NH_2 creates two new states near the Fermi level (Figure 17.12), resulting in degenerate p-type

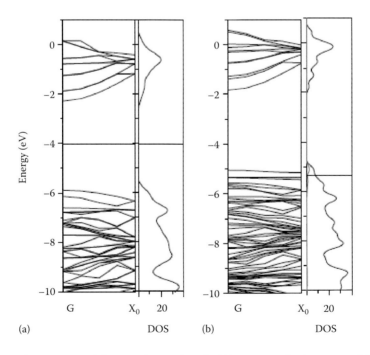

Figure 17.12   Band structure and density of states (DOS) of (a) a pristine (8, 0) BNNT and (b) a functionalized BNNT with an attached NH_2 radical. (Cao, F.L. et al., *Nanotechnology* 20(14), 145703, 2009.)

semiconducting behavior and dramatically increased electrical conductivity. So the NH_3 plasma–treated BNNTs are potentially more suitable for electronic applications.

Oxygen doping and nitrogen functionalization of BNNTs were achieved using oxygen (O_2) and nitrogen with 15% hydrogen ($N_2 + H_2$) plasma treatments (Dai et al. 2011). In this study, the BNNTs were produced by a boron ink method (Li et al. 2010b), and the treatments were carried out in a custom-built inductively coupled RF plasma reactor. Both the O_2 and $N_2 + 15\%$ H_2 plasma treatments were conducted at 100 W without additional bias, and the electron energy distribution function in an oxygen plasma has a modest high-energy tail (>10 eV). This makes it possible to introduce functionalization of BNNTs but with limited surface damage. Before the O_2 and $N_2 + H_2$ plasma treatments, the BNNTs were pretreated by argon (Ar) plasma with a power of 100 W for two purposes. One purpose is to remove possible amorphous BN contamination layer on the BNNTs; the other is to activate the surface of the BNNTs for further modification and functionalization. It was found that the O_2 plasma treatment was efficient in creating N vacancies but had little impact on the B atoms. The N vacancies were healed by the O radicals created by the plasma, giving rise to O doping in the BNNTs. Such O doping can change the optical, electrical, and magnetic properties of the BNNTs. In addition, oxygen functional groups such as carboxyl C=O were also found on the surface of the O_2 plasma–treated BNNTs. In terms of the introduction of nitrogen functional groups, the $N_2 + H_2$ plasma treatment on the BNNTs is of more interest. Amide groups were found on the $N_2 + H_2$ plasma–treated BNNTs under plasma conditions. The O content in the BNNTs was also increased after the plasma treatment, which might be due to the exposure of the sample to air. It should be noted that the surface damage in both the O_2 and $N_2 + H_2$ plasma treatments was much less than that of the aggressive NH_3 plasma treatment and could be well controlled via treatment time (Figure 17.13).

Recently, $N_2 + H_2$ plasma in either CW mode or a combination of continuous wave and pulse mode (CW + P) was used to modify wettability of BNNT films and increase the cell proliferation (Li et al. 2012). The BNNT films on steel substrate were also produced by the B ink painting method (Li et al. 2010a). The pristine BNNT films are highly nonwettable to water with an average water contact angle of 158.1° ± 3.6° due to the nanoroughness-enabled Cassie state (Li and Chen 2010). After the $N_2 + H_2$ plasma of 100 W in CW mode for 5 min with an Ar plasma pretreatment for 1 min, the contact angle of the BNNT films reduced to 65.2° ± 4.2°. After the same plasma treatment for 10 min, the BNNT films became superhydrophilic with contact angles <5°. Similar superhydrophilicity can be achieved using CW + P mode plasma (CW plasma for 5 min and P plasma for 10 min with a duty cycle of 10% and each duty cycle lasting 1 ms) with less energy input (Figure 17.14). The (super)hydrophilic patterns on highly hydrophobic BNNT films were produced for the first time (Figure 17.15). The pattern was achieved by the $N_2 + H_2$ plasma treatment using a patterned mask on top of the BNNT film.

It is found that the BNNT films after the $N_2 + H_2$ plasma treatment of the maximum energy input kept the original appearance and roughness. Furthermore, the surface structure of individual BNNTs was basically retained. So the wettability modification of the BNNTs after the plasma treatment was found mainly due to the chemical functionalization rather than physical morphology change. This was confirmed by XPS analyses that amide, amine, and imine (nitrile) groups are attached to the plasma-treated BNNT films (Figure 17.16). It should be noted that the $N_2 + H_2$ plasma treatment using CW + P mode is able to introduce

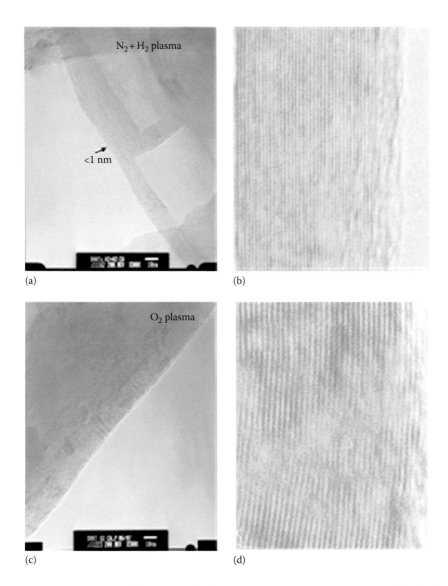

Figure 17.13   HRTEM images of the BNNTs (a, b) after N_2 + H_2 plasma treatment, and (c, d) after the O_2 plasma treatment. (Dai, X.J. et al., *Nanotechnology* 22(24), 245301, 2011.)

amine more effectively. The content of primary amines on N_2 + H_2 plasma in CW + P mode (CW mode for 5 min and P mode for 10 min)–treated BNNT film was more than 2%, which is much larger than the plasma in CW mode for 10 min. The difference can be attributed to the fact that the CW-mode plasma with continuous and higher-energy inputs supports the dissociation and further reaction of freshly formed amine to form amide and imine with the presence of O and C impurities, while the P mode with much less energy input and off-time

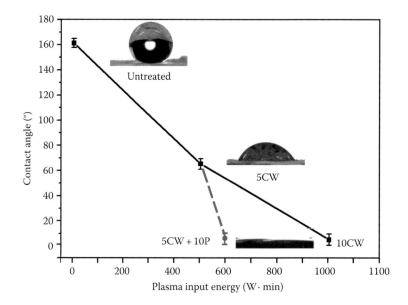

Figure 17.14   Contact angle changes of the BNNT films after $N_2 + H_2$ plasma treatments with different energy inputs (time) and modes (CW or CW + P). (Li, L. et al., *Journal of Physical Chemistry C* 116(34), 18334, 2012.)

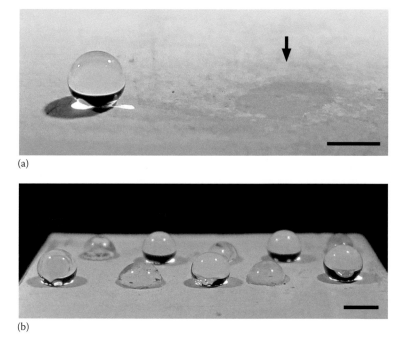

Figure 17.15   Photos of (super)hydrophilic patterns on hydrophobic BNNT films created by masked $N_2 + H_2$ plasma treatments: (a) in CW mode for 5 min and P mode for 10 min (with a superhydrophilic spot arrowed) and (b) in CW mode for 5 min. Both scale bars are 2 mm. (Li, L. et al., *Journal of Physical Chemistry C* 116(34), 18334, 2012.)

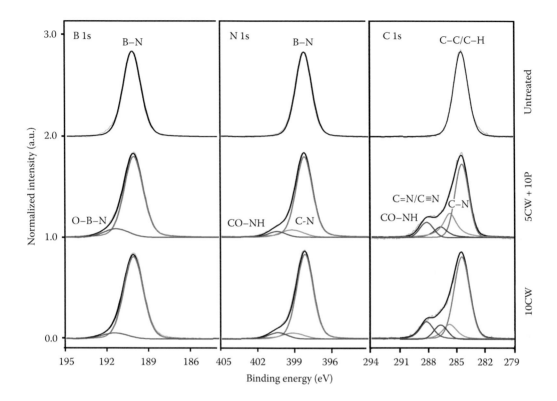

Figure 17.16 B 1s, N 1s, and C 1s regions of the XPS spectra of BNNT films before plasma treatment (upper row) and after plasma treatments in CW + P mode (CW mode for 5 min and P mode for 10 min) (middle row) and CW mode for 10 min (lower row). In the fittings: light gray, BN_xO_y; dark gray, amide; black, amine; green, imine (nitrile). (Li, L. et al., *Journal of Physical Chemistry C* 116(34), 18334, 2012.)

can retain more amine. Besides, the functional groups introduced by the CW + P mode are more stable on the BNNTs than those by the CW mode.

Cell proliferation on the pristine and $N_2 + H_2$ plasma–treated BNNT films was tested using two representative cell lines: human primary mammary fibroblasts and a transformed cell line TXP RFP3. It is found that both the untreated and plasma-treated BNNT films could support the growth of both types of cells, but the plasma treatment greatly enhanced the number of cells (up to six times). The increases in the cell numbers are due to the change of the BNNT films from hydrophobic to superhydrophilic after the plasma treatment and the attached functional groups, such as amine and amide, which have the ability to covalently couple cells. The morphologies of the cells on the plasma-treated BNNT films showed more prolonged extension of membrane projections for both cell lines (Figure 17.17), indicating that the plasma treatment increased the cell proliferation and viability. It is interesting that some cells showed a capacity to bundle the BNNTs beneath together to form a *yarn*, suggesting a strong adhesion between the cell and the plasma-treated BNNTs. The wettability-modified BNNTs by the plasma are also suitable for many other applications that require wettable BNNTs, such as nanocomposite and desalination.

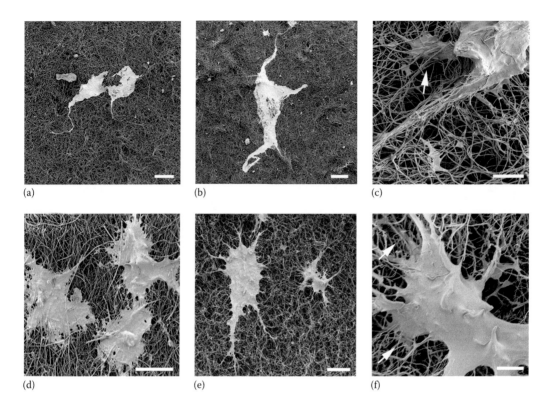

Figure 17.17   SEM images of fibroblasts on (a) untreated and (b, c) plasma-treated BNNT films, and TXP RFP3 cells on (d) untreated and (e, f) plasma-treated BNNT films. Scale bars: (a, b) 50 μm; (c) 5 μm; (d, e) 20 μm; (f) 5 μm. (Li, L. et al., *Journal of Physical Chemistry C* 116(34), 18334, 2012.)

17.5   SUMMARY

This chapter provides an overview of the research on plasma surface functionalization of CNTs with amine groups as well as the mechanical properties of CNT-reinforced epoxy resins. The plasma functionalization of BNNTs also has been reviewed. It has demonstrated that amine groups, especially primary amines produced by the plasma (both N-containing gas and monomer) on the nanotube surface, can directly react with epoxy resin by chemical bonding as well as improve their dispersion in the epoxy matrix, therefore enhancing mechanical properties of nanocomposites. Moreover, plasma functionalization of BNNTs can be used in biomedical application. Furthermore, O doping of BNNTs using O_2 plasma was achieved, where N vacancies were created and healed by the O radicals generated by O_2 plasma. Plasma techniques provide a nondestructive and more environmentally friendly alternative for the functionalization of nanotubes for many applications. Although, to achieve controllable and selectable functionalities, large-scale and uniform plasma treatment of nanomaterials is still a challenging task, our work has taken some forward steps: (1) a combination of CW pulsed plasma methods can give a more selectable and higher level of primary amine groups on the nanotubes surface and (2) a stirring plasma system can provide a practical approach for uniform functionalization of larger amounts of nanomaterial in a low-pressure plasma system.

ACKNOWLEDGMENTS

We very much appreciate the technical support provided by R. Lovett, M. Wright, R. Pow, A. Deveth, and L. Kviz, valuable discussions on composites from K. Magniez and J. Bungur, and valuable scientific discussions and advice from P. R. Lamb. We would like to acknowledge assistance with AFM measurements from D. R. de Celis Leal. We thank J. du Plessis (RMIT University) for valuable discussions on XPS, J. Peng and M. Field (RMIT University) for conducting TEM, and RMIT University for access to their XPS and TEM facilities.

REFERENCES

Abou Rich, S., M. Yedji, J. Amadou, G. Terwagne, A. Felten, L. Avril, and J. J. Pireaux. 2012. Polymer coatings to functionalize carbon nanotubes. *Physica E: Low-Dimensional Systems and Nanostructures* 44 (6):1012–1020. doi:10.1016/j.physe.2010.11.028.

Ariga, K., Q. Ji, J. P. Hill, Y. Bando, and M. Aono. 2012. Forming nanomaterials as layered functional structures toward materials nanoarchitectonics. *NPG Asia Materials* 4:e17.

Baughman, R. H., A. A. Zakhidov, and W. A. de Heer. 2002. Carbon nanotubes—The route toward applications. *Science* 297 (5582):787–792. doi:10.1126/science.1060928.

Bubert, H., S. Haiber, W. Brandl, G. Marginean, M. Heintze, and V. Brüser. 2003. Characterization of the uppermost layer of plasma-treated carbon nanotubes. *Diamond and Related Materials* 12 (3–7):811–815.

Byrne, M. T. and Y. K. Gun'ko. 2010. Recent advances in research on carbon nanotube–polymer composites. *Advanced Materials* 22 (15):1672–1688. doi:10.1002/adma.200901545.

Cantini, M., P. Rico, D. Moratal, and M. Salmeron-Sanchez. 2012. Controlled wettability, same chemistry: Biological activity of plasma-polymerized coatings. *Soft Matter* 8 (20):5575–5584.

Cao, F. L., W. Ren, Y. M. Ji, and C. Y. Zhao. 2009. The structural and electronic properties of amine-functionalized boron nitride nanotubes via ammonia plasmas: A density functional theory study. *Nanotechnology* 20 (14):145703. doi:10.1088/0957-4484/20/14/145703.

Chan, C. M., T. M. Ko, and H. Hiraoka. 1996. Polymer surface modification by plasmas and photons. *Surface Science Reports* 24 (1–2):1–54. doi:10.1016/0167-5729(96)80003-3.

Chen, C., B. Liang, D. Lu, A. Ogino, X. Wang, and M. Nagatsu. 2010a. Amino group introduction onto multiwall carbon nanotubes by NH_3/Ar plasma treatment. *Carbon* 48 (4):939–948.

Chen, C., B. Liang, A. Ogino, X. Wang, and M. Nagatsu. 2009. Oxygen functionalization of multiwall carbon nanotubes by microwave-excited surface-wave plasma treatment. *Journal of Physical Chemistry C* 113 (18):7659–7665. doi:10.1021/jp9012015.

Chen, C., A. Ogino, X. Wang, and M. Nagatsu. 2010b. Plasma treatment of multiwall carbon nanotubes for dispersion improvement in water. *Applied Physics Letters* 96 (13):131504.

Chen, C., A. Ogino, X. Wang, and M. Nagatsu. 2011. Oxygen functionalization of multiwall carbon nanotubes by Ar/H_2O plasma treatment. *Diamond and Related Materials* 20 (2):153–156. doi:10.1016/j.diamond.2010.11.018.

Chen, I. H., C. C. Wang, and C. Y. Chen. 2010c. Preparation of carbon nanotube (CNT) composites by polymer functionalized CNT under plasma treatment. *Plasma Processes and Polymers* 7 (1):59–63.

Chen, K. S., S. C. Chen, W. C. Lien, J. C. Tsai, Y. A. Ku, H. R. Lin, F. H. Lin et al. 2007. Surface modification of materials by plasma process and UV-induced grafted polymerization for biomedical applications. *Journal of the Vacuum Society of Japan* 50 (10):609–614.

Chen, X., J. Wang, M. Lin, W. Zhong, T. Feng, X. Chen, J. Chen, and F. Xue. 2008. Mechanical and thermal properties of epoxy nanocomposites reinforced with amino-functionalized multi-walled carbon nanotubes. *Materials Science and Engineering A* 492 (1–2):236–242. doi:10.1016/j.msea.2008.04.044.

Chen, Z. Q., X. J. Dai, P. R. Lamb, D. R. de Celis Leal, B. L. Fox, Y. Chen, J. du Plessis, M. Field, and X. Wang. 2012. Practical amine functionalization of multi-walled carbon nanotubes for effective interfacial bonding. *Plasma Processes and Polymers* 9 (7):733–741. doi:10.1002/ppap.201100203.

Chen, Z. Q., X. J. Dai, K. Magniez, P. R. Lamb, D. R. de Celis Leal, B. L. Fox, and X. Wang. 2013. Improving the mechanical properties of epoxy using multiwalled carbon nanotubes functionalized by a novel plasma treatment. *Composites Part A: Applied Science and Manufacturing* 45:145–152. doi:10.1016/j.compositesa.2012.09.005.

Chen, Z., X. J. Dai, K. Magniez, P. R. Lamb, B. L. Fox, and X. Wang. 2014. Improving the mechanical properties of multiwalled carbon nanotube/epoxy nanocomposites using polymerization in a stirring plasma system. *Composites Part A: Applied Science and Manufacturing* 56:172–180. doi: http://dx.doi.org/10.1016/j.compositesa.2013.10.009.

Chirila, V., G. Marginean, and W. Brandl. 2005. Effect of the oxygen plasma treatment parameters on the carbon nanotubes surface properties. *Surface and Coatings Technology* 200 (1–4):548–551.

Choukourov, A., J. Kousal, D. Slavínská, H. Biederman, E. R. Fuoco, S. Tepavcevic, J. Saucedo, and L. Hanley. 2004. Growth of primary and secondary amine films from polyatomic ion deposition. *Vacuum* 75 (3):195–205. doi: http://dx.doi.org/10.1016/j.vacuum.2004.02.006.

Coleman, J. N., U. Khan, W. J. Blau, and Y. K. Gun'ko. 2006. Small but strong: A review of the mechanical properties of carbon nanotube-polymer composites. *Carbon* 44 (9):1624–1652.

Cooper, C. A., S. R. Cohen, A. H. Barber, and H. D. Wagner. 2002. Detachment of nanotubes from a polymer matrix. *Applied Physics Letters* 81 (20):3873–3875.

Dai, X. J., J. du Plessis, I. L. Kyratzis, G. Maurdev, M. G. Huson, and C. Coombs. 2009. Controlled amine functionalization and hydrophilicity of a poly(lactic acid) fabric. *Plasma Processes and Polymers* 6 (8):490–497.

Dai, X. J. J., Y. Chen, Z. Q. Chen, P. R. Lamb, L. H. Li, J. du Plessis, D. G. McCulloch, and X. G. Wang. 2011. Controlled surface modification of boron nitride nanotubes. *Nanotechnology* 22 (24):245301. doi: 10.1088/0957-4484/22/24/245301.

Delpeux, S., K. Metenier, R. Benoit, F. Vivet, L. Boufendi, S. Bonnamy, and F. Beguin. 1999. Functionalisation of carbon nanotubes for composites. *AIP Conference Proceedings* 486 (1):470–473.

Denes, F. S. and S. Manolache. 2004. Macromolecular plasma-chemistry: An emerging field of polymer science. *Progress in Polymer Science* 29 (8):815–885.

Denis, L., P. Marsal, Y. Olivier, T. Godfroid, R. Lazzaroni, M. Hecq, J. Cornil, and R. Snyders. 2010. Deposition of functional organic thin films by pulsed plasma polymerization: A joint theoretical and experimental study. *Plasma Processes and Polymers* 7 (2):172–181.

Favia, P., M. V. Stendardo, and R. d'Agostino. 1996. Selective grafting of amine groups on polyethylene by means of NH_3–H_2 RF glow discharges. *Plasmas and Polymers* 1 (2):91–112. doi: 10.1007/bf02532821.

Felten, A., C. Bittencourt, J. J. Pireaux, G. Van Lier, and J. C. Charlier. 2005. Radio-frequency plasma functionalization of carbon nanotubes surface O_2, NH_3, and CF_4 treatments. *Journal of Applied Physics* 98 (7):1–9.

Förch, R., Z. Zhang, and W. Knoll. 2005. Soft plasma treated surfaces: Tailoring of structure and properties for biomaterial applications. *Plasma Processes and Polymers* 2 (5):351–372. doi: 10.1002/ppap.200400083.

Frankland, S. J. V., A. Caglar, D. W. Brenner, and M. Griebel. 2002. Molecular simulation of the influence of chemical cross-links on the shear strength of carbon nanotube-polymer interfaces. *Journal of Physical Chemistry B* 106 (12):3046–3048.

Fridman, G., G. Friedman, A. Gutsol, A. B. Shekhter, V. N. Vasilets, and A. Fridman. 2008. Applied plasma medicine. *Plasma Processes and Polymers* 5 (6):503–533.

Garg, P., B. P. Singh, G. Kumar, T. Gupta, I. Pandey, R. K. Seth, R. P. Tandon, and R. B. Mathur. 2010. Effect of dispersion conditions on the mechanical properties of multi-walled carbon nanotubes based epoxy resin composites. *Journal of Polymer Research* 18 (6):1397–1407. doi: 10.1007/s10965-010-9544-8.

Geng, Y., M. Y. Liu, J. Li, X. M. Shi, and J. K. Kim. 2008. Effects of surfactant treatment on mechanical and electrical properties of CNT/epoxy nanocomposites. *Composites Part A: Applied Science and Manufacturing* 39 (12):1876–1883. doi: 10.1016/j.compositesa.2008.09.009.

Gohel, A., K. C. Chin, Y. W. Zhu, C. H. Sow, and A. T. S. Wee. 2005. Field emission properties of N_2 and Ar plasma-treated multi-wall carbon nanotubes. *Carbon* 43 (12):2530–2535.

Gojny, F. H., J. Nastalczyk, Z. Roslaniec, and K. Schulte. 2003. Surface modified multi-walled carbon nanotubes in CNT/epoxy-composites. *Chemical Physics Letters* 370 (5–6):820–824.

Gojny, F. H. and K. Schulte. 2004. Functionalisation effect on the thermo-mechanical behaviour of multi-wall carbon nanotube/epoxy-composites. *Composites Science and Technology* 64 (15 Spec. Iss.):2303–2308.

Gojny, F. H., M. H. G. Wichmann, B. Fiedler, and K. Schulte. 2005. Influence of different carbon nanotubes on the mechanical properties of epoxy matrix composites—A comparative study. *Composites Science and Technology* 65 (15–16):2300–2313. doi: 10.1016/j.compscitech.2005.04.021.

Gojny, F. H., M. H. G. Wichmann, U. Köpke, B. Fiedler, and K. Schulte. 2004. Carbon nanotube-reinforced epoxy-composites: Enhanced stiffness and fracture toughness at low nanotube content. *Composites Science and Technology* 64 (15):2363–2371.

Gomathi, N., A. Sureshkumar, and S. Neogi. 2008. RF plasma-treated polymers for biomedical applications. *Current Science* 94 (11):1478–1486.

Gomez, E., D. A. Rani, C. R. Cheeseman, D. Deegan, M. Wise, and A. R. Boccaccini. 2009. Thermal plasma technology for the treatment of wastes: A critical review. *Journal of Hazardous Materials* 161 (2–3):614–626. doi: 10.1016/j.jhazmat.2008.04.017.

González-Domínguez, J. M., A. Ansón-Casaos, A. M. Díez-Pascual, B. Ashrafi, M. Naffakh, D. Backman, H. Stadler, A. Johnston, M. Gómez, and M. T. Martínez. 2011. Solvent-free preparation of high-toughness epoxy–SWNT composite materials. *ACS Applied Materials & Interfaces* 3 (5):1441–50. doi: 10.1021/am101260a.

Grace, J. and L. Gerenser. 2003. Plasma treatment of polymers. *Journal of Dispersion Science and Technology* 24 (3/4):305.

Gupta, A. K., R. R. Naregalkar, V. D. Vaidya, and M. Gupta. 2007. Recent advances on surface engineering of magnetic iron oxide nanoparticles and their biomedical applications. *Nanomedicine* 2 (1):23–39.

Hosur, M. V., T. Rahman, S. Brundidge-Young, and S. Jeelani. 2010. Mechanical and thermal properties of amine functionalized multi-walled carbon nanotubes epoxy-based nanocomposite. *Composite Interfaces* 17:197–215.

Hu, L., D. S. Hecht, and G. Grüner. 2010. Carbon nanotube thin films: Fabrication, properties, and applications. *Chemical Reviews* 110 (10):5790–5844. doi: 10.1021/cr9002962.

Huang, S. and L. Dai. 2002. Plasma etching for purification and controlled opening of aligned carbon nanotubes. *Journal of Physical Chemistry B* 106 (14):3543–3545. doi: 10.1021/jp014047y.

Hussain, F., M. Hojjati, M. Okamoto, and R. E. Gorga. 2006. Review article: Polymer-matrix nanocomposites, processing, manufacturing, and application: An overview. *Journal of Composite Materials* 40 (17):1511–1575. doi: 10.1177/0021998306067321.

Ikuno, T., T. Sainsbury, D. Okawa, J. M. J. Frehet, and A. Zettl. 2007. Amine-functionalized boron nitride nanotubes. *Solid State Communications* 142 (11):643–646. doi: 10.1016/j.ssc.2007.04.010.

Inagaki, N., S. Tasaka, and H. Abe. 1992. Surface modification of polyethylene powder using plasma reactor with fluidized bed. *Journal of Applied Polymer Science* 46 (4):595–601.

Inagaki, N., S. Tasaka, and K. Ishii. 1993. Surface modification of polyethylene and magnetite powders by combination of fluidization and plasma polymerization. *Journal of Applied Polymer Science* 48 (8):1433–1440.

Jones, J. G., A. R. Waite, C. Muratore, and A. A. Voevodin. 2008. Nitrogen and hydrogen plasma treatments of multiwalled carbon nanotubes. *Journal of Vacuum Science & Technology: Part B* 26 (3):995–1000. doi: 10.1116/1.2917068.

Kale, K. H. and A. N. Desai. 2011. Atmospheric pressure plasma treatment of textiles using non-polymerising gases. *Indian Journal of Fibre & Textile Research* 36 (3):289–299.

Kalita, G., S. Adhikari, H. R. Aryal, R. Afre, T. Soga, M. Sharon, and M. Umeno. 2009. Functionalization of multi-walled carbon nanotubes (MWCNTs) with nitrogen plasma for photovoltaic device application. *Current Applied Physics* 9 (2):346–351.

Kang, E. T. and K. G. Neoh. 2009. Surface modification of polymers. In *Encyclopedia of Polymer Science and Technology*. ed. Mark. H. F., John Wiley & Sons, Inc., New York.

Karousis, N., N. Tagmatarchis, and D. Tasis. 2010. Current progress on the chemical modification of carbon nanotubes. *Chemical Reviews* 110 (9):5366–5397.

Keledi, G., J. Hari, and B. Pukanszky. 2012. Polymer nanocomposites: Structure, interaction, and functionality. *Nanoscale* 4 (6):1919–1938.

Khare, B., P. Wilhite, B. Tran, E. Teixeira, K. Fresquez, D. N. Mvondo, C. Bauschlicher, and M. Meyyappan. 2005. Functionalization of carbon nanotubes via nitrogen glow discharge. *Journal of Physical Chemistry B* 109 (49):23466–23472. doi: 10.1021/jp0537254.

Khare, B., P. Wilhite, R. C. Quinn, B. Chen, R. H. Schingler, B. Tran, H. Imanaka, C. R. So, C. W. Bauschlicher, and M. Meyyappan. 2004. Functionalization of carbon nanotubes by ammonia glow-discharge: Experiments and modeling. *Journal of Physical Chemistry B* 108 (24):8166–8172. doi: 10.1021/jp049359q.

Kuzuya, M., Y. Sasai, Y. Yamauchi, and S. I. Kondo. 2008. Pharmaceutical and biomedical engineering by plasma techniques. *Journal of Photopolymer Science and Technology* 21 (6):785–798.

Lachman, N. and H. Daniel Wagner. 2010. Correlation between interfacial molecular structure and mechanics in CNT/epoxy nano-composites. *Composites Part A: Applied Science and Manufacturing* 41 (9):1093–1098.

Lee, H. J., J. Park, O. J. Yoon, H. W. Kim, D. Y. Lee, D. H. Kim, W. B. Lee, N.-E. Lee, J. V. Bonventre, and S. S. Kim. 2011. Amine-modified single-walled carbon nanotubes protect neurons from injury in a rat stroke model. *Nature Nanotechnology* 6 (2):121–125. doi: http://www.nature.com/nnano/journal/v6/n2/abs/nnano.2010.281.html#supplementary-information.

Li, L., X. J. Dai, H. S. Xu, J. H. Zhao, P. Yang, G. Maurdev, J. du Plessis, P. R. Lamb, B. L. Fox, and W. P. Michalski. 2009. Combined continuous wave and pulsed plasma modes: For more stable interfaces with higher functionality on metal and semiconductor surfaces. *Plasma Processes and Polymers* 6 (10):615–619.

Li, L., L. H. Li, S. Ramakrishnan, X. J. J. Dai, K. Nicholas, Y. Chen, Z. Q. Chen, and X. W. Liu. 2012. Controlling wettability of boron nitride nanotube films and improved cell proliferation. *Journal of Physical Chemistry C* 116 (34):18334–18339. doi: 10.1021/Jp306148e.

Li, L. H. and Y. Chen. 2010. Superhydrophobic properties of nonaligned boron nitride nanotube films. *Langmuir* 26 (7):5135–5140. doi: 10.1021/La903604w.

Li, L. H., Y. Chen, and A. M. Glushenkov. 2010a. Boron nitride nanotube films grown from boron ink painting. *Journal of Materials Chemistry* 20 (43):9679–9683. doi: 10.1039/C0jm01414a.

Li, L. H., Y. Chen, and A. M. Glushenkov. 2010b. Synthesis of boron nitride nanotubes by boron ink annealing. *Nanotechnology* 21 (10):105601. doi: 10.1088/0957-4484/21/10/105601.

Li, X., H. Gao, W. A. Scrivens, D. Fei, X. Xu, M. A. Sutton, A. P. Reynolds, and M. L. Myrick. 2004. Nanomechanical characterization of single-walled carbon nanotube reinforced epoxy composites. *Nanotechnology* 15 (11):1416.

Li, Y. B., B. Q. Wei, J. Liang, Q. Yu, and D. H. Wu. 1999. Transformation of carbon nanotubes to nanoparticles by ball milling process. *Carbon* 37 (3):493–497. doi: 10.1016/s0008-6223(98)00218-8.

Liu, C. K., J. M. Wu, and H. C. Shih. 2010. Application of plasma modified multi-wall carbon nanotubes to ethanol vapor detection. *Sensors and Actuators B: Chemical* 150 (2):641–648. doi: 10.1016/j.snb.2010.08.026.

Loh, I. H., M. Klausner, R. F. Baddour, and R. E. Cohen. 1987. Surface modifications of polymers with fluorine-containing plasmas: Deposition versus replacement reactions. *Polymer Engineering and Science* 27 (11):861–868. doi: 10.1002/pen.760271115.

Lu, F., L. Gu, M. J. Meziani, X. Wang, P. G. Luo, L. M. Veca, L. Cao, and Y.-P. Sun. 2009. Advances in bioapplications of carbon nanotubes. *Advanced Materials* 21 (2):139–152. doi: 10.1002/adma.200801491.

Luais, E., C. Thobie-Gautier, A. Tailleur, M. A. Djouadi, A. Granier, P. Y. Tessier, D. Debarnot, F. Poncin-Epaillard, and M. Boujtita. 2010. Preparation and modification of carbon nanotubes electrodes by cold plasmas processes toward the preparation of amperometric biosensors. *Electrochimica Acta* 55 (27):7916–7922.

Ma, P. C., N. A. Siddiqui, G. Marom, and J. K. Kim. 2010. Dispersion and functionalization of carbon nanotubes for polymer-based nanocomposites: A review. *Composites Part A: Applied Science and Manufacturing* 41 (10):1345–1367.

Meyyappan, M. 2011. Plasma nanotechnology: Past, present and future. *Journal of Physics D: Applied Physics* 44 (17):174002.

Milella, A. 2008. Plasma processing of polymers. In *Encyclopedia of Polymer Science and Technology*, ed. Mark, H. F., John Wiley & Sons, Inc., New York.

Moreau, M., N. Orange, and M. G. J. Feuilloley. 2008. Non-thermal plasma technologies: New tools for biodecontamination. *Biotechnology Advances* 26 (6):610–617. doi: 10.1016/j.biotechadv.2008.08.001.

Oliver, W. C. and G. M. Pharr. 1992. An improved technique for determining hardness and elastic modulus using load and displacement sensing indentation experiments. *Journal of Materials Research* 7:1564–1583. doi: 10.1557/JMR.1992.1564.

Perucca, M. 2010. Introduction to plasma and plasma technology. In *Plasma Technology for Hyperfunctional Surfaces*, ed. Rauscher, H., Perucca, M., Buyle, G., pp. 1–32. Wiley-VCH Verlag GmbH & Co. KGaA, Weinheim, Germany.

Ramanathan, T., F. T. Fisher, R. S. Ruoff, and L. C. Brinson. 2005. Amino-functionalized carbon nanotubes for binding to polymers and biological systems. *Chemistry of Materials* 17 (6):1290–1295. doi: 10.1021/cm048357f.

Rana, S., R. Alagirusamy, and M. Joshi. 2009. A review on carbon epoxy nanocomposites. *Journal of Reinforced Plastics and Composites* 28 (4):461–487. doi: 10.1177/0731684407085417.

Roy, S. S., P. Papakonstantinou, T. I. T. Okpalugo, and H. Murphy. 2006. Temperature dependent evolution of the local electronic structure of atmospheric plasma treated carbon nanotubes: Near edge x-ray absorption fine structure study. *Journal of Applied Physics* 100 (5):053703.

Ruelle, B., A. Felten, J. Ghijsen, W. Drube, R. L. Johnson, D. Liang, R. Erni et al. 2009. Functionalization of MWCNTs with atomic nitrogen. *Micron* 40 (1):85–88.

Ruelle, B., S. Peeterbroeck, R. Gouttebaron, T. Godfroid, F. Monteverde, J. P. Dauchot, M. Alexandre, M. Hecq, and P. Dubois. 2007. Functionalization of carbon nanotubes by atomic nitrogen formed in a microwave plasma Ar + N_2 and subsequent poly([varepsilon]-caprolactone) grafting. *Journal of Materials Chemistry* 17 (2):157–159.

Scaffaro, R., A. Maio, S. Agnello, and A. Glisenti. 2012. Plasma functionalization of multiwalled carbon nanotubes and their use in the preparation of nylon 6-based nanohybrids. *Plasma Processes and Polymers* 9 (5):503–512. doi: 10.1002/ppap.201100140.

Shao, D., J. Hu, and X. Wang. 2010. Plasma induced grafting multiwalled carbon nanotube with chitosan and its application for removal of UO_2^{2+}, Cu^{2+}, and Pb^{2+} from aqueous solutions. *Plasma Processes and Polymers* 7 (12):977–985. doi: 10.1002/ppap.201000062.

Shen, J., W. Huang, L. Wu, Y. Hu, and M. Ye. 2007. The reinforcement role of different amino-functionalized multi-walled carbon nanotubes in epoxy nanocomposites. *Composites Science and Technology* 67 (15–16):3041–3050. doi: 10.1016/j.compscitech.2007.04.025.

Shishoo, R. ed. 2007. *Plasma Technologies for Textiles*. Woodhead Publishing Limited/CRC Press LLC, Abington, MA.

Siow, K. S., L. Britcher, S. Kumar, and H. J. Griesser. 2006. Plasma methods for the generation of chemically reactive surfaces for biomolecule immobilization and cell colonization—A review. *Plasma Processes and Polymers* 3 (6–7):392–418.

Sreenivasan, R. and K. K. Gleason. 2009. Overview of strategies for the CVD of organic films and functional polymer layers. *Chemical Vapor Deposition* 15 (4–6):77–90.

Tao, X., M. Bai, X. Li, H. Long, S. Shang, Y. Yin, and X. Dai. 2011. CH_4–CO_2 reforming by plasma—Challenges and opportunities. *Progress in Energy and Combustion Science* 37 (2):113–124. doi: 10.1016/j.pecs.2010.05.001.

Thostenson, E. T., C. Li, and T.-W. Chou. 2005. Nanocomposites in context. *Composites Science and Technology* 65 (3–4):491–516. doi: http://dx.doi.org/10.1016/j.compscitech.2004.11.003.

Tjong, S. C. 2006. Structural and mechanical properties of polymer nanocomposites. *Materials Science and Engineering R: Reports* 53 (3–4):73–197. doi: 10.1016/j.mser.2006.06.001.

Tseng, C. H., C. C. Wang, and C. Y. Chen. 2007. Functionalizing carbon nanotubes by plasma modification for the preparation of covalent-integrated epoxy composites. *Chemistry of Materials* 19 (2):308–315. doi: 10.1021/cm062277p.

Vandenbroucke, A. M., R. Morent, N. De Geyter, and C. Leys. 2011. Non-thermal plasmas for non-catalytic and catalytic VOC abatement. *Journal of Hazardous Materials* 195:30–54. doi: 10.1016/j.jhazmat.2011.08.060.

Wang, J., Z. Fang, A. Gu, L. Xu, and F. Liu. 2006. Effect of amino-functionalization of multi-walled carbon nanotubes on the dispersion with epoxy resin matrix. *Journal of Applied Polymer Science* 100 (1):97–104. doi: 10.1002/app.22647.

Wertheimer, M. R., A. C. Fozza, and A. Hollander. 1999. Industrial processing of polymers by low-pressure plasmas: The role of VUV radiation. *Nuclear Instruments and Methods in Physics Research B* 151 (1–4):65–75.

Xie, S. Y., W. Wang, K. A. S. Fernando, X. Wang, Y. Lin, and Y. P. Sun. 2005. Solubilization of boron nitride nanotubes. *Chemical Communications* (29):3670–3672. doi: 10.1039/B505330g.

Yan, Y. H., M. B. Chan-Park, Q. Zhou, C. M. Li, and C. Y. Yue. 2005. Functionalization of carbon nanotubes by argon plasma-assisted ultraviolet grafting. *Applied Physics Letters* 87 (21):1–3.

Yang, K., M. Gu, Y. Guo, X. Pan, and G. Mu. 2009. Effects of carbon nanotube functionalization on the mechanical and thermal properties of epoxy composites. *Carbon* 47 (7):1723–1737. doi: 10.1016/j.carbon.2009.02.029.

Yook, J. Y., J. Jun, and S. Kwak. 2010. Amino functionalization of carbon nanotube surfaces with NH_3 plasma treatment. *Applied Surface Science* 256 (23):6941–6944. doi: 10.1016/j.apsusc.2010.04.075.

Yoon, O. J., H. W. Kim, D. J. Kim, H. J. Lee, J. Y. Yun, Y. H. Noh, D. Y. Lee, D. H. Kim, S. S. Kim, and N. E. Lee. 2009. Nanocomposites of electrospun poly[(D,L-lactic)-co-(glycolic acid)] and plasma-functionalized single-walled carbon nanotubes for biomedical applications. *Plasma Processes and Polymers* 6 (2):101–109. doi: 10.1002/ppap.200800081.

Yoon, O. J., H. J. Lee, Y. M. Jang, H. W. Kim, W. B. Lee, S. S. Kim, and N.-E. Lee. 2011. Effects of O_2 and N_2/H_2 plasma treatments on the neuronal cell growth on single-walled carbon nanotube paper scaffolds. *Applied Surface Science* 257 (20):8535–8541. doi: 10.1016/j.apsusc.2011.05.009.

Zhang, C., F. Zhu, Z. Wang, L. Meng, and Y. Liu. 2012. Amino functionalization of multiwalled carbon nanotubes by gamma ray irradiation and its epoxy composites. *Polymer Composites* 33 (2):267–274. doi: 10.1002/pc.22144.

Zhi, C. Y., Y. Bando, C. Tang, and D. Golberg. 2010. Boron nitride nanotubes. *Materials Science and Engineering R: Reports* 70 (3–6):92–111. doi: 10.1016/j.mser.2010.06.004.

Zhi, C. Y., Y. Bando, C. C. Tang, S. Honda, K. Sato, H. Kuwahara, and D. Golberg. 2005. Covalent functionalization: Towards soluble multiwalled boron nitride nanotubes. *Angewandte Chemie International Edition* 44 (48):7932–7935. doi: 10.1002/anie.200502846.

Applications

Chapter 18

Boron Nitride Nanotubes as Nanofillers/ Reinforcement for Polymer, Ceramic, and Metal Matrix Composites

Debrupa Lahiri and Arvind Agarwal

CONTENTS

18.1 INTRODUCTION

Boron nitride nanotube (BNNT) is a potential reinforcement for structural composites owing to its (1) excellent elastic modulus (750–1200 GPa; Chopra and Zetll 1998, Suryavanshi 2004), (2) tensile strength (up to 61 GPa; Arenal 2011), (3) high bending flexibility (Golberg 2007), (4) high fracture strain (Ghassemi 2011), and (5) high-temperature inertness (oxidation starts >950°C; Golberg 2001, 2007). While discussing BNNT as reinforcement to structural composites, an obvious comparison with carbon nanotubes (CNTs) is made. This is because these two types of nanotubes, BNNT and CNT, are not only structural analogues, but their elastic modulus and tensile strength are also similar. However, BNNTs show higher oxidation temperature as compared to CNT, which could be beneficial for processing composites at higher temperature, especially the ceramic and metallic matrix ones, without damaging the reinforcement phase.

The structure and properties of BNNTs are already described in detail in other chapters of this book. Thus, this chapter would briefly mention about the BNNT structure and properties with an emphasis as reinforcement for composite materials. BNNTs generally exist as a multiwalled tubular structure, which is made by rolling hexagonal boron nitride (h-BN) sheets. BN tubular shells are separated by an average intershell distance of 0.33–0.34 nm (similar to multiwalled CNTs), which is a characteristic of d_{0002} spacing in a hexagonal BN. Individual tubes within the bundles interact through weak van der Waals forces and are packed in a honeycomb-like array in cross section. B and N atomic planes in h-BN sheet are slightly shifted in a direction perpendicular to the tube axis. The partially ionic character of B–N bonding in a BN layered material may stabilize double- or multilayered morphologies owing to strong *lip–lip* interactions between adjacent layers. These interactions favor the placement of B atoms strictly above or below N atom (Golberg et al. 2007, 2010). Table 18.1 presents some of the unique mechanical, thermal, and electrical properties of BNNT.

The potential of BNNTs has not been explored as immensely as CNTs, largely due to challenges in BNNT synthesis, which lead to very limited availability of BNNT. But an increasing interest in BNNT and its composites has opened up possibilities of several applications. The insulating nature of BNNT could be taken advantage for electrically insulated nanocables with embedded metallic or semiconducting nanowires. Such cables may be utilized in downsized electrical devices and complex multicable circuits, where each cable should perform independently, without current leakage between them (Golberg et al. 2007). Carbon-doped BNNTs are suitable for field emitters with better environmental stability (Golberg et al. 2007). BNNTs possess piezoelectric characteristics that could be used in precision piezoelectric devices to measure or apply force at high resolution (Bai et al. 2007). BNNT also has bright prospect for nonlinear optical and optoelectronic applications. BNNTs may be ideal candidates for optical devices working in the UV regime (Wu et al. 2004). Gas adsorption ability of BNNT may also be used for hydrogen storage and thus offering solution to current environmental pollution

TABLE 18.1   MECHANICAL AND PHYSICAL PROPERTIES OF BNNT IN COMPARISON TO CNT

Property	BNNT	CNT
Elastic modulus (GPa)	750–1200 (Chopra and Zetll 1998, Suryavanshi et al. 2004)	270–950 GPa (Yu et al. 2000)
Tensile strength (GPa)	≥61 (Arenal et al. 2011)	11–63 GPa (Yu et al. 2000)
Specific heat capacity (at 300 K), $J\,kg^{-1}\,K^{-1}$	~1000 (Xiao et al. 2004)	~480 (Yi et al. 1999)
Thermal conductivity at RT, $W\,m^{-1}\,K^{-1}$	200–300 (Zhi et al. 2009, Terao et al. 2010)	3000 (Kim et al. 2001)
Electrical conductivity, $S\,cm^{-1}$	3.3×10^{-3} (Tang et al. 2006)	1850 (Bando et al. 1999)
Band gap (eV)	5–6 (Golberg et al. 2007)	0.2–2 eV (Zhao and Dai 2004)
Oxidation start temperature (K)	1223 (Golberg et al. 2007)	773 (Golberg et al. 2007)

Number of publications

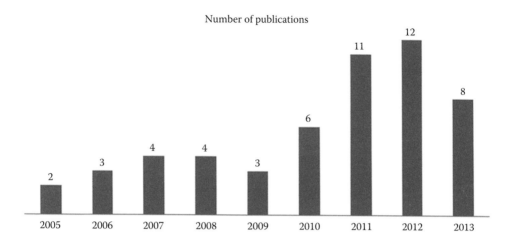

Figure 18.1 Year-wise publication plot for BNNT-reinforced composites. (From Scopus.com. Data Collected in August, 2013.)

(Ma et al. 2002). Apart from these listed applications, BNNT is gaining popularity as reinforcement in polymer and ceramic matrix composites due to its excellent mechanical and thermal properties.

Figure 18.1 presents year-wise publication plot for BNNT-reinforced composites, which reveals its early stage of development. At the same time, the plot shows a trend of growing research interest in the field, which would establish BNNT as a very potential reinforcement to various material systems (polymer/metal/ceramic). Studies on BNNT-based composites started with polymer, as these composites are relatively easy to fabricate and often do not need exposure to high temperature and pressure. Researchers in the field have found that randomly and aligned BNNTs reinforced in polymer matrix improve the thermal, mechanical, and optical properties of polymers (Zhi et al. 2005, 2006, 2009, Ravichandran et al. 2008, Lahiri et al. 2010, Terao et al. 2010, Huang et al. 2013, Li et al. 2013, Nurul and Mariatti 2013). Elastic modulus increases by as high as 1370% with 2 wt.% BNNT addition (Lahiri et al. 2010), whereas 20 times increase in thermal conductivity is noted with a high load of BNNT addition (up to 37 wt.%) (Zhi et al. 2009). Alignment of BNNTs further augments the thermal conductivity of the polymer matrix composite (Terao et al. 2010). Other studies have reported better thermal stability and lower coefficient of thermal expansion (Zhi et al. 2009, Terao et al. 2010). The use of BNNT as a reinforcement to biodegradable polymer scaffold in orthopedic application has also been established by a few recent studies (Lahiri et al. 2010).

Ceramic-based composites are fewer in number as compared to polymer matrix composites. The main aim of adding BNNT to ceramic is to increase the fracture toughness and strength. BNNTs are used as reinforcement to different ceramics, namely, Al_2O_3, Si_3N_4, AlN, ZrB_2–SiC, ZrO_2, SiO_2, and barium calcium aluminosilicate glass (Bansal et al. 2006, Choi et al. 2007, Huang et al. 2007, Li et al. 2010, Du et al. 2011, 2012, Xu et al. 2012, Yue et al. 2013). BNNT is found to induce high-temperature superplasticity in the ceramics by controlled dynamic grain growth and energy absorption mechanism (Huang et al. 2007). Increase in elastic modulus, fracture toughness, and strength is reported for BNNT-reinforced ceramics (Bansal et al. 2006, Choi et al. 2007, Du et al. 2011, 2012, Xu et al. 2012, Yue et al. 2013). In case of ceramic matrix composite also, the suitability of BNNT reinforcement in orthopedic application has been probed by studying hydroxyapatite (HA)–BNNT composite material system (Lahiri et al. 2011a).

The BNNT-reinforced *metal* matrix composite is the least studied material system in this category. Only a few studies are existing in this field, and most of them are on aluminum-based composites (Singhal et al. 2011, Lahiri et al. 2013, Yamaguchi et al. 2012, 2013), due to the immense interest in the development of high-strength low-weight composites. BNNTs are found to have constructive influence of strengthening the metal matrix composites (Singhal et al. 2011, Yamaguchi et al. 2012, 2013, Lahiri et al. 2013).

In this chapter, BNNT-reinforced polymer, ceramic, and metal matrix composites are discussed in detail. The role of BNNT in strengthening the structural composites is analyzed. The projected applications of these composites are significantly different and thus their functionalities are also discussed.

18.2 POLYMER–BNNT COMPOSITES

A copolymer of polylactic acid and polycaprolactone (PLC) is a popular choice for orthopedic scaffold material, as this copolymer gives the freedom of tailoring the strength, deformability, and degradation rate, suitable for the intended application. However, the elastic modulus and strength of this copolymer demand further enhancement to match with hard bone tissue for their application in orthopedic scaffold. Different reinforcements have been investigated for improving the mechanical properties of this copolymers, for example, HA (Rizzi et al. 2001, Causa et al. 2006, Kim 2007); organic phases like polysaccharides, starch, and cellulose; silk fibers (Cheung et al. 2008, Li et al. 2008); and CNTs (Kim et al. 2008, Xu 2008, Lahiri et al. 2009). Multiwalled BNNTs were chosen for this purpose owing to their excellent elastic modulus and strength (Lahiri et al. 2010).

18.2.1 Synthesis of PLC–BNNT Composite

A colloidal suspension of PLC was prepared with acetone as the solvent. BNNTs were added to the colloidal suspension by ultrasonication, followed by casting and curing at room temperature. The PLC–BNNT composites were synthesized with 2 and 5 wt.% BNNT reinforcement content, referred to as PLC–2BNNT and PLC–5BNNT hereafter. The cured suspension was peeled off from the Petri dish to form a freestanding film with 200–500 μm thickness. An interesting feature of the composite was the increase in density with BNNT content. The density values noted are 0.71, 1.15, and 1.33 g cm^{-3} for PLC, PLC–2BNNT, and PLC–5BNNT composite films, respectively. The theoretical density of the composites calculated using the as-measured density of PLC shows 55% and 71% higher density for PLC–2BNNT and PLC–5BNNT composite films, respectively. This difference is due to the significantly higher amount of porosity content in the PLC film as compared to BNNT, as observed in SEM images (Figure 18.2). Processing conditions of all the compositions being same, the decreasing porosity of the composite films is attributed to the presence of BNNT in the PLC matrix. The role of BNNT on the porosity content of the composite film is explained in terms of interaction between PLC and BNNT in solution. During curing operation at room temperature, PLC coagulates from the colloidal solution and forms separated agglomerates. These agglomerates are responsible for the porous structure upon drying due to shrinkage,

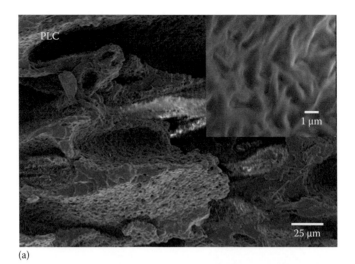

(a)

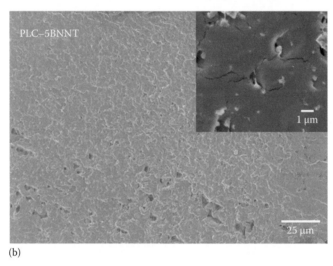

(b)

Figure 18.2 Cross section of (a) PLC and (b) PLC–5BNNT composite films revealing the porosity content. (Reprinted from *Acta Biomaterialia*, 6, Lahiri, D., Rouzaud, F., Richard, T., Keshri, A.K., Bakshi, S.R., Kos, L., and Agarwal, A., Boron nitride nanotube reinforced polylactide–polycaprolactone copolymer composite: Mechanical properties and cytocompatibility with osteoblasts and macrophages in vitro, 3524–3533, Copyright 2010a, with permission from Elsevier.)

and in turn, it creates a rough surface also. On the other hand, BNNTs form a uniform suspension with the acetone resulting in good wetting and strong interaction with the copolymer matrix. The good wetting and bonding of BNNT with the polymer matrix are attributed to the helical structure of the polymer that tends to form a coil, thus wrapping BNNTs and leading to a strong p–p interaction (Zhi et al. 2005, 2006). The uniform dispersion of BNNT in the colloidal solution leads to excellent wetting, and thus, the uniform distribution of PLC is maintained in the solution during drying resulting in a denser structure as compared to 100% PLC film.

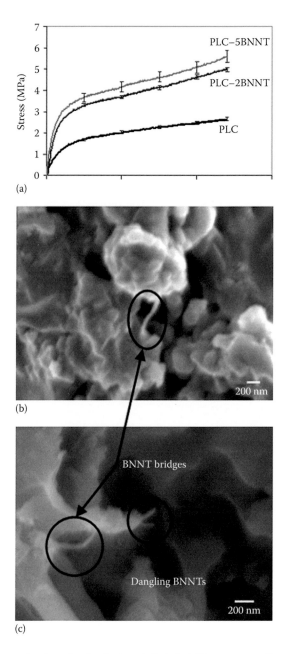

(a)

(b)

(c)

Figure 18.3 (a) Stress–strain plots for the PLC and PLC–BNNT composite films obtained through tensile testing; (b, c) SEM micrographs of fracture surface of PLC–5BNNT composite revealing a strong interface and BNNT bridges in the PLC matrix. (Reprinted from *Acta Biomaterialia*, 6, Lahiri, D., Rouzaud, F., Richard, T., Keshri, A.K., Bakshi, S.R., Kos, L., and Agarwal, A., Boron nitride nanotube reinforced polylactide–polycaprolactone copolymer composite: Mechanical properties and cytocompatibility with osteoblasts and macrophages in vitro, 3524–3533, Copyright 2010a, with permission from Elsevier.)

18.2.2 Tensile Properties and Strengthening Mechanism of PLC–BNNT Composite

One of the aims of adding BNNT to PLC matrix was to enhance the tensile strength. The strengthening effect is evaluated through uniaxial tensile test of the composite film. Figure 18.3a shows the engineering stress–strain behavior for PLC and PLC–BNNT composite films. Addition of BNNT improves the tensile strength from 2.67 MPa in PLC to 4.98 and 5.59 MPa for PLC–2BNNT and PLC–5BNNT, which is an increase by 87% and 109%, respectively. Tensile samples from the three films did not fail and retained their original shape, upon unloading, without any visible deformation after a maximum strain of 2.4 was achieved. Generally, addition of a strong second phase to a flexible matrix often shows a negative effect on its deformability. This negative effect is more prominent in case of a flexible matrix, for example, polymer. But in case of the PLC–BNNT composite, the deformability of the PLC is not affected up to 240% elongation. Such behavior is attributed to the high flexibility of BNNTs (Golberg et al. 2007). The strength of the composite structure with nanosized reinforcement is achieved by effective load transfer through the large interface area between matrix and the reinforcement. Such load transfer becomes effective only when interfacial bond is good enough to transfer the load without getting disturbed/damaged. Further, the fiber-type geometry of the reinforcement makes them more effective in increasing the tensile strength of the matrix. The evidence of BNNT strengthening of the PLC matrix is available in the SEM images of the fracture surface of PLC–5BNNT (Figure 18.3b,c) revealing that the BNNTs bridges within the PLC matrix. Dangling BNNTs with the one end fully embedded in the polymer matrix are also signs of strong interfacial bonding. BNNTs behave as rigid reinforcements and provide benefits of fiber strengthening.

The elastic modulus of the PLC–BNNT composites, evaluated using the nanoindentation technique, shows an improvement of 100% with 2 wt.% BNNT and 1370% with 5 wt.% BNNT additions. This is due to the strengthening effect achieved through BNNT reinforcement that resists the elastic deformation of the composites.

18.2.3 Biocompatibility of PLC–BNNT Composite for Orthopedic Application

The intended application of the PLC–BNNT composite in orthopedic scaffold demands the assessment of biocompatibility, as BNNT is new to the field of biomaterials and biomedicine. BNNT has been studied for the drug delivery system for the past few years (Ciofani et al. 2009, 2010). The cytotoxicity studies on BNNT have mostly reported no negative effect on different cell types (Ciofani et al. 2008, 2009, 2010, Chen et al. 2009, Lahiri et al. 2010). Gene expression studies have indicated that BNNT influence the accelerated differentiation of osteoblast cells (Lahiri et al. 2010). BNNTs are also found highly internalized by mouse myoblast (muscle) cells, with neither adversely affecting its viability nor significantly interfering with myotube formation (Ciofani et al. 2010). Chen et al. (2009) have shown BNNTs to be noncytotoxic to human embryonic kidney cells and reported that BNNTs do not inhibit cell proliferation even after 4 days. Ciofani et al. (2008) demonstrated good cytocompatibility and cellular uptake of polyethyleneimine-coated BNNTs in a human neuroblastoma cell line. In addition, BNNTs also favor attachment of protein on their surface and thus are potentially suitable for nanobiological applications (Zhi et al. 2005). All these studies indicate a safe and bright future for BNNTs in bioapplication.

These studies encouraged further evaluation of BNNT, specifically for orthopedic application. The application of BNNTs in orthopedics necessitates evaluation of their cytotoxic behavior with bone cells, for example, osteoblasts. Further, the PLC matrix being biodegradable, BNNTs may get exposed to the bloodstream after degradation of the scaffold. BNNTs, exposed in the bloodstream, interact first with macrophages, as these blood cells are supposed to internalize the foreign elements entering the bloodstream to prevent any harmful reaction. Thus, the compatibility of BNNT with macrophages also seems important. Cytotoxicity assay for bare BNNTs can measure the population of dead cells, when incubated in medium with BNNT and expressed in comparative mode with respect to the cells incubated in medium without BNNTs. The cytotoxicity assay outcomes revealed a nonsignificant cytotoxicity induced by the BNNT for both osteoblast and macrophage cell lines (Lahiri et al. 2010). Such behavior is primarily attributed to the chemical inertness and structural stability of BNNT.

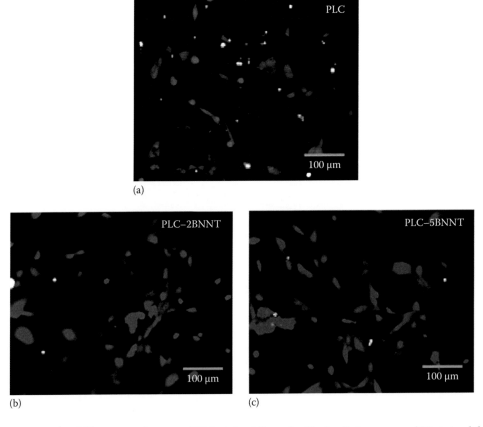

Figure 18.4 (a–c) Fluorescent images of FDA-stained live osteoblast cells in green and PI-stained dead cells in red obtained on PLC and PLC–BNNT composite surfaces after incubation for 2.5 days. (Reprinted from *Acta Biomaterialia*, 6, Lahiri, D., Rouzaud, F., Richard, T., Keshri, A.K., Bakshi, S.R., Kos, L., and Agarwal, A., Boron nitride nanotube reinforced polylactide–polycaprolactone copolymer composite: Mechanical properties and cytocompatibility with osteoblasts and macrophages in vitro, 3524–3533, Copyright 2010a, with permission from Elsevier.)

Once the cytotoxicity of BNNT has been checked, the next step should be the compatibility of the composite surface to bone cells. The viability of osteoblast cells, which is the ratio of live to dead cells, is used to understand the biocompatibility of PLC-based composite surfaces. The viability ratio is noted as 90% ± 2 and 91% ± 4 for PLC–2BNNT and PLC–5BNNT composites, which is much higher than 59% ± 4 as observed on PLC (p value <0.05). Figure 18.4 is a representative picture of live and dead cells on the three PLC composite films, captured after 2.5 days of incubation. A typical lens shape of osteoblasts suggests the presence of normal and healthy cell behavior on PLC, PLC–2BNNT, and PLC–5BNNT composite films with no significant difference in the cellular morphology. These results indicate that the presence of BNNTs improve the biocompatibility of the PLC copolymer in terms of osteoblast cell viability.

In addition, the effect of PLC–BNNT composites on the differentiation state of osteoblasts is evaluated by assessing the expression of the transcription factor *Runx2* for the osteoblasts incubated on these surfaces. *Runx2* (also known as *Cbfa*) is a master regulator of osteoblastogenesis and coordinates the integration of signaling events and other transcription factors involved in this process (Marie 2008). A fourfold and a sevenfold increase in the levels of expression of *Runx2* in osteoblasts grown on PLC–2BNNT and PLC–5BNNT, respectively, when compared to PLC alone, indicates the positive effect of the presence of BNNTs on accelerated osteoblast differentiation and growth. These results are attributed to the natural affinity of protein to BNNTs (Zhi et al. 2005). Regulation of osteoblast differentiation is mediated by the presence of bone morphogenetic proteins (BMPs). Hence, strong interaction and immobilization of BMPs on the BNNT surface can enhance the osteoblast cell differentiation in PLC–BNNT composites. Thus, PLC–BNNT composites are found to be suitable to be used as a biodegradable orthopedic scaffold.

18.3   CERAMIC–BNNT COMPOSITES

BNNTs are used as reinforcement to HA ($Ca_{10}(PO_4)_6(OH)_2$) also, which is a calcium phosphate–based ceramic used in orthopedic implant and scaffold (Lahiri et al. 2011a). Chemical composition, crystal structure, and Ca:P ratio (1.67) of HA, which is similar to apatite found in human skeleton (Gu et al. 2002, Khor et al. 2003, White et al. 2007), led to its clinical use in orthopedic implant/scaffold material. However, the major shortcomings of HA in load-bearing applications are its poor fracture toughness and wear resistance. A second-phase reinforcement is often used to enhance these performances of a brittle ceramic, and fiber-shaped reinforcements are found most suitable for fracture toughening. In this connection, CNT has appeared as a strong contender as a reinforcement to HA due to its excellent elastic modulus and fiber-like morphology, which plays an active role in improving fracture toughness through crack bridging (Balani et al. 2007, Sarkar et al. 2007, Lahiri et al. 2010). It has been studied that CNT acts as grain refiner for sintered HA (Lahiri et al. 2010), further aiding the toughening. Similarity of BNNT to CNT in structure and mechanical properties has led to the evaluation of potential of BNNT as reinforcement to HA.

18.3.1   Synthesis of HA–BNNT Composite

HA–BNNT composite with 4 wt.% BNNT reinforcement was successfully synthesized through the powder processing route using spark plasma sintering (SPS) as a consolidation method (Lahiri et al. 2011a). Effective and homogeneous dispersion of BNNT obtained through the wet mixing method was retained in the final consolidated structure after SPS at 1373 K and 70 MPa for 5 min. Uniform distribution of the reinforcement phase in the matrix is essential for the strengthening or toughening of any composite structure. Another important factor is the retention of undamaged

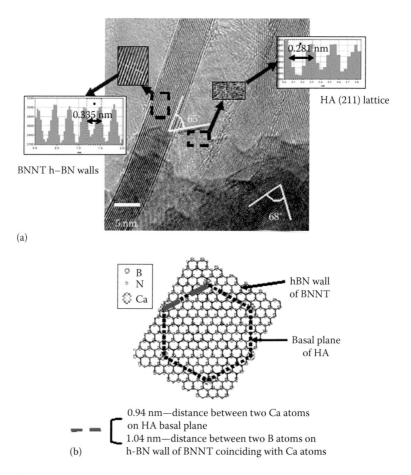

(a)

(b)

Figure 18.5 (a) High-resolution TEM image of the HA–BNNT interface revealing the lattice fringes for (211) planes of HA and wall spacing of BNNT; (b) schematic of atomic arrangement across the interface. (Reprinted from *J. Mech. Behav. Biomed. Mater.*, 4, Lahiri, D., Singh, V., Benaduce, A.P., Seal, S., Kos, L., and Agarwal, A., Boron nitride nanotube reinforced hydroxyapatite composite: Mechanical and tribological performance and in-vitro biocompatibility to osteoblasts, 44–56, Copyright 2011, with permission from Elsevier.)

nanoscale reinforcement phase in the matrix. This is especially crucial for ceramic and metal matrix composites, as the exposure to high temperature and/or pressure required for consolidation can have detrimental effect on the retention of unaltered structures of reinforcement phase. High flexibility and capability of withstanding a large amount of stress without any permanent deformation (Golberg et al. 2007) are unique properties of BNNT that have led to the presence of undamaged BNNT in consolidated HA matrix after high-temperature and high-pressure exposure in SPS, as observed in high resolution transmission electron micrograph (HRTEM) micrograph (Figure 18.5a). BNNTs are responsible for increasing the density of SPS-consolidated HA from 92% TD to 97% TD. Higher densification of composite is attributed to better thermal conductivity of BNNT (Gaona et al. 2007, Zhi et al. 2009, Terao et al. 2010), which causes more uniform thermal gradient throughout the pellet thickness and thus better consolidation. Further, BNNTs are instrumental in refining the grain size of sintered HA pellets by three times as compared to HA, through grain boundary pinning. Finer grain size leads to higher fracture toughness in ceramics.

18.3.2 HA–BNNT Interface

The matrix-reinforcement interface dictates the mechanical performance of any composite structure. Thus, the type of bonding and strength of HA–BNNT interface is very important toward its application in load-bearing orthopedic implants. Both HA and BNNT are chemically nonreacting, which is also confirmed by the absence of an interfacial reaction product and clean interface in HRTEM (Figure 18.5a). As a result, ionic or covalent bonds cannot dominate the HA–BNNT interface, and van der Waals bond is the most probable. Hence, this interfacial strength is mainly governed by the work of adhesion, which is dictated by the lattice arrangement at the interface. Work of adhesion becomes higher when the lattice mismatch is minimal, resulting in reduced lattice strain. A higher lattice mismatch, $\delta > 0.25$, leads to an incoherent interface and poor bonding (Porter and Easterling 2001). Analysis of the relationship between lattice fringes of HA and BNNT at the interface, as obtained in HRTEM micrograph (Figure 18.5a), gives an idea about the crystallographic arrangement along the interface. The (211) planes of HA are arranged at an angular range of 65°–68° to the outer wall of BNNT. Similar angular relationship is found between (211) planes and the basal plane in hexagonal. Thus, the basal planes of HA should be parallel to the BNNT surface, which is the h-BN sheet. Figure 18.5b presents a schematic of the basal plane of HA crystal with Ca atoms at each corner superimposed on the h-BN wall, with appropriate lattice spacing. The mismatch between two superimposed pair of Ca (in HA) and B (in h-BN) atoms, calculated as per Figure 18.5b, is $\delta \sim 0.11$, which is much lower than the incoherence limit of 0.25. Thus, the preferential alignment of HA crystals on the BNNT surface, in the HA–BNNT composite structure, suggests a strong coherent interfacial bond with minimal lattice strain. Similar observation is reported at the interface of HA and CNT of SPS-processed composite structure (Lahiri et al. 2010).

The strength of the HA–BNNT interface can also be estimated based on the model proposed by Chen et al. for the Al_2O_3–CNT system (Chen et al. 2008). The details of this estimation are available in a prior publication of the present authors (Lahiri et al. 2011a). The calculations reveal that the minimum energy required for pulling out one BNNT from the HA matrix is $2 \, J \, m^{-2}$, which is double as compared to the fracture energy of monolithic HA of $1 \, J \, m^{-2}$ (Nakira and Eguchi 2001). Thus, the presence of BNNT reinforcement in the HA matrix is bound to increase the fracture toughness of the latter.

18.3.3 Fracture Toughening of HA–BNNT

The theoretical calculations show that the presence of BNNT reinforcement in the HA matrix should increase the fracture toughness of the latter (Lahiri et al. 2011a). Elastic modulus, measured through nanoindentation, reveals the elastic modulus of HA–BNNT as 205 ± 15 GPa as compared to 93 ± 9 GPa for HA, which is an increase by 120% with BNNT addition. The improvement in elastic modulus with BNNT reinforcement can be explained as a result of two major factors: (1) higher elastic modulus of BNNT and (2) strong bonding at the HA–BNNT interface. Further, retention of the defect-free BNNT structure during SPS makes its contribution more effective toward enhancing the elastic modulus.

A very effective method of measuring the fracture toughness of brittle ceramic is using the length of radial crack from Vickers' indentation and using Anstis's relationship (Anstis et al. 1981). The fracture toughness thus calculated shows 86% improvement with BNNT addition, with the values being $1.6 \, (\pm 0.3) \, MPa.m^{0.5}$ for HA–BNNT and $0.85 \, (\pm 0.3) \, MPa.m^{0.5}$ for HA (Lahiri et al. 2011a). Shorter length of radial crack, which is responsible for improved fracture toughness in HA–BNNT,

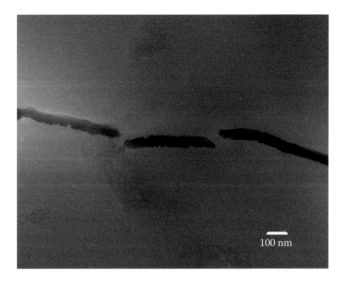

Figure 18.6 BNNT bridge across a crack generated from the corners of a microindent impression. (Reprinted from *J. Mech. Behav. Biomed. Mater.*, 4, Lahiri, D., Singh, V., Benaduce, A.P., Seal, S., Kos, L., and Agarwal, A., Boron nitride nanotube reinforced hydroxyapatite composite: Mechanical and tribological performance and in-vitro biocompatibility to osteoblasts, 44–56, Copyright 2011, with permission from Elsevier.)

is attributed to two major factors—(1) grain size refinement and (2) crack bridging by BNNTs. A study by Wang and Shaw has reported simultaneous improvement in hardness and toughness in sintered HA pellet due to refinement in grain size (Wang and Shaw 2009). Deflection of crack and transition of cracking mode from transgranular to intergranular lead to the improvement in fracture toughness of HA with refined grain size. BNNTs are found to induce refinement of grains in HA matrix by three times. In addition, higher pull-out energy of BNNT from HA matrix (refer to Section 18.3.2) restricts the propagation of cracks in the vicinity of BNNT, as more energy is required for interface debonding. BNNT bridges in a radial crack generated by the indentation are found to stop the crack propagation (Figure 18.6). Deflection in the crack path by BNNT bridges leads to absorption of fracture energy and thus increases fracture toughness.

18.3.4 Tribological Behavior of HA–BNNT

BNNT reinforcement shows significant improvement in the wear resistance of HA matrix, when evaluated using ball-on-disk test. The coefficient of friction (CoF) increases by ~25% with BNNT reinforcement in HA (Lahiri et al. 2011a). Higher fracture toughness and elastic modulus of the HA–BNNT composite cause more resistance to mass removal and as a result increase in the lateral (transverse) force and thus CoF. However, the h-BN sheet is otherwise used as a good lubricator and decreases the CoF of the system (Pawlak et al. 2008, 2009). Hence, higher CoF in HA–BNNT wear track must be due to the absence of peeled-off h-BN sheet. This is possible due to the capability of BNNT to withstand high amount of deformation without getting damaged as discussed earlier (Golberg et al. 2007).

Presence of BNNT is found to increase the wear resistance of HA matrix by 75% (Lahiri et al. 2011a). The increase in the wear resistance of HA–BNNT is the result of its improved elastic modulus and fracture toughness. Toughened matrix is supposed to restrict loss of mass on wear track due to fracture and chipping. The morphology of wear tracks (Figure 18.7) reveals further details of wear mechanism in the HA–BNNT composite. Figure 18.7a and c shows a flat morphology in the HA wear track, which is an indicator of abrasive wear mechanism and total detachment of mass from the wear track. On the contrary, the wear track on HA–BNNT (Figure 18.7b, d) shows shear displacement of mass, resulting in pileup, which is not expected in brittle ceramics like HA. Huang et al. have also reported superplasticity introduced in ceramics as a result of BNNT reinforcement (Huang et al. 2007), due to the obstacle in dynamic grain growth at higher temperature and *sword in sheathe* phenomenon of load transfer in BNNT. The sword in sheathe phenomenon can effectively transfer load from the matrix to the outermost wall of BNNT, followed by inner layers gradually. The gradual sliding of BNNT layers transforms the applied force to strain energy. This energy absorption mechanism causes the dislodgement of mass toward the periphery of the wear track, still being held together with BNNT bridges (Figure 18.7e) and thus creating the pileup. The earlier discussion shows the modification of wear mechanism in HA by the presence of BNNT, which in turn provides improved wear resistance of the composite aiding its orthopedic application.

18.3.5   Biocompatibility of HA–BNNT Composite

Compatibility of the HA–BNNT composite to bone cells is again a critical issue for its intended orthopedic application. Section 18.2.2 has already discussed about cytotoxicity evaluation of bare BNNTs with bone cells and macrophages, which are most relevant to the orthopedic application, and the findings of in vitro studies established no negative effect of BNNTs on these cells (Lahiri et al. 2010). Proliferation and viability assessment of osteoblast cells on HA and HA–BNNT surfaces after in vitro culturing for 1 and 3 days reveals an increase in cell population with number of days on both the surfaces. Figure 18.8 presents the fluorescent images of cells on both surfaces after 1 and 3 days of incubation. The cells exhibit typical lens shape suggesting the normal cell growth behavior. Population of osteoblast cells after 3 days of incubation is found marginally higher on HA–BNNT (Lahiri et al. 2011a). Calculation of % live cells on these surfaces over different incubation periods reveals comparable viability on HA and HA–BNNT surfaces.

Apart from biocompatibility to bone cells, another quality of an ideal orthopedic implant surface is to get attached with newly integrated bone. This phenomenon depends on the affinity of the surface for apatite precipitation. In case of the HA–BNNT composite, the matrix itself possesses a similar chemical composition to the mineral phase of human bone and thus inducts apatite precipitation quite obviously. Thus, the apatite formability of the BNNT surface becomes the deciding factor about the suitability of the HA–BNNT surface being integrated to neobone while implanted. An investigation was carried out on apatite formation ability of BNNT, immersed in simulated body fluid environment for a period of 7, 14, and 28 days (Lahiri et al. 2011b). When soaked in the simulated body fluid, BNNTs are found to induce amorphous flakes and crystalline needlelike HA precipitates on their surface (Figure 18.9). The precipitation process is found to have a threshold period of ~6.2 days. The mass of HA precipitate increases gradually with the soaking time. Lattice fringe analysis from high-resolution TEM study reveals a hexagonal crystal structure of HA needles, which does not show any specific crystallographic orientation

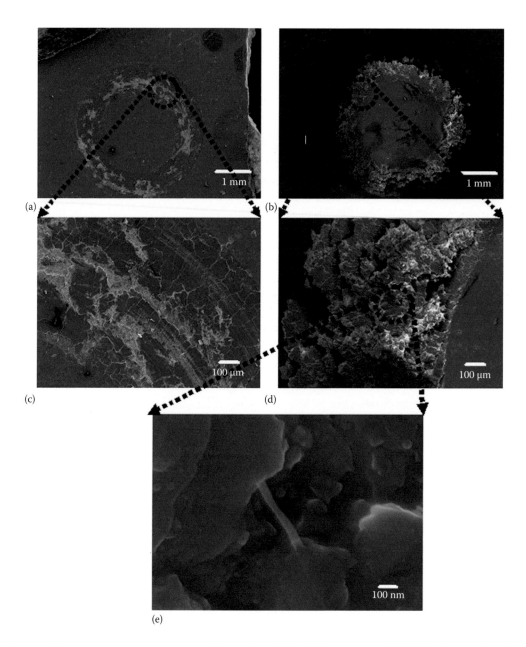

Figure 18.7   Macro wear tracks on (a, c) HA and (b, d) HA–BNNT surfaces; (e) BNNT bridge holding the loosened mass on the wear track of HA–BNNT. (Reprinted from *J. Mech. Behav. Biomed. Mater.*, 4, Lahiri, D., Singh, V., Benaduce, A.P., Seal, S., Kos, L., and Agarwal, A., Boron nitride nanotube reinforced hydroxy-apatite composite: Mechanical and tribological performance and in-vitro biocompatibility to osteoblasts, 44–56, Copyright 2011, with permission from Elsevier.)

1 day 3 days

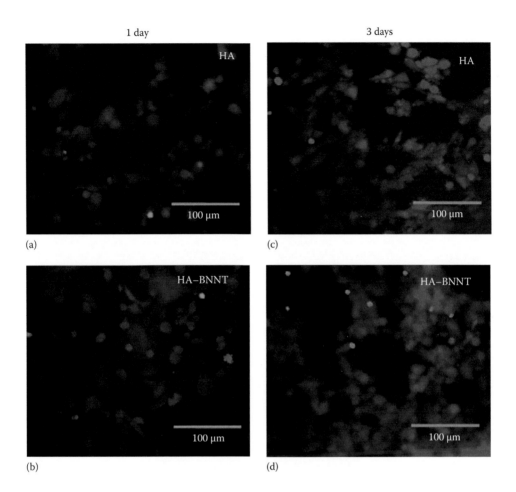

(a) (c)

(b) (d)

Figure 18.8 Fluorescent images of FDA–PI stained live (gray) and dead (black) osteoblast cells incubated on HA and HA–BNNT surfaces for 1 and 3 days. (Reprinted from *J. Mech. Behav. Biomed. Mater.*, 4, Lahiri, D., Singh, V., Benaduce, A.P., Seal, S., Kos, L., and Agarwal, A., Boron nitride nanotube reinforced hydroxyapatite composite: Mechanical and tribological performance and in-vitro biocompatibility to osteoblasts, 44–56, Copyright 2011, with permission from Elsevier.)

relationship with the BNNT surface. This could be due to the presence of a thin amorphous HA layer on the BNNT surface that disturbs definite orientation relationship.

Compatibility of bone cells to the HA–BNNT surface and apatite formability of BNNT establishes the biological suitability of this composite for orthopedic applications. The toughening and tribological effects of BNNT reinforcement were already encouraging for the HA–BNNT composite for structural application.

18.4 METAL–BNNT COMPOSITES

BNNT-reinforced metal matrix composites are most challenging to synthesize due to the high reactivity of most of the metals at consolidation temperature, which causes formation of reaction product at the interface. The challenge becomes more several in case of nanosized fillers,

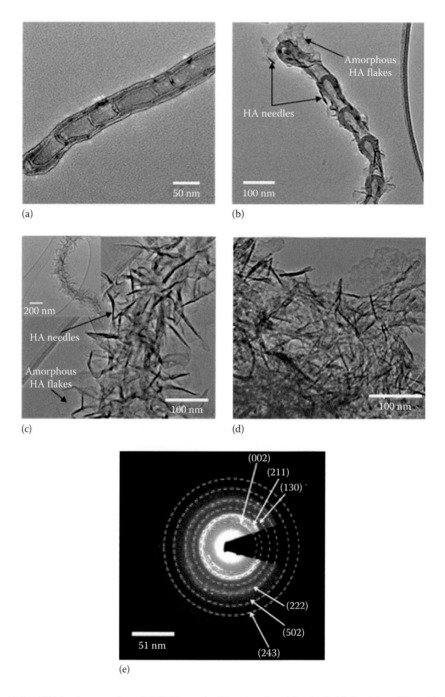

Figure 18.9   TEM micrographs of BNNTs soaked in simulated body fluid for (a) 0, (b) 7, (c) 14, and (d) 28 days; (e) selected area diffraction (SAD) pattern BNNTs soaked for 28 days revealing crystalline HA. (Reprinted with permission from Lahiri, D., Singh, V., Keshri, A.K., Seal, S., and Agarwal, A., Apatite formability of boron nitride nanotubes, *Nanotechnology*, 2011, 22, 205601, with permission from Institute of Physics.)

as reaction product formation can cause significant loss in their volume and often destruction of the structure causing the reinforcement ineffective. Thus, considerations must be given toward the reaction thermodynamics and kinetics at the interface while choosing a nanofiller for any metal matrix composite.

Aluminum–BNNT composites seem to be potentially attractive due to their projected structural applications requiring light weight, high specific strength, good corrosion resistance, and high thermal and electrical conductivity. CNT, the structural analogue to BNNT, has been researched as a structural reinforcement to aluminum for more than a decade for this purpose (Laha et al. 2004, Deng et al. 2007, He et al. 2007, Laha et al. 2007, Choi et al. 2009, Kwon et al. 2009, Perez-Bustamante et al. 2009, Agarwal et al. 2010, Bakshi et al. 2010). However, reactivity and formation of brittle carbide at the Al/CNT interface make it weak, which has restricted the Al–CNT composite system from achieving expected mechanical properties (Agarwal et al. 2010, Bakshi et al. 2010, Bakshi and Agarwal 2011). BNNT has the potential to be a suitable alternative to CNT for Al-matrix composites. In addition to comparable elastic modulus and strength to CNT, BNNT is more flexible and capable of withstanding heavy deformation without getting damaged (Golberg et al. 2007) and higher fracture strain than CNT (Ghassemi et al. 2011). More interestingly, BNNTs do not oxidize up to 950°C, as compared to CNT, which starts oxidizing at 500°C (Golberg et al. 2001, 2007). The oxidation temperature of BNNT being much higher than the melting point of aluminum (660°C), the consolidation process of the composite reduces the chance of chemical reaction on the BNNT surface. Keeping these positive aspects in mind, the first investigation was required on reactivity between Al and BNNT at interface.

18.4.1 Interfacial Reaction for Aluminum–BNNT System

The experimental evaluation of reactivity between Al and BNNT was performed by heat-treating the Al–BNNT composite powder at 650°C for different times up to 120 min (Lahiri et al. 2012). Figure 18.10 presents the high-resolution TEM micrographs of composite powder heat-treated at different time periods. An exposure of 10 min at 650°C results in the nucleation of a discontinuous reaction product at few locations on the BNNT surface (Figure 18.10b). The thickness of the reaction layer increases with time and leads to a thin, uneven, and semicontinuous layer after 60 min of soaking (Figure 18.10c). However, a significant change in the volume and morphology of the reaction product is noticed after 120 min exposure, resulting in much distinct globular crystals (Figure 18.10d). Reaction product layer thickness is found to be 0.35–2.3 nm (average: 1.5 nm) at 60 min, and the same increases to 3.8–9 nm (average: 7.3 nm) at 120 min. Identity of the reaction products was established from the lattice fringes in high-resolution TEM image (Figure 18.11) of the BNNT surface after heat treatment at 650°C for 10 min, with traces of discretely nucleating crystals. This analysis reveals nucleation of AlN and AlB_2 crystallites at the Al–BNNT interface. Based on the spatial distribution of these crystallites on the BNNT surface and their crystallographic orientation, the reaction mechanism at the Al–BNNT interface was predicted.

AlN nucleates as nearly spherical crystals of ~7 nm size, whereas AlB_2 grows into an elongated and big crystal with ~35 nm length five times larger than AlN. The size, shape, and spatial distribution through the interface can be explained in terms of the growth mechanism of AlN and AlB_2, their orientation with BNNT wall, and diffusion of Al, B, and N in each other. Figure 18.11 reveals the (100) planes of AlN aligned at 23° with BNNT wall. This indicates that the c-axis of hexagonal wurtzite-type crystals of AlN will be aligned at 23° to BNNT walls, which happens to be the

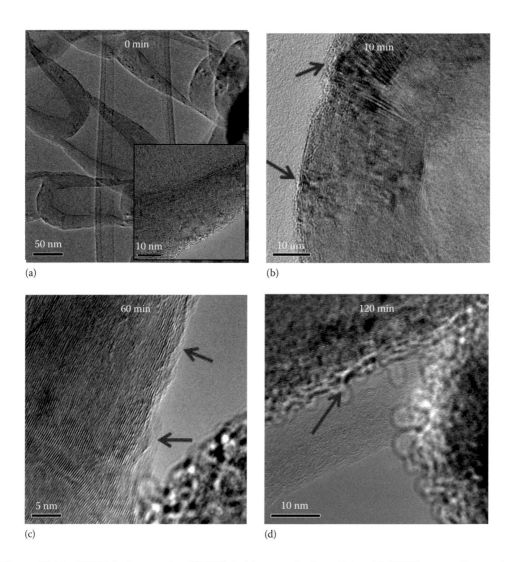

Figure 18.10 HRTEM micrographs of BNNT in (a) as-received condition; Al–BNNT composite powder exposed in argon atmosphere at 650°C for (b) 10 min, (c) 60 min, and (d) 120 min. Reaction products are marked by arrows in the micrographs. (Reproduced from Lahiri, D. et al., *J. Mater. Res.*, 27, 2760. With permission from MRS.)

favorable growth direction of AlN also. As a result, numerous AlN crystals, nucleated on the BNNT surface, would hinder the growth of each other due to their low-angle (23°) alignment, resulting in smaller crystallite size. AlB_2 crystals are oriented with their (101) set of planes aligned perpendicular to the BNNT surface (Figure 18.11). Considering the hexagonal structure of AlB_2 with alternate arrangement layers of Al and B, the basal plane should make 38.6° angles with BNNT wall. The preferable growth direction for AlB_2 being along their basal plane, the angular alignment of growth direction with BNNT surface (38.6°) is higher. This allows the growth of AlB_2 to larger size as compared to AlN. In addition, the natural tendency of AlB_2 for growing into large crystals at the expense of smaller crystals helps in the formation of five times larger crystals than AlN.

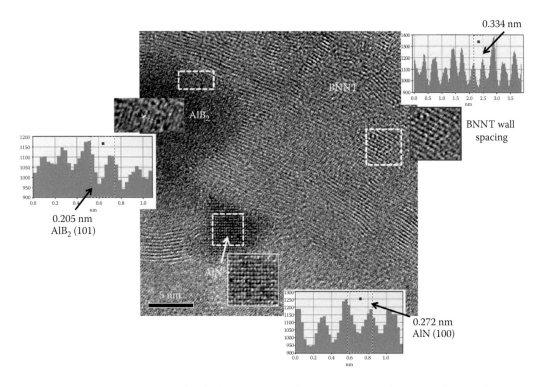

Figure 18.11   HRTEM micrographs of Al–BNNT exposed in argon at 650°C for 10 min show nucleation of AlN and AlB$_2$ crystals. (Reproduced from Lahiri, D. et al., *J. Mater. Res.*, 27, 2760. With permission from MRS.)

The spatial distribution of AlN and AlB$_2$ across the Al–BNNT interface is dictated by inter-diffusion of Al, B, and N in each other. Smaller N atoms (atomic radius, 65 pm) are able to diffuse faster in Al than relatively larger B atoms (atomic radius, 85 pm), resulting in more AlN formation toward the Al side of the interface. As a result, higher concentration of B atoms is left toward the BN side of the interface. The heat treatment temperature (650°C) is high enough for Al to be highly thermally active to diffuse fast through BN surface to form AlB$_2$. Confirming this hypothesis, AlN is found nucleating on the outer surface of BNNT and AlB$_2$ in the grooves of surfaces at the BNNT wall (Figure 18.11). In addition to these discussed points, the reactivity at the Al–BNNT interface is also influenced by morphological features and defects present. The damaged or broken sites on the surface of BNNT, knots of bamboo-shaped BNNTs, etc., aggravate the rate of reaction due to accelerated diffusion kinetics at disturbed lattices (Lahiri et al. 2012).

The reaction mechanism at the Al–BNNT interface discussed till now can predict the behavior of the Al–BNNT interface in the macroscale composite structure. The two most important factors that dictate the interfacial strength is the thickness of the reaction product layer formation at the interface and coherency of the reaction products along interface. In the study discussed earlier (Lahiri et al. 2012), BNNTs are found forming a very thin reaction layer upon exposure of the composite powder at 650°C for 120 min. The average thickness of the reaction products formed between the Al and BNNT interface is very thin (1.5 nm for 60 min treatment), compared to the Al–CNT interface at much less severe conditions and very short exposure (Ci et al. 2006, Bakshi et al. 2009, Kwon et al. 2009). Further, carbide formed at the Al–CNT interface (Al_4C_3) is deleterious to the mechanical performance of the composite, due to the inherent brittleness and hygroscopic nature of the carbide (Lahiri et al. 2012). In comparison, the reaction products of Al and BNNT (AlB_2 and AlN) are not so detrimental, as AlN reinforcement is found to enhance the hardness and flexural strength of Al matrix and AlB_2 is also least detrimental to the fracture toughness as compared to Al_4C_3, when formed simultaneously in a composite interface (Lahiri et al. 2012). Thus, a thin layer of reaction product at the BNNT interface is actually not deleterious but can improve interfacial bonding and promote strengthening of the composite structure.

The interfacial orientation relationship, as presented in the schematic in Figure 18.12, is also important to determine the interfacial strength of the Al–BNNT composite. AlN forms a stable interface along Al (111) and AlN (002) crystallographic orientation as these surfaces form the lowest energy coherent interface (Arslan et al. 2003, Montesa et al. 2011). This could be the reason for AlN crystals being equiaxed, which denotes coherency of second phase in the matrix with less misfit strain. On the contrary, the interfacial energy of the Al/AlB_2 interface is much higher than that as they only one set of partially coherent planes between the two (Xioa et al. 2006, Han et al. 2011), which makes the Al–AlB_2 interface weaker. Though the shape of AlB_2 is elongated, which means a semicoherent boundary, the larger edge of AlB_2 indicates less misfit. Hence, most part of AlB_2 also offers better bonding with the matrix. The spatial distribution of reaction products along the interface, as discussed earlier, reveals that AlN crystals will mostly cover the surface of BNNTs with embedded AlB_2 at few locations. Since AlN forms a more stable interface, it is

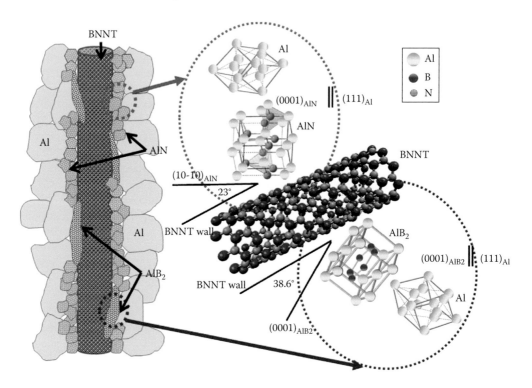

Figure 18.12 Schematic of the Al–BNNT interface presenting the spatial distribution crystallographic orientation of reaction products across the interface. (Reproduced from Lahiri, D. et al., *J. Mater. Res.*, 27, 2760. With permission from MRS.)

expected to offer a strong interface with BNNT reinforcement in the Al matrix. These findings (Lahiri et al. 2012) encourage the fabrication of the bulk Al–BNNT composite, which would be described in the following section.

18.4.2   Synthesis of Aluminum–BNNT Composite

Bulk Al–BNNT composites were synthesized by the powder metallurgy route (Lahiri et al. 2013). BNNTs were synthesized by ball milling and annealing. Ultrasonication is found to be a very effective technique for uniformly dispersing BNNTs in aluminum. SPS at 500°C and 80 MPa for 1 h in vacuum resulted in 97.8% TD for aluminum and 2 vol.% for the BNNT composite. BNNTs were not damaged during the high-temperature and high-pressure exposure of SPS. No significant reaction at the Al–BNNT interface is also reported (Lahiri et al. 2013), which can be attributed to high inertness of BNNT (Golberg et al. 2001, 2007), lower mass of reaction product formation, and lower thermodynamic probability of reaction with Al (Lahiri et al. 2012). Observation of the fracture surface of Al and Al–5 vol.% BNNT structure reveals signs of ductile fracture dominating in both cases (Figure 18.13a,b). Highly localized necking during fracture leads to the formation of thin high ridges on the fracture surface. High-magnification micrograph of composite fracture surface reveals protruding undamaged cylindrical BNNTs (Figure 18.13c), which become effective during strengthening.

18.4.3   Strengthening in Aluminum–BNNT Composite

BNNTs are expected to strengthen the Al matrix due to their excellent elastic modulus, tensile strength, and high flexibility. Micropillar compression testing was carried out to evaluate the strengthening in bulk Al–BNNT composite (Lahiri et al. 2013). Figures 18.14a and 18.15c show the SEM images of a representative focused ion beam (FIB) fabricated Al–BNNT micropillar before and after compression testing through the nanoindentation probe. Figure 18.14c shows a representative stress–strain curve for Al and Al–BNNT, derived from the load–displacement plot obtained during compression test. Initial linear part of the stress–strain curve shows higher elevation for Al–BNNT, indicating improvement in elastic modulus. Further, yield strength (YS) and compressive strength (CS) are also found enhanced in Al–BNNT. YS shows 54% improvement over Al, with a noted value of 88 MPa in composite structure. Similar improvement in CS is also reported with 142 and 216 MPa for Al and Al–5BNNT, respectively.

YS is related to the elastic deformation of the material, whereas CS is a combination of both elastic and plastic. Thus, BNNT reinforcement is found to manipulate both elastic and plastic deformation behavior of sintered Al matrix. Fracture surface makes an angle of ~45° with the loading direction (Figure 18.14c), which is the maximum shear stress direction in a polycrystalline structure. Thus, the deformation in Al–BNNT is dominated by slipping under shear stress, which indicates ductile behavior. The high-magnification micrographs of composite micropillar fracture surface shows protruded BNNT (Figure 18.14d) and BNNT bridge across a crack (Figure 18.14e). These features are indicators of load-bearing ability of BNNT and strengthening of the Al matrix. The *sword in sheathe* deformation of BNNT has also been observed as an additional strengthening mechanism by other researchers (Huang et al. 2007, Lahiri et al. 2011a, 2013, Yamaguchi et al. 2012). Effective sharing of load by the reinforcement restricts both elastic and plastic deformation of the matrix, leading to improvement in YS and CS. In addition, reaction products at the Al–BNNT interface are also expected to provide interfacial bonding, which helps in effective load transfer from matrix to reinforcement.

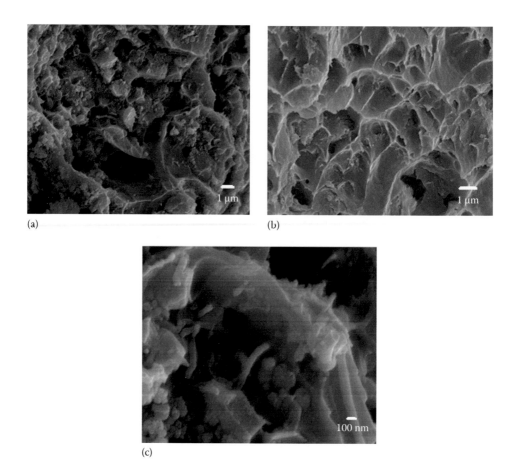

Figure 18.13 SEM micrograph of the fracture surfaces of (a) Al and (b) Al–5 vol.% BNNT indicating ductile failure. (c) Al–BNNT fracture surface shows traces of tubular-shaped BNNTs protruding out of the Al matrix. (Reprinted from *Mater. Sci. Eng. A*, 574, Lahiri, D., Hadjikhani, A., Zhang, C., Xing, T., Li, L.H., Chen, Y., and Agarwal, A., Boron nitride nanotubes reinforced aluminum composites prepared by spark plasma sintering: Microstructure, mechanical properties and deformation behavior, 149–156, Copyright 2013, with permission from Elsevier.)

18.4.4 Formability of Al–BNNT Composite

Formability is very important for any structural material to fabricate it to the desired intricate shape without forming cracks and defects in the structure. However, strengthening by different methods, including adding a reinforcement phase, often makes metallic structures less ductile. Thus, for BNNT to be a successful reinforcement to Al, the composite should not pose brittleness to the matrix. In order to evaluate the effect on ductility, Al and Al–BNNT pellets were subjected to 75% reduction in thickness via warm rolling (80°C) up to a thickness of ~150 μm (Figure 18.15a). Such heavy deformation at low temperature did not introduce any major crack or disintegration, revealing the ductility of the composite structure. Fracture surfaces perpendicular to the rolling direction show highly dense structure and signs of ductile failure (Figure 18.15b, c). High-magnification micrograph of composite fracture surface also

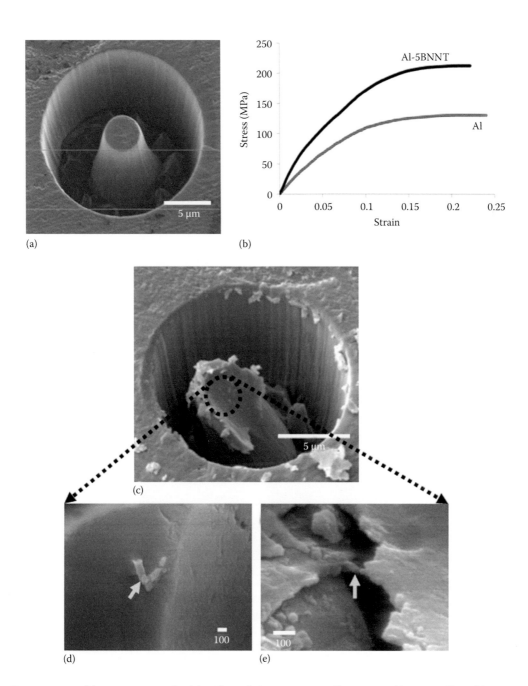

Figure 18.14 (a) SEM micrograph of the Al–5 vol.% BNNT micropillar prepared by FIB milling, (b) stress–strain plot obtained through compression tests of micropillar, and (c) micrograph of broken Al–5 vol.% BNNT micropillar. High-magnification micrographs of the same broken pillar show BNNTs (d) protruding out of the Al matrix and (e) forming a bridge across the crack. (Reprinted from *Mater. Sci. Eng. A*, 574, Lahiri, D., Hadjikhani, A., Zhang, C., Xing, T., Li, L.H., Chen, Y., and Agarwal, A., Boron nitride nanotubes reinforced aluminum composites prepared by spark plasma sintering: Microstructure, mechanical properties and deformation behavior, 149–156, Copyright 2013, with permission from Elsevier.)

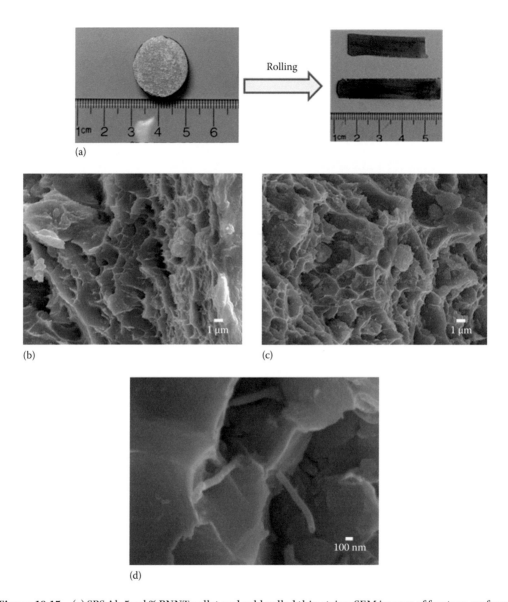

Figure 18.15 (a) SPS Al–5 vol.% BNNT pellet and cold-rolled thin strips. SEM images of fracture surfaces from the rolled (b) Al and (c, d) Al–5vol.% BNNT thin strips. The high-magnification image (d) reveals survival of BNNT even after heavy deformation. (Reprinted from *Mater. Sci. Eng. A*, 574, Lahiri, D., Hadjikhani, A., Zhang, C., Xing, T., Li, L.H., Chen, Y., and Agarwal, A., Boron nitride nanotubes reinforced aluminum composites prepared by spark plasma sintering: Microstructure, mechanical properties and deformation behavior, 149–156, Copyright 2013, with permission from Elsevier.)

reveals the presence of undamaged tubular BNNTs protruding from the matrix, which is possible due to high flexibility and fracture strain of BNNT (Golberg et al. 2007, Ghassemi et al. 2011). The survival of undamaged BNNTs in the rolled structure ensures retaining of strengthening capability of the composite after fabrication to final shape also. In addition, the rolling-induced deformation increases the hardness of Al–BNNT by 59%, as compared to 21% in Al.

This differential improvement in the hardness of deformed Al matrix with BNNT reinforcement is due to improved densification and bonding of BNNT with the Al matrix, which causes further strain hardening.

18.5 SUMMARY

BNNT is found to be an effective reinforcement for a wide variety of materials including polymer (PLC), ceramics (HA), and metals (aluminum). Excellent elastic modulus and strength of BNNTs can strengthen the polymer matrix composites, still retaining their ductility. On the other hand, fiber-type BNNT reinforcement toughens the ceramic base structural composites by restricting the propagation of crack, in addition to improvement in elastic modulus and wear resistance. The inert nature and high flexibility of BNNT allow high-temperature and high-pressure consolidation of ceramic and metal matrix composites. The inertness of BNNT also prevents excessive interfacial reaction with the matrix, which is critical for metal-based composites. BNNT-reinforced aluminum composite shows excellent formability indicating ease of manufacturing. The biocompatibility of BNNT also encourages its application as biomaterial-based composites for structural applications such as load-bearing implants and scaffolds.

REFERENCES

Agarwal, A., Bakshi, S.R., Lahiri, D. (2010) *Carbon Nanotubes Reinforced Metal Matrix Composites.* Boca Raton, FL: CRC Press, ISBN: 9781439811498.

Anstis, G.R., Chantikul, P.B., Lawn, R., Marshall, D.B. (1981) A critical evaluation of indentation techniques for measuring fracture toughness: I, Direct crack measurements. *Journal of the American Ceramic Society* 64: 533–538.

Arenal, R., Wang, M.-S., Xu, Z., Loiseau, A., Golberg, D. (2011) Young modulus, mechanical and electrical properties of isolated individual and bundled single-walled boron nitride nanotubes. *Nanotechnology* 22: 265704.

Arslan, G., Kara, F., Turan, S. (2003) Quantitative X-ray diffraction analysis of reactive infiltrated boron carbide–aluminium composites. *Journal of the European Ceramic Society* 23: 1243–1255.

Bai, X., Golberg, D., Bando, Y., Zhi, C., Tang, C., Mitome, M., Kurashima, K. (2007) Deformation-driven electrical transport of individual boron nitride nanotubes. *Nano Letters* 7: 632–637.

Bakshi, S.R., Agarwal, A. (2011) An analysis on the factors affecting strengthening in carbon nanotube reinforced aluminum composites. *Carbon* 49: 533–544.

Bakshi, S.R., Keshri, A.K., Singh, V., Seal, S., Agarwal, A. (2009) Interface in carbon nanotube reinforced aluminum silicon composites: Thermodynamic analysis and experimental verification. *Journal of Alloys and Compounds* 481: 207–213.

Bakshi, S.R., Lahiri, D., Agarwal, A. (2010) Carbon nanotube reinforced metal matrix composites—A review. *International Materials Review* 55: 41–64.

Balani, K., Anderson, R., Laha, T., Andara, M., Tercero, J., Crumpler, E., Agarwal, A. (2007) Plasma-sprayed carbon nanotube reinforced hydroxyapatite coatings and their interaction with human osteoblasts in vitro. *Biomaterials* 28: 618–624.

Bando, Y., Zhao, X., Shimoyama, H., Sakai, G., Kaneto, K. (1999) Physical properties of multiwalled carbon nanotubes. *International Journal of Inorganic Materials* 1: 77–82.

Bansal, N.P., Hurst, J.B., Choi, S.R. (2006) Boron nitride nanotubes-reinforced glass composites. *Journal of American Ceramic Society* 89: 388–390.

Deng, C.F., Wang, D.F., Zhang, X.X., Li, A.B. (2007) Processing and properties of carbon nanotubes reinforced aluminum composites. *Materials Science and Engineering A* 444: 138–145.

Causa, F., Netti, P.A., Ambrosi, L., Ciapetti, G., Baldini, N., Pagani, S., Martini, D., Giunti, A. (2006) Poly-ϵ-caprolactone/hydroxyapatite composites for bone regeneration: In vitro characterization and human osteoblast response. *Journal of Biomedical Materials Research A* 76A: 151–162.

Chen, X., Wu, P., Rousseas, M., Okawa, D., Gartner, Z., Zetll, Z.A., Bertozzi, C.R. (2009) Boron nitride nanotubes are noncytotoxic and can be functionalized for interaction with proteins and cells. *Journal of American Chemical Society Communication* 131: 890–891.

Chen, Y., Balani, K., Agarwal, A. (2008) Analytical model to evaluate interface characteristics of carbon nanotube reinforced aluminum oxide nanocomposites. *Applied Physics Letters* 92: 011916.

Cheung, H.Y., Lau, K.Y.T., Tao, X.M., Hui, D. (2008) A potential material for tissue engineering: Silkworm silk/PLA biocomposite. *Composites* B39: 1026–1033.

Choi, H., Shin, J., Min, B., Park, J., Bae, D. (2009) Reinforcing effects of carbon nanotubes in structural aluminum matrix nanocomposites. *Journal of Materials Research* 24: 2610–2616.

Choi, S.R., Bansal, N.P., Garg, A. (2007) Mechanical and microstructural characterization of boron nitride nanotubes-reinforced SOFC seal glass composite. *Materials Science and Engineering A* 460–446: 509–515.

Chopra, N.G., Zetll, A. (1998) Measurement of the elastic modulus of a multi walled boron nitride nanotubes. *Solid State Communication* 105(5) 297–300.

Ci, L., Ryu, Z., Jin-Phillipp, N.Y., Ruhle, M. (2006) Investigation of the interfacial reaction between multi-walled carbon nanotubes and aluminum. *Acta Materialia* 54: 5367–5375.

Ciofani, G. (2010) Potential applications of boron nitride nanotube as drug delivery systems. *Expert Opinions in Drug Delivery* 7: 889–893.

Ciofani, G., Raffa, V., Menciassi, A., Cuschieri, A. (2008) Cytocompatibility, interactions, and uptake of poly-ethyleneimine-coated boron nitride nanotubes by living cells: Confirmation of their potential for biomedical application. *Biotechnology and Bioengineering* 101: 850–858.

Ciofani, G., Raffa, V., Mencissia, A., Cuschieria, A. (2009) Boron nitride nanotubes: An innovative tool for nanomedicine. *Nanotoday* 4: 8–10.

Ciofani, G., Ricotti, L., Danti, S., Moscato, S., Nesti, C., D'Alessandro, D., Dinucci, D. et al. (2010) Investigation of interactions between poly-l-lysine-coated boron nitride nanotubes and C2C12 cells: up-take, cytocompatibility, and differentiation. *International Journal of Nanomedicine* 5: 285–298.

Du, M., Bi, J.-Q., Wang, W.-L., Sun, N.-N, Long, X.-L. (2012) Influence of sintering temperature on microstructure and properties of SiO_2 ceramic incorporated with boron nitride nanotubes. *Materials Science and Engineering A* 543: 271–276.

Du, M., Bi, J.-Q., Wang, W.-L., Sun, X.-L., Long, N.-N., Bi, Y.-J. (2011) Fabrication and mechanical properties of SiO_2–Al_2O_3–BNNPs and SiO_2–Al_2O_3–BNNTs composites. *Materials Science and Engineering A* 530: 669–674.

Gaona, M., Lima, R.S., Marple, B.R. (2007) Nanostructured titania/hydroxyapatite composite coatings deposited by high velocity oxy-fuel (HVOF) spraying. *Materials Science and Engineering A* 458: 141–149.

Ghassemi, H.M., Lee, C.H., Yap, Y.K., Yassar, R.S. (2011) In situ observation of reversible rippling in multi-walled boron nitride nanotubes. *Nanotechnology* 22: 115702.

Golberg, D., Bando, Y., Huang, Y., Terao, T., Mitome, M., Tang, C., Zhi, C. (2010) Boron nitride nanotubes and nanosheets. *ACS Nano* 4: 2979–2993.

Golberg, D., Bando, Y., Kurashima, K., Sato, T. (2001) Synthesis and characterization of ropes made of BN multiwalled nanotubes. *Scripta Materialia* 44: 1561–1565.

Golberg, D., Bando, Y., Tang, C., Zhi, C. (2007) Boron nitride nanotubes. *Advanced Materials* 19: 2413–2432.

Gu, Y.W., Loha, N.H., Khor, K.A., Tor, S.B., Cheang, P. (2002) Spark plasma sintering of hydroxyapatite powder. *Biomaterials* 23: 37–43.

Han, Y.F., Dai, Y.B., Wang, J., Shu, D., Sun, B.D. (2011) First-principles calculations on Al/AlB2 interfaces. *Applied Surface Science* 257: 7831–7836.

He, C., Zhao, N., Shi, C., Du, X., Li, J., Li, H., Cui, Q. (2007) An approach to obtaining homogeneously dispersed carbon nanotubes in Al powders for preparing reinforced Al-matrix composites. *Advanced Materials* 19: 1128–1132.

Huang, Q., Bando, Y., Xu, T., Nishimura, X., Zhang, C., Tang, C., Xu, F., Gao, L., Golberg, D. (2007) Enhancing superplasticity of engineering ceramics by introducing BN nanotubes. *Nanotechnology* 18: 485706.

Huang, X., Zhi, C., Jiang, P., Golberg, D., Bando, Y., Tanaka, T. (2013) Polyhedral oligosilsesquioxane-modified boron nitride nanotube based epoxy nanocomposites: An ideal dielectric material with high thermal conductivity. *Advanced Functional Materials* 23: 1824–1831.

Khor, K.A., Li, H., Cheang, P. (2003) Effect of spark plasma sintering on the microstructure and in vitro behavior of plasma sprayed HA coatings. *Biomaterials* 24: 2695–2705.

Kim, H.S., Chae, Y.S., Choi, J.H., Yoon, J.S., Jin, H.J. (2008) Thermal properties of poly(ecaprolactone)/multi-walled carbon nanotubes composites. *Advanced Composite Materials* 17: 157–166.

Kim, H.-W. (2007) Biomedical nanocomposites of hydroxyapatite/polycaprolactone obtained by surfactant mediation. *Journal of Biomedical Materials Research A* 83A: 169–177.

Kim, P., Shi, L., Majumdar, A., McEuen, P.L. (2001) Thermal transport measurements of individual multiwalled nanotubes. *Physical Review Letters* 87: 215502.

Kwon, H., Estili, M., Takagi, K., Miyazaki, T., Kawasaki, A. (2009) Combination of hot extrusion and spark plasma sintering for producing carbon nanotube reinforced aluminum matrix composites. *Carbon* 47: 570–577.

Laha, T., Agarwal, A., McKechnie, T., Seal, S. (2004) Synthesis and characterization of plasma spray formed carbon nanotube reinforced aluminum composite. *Materials Science and Engineering A* 381: 249–258.

Laha, T., Kuchibatla, S., Seal, S., Li, W., Agarwal, A. (2007) Interfacial phenomena in thermally sprayed multi-walled carbon nanotube reinforced aluminum nanocomposite. *Acta Materialia* 55: 1059–1066.

Lahiri, D., Hadjikhani, A., Zhang, C., Xing, T., Li, L.H., Chen, Y., Agarwal, A. (2013) Boron nitride nanotubes reinforced aluminum composites prepared by spark plasma sintering: Microstructure, mechanical properties and deformation behavior. *Materials Science and Engineering A* 574: 149–156.

Lahiri, D., Rouzaud, F., Namin, S., Keshri, A.K., Valdes, J.J., Kos, L., Agarwal, A. (2009) Carbon nanotube reinforced polylactide–caprolactone copolymer: Mechanical strengthening and interaction with human osteoblasts in vitro. *Journal of Applied Materials and Interfaces* 1: 2470–2476.

Lahiri, D., Rouzaud, F., Richard, T., Keshri, A.K., Bakshi, S.R., Kos, L., Agarwal, A. (2010a) Boron nitride nanotube reinforced polylactide–polycaprolactone copolymer composite: Mechanical properties and cytocompatibility with osteoblasts and macrophages in vitro. *Acta Biomaterialia* 6: 3524–3533.

Lahiri, D., Singh, V., Benaduce, A.P., Seal, S., Kos, L., Agarwal, A. (2011a) Boron nitride nanotube reinforced hydroxyapatite composite: Mechanical and tribological performance and in-vitro biocompatibility to osteoblasts. *Journal of the Mechanical Behavior of Biomedical Materials* 4: 44–56.

Lahiri, D., Singh, V., Keshri, A.K., Seal, S., Agarwal, A. (2010b) Carbon nanotube toughened hydroxyapatite by spark plasma sintering: Microstructural evolution and multiscale tribological properties. *Carbon* 48: 3103–3120.

Lahiri, D., Singh, V., Keshri, A.K., Seal, S., Agarwal, A. (2011b) Apatite formability of boron nitride nanotubes. *Nanotechnology* 22: 205601.

Lahiri, D., Singh, V., Li, L.H., Xing, T., Seal, S., Chen, Y., Agarwal, A. (2012) Insight into reactions and interfaces between boron nitride nanotube and aluminum. *Journal of Materials Research* 27: 2760–2770.

Li, L., Chen, Y., Stachurski, H. (2013) Boron nitride nanotube reinforced polyurethane composites. *Progress in Natural Sciences: Materials International* 23: 170–173.

Li, W., Qiao, X., Sun, K., Chen, X. (2008) Mechanical and viscoelastic properties of novel silk fibroin fiber/poly(e-caprolactone) biocomposites. *Journal of Applied Polymer Science* 110: 134–139.

Li, Y., Liang, H., Zhang, J. (2010) Fabrication and properties of AlN/carbon-doped boron nitride nanotubes composites. *Journal of Chinese Ceramic Society* 38: 1440–1444.

Ma, R., Bando, Y., Zhu, H., Saito, T., Xu, C., Wu, D. (2002) Hydrogen uptake in boron nitride nanotubes at room temperature. *Journal of American Chemical Society* 124: 7672–7623.

Marie, P.J. (2008) Transcription factors controlling osteoblastogenesis. *Archives of Biochemistry and Biophysics* 473: 98–105.

Montesa, C.M., Shibata, N., Tohei, T., Ikuhara, Y. (2011) TEM observation of liquid-phase bonded aluminum–silicon/aluminum nitride hetero interface. *Journal of Materials Science* 46: 4392–4396.

Nakahira, A., Eguchi, K. (2001) Evaluation of microstructure and some properties of hydroxyapatite/Ti composites. *Journal of Ceramic Processing Research* 2: 108–112.

Nurul, M.S., Mariatti, M. (2013) Effect of hybrid nanofillers on the thermal, mechanical, and physical properties of polypropylene composites. *Polymer Bulletin* 70: 871–884.

Pawlak, Z., Kaldonski, T., Pai, R., Bayraktar, E., Oloyede, A. (2009) A comparative study on the tribological behaviour of hexagonal boron nitride (h-BN) as lubricating microparticles—An additive in porous sliding bearings for a car clutch. *Wear* 267: 1198–1202.

Pawlak, Z., Pai, R., Bayraktar, E., Kaldonski, T., Oloyede, A. (2008) Lamellar lubrication in vivo and vitro: Friction testing of hexagonal boron nitride. *BioSystems* 94: 202–208.

Pérez-Bustamante, R., Gómez-Esparza, C.D., Estrada-Guel, I., Miki-Yoshida, M., Licea-Jiménez, L., Pérez-García, S.A., Martínez-Sánchez, R. (2009) Microstructural and mechanical characterization of Al–MWCNT composites produced by mechanical milling. *Materials Science and Engineering A* 502: 159–163.

Porter, D.A., Easterling, K.E. (2001) *Phase Transformation in Metals and Alloys*, 2nd edn. Cheltenham, U.K.: CRC Press.

Ravichandran, J., Manoj, A.G., Liu, J., Manna, I., Carroll, D.L. (2008) A novel polymer nanotube composite for photovoltaic packaging applications. *Nanotechnology* 19: 085712.

Rizzi, S.C., Heath, D.J., Coombes, A.G.A., Bock, N., Textor, M., Downes, S. (2001) Biodegradable polymer/hydroxyapatite composites: Surface analysis and initial attachment of human osteoblasts. *Journal of Biomedical Materials Research* 55: 475–486.

Sarkar, S.K., Youn, M.H., Oh, I.H., Lee, B.T. (2007) Fabrication of CNT reinforced HAp composites by spark plasma sintering. *Materials Science Forum* 534–536: 893–896.

Singhal, S.K., Srivastava, A.K., Pasricha, R., Mathur, R.B. (2011) Fabrication of Al-matrix composites reinforced with amino functionalized boron nitride nanotubes. *Journal of Nanoscience and Nanotechnology* 11: 5179–5186.

Suryavanshi, A.P., Yu, M.F., Wen, J., Tang, C., Bando, Y. (2004) Elastic modulus and resonance behavior of boron nitride nanotubes. *Applied Physics Letters* 84: 2527–2529.

Tang, C., Bando, Y., Huang, Y., Yue, S., Gu, C., Xu, F., Golberg, D. (2006) Fluorination and electrical conductivity of BN nanotubes. *Journal of American Chemical Society* 127: 6552–6553.

Terao, T., Zhi, C., Bando, Y., Mitome, M., Tang, C., Golberg, D. (2010) Alignment of boron nitride nanotubes in polymeric composite films for thermal conductivity improvement. *Journal of Physical Chemistry* 114: 4340–4344.

Wang, J., Shaw, L.L. (2009) Nanocrystalline hydroxyapatite with simultaneous enhancements in hardness and toughness. *Biomaterials* 30: 6565–6572.

White, A.A., Best, S.M., Kinloch, I.A. (2007) Hydroxyapatite-carbon nanotube composites for biomedical applications: A review. *International Journal of Applied Ceramic Technology* 4: 1–13.

Wu, J., Han, W.Q., Walukiewicz, W., Ager III, J.W., Shan, W., Haller, E.E., Zetll, A. (2004) Raman spectroscopy and time-resolved photoluminescence of BN and BxCyNz nanotubes. *Nano Letters* 4: 647–650.

Xiao, X.L., McCulloch, D.G., Mckenzie, D.R., Bilek, M.M.M. (2006) The microstructure and stability of Al/AlN multilayered films. *Journal of Applied Physics* 100: 013504.

Xiao, Y., Yan, X.H., Xiang, J., Mao, Y.L., Zhang, Y., Cao, J.X., Ding, J.W. (2004) Specific heat of single-walled boron nitride nanotubes. *Applied Physics Letters* 84: 4626–4628.

Xu, G., Du, L., Wang, H., Xia, R., Ming, X., Zhu, Q. (2008) Nonisothermal crystallization kinetics and thermomechanical properties of multiwalled carbon nanotubereinforced poly(e-caprolactone) composites. *Polymer International* 57: 1052–1066.

Xu, J.-J., Bai, Y.-J., Wang, W.-L., Wang, S.-R., Han, F.-D., Qi, Y.-X., Bi, J.-Q. (2012) Toughening and reinforcing zirconia ceramics by introducing boron nitride nanotubes. *Materials Science and Engineering A* 546: 301–306.

Yamaguchi, M., Pakdel, A., Zhi, C., Bando, Y., Tang, D.-M., Faerstein, K., Shtansky, D., Golberg, D. (2013) Utilization of multiwalled boron nitride nanotubes for the reinforcement of lightweight aluminum ribbons. *Nanoscale Research Letters* 8: 36.

Yamaguchi, M., Tang, D.-M., Zhi, C., Bando, Y., Sntansky, D., Golberg. D. (2012) Synthesis, structural analysis and in situ transmission electron microscopy mechanical tests on individual aluminum matrix/boron nitride nanotube nanohybrids. *Acta Materialia* 60: 6213–6222.

Yi, W., Lu, L., Zhang, D.L., Pan, Z.W., Xie, S.S. (1999) Linear specific heat of carbon nanotubes. *Physical Review B* 59: 9015–9018.

Yu, M.F., Laurie, O., Dyer, M.J., Moloni, K., Kelly, T.F., Rouff, R.S. (2000) Strength and breaking mechanism of multiwalled carbon nanotubes. *Science* 287: 637–640.

Yue, C., Liu, W., Zhang, L., Zhang, T., Chen, Y. (2013) Fracture toughness and toughening mechanisms in a (ZrB2–SiC) composite reinforced with boron nitride nanotubes and boron nitride nanoplatelets. *Scripta Materialia* 68: 579–582.

Zhao, J.X., Dai, B.Q. (2004) DFT studies of electro-conductivity of carbon-doped boron nitride nanotubes. *Materials Chemistry and Physics* 88: 244–249.

Zhi, C., Bando, Y., Tang, C. (2006) Boron nitride nanotube/polystyrene composites. *Journal of Materials Research* 21: 2794–2800.

Zhi, C., Bando, Y., Tang, C., Honda, S., Sato, K., Kuwahara, H., Golberg, D. (2005) Characteristics of boron nitride nanotube–polyaniline composites. *Angewandte Chemie* 44 7929–7932.

Zhi, C., Bando, Y., Terao, T., Tang, C., Kuwahara, H., Golberg, D. (2009) Towards thermoconductive, electrically insulating polymeric composites with boron nitride nanotubes as fillers. *Advanced Functional Materials* 19: 1857–1862.

Chapter 19

Preparation and Application of Long Boron Nitride Nanotubes

Yuanlie Yu, Hua Chen, Yun Liu, Lu Hua Li, and Ying (Ian) Chen

CONTENTS

19.1 INTRODUCTION

Since the discovery of the carbon nanotubes (CNTs) in 1991 (Iijima 1991), a rapid development has been achieved in the fabrication and application of CNTs. In 2004, single-wall CNTs with a length of 4 cm were successfully synthesized at a high growth rate of 11 μm/s by a catalytic chemical vapor deposition approach (Zheng et al. 2004). Such long nanotubes have many advantages and enable many new applications (Baughman et al. 2002). For example, long CNTs can be spun into meter-long fibers of a high Young modulus (Jiang et al. 2002). Long metallic nanotubes can be readily assembled into a microelectromechanical system (MEMS) or nanoscale semiconductor devices (Zhu et al. 2002). Structurally similar to CNTs, boron nitride nanotubes (BNNTs) exhibit excellent mechanical (Chopra and Zettl 1998; Suryavanshi et al. 2004; Golberg et al. 2007) and thermal (Sanchez-Portal and Hernandez 2002; Popov 2003; Wirtz and Rubio 2003) properties comparable to CNTs. On the other hand, BNNTs are electrically insulating and exhibit better chemical stability and higher resistance to oxidation at high temperatures (Rubio et al. 1994; Chen et al. 2004) than CNTs. These unique properties make BNNTs very attractive for reinforcing composites and serving as insulator parts in 3D CNT-based MEMS. In this chapter, we brief the fabrication of over 1.0 mm long BNNTs through an optimized ball milling and annealing process, and demonstrate their potential applications in humidity-sensing unit, field-emission devices, and the treatment of oily wastewater.

19.2   FABRICATION AND CHARACTERIZATION OF LONG BNNTs

19.2.1   Fabrication of Long BNNTs

Long BNNTs were synthesized using an optimized ball milling and annealing process. In a typical process, amorphous boron (B) powder was first loaded into a stainless steel milling jar with steel balls (25.4 mm in diameter AISI 420) under NH_3 atmosphere at 300 kPa. The weight ratio of the steel balls to B powder is 50:1. The ball milling process was performed for 50 h at a rotation speed of 300 rpm. This milling process introduces about 1.5 at% iron nanoparticles homogeneously to the B powder. The milled powder was then annealed at 1100°C in a tube furnace in an N_2 flow of 100 mL/min. After annealing for 15 h, a large amount of long BNNTs was produced.

19.2.2   Characterization

The structure of the as-synthesized samples was analyzed using x-ray powder diffractometer (XRD) (Philips 3020, Co target, 40 kV, 30 mA). Field-emission scanning electron microscopy (FESEM, Hitachi 4500) was used to characterize the sample morphology. Transmission electron microscope (TEM, Philips CM300) was employed to examine individual structures. Thermal gravimetric analyzer (TGA) (Shimadzu TGA-50, N_2, 20°C/min) was used to monitor nitriding reactions.

19.2.3   Results and Discussion

The product after the annealing contained a high density of BNNTs with diameters in the range of 50–200 nm, as shown in the FESEM image in Figure 19.1a. Due to their great length, the nanotubes tended to interweave together. In order to reveal the length of the nanotubes, an individual nanotube was pulled out from the product using very sharp tweezers and placed on the surface of a silicon substrate under a stereoscopic microscope. As shown in the FESEM images in Figure 19.1c, the BNNT was found to be over 1.0 mm long. It should be mentioned that the original length of this tube might be even longer, as it is possible that the BNNT was broken during the pulling-out process. The interior structure of the as-synthesized BNNTs was determined by TEM. The TEM image in Figure 19.1b reveals a bamboo-like structure. According to the high-resolution TEM (HRTEM) image inserted in Figure 19.1b, the intrinsic plane spacing of the BNNT walls is calculated to be about 0.34 nm, corresponding to that of (002) plane in h-BN. It is noteworthy that Fe particles, acting as catalyst, were found embedded in the tips of the nanotubes (Chen et al. 1999; Yuan et al. 2003). The existence of Fe was also confirmed by XRD analyses. The XRD pattern in Figure 19.2 shows six diffraction peaks associated to (002), (100), and (004) planes of h-BN, (110) and (200) planes of α-Fe, as well as (111) plane of γ-Fe phases. The width of the (002) diffraction peak is broader than that of the bulk h-BN particles owing to the small size effect of the BNNTs.

The long annealing or growth time is very critical for the formation of long BNNTs, which was revealed by the parallel experiments conducted. The growth time and the corresponding length of the BNNTs at 1100°C are outlined in Table 19.1. The growth rate at different growth times (1, 2, 4, and 15 h) can be estimated. At the beginning of the growth stage (first 4 h), the growth rate is about 34.7 nm/s, which is higher than that of the further growth stage of up to 15 h. In other words, the growth rate decreases with increasing growth time. This interesting phenomenon therefore brings about the following question: what are the critical parameters controlling the formation of long BNNTs?

To answer this question, the growth mechanism needs to be examined first. The special bamboo-structured BNNTs are formed via metal catalytic growth mechanism (Zhang et al. 2001; Chadderton and Chen 2002; Ma et al. 2002; Tang et al. 2002; Velazquez-Salazar et al. 2005;

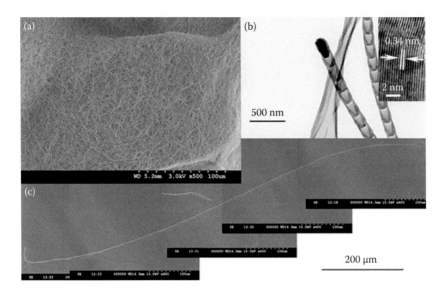

Figure 19.1   (a) FESEM image of as-synthesized BNNTs with a high yield, (b) TEM image revealing a bamboo-type structure (inset is the corresponding HRTEM image), and (c) FESEM image showing an individual BNNT over 1.0 mm long. (Reprinted from *Chem. Phys. Lett.*, 463, Chen, H. et al., Over 1.0 mm-long boron nitride nanotubes, 130–133, Copyright 2008, with permission from Elsevier.)

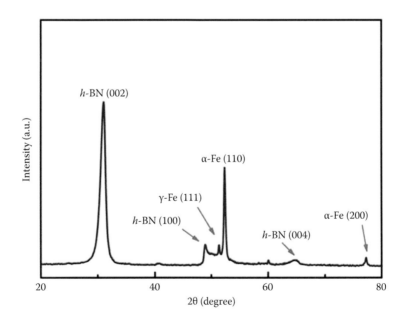

Figure 19.2   X-ray powder diffraction pattern of the produced BNNTs. (Reprinted from *Chem. Phys. Lett.*, 463, Chen, H. et al., Over 1.0 mm-long boron nitride nanotubes, 130–133, Copyright 2008, with permission from Elsevier.)

TABLE 19.1   THE GROWTH TIME AND CORRESPONDING LENGTH OF THE AS-SYNTHESIZED BNNTs

Sample No.	Growth Time (h)	Length (mm)
1	1	0.13
2	2	0.31
3	4	0.50
4	15	1.05

Oku et al. 2007), in which the nanotubes grow out from the milled B powders via a nitriding reaction to convert B to BN during the annealing process (Yu et al. 2005). In detail, the B atoms firstly diffuse into the Fe particles; simultaneously the N_2 is decomposed to N atoms on the surface of the Fe particles and diffuse into them. Then, the diffused B and N react inside the Fe nanoparticles forming h-BN. Because the austenite iron can accommodate only limited amounts of B (0.06 at%) (Villars et al. 1995) within the annealing temperature range, the supersaturated BN starts to precipitate layer by layer on the surfaces of Fe nanoparticles to form the BNNTs. Such BN dissolution and precipitation catalytic process to grow BNNTs is supported by the high-purity BNNT product as shown in Figure 19.1a, because direct reaction between B and N_2 normally requires a much higher temperature (Grieveson and Turkdogan 1964).

To further prove this growth mechanism, TGA analyses have been performed to show the nitriding reaction of the ball-milled B (15.20 mg) powder under N_2 atmosphere by monitoring the sample weight changes as a function of temperature (Figure 19.3). The weight increases with the increased temperature can be attributed to three processes: the absorption of N_2, the dissolution of N in the Fe nanoparticles, and the formation of BN. Additionally, the fastest reaction rate is found at around 1050°C, as indicated by the peak in the differential curve (dashed line). Isothermal annealing in this temperature range guarantees a fast formation of BN phase, which is an important factor for the growth of long BNNTs.

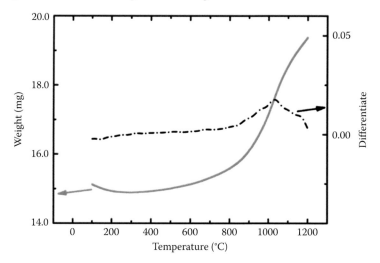

Figure 19.3   TGA and associated differential curves, presenting the weight increase as a function of temperature. (Reprinted from *Chem. Phys. Lett.*, 463, Chen, H. et al., Over 1.0 mm-long boron nitride nanotubes, 130–133, Copyright 2008, with permission from Elsevier.)

The ratio of B and N (B/N) dissolved in Fe nanoparticles increases dramatically with the increase in temperature, according to the B–Fe–N phase diagrams at 950°C, 1050°C, and 1150°C (Villars et al. 1995). At 1130°C, the B/N ratio is close to one; however, if the temperature is too high (i.e., 1300°C), the nanosized Fe particles would melt and merge together to form large particles, which impede the nanotube growth. In addition, the B particles would also react directly with N_2 to form BN particles instead of BNNTs at this temperature (Yu et al. 2006). On the other hand, if the temperature is too low (i.e., 900°C), the low diffusion rate of the B and N in Fe would slow down the precipitation rate of BN layers and thus lower growth rate of BNNTs. Therefore, the selected annealing temperature of 1100°C provides two essential parameters for the formation of long BNNTs: fast N dissolution rate and suitable B/N ratio (about 1) in Fe catalytic particles ensuring a high formation rate of BN phase. The TGA analyses show a similar decreased growth rate with the increase in annealing time. At 1100°C, the sample shows a very high weight increase rate of 9.8×10^{-3} mg/s, which could be assigned to the fast growth rate of nanotubes in the beginning stage of growth. Lower growth rates observed in further growth indicates that the growth process of the long BNNTs is certainly not limited to the reaction rate but might be related to the slow precipitation rate of BN layers to form the special bamboo-like structure. Noticeably, the diameter distribution of the as-synthesized BNNTs almost remains unchanged during the longitudinally growth of BNNTs. The growth process of the long BNNTs is therefore most likely limited by the slow formation process of BN layers on the surface of Fe nanoparticles. With the increase in the curvature of BN layers, the Fe nanoparticles were sucked into the nanotube by capillary effect (Chadderton and Chen 2002; Chen et al. 2006), as shown in the TEM image in Figure 19.1b. As a result, the contact area among Fe, B, and N decreases. The long bamboo-like BNNTs are formed through repeating the earlier processes many times, which obviously requires long growth/annealing time.

19.3 ELECTRICAL CONDUCTIVITY OF BNNTs

The successful synthesis of long BNNTs (over 1 mm) makes it possible to manipulate and assemble individual BNNTs into devices under an optical microscope. This section reviews prototype BNNT devices assembled under an optical microscope and measurement of electrical conductivity of the long BNNTs with or without gold decoration.

A testing system has been designed to investigate the conductivity of a long bamboo-like BNNT, as shown in Figure 19.4. A single long BNNT is fixed between two nickel (Ni) electrodes with silver paste. The electric conductance of the BNNT is measured using the two-point configuration.

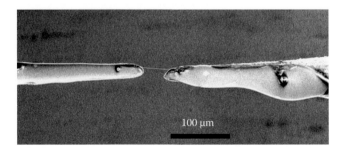

100 μm

Figure 19.4   FESEM image of a single BNNT connected on a pair of Ni electrodes. (Reprinted from *Chem. Phys. Lett.*, 463, Chen, H. et al., Over 1.0 mm-long boron nitride nanotubes, 130–133, Copyright 2008, with permission from Elsevier.)

It is well known that bulk h-BN has a very high resistivity in the order of 10^{14} $\Omega \cdot cm$, but BNNTs possess a relatively low resistivity of 300 $\Omega \cdot cm$ (Tang et al. 2005). The big difference between them in electrical conducting needs to be clarified. Traditional testing methods performed by two- or four-probe setup on the SiO_2 substrate work well on the measurement of electrical properties of CNTs (Yao et al. 1999), but they are not suitable for insulating BNNTs because influence of the SiO_2 substrate to the measurement of resistance is not negligible as BNNTs have comparable insulating property with the SiO_2 substrate. Therefore, a new setup without touching SiO_2 substrate is assembled using a long BNNT to measure the electrical resistance of individual BNNTs (Figure 19.4). Two Ni electrodes were mounted onto an alumina ceramic plate with a gap of 10 mm for a good insulation. The tips of the Ni electrodes were bent close to each other to avoid the influence of the resistance from the holder for reliable measurements. The individual BNNTs were fixed between the two tips of electrodes by conductive silver paste (DuPont), and the electrical contact was further improved by heat treatment of the system at 400°C in the air for 10 min, ensuring an ohmic contact between the BNNTs and the Ni electrodes (Kimura et al. 2001). As shown in Figure 19.4, the length of the mounted BNNT between two electrodes is about 47 µm, and the diameter is around 157 nm. $I–V$ characteristics of the nanotubes were measured utilizing a pA meter/DC voltage source (HP 4140B). Figure 19.5 shows the $I–V$ characteristic of the background (open circuit or the signal from the sample holder) and the individual BNNT. The resistance of this BNNT is 1.9×10^{12} Ω, two orders of magnitude smaller than that of the background (1.3×10^{14} Ω). The corresponding electrical resistivity of the BNNT can be calculated to be about 7.8×10^{4} $\Omega \cdot cm$. Additionally, the $I–V$ curve of the BNNT presents a linear relation, confirming an ohmic contact between the BN and electrodes. The average value of the measured resistivity from eight BNNTs was $(7.1 \pm 0.9) \times 10^{4}$ Ω cm.

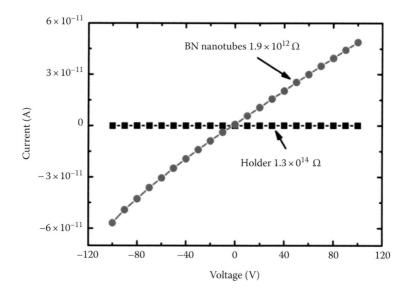

Figure 19.5 The $I–V$ curves of the sample holder and an individual BNNT. (Reprinted from *Chem. Phys. Lett.*, 463, Chen, H. et al., Over 1.0 mm-long boron nitride nanotubes, 130–133, Copyright 2008, with permission from Elsevier.)

The control over the electrical conductivity of the BNNTs is of great importance for various applications. Previous work has demonstrated that the resistivity of the BNNTs could decrease to 0.2–0.6 $\Omega \cdot$ cm by fluorine doping (Tang et al. 2005). We investigated the influence of surface modification on the electrical conductivity of the BNNTs by sputtering Au nanoparticles on the surface of the BNNTs by a DC magnetron sputter (Emitech K575X) in Ar atmosphere (10 Pa) with a sputtering current of 5 mA for a short time. The ceramic holder was fully covered during sputtering to avoid Au deposition. Figure 19.6 shows the TEM images of the Au-decorated BNNTs (Au-BNNTs) with different sputtering times of 20, 40, 80, and 120 s, referring to Au20, Au40, Au80, and Au120, respectively. Figure 19.6 clearly shows that the size of Au spots increases with the increase in sputtering time. The Au particles on Au20 and Au40 are separated from each other. The adjacent Au spots of Au80 start to connect and form islands, as revealed in Figure 19.6c. When the sputtering time further increases, the Au islands further expand, and many of them are interlinked (see Figure 19.6d) to form larger islands. The inset in Figure 19.6b is an HRTEM image of an Au spot showing an interplanar spacing of approximately 0.24 nm, which is identical to the interplanar spacing of Au (111) planes.

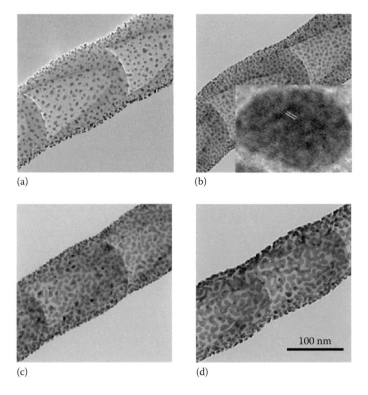

(a)   (b)

(c)   (d)

Figure 19.6   TEM images of Au-BNNTs under different sputtering time. (a) Small Au spots were obtained at 20 s. (b) The size of the Au spots increased at 40 s; inset TEM image shows an Au spot with the interplanar spacing of 0.24 nm referring to Au (111) planes. (c) Some Au spots combined together to form Au islands at 80 s. (d) Au islands were connected to form quasi-continuous Au coating at 120 s. (Reprinted with permission from Chen, H. et al., Nano Au-decorated boron nitride nanotubes: Conductance modification and field-emission enhancement, *Appl. Phys. Lett.*, 92, 243105, 2008, Copyright 2008, American Institute of Physics.)

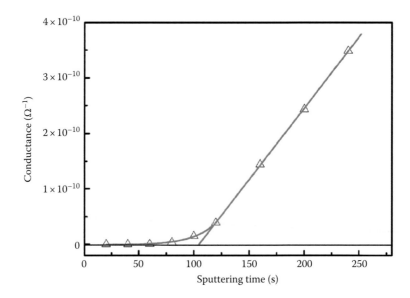

Figure 19.7 Conductance of the Au-BNNTs with respect to different sputtering times. After 120 s, there is a linear relationship between the conductance and sputtering time. (Reprinted with permission from Chen, H. et al., Nano Au-decorated boron nitride nanotubes: Conductance modification and field-emission enhancement, *Appl. Phys. Lett.*, 92, 243105, 2008, Copyright 2008, American Institute of Physics.)

Figure 19.7 shows the conductance of Au-BNNTs (Ω^{-1}) with respect to the sputtering time (seconds), which clearly shows that the relationship between the conductance and the sputtering time can be fitted perfectly with an exponential function at the sputtering time less than 120 s. However, once the sputtering time is over 120 s, the conductance linearly increases with the sputtering time. Figure 19.8 shows the *I–V* curves of Au40, Au80, and Au120 in conductance measurements in which the linear relationship is obtained for all of the samples suggesting an ohmic contact between the individual nanotube and electrodes. As the sputtering time exceeded 120 s, a quasi-continuous Au layer formed. Consequently, the metallic conductance of the decorated Au layer became dominant. Therefore, after the formation of the continuous Au layers, the prolongation of the sputtering time can lead only to a linear rise in conductance. However, if the Au coverage is below a critical fraction, the conductance of the Au-BNNTs is strongly dependent on that of the pure BNNTs based on the percolation theory (Sykes and Essam 1964). Here, we also calculate the critical fraction according to the same theory that relates to the Au coverage sputtered around 105 s.

19.4   Au-BNNTs AS FIELD EMITTERS

As an important application, reliable field-emission characteristics have been generated from BN thin films (Sugino et al. 2003; Luo et al. 2004), BNNTs (Cumings and Zettl 2004), BN nanorods (Zhang et al. 2007), and BN-coated Si tip (Sugino et al. 1997). The tubular shape and the surface negative electron affinity (NEA) (Powers et al. 1995; Loh et al. 1999) of BNNTs make them promising candidates as field emitters due to the increased field enhancement factor.

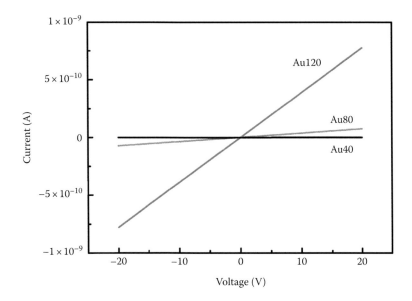

Figure 19.8   *I–V* curves of Au40, Au80, and Au120. The linear relationship suggests ohmic contact between the single nanotube and electrodes for all of samples. (Reprinted with permission from Chen, H. et al., Nano Au-decorated boron nitride nanotubes: Conductance modification and field-emission enhancement, *Appl. Phys. Lett.*, 92, 243105, 2008, Copyright 2008, American Institute of Physics.)

Carbon-doped BNNTs have been illustrated in the prospect to lower the turn-on fields and to subsequently increase the field-emission current densities. In Section 19.3, we describe that the decoration of Au on the surface of BNNTs can effectively improve the electrical conductivity of the BNNTs, and the improved electrical conductivity of BNNTs may lead to enhanced field-emission properties.

To investigate the field-emission properties of the pure and Au-BNNTs (Au40 and Au80), the long bamboo-like BNNTs with and without Au coating were randomly disposed on an Si substrate as the cathode. A metal plate was utilized as the anode 100 μm away from the cathode. The measurement was carried out in a vacuum chamber with an ultimate pressure of 5×10^{-7} Pa. The emission current was measured by a Keithley 485 picoammeter.

Figure 19.9 outlines the emission current density (*J*) of the pure and Au-BNNTs as a function of the macroscopic electric field (E_{mac}) applied on the samples. Here, $E_{mac} = V/d$, where *V* and *d* stand for the cathode–anode voltage and distance, respectively. The turn-on fields of pure BNNTs, Au40, and Au80 (arbitrarily defined as the electrical field resulting in an emission current density of 10 nA/cm²) were 18, 11, and 3.9 V/μm, respectively. The results well demonstrate that the Au decoration can significantly decrease the turn-on field of the BNNT emitters. Under the same macroscopic electrical field, the field-emission current density of Au40 was two orders higher than that of the pure BNNTs. Similarly, the emission current density of Au80 was about two orders higher than that of Au40. These results indicate that the field-emission properties of BNNTs can be dramatically improved by the Au decoration.

Fowler–Nordheim (FN) theory (Fowler and Nordheim 1928), which describes the electron tunneling through an interface barrier, has been widely used to interpret the mechanism of electron

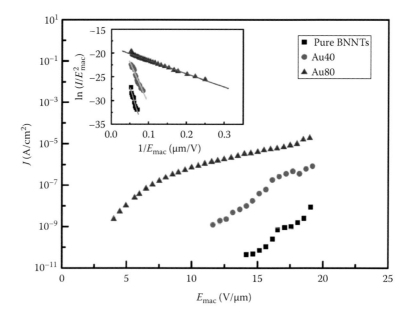

Figure 19.9 Emission current densities of pure and Au-decorated BNNTs as a function of E_{mac}. The emission current density of the pure BNNTs is unsaturated at high field. The emission current density of Au40 increases by almost two orders, and it remains in the unsaturated state under the same macroscopic electrical field. For Au80, the current density saturation appears at the high electrical field range. The inset shows corresponding FN plots in which the Au80 has a significant variation in the slope. (Reprinted with permission from Chen, H. et al., Nano Au-decorated boron nitride nanotubes: Conductance modification and field-emission enhancement, *Appl. Phys. Lett.*, 92, 243105, 2008, Copyright 2008, American Institute of Physics.)

field emission. In the FN model, the J of the emitter is expressed as a function of the tip work function (Φ) and the local electrical field at the emitter surface (E_{loc}) (Collins and Zettl 1997):

$$J \propto E_{loc}^2 \exp(-6.8 \times 10^{10} \Phi^{3/2}/E_{loc}) \tag{19.1}$$

where E_{loc}, Φ, and J have the units of V/μm, eV, and A/cm^2, respectively. Marcus et al. (1990) investigated the relationship between E_{mac} and E_{loc} for an emitter of hemispherical morphology. When the distance between the emitter and the anode d is much greater than the radius of the emitter tip (r_{tip}), E_{loc} is reciprocally proportional to r_{tip}, ($E_{loc} = V/r_{rip}$). A factor α (≥ 1) is introduced where $d \gg r_{tip}$ is ungratified. In this case, E_{loc} could be expressed as

$$E_{loc} = V/(\alpha r_{tip}) = d/\alpha r_{tip} \times E_{mac} \tag{19.2}$$

Modifying Equation 19.1 by introducing Equation 19.2 and changing current density J to current I, the FN equation can be expressed as

$$\ln(I/E_{mac}^2) = (1/E_{mac})[-6.8 \times 10^{10} (\alpha r_{tip}/d) \Phi^{3/2}] + \text{offset} \tag{19.3}$$

This equation predicts a linear relation between $\ln(I/E_{mac}^2)$ and $1/E_{mac}$, that is, FN plot. For our particular configuration and measurement condition, r_{tip} and d are ensemble averages of randomly oriented BNNTs (Zhang et al. 2007). Because of the large quantity of nanotubes and the high uniformity of the morphology, it can be safely assumed that the values of r_{tip} and d of the measured samples are very close to each other. Hence, variations in the work function of the samples are evaluated through the slopes (S_{FN}) of the FN plots. $S_{FN} = -6.8 \times 10^{10} \, (\alpha r_{tip}/d)\Phi^{3/2} = -A\Phi^{3/2}$, A is a geometric factor and keeps constant for all our samples. The inset of Figure 19.9 is the FN plots of our samples. The work function Φ (~6 eV) and the S_{FN} of pure BNNTs can be used to acquire the constant A. Then, the work functions Φ of Au40 and Au80 are estimated to be 4.8 and 1.4 eV. The work function of Au40 approaches the value of pure gold (4.8 eV) (Anderson 1959). For Au80, the smaller work function suggests formation of a much smaller surface barrier. This indicates that Au possibly diffused into the BNNTs during the deposition so that a surface of NEA might be formed in Au80 as observed on the Au–Al_2O_3 interface (Banan-Sadeghian et al. 2008). However, a detailed investigation is required to further clarify the mechanism of the dramatic decrease in the work function of Au80. Nevertheless, the results suggest that Au decoration can significantly improve the field-emission characteristics of BNNTs by decreasing the work function.

Unlike the field emission of CNTs, where the current saturation happens under high electrical fields (>~5 V/μm) (Klinke et al. 2005), the emission current density of pure BNNTs did not saturate even at the much higher electrical field (see Figure 19.9). For CNT emitters, Zhong et al. (2002) addressed that the current saturation occurred at a large emission current because the contact resistance between the substrate and CNTs obstructed the electron transport to the emitting sites. The current saturation of Au80 may arise from a similar mechanism due to a reduced contact resistance. It can be concluded from the field-emission tests that the conductance of Au-BNNTs increases with the increase in sputtering time. More importantly, the Au decoration can modify the work function of the BNNTs, and as a consequence, the field-emission current densities of Au-BNNTs are significantly enhanced. The turn-on field of such Au-BNNTs is reduced to one-third, and the emission current density is increased by four orders compared to pure BNNTs.

19.5   LONG BNNTs AS HUMIDITY-SENSING UNIT

Humidity monitoring and measurement play a more and more important role in our lives, because many of civil and industrial fields, such as quality management, electronics, plant cultivation, household appliances, drug manufacturing, agriculture, weather forecast, and automobiles, need to control the environmental humidity (Yamazoe and Shimizu 1986; Kuang et al. 2007; Zhang et al. 2010). Recently, many 1D nanomaterials have been selected to assemble humidity sensors, such as SnO_2 nanowires (Kuang et al. 2007), Al_2O_3 nanowires (Cheng et al. 2011), CNTs (Huang et al. 2007), TiO_2 nanotubes (Zhang et al. 2008), ZnO_2 nanowires (Wang et al. 2004), and polymer nanofibers (Lin et al. 2012) due to their relatively large specific surface area and exceptional surface properties. Because of the complexity and multiplicity of the environment, it is then a general requirement to develop different humidity-sensing materials that can be used in a wild environment. BNNTs are attractive due to their special chemical and physical stability. In particular, the successful synthesis of long BNNTs (over 1 mm) as well as the improvement of electrical conductivity of BNNTs by Au decoration makes it possible to manipulate and assemble individual BNNTs into devices under an optical microscope (Chen et al. 2004).

In this section, we will discuss the humidity-sensing performance of an individual long BNNT and Au-decorated BNNT through measuring their resistance relative to the relative humidity (RH), including the analysis of corresponding response and recovery characteristics. In addition, based on the analyses of the conductance change of BNNTs and Au-BNNTs at different RH levels, the sensing mechanisms will also be discussed.

The testing system for the humidity-sensing measurement is similar to that described in Section 19.3, which was initially designed to measure the conductance of bamboo-like BNNTs (Figure 19.4). Au nanoparticles were deposited on the surface of the BNNT by a DC magnetron sputter (Emitech K575X) in Ar atmosphere (10 Pa) with a sputtering current of 5 mA and a sputtering time of 20 s. In order to avoid Au depositing on the ceramic holder, the holder was fully covered during the sputtering. The electrical features of the individual BNNT with and without Au decoration were tested as a function of relative RH with a Keithley 6517 LCR analyzer in a home-built testing chamber illustrated in Figure 19.10.

In the humidity tests, saturated aqueous solutions of $ZnBr_2$, $MgCl_2$, NaBr, KI, NH_4Cl, KCl, and $Pb(NO_3)_2$ were selected to control the RH level at room temperature. Because the humidity of the saturated aqueous solutions depended on the environmental temperature, the RH values of saturated aqueous solutions were double-checked with a humidity meter (Lutron HT-3015HA) before measurements. The BNNT and Au-BNNT were suspended in the sealed chamber without contacting the solution.

Figure 19.11a and c show the SEM images of individual BNNT and Au-BNNT in as-assembled test units whereat the BNNTs are hung over two Ni electrodes. The closed views of the corresponding BNNT (Figure 19.11b) and Au-BNNT (Figure 19.11d) show that the diameters of both BNNTs are about 200 nm. The TEM and AFM images inserted in Figure 19.11d indicate that isolated Au nanoparticles with a diameter in the range of 5–20 nm were deposited onto the surface of the BNNT.

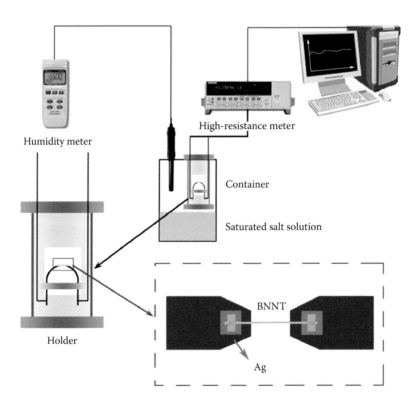

Figure 19.10   A schematic drawing of the humidity-sensing test system and of BNNT test unit. (Reprinted from *Electrochem. Commun.*, 30, Yu, Y. et al., Humidity sensing properties of single Au-decorated boron nitride nanotubes, 29–33, Copyright 2013, with permission from Elsevier.)

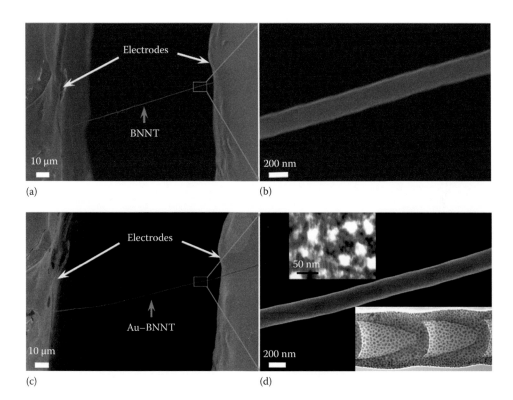

Figure 19.11   (a, c) The SEM images of typical test units with n individual BNNT and Au-BNNT. (b, d) The corresponding higher-magnification SEM images in (a) and (c) marked with square [insets in (d): AFM (up) and TEM (down) images of the Au-BNNT]. (Reprinted from *Electrochem. Commun.*, 30, Yu, Y. et al., Humidity sensing properties of single Au-decorated boron nitride nanotubes, 29–33, Copyright 2013, with permission from Elsevier.)

Figure 19.12a shows the typical current–voltage (*I–V)* curves of the container, the holder, and the test units in the relatively static RH = 85.0% atmosphere at room temperature. Noticeably, ohmic contacts are obtained for both BNNT and Au-BNNT test units in the voltage range of –10 to +10 V. The resistance of the Au-BNNT (3.3×10^7 Ω) is much smaller than that of the intrinsic BNNT (1.0×10^8 Ω), container (4.0×10^{10} Ω), and holder (2.0×10^8 Ω), indicating an improvement of the conductance of BNNTs by Au decoration. Lin et al. (2011) have demonstrated that water could destabilize the B–N bonds of bulk hexagonal boron nitride (*h*-BN) through O atoms attacking B–N and accelerating their hydrolysis to form layered *h*-BN nanosheets under sonication without surfactant or organic functionalization. Similar B–N bonds also envelope the outer surface of BNNTs. So it is reasonable to assume that H_2O molecules could easily absorb on the surface of BNNTs through B–O bonds to form a H_2O absorption layer (Figure 19.13a). Especially, the bamboo-like BNNTs have a lot of suspending bonds on the surface or open edges of the basal planes along the tube, making the nucleophilic attacking much easier to occur. Once the water molecules were adsorbed on the surface of BNNTs, the electrical conductance can be improved due to the polar structure of H_2O molecules (Lin et al. 2011). Therefore, based on the aforementioned principle, the BNNTs can be used as humidity sensors. Figure 19.12b outlined the humidity-sensing performances of

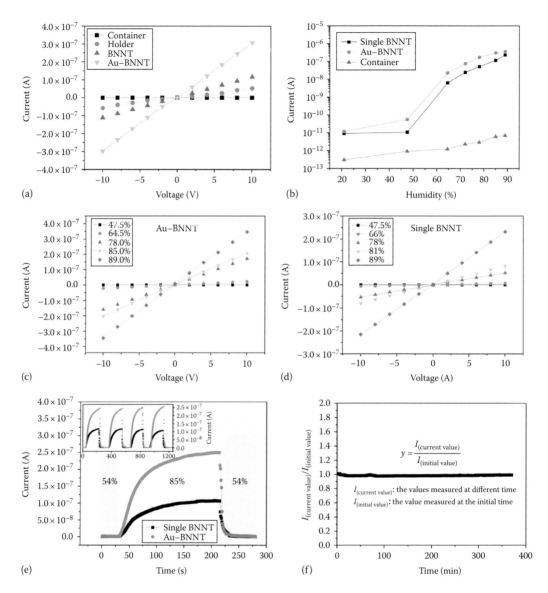

Figure 19.12   (a) The *I–V* curves of the container, holder, a single BNNT, and Au-BNNT in the static RH = 85.0% atmosphere at room temperature. (b) The *I–H* curves of the container, a single BNNT, and Au-BNNT. Bias voltage of 10 V is applied during the test. (c) *I–V* curves of a single Au-BNNT in different RH levels from 47.5% to 89.0% RH measured at room temperature. (d) *I–V* curves of a single BNNT in different static RH levels from 47.5% to 89.0% RH measured at room temperature. (e) Response and recovery characteristics of a single BNNT- and Au-BNNT-based humidity-sensing units measured by switching the RH levels between RH = 54.0% and 85.0% (Inset is four cycles). (f) The curve of $I_{\text{(current value)}}/I_{\text{(initial value)}}$ of Au-BNNT at 90.0% RH measured at room temperature (Applied voltage: 10 V). (Reprinted from *Electrochem. Commun.*, 30, Yu, Y. et al., Humidity sensing properties of single Au-decorated boron nitride nanotubes, 29–33, Copyright 2013, with permission from Elsevier.)

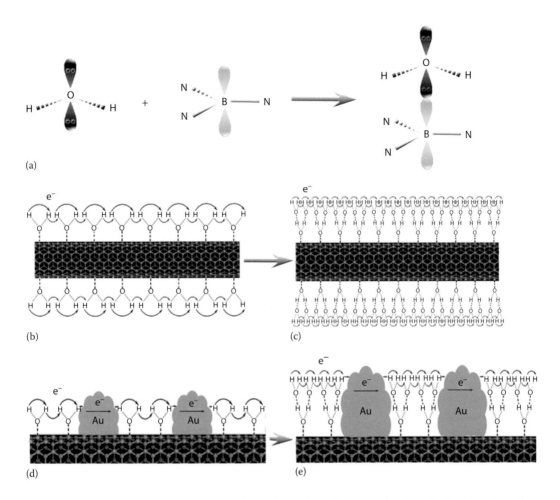

Figure 19.13 Schematic of electrons transfer on the surface of water molecules adsorbed BNNTs and Au decorated BNNTs: (a) Interaction of H_2O molecules and B atoms in BNNTs, (b) and (c) transfer of electrons on the surface of water molecules adsorbed BNNTs, (d) and (e) transfer of electrons on the surface of water molecules adsorbed Au decorated BNNTs.

the BNNT and Au-BNNT. It can be seen that the conductance for both BNNT and Au-BNNT varied slightly in the lower RH levels from 20% to 50%. This can be ascribed to a low degree of adsorption of water molecules as shown in Figure 19.13b, because only a small amount of water molecules can reach the surface of BNNTs at low RH. As shown in Figure 19.13c, more H_2O molecules can reach the surface of BNNTs with the increase in humidity levels, and additional H_2O molecules can be further adsorbed through a hydrogen bond on the surface of the chemisorbed layer (Lin et al. 2011). Correspondingly, the electrical current response was enhanced due to the existence of such hydrogen bonds, as shown in Figure 19.12b, that the conductance response increased significantly between 50% and 90% RH.

Because of the better conductance of Au-BNNTs than that of the pristine BNNTs, the Au-BNNTs are expected to have a better performance in environmental RH sensing. Figure 19.12c and d compares the electrical response properties of the BNNT- and Au-BNNT-based sensing elements in different static RH atmospheres of 47.5%–89.0% at room temperature. Because the electrical conductivity of Au is much better than that of H_2O, the resistance of Au can be ignored compared with that of H_2O in the current case. So the Au nanoparticles can enormously enhance the electron transport and thus improve the electrical conductivity (see Figure 19.12c, d). For Au-BNNT, the I–V curves exhibit a good linear behavior at different humidity atmospheres. The resistance of the Au-BNNT decreases promptly with the increase in RH value. The resistance of the Au-BNNT in 44.0% RH is calculated to be about 5.0×10^{11} Ω, which is about four and seven orders of magnitude larger than those in 78.0% (5.8×10^7 Ω) and 89.0% (2.8×10^7 Ω) RH, respectively. It is clear that the humidity has a strong influence on the conductance of the Au-BNNT, indicating the effectiveness of Au-BNNTs in humidity sensing. The response and recovery behaviors are two of the most important parameters to evaluate the ability of a humidity material's response to adsorption and desorption of water molecules (Cheng et al. 2011). These two parameters were studied by monitoring the current change (90% of the maximum value and at a voltage of 10 V) of the BNNTs alternately exposed to 54.0% (air) and 85.0% RH. The response time of the Au-BNNT is faster than that of the pristine BNNT: The response time is about 100 s for the Au-BNNT and 110 s for the pristine BNNT (see Figure 19.12e). More importantly, the current change is very small during the testing period of 6 h at a high humidity of 90% RH (Figure 19.12f). These demonstrate that the Au-BNNT-based humidity-sensing unit is relatively stable and potentially useful in detecting environmental humidity.

19.6   LONG BNNT COATING TO SEPARATE OIL AND WATER

Nowadays, the separation of oil from oil-polluted water generated from many industrial manufactures, such as petrochemical, food, textiles, leather manufactures, and metallurgy, is a challenging task (Kajitvichyanukul et al. 2008). Many approaches have been traditionally employed to separate oil from water, such as oil skimmers, centrifuges, coalesces, settling tanks, depth filters, magnetic separations, flotation technologies, oil-absorbing materials, and burning (Bayat et al. 2005; Adebajo et al. 2009; Lei et al. 2013; Zhang et al. 2013). Since Feng et al. (2004) first reported a superhydrophobic and superoleophilic coating deposited on a stainless steel mesh for oil/water separation, the superhydrophobicity/superoleophilicity or superhydrophilicity/superoleophobicity feature has attracted much attention. Recently, a series of materials with such feature have been developed, such as TiO_2-coated mesh (Gao et al. 2013), marshmallow-like gel (Hayase et al. 2013), PVDF membrane (Zhang et al. 2013), polytetrafluoroethylene-coated mesh (Feng et al. 2004), trichloromethylsilane-coated polyester textile (Zhang and Seeger 2011), nanoporous polydivinylbenzene materials (Zhang et al. 2009), and cross-linked oil-absorbing polymer gels (Sonmez and Wudl 2005; Ono et al. 2007). Despite these pioneering works, it is still of great importance to develop novel materials for oil/water separation with higher separation efficiency and better recyclability. In addition, our previous work has also demonstrated that the as-fabricated BNNT films are nonwettable to water due to their nanometer-scale roughness as well as adsorbed hydrocarbon contaminations (Li et al. 2010a,b, 2012). Such a character can be further enhanced if BNNTs are partially and vertically aligned (Lee et al. 2009). With additional properties such as superior mechanical performance, high thermal conductivity, outstanding electrical insulation, as well as excellent chemical stability and superb oxidation resistance (Rubio et al. 1994; Sanchez-Portal and Hernandez 2002; Popov 2003; Wirtz and Rubio 2003; Chen et al. 2004; Suryavanshi et al. 2004; Golberg et al. 2007), BNNTs become attractive candidate as a superhydrophobic surface in harsh chemical and thermal conditions. In the next section, the superhydrophobicity and superoleophobicity of stainless steel meshes coated by long BNNTs as well as their potential application in oily water treatment are reviewed.

The BNNT-coated stainless steel mesh is produced via an in situ vapor–liquid–solid (VLS) growth process. The milled B powder (see Section 19.2.1) was loaded into an iron boat and covered with a cleaned 2×2 cm stainless steel mesh (without touching the B powder). The mesh has a pore size of approximately 150 µm (100 mesh). The resultant iron boat with this mesh cover was placed in a tube furnace and annealed at 1100°C for 15 h under a H_2/N_2 (5% H_2) gas flow (100 mL/min). During the annealing, the B was partially evaporated and converted to BNNTs covering the mesh surface via the VLS growth mechanism. The Fe in the stainless steel mesh acted as catalysts.

Figure 19.14a and b displays typical FESEM images of the mesh with and without BNNT coating. It can be seen that BNNTs are homogeneously formed on the mesh surface. Compared to the 150 µm hole size of the uncoated mesh, the hole size decreases to ~50–100 µm after BNNT coating. Correspondingly, the gird width increases from ~70 µm before BNNT coating to over 100 µm after BNNT coating. Most importantly, the BNNTs coating does not completely block the holes, ensuring that liquid wettable to BNNTs can quickly pass through the mesh. The enlarged FESEM images (Figure 19.14c, d) clearly show that a large amount of BNNTs with diameter in the range of 100–400 nm are uniformly and randomly distributed on the surface of the grids. The BNNTs are robustly attached to the mesh surface, because the stainless steel mesh acted as catalyst or root for BNNT growth. The energy dispersive x-ray spectrum (EDS) inserted in Figure 19.14d confirms the BN composition in the as-synthesized nanotubes. The typical XRD pattern (Figure 19.14e) of the BNNTs shows two reflections with the d-spacing of 0.336 and 0.215 nm, corresponding to (002) and (100) planes of h-BN structure (JCPDF card No. 73-2095) (Wang et al. 2011). The TEM image (Figure 19.14f) demonstrates that the as-synthesized BNNTs also have a bamboo-like structure. According to the HRTEM image (inset in Figure 19.14f), the intrinsic plane spacing of the BNNTs walls is about 0.34 nm corresponding to that of h-BN (002) plane.

The BNNT-coated mesh is superhydrophobic and superoleophilic. Figure 19.15 shows the wetting behaviors of water and oil on the surface of the BNNT-coated and uncoated meshes. Interestingly, as shown in Figure 19.15a, water droplets can steadily rest on the BNNT-coated mesh with a high contact angle (CA); however, oil droplets can rapidly spread and penetrate through the mesh (Figure 19.15b) as long as they touch the surface of the coated mesh, which is typical for a superoleophilic surface. Noticeably, the uncoated mesh is also hydrophobic with the water CA of approximately 115°, but the water droplets are adsorbed by the uncoated mesh once they touch the surface. Figure 19.15c exhibits the shape of a water droplet on the uncoated mesh under a tilted angle of 0° and 180°. The water droplet can hang on the uncoated mesh and maintain the semispherical shape without obvious distortion, even if the uncoated mesh is gradually tilted to 180°. Figure 19.15d and e shows the shapes of a water droplet and an oil droplet on the surface of the BNNT-coated mesh, respectively. The BNNT-coated mesh shows a high water CA, for example, 156° (Figure 19.15d) and a very small oil CA of less than 1°. Additionally, the BNNT-coated mesh has a small water CA hysteresis so that water droplets can easily roll off once the mesh is slightly tilted. As an example, Figure 19.16 exhibits the process of a 10 µL water droplet that was dropped and immediately rolled off the slightly tilted BNNT-coated mesh. It can be concluded that the BNNT coating improves the hydrophobicity of the mesh.

Generally, the wettability of a solid surface is mainly dominated by its surface chemistry and geometrical roughness (Feng et al. 2002). Lee et al. (2009) have reported that the BNNT arrangement is very important for water wettability. That is, partially vertically aligned BNNTs exhibit excellent water repellency with water CAs over 150°. The enlarged view of the BNNT-coated mesh (Figure 19.14c) shows that the as-synthesized BNNTs are randomly but partially vertically

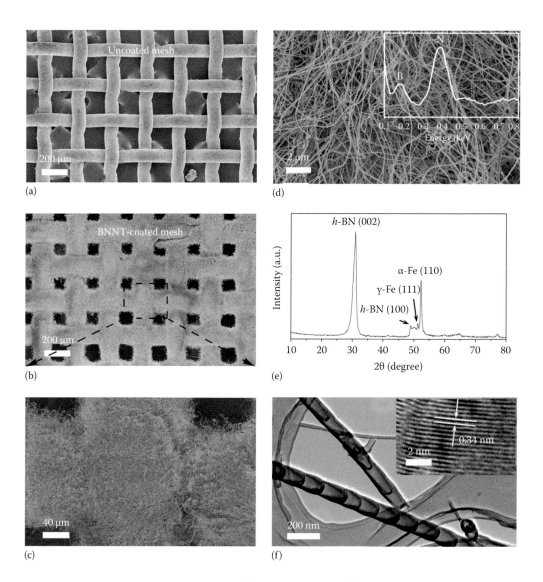

Figure 19.14   FESEM images of the uncoated (a) and BNNT-coated (b) stainless steel meshes, the BNNTs on the mesh surface (c), and corresponding peel-off BNNTs (d), as well as the x-ray diffraction pattern (e) and TEM image (f) of the BNNTs. The insets in (d) and (f) are the EDS and high-resolution image of the BNNTs, respectively. (Reprinted from *Adv. Mater. Interfaces*, 1, Yu, Y. et al., Superhydrophobic and superoleophilic boron nitride nanotube-coated stainless steel meshes for oil and water separation, 1300002, 1–5, Copyright 2014, with permission from Willey.)

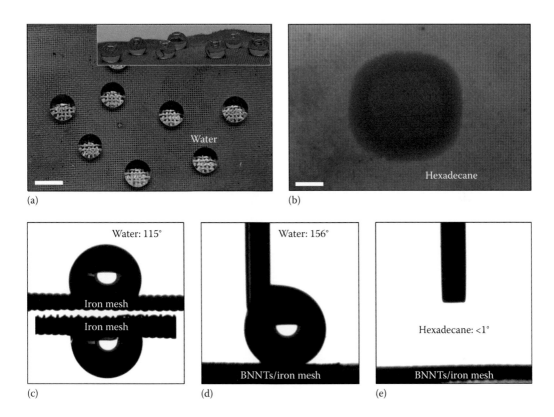

Figure 19.15 (a) A photograph of water droplets on the surface of the BNNT-coated mesh (inset: a side view of a water droplet), (b) a photograph of an oil droplet completely spread on the coated mesh surface, (c, d) the shapes of a water droplet (6 μL) staying on the uncoated (up: no tilting, down: after tilting 180°) and coated mesh with a water CA of 115° and 156°, respectively. (e) A photograph of the hexadecane droplet on the coated mesh surface with a CA of nearly zero. (Reprinted from *Adv. Mater. Interfaces*, 1, Yu, Y. et al., Superhydrophobic and superoleophilic boron nitride nanotube-coated stainless steel meshes for oil and water separation, 1300002, 1–5, Copyright 2014, with permission from Willey.)

aligned on the surface of the mesh, which can enhance the superhydrophobicity. In addition, Cassie equation (Cassie and Baxter 1944) can also be used to interpret the CA of the heterogeneous interface composed of air and a hydrophobic solid:

$$\cos \theta r = f_1 \cos \theta - f_2$$

where

  θr is the CA of the as-fabricated superhydrophobic surface
  θ is the original CA of the material of a smooth surface
  f_1 and f_2 ($f_1 + f_2 = 1$) are the fractional areas of the solid and air in the surface microstructure of the compound surface, respectively

This equation suggests that the increase in the fraction of air (f_2) will increase the CA of as-fabricated superhydrophobic surface (θr). As shown in Figure 19.14d, the as-synthesized BNNTs exhibit a highly porous structure with many interspaces among the BNNTs, which can further improve the hydrophobicity.

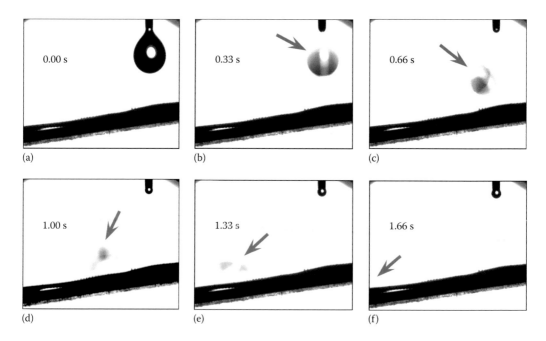

Figure 19.16   The process of a 10 µL water droplet dropped on the surface and rolled down rapidly.

The BNNT-coated mesh with such a unique superhydrophobic and superoleophilic feature is expected to enable the separation of insoluble oil from water. In order to prove this concept, an oil/water separation setup was constructed, as shown in Figure 19.17. The uncoated and BNNT-coated mesh acting as the separation membranes were placed between two glass tubes that were clamped together. Hexane and hexadecane as well as their water mixtures (hexane/water and hexadecane/water) were selected as model oils and oil/water mixtures to demonstrate separating effects. Figure 19.17a through c shows that water (Figure 19.17a), hexane (Figure 19.17b), and hexadecane (Figure 19.17c) can all easily pass through the uncoated mesh and flow down quickly. However, for the BNNT-coated mesh, the water stayed steadily above the mesh (Figure 19.17d), while the hexane (Figure 19.17e) and hexadecane (Figure 19.17f) easily passed through the mesh and flew down quickly. Figure 19.17g through j exhibits the separation process of insoluble oil and water, where the glass tube is tilted in order to make sure that the insoluble oil can contact the mesh. After the mixtures of hexane/water and hexadecane/water were poured into the glass tube, the hexane and hexadecane quickly penetrated through the coated mesh and dropped into the beaker below due to the superoleophilicity of the BNNT-coated mesh while water retained above the mesh because of its superhydrophobicity. As previously discussed, BNNTs have a strong adhesion to the stainless steel mesh because BNNTs have one end attached to the surface of the mesh (Li et al. 2010). The efficiency of the oil/water separation of the BNNT-coated mesh remained the same after over 100 cycles without any visible deterioration. This suggests that the BNNT-coated mesh not only presents superior hydrophobic and oleophilic properties but also has the virtue of stability, durability, and repeatability. These, in conjunction with excellent mechanical and chemical stability of the BNNTs, make the BNNT-coated stainless steel mesh a promising candidate for oil-polluted water treatments.

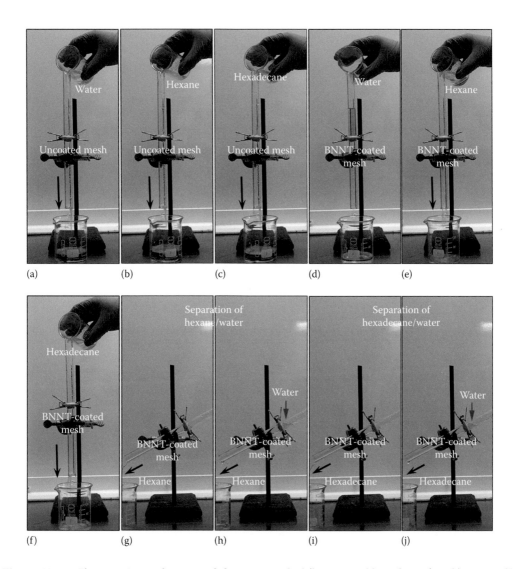

Figure 19.17 The experimental setup and the process of oil (hexane and hexadecane) and/or water filtering/separation: (a–c) are for an uncoated mesh, (d–j) are for a BNNT-coated mesh. Water, hexane, and hexadecane can penetrate through the uncoated mesh and flow down rapidly. Water stays over the coated mesh, and hexane and hexadecane still go through the coated mesh after filtering or separation.

19.7 CONCLUSIONS

BNNTs with a length over 1.0 mm can be synthesized by an optimized ball milling and annealing process at an annealing temperature of 1100°C. Such long BNNTs ensure the reliable setup based on a single tube configuration for accurate measuring electrical property. The average resistivity of the as-synthesized long BNNTs is found to be $(7.1 \pm 0.9) \times 10^4$ $\Omega \cdot$cm. The conductance of such long BNNTs can be enhanced by Au decoration. The introduction of Au nanoparticles on BNNT surface modifies the work function of the BNNTs, and as a consequence, the field-emission current densities of Au-BNNTs are significantly enhanced. Correspondingly, the turn-on field of such Au-BNNTs is reduced to one-third, and the emission current density is increased by four orders in contrast to pure BNNTs. In addition, the resistance of both individual BNNT and Au-BNNT decreases with the increase in humidity at room temperature. The adsorption/desorption tests show that the individual Au-BNNT has a rapid response and recovery time of 100 s/15 s compared to 110 s/15 s of pristine BNNT. More importantly, the as-synthesized long BNNTs are superhydrophobic and superoleophilic without any modification. A new kind of BNNT-coated mesh with superhydrophobicity and superoleophilicity can be constructed. The BNNTs with diameters in the range of 100–400 nm are partially, vertically aligned on the surface of the stainless steel mesh, altering the surface properties of the stainless steel mesh from hydrophobicity with a high water CA hysteresis to superhydrophobicity and superoleophilicity without hysteresis. The BNNT-coated mesh developed in this work can therefore be practically used as a highly efficient filtration membrane for the separation of oil from water in oil-polluted water, benefiting the environment and human health. BNNTs exhibit excellent optical, mechanical, and thermal properties, making the BNNTs very attractive in many practical applications. In the future, we think that there are several aspects that can be further improved:

1. The conductance of BNNTs can be improved by decoration/doping with other elements to improve their application in electric devices.
2. Different kinds of BNNT-coated meshes/BNNT-based films can be constructed to realize their application in practical waste water treatments.

REFERENCES

Adebajo, M.O., Frost, R.L., Kloprogge, J.T., Carmody, O., and Kokot, S. (2003) Porous materials for oil spill cleanup: A review of synthesis and absorbing properties, *Journal of Porous Materials* 10(3), 159–170.

Anderson, P.A. (1959) Work function of gold, *Physics Review* 115(3), 553–554.

Banan-Sadeghian, R., Montreal, C.U., Badilescu, S., and Djaoued, Y. (2008) Ultra-low-voltage Schottky-barrier field-enhanced electron emission from gold nanowires electrochemically grown in modified porous alumina templates, *IEEE Electron Device Letters* 29(4), 312–314.

Baughman, R.H., Zakhidov, A.A., and de Heer, W.A. (2002) Carbon nanotubes—The route toward applications, *Science* 297(5582), 787–792.

Bayat, A., Aghamiri, S.F., Moheb, A., and Vakili-Nezhaad, G.R. (2005) Oil spill cleanup from sea water by sorbent materials, *Chemical Engineer & Technology* 28(12), 1525–1528.

Cassie, A.B.D. and Baxter, S. (1944) Wettability of porous surface, *Transactions of the Faraday Society* 40, 546–551.

Chadderton, L.T. and Chen, Y. (2002) A model for the growth of bamboo and skeletal nanotubes: Catalytic capillarity, *Journal of Crystal Growth* 240(1–2), 164–169.

Chen, H., Chen, Y., Liu, Y., Fu, L., Huang, C., and Llewellyn, D. (2008a) Over 1.0 mm-long boron nitride nanotubes, *Chemical Physics Letters* 463, 130–133.

Chen, H., Zhang, H., Fu, L., Chen, Y., Williams, J., Yu, C., and Yu, D. (2008b) Nano Au-decorated boron nitride nanotubes: Conductance modification and field-emission enhancement, *Applied Physics Letters* 92, 243105.

Chen, H., Chen, Y., Li, B.C.P., Williams, J.S., and Ringer, S. (2006) High-yield boron nitride bamboo nanotubes, in *30th Annual Condensed Matter and Materials Meeting*, Wagga Wagga, New South Wales, Australia.

Chen, Y., Chadderton, L.T., Gerald, J.F., and Williams, J.S. (1999) A solid-state process for formation of boron nitride nanotubes, *Applied Physics Letters* 74(20), 2960–2962.

Chen, Y., Zou, J., Campbell, S.J., and Le Caer, G. (2004) Boron nitride nanotubes: Pronounced resistance to oxidation, *Applied Physics Letters* 84(13), 2430–2432.

Cheng, B.C., Tian, B.X., Xie, C.C., Xiao, Y.H., and Lei, S.J. (2011) Highly sensitive humidity sensor based on amorphous Al_2O_3 nanotubes, *Journal of Materials Chemistry* 21(1), 1907–1912.

Chopra, N.G. and Zettl, A. (1998) Measurement of the elastic modulus of a multi-wall boron nitride nanotube, *Solid State Communications* 105(5), 297–300.

Collins, P.G. and Zettl, A. (1997) Unique characteristics of cold cathode carbon-nanotube-matrix field emitters, *Physics Review B: Condensed Matter and Materials Physics* 55(15), 9391–9399.

Cumings, J. and Zettl, A. (2004) Field emission and current-voltage properties of boron nitride nanotubes, *Solid State Communications* 129(10), 661–664.

Feng, L., Li, S., Li, Y., Li, H., Zhang, L., Zhai, J., Song, W., Liu, B., Jiang, L., and Zhu, D. (2002) Super-hydrophobic surfaces: From natural to artificial, *Advanced Materials* 14(24), 1857–1860.

Feng, L., Zhang, Z.Y., Mai, Z.Y., Ma, Y.M., Liu, B.Q., Jiang, L., and Zhu, D.B. (2004) A super-hydrophobic and super-oleophilic coating mesh film for the separation of oil and water, *Angewandte Chemie* 116(15), 2046 –2048.

Fowler, R.H. and Nordheim, L. (1928) Electron emission in intense electric fields, *Proceedings of the Royal Society of London. Series A: Containing Papers of a Mathematical and Physical Character* 119(781), 173–181.

Gao, C.G., Sun, Z.X., Li, K., Chen, Y.N., Cao, Y.Z., Zhang, S.Y., and Feng, L. (2013) Integrated oil separation and water purification by a double-layer TiO_2-based mesh, *Energy and Environmental Science* 6(4), 1147–1151.

Golberg, D., Costa, P.M.F.J., Lourie, O., Mitome, M., Bai, X.D., Kurashima, K.J., Zhi, C.Y., Tang, C.C., and Bando, Y. (2007) Direct force measurements and kinking under elastic deformation of individual multiwalled boron nitride nanotubes, *Nano Letters* 7(7), 2146–2151.

Grieveson, P. and Turkdogan, E.T. (1964) Kinetics of reaction of gaseous nitrogen with iron part I: Kinetics of nitrogen solution in gamma iron, *Transactions of the Metallurgical Society of AIME* 230, 1604–1609.

Hayase, G., Kanamori, K., Fukuchi, M., Kaji, H., and Nakanishi, K. (2013) Facile synthesis of marshmallow-like macroporous gels usable under harsh conditions for the separation of oil and water, *Angewandte Chemie International Edition* 52(7), 1986–1989.

Huang, J.R., Li, M.Q., Huang, Z.Y., and Liu, J.H. (2007) A novel conductive humidity sensor based on field ionization from carbon nanotubes, *Sensors Actuators A: Physical* 133(2), 467–471.

Iijima, S. (1991) Helical microtubules of graphitic carbon, *Nature* 354, 56–58.

Jiang, K.L., Li, Q.Q., and Fan, S.S. (2002) Spinning continuous carbon nanotube yarns, *Nature* 419(6909), 801.

Kajitvichyanukul, P., Hung, Y.T., and Wang, L.K. (2008) *Handbook of Environmental Engineering, Membrane and Desalination Technologies*, vol. 13, pp. 639–668. New York: Humana Press.

Kimura, C., Yamamoto, T., and Sugino, T. (2001) Study on electrical characteristics of metal/boron nitride/metal and boron nitride/silicon structures, *Diamond and Related Materials* 10(3–7), 1404–1407.

Klinke, C., Delvigne, E., Barth, J.V., and Kern, K. (2005) Enhanced field emission from multiwall carbon nanotube films by secondary growth, *Journal of Physics Chemistry B:* 109(46), 21677–21680.

Kuang, Q., Lao, C.S., Wang, Z.L., Xie, Z.X., and Zheng, L.S. (2007) High-sensitivity humidity sensor based on a single SnO_2 nanowire, *Journal of American Chemical Society* 129(19), 6070–6071.

Lee, C.H., Drelich, J., and Yap, Y.K. (2009) Superhydrophobicity of boron nitride nanotubes grown on silicon substrates, *Langmuir* 25(9), 4853–4860.

Lei, W.W., Portehault, D., Liu, D., Qin, S., and Chen, Y. (2013) Porous boron nitride nanosheets for effective water cleaning, *Nature Communications* 4, 1777.

Li, L., Li, L.H., Ramakrishnan, S., Dai, X.J., Nicholas, K., Chen, Y., Chen, Z.Q., and Liu, X.W. (2012) Controlling wettability of boron nitride nanotube films and improved cell proliferation, *Journal of Physics Chemistry C: Nanomaterials and Nanostructures* 116(34), 18334–18339.

Li, L.H. and Chen, Y. (2010) Superhydrophobic properties of nonaligned boron nitride nanotube films, *Langmuir* 26(7), 5135–5140.

Li, L.H., Chen, Y., and Glushenkov, A.M. (2010a) Synthesis of boron nitride nanotubes by boron ink annealing, *Nanotechnology* 21(10), 105601.

Li, L.H., Chen, Y., and Glushenkov, A.M. (2010b) Boron nitride nanotube films grown from boron ink painting, *Journal of Materials Chemistry* 20, 9679–9683.

Lin, Q.Q., Li, Y., and Yang, M.J. (2012) Polyaniline nanofiber humidity sensor prepared by electrospinning, *Sensors Actuators B: Chemical* 161(1), 967–972.

Lin, Y., Williams, T.V., Xu, T.B., Cao, W., Elsayed-Ali, H.E., and Connell, J.W. (2011) Aqueous dispersions of few-layered and monolayered hexagonal boron nitride nanosheets from sonication-assisted hydrolysis: Critical role of water, *Journal of Physics Chemistry C: Nanomaterials and Nanostructures* 115(6), 2679–2685.

Loh, K.P., Sakaguchi, I., Gamo, M.N., Tagawa, S., Sugino, T., and Ando, T. (1999) Surface conditioning of chemical vapor deposited hexagonal boron nitride film for negative electron affinity, *Applied Physics Letters* 74(1), 28–30.

Luo, H.T., Funakawa, S., Shen, W.Z., and Sugino, T. (2004) Field emission characteristics of BN nanofilms grown on GaN substrates, *Journal of Vacuum Science and Technology B: Microelectronics and Nanometer Structures* 22(4), 1958–1963.

Ma, R.Z., Bando, Y., and Sato, T. (2002) Bamboo-like boron nitride nanotubes, *Journal of Electron Microscopy* 51(suppl. 1), S259–S263.

Marcus, R.B., Chin, K.K., Yuan, Y., Wang, H., and Carr, W.N. (1990) Simulation and design of field emitters, *IEEE Transactions Electron Devices* 37(6), 1545–1550.

Oku, T., Koi, N., Narita, I., Suganuma, K., and Nishijima, M. (2007) Formation and atomic structures of boron nitride nanotubes with cup-stacked and Fe nanowire encapsulated structures, *Materials Transactions* 48(4), 722–729.

Ono, T., Sugimoto, T., Shinkai, S., and Sada, K. (2007) Lipophilic polyelectrolyte gels as super-absorbent polymers for nonpolar organic solvents, *Nature Materials* 6(6), 429–433.

Popov, V.N. (2003) Lattice dynamics of single-walled boron nitride nanotubes, *Physics Review B: Condensed Matter and Materials Physics* 67(8), 085408.

Powers, M.J., Benjamin, M.C., Porter, L.M., Nemanich, R.J., Davis, R.F., Cuomo, J.J., Doll, G.L., and Harris, S.J. (1995) Observation of a negative electron affinity for boron nitride, *Applied Physics Letters* 67(26), 3912–3914.

Rubio, A., Corkoll, J.F., and Cohen, M.L. (1994) Theory of graphitic boron nitride nanotubes, *Physics Review B: Condensed Matter and Materials Physics* 49(7), 5081–5084.

Sanchez-Portal, D. and Hernandez, E. (2002) Vibrational properties of single-wall nanotubes and monolayers of hexagonal BN, *Physics Review B: Condensed Matter and Materials Physics* 66(23), 235415.

Sonmez, H.B. and Wudl, F. (2005) Cross-linked poly(orthocarbonate)s as organic solvent Sorbents, *Macromolecules* 38(5), 1623–1626.

Sugino, T., Kawasaki, S., Tanioka, K., and Shirafuji, J. (1997) Electron emission from boron nitride coated Si field emitters, *Applied Physics Letters* 71(18), 2704–2706.

Sugino, T., Kimura, C., Yamamoto, T., and Funakawa, S. (2003) Tunneling controlled field emission of boron nitride nanofilm, *Diamond and Related Materials* 12(3–7), 464–468.

Suryavanshi, A.P., Yu, M.F., Wen, J.G., and Tang, C.C. (2004) Elastic modulus and resonance behavior of boron nitride nanotubes, *Applied Physics Letters* 84(14), 2527–2529.

Sykes, M.F. and Essam, J.W. (1964) Critical percolation probabilities by series methods, *Physics Review* 133(1), A310–A315.

Tang, C.C., Bando, Y., Huang, Y., Yue, S.G., Gu, C.G. Xu, F.F., and Golberg, D. (2005) Fluorination and electrical conductivity of BN nanotubes, *Journal of American Chemical Society* 127(18), 6552–6553.

Tang, C.C., Bando, Y., and Sato, T. (2002) Catalytic growth of boron nitride nanotubes, *Chemistry Physics Letters* 362(3–4), 185–189.

Velazquez-Salazar, J.J., Munoz-Sandoval, E., Romo-Herrera, J.M., Lupo, F., Rühle, M., Terrones, H., and Terrones, M. (2005) Synthesis and state of art characterization of BN bamboo-like nanotubes: Evidence of a root growth mechanism catalyzed by Fe, *Chemical Physics Letters* 416(4–6), 342–348.

Villars, P., Prince, A., and Okamoto, H. (1995) *Handbook of Ternary Alloy Phase Diagrams*. Geauga County, OH: ASM International.

Wan, Q., Li, Q.H., Chen, Y.J., and Wang, T.H. (2004) Positive temperature coefficient resistance and humidity sensing properties of Cd-doped ZnO nanowires, *Applied Physics Letters* 84(16), 3085–3087.

Wang, J.L., Zhang, L.P., Zhao, G.W., Gu, Y.L., Zhang, Z.H., Zhang, F., and Wang, W.M. (2011) Selective synthesis of boron nitride nanotubes by self-propagation high-temperature synthesis and annealing process, *Journal of Solid State Chemistry* 184(9), 2478–2484.

Wirtz, L. and Rubio, A. (2003) *Ab initio* calculations of the lattice dynamics of boron nitride nanotubes, *Physics Review B: Condensed Matter and Materials Physics* 68(4), 045425.

Yamazoe, N. and Shimizu, Y. (1986) Humidity sensors: Principles and applications, *Sensors and Actuators* 10(3–4), 379–398.

Yao, Z., Postma, H.W.C., Balents, L., and Dekker, C. (1999) Carbon nanotube intramolecular junctions, *Nature* 402, 273–276.

Yu, J., Chen, Y., Elliman, R.G., and Petravic, M. (2006) Isotopically enriched [10]BN nanotubes, *Advanced Materials* 18(16), 2157–2160.

Yu, J., Chen, Y., Wuhrer, R., Liu, Z.W., and Ringer, S.P. (2005) In situ formation of BN nanotubes during nitriding reactions, *Chemical Materials* 17(20), 5172–5176.

Yu, Y., Chen, H., Liu, Y., Li, L.H., and Chen, Y. (2013) Humidity sensing properties of single Au-decorated boron nitride nanotubes, *Electrochemistry Communications* 30, 29–33.

Yu, Y., Chen, H., Liu, Y., Craig, V., Li, L.H., and Chen. Y. (2014) Superhydrophobic and superoleophilic boron nitride nanotube-coated stainless steel meshes for oil and water separation, *Advanced Materials Interfaces* 1, 1300002.

Yuan, S.D., Ding, X.X., Zhang, Z.X., Huang, X.T., Gan, Z.W., Tang, C., and Qi, S.R. (2003) Synthesis of BN nano-bamboos and nanotubes from barium metaborate, *Journal of Crystal Growth* 256(1–2), 67–72.

Zhang, H.Z., Zhao, Q., Yu, J., Yu, D.P., and Chen, Y. (2007) Field-emission characteristics of conical boron nitride nanorods, *Journal of Physics D: Applied Physics* 40(1), 144–147.

Zhang, J.P. and Seeger, S. (2011) Polyester materials with superwetting silicone nanofilaments for oil/water separation and selective oil absorption. *Advanced Functional Materials* 21(24), 4699–4704.

Zhang, W.B., Shi, Z., Zhang, F., Liu, X., Jin, J., and Jiang, L. (2013) Superhydrophobic and superoleophilic PVDF membranes for effective separation of water-in-oil emulsions with high flux, *Advanced Materials* 25(14), 2071–2076.

Zhang, X.X., Li, Z.Q., Wen, G.H., Fung, K.K., Chen, J.L., and Li, Y.D. (2001) Microstructure and growth of bamboo-shaped carbon nanotubes, *Chemical Physics Letters* 333(6), 509–514.

Zhang, Y., Zheng, X.J., Zhang, T., Gong, L.J., Dai, S.H., and Chen, Y.Q. (2010) Humidity sensing properties of the sensor based on $Bi_{0.5}K_{0.5}TiO_3$ powder, *Sensors Actuators B: Chemical* 147(1), 180–184.

Zhang, Y.L., Wei, S., Liu, F.J., Du, Y.C., Liu, S., Ji, Y.Y., Yokoi, T., Tatsumi, T., and Xiao, F.S. (2009) Superhydrophobic nanoporous polymers as efficient adsorbents for organic compounds, *NanoToday* 4(2), 135–142.

Zhang, Y.Y., Fu, W.Y., Yang, H.B., Qi, Q., Zeng, Y., Zhang, T., Ge, R.X., and Zou, G.T. (2008) Synthesis and characterization of TiO_2 nanotubes for humidity sensing, *Applied Surface Science* 254(17), 5545–5547.

Zheng, L.X., O'Connell, M.J., Doorn, S.K., Liao, X.Z., Zhao, Y.H., Akhadov, E.A., Hoffbauer, M.A. et al. (2004) Ultralong single-wall carbon nanotubes, *Nature Materials* 3(10), 673–676.

Zhong, D.Y., Zhang, G.Y., Liu, S., Sakurai, T., and Wang, E.G. (2002) Universal field-emission model for carbon nanotubes on a metal tip, *Applied Physics Letters* 80(3), 506–508.

Zhu, H.W., Xu, C.L., Wei, B.Q., Vajtai, R., and Ajayan, P.M. (2002) Direct synthesis of long single-walled carbon nanotube strands, *Science* 296(5569), 884–886.

Chapter 20

Controlled Synthesis of Functional Boron Nitride Nanostructures for Applications

Boyi Hao, Chee Huei Lee, Jiesheng Wang, Anjana Asthana,
Dustin Winslow, Dongyan Zhang, and Yoke Khin Yap

CONTENTS

20.1 INTRODUCTION TO BORON NITRIDE NANOSTRUCTURES

Boron nitride (BN) nanostructures can have varied structures. They can simply be nanoscale particles of various bulk BN materials, including hexagonal-phase BN (h-BN) and cubic-phase BN (c-BN). However, many of the BN nanostructures of popular interest were discovered in the past 20 years and are unique in their structures and properties. These include BN nanotubes (BNNTs), BN nanosheets (BNNSs), and BN nanoribbons (BNNRs). Furthermore, some new functional BN nanostructures were recently demonstrated by the authors including quantum dots functionalized BNNTs (QDs-BNNTs) and branching CNT-BNNT heterojunctions. The controlled growth and interesting properties of these nanostructures will be described in this chapter. Prior to the detailed discussions in subsequent subsections, a brief introduction of bulk BN phases is described here as a reference and comparison for the later discussed BN nanostructures.

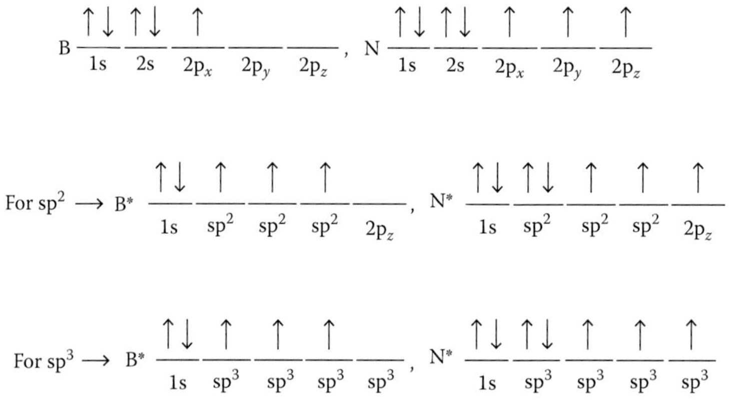

$$\text{B} \quad \underset{1s}{\uparrow\downarrow} \ \underset{2s}{\uparrow\downarrow} \ \underset{2p_x}{\uparrow} \ \underset{2p_y}{\quad} \ \underset{2p_z}{\quad} \ , \quad \text{N} \quad \underset{1s}{\uparrow\downarrow} \ \underset{2s}{\uparrow\downarrow} \ \underset{2p_x}{\uparrow} \ \underset{2p_y}{\uparrow} \ \underset{2p_z}{\uparrow}$$

$$\text{For sp}^2 \rightarrow \text{B}^* \quad \underset{1s}{\uparrow\downarrow} \ \underset{\text{sp}^2}{\uparrow} \ \underset{\text{sp}^2}{\uparrow} \ \underset{\text{sp}^2}{\uparrow} \ \underset{2p_z}{\quad} \ , \quad \text{N}^* \quad \underset{1s}{\uparrow\downarrow} \ \underset{\text{sp}^2}{\uparrow\downarrow} \ \underset{\text{sp}^2}{\uparrow} \ \underset{\text{sp}^2}{\uparrow} \ \underset{2p_z}{\uparrow}$$

$$\text{For sp}^3 \rightarrow \text{B}^* \quad \underset{1s}{\uparrow\downarrow} \ \underset{\text{sp}^3}{\uparrow} \ \underset{\text{sp}^3}{\uparrow} \ \underset{\text{sp}^3}{\uparrow} \ \underset{\text{sp}^3}{\quad} \ , \quad \text{N}^* \quad \underset{1s}{\uparrow\downarrow} \ \underset{\text{sp}^3}{\uparrow\downarrow} \ \underset{\text{sp}^3}{\uparrow} \ \underset{\text{sp}^3}{\uparrow} \ \underset{\text{sp}^3}{\uparrow}$$

Figure 20.1 Ground state electronic configurations of B and N atoms and hybridized electronic configurations of BN bonds. (With kind permission from Springer Science+Business Media: *B–C–N Nanotubes and Related Nanostructures*, Introduction to B–C–N materials, 2009, Y. K. Yap (Ed.), p. 1, Lee, C.H., Kayastha, V., Wang, J. et al.)

BN materials are constructed by boron (B) and nitrogen (N) atoms, the group IIIA and VA elements, respectively, which are on either side of the group IVA element carbon (C) in the *Periodic Table of Elements*. Since B–N bonds and C–C bonds are isoelectronic (having the same number of valence electrons), BN materials form covalent structures similar to the carbon allotropes. The electronic configurations of ground state B and N atoms and hybridized B–N bonds are shown in Figure 20.1 (Lee et al. 2009b).

As shown, one can observe that an additional electron is localized at the nitrogen atoms for both the sp^2 and sp^3 hybridized bonds. Although all the electrons in the hybridized orbitals will redistribute to form the necessary bonds, these covalent bonds are ionic in nature and are more electronegative at the nitrogen site. In fact, BN can appear in the *hexagonal* phase (*h*-BN), the *cubic* phase (*c*-BN), the *rhombohedral* phase (*r*-BN), and the *wurtzite* phase (*w*-BN) as shown in Figure 20.2 (Medlin et al. 1994; Lee et al. 2009b). These phases are similar to *hexagonal* graphite (*ABAB...*), *cubic* diamonds, *rhombohedral* graphite (*ABCABC...*), and *hexagonal* diamonds (*Lonsdaleite*). As shown, *h*-BN and *r*-BN are sp^2 hybridized, while the hybridization of *c*-BN and *w*-BN are sp^3. Thin layers of *h*-BN have recently gained increasing attention due to their structural similarity to graphene sheets. Mono- and few-layer *h*-BN sheets are often referred to as *white graphene* or BNNSs. BNNSs with high aspect ratios (say with narrow width less than 100 nm) are referred to as BNNRs.

20.2 BORON NITRIDE NANOTUBES

BNNTs are most easily understood as sheets of *h*-BN that have been rolled up into seamless cylindrical tubes. As shown in Figure 20.3 (Wang et al. 2010), BNNTs can be classified by vectors (n, m) according to the rolling/chiral angles. In fact, BNNTs were first theoretically predicted in 1994 (Blase et al. 1994; Rubio et al. 1994) and were experimentally produced in the following year (Chopra et al. 1995). Depending on the number of layers of *h*-BN sheets involved, BNNTs can be single-walled (SW) in which case of a single layer of *h*-BN sheet is rolled into a tube, or multi-walled (MW) in which several stacked *h*-BN sheets are rolled into a single tube. SW-BNNTs in the $(n, 0)$ configuration represents the zigzag nanotube structure, SW-BNNTs in the (n, n)

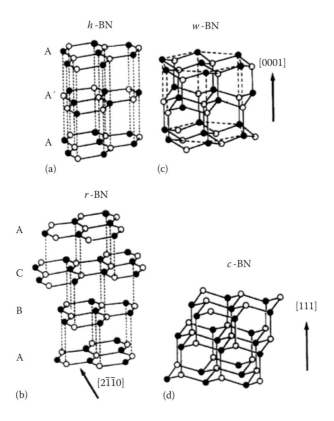

Figure 20.2 Graphical representations of (a) *h*-BN, (b) *r*-BN, (c) *w*-BN, and (d) *c*-BN phases. (Reprinted with permission from Medlin, D., Friedmann, T., Mirkarimi, P. et al., Evidence for rhombohedral boron nitride in cubic boron nitride films grown by ion-assisted deposition, *Phys. Rev. B*, 50(11), 7884, 1994. Copyright 1994, American Physical Society.)

configuration are known as armchair nanotubes, and all other configurations are chiral nanotubes. As examples, the atomic arrangements of (a) zigzag (10, 0), (b) armchair (6, 6), and (c) chiral (7, 5) SW-BNNTs are illustrated in Figure 20.4 (Lee et al. 2009b).

BNNTs are structurally similar to carbon nanotubes (CNTs) and exhibit some similar properties, particularly for their extraordinary mechanical properties (Hernandez et al. 1998; Kudin et al. 2001; Suryavanshi et al. 2004). Despite these similarities, BNNTs differ from CNTs with respect to some very important properties. Unlike CNTs, BNNTs possess uniform electronic properties that are insensitive to their diameters and chiralities (Blase et al. 1994; Rubio et al. 1994). Theoretically, their band gaps are tunable and can even be eliminated by the application of transverse electric fields through the giant dc Stark effect (Chen et al. 2004a; Khoo and Louie 2004; Ishigami et al. 2005). In addition, BNNTs demonstrate high oxidation resistance above 900°C (Golberg et al. 2001; Chen et al. 2004b) and excellent piezoelectricity (Mele and Král 2002; Nakhmanson et al. 2003), and present a possible application for room-temperature hydrogen storage (Jhi and Kwon 2004). Superlattices or isolated CNT/BNNT junctions (Blase et al. 1997) are predicted to produce itinerant ferromagnetism and spin polarization (Choi et al. 2003). The unique combination of these properties makes BNNTs very attractive for innovative applications in various branches of science and industrial sectors.

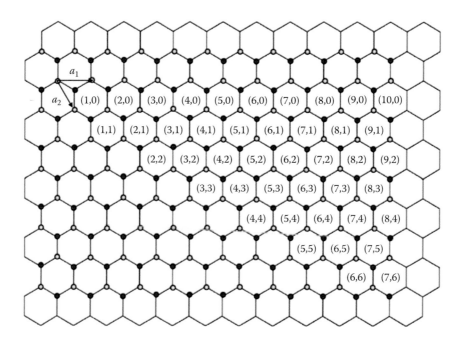

Figure 20.3   Vector (n, m) of SW-BNNTs on an h-BN sheet. (Reprinted with permission from Wang, J. S., C. H. Lee, and Y. K. Yap. Recent advancements in boron nitride nanotubes. *Nanoscale* 2(10) (2010): 2028–2034. Copyright 2010 by Royal Society of Chemistry.)

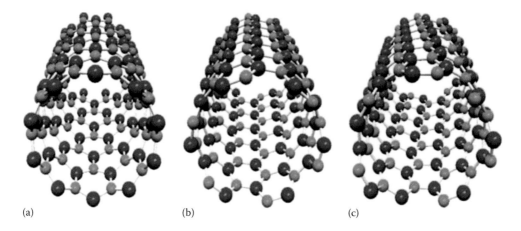

(a)   (b)   (c)

Figure 20.4   (a) Zigzag (10, 0), (b) armchair (6, 6), and (c) chiral (7, 5) BNNTs. (With kind permission from Springer Science+Business Media: *B–C–N Nanotubes and Related Nanostructures*, Introduction to B–C–N materials, 2009, p. 1, Lee, C.H., Kayastha, V., Wang, J. et al.)

The growth of BNNTs is extremely challenging. Various mechanisms have been developed for the growth of BNNTs including arc discharge (Loiseau et al. 1996; Terrones et al. 1996; Chopra and Zettl 1998), laser ablation (Zhang et al. 1998; Lee et al. 2001), substitution reactions from CNTs (Han et al. 1998), ball-milling (Chen et al. 1999), and chemical vapor deposition (CVD) using borazine (Lourie et al. 2000). These BNNTs were contaminated by numerous impurities including amorphous boron nitride (a-BN) powders and other solid-state by-products. Also, it is impossible to use these techniques to directly grow BNNTs on substrates for device fabrication. High growth temperatures, low production yield, and contamination have prevented large-scale synthesis and the development of applications for BNNTs.

In 2005, we reported the growth of BNNTs directly on substrates at 600°C by a plasma-enhanced pulsed-laser deposition (PE-PLD) technique (Wang et al. 2005). A negative substrate bias voltage induced by a nitrogen RF-plasma (Yap et al. 1998) generates the reactive condition for growing BNNTs. The substrate bias accelerates the positive ions in the RF plasma and the BN vapor to bombard on the substrate surface. When the kinetic energies of these ions are sufficient, the deposition rate of BN films is balanced by the rate of re-sputtering and results in total re-sputtering. The growth of BNNTs is highly reproducible under this condition, which happened in our experiments when the substrate bias is between −360 and −450 V.

Scanning electron microscopy (SEM) indicates that multiple BNNTs grown from adjacent Fe catalyst particles tend to form vertical bundles. BNNTs grown at −380 V appear in conical features as shown in Figure 20.5a. For samples grown at higher substrate bias (−450 V), individual BNNTs inside the bundles can clearly be resolved (Figure 20.5b). This is because as the plasma heating is enhanced, then the size of the Fe nanoparticles increase, which in turn causes an

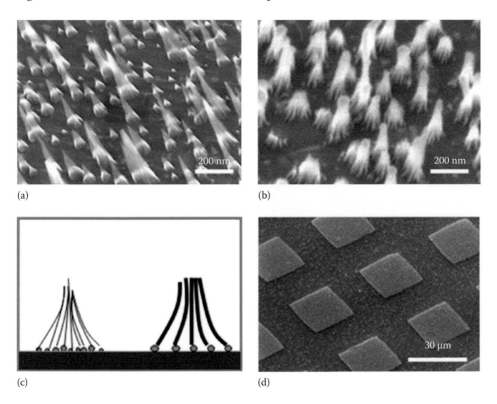

(a)

(b)

(c)

(d)

Figure 20.5   SEM images of BNNT bundles grown at a substrate bias of (a) −380 V, (b) −450 V, and (c) their corresponding bundling configurations (left and right, respectively). (d) Patterned growth of BNNTs. (Reprinted with permission from Wang, J., Kayastha, V.K., Yap, Y.K. et al., Low temperature growth of boron nitride nanotubes on substrates, *Nano Lett.*, 5(12), 2528–2532, 2005. Copyright 2005, American Chemical Society.)

increase in the grown BNNTs (~20 nm). The bundled configurations of BNNTs with both small diameters and big diameters are shown in Figure 20.5c (left and right, respectively). These BNNT bundles can be grown in regular patterned arrays (Figure 20.5d) with a patterned Fe film created by using shadow mask during the deposition of the Fe catalyst. These results indicate that the growth location of BNNTs is controllable by controlled patterning of the Fe nanoparticle catalysts, which is similar to the catalytic patterned growth of CNTs by CVD technique.

20.2.1 Controlled Growth of BNNTs by Growth Vapor Trapping Method

Convenient growth techniques for BNNTs are required to accelerate future investigation of the properties and the applications of BNNTs. The most prospective approach is by thermal CVD, which has gained popularity for the growth of CNTs and various ZnO nanostructures. The boric oxide CVD (BOCVD) method is considered to be a significant advancement for the synthesis of BNNTs (Tang et al. 2002; Zhi et al. 2005a). In the BOCVD technique, B, MgO, and various other metal oxide catalysts are used as precursors. A growth temperature >1300°C is needed to activate the precursors. Volatile B_xO_y vapors are generated by the reaction of the heated precursors and are carried by an Ar gas flow to interact with anhydrous ammonia gas (NH_3) to form BNNTs. This technique requires a specially designed vertical induction furnace with rapid heating and a large temperature gradient.

In 2008 (Lee et al. 2008), we demonstrate a simple approach for the growth of BNNTs in a conventional horizontal tube furnace consisting of a quartz tube vacuum chamber, commonly used for the synthesis of CNTs (Kayastha et al. 2004, 2007) and various ZnO nanostructures (Mensah et al. 2007). Therefore, this approach can be easily reproduced by researchers working on the growth of CNTs and nanowires (NWs) and allow them to also begin research activity focused on BNNTs. The essential accessory required for this experimental setup is a quartz test tube, which is 60 cm long and 2 cm in diameter (Figure 20.6a). A mixture of B, MgO, and FeO (or either Fe_2O_3 or Fe_3O_4) powders is used as the precursor material. The molar ratio of B:MgO:FeO (Fe_2O_3, Fe_3O_4) is fixed at 2:1:1 for all experiments. A total of 100 mg of these powders is placed in an alumina combustion boat. The boat is placed at the closed end of the quartz test tube, and the quartz test tube is loaded into the quartz vacuum chamber so that the closed end of the test tube is located at the center of the heating zone as shown in Figure 20.6a. Several Si substrates can be placed on top of the alumina boat to increase the surface area available for the deposition of BNNTs. The quartz tube chamber is evacuated to about ~30 mTorr before 200 sccm of NH_3 gas is flowed through the evacuated chamber. The precursors are heated to 1200°C and held for 1 h. At this temperature, reactive B_xO_y vapors will be generated and react with NH_3 gas to form BNNTs. It is believed that FeO (Fe_2O_3, Fe_3O_4) and MgO are serving as catalysts for the creation of B_xO_y. Upon removal of the alumina boat and substrates, white coating (BNNTs) will be observed on the precursor materials, on the test tube, around the inner side walls of the alumina boat, and on the Si substrates (Figure 20.6b).

The BNNTs produced through the use of the aforementioned thermal CVD technique have been studied to identify their purity and uniformity. Figure 20.7a is an SEM image, which shows the morphologies of the as-grown BNNTs. Clean and long BNNTs can be clearly observed on the as-grown products. The typical diameter of these tubes is between 10 and 100 nm, depending on the experimental parameters. The as-grown BNNTs are estimated to be longer than 10 μm. Transmission electron microscopy (TEM) indicates that these BNNTs have high-ordered tubular structures (Figure 20.7b, c), with some amorphous BN coating on the side walls. Energy-filtered imaging was used to identify the compositional property of these BNNTs. As shown in Figure 20.7d, mapping of boron and nitrogen atoms clearly indicates that the as-grown nanotube is a BNNT. Carbon is not detected since no carbon source was used in the growth. This result

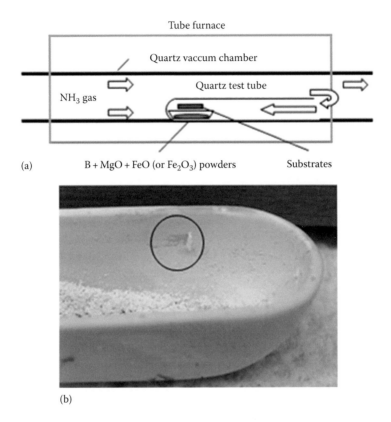

(a)

(b)

Figure 20.6   (a) Experimental setup for the growth of BNNTs in a horizontal tube furnace. (b) A photo-graph of the precursor materials in the alumina boat coated with a white layer of BNNTs. These BNNTs can be easily collected by mechanical scratching (marked by a circle). (Reprinted with permission from Lee, C.H., Wang, J., Kayatsha, V.K. et al., Effective growth of boron nitride nanotubes by thermal chemical vapor deposition, *Nanotechnology*, 19(45), 455605, 2008. Copyright 2008, Institute of Physics.)

has been confirmed through the use of electron energy loss spectroscopy (EELS). As shown in Figure 20.7e, sharp boron and nitrogen K-edge bands were detected, which are composed of both the π^* and σ^* peaks that are typical for sp^2-bonded BN networks.

In 2010, we reported that MgO, Ni, and Fe were active catalysts, which enabled patterned growth of BNNTs directly on Si substrates using the Growth Vapor Trapping (GVT) approach (Lee et al. 2010). To prepare the substrate for patterned growth, an Al_2O_3 under layer (30 nm) was initially coated on an Si or SiO_2 substrate by PLD. Subsequently, a thin film layer of MgO (15–30 nm), Ni, or Fe (10 nm) was deposited on top of the Al_2O_3 layer. In order to demonstrate the patterned growth of BNNTs, shadow masks (150 and 200 mesh) were used to define the location of the catalysts. These substrates were then placed on the top of an alumina combustion boat with the catalyzed films facing upward, and the alumina boats were filled with the aforementioned precursors: B, MgO, and FeO. As before, the setup was loaded into a closed-end quartz tube in a horizontal tube furnace. The precursors and substrates were then heated up to 1100°C–1200°C with an NH_3 flow rate between 200 and 350 sccm, and the temperature and flow rate were kept for ~30 min.

The morphology of the as-grown BNNT film was examined by SEM. A representative image is shown in Figure 20.8. These BNNTs have a typical diameter of 60 nm. The average length of the BNNTs

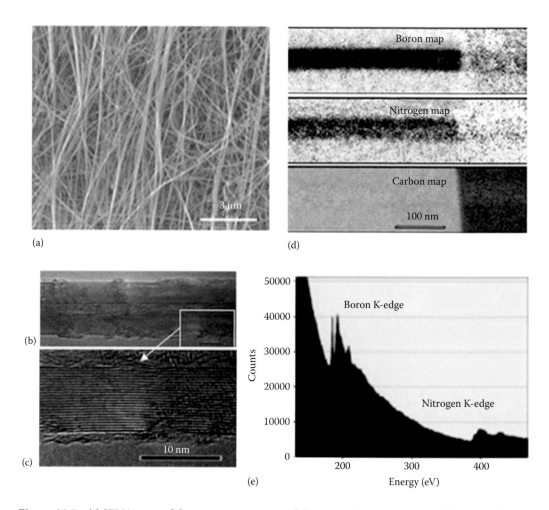

Figure 20.7   (a) SEM images of the as-grown BNNTs and their typical TEM images at (b) low and (c) high magnification. (d) Energy-filtered image shown that these nanotubes consist of boron and nitrogen atoms but not carbon. This is confirmed by (e) electron energy loss spectroscopy (EELS). (Reprinted with permission from Lee, C.H., Wang, J., Kayatsha, V.K. et al., Effective growth of boron nitride nanotubes by thermal chemical vapor deposition, *Nanotechnology*, 19(45), 455605, 2008. Copyright 2008, Institute of Physics.)

is greater than 10 μm. The patterned growth of BNNTs are preferentially grow along the direction perpendicular to the substrate surface. It is of particular interest that these vertically aligned BNNTs exhibit a superhydrophobic behavior and may be applicable as transparent, self-cleaning, and anti-corrosive coatings (Lee et al. 2009b). As shown, well-defined patterns of BNNTs can be deposited using this approach. This implies that the chemical reaction between the spatially predefined catalysts (MgO, Ni, or Fe) and the reactive growth species from the combustion boat (B_2O_2 or BN vapors) plays an important role in the growth of BNNTs. The proposed chemical process is as follows:

$$B_2O_2 (g) + MgO (s) + 2NH_3 (g) \rightarrow 2BN (BNNTs) + MgO(s) + 2H_2O (g) + H_2 (g) \qquad (20.1)$$

or

$$B_2O_2 (g) + Ni/Fe (s) + 2NH_3 (g) \rightarrow 2BN (BNNTs) + Ni/Fe(s) + 2H_2O (g) + H_2 (g) \qquad (20.2)$$

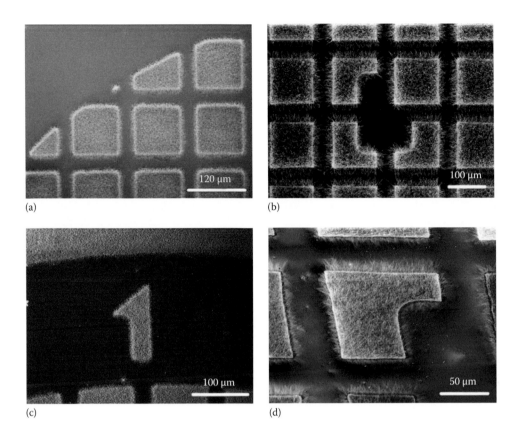

(a)

(b)

(c)

(d)

Figure 20.8   (a to d) Demonstration of well defined patterned growth of BNNTs on a substrate. (Artificial color was added to the SEM images for clarity.) (Reprinted with permission from Lee, C. H., M. Xie, V. Kayastha et al. Patterned growth of boron nitride nanotubes by catalytic chemical vapor deposition. *Chemistry of Materials* 22(5) (2010): 1782–1787. Copyright 2010 by American Chemical Society.)

The generation of B_2O_2 vapor from the combustion boat is described by

$$2B\ (s) + 2MgO\ (s) + FeO\ (s) \rightarrow B_2O_2\ (g) + 2Mg\ (s) + Fe\ (s) \tag{20.3}$$

20.2.2 Functional BNNTs for Applications

Since BNNTs have several properties that are not typically exhibited by a single material, they offer unique solutions for various technological challenges. However, the successful deployment of BNNTs into technological applications will strongly depend on two factors (Wang et al. 2010): (1) the availability of high-quality BNNTs and (2) the demonstration of prospective applications. As was recently discussed, the synthesis of high-quality BNNTs in industrial volumes is still not forthcoming. However, recently, two companies have begun to commercialize larger quantities of BNNTs: *BNNT, LLC* (www.BNNT.com) and *Nano Innovations, LLC* (www.nano-innov.com). The BNNTs produced by *BNNT, LLC* are produced by a pressurized vapor/condenser method (Smith et al. 2009), while the high-quality BNNTs being produced by *Nano Innovations, LLC* are grown with the use of a new CVD technology. *Nano Innovations,*

LLC is also working on the commercialization of several other BN nanostructures, including BNNSs, and QDs-BNNTs. It should also be noted that Professor Ying Chen has been producing small quantities of BNNTs for commercial and research purposes for many years (see http://www.deakin.edu.au/research/ifm/research/nanotechnology/services.php).

As mentioned, the second barrier to the adoption of BNNTs for industrial applications is the dearth of any demonstration showing successful use of their novel properties in an application. With several companies beginning to produce large quantities of BNNTs, it is hoped that they will become more attractive to academic and industry researchers.

Due to the low reactivity and superhydrophobicity of BNNTs, it is difficult to disperse them for easy use in an industrial or commercial setting. Effective functionalization of BNNTs is a vital step that is necessary to incorporate the BNNTs into organic or aqueous media. The incorporation of BNNTs into organic solvents is important for adoption into industries that use composite materials including possible use in high-strength, lightweight reinforcement applications. On the other hand, the inclusion of BNNTs in aqueous media will allow water-based applications to be developed including those in the biomedical and biosensing industries. These BNNT functionalization studies have already begun in an academic setting in both organic (Velayudham et al. 2010) and aqueous media (Lee at al. 2012). In fact, these initial functionalization studies led to the first observation of the superhydrophobicity of BNNTs (Lee et al. 2009b; Boinovich et al. 2011), which shows the potential for BNNTs to be used as mechanical and chemical robust self-cleaning coatings that are transparent to UV–visible light.

20.2.2.1 Functional BNNTs in Organic Solvents

Noncovalent functionalization of BNNTs has proved to be advantageous over covalent functionalization approaches since the functionalized BNNTs can keep the intrinsic properties of the as-grown BNNTs, such as high mechanical strength, without creating defects in the BNNT lattice. BNNTs have been functionalized and solubilized in chloroform by using poly[*m*-phenylenevinylene-co-(2,5-dioctoxypphenylenevinylene)] (Zhi et al. 2005b). This has allowed for purification of the BNNTs via wrapping due to the π–π interactions between the polymer and BNNTs (Zhi et al. 2006). Covalent linking of diamine-terminated polyethylene glycol (PEG) (Xie et al. 2005) to the surface of BNNTs has been demonstrated and used to tune the band gap of BNNTs (Zhi et al. 2006). Additionally, self-organized composite films of BNNTs and polyaniline were prepared by sonicating a mixture of BNNTs and polyaniline in *N,N*-dimethylformamide through efficient π–π interaction between the polyaniline and the BNNTs (Zhi et al. 2005b). In 2010, noncovalent functionalization of BNNTs was demonstrated by using poly(*p*-phenylene-ethynylene)s (PPEs). This approach was compared to a similar approach using polythiophene (Velayudham et al. 2010).

As shown in Figure 20.9a, high-resolution TEM imaging of BNNT functionalized with PPE shows a layer of the polymer that is less than 1 nm thick on the walls of BNNTs. These functionalized BNNTs have well-ordered tubular structures with wall thicknesses of ~30 nm, indicating that BNNTs still retain their crystalline morphology after functionalization. An SEM image of the composite material formed by PPEs and BNNTs shows that BNNTs are well dispersed and aligned along the flow direction of the solution (Figure 20.9b). The molecular model of the composite material composed of BNNTs and PPEs obtained by using Hyperchem software shows that the PPE backbone attaches to the BNNT surface through π–π stacking interactions (Figure 20.9c). This helps enhance π-conjugated of PPE while the polymer side chains are flexible, which enhances solvation of BNNTs and solubilizes BNNTs in a series of organic solvents.

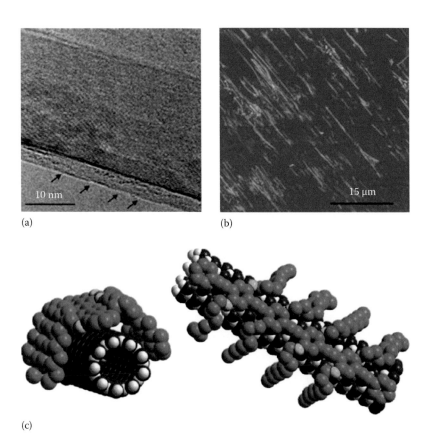

(a)   (b)

(c)

Figure 20.9   (a) High-resolution TEM image at the BNNT side walls functionalized with PPE (polymer A). A layer of PPE can be clearly seen as indicated by arrows. (b) SEM image of aligned BNNTs functionalized with polymer A. (c) Schematic models of composite material of BNNTs and polymer A at various viewing angles. (Reprinted with permission from Velayudham, S., Lee, C.H., Xie, M. et al., Noncovalent functionalization of boron nitride nanotubes with poly(p-phenylene-ethynylene)s and polythiophene, *ACS Appl. Mater. Interfaces*, 2(1), 104–110, 2010. Copyright 2010, American Chemical Society.)

20.2.2.2   Functional BNNTs in Water

Noncovalent functionalization of BNNTs in water is mostly unexplored (Chen et al. 2009; Ciofani et al. 2009, 2010). In particular, the stability of the dispersed BNNTs in water has not been thoroughly investigated. Specifically, the effect of dispersed BNNTs on biological systems has not been well studied. Recently, the toxicity of BNNTs on human embryonic kidney (HEK293) cells was found (Horvath et al. 2011), which is contradictory to the data released in an earlier study (Chen et al. 2009). Since short nanotube length was reported as an essential factor for CNTs to be biological compatible without causing accumulation and toxicity in test animals (Liu et al. 2006; Chen et al. 2009), it is therefore important to establish the mechanism behind functionalization of BNNTs in water and their dispersion stability. Similarly, it is important to explore possible length-shortening processes in the event that the biological compatibility is dependent on BNNT length.

In 2012, the first success in functionalized MW-BNNTs in water, by the use of methoxy-poly(ethylene glycol)-1,2-distearoyl-*sn*-glycero-3-phosphoethanolamine-N conjugates (mPEG-DSPE),

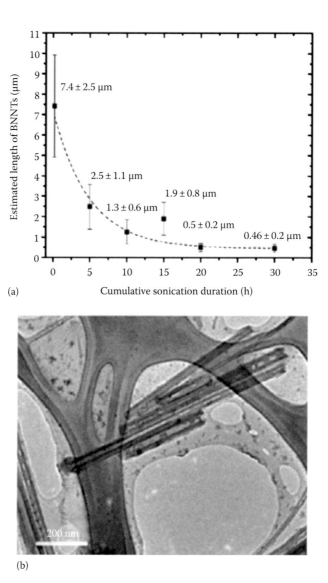

(a)

(b)

Figure 20.10   (a) Measurement and statistical analysis of nanotube length after various cumulative soni-cation processing and (b) TEM image of the shortened BNNTs. (Reprinted with permission from Lee, C.H., Zhang, D., and Yap, Y.K., Functionalization, dispersion, and cutting of boron nitride nanotubes in water, *J. Phys. Chem. C*, 116(2), 1798–1804, 2011. Copyright 2011, American Chemical Society.)

was demonstrated (Lee et al. 2012). This PEGylated phospholipid is biocompatible. Once dis-persed, these mPEG-DSPE functionalized BNNTs (mPEG-DSPE/BNNTs) are stable for more than 3 months without noticeable aggregation. Furthermore, it has been shown that sonication of BNNTs in solution for an hour will shorten the length of as-grown BNNTs (>10 μm) to less than 500 nm, as shown in Figure 20.10. It is believed that the combination of both the ability to sig-nificantly shorten the BNNTs and the ability to keep them dispersed in suspension will allow for implementation in biomedical applications.

20.2.2.3 Superhydrophobicity of BNNTs

Highly hydrophobic surfaces and coatings have gained significant research interest due to their water-repelling, self-cleaning, and antifouling properties (Feng et al. 2002; Blossey 2003). A material is said to be superhydrophobic if the water contact angle (CA) is larger than 150° and if the CA hysteresis (or sliding angle) is small, less than 5°–10° (Sun et al. 2005). In some cases, surfaces with high water CAs can be prepared by a careful manipulation of surface topography and structure at micrometer- and nanometer-size scales (Ma and Hill 2006). Historically, hydrophobic organic materials such as fluorinated polymers (Morra et al. 1989), olefins, fluorohydrocarbons (Kim et al. 2005), and silicon-based hydrocarbons (Jin et al. 2005; Khorasani et al. 2005) were frequently used when a hydrophobic surface was required. However, most of these organic materials are not thermally stable enough for high-temperature applications. Recently, it was shown that superhydrophobic coatings and films can be prepared using small diameter fibers, wires, rods, and tubes (Feng et al. 2003; Feng and Jiang 2006; Ma et al. 2008). Several research groups used CNTs (Lau et al. 2003; Huang et al. 2005; Zhu et al. 2005; Kakade et al. 2008), TiO_2 (Sun et al. 2005; Lai et al. 2008), and ZnO (Feng et al. 2004, 2005) NWs or nanorods as building blocks to prepare coatings with controllable wetting characteristics.

In 2009, the superhydrophobicity of BNNTs was discovered (Lee et al. 2009a). As was briefly discussed earlier, BNNTs are structurally similar to CNTs. They possess the same super strong mechanical properties as CNTs (Rubio et al. 1994; Chopra and Zettl 1998), due to the strong sp^2 bonding within their hexagonal networks. Unlike CNTs, which can be semimetallic or semiconducting, BNNTs are always insulating (Blase et al. 1994; Rubio et al. 1994) with an electronic band gap of ~5.9 eV (Lee et al. 2008). Moreover, BNNTs are chemically more stable, compared to CNTs (Chen et al. 2004b). Additionally, they are excellent thermal conductors (Chang et al. 2005) with high resistance to oxidation at elevated temperatures, above 900°C (Chen et al. 2004b).

While h-BN films are partially wetted by water with an advancing CA of about 50°, randomly aligned BNNTs can achieve a superhydrophobic state with an advancing water CA that exceeds 150° as shown in Figure 20.11. Our results show that the pH value of water does not affect the wetting characteristics of BNNTs (Lee et al. 2009a). Since BN is chemically inert, resistive to

(a)

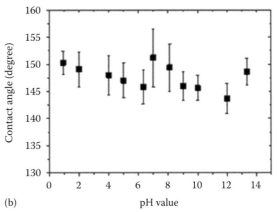
(b)

Figure 20.11 (a) Images of water droplets on a BNNT film coated on an Si substrate. (b) Advancing contact angles measured for water of varying pH on the BNNT film. (Reprinted with permission from Lee, C.H., Drelich, J., and Yap, Y.K., Superhydrophobicity of boron nitride nanotubes grown on silicon substrates, *Langmuir*, 25(9), 4853–4860, 2009. Copyright 2009, American Chemical Society.)

oxidation up to 900°C, and transparent to visible–UV light, BNNTs are an attractive option for self-cleaning, transparent (up to 5.9 eV), insulating, anticorrosive coatings that need to perform under rigorous chemical and thermal conditions.

20.3   OTHER FUNCTIONAL BORON NITRIDE NANOSTRUCTURES

It should now be clear that many of the potential applications for BNNTs are based on their unique combination of mechanical, biomedical, and surface chemical properties. However, the careful reader may have noticed that there has been a lack of discussion concerning the possible electronic applications for BNNTs. We will now discuss some of the electronic possibilities for functionalized BNNTs as well as the exciting possibilities that will be available due to the recent discovery of several new functional BN nanostructures. Some of these latest results are briefly described here.

20.3.1   Quantum Dots Functionalized BNNTs

Recall that although BNNTs are structurally similar to CNTs, their electronic properties are significantly different. Specifically, BNNTs have a wide band gap (~6 eV), and their band gap is not sensitive to differences in the BNNT's diameter, chirality, or number of tubular shells. At first glance, it would appear that BNNTs do not show much promise for use in electronic devices and sensors. However, their band gap is tunable under certain special conditions. These include lattice doping, through exploitation of the giant Stark effect, and structural deformation of the BNNT lattice (Golberg et al. 2007, 2010; Wang et al. 2009a). Unfortunately, experiments focused on doping BNNTs with carbon (Suenaga et al. 1997) and fluorine (Tang et al. 2005) have led to phase segregation and defective BNNTs, respectively. Obviously, due to the defects introduced by doping, this particular method has failed to produce a viable BNNT semiconductor. More recently, bent BNNTs have been shown to conduct a minute current (Bai et al. 2007; Ghassemi et al. 2012), which is attributed to the radial deformation, which was predicted to result in a reduced band gap. Although this small amount of current is academically interesting, it is not enough to allow for BNNTs to be used as conduction channel in electronic devices. Nevertheless, there may be a very exciting possible market opening up for functionalized BNNT production due to recent developments.

Semiconductors are the basic building block on which our entire modern technological world is built. They are indispensable for electronic devices including field-effect transistors (FETs). However, future FETs, based on traditional semiconductor technology, have begun to encounter various fundamental limitations (Theis and Solomon 2010; Morton et al. 2011), including (1) high power consumption due to leakage in the semiconducting channels, (2) short channel effects as the channel length approaches the scale of the depletion layer width, and (3) high contact resistance at the semiconducting channels. Novel device concepts have been proposed to overcome these issues including the exploration of alternative gating architecture (Ferain et al. 2011), tunnel FETs (TFETs) (Ionescu and Riel 2011), and spintronic devices (Morton et al. 2011). In particular, TFETs with low turn-on voltage and high sub-threshold slope are important for next generation very-large-scale integration technology. Nevertheless, all these devices are still based on standard semiconductors.

In 2013, a novel method was demonstrated that uses BNNTs to create room-temperature tunneling FETs without semiconductors (Lee et al. 2013). These devices are based on quantum tunneling of electrons between gold QDs deposited on the insulating BNNTs (QDs-BNNTs). The QDs-BNNTs are insulating at low bias voltages, but allow transport of electrons by way of tunneling through the array of QDs only when sufficient potential is applied. Since the switching behaviors are based on quantum tunneling, these FETs have significantly reduced leakage current and contact resistance. In addition, the performance of the QDs-BNNTs FETs increases as the tunneling channel is decreased, in contrast to the reduced performance due to short channel effects in standard Si devices. Therefore, QDs-BNNTs may offer a viable solution to fundamental limitations that plague current semiconductor technology.

As shown in Figure 20.12a, an isolated QD-BNNT is in contact with two STM probes at a defined channel length (L). At low voltages (*off* state, current $\sim10^{-11}$ A), this QD-BNNT is as insulating as a pure BNNT (Figure 20.12b,c). When an increased bias voltage is applied, the QD-BNNT switches to a conducting state (*on* state) with a current level as high as $\sim10^{-7}$ A. Current–voltage (I–V) curves in Figure 20.12c show that the turn-on voltages (V_{on}) of this QD-BNNT decrease from about 34 to 2.0 V as the nanotube channel length (L) is reduced from 2.37 to 1.29 μm. For this QD-BNNT, under these operating parameters, the ratio between the on and off states is estimated to be on the order of 10^4.

A theoretical simulation of the QD-BNNT indicates that the system shows phenomenon consistent with Coulomb blockade behavior. The QD-BNNT is modeled as localized states located on each gold QD with a finite barrier between each localized state. The barriers between the gold QDs are represented in Figure 20.12d and e. By applying an electric potential (V_{sd}) across the QDs with the STM probes acting as source and drain electrodes, a potential gradient is established across the QD array, as illustrated schematically in Figure 20.12f. As V_{sd} increases, there is also an accompanying band bending between QDs. In the *on* state, the I–V characteristics depend on the nature of the tunneling process across the junctions. Additionally, V_{on} can be reduced close to zero, and the effect of gate potential has been explored. See reference for more details.

More recently, QDs-BNNTs have been created by using other metallic nanoparticles, Fe in this case (Boyi Hao et al. in preparation). As shown in Figure 20.13a, the Fe QDs are randomly formed on the surface of a relatively thick BNNT ($\sim$120 nm in diameter). By the use of an STM–TEM stage (inset in Figure 20.13b), I–V data are collected in real-time observation of TEM. The I–V curve in Figure 20.13b shows that this QD-BNNT behaves as a rectifying channel as was previously discussed for gold QDs-BNNTs. This result suggests that the random distribution of Fe QDs within a channel width of $\sim$120 nm did not prevent electron tunneling across the QDs-BNNTs.

20.3.2 Heterojunctions of Carbon Nanotubes and Boron Nitride Nanotubes

Theoretical studies indicate that BNNT/CNT junctions (Blase et al. 1997; Choi et al. 2003) and *h*-BN/graphene sheets (Okada et al. 2000; Okada and Oshiyama 2001) are energetically stable. Various configurations of such BNNT/CNT nanotubular junctions have been evaluated. Some of the conclusions from the theoretical studies are summarized as follows:

1. BNNT/CNT junctions are energetically stable if they are in either the armchair or the zigzag configuration.
2. For the zigzag BNNT/CNT, the junction is an insulator/semiconductor interface. They possess flat band structures with tunable direct band gaps ($\sim$0.5–2.0 eV). Thus, zigzag BNNT/CNT junctions are applicable for nanoferromagnetic, spintronic, and tunable photonic/optical devices. These properties are achievable only for single-wall BNNT/CNT junctions, because multi-wall CNTs are not semiconductors.
3. Armchair BNNT/CNT junctions are insulator/semimetal interfaces with a direct band gap. They can form Schottky barrier devices, diodes, and QDs. These properties are available in both single- and multi-wall BNNT/CNT junctions that have semimetallic CNT segments.

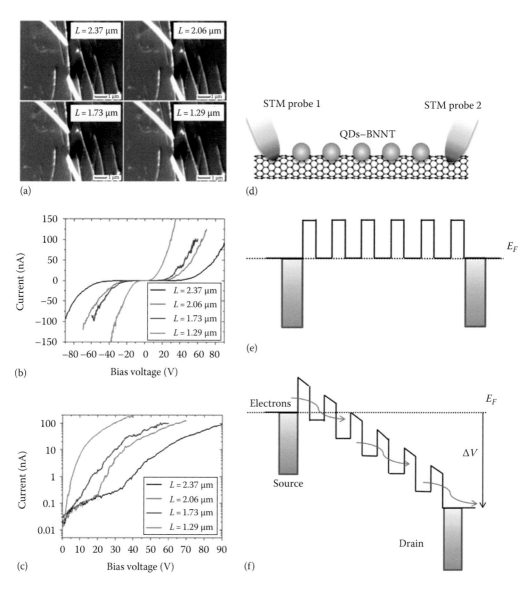

Figure 20.12   (a) SEM images of a QD-BNNT nanostructure as contacted by two STM probes at various conduction channel lengths (L) and the corresponding I–V curves measured at room temperature (b, c). (d) Schematic of a QD-BNNT in contact with two STP probes and the corresponding energy diagram (e). (f) Electron tunneling across the QDs due the applied electric field between the probes. (Lee, C.H., Qin, S., Savaikar, M.A. et al.: Room-temperature tunneling behavior of boron nitride nanotubes functionalized with gold quantum dots. *Adv. Mater.* 2013. 25(33), 4544–4548. Copyright Wiley-VCH Verlag GmbH & Co. KGaA. Reproduced with permission.)

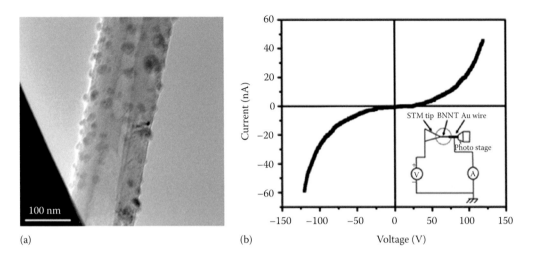

(a)   (b)

Figure 20.13   (a) TEM image of an Fe QD-BNNT in contact with an STM tip (inset) and (b) the corresponding *I–V* characters measured at room temperature.

In 2008 (Yap 2008) and 2011(Yap 2011), two types of heterojunctions were reported: (1) coaxial junctions and (2) branching junctions. The coaxial junctions require the use of a common catalyst for the growth of BNNT and CNT segments in a coaxial manner. The branching junctions were prepared by decorating catalytic nanoparticles onto BNNTs followed by the growth of branching CNTs. The electronic properties of the branching junctions have also been reported (Wang et al. 2011; Yap 2012). As shown in Figure 20.14, a branching CNT–BNNT heterojunction, as probed by two STM tips, produces a rectifying *I–V* characteristic. As a low bias voltage is applied between the STM probes, the branching CNT–BNNT heterojunction is insulating. At a higher bias voltage, the applied potential is sufficient to overcome the Schottky barrier at the junction and allow electron tunneling across the junctions (Boyi Hao et al. in preparation).

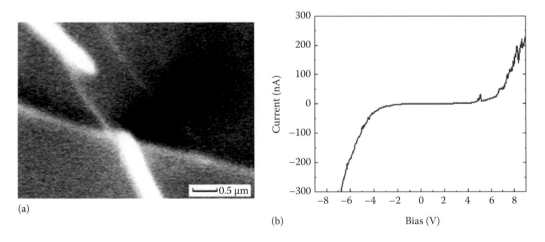

(a)   (b)

Figure 20.14   (a) A branching CNT–BNNT junction probed by two STM probes and (b) the corresponding *I–V* curve.

20.3.3   Boron Nitride Nanosheets and Boron Nitride Nanoribbons

BNNSs and BNNRs are analogous to graphene sheets (Novoselov et al. 2004; Zhang et al. 2005) and graphene nanoribbons (GNRs) (Barone et al. 2006; Han et al. 2007; Li et al. 2008). A GNR and a BNNR are illustrated in Figure 20.15 for comparison (Yap 2011). The growth of BNNSs by CVD has been reported (Shi et al. 2010; Song et al. 2010). Recently, a new class of BNNSs was also reported (Yu et al. 2010; Pakdel et al. 2011). These BNNSs have curly morphologies and were shown to be superhydrophobic. On the other hand, production of BNNRs has also been reported. The production methodology is best described by unzipping as-grown BNNTs (Zeng et al. 2010; Erickson et al. 2011). However, it should be noted that the growth of BNNRs by CVD was reported as far back as 2007 (Yap 2007; Wang et al. 2009b).

In 2007, we discovered that BNNRs can be grown at 800°C, significantly lower than the growth temperatures of our BNNTs (Yap 2007). These BNNRs were named as single crystalline BNNWs

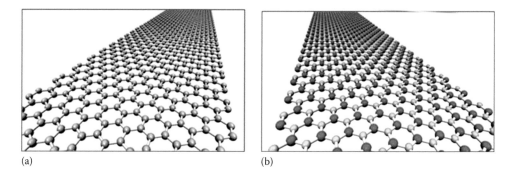

(a)   (b)

Figure 20.15   Schematic representation of a zigzag (a) GNR and (b) BNNR.

Figure 20.16   SEM image of BNNRs grown in arrays of square patterns and TEM image of the individual BNNR (Yap, Y. K. CAREER: Synthesis, characterization and discovery of frontier carbon materials. Solid state and materials chemistry highlights of National Science Foundation, 2007.)

due to their multilayered structures. As shown in Figure 20.16, the growth locations of these BNNRs are fully controllable by the patterns of the Fe catalyst. TEM images indicate that the width of these BNNRs is ~12 nm with single crystallinity. The high-resolution TEM images show symmetric hexagonal atomic structures with lattice constants of *h*-BN.

20.4 SUMMARY

Although BN nanomaterials have been available in the form of BNNTs, for nearly two decades, they have remained difficult to produce, and therefore, they have not seen the same interest that carbon nanomaterials have seen. However, recent studies have begun to probe the properties of BN nanomaterials, and many studies show evidence that BN nanomaterials possess a unique set of properties not typically seen together in nature. This new excitement on BN nanomaterials has led to the discovery and synthesis of other BN configurations including BNNRs and BNNSs. New work in the functionalization of BN nanomaterials has paved the way for application studies using these materials. Along with functionalization, exciting work has shown that exotic chimerical BN nanomaterials are a real possibility. These include the production of FETs with QDs-BNNTs, biomedical application of functionalized BNNTs, and the development of CNT/BNNT heterojunctions. With the approaching commercial availability of BN nanomaterials, the future study and application of these fantastic nanomaterials look very bright.

REFERENCES

Bai, X., D. Golberg, Y. Bando et al. Deformation-driven electrical transport of individual boron nitride nanotubes. *Nano Letters* 7(3) (2007): 632–637.

Barone, V., O. Hod, and G. E. Scuseria. Electronic structure and stability of semiconducting graphene nanoribbons. *Nano Letters* 6(12) (2006): 2748–2754.

Blase, X., J.-C. Charlier, A. De Vita et al. Theory of composite BCN nanotube heterojunctions. *Applied Physics Letters* 70 (1997): 197.

Blase, X., A. Rubio, S. Louie et al. Stability and band gap constancy of boron nitride nanotubes. *Europhysics Letters* 28(5) (1994): 335.

Blossey, R. Self-cleaning surfaces-virtual realities. *Nature Materials* 2(5) (2003): 301–306.

Boinovich, L. B., A. M. Emelyanenko, A. S. Pashinin et al. Origins of thermodynamically stable superhydrophobicity of boron nitride nanotubes coatings. *Langmuir* 28(2) (2011): 1206–1216.

Chang, C. W., W.-Q. Han, and A. Zettl. Thermal conductivity of B–C–N and BN nanotubes. *Applied Physics Letters* 86(17) (2005): 173102.

Chen, C.-W., M.-H. Lee, and S. Clark. Band gap modification of single-walled carbon nanotube and boron nitride nanotube under a transverse electric field. *Nanotechnology* 15(12) (2004a): 1837.

Chen, X., P. Wu, M. Rousseas et al. Boron nitride nanotubes are noncytotoxic and can be functionalized for interaction with proteins and cells. *Journal of the American Chemical Society* 131(3) (2009): 890–891.

Chen, Y., L. T. Chadderton, J. F. Gerald et al. A solid-state process for formation of boron nitride nanotubes. *Applied Physics Letters* 74(20) (1999): 2960–2962.

Chen, Y., J. Zou, S. J. Campbell et al. Boron nitride nanotubes: Pronounced resistance to oxidation. *Applied Physics Letters* 84(13) (2004b): 2430–2432.

Choi, J., Y.-H. Kim, K. Chang et al. Itinerant ferromagnetism in heterostructured C/BN nanotubes. *Physical Review B* 67(12) (2003): 125421.

Chopra, N. G., R. Luyken, K. Cherrey et al. Boron nitride nanotubes. *Science* 269(5226) (1995): 966–967.

Chopra, N. G. and A. Zettl. Measurement of the elastic modulus of a multi-wall boron nitride nanotube. *Solid State Communications* 105(5) (1998): 297–300.

Ciofani, G., S. Danti, D. D'Alessandro et al. Enhancement of neurite outgrowth in neuronal-like cells following boron nitride nanotube-mediated stimulation. *ACS Nano* 4(10) (2010): 6267–6277.

Ciofani, G., V. Raffa, A. Menciassi et al. Folate functionalized boron nitride nanotubes and their selective uptake by glioblastoma multiforme cells: Implications for their use as boron carriers in clinical boron neutron capture therapy. *Nanoscale Research Letters* 4(2) (2009): 113–121.

Erickson, K. J., A. L. Gibb, A. Sinitskii et al. Longitudinal splitting of boron nitride nanotubes for the facile synthesis of high quality boron nitride nanoribbons. *Nano Letters* 11(8) (2011): 3221–3226.

Feng, L., S. Li, Y. Li et al. Super-hydrophobic surfaces: From natural to artificial. *Advanced Materials* 14(24) (2002): 1857–1860.

Feng, L., Z. Yang, J. Zhai et al. Superhydrophobicity of nanostructured carbon films in a wide range of pH values 13. *Angewandte Chemie International Edition* 42(35) (2003): 4217–4220.

Feng, X., L. Feng, M. Jin et al. Reversible super-hydrophobicity to super-hydrophilicity transition of aligned ZnO nanorod films. *Journal of American Chemical Society* 126(1) (2004): 62–63.

Feng, X. and L. Jiang. Design and creation of superwetting/antiwetting surfaces. *Advanced Materials* 18(23) (2006): 3063–3078.

Feng, X. J., J. Zhai, and L. Jiang. The fabrication and switchable superhydrophobicity of TiO$_2$ nanorod films. *Angewandte Chemie International Edition* 44(32) (2005): 5115–5118.

Ferain, I., C. A. Colinge, and J.-P. Colinge. Multigate transistors as the future of classical metal-oxide-semiconductor field-effect transistors. *Nature* 479(7373) (2011): 310–316.

Ghassemi, H. M., C. H. Lee, Y. K. Yap et al. Field emission and strain engineering of electronic properties in boron nitride nanotubes. *Nanotechnology* 23(10) (2012): 105702.

Golberg, D., Y. Bando, Y. Huang et al. Boron nitride nanotubes and nanosheets. *ACS Nano* 4(6) (2010): 2979–2993.

Golberg, D., Y. Bando, K. Kurashima et al. Synthesis and characterization of ropes made of BN multiwalled nanotubes. *Scripta Materialia* 44(8–9) (2001): 1561–1565.

Golberg, D., Y. Bando, C. Tang et al. Boron nitride nanotubes. *Advanced Materials* 19(18) (2007): 2413–2432.

Han, M. Y., B. Özyilmaz, Y. Zhang et al. Energy band-gap engineering of graphene nanoribbons. *Physical Review Letters* 98(20) (2007): 206805.

Han, W., Y. Bando, K. Kurashima et al. Synthesis of boron nitride nanotubes from carbon nanotubes by a substitution reaction. *Applied Physics Letters* 73(21) (1998): 3085–3087.

Hernandez, E., C. Goze, P. Bernier et al. Elastic properties of C and B$_x$C$_y$N$_z$ composite nanotubes. *Physical Review Letters* 80(20) (1998): 4502.

Horvath, L., A. Magrez, D. Golberg et al. In vitro investigation of the cellular toxicity of boron nitride nanotubes. *ACS Nano* 5(5) (2011): 3800–3810.

Huang, L., S. P. Lau, H. Y. Yang et al. Stable superhydrophobic surface via carbon nanotubes coated with a ZnO thin film. *Journal of Physical Chemistry B* 109(16) (2005): 7746–7748.

Ionescu, A. M. and H. Riel. Tunnel field-effect transistors as energy-efficient electronic switches. *Nature* 479(7373) (2011): 329–337.

Ishigami, M., J. D. Sau, S. Aloni et al. Observation of the giant stark effect in boron-nitride nanotubes. *Physical Review Letters* 94(5) (2005): 46.

Jhi, S.-H. and Y.-K. Kwon. Hydrogen adsorption on boron nitride nanotubes: A path to room-temperature hydrogen storage. *Physical Review B* 69(24) (2004): 245407.

Jin, M., X. Feng, J. Xi et al. Super-hydrophobic PDMS surface with ultra-low adhesive force. *Macromolecular Rapid Communications* 26(22) (2005): 1805–1809.

Kakade, B., R. Mehta, A. Durge et al. Electric field induced, superhydrophobic to superhydrophilic switching in multiwalled carbon nanotube papers. *Nano Letters* 8(9) (2008): 2693–2696.

Kayastha, V., Y. K. Yap, S. Dimovski et al. Controlling dissociative adsorption for effective growth of carbon nanotubes. *Applied Physics Letters* 85(15) (2004): 3265–3267.

Kayastha, V. K., S. Wu, J. Moscatello et al. Synthesis of vertically aligned single- and double-walled carbon nanotubes without etching agents. *Journal of Physical Chemistry C* 111(28) (2007): 10158–10161.

Khoo, K. and S. G. Louie. Tuning the electronic properties of boron nitride nanotubes with transverse electric fields: A giant dc Stark effect. *Physical Review B* 69(20) (2004): 201401.

Khorasani, M. T., H. Mirzadeh, and Z. Kermani. Wettability of porous polydimethylsiloxane surface: Morphology study. *Applied Surface Science* 242(3–4) (2005): 339–345.

Kim, S. H., J.-H. Kim, B.-K. Kang et al. Superhydrophobic CF_x coating via in-line atmospheric RF plasma of He–CF_4–H_2. *Langmuir* 21(26) (2005): 12213–12217.

Kudin, K. N., G. E. Scuseria, and B. I. Yakobson. C_2F, BN, and C nanoshell elasticity from ab initio computations. *Physical Review B* 64(23) (2001): 235406.

Lai, Y., C. Lin, H. Wang et al. Superhydrophilic–superhydrophobic micropattern on TiO_2 nanotube films by photocatalytic lithography. *Electrochemistry Communications* 10(3) (2008): 387–391.

Lau, K. K. S., J. Bico, K. B. K. Teo et al. Superhydrophobic carbon nanotube forests. *Nano Letters* 3(12) (2003): 1701–1705.

Lee, C. H., J. Drelich, and Y. K. Yap. Superhydrophobicity of boron nitride nanotubes grown on silicon substrates. *Langmuir* 25(9) (2009a): 4853–4860.

Lee, C. H., V. Kayastha, J. Wang et al. Introduction to B–C–N materials. In *B–C–N Nanotubes and Related Nanostructures*, Y. K. Yap (Ed.), Springer, New York, pp. 1–22, 2009b.

Lee, C. H., S. Qin, M. A. Savaikar et al. Room-temperature tunneling behavior of boron nitride nanotubes functionalized with gold quantum dots. *Advanced Materials* 25(33) (2013): 4544–4548.

Lee, C. H., J. Wang, V. K. Kayatsha et al. Effective growth of boron nitride nanotubes by thermal chemical vapor deposition. *Nanotechnology* 19(45) (2008): 455605.

Lee, C. H., M. Xie, V. Kayastha et al. Patterned growth of boron nitride nanotubes by catalytic chemical vapor deposition. *Chemistry of Materials* 22(5) (2010): 1782–1787.

Lee, C. H., D. Zhang, and Y. K. Yap. Functionalization, dispersion, and cutting of boron nitride nanotubes in water. *Journal of Physical Chemistry C* 116(2) (2012): 1798–1804.

Lee, R., J. Gavillet, M. L. de La Chapelle et al. Catalyst-free synthesis of boron nitride single-wall nanotubes with a preferred zig-zag configuration. *Physical Review B* 64(12) (2001): 121405.

Li, X., X. Wang, L. Zhang et al. Chemically derived, ultrasmooth graphene nanoribbon semiconductors. *Science* 319(5867) (2008): 1229–1232.

Liu, Z., W. Cai, L. He et al. In vivo biodistribution and highly efficient tumour targeting of carbon nanotubes in mice. *Nature Nanotechnology* 2(1) (2006): 47–52.

Loiseau, A., F. Willaime, N. Demoncy et al. Boron nitride nanotubes with reduced numbers of layers synthesized by arc discharge. *Physical Review Letters* 76(25) (1996): 4737.

Lourie, O. R., C. R. Jones, B. M. Bartlett et al. CVD growth of boron nitride nanotubes. *Chemistry of Materials* 12(7) (2000): 1808–1810.

Ma, M. and R. M. Hill. Superhydrophobic surfaces. *Current Opinion in Colloid & Interface Science* 11(4) (2006): 193–202.

Ma, M., R. M. Hill, and G. C. Rutledge. A review of recent results on superhydrophobic materials based on micro- and nanofibers. *Journal of Adhesion Science and Technology* 22 (2008): 1799–1817.

Medlin, D., T. Friedmann, P. Mirkarimi et al. Evidence for rhombohedral boron nitride in cubic boron nitride films grown by ion-assisted deposition. *Physical Review B* 50(11) (1994): 7884.

Mele, E. and P. Král. Electric polarization of heteropolar nanotubes as a geometric phase. *Physical Review Letters* 88(5) (2002): 056803.

Mensah, S. L., V. K. Kayastha, I. N. Ivanov et al. Formation of single crystalline ZnO nanotubes without catalysts and templates. *Applied Physics Letters* 90(11) (2007): 113108–113103.

Morra, M., E. Occhiello, and F. Garbassi. Contact angle hysteresis in oxygen plasma treated poly(tetrafluoroethylene). *Langmuir* 5(3) (1989): 872–876.

Morton, J. J., D. R. McCamey, M. A. Eriksson et al. Embracing the quantum limit in silicon computing. *Nature* 479(7373) (2011): 345–353.

Nakhmanson, S., A. Calzolari, V. Meunier et al. Spontaneous polarization and piezoelectricity in boron nitride nanotubes. *Physical Review B* 67(23) (2003): 235406.

Novoselov, K. S., A. K. Geim, S. Morozov et al. Electric field effect in atomically thin carbon films. *Science* 306(5696) (2004): 666–669.

Okada, S., M. Igami, K. Nakada et al. Border states in heterosheets with hexagonal symmetry. *Physical Review B* 62(15) (2000): 9896.

Okada, S. and A. Oshiyama. Magnetic ordering in hexagonally bonded sheets with first-row elements. *Physical Review Letters* 87(14) (2001): 146803.

Pakdel, A., C. Zhi, Y. Bando et al. Boron nitride nanosheet coatings with controllable water repellency. *ACS Nano* 5(8) (2011): 6507–6515.

Rubio, A., J. L. Corkill, and M. L. Cohen. Theory of graphitic boron nitride nanotubes. *Physical Review B* 49(7) (1994): 5081–5084.

Shi, Y., C. Hamsen, X. Jia et al. Synthesis of few-layer hexagonal boron nitride thin film by chemical vapor deposition. *Nano Letters* 10(10) (2010): 4134–4139.

Smith, M. W., K. C. Jordan, C. Park et al. Very long single-and few-walled boron nitride nanotubes via the pressurized vapor/condenser method. *Nanotechnology* 20(50) (2009): 505604.

Song, L., L. Ci, H. Lu et al. Large scale growth and characterization of atomic hexagonal boron nitride layers. *Nano Letters* 10(8) (2010): 3209–3215.

Suenaga, K., C. Colliex, N. Demoncy et al. Synthesis of nanoparticles and nanotubes with well-separated layers of boron nitride and carbon. *Science* 278(5338) (1997): 653–655.

Sun, T., L. Feng, X. Gao et al. Bioinspired surfaces with special wettability. *Accounts of Chemical Research* 38(8) (2005): 644–652.

Suryavanshi, A. P., M.-F. Yu, J. Wen et al. Elastic modulus and resonance behavior of boron nitride nanotubes. *Applied Physics Letters* 84(14) (2004): 2527–2529.

Tang, C., Y. Bando, Y. Huang et al. Fluorination and electrical conductivity of BN nanotubes. *Journal of the American Chemical Society* 127(18) (2005): 6552–6553.

Tang, C., Y. Bando, T. Sato et al. A novel precursor for synthesis of pure boron nitride nanotubes. *Chemical Communications* 12 (2002): 1290–1291.

Terrones, M., W. Hsu, H. Terrones et al. Metal particle catalysed production of nanoscale BN structures. *Chemical Physics Letters* 259(5) (1996): 568–573.

Theis, T. N. and P. M. Solomon. It's time to reinvent the transistor! *Science* 327(5973) (2010): 1600–1601.

Velayudham, S., C. H. Lee, M. Xie et al. Noncovalent functionalization of boron nitride nanotubes with poly (p-phenylene-ethynylene) s and polythiophene. *ACS Applied Materials and Interfaces* 2(1) (2010): 104–110.

Wang, J., V. K. Kayastha, Y. K. Yap et al. Low temperature growth of boron nitride nanotubes on substrates. *Nano Letters* 5(12) (2005): 2528–2532.

Wang, J., C. H. Lee, Y. Bando et al. Multiwalled boron nitride nanotubes: Growth, properties, and applications. In *B–C–N Nanotubes and Related Nanostructures*, pp. 23–44, 2009a.

Wang, J. S., C. H. Lee, V. K. Kayastha et al. First success in the synthesis of boron nitride nanoribbons and hetero-junctions of boron nitride nanotubes and carbon nanotubes. Paper K17.6 presented at *2009 Materials Research Society Fall Meeting*, Boston, MA, November 30–December 4, 2009b.

Wang, J. S., C. H. Lee, and Y. K. Yap. Recent advancements in boron nitride nanotubes. *Nanoscale* 2(10) (2010): 2028–2034.

Wang, J. S., C. H. Lee and Y. K. Yap. Hetero-junctions of carbon nanotubes and boron nitride nanotubes. Paper CT38 at the *12th International Conference on the Science and Application of Nanotubes*, Cambridge, U.K., July 10–16, 2011.

Xie, S.-Y., W. Wang, K. S. Fernando et al. Solubilization of boron nitride nanotubes. *Chemical Communications* 29 (2005): 3670–3672.

Yap, Y., S. Kida, T. Aoyama et al. Influence of negative dc bias voltage on structural transformation of carbon nitride at 600°C. *Applied Physics Letters* 73(7) (1998): 915–917.

Yap, Y. K. CAREER: Synthesis, characterization and discovery of frontier carbon materials. Solid state and materials chemistry highlights of National Science Foundation, 2007. http://www.nsf.gov/mps/dmr/highlights/07highlights/ssmc.jsp.

Yap, Y. K. Hetero-junctions of boron nitride and carbon nanotubes: Synthesis and characterization. Paper presented at *DOE-BES Contractors Meeting*, Warrenton, VA, March 16–19, 2008.

Yap, Y. K. B–C–N nanotubes, nanosheets, nanoribbons, and related nanostructures. In the AZoNano.com "Nanotechnology Thought Leaders" Series, April 20, 2011. Available at: http://www.azonano.com/article.aspx?ArticleID=2847.

Yap, Y. K. Functional boron nitride nanotubes: Controlled growth and applications. In *13th International Conference on the Science and Application of Nanotubes*, Brisbane, Queensland, Australia, June 24–29, 2012.

Yu, J., L. Qin, Y. Hao et al. Vertically aligned boron nitride nanosheets: Chemical vapor synthesis, ultraviolet light emission, and superhydrophobicity. *ACS Nano* 4(1) (2010): 414–422.

Zeng, H., C. Zhi, Z. Zhang et al. "White graphenes": Boron nitride nanoribbons via boron nitride nanotube unwrapping. *Nano Letters* 10(12) (2010): 5049–5055.

Zhang, Y., Y.-W. Tan, H. L. Stormer et al. Experimental observation of the quantum Hall effect and Berry's phase in graphene. *Nature* 438(7065) (2005): 201–204.

Zhang, Y., Y. Tang, N. Wang et al. Silicon nanowires prepared by laser ablation at high temperature. *Applied Physics Letters* 72 (1998): 1835.

Zhi, C., Y. Bando, C. Tan et al. Effective precursor for high yield synthesis of pure BN nanotubes. *Solid State Communications* 135(1–2) (2005a): 67–70.

Zhi, C., Y. Bando, C. Tang et al. Perfectly dissolved boron nitride nanotubes due to polymer wrapping. *Journal of the American Chemical Society* 127(46) (2005b): 15996–15997.

Zhi, C., Y. Bando, C. Tang et al. Purification of boron nitride nanotubes through polymer wrapping. *Journal of Physical Chemistry B* 110(4) (2006): 1525–1528.

Zhu, L., Y. Xiu, J. Xu et al. Superhydrophobicity on two-tier rough surfaces fabricated by controlled growth of aligned carbon nanotube arrays coated with fluorocarbon. *Langmuir* 21(24) (2005): 11208–11212.

Recent Advancements in Boron Nitride Nanotube Biomedical Research

Gianni Ciofani, Barbara Mazzolai, and Virgilio Mattoli

CONTENTS

21.1 INTRODUCTION

During the past decades, nanostructured materials have been widely investigated for a huge spectrum of advanced technological applications, ranging from microelectronics to solar cells (Narayan 2012; Ferric et al. 2013; Kujawa and Winnik 2013). In recent years, multifunctional nanoparticle systems have been developed owning specific features that render them particularly suitable in biomedicine, for applications like medical diagnosis, drug delivery, and prosthetics (Gao and Xu 2009). Indeed, the exploitation of nanotechnology tools toward the development of biomedical devices has been found to allow for the control of the interactions between material surfaces and biological entities down to the molecular level. In this way, nanosurfaces, nanostructures, and, more generally, nanomaterials have been used in order to mimic the biological microenvironment and thus to promote specific cellular functions, like adhesion, mobility, and differentiation (Ferreira 2009).

Among nanomaterials, boron nitride nanotubes (BNNTs) represent an innovative and extremely intriguing example (Golberg et al. 2007; Terrones et al. 2007). Thanks to their impressive chemical and physical properties, they have found plenty of applications in the nanotechnology field, and studies about a possible exploitation in biomedicine as nanotransducers (Ciofani et al. 2010a) and drug nanocarriers (Ciofani 2010) have recently been proposed.

A BNNT presents an analogue structure of a carbon nanotube (CNT), with B and N atoms that entirely substitute C atoms in a graphitic-like sheet, with almost no change in atomic spacing. However, the significant ionic component of the B–N bonds strongly affects their features, which include an extremely high Young's modulus (Chopra and Zettl 1998), superior chemical

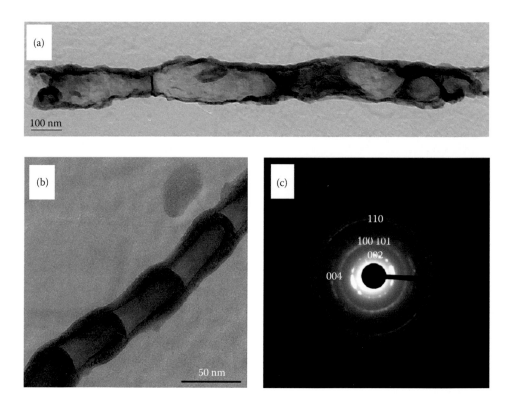

Figure 21.1   BNNT ultrastructural analysis. TEM images of BNNTs characterized by noncontinuous walls and absence of regular stacking of single units (a), and showing the well-known bamboo-like shape (b). Electron diffraction pattern of a bamboo-shaped BNNT: diffraction rings demonstrate the hexagonal structure of boron nitride (c). (Reproduced from *Colloid. Surf. B*, 111, Del Turco, S., Ciofani, G., Cappello, V. et al., Cytocompatibility evaluation of glycol-chitosan coated boron nitride nanotubes in human endothelial cells, 142–149, Copyright 2013, with permission from Elsevier.)

and thermal stability with respect to CNTs (Chen et al. 2004), a constant band gap of about 5.5 eV (Blasè et al. 1995), and very intriguing piezoelectric properties (Mele and Kral 2002; Nakhmanson et al. 2003).

Figure 21.1 shows, as an example, transmission electron microscopy (TEM) images of typical BNNTs obtained by using an annealing method from boron-containing precursors (Wang et al. 2010). We can observe as they are organized in shape characterized by a noncontinuous wall and by the absence of regular stacking (Figure 21.1a), as well by the well-known bamboo-like structure formed by a regular stacking of repeated units (Figure 21.1b). The electron diffraction pattern shows diffraction rings that can be indexed according to the hexagonal structure of boron nitride, thus suggesting the crystalline nature of the nanotube walls (Figure 21.1c).

In our recent review on the journal *Small* (Ciofani et al. 2013a), we have summarized and discussed the most recent advancements in biological and biomedical applications of BNNTs, with particular attention to those researches that seemed to be closest to an actual translational exploitation. In this chapter, we will go in detail about studies of biocompatibility or other significant results that emerged in the past 2 years, demonstrating once more the impressive interest of the bio-related research community toward BNNTs.

21.2 PROGRESS IN BIOCOMPATIBILITY ASSESSMENT

Dispersion of BNNTs in aqueous solutions represents a mandatory step before any further biological investigation. Several examples of BNNT dispersions with different surfactants or stabilizing agents can be found in the literature; among these, we have to mention solutions based on polyoxyethylene sorbitan monooleate (Tween 80; Horváth et al. 2011), glycol-chitosan (GC; Ciofani et al. 2010b), poly-L-lysine (Ciofani et al. 2010c), polyethyleneimine (Ciofani et al. 2008), dendrimers (Chen et al. 2009), PEGylated phospholipids (Lee et al. 2012), and many others reviewed in detail in Ciofani et al. (2013a).

Here, it is worth to mention an innovative approach proposed by Gao et al. (2012), involving the use of gum Arabic (GA), a natural polysaccharide extracted from exudates of *Arabic senegal* and *Acacia seyal* trees. Its chemical structure, not yet fully understood (Islam et al. 1997), is exploited to obtain stable and highly biocompatible dispersions of nanoparticles, including quantum dots (Wu and Chen 2010), magnetic nanoparticles (Batalha et al. 2010), and CNTs (Bandyopadhyaya et al. 2002).

GA was thus efficiently used for an easy disentanglement of BNNTs from bundles, thanks to a sonication procedure of BNNTs in GA water solutions, obtaining highly concentrated and stable BNNT dispersions in aqueous environments. Fourier transform infrared spectroscopy and fluorescence analysis confirmed a strong interaction between GA and BNNTs. Indeed, UV/visible spectra revealed a red shift of the maximum absorption peak, suggesting a modification of the electronic structure of BNNTs following GA noncovalent functionalization.

Finally, several other proteins (including streptavidin, bovine serum albumin, lysozyme, and immunoglobulin G) could be immobilized on GA-functionalized BNNTs, thanks to strong electrostatic interactions, as demonstrated by atomic force microscopy. This approach resulted extremely straightforward and efficient, opening interesting perspectives in the applications of BNNTs as vectors for the delivery of biomolecules and in other bio-related fields.

The actual exploitation of innovative inorganic nanoparticles in biomedicine has however to be derived from a previous deep and accurate investigation of their biocompatibility, both in vitro and in vivo. As we have recently reviewed (Ciofani et al. 2013a), in the past years, we have assisted to an increasing interest toward BNNT investigation in biological context. Many studies have pointed out favorable interactions between BNNTs and living matter, describing cytocompatibility of these innovative nanovectors on different cell types, including human embryonic kidney cells HEK 293 and Chinese hamster ovary cells CHO (Chen et al. 2009), human osteoblasts and mouse macrophages (Lahiri et al. 2010), human neuroblastoma cells SH-SY5Y (Ciofani et al. 2010b), mouse myoblasts C2C12 (Ciofani et al. 2010c), and neuronal-like PC12 cells (Ciofani et al. 2010a). Researches reporting on negative cellular effects are limited (Horváth et al. 2011) and most probably due to specific physical properties of the tested BNNTs (specifically, very long nanotubes with a huge aspect ratio). Here, we summarize some recent findings that further corroborate the hypothesis of BNNTs as safe nanomaterials.

The first barrier that injected nanoparticles find on their pathway toward a desired site is represented by the vascular endothelium (Couvreur and Vauthier 2006). Moreover, by virtue of its strategic location and important role in cardiovascular disease, endothelium itself represents an important target for drug or gene therapy (Yang et al. 2013). For all these reasons, an evaluation of the interactions between endothelial cells and nanoparticles can give important information about their potential, early stage in vivo behavior. At this aim, glycol-chitosan coated BNNTs (GC-BNNTs) have been tested by our group on human umbilical vein endothelial cells (HUVECs; Del Turco et al. 2013). Observation under microscopy of cell cultures at the endpoint of incubation (in the range of 0–100 µg/mL) did not show alteration of cell morphology and of cell monolayer integrity. Amido-black assay showed absence of cell necrosis up to 100 µg/mL of concentration, and the surface expression of E 1/1 constitutive antigen was confirmed. All these results were corroborated by several other complimentary assays, including evaluation of DNA synthesis and damage, analysis of cytoskeleton and focal adhesion organization, and specific assays aiming at evaluating activation of endothelium (surface expression of adhesion molecules, such as VCAM-1 and ICAM-1). Taken together, all these results clearly showed excellent cell viability following GC-BNNT treatment, with an optimal concentration range to exclude cytotoxicity, inhibition of proliferation, and cell activation, between 5 and 50 µg/mL.

Internalization was finally assessed through TEM analysis, which demonstrated absence of alterations in terms of cytosolic organelle ultrastructure. BNNTs appear internalized by HUVECs, without any preferential localization in the cellular cytoplasm or in organelles, but resulted enveloped by cytoskeleton, the structures of which appeared particularly enriched in the regions surrounding BNNTs.

Evaluation of BNNT biocompatibility was also performed by culturing cells *on* them, rather than allowing the incubation with BNNT-doped media. In a recent work, Li et al. (2012) prepared and characterized BNNT films, with special attention to bioapplications. These nanostructures present huge roughness that could be positive for cell adhesion and proliferation, and present a 3D nanoporous aspect that could be exploited as scaffold for tissue engineering applications. However, wettability of BNNT films is a key point to be addressed.

Films of high-density and pure BNNTs were grown on steel substrates by using the boron ink approach (Li and Chen 2010; Li et al. 2010). Thereafter, a room-temperature nitrogen/hydrogen ($N_2 + 15\%\ H_2$) mixture gas plasma treatment, by using continuous wave (CW) and a combination of continuous and pulsed (CW+P) modes, was used to change surface features of BNNT film from highly hydrophobic to hydrophilic. The followed protocols introduced negligible defects in BNNTs, and instead allowed their functionalization with amine, amide, and imine groups. These substrates were tested as surfaces for cell culturing on human fibroblasts and on a transformed mammary cell line (TXP RFP3). No evident sign of cytotoxicity was observed, even if cell proliferation was much more enhanced on plasma-treated surfaces. This increment in terms of cell number is due to the change of the BNNT films from hydrophobic to hydrophilic, being cell attachment more difficult on hydrophobic surfaces. However, it is not possible to exclude that also surface functionalization of BNNTs can have beneficial effects on cell proliferation, because of the introduction of positively charged groups that help cell adhesion and spreading.

Cell morphology was investigated by SEM analysis (Figure 21.2, see caption for details) and revealed that plasma-treated BNNT films provide a better environment for cell adhesion and viability: most cells appear attached and well spread on the BNNTs, suggesting a strong interaction with the substrate. Moreover, some cell projections are directed inside the BNNT films, forming a 3D architecture (indicated by arrows in Figure 21.2c, f), thus opening interesting opportunities in the exploitation of these nanomaterials for tissue engineering applications.

Cytocompatibility is the first necessary step to be accomplished for any toxicological evaluation of innovative nanomaterials, in particular concerning inorganic nanoparticles. However, since in vitro studies can highlight any adverse effect only at the cellular level, passage to in vivo investigation is mandatory before any realistic biomedical exploitation.

At this regard, very recently, our group started a systematic in vivo investigation of BNNT toxicology in rabbits (Ciofani et al. 2012a, 2013b), monitoring hematological values up to 7 days following injections of BNNT doses up to 10 mg/kg *per* animal. Moreover, BNNT half-life (HL) in blood was calculated, detecting, for the first time in the literature, this relevant pharmacokinetic value.

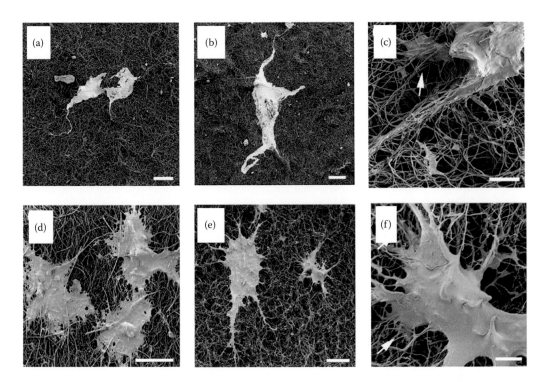

Figure 21.2   SEM images of fibroblasts on untreated (a) and plasma-treated (b, c) BNNT films, and of TXP RFP3 cells on untreated (d) and plasma-treated (e, f) BNNT films. Scale bars: (a, b) 50 μm, (c) 5 μm, (d, e) 20 μm, (f) 5 μm. (Reproduced with permission from Li, L., Li, L.H., Ramakrishnan, S. et al., Controlling wettability of boron nitride nanotube films and improved cell proliferation, *J. Phys. Chem. C*, 116, 18334–18339, 2012. Copyright 2012, American Chemical Society.)

Typical blood values (number of white cells, red cells, platelets, etc.) and blood biochemical parameters quantifying both renal and hepatic functions resulted within the tabulated ranges for rabbit, and/or not significantly altered before and during all the treatment (at 1, 3, and 7 days since injection), for all the experimental trials (5, 10, and 5 mg/kg once per day for 3 days). A terminal HL of about 1.5 h was found, a value similar to other inorganic nanoparticles, like CNTs (Cherukuri et al. 2006). Moreover, the absence of multiexponential kinetic components in the BNNT blood concentration trend indicates no significant temporary accumulation of the nanoparticles in the tissues, which could theoretically behave as reversible reservoirs (Neubauer et al. 2008); these data suggest a relatively high clearance of BNNT form the blood and a quick distribution in the organism and/or excretion. These results, even if preliminary, are extremely encouraging and confirm suitability of BNNTs as innovative and promising nanomaterials for biomedical applications.

21.3   PROGRESS IN BIOMEDICAL APPLICATIONS

Besides their use as nanovectors for drug delivery, BNNTs can be moreover considered as *drug* themselves, if we think about their exploitation in the boron neutron capture therapy (BNCT). This is in fact a clinical approach (often the unique) of increasing use in the treatment of several aggressive cancers, including cerebral *glioblastoma multiforme* (Capala et al. 2003). However, the essential requirement for this therapy is a selective targeting of tumor cells by sufficient quantities of ^{10}B atoms, able to capture low-energy thermal neutrons and immediately to decay into highly energetic, cell-disrupting radiation (^{4}He and ^{7}Li). The low selective accumulation of boron species in *glioblastoma multiforme* cells often represents the main problem in BNCT.

If a systemic administration of the boron-containing drug is envisaged, the problem of blood–brain barrier (BBB) crossing has to be considered. The BBB represents an extremely exclusive biological barrier, which restricts the access to most of the diagnostic and therapeutic agents, including nanoparticles (Abbott et al. 2010). The functionalization of the latter with specific moieties, which enable their receptor-mediated transcytosis through the BBB, is therefore mandatory. Among these, transferrin has been widely used as a successful targeting ligand (Ulbrich et al. 2009), as its receptors are highly expressed by BBB capillaries to mediate the delivery of iron to the brain.

Transferrin-conjugated BNNTs (tf-BNNTs) have been prepared by our group (Ciofani et al. 2012b) following a covalent functionalization approach where BNNTs are firstly strongly oxidized with nitric acid and thereafter grafted with 3-aminopropyl-triethoxysilane (APTES) in order to expose amino groups on their walls. Thereafter, green fluorescent-labeled (to allow BNNT tracking) transferrin is covalently bonded through the protein carboxylic residues.

The obtained tf-BNNTs were characterized and thereafter in vitro tested on HUVECs. Confocal microscopy highlighted a high internalization of tf-BNNTs by the cells, as evidenced by the presence of highly fluorescent green spots, which display a preferential positioning in the perinuclear area (Figure 21.3: cytoskeletal f-actin stained in red with TRITC-phalloidin, nuclei counterstained in blue with DAPI). Moreover, the tf-BNNT uptake by the endothelial cells resulted primarily mediated by the transferrin–transferrin receptor interaction, as demonstrated by internalization assays performed after saturation of transferrin receptors with *free* transferrin in the culture medium: in this case, fluorescence analysis revealed a significantly reduced uptake of tf-BNNTs. These results confirm the possibility to enhance and to target the uptake of BNNTs toward specific cells after an appropriate functionalization, thus allowing their use as selective nanovectors for cancer therapy and drug delivery.

The development of BNNTs as vectors for pharmaceutical applications would strongly benefit from the possibility of their noninvasive tracking through standard medical imaging approaches. A first attempt in this direction was performed by Soares and colleagues, who, in a recent work, injected ^{99m}Tc radiolabeled BNNTs in mice (Soares et al. 2012), evaluating their biodistribution by using scintigraphic imaging. However, use of nonionizing radiation should be privileged, if clinical applications are envisaged. At his aim, our group elaborated a straightforward strategy to obtain Gd-loaded BNNTs (Gd@BNNTs), in order to obtain BNNT tracking capability through magnetic resonance imaging (MRI). Gd@BNNTs were obtained through a simple sonication of defects-enriched BNNTs in a $GdCl_3$ solution, followed by their purification and stabilization (Ciofani et al. 2013c).

The relaxometric characterization at 7 T indicated that Gd@BNNTs have a longitudinal r_1 value comparable to clinically used Gd-based contrast agents at high fields. Instead, transversal relaxivity r_2 resulted largely exceeding the values of commercially available Gd-based contrast agents, and even higher than those found for T_2 contrast agents based on superparamagnetic iron oxide nanoparticles. A high r_2/r_1 ratio was determined for Gd@BNNTs (about 110), thus suggesting their use as negative MRI contrast agents at high fields. The unexpectedly high r_2

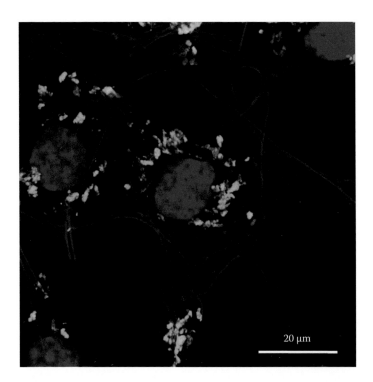

20 µm

Figure 21.3 Transferrin-grafted boron nitride nanotubes (green fluorescent) are actively internalized by human umbilical vein endothelial cells (in red, f-actin staining; in blue, nucleus staining). (Reproduced from *Int. J. Pharm.*, 436, Ciofani, G., Del Turco, S., Genchi, G.G., D'Alessandro, D., Basta, G., and Mattoli, V., Transferrin-conjugated boron nitride nanotubes: Protein grafting, characterization, and interaction with human endothelial cells, 444–453, Copyright 2012, with permission from Elsevier.)

value is probably ascribable to the confinement of Gd ions within the nanostructure, but further studies are in progress to address this point.

Obtained Gd@BNNTs resulted fully cytocompatible even at high concentrations (up to 100 µg/mL) and were able to label cells in MRI in vitro experiments. Indeed, the observed relaxometric properties and the in vitro results suggest as Gd@BNNTs hold promise for their future exploitation in theranostic applications, particularly in cancer therapy (Kim et al. 2013).

An interesting example of BNNTs as vectors in chemotherapy treatment is offered by the work of Li et al. (2013). Here, BNNTs–mesoporous silica hybrids (BNNTs@MS) with tunable surface Z-potential were synthesized, characterized, and in vitro tested for intracellular delivery of doxorubicin.

The functionalization of BNNTs with silica, a widely used material with optimal biocompatibility and easy to be functionalized with different chemical groups, allowed the improvement of the BNNT water suspension, the possibility to control the surface charge (from negative to positive, thus enhancing the cellular up-take), and finally to tailor the intracellular drug delivery. Doxorubicin was loaded through π-stacking on BNNTs, BNNTs@MS, and BNNTs@MS-NH$_2$ (the latter obtained with a treatment with APTES). Results indicated a high drug-loading capability of the nanostructures and a substantial increment of drug uptake by LNcap prostate cancer cells, with a consequent higher ability to kill cancer cells with respect to the free drug. Interestingly, better drug uptake and better results in terms of efficiency were achieved with BNNTs@MS-NH$_2$, because of the positive Z-potential that allows for a better cellular internalization. Overall, the enhanced uptake of doxorubicin through binding to BNNTs@MS is supposed to be due to the different uptake mechanisms, that is, simple diffusion in the case of *free* drug, and via endocytosis for doxorubicin loaded onto the nanotubes.

The exploration of BNNTs as anticancer drug delivery system could provide a multicomponent vector by combining chemotherapy with a physical approach, thanks to the peculiar properties of BNNTs. It has been found, for example, that piezoelectricity of BNNTs (Bai et al. 2007; Dai et al. 2009) can be exploited in biological context, eliciting specific functions in the cells. After preliminary results on differentiation of neuronal-like PC12 cells (Ciofani et al. 2010a), very recently, we have addressed our attention on complex in vitro systems.

Exciting results have been obtained by using BNNT-mediated stimulation on a coculture of human dermal fibroblasts and murine skeletal muscle C2C12 on flexible flat (F) or microgrooved (µG) polyacrylamide gels (Ricotti et al. 2013). The fibroblasts were used as feeder layer for the differentiation of skeletal myotubes, and the coculture was treated for several days with low-serum culture medium supplemented with 10 µg/mL BNNTs, and stimulated with low-frequency ultrasound (US, 10 s/day, 20 W, 40 kHz).

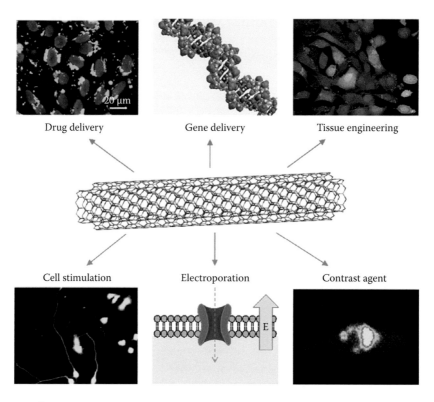

Figure 21.4 Schematic representation of potential applications of BNNTs in biomedicine. (From Ciofani, G., Danti, S., Genchi, G.G., Mazzolai, B., and Mattoli, V.: Boron nitride nanotubes: Biocompatibility and potential spill-over in nanomedicine. *Small*. 2013a. 9. 1672–1685. Copyright Wiley-VCH Verlag GmbH & Co. KGaA, Weinheim. Reproduced with permission.)

An isotropic cell distribution was clearly visible on F samples, while a strong preferential alignment of cells along the groove axis was found on μG samples. C2C12 differentiation was dramatically affected by this culturing protocol, with a significant increment both in fusion index and in some key-genes expression, like myogenin, muscle LIM protein, and MHC-IIa, with respect to cultures on F samples and not treated with BNNT+US stimulation. After 7 days of differentiation, analysis of immunofluorescence images confirmed that cells cultured on μG and treated with the BNNT+US stimulation formed longer and wider multinucleated myotubes with respect to other control samples.

From all the aforementioned examples, we can easily deduce that as interest toward bioapplications of BNNTs is exponentially increasing, they will be soon exploited in a wide range of fields, including cell/tissue stimulation, tissue engineering, and drug/gene delivery for cancer therapy (Ciofani et al. 2013a; Figure 21.4).

21.4 CONCLUSIONS AND PERSPECTIVES

In nanomedicine, just recently, scientists have started to exploit intrinsic properties of nanomaterials, rather than simply using them as carriers for medications. In this way, the plain nanomaterial represents a really active device, able to respond to external stimuli with intrinsic modifications of its chemical and/or physical characteristics. Given that any nanomaterial undergoing a dynamic and reversible change in its critical properties can be considered as a *smart* nanomaterial, it is easy to understand that smart nanomaterials could represent election tools to interact with biological systems and eventually to foster cell and tissue functions (Greco and Mattoli 2012).

In this regard, BNNTs represent actual smart nanomaterials, behaving as multifunctional nanovectors able to carry both chemical and physical stimuli in a living system. Recent advancements, as demonstrated in this chapter, are highly promising, and we are fully convinced that BNNTs will soon find tangible and concrete applications even in the preclinical field.

REFERENCES

Abbott, N.J., Patabendige, A.A., Dolman, D.E., Yusof, S.R., and Begley, D.J. 2010. Structure and function of the blood–brain barrier. *Neurobiol. Dis.* 37: 13–25.

Bai, X., Golberg, D., Bando, Y., Zhi, C.Y., Tang, C., and Mitome, M. 2007. Deformation-driven electrical transport of individual boron nitride nanotubes. *Nano Lett.* 7: 632–637.

Bandyopadhyaya, R., Nativ-Roth, E., Regev, O., and Yerushalmi-Rozen, R. 2002. Stabilization of individual carbon nanotubes in aqueous solutions. *Nano Lett.* 2: 25–28.

Batalha, L., Hussain, A., and Roque, A.C.A. 2010. Gum arabic coated magnetic nanoparticles with affinity ligands specific for antibodies. *J. Mol. Recognit.* 23: 462–471.

Blasé, X., Rubio, A., Louie, S.G., and Cohen, M.L. 1995. Quasiparticle band structure of bulk hexagonal boron nitride and related systems. *Phys. Rev. B* 51: 6868–6875.

Capala, J., Stenstam, B.H., Sköld, K. et al. 2003. Boron neutron capture therapy for glioblastoma multiforme: Clinical studies in Sweden. *J. Neurooncol.* 62: 135–144.

Chen, X., Wu, P., Rousseas, M. et al. 2009. Boron nitride nanotubes are noncytotoxic and can be functionalized for interaction with proteins and cells. *J. Am. Chem. Soc.* 131: 890–891.

Chen, Y., Zou, J., Campbell, S.J., and Le Caer, G. 2004. Boron nitride nanotubes: Pronounced resistance to oxidation. *Appl. Phys. Lett.* 84: 2430–2432.

Cherukuri, P., Gannon, C.J., Leeuw, T.K. et al. 2006. Mammalian pharmacokinetics of carbon nanotubes using intrinsic near-infrared fluorescence. *Proc. Natl. Acad. Sci. USA* 103: 18882–18886.

Chopra, N.G. and Zettl, A. 1998. Measurement of the elastic modulus of a multi-wall boron nitride nanotube. *Solid State Commun.* 105: 297–300.

Ciofani, G. 2010. Potential applications of boron nitride nanotubes as drug delivery systems. *Expert Opin. Drug Deliv.* 7: 889–893.

Ciofani, G., Boni, A., Calucci, L. et al. 2013c. Gd-doped BNNTs as T_2-weighted MRI contrast agents. *Nanotechnology* 24: 315101(1–7).

Ciofani, G., Danti, S., D'Alessandro, D. et al. 2010a. Enhancement of neurite outgrowth in neuronal-like cells following boron nitride nanotube-mediated stimulation. *ACS Nano* 4: 6267–6277.

Ciofani, G., Danti, S., D'Alessandro, D., Moscato, S., and Menciassi, A. 2010b. Assessing cytotoxicity of boron nitride nanotubes: Interference with the MTT assay. *Biochem. Biophys. Res. Commun.* 394: 405–411.

Ciofani, G., Danti, S., Genchi, G.G. et al. 2012a. Pilot in vivo toxicological investigation of boron nitride nanotubes. *Int. J. Nanomed.* 7: 19–24.

Ciofani, G., Danti, S., Genchi, G.G., Mazzolai, B., and Mattoli, V. 2013a. Boron nitride nanotubes: Biocompatibility and potential spill-over in nanomedicine. *Small* 9: 1672–1685.

Ciofani, G., Danti, S., Nitti, S., Mazzolai, B., Mattoli, V., and Giorgi, M. 2013b. Biocompatibility of boron nitride nanotubes: An up-date of in vivo toxicological investigation. *Int. J. Pharm.* 444: 85–88.

Ciofani, G., Del Turco, S., Genchi, G.G., D'Alessandro, D., Basta, G., and Mattoli, V. 2012b. Transferrin-conjugated boron nitride nanotubes: Protein grafting, characterization, and interaction with human endothelial cells. *Int. J. Pharm.* 436: 444–453.

Ciofani, G., Raffa, V., Menciassi, A., and Cuschieri, A. 2008. Cytocompatibility, interactions and uptake of polyethyleneimine-coated boron nitride nanotubes by living cells: Confirmation of their potential for biomedical applications. *Biotechnol. Bioeng.* 101: 850–858.

Ciofani, G., Ricotti, L., Danti, S. et al. 2010c. Investigation of interactions between poly-L-lysine coated boron nitride nanotubes and C2C12 cells: Up-take, cytocompatibility and differentiation. *Int. J. Nanomed.* 5: 285–298.

Couvreur, P. and Vauthier, C. 2006. Nanotechnology: Intelligent design to treat complex disease. *Pharm. Res.* 23: 1417–1450.

Dai, Y., Guo, W., Zhang, Z., Zhou, B., and Tang, C. 2009. Electric-field-induced deformation in boron nitride nanotubes. *J. Phys. D: Appl. Phys.* 42: 1–4.

Del Turco, S., Ciofani, G., Cappello, V. et al. 2013. Cytocompatibility evaluation of glycol-chitosan coated boron nitride nanotubes in human endothelial cells. *Colloids Surf. B* 111: 142–149.

Ferreira, L. 2009. Nanoparticles as tools to study and control stem cells. *J. Cell. Biochem.* 108: 746–752.

Ferric, C., Selly, E., Adityawarman, D., and Indarto, A. 2013. Application of nanotechnologies in the energy sector: A brief and short review. *Front. Energy* 7: 6–18.

Gao, J. and Xu, B. 2009. Applications of nanomaterials inside cells. *Nano Today* 4: 37–51.

Gao, Z., Zhi, C., Bando, Y., Golberg, D., Komiyama, M., and Serizawa, T. 2012. Efficient disentanglement of boron nitride nanotubes using water-soluble polysaccharides for protein immobilization. *RSC Adv.* 2: 6200–6208.

Golberg, D., Bando, Y., Tang, C., and Zhi, C. 2007. Boron nitride nanotubes. *Adv. Mater.* 19: 2413–2432.

Greco, F. and Mattoli, V. 2012. Introduction to active smart materials for biomedical applications. In *Piezoelectric Nanomaterials for Biomedical Applications*, G. Ciofani and A. Menciassi (eds.), pp. 239–245. Springer, Berlin, Germany.

Horváth, L., Magrez, A., Golberg, D. et al. 2011. In vitro investigation of the cellular toxicity of boron nitride nanotubes. *ACS Nano* 5: 3800–3810.

Islam, A.M., Phillips, G.O., Sljivo, A., Snowden, M.J., and Williams, P.A. 1997. A review of recent developments on the regulatory, structural and functional aspects of gum arabic. *Food Hydrocoll.* 11: 493–505.

Kim, T.J., Chae, K.S., Chang, Y., and Lee, G.H. 2013. Gadolinium oxide nanoparticles as potential multimodal imaging and therapeutic agents. *Curr. Top. Med. Chem.* 13: 422–433.

Kujawa, P. and Winnik, F.M. 2013. Innovation in nanomedicine through materials nanoarchitectonics. *Langmuir* 29: 7354–7361.

Lahiri, D., Rouzaud, F., Richard, T. et al. 2010. Boron nitride nanotube reinforced polylactide–polycaprolactone copolymer composite: Mechanical properties and cytocompatibility with osteoblasts and macrophages *in vitro*. *Acta Biomater.* 6: 3524–3533.

Lee, C.H., Zhang, D., and Yap, Y.K. 2012. Functionalization, dispersion, and cutting of boron nitride nanotubes in water. *J. Phys. Chem. C* 116: 1798–1804.

Li, L., Li, L.H., Ramakrishnan, S. et al. 2012. Controlling wettability of boron nitride nanotube films and improved cell proliferation. *J. Phys. Chem. C* 116: 18334–18339.

Li, L.H. and Chen, Y. 2010. Superhydrophobic properties of nonaligned boron nitride nanotube films. *Langmuir* 26: 5135–5140.

Li, L.H., Chen, Y., and Glushenkov, A.M. 2010. Boron nitride nanotube films grown from boron ink painting. *J. Mater. Chem.* 20: 9679–9683.

Li, X., Zhi, C., Hanagata, N., Yamaguchi, M., Bando, Y., and Golberg, D. 2013. Boron nitride nanotubes functionalized with mesoporous silica for intracellular delivery of chemotherapy drug. *Chem. Commun.* 49: 7337–7339. doi:10.1039/C3CC42743A.

Mele, E.J. and Kral, P. 2002. Electric polarization of heteropolar nanotubes as a geometric phase. *Phys. Rev. Lett.* 88: 056803(1–4).

Nakhmanson, S.M., Calzolari, A., Meunier, V., Bernholc, J., and Buongiorno Nardelli, M. 2003. Spontaneous polarization and piezoelectricity in boron nitride nanotubes. *Phys. Rev. B* 67: 2354061–2354065.

Narayan, J. 2012. Nanoscience to nanotechnology to manufacturing transition. *Int. J. Nanotechnol.* 9: 914–941.

Neubauer, A.M., Sim, H., Winter, P.M. et al. 2008. Nanoparticle pharmacokinetic profiling in vivo using magnetic resonance imaging. *Magn. Reson. Med.* 60: 1353–1361.

Ricotti, L., Fujie, T., Vazão, H. et al. 2013. Boron nitride nanotube-mediated stimulation of cell co-culture on micro-engineered hydrogels. *PLoS One* 8(8): e71707.

Soares, D.C.F., Ferreira, T.H., Ferreira, C.D.A., Cardoso, V.N., and De Sousa, E.M.B. 2012. Boron nitride nanotubes radiolabeled with [99m]Tc: Preparation, physicochemical characterization, biodistribution study, and scintigraphic imaging in Swiss mice. *Int. J. Pharm.* 423: 489–495.

Terrones, M., Romo-Herrera, J.M., Cruz-Silva, E. et al. 2007. Pure and doped boron nitride nanotubes. *Mater. Today* 10: 30–38.

Ulbrich, K., Hekmatara, T., Herbert, E., and Kreuter, J. 2009. Transferrin- and transferrin-receptor-antibody-modified nanoparticles enable drug delivery across the blood–brain barrier (BBB). *Eur. J. Pharm. Biopharm.* 71: 251–256.

Wang, J., Gu, Y., Zhang, L., Zhao, G., and Zhang, Z. 2010. Synthesis of boron nitride nanotubes by self-propagation high-temperature synthesis and annealing method. *J. Nanomater.* 540456(1–6).

Wu, C. and Chen, D. 2010. Facile green synthesis of gold nanoparticles with gum arabic as a stabilizing agent and reducing agent. *Gold Bull.* 43: 234–240.

Yang, H., Zhao, F., Li, Y. et al. 2013. VCAM-1-targeted core/shell nanoparticles for selective adhesion and delivery to endothelial cells with lipopolysaccharide-induced inflammation under shear flow and cellular magnetic resonance imaging *in vitro*. *Int. J. Nanomed.* 8: 1897–1906.

Chapter 22

Titanium-Based Nanorods and Nanosheets as Efficient Electrode Materials

Tao Tao

CONTENTS

22.1 INTRODUCTION

Titanium-based material is one of the most studied materials due to its unique properties and wide potential application in many fields (Chen and Mao 2007; Roy et al. 2011; Chen et al. 2013). For many of these applications, it is very important to manipulate their morphology, crystalline texture, and surface characteristics for better control of the properties. Especially, 1D and 2D titanium-based nanostructures, such as nanotubes, nanorods, and nanosheets, which have high surface-to-volume ratio and excellent electronic transport property, are of great interest. The detailed introduction to 1D and 2D titanium-based nanostructures can be found in review articles (Chen et al. 2011; Léonard and Talin 2011; Liu and Liu 2012).

In general, the fabrication of nanostructured electrodes seems to be one of the most promising tracks for improving the performances of power sources (Yang et al. 2009; Zhang et al. 2013). One-dimensional and two-dimensional titanium-based nanostructures have been investigated as key materials for fundamental research and technological applications in the fields of electrochemical energy storage due to their capacity to offer high surface area and greatly improved electron transfer pathways, and their advantages in terms of cost, safety, and rate capability (Djenizian et al. 2011; Zhu et al. 2012). For example, it has been consistently reported that 1D and 2D TiO_2 nanostructures can be an ideal host for reversible lithium insertion/removal and, hence, are promising anode materials for lithium ion batteries (Hu et al. 2006; Liu et al. 2012). It has also been demonstrated that TiO_2 nanotubes and $FeTiO_3$ nanoflowers show functionality as an electrode material for supercapacitors (Tao et al. 2011; Lu et al. 2012). Some main information on TiO_2 and $FeTiO_3$ nanostructures relevant to the work described in this chapter is given next.

Energy conversion and storage usually involve chemical reaction and/or physical interaction at the surface or interface. Lithium ion batteries and electrochemical capacitors, as important energy storage devices, have attracted growing interest in recent years (Lee and Cho 2011; Wang et al. 2012). They consist of two electrodes immersed in an electrolyte. Lithium ions move from the anode to cathode during discharge and back when charging, which forms the working principle of the lithium ion battery, is influenced by transport of lithium ions in the electrolyte. Electrochemical supercapacitors now are a new type of capacitors with complete different energy storage principles. Their capacitance value is determined by two storage principles including electrostatic and electrochemical charge storage. Electrostatic storage of the electrical energy is achieved by charge separation in an electrical double layer at the interfaces of electrodes with electrolyte. Electrochemical storage of the electrical energy is achieved by fast surface redox reactions with specifically adsorbed ions from the electrolyte. Several excellent review articles focused on lithium ion batteries and supercapacitors have been published (Simon and Gogotsi 2008; Etacheri et al. 2011; Marom et al. 2011).

22.1.1   Titanium Dioxide

Titanium dioxide (TiO_2) with a large band gap (3.1 eV) is a chemically stable nontoxic transition-metal oxide. It has three common crystalline polymorphs: rutile (space group $P4_2/mnm$), anatase ($I4_1/amd$), and brookite (*Pbca*). Rutile is the thermodynamically most stable modification of TiO_2. The metastable anatase and brookite phases convert to rutile at a high temperature (Ovenstone and Yanagisawa 1999). The properties of TiO_2 are known to improve greatly when this material is in a nanostructure form. Different nanostructures including 0D, 1D, and 2D exhibit their unique performance based on surface and structural properties respectively.

Various approaches have been suggested to fabricate TiO_2 nanostructures, such as a template-assisted method (Jung et al. 2002), sol–gel process (Tsai et al. 2009), electrochemical anodic oxidation (Gong et al. 2001), chemical vapor deposition (Pradhan et al. 2003), and hydrothermal process (Kasuga et al. 1999). TiO_2 nanowires and nanotubes have been synthesized by a simple hydrothermal reaction between NaOH and TiO_2 under controlled temperature, time, and/or pressure (Armstrong et al. 2004; Zhou et al. 2010; Tan et al. 2012). TiO_2 nanorods have been reported by Zhang et al. (Zhang and Gao 2003) via treating a dilute $TiCl_4$ solution at 60°C–150°C for 12 h in the presence of acid or inorganic salts. Sheet-like anatase TiO_2 has been produced via a simple hydrothermal method using tetrabutyl titanate, $Ti(OBu)_4$, as a source and 47% hydrofluoric acid solution as the solvent (Han et al. 2009).

22.1.2   Iron (II) Titanium

Iron (II) titanium ($FeTiO_3$), namely, ilmenite, is a weak magnetic titanium–iron oxide mineral, which is iron-black or steel-gray (Tao et al. 2013). It is a semiconductor with a band gap of about 2.54–2.58 eV. Ilmenite is available in large quantities (world's total reserves are in excess of 680 million tons), cheap (80–107 USD/metric tons in 2004–2008 with a recent peak of 250–350 USD/metric tons in 2012), and can be found in various geographical locations—America, Australia, Europe, Asia, and Africa. According to 2011 statistics, Australia is the world's largest ilmenite ore producer, with about 1.3 million ton of production, followed by South Africa, Canada, Mozambique, India, China, Vietnam, Ukraine, Norway, Madagascar, and the United States. Additionally, the same phase of $FeTiO_3$ can be prepared in the laboratory. Aggregates of $FeTiO_3$ nanoparticles have been mechanochemically synthesized by the milling of TiO_2 in steel-milling equipment (Ohara et al. 2010). Single crystalline $FeTiO_3$ nanodisks have been prepared by a hydrothermal reaction between $FeSO_4 \cdot 7H_2O$ and titanium isopropoxide at 220°C in aqueous tetrabutyl-ammonium hydroxide (Kim et al. 2009).

In this chapter, combination methods for synthesizing TiO_2 nanorods from ilmenite have been developed in an Australian lab (Yu et al. 2009, 2010; Tao et al. 2013; Tao and Chen 2013). These methods consist of two steps: ball milling as a first step, and annealing or wet chemical treatment as a second step. The obtained TiO_2 nanorods exhibit an excellent electrochemical performance, when they are used in an anode for lithium ion batteries.

Nanostructured ilmenite $FeTiO_3$ can be obtained by high-energy ball milling of natural ilmenite and subsequent mild hydrothermal treatment in 2 M NaOH aqueous solution. The nanostructured $FeTiO_3$ forms in a special morphology of nanosheets (Tao et al. 2011). $FeTiO_3$ nanosheets are evaluated as electrode materials for electrochemical supercapacitors.

22.2   EXPERIMENTAL CONCEPTS AND METHODS

22.2.1   Ball Milling

Ball milling experiments are usually conducted in cylindrical vials containing balls. The vials are occasionally equipped with gas inlets in order to vary the grinding atmosphere (Ar, O_2, H_2, N_2, or vacuum), and the grinding is done by placing the containers on various types of grinders, vibratory or planetary, to obtain powders with highly controlled physical characteristics. The repeated grindings (e.g., welding and fracturing of powders) enable us to successively create the defects and new interfaces, therefore leading to a decrease in crystallite size, a greater increase in the surface area of powders, changes in chemical reactivity and volatility of the materials, and highly homogeneous mixing of components. The obtained powders are generally very different from those obtained via thermal treatment as far as their structure and texture are concerned. More detailed description of ball milling can be found elsewhere (Chen 1997; Chen et al. 1997, 1999; Park et al. 2001; Li et al. 2004; Cheng et al. 2005;

Glushenkov et al. 2008; Cao et al. 2009 ; Luo et al. 2009; Ren and Gao 2010; Sarkar et al. 2010; Tao et al. 2010; Hou et al. 2011; Jin et al. 2011; Sun et al. 2011).

22.2.2   Annealing

Annealing treatments are largely characterized by induced microstructural changes that are ultimately responsible for altering the material's mechanical properties. Furthermore, an efficient annealing technology can well control thermal budgets and possibly improve the crystal structure and crystal quality of material. The milled precursor will be annealed at an elevated temperature to yield the desired product.

A naturally occurring iron titanate ($FeTiO_3$) was used as a starting material for the preparation of titanium-based nanostructures. Ilmenite (99% purity) was provided by Consolidated Rutile Limited located in Australia, and its Chemical composition (wt%) is TiO_2 (dry basis) 49.6, iron (total) 35.1, FeO 32.8, Fe_2O_3 13.7, Al_2O_3 0.47, Cr_2O_3 0.25, SiO_2 0.45.

TiO_2 nanorods were synthesized by ball milling and annealing. Several grams of the mixture of ilmenite ($FeTiO_3$) and active carbon (weight ratio of 4:1) were milled in vacuum atmosphere in a Fritsch planetary ball mill with 10 steel balls (diameter 1 cm) for 50 h at room temperature. Isothermal annealing of the mixture of ilmenite and carbon was conducted in a horizontal tube furnace in different atmospheres (Ar or N_2–5% H_2) at a flow rate of 100 mL min^{-1} and at different temperatures (700°C–1200°C).

22.2.3   Wet Chemical Treatment

The wet chemical treatment is almost identical to the hydrothermal method except that the temperature used here is much lower. The milled ilmenite powder will be used as a precursor for the growth of nanostructured materials during wet chemical treatment process. In all of the wet chemical experiments, the powders, stirring rods, and their corresponding aqueous solutions are placed in SchÖtt bottles. Magnetic stirring is used to stir the solutions. A paraffin-oil bath on top of a hot plate is used for heating the immersed bottles.

Ilmenite powder (10 g) and four hardened steel balls (diameter: 25.4 mm) were loaded into the stainless-steel container of a magneto-ball mill. Ball milling was conducted for 150 h at room temperature under an argon atmosphere (100 kPa). The magnet was located on the bottom of the mill at a 45° angle relative to the vertical direction, and the rotation speed was 160 rpm.

$FeTiO_3$ nanosheets were prepared from milled ilmenite by the following mild hydrothermal treatment. The milled ilmenite (3 g) was treated with a 2 M aqueous solution of NaOH (300 mL) at 120°C for 2 h. The filtered samples were washed and dried at 90°C for 4 h.

The rutile TiO_2 nanorods were prepared by leaching an NaOH-treated sample (1 g) in 4 M HCl (100 mL) at 90°C for 4 h. The suspension was filtered after leaching, and the leached samples were washed and dried at 90°C for 4 h.

22.2.4   Characterization of Materials

A set of characterization methods were applied to the assessment of nanostructured materials including scanning electron microscopy (SEM), transmission electron microscopy (TEM), x-ray diffraction (XRD), x-ray energy dispersive spectroscopy (EDS) system, low-temperature N$_2$ adsorption, backscattered electron imaging (BSE), and x-ray photoelectron spectroscopy.

22.2.5 Electrochemical Testing

The electrochemical properties of nanostructured electrode materials were tested using an electrochemical workstation from Solartron Analytical (an eight-channel 1470E potentiostat/ galvanostat with an attached 1255B frequency response analyzer) and Ivium Technologies (an eight-channel Ivium-n-stat potentiostat–galvanostat fitted with impedance capability on each channel).

The electrodes for supercapacitors were assessed in aqueous electrolytes using three-electrode electrochemical cells. Platinum wire was used as a counter electrode (needed for passing current), and a suitable electrode (Ag/AgCl or Hg/HgO, depending on the electrolyte) was used as a reference electrode. The slurry for the working electrodes was prepared by mixing 70 wt% FeTiO$_3$ nanostructures with 20 wt% carbon black (Aldrich no. 699633) and 10 wt% polyvinylidene difluoride (PVDF) in N-methyl-2-pyrrolidinone (NMP). Aqueous electrolytes of basic (KOH), acidic (H$_2$SO$_4$), and neutral (NaCl) types were used for the assessment of capacitive properties in the corresponding media.

To test the electrochemical performance of TiO$_2$ nanorods, the active material was mixed with carbon black and a binder, PVDF, in a weight ratio of 80:10:10 in a solvent (NMP). The slurry was uniformly pasted onto Cu foil substrates, and these coated electrodes were dried in a vacuum oven at 100°C for 12 h. Coin cells (CR2032 type) were fabricated using Li foil as the counter electrode, a porous polyethene film as the separator, and LiPF$_6$ (1 M) in ethylene carbonate/dimethyl carbonate/diethyl carbonate (EC/DMC/DEC, 1:1:1 vol%) as the electrolyte in an Ar-filled glove box (Innovative Technology, USA). The cells were tested over a voltage range of 1–3 V vs Li/Li$^+$.

Cyclic voltammetry (CV), galvanostatic charge/discharge (GCD), and impedance spectroscopy were primary methods for the testing of electrochemical properties. GCD was performed to assess charge–discharge behavior and rate capability at various current rates. Cyclic stability was monitored by repeating the charging/discharging routine for numerous cycles.

22.3 PREPARATION OF TIO$_2$ NANORODS FROM THE NATURAL MINERAL ILMENITE

One-dimensional TiO$_2$ nanostructures, such as nanorods, nanotubes, and nanowires, have presented highly interesting properties for possible application in lithium ion batteries (Armstrong et al. 2006; Ortiz et al. 2009; Qiu et al. 2010). The synthesis methods as well as the precursors chosen for preparing these nanostructures play a key role. We demonstrate two different combination methods for the mass production of TiO$_2$ nanorods from mineral ilmenite, and they are presented in this section. One method that is a solid-state process includes ball milling and two sequential annealing. First high-temperature annealing produced metastable titanium oxide phases, and subsequent second low-temperature annealing in N$_2$–5% H$_2$ activates the growth of rutile nanorods. The other method for the synthesis of TiO$_2$ nanorods from ilmenite includes ball milling and two sequential wet chemical treatments in (a) aqueous NaOH solution (an optional step is used for the removal of SiO$_2$ impurities from ball-milled ilmenite) and (b) aqueous solution of HCl.

22.3.1 Synthesis of TiO$_2$ Nanorods via a Solid-State Ball Milling and Annealing Process

Figure 22.1a shows an SEM image of starting ilmenite, which consists of large particles in the range of about 100 μm. Ball milling of the mixture of ilmenite and activated carbon leads to significant changes in morphology. The changes are depicted in Figure 22.1b. Both FeTiO$_3$ and C have been reduced down to small particles of about 100 nm. The XRD pattern of the milled mixture (Figure 22.1c) shows the peaks of FeTiO$_3$ phase (Powder Diffraction file 29-733). The peak around 21° is associated with the amorphous structure of activated carbon. The relatively broadened peak shapes are caused by the crystal size reduction induced by ball milling treatment.

The product of first annealing of the milled mixture at 1200°C for 8 h in Ar–5% H$_2$ gases flowing at 100 mL min^{-1} to induce carbothermic reductions, which is shown by the XRD diffraction pattern in Figure 22.2a. It reveals the presence of monoclinic Ti$_3$O$_5$, Ti$_2$O$_3$, and α-Fe phases, indicating the partial reduction of FeTiO$_3$ by C and H$_2$. The BSE image of the reduced sample in Figure 22.2b shows a mixture of small bright Fe-dominant particles and large gray Ti-containing particles. The different contrasts are due to different atomic masses of Ti and Fe. EDS mapping was used to examine chemical compositions of the mixture. The Ti and Fe elemental mapping images are shown in Figure 22.2c and d, which can confirm the chemical nature of the Ti and Fe particles in the BSE image. These results indicate that FeTiO$_3$ has been reduced into large titanium oxide particles and small iron particles.

The reduced sample was further annealed for 4 h in N$_2$–5% H$_2$ at a lower temperature of 700°C to activate the 1D growth of rutile structure. The SEM image of the final product in Figure 22.3a

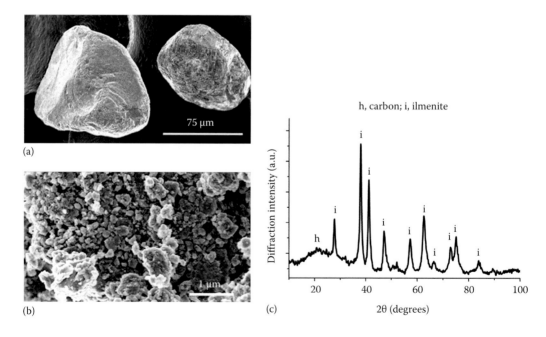

(a)

(b)

(c)

Figure 22.1 SEM images of ilmenite (a) and ball-milled mixture of ilmenite and carbon (b), and (c) XRD pattern of the milled mixture. (Reprinted with permission from Yu, J., Chen, Y., and Glushenkov, A.M., Titanium oxide nanorods extracted from ilmenite sands, *Cryst. Growth Design*, 9, 1240–1244, 2009. Copyright 2009, American Chemical Society.)

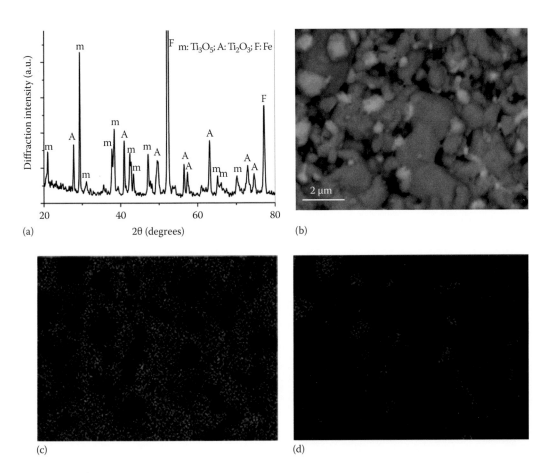

m: Ti$_3$O$_5$; A: Ti$_2$O$_3$; F: Fe

(a)   2θ (degrees)

(b)

(c)   (d)

Figure 22.2   (a) XRD pattern, (b) BSE image, (c) Ti mapping, and (d) Fe mapping of the sample after the first annealing at 1200°C. (Reprinted with permission from Yu, J., Chen, Y., and Glushenkov, A.M., Titanium oxide nanorods extracted from ilmenite sands, *Cryst. Growth Design*, 9, 1240–1244, 2009. Copyright 2009, American Chemical Society.)

shows a layer of nanorods of around 100 nm in diameter and a few micrometers in length cover on all titanium oxide surfaces. Typical rectangular cross sections and sizes of the nanorods can be seen clearly in Figure 22.3b. Figure 22.3c shows a bright-field TEM image of a nanorod, and its corresponding SAED pattern is displayed in Figure 22.3d. Electron diffraction pattern consists of a regular periodic array of dots and indicates that the nanorod has a single crystalline structure. The [002] direction of elongation was concluded from the SAED pattern (its orientation was carefully corrected for the rotation induced by the lenses of the microscope). The nanorod has a nonuniform contrast in a bright-field image, which is related to the complex shape of the cross section of the nanorods. The darker area in the center of the nanorod corresponds, most likely, to a larger thickness of its crystal in this area. Figure 22.4e shows a high-resolution image taken from the edge of the same nanorod. Two types of lattice planes are resolved with the distances of 0.23 and 0.25 nm corresponding to (200) and (101) crystal planes of rutile TiO$_2$, respectively. TEM analysis reveals that the nanorod growth direction is parallel to the c axis of tetragonal cell of rutile structure, and the nanorod side walls of (002) and (101) planes actually construct unit cells

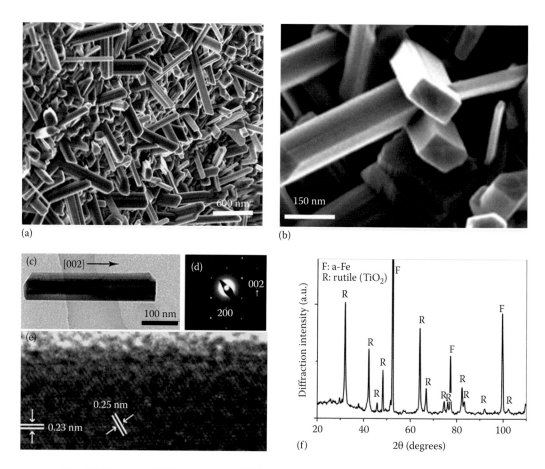

Figure 22.3   SEM images, TEM images, and XRD patterns showing sample morphology changes during the second annealing at 700°C in N_2–5% H_2. SEM images of nanorods: (a) low magnification and (b) high magnification; TEM image of a nanorod: (c) low magnification, (d) SAED pattern, and (e) high-magnification image taken from the edge of the rod; (f) XRD pattern of the sample after second annealing. (Reprinted with permission from Yu, J., Chen, Y., and Glushenkov, A.M., Titanium oxide nanorods extracted from ilmenite sands, *Cryst. Growth Design*, 9, 1240–1244, 2009. Copyright 2009, American Chemical Society.)

of tetragonal rutile. This can explain the rectangular cross sections observed in Figure 22.3b. The XRD pattern of the final sample in Figure 22.3f shows only TiO_2 rutile and Fe phases. Ti_3O_5 and Ti_2O_3 phases are no longer detected. A selective chemical leaching treatment using 3 M HCl solution was found to effectively remove most Fe.

The formation of special nanorod structure may be because of one-directional growth of the rutile crystals in the second annealing treatment. A number of nanorods formed at different stages found that nanorods with large cross sections seem to be etched during extended annealing in hydrogen-containing gas. One large nanorod became several nanorods with smaller diameters. This phenomena might be caused by hydrogen gas reduction along [002] directions. Without hydrogen gas, rutile particles can be produced from ilmenite but not in nanorod form. Therefore, we believe that hydrogen gas has played important roles in the two annealing processes. The morphological transformation and the formation mechanism of these TiO_2 nanorods have been previously discussed in detail elsewhere (Yu et al. 2009; Yu and Chen 2010).

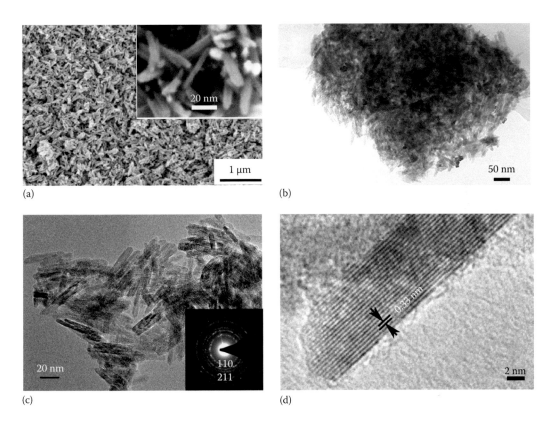

Figure 22.4 SEM images of (a) nanorods obtained by HCl leaching of the milled and NaOH-treated ilmenite, and TEM images of the obtained nanorods (b–d): (b, c) low-resolution images, the inset of (c) is the SAED pattern, and (d) high-resolution image. (Tao, T., Chen, Y., Zhou, D., Zhang, H.Z., Liu, S., Amal, R., Sharma, N., and Glushenkov, A.M.: Expanding the applications of the ilmenite mineral to the preparation of nanostructures: TiO₂ nanorods and their photocatalytic properties in the degradation of oxalic acid. *Chem. Eur. J.* 2013. 19. 1091–1096. Copyright Wiley-VCH Verlag GmbH & Co. KGaA. Reproduced with permission.)

In summary, a novel solid-state method for producing rutile TiO₂ nanorods from ilmenite sands has been demonstrated. This method includes three steps: ball milling of a mixture of ilmenite and activated carbon in a planetary milling device, first annealing of the ball milled mixture at 1200°C in Ar–5% H₂ results in the reduction of ilmenite into metastable titanium oxides and catalytic iron, and the second annealing at 700°C in N₂–5% H₂ atmosphere to activate the growth of nanosized rutile rods. The electron microscopy study has demonstrated that the growth direction of the obtained nanorod is parallel to the c axis of tetragonal cell of rutile structure, and the nanorod side walls of (002) and (101) planes actually construct unit cells of tetragonal rutile.

22.3.2 Synthesis of TiO₂ Nanorods via Ball Milling and Wet Chemical Treatment Processes

An SEM image of the material after leaching in an aqueous solution of HCl is shown in Figure 22.4a. As a result of leaching, ilmenite changes its morphology from irregular aggregates of submicrometer particles into a rod-like nanoarchitecture. The typical length of the rods is in the range 50–100 nm, their width is between 5 and 20 nm, and their thickness is in the range 2–5 nm. The nanorods were also evaluated by TEM; low-magnification images are shown in Figure 22.4b and c. Aggregates of short nanorods are visible, and their electron diffraction pattern (the inset of Figure 22.4c) matches the expected pattern of a rutile TiO_2 phase. The TiO_2 nanorods appear to be single crystals, and the single-crystalline nature of a TiO_2 nanorod is demonstrated in Figure 22.4d. During the TEM analysis, we observed that the nanorods were sensitive to beam damage and easily broke into tiny nanocrystallites under the electron beam.

XRD pattern of the samples after the wet chemical treatments in a 4 M aqueous solution of HCl for 8 h is presented in Figure 22.5. It matched the standard pattern of tetragonal rutile TiO_2 (JCPDS No. 01-076-1939). Thus, we conclude that rutile TiO_2 nanorods can be prepared from ilmenite by ball milling, treatment in an alkaline solution (optional), and leaching in dilute HCl.

The formation mechanism of the TiO_2 nanorods during the acid leaching of ilmenite can be interpreted as a dissolution/hydrolysis/precipitation process. Both iron and titanium migrate into the solution to form $FeCl_2$ and $TiOCl_2$, when the milled ilmenite comes into contact with the HCl solution. $TiOCl_2$ may be subsequently hydrolyzed and precipitate in the form of tiny TiO_2 crystals that continue their growth in a 1D fashion. The mechanism has been discussed in detail elsewhere (Tao et al. 2013).

In summary, another new method for the synthesis of single-crystal TiO_2 nanorods from natural ilmenite has also been demonstrated. This method includes ball milling and two sequential wet chemical treatments in (1) aqueous NaOH (optional) and (2) aqueous HCl.

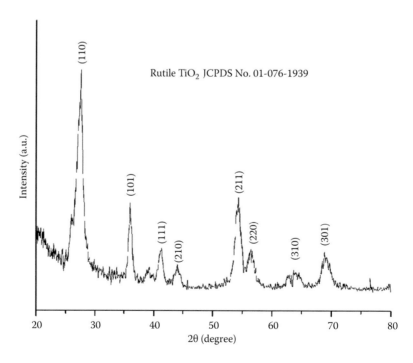

Figure 22.5 XRD patterns of the materials that were obtained from ilmenite after leaching of the milled and the NaOH-treated sample in HCl for 8 h.

Dissolution–hydrolysis–precipitation is proposed as the growth mechanism. Electron microscopy demonstrated that the nanorods had good crystallinity, and their dimensions were 50–100 nm (length) by 5–20 nm (width) by 2–5 nm (thickness).

22.4   GROWTH OF ILMENITE (FeTiO$_3$) NANOSHEETS FROM THE MILLED ILMENITE POWDER

Ilmenite FeTiO$_3$ is a semiconductor with a band gap of about 2.54–2.58 eV and distinct magnetic properties. Nanostructured ilmenite may possess attractive properties, opening novel applications for FeTiO$_3$. For example, heterojunctions of FeTiO$_3$ nanodisks and TiO$_2$ nanoparticles have been shown to act as a photocatalyst with enhanced photocatalytic activity. Thus, it is important to find an efficient way for producing this nanostructured material.

We have developed a combination method capable of producing large quantities of FeTiO$_3$ sheet–like nanostructures, and it is presented in this section. An ilmenite powder is first milled in a ball mill to prepare a nanocrystalline precursor and then treated in 2 M aqueous NaOH solution to produce nanosheets.

According to our results, the hydrothermal treatment of the ball-milled ilmenite in 2 M NaOH aqueous solution at 120°C leads to noticeable changes in the material's morphology. The SEM images of the product after such a treatment are shown in Figure 22.6a and b. This material is composed of uniform flowerlike architectures of about 1–2 µmin diameter. Each nanoflower consists of a number of petals with smooth surfaces (Figure 22.6b). Each petal is 5–20 nm thick and 100–200 nm wide, and different petals are interconnected. The quantitative EDS analysis reveals that the elemental composition (wt%) of nanoflowers is nearly identical to that of the original ilmenite FeTiO$_3$. Figure 22.6c through g shows the results of the TEM characterization. A typical nanosheet (petal) of nanoflowers oriented in a convenient way (with its side wall being perpendicular to the electron beam) is depicted in Figure 22.6c. An inverse FFT image of the petal is shown in Figure 22.6d, and the characteristic angle between (011) and (003) planes has a good fit to the ideal atomic arrangement in the FeTiO$_3$ crystal (Figure 22.6f). A fast Fourier transform (FFT) pattern derived from the plate shown in Figure 22.6c is presented in Figure 22.6e. It represents a periodic array of spots, which are slightly distorted into small arcs. Such a pattern indicates that the plate is either a single crystal with the presence of some defects (possibly dislocations) introducing minor rotations or can be described as a highly textured polycrystal in which nanocrystalline grains are slightly disoriented. The pattern can be indexed as the same phase of ilmenite FeTiO$_3$, as shown by XRD (Figure 22.7). The incident electron beam is close to the normal of the plate, which is, therefore, parallel to the [100] crystallographic direction (Figure 22.6g). The exposure plane of the nanosheets can be identified as (2$\bar{1}$0).

The XRD pattern of the sheet-like nanostructures (Figure 22.7) is very close to that of the milled ilmenite and agrees well with the standard XRD pattern of ilmenite FeTiO$_3$ (JCPDS 01-075-1211). The diffraction peaks of them are weak and broad. The broadening of the peaks is due to the small crystallite size.

We believe that the milled FeTiO$_3$ powders can react with NaOH under hydrothermal conditions. According to the fact that the presence of NaOH is critical for the formation of nanosheets, this reaction is particularly important. Sodium titanate or another phase that forms as a result of chemical reaction between FeTiO$_3$ and NaOH must be soluble in water in order for the dissolution–precipitation mechanism to happen, as it has been discussed in more detail elsewhere (Tao et al. 2011).

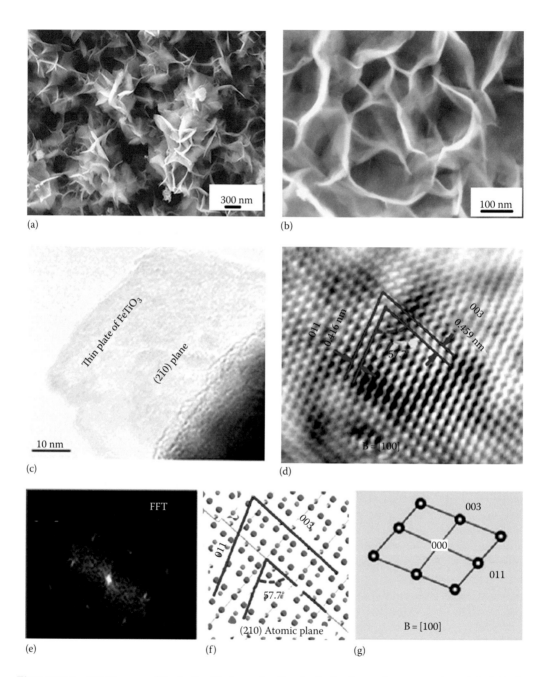

Figure 22.6   SEM images of the ball-milled sample after the hydrothermal treatment in 2 M NaOH solution for 2 h (a, b), and TEM characterization of nanosheets: (c) A bright-field image showing a typical nanopetal, (d) an inverse FFT image showing (011) and (003) lattice fringes, (e) the FFT pattern derived from (c), (f) top view of the $(2\bar{1}0)$ atomic plane, and (g) an indexed pattern corresponding to the zone axis of [100]. (Reprinted with permission from Tao, T., Glushenkov, A.M., Liu, H.W., Liu, Z.W., Dai, X.J.J., Chen, H., Ringer, S.P., and Chen, Y., Ilmenite $FeTiO_3$ nanoflowers and their pseudocapacitance, *J. Phys. Chem. C*, 115, 17297–17302, 2011. Copyright 2011, American Chemical Society.)

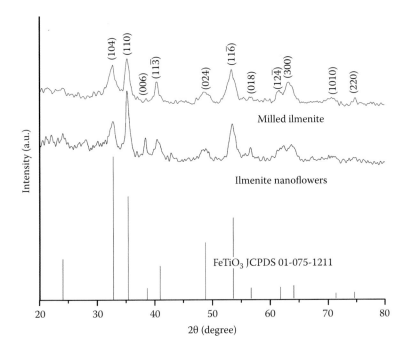

Figure 22.7   XRD patterns for the milled FeTiO$_3$ and the resulting nanosheets (strong lines from the JCPDS 01-075-1211 card are also shown at the bottom of the graph). (Reprinted with permission from Tao, T., Glushenkov, A.M., Liu, H.W., Liu, Z.W., Dai, X.J.J., Chen, H., Ringer, S.P., and Chen, Y., Ilmenite FeTiO$_3$ nanoflowers and their pseudocapacitance, *J. Phys. Chem. C*, 115, 17297–17302, 2011. Copyright 2011, American Chemical Society.)

In summary, we have demonstrated that sheet-like FeTiO$_3$ nanostructures can be prepared via high-energy ball milling of ilmenite and subsequent mild hydrothermal treatment in 2 M NaOH aqueous solution. The sheet-like nanostructures are composed of a number of petals with smooth surfaces, and each petal is 5–20 nm thick and 100–200 nm wide. It is demonstrated that the presence of NaOH in water is necessary for the formation of nanosheets. A dissolution–precipitation mechanism is proposed to explain the formation of the FeTiO$_3$ nanostructures.

22.5   APPLICATIONS IN ELECTROCHEMICAL SUPERCAPACITORS AND LITHIUM ION BATTERIES

22.5.1   TiO$_2$ Nanorods as Anode for Lithium Ion Batteries

TiO$_2$ has been investigated as a possible candidate of anode materials for lithium ion batteries for a long time. It is notable that the performance of TiO$_2$ depends greatly on its crystalline phase, size, surface state, and microstructures (Yue et al. 2009; Shim et al. 2010). The electrochemical performances of TiO$_2$ nanorods prepared by leaching the ball-milled ilmenite have been evaluated for their application in lithium ion batteries and are reviewed next.

Figure 22.8a shows a cyclic voltammogram (CV) profile of TiO$_2$ nanorods prepared by treating the milled ilmenite powder in 2 M NaOH solution. The nanorods depict typical electrochemical

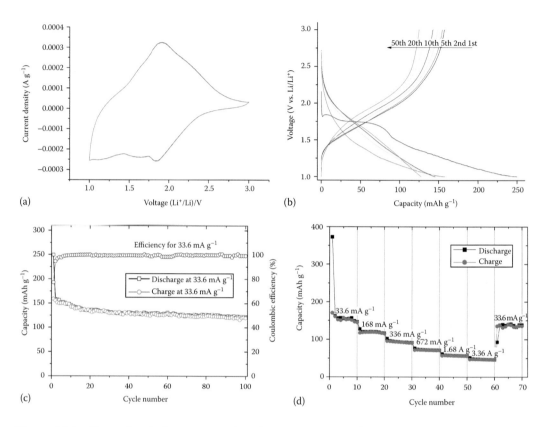

Figure 22.8 Electrochemical properties of rutile TiO_2 nanorods: (a) CV at a scan rate of 0.2 mV s^{-1} between 1.0 and 3.0 V; (b) selected galvanostatic discharge/charge voltage curves at a constant current rate of 33.6 mA g^{-1}; (c) variation of charge–discharge vs cycle number and coulombic efficiency at a constant current rate of 33.6 mA g^{-1}; (d) rate capability at 33.6 mA g^{-1}–3.36 A g^{-1}. The cycling tests are conducted with a voltage window of 1.0–3.0 V. (Reprinted from *Mater. Lett.*, 98, Tao, T. and Chen, Y., Direct synthesis of rutile TiO_2 nanorods with improved electrochemical lithium ion storage properties, 112–115, Copyright 2013, with permission from Elsevier.)

characteristics of rutile TiO_2, exhibiting cathodic insertion of lithium at 1.84 V and anodic extraction of lithium at 1.91 V vs Li/Li$^+$. The galvanostatic discharge/charge of the nanorods in the voltage range of 3.0–1.0 V (vs Li/Li$^+$) at a current density of 33.6 mA g^{-1} up to 50 cycles is presented in Figure 22.8b. The initial discharge and charge capacities are 249.9 and 157.5 mAhg^{-1}, respectively, with an irreversible capacity of 92.4 mAhg^{-1}. The irreversible capacity loss during the first charge and discharge was also observed for other TiO_2 polymorphs and nanosized rutile. Despite the capacity decay in the first several cycles, these nanorods exhibit good capacity retention over extended cycling and are able to deliver a reversible capacity of about 121 mAhg^{-1} after 100 cycles (Figure 22.8c). The high coulombic efficiency of over 98% for TiO_2 nanorod electrode can be remained after the sixth cycle. Additionally, the cycling response at continuously variable rates on the nanorods was also evaluated, as presented in Figure 22.8d. While cycling at higher current rates of 168 mA g^{-1} to 3.36 A g^{-1}, comparable capacities of 52–127 mAhg^{-1} can be delivered. After deep cycling at high rates, a capacity of about 140 mAhg^{-1} was detected in the 65th cycle when the current rate was returned to the value of 33.6 mA g^{-1}.

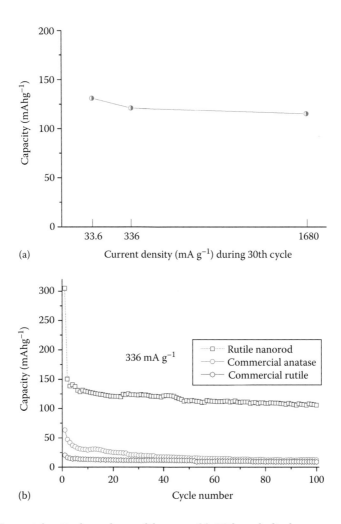

Figure 22.9   (a) Current density dependence of the reversible 30th-cycle discharge capacity for rutile TiO_2 nanorods; (b) cycling performances of rutile TiO_2 nanorods, commercial anatase TiO_2, and commercial rutile TiO_2. (Reprinted from *Mater. Lett.*, 98, Tao, T. and Chen, Y., Direct synthesis of rutile TiO_2 nanorods with improved electrochemical lithium ion storage properties, 112–115, Copyright 2013, with permission from Elsevier.)

Figure 22.9a shows high-rate capability of rutile TiO_2 nanorods. No severe capacity decay (12.2%) was observed with increasing constant current density from 33.6 to 1680 mA g^{-1} in the rutile nanorod electrode. The curves of discharge capacity vs cycle number for the rutile TiO_2 nanorods, the commercial anatase TiO_2, and commercial rutile TiO_2 electrode at a current density of 336 mA g^{-1} are exhibited in Figure 22.9b. The results indicate that the specific capacity of the nanorod electrode is much higher than that of the commercial TiO_2 electrodes, such as commercial anatase (Sigma Aldrich) and commercial rutile TiO_2 (Sigma Aldrich). The enhanced capacity of the rutile TiO_2 nanorods is ascribed to their unique structural characteristics.

Lithium insertion/extraction into/from rutile is highly anisotropic and proceeds mainly through fast diffusion along the *c*-axis channels while Li diffusion in the *ab*-planes is very slow

(Deng et al. 2009; Pfanzelt et al. 2011). The small nanometric size in the *ab*-plane drastically can decrease the diffusion length in comparison to microsized materials and overcomes the low diffusion coefficient D_{ab}. Additionally, a lower crystallite size opens the lattice leading to higher lithium diffusion rates and less strain due to lattice expansion during the insertion process. On the other hand, the high specific surface area of nanocrystalline rutile provides more active sites for Li insertion and also for surface Li-ion storage. Consequently, Li insertion occurs mostly on a thin surface layer rather than in the bulk. Thus, it is believed that nanosizing increases both the capacity and the rate capability of rutile TiO_2.

22.5.2 $FeTiO_3$ Nanosheets as Electrode Material for Electrochemical Supercapacitors

Nanostructured $FeTiO_3$ is an interesting representative of a ternary oxide, and its potential requires to be explored carefully. We have recently investigated the pseudocapacitive behavior of $FeTiO_3$ nanosheets. Their properties were evaluated in several types of aqueous solutions. The motivation for these measurements is that a number of metal oxides have been previously found to have the so-called pseudocapacitive behavior. These metals are normally multivalent metals, and most of the systems tested are oxides involving a single type of metal atoms. It is of high interest to evaluate ternary oxides because it is expected that they may provide capacitances over larger potential windows and involve redox chemical reactions over two types of metal cations.

CV curves of the $FeTiO_3$ nanosheets in three aqueous electrolytes—3 M KCl, 1 M H_2SO_4, and 1 M KOH—at sweep rates of 5 and 50 mV s^{-1} are shown in Figure 22.10a through c, and pseudocapacitive behavior is evident in all cases. The potential windows are –1.2 to –0.4 V vs Ag/AgCl in 3 M KCl, 0–0.9 V vs Ag/AgCl in 1 M H_2SO_4, and –1.2 to –0.4 V vs Hg/HgO in 1 M KOH aqueous electrolyte. The CV curves taken at 5 mV s^{-1} are closer to the rectangular ones in 3 M KCl, whereas the CV in 1 M H_2SO_4 shows a pair of distinct redox peaks. The area enclosed by the CV curve is much smaller for the working electrode in 1 M H_2SO_4 electrolyte, suggesting that the charge storage ability of $FeTiO_3$ in this electrolyte is lower than that in 3 M KCl or 1 M KOH. The CV curves taken at the sweep rate of 50 mV s^{-1} are elliptical in shape, which may be an evidence of limitations in the conductivity of the electrode. Electrochemical properties of $FeTiO_3$ nanosheets were also assessed by GCD experiments, as shown in Figure 22.10d through f. The testing was conducted in 1 M KOH electrolyte in the potential range of between –1.2 and –0.4 V vs Hg/HgO reference electrode at current densities ranging from 0.5 to 5 A g^{-1}. The GCD curves of the $FeTiO_3$ nanosheets at current rates of 0.5 and 2 A g^{-1} are shown in Figure 22.10d. The capacitance measured at 0.5 A g^{-1} is 122 ± 14.5 F g^{-1}. The long-term cycle stability was investigated at a constant current density of 0.5 A g^{-1} over 1000 cycles and the graph of the capacitance as a function of the cycle number (Figure 22.10e). No noticeable decay in the specific capacitance was found after 1000 cycles. The $FeTiO_3$ nanosheets are able to retain the capacitance of 50 ± 6 F g^{-1} at a high current rate of 5 A g^{-1} (Figure 22.10f). We should note that GCD experiments conducted at slow current rates (50 and 100 mA g^{-1}) showed some deviation from the behavior. Particularly, the coulombic efficiency deviates from being close to 100%, and charge or discharge branches of the profile may become extended.

A good electrochemical behavior of the sheet-like $FeTiO_3$ nanostructures can be attributed to their high surface area (26 m^2 g^{-1}) and interconnected hierarchical structure of petals. The sheet-like morphology is expected to provide a larger electrode–electrolyte contact area, good penetration of electrolyte throughout the electrode, and a convenient conduction path for electrons traveling in the active component of the electrode.

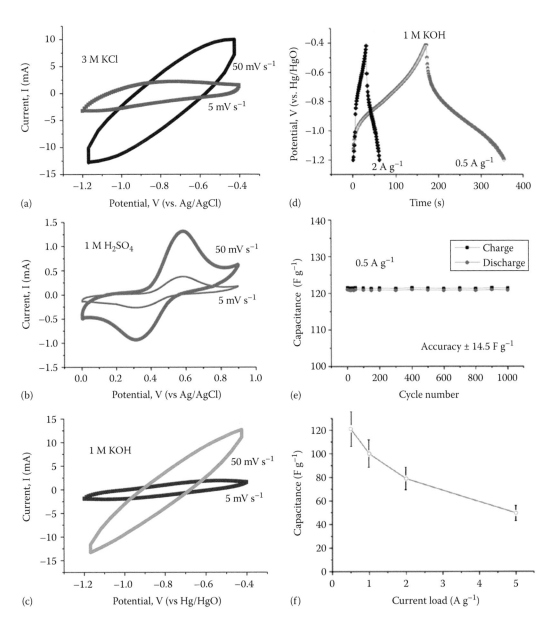

Figure 22.10   Cyclic voltammetry curves of the FeTiO₃ nanosheets recorded at sweep rates of 5 and 50 mV s⁻¹ in 3 M KCl (a), 1 M H₂SO₄ (b), and 1 M KOH (c) aqueous electrolytes, and galvanostatic charge–discharge measurements: (d) charge–discharge profiles measured at current rates of 0.5 and 2 A g⁻¹, (e) cyclic stability at 0.5 A g⁻¹, and (f) capacitance retention at high current loads. (Reprinted with permission from Tao, T., Glushenkov, A.M., Liu, H.W., Liu, Z.W., Dai, X.J.J., Chen, H., Ringer, S.P., and Chen, Y., Ilmenite FeTiO₃ nanoflowers and their pseudocapacitance, *J. Phys. Chem. C*, 115, 17297–17302, 2011. Copyright 2011, American Chemical Society.)

22.6 CONCLUSIONS

Combination methods for the synthesis of titanium-based nanostructures including TiO_2 nanorods and $FeTiO_3$ nanosheets from natural ilmenite have been presented. In general, these methods consist of two steps: ball milling as a first step, and annealing or wet chemical treatment as a second step.

The growth direction of the rutile TiO_2 nanorod obtained by annealing the milled ilmenite powders is parallel to the c axis of tetragonal cell of rutile structure, and the nanorod side walls of (002) and (101) planes actually construct unit cells of tetragonal rutile. The single-crystal nanorods prepared by leaching the milled ilmenite powders in aqueous HCl have good crystallinity, and their dimensions were 50–100 nm (length) by 5–20 nm (width) by 2–5 nm (thickness). The sheet-like $FeTiO_3$ nanostructures obtained by treating the milled ilmenite in 1 M NaOH aqueous solution are composed of a number of petals with smooth surfaces, and each petal is 5–20 nm thick and 100–200 nm wide. Thus, we believe that mineral ilmenite would be a perfect precursor for producing the titanium-based nanostructures if a suitable synthesis method was developed.

The rutile TiO_2 nanorods obtained by leaching the milled ilmenite in aqueous HCl as anode for lithium ion batteries show excellent electrochemical properties, such as large reversible charge–discharge capacity, good cycling stability, and high rate performance. Moreover, their electrochemical performance is better than that of two commercial samples including commercial anatase and rutile TiO_2. $FeTiO_3$ nanosheets synthesized by treating the milled ilmenite in NaOH aqueous solution have good electrochemical properties in aqueous electrolytes, and the capacitance of 122 ± 14.5 F g^{-1} is measured in 1 M KOH aqueous electrolyte at the current rate of 500 mA g^{-1}, and 50 ± 6 F g^{-1} is retained at 5 A g^{-1}. The material has a good long-term cycling stability. According to our data, $FeTiO_3$ nanosheets show functionality as an electrode material for supercapacitors.

The research presented in this chapter is a new finding in the synthesis of TiO_2 nanorods and $FeTiO_3$ nanosheets. It is expected that the technology demonstrated for synthesizing Ti-based nanomaterials can be extended to produce other nanostructures such as TiO_2 nanosheets, $FeTiO_3$ nanotubes, and $Li_4Ti_5O_{12}$ nanomaterials, and the continuing breakthroughs in the synthesis and modifications of Ti-based nanomaterials can bring new properties and new applications with improved performance.

ACKNOWLEDGMENTS

I gratefully acknowledge financial support from the Australian Research Council under the Centre of Excellence program and the China Scholarship Council (CSC) for providing the scholarship. I also acknowledge the help of Dr. Alexey M. Glushenkov (Deakin University) for testing of electrode materials for lithium ion batteries.

Dr. Hongwei Liu (University of Sydney), Prof. Zongwen Liu (University of Sydney), Dr. Hongzhou Zhang (Trinity College Dublin, Republic of Ireland), and Miss Dan Zhou (Trinity College Dublin, Republic of Ireland) are acknowledged for TEM support. Dr. Hua Chen (The Australian National University) and Dr. Xiujuan J. Dai (Deakin University) are also acknowledged for their scientific and technical assistance.

REFERENCES

Armstrong, A. R., Armstrong, G., Canales, J., and Bruce, P. G. (2004) TiO_2-B nanowires, *Angewandte Chemie International Edition 43*(17), 2286–2288.

Armstrong, G., Armstrong, A. R., Bruce, P. G., Reale, P., and Scrosati, B. (2006) TiO_2(B) nanowires as an improved anode material for lithium-ion batteries containing $LiFePO_4$ or $LiNi_{0.5}Mn_{1.5}O_4$ cathodes and a polymer electrolyte, *Advanced Materials 18*(19), 2597–2600.

Cao, F., Hu, W., Zhou, L., Shi, W. D., Song, S. Y., Lei, Y. Q., Wang, S., and Zhang, H. J. (2009) 3D Fe_3S_4 flower-like microspheres: High-yield synthesis via a biomolecule-assisted solution approach, their electrical, magnetic and electrochemical hydrogen storage properties, *Dalton Transactions 42*, 9246–9252.

Chen, J. S., Archer, L. A., and Lou, X. W. (2011) SnO_2 hollow structures and TiO_2 nanosheets for lithium-ion batteries, *Journal of Materials Chemistry 21*(27), 9912–9924.

Chen, X. B. and Mao, S. S. (2007) Titanium dioxide nanomaterials: Synthesis, properties, modifications, and applications, *Chemical Reviews 107*(7), 2891–2959.

Chen, Y. (1997) Low-temperature oxidation of ilmenite ($FeTiO_3$) induced by high energy ball milling at room temperature, *Journal of Alloys and Compounds 257*(1–2), 156–160.

Chen, Y., Hwang, T. H., and Williams, J. S. (1997) Mechanically activated carbothermic reduction of ilmenite, *Metallurgical and Materials Transactions A 28*(5), 1115–1121.

Cheng, Y., Wang, Y. S., Zheng, Y. H., and Qin, Y. (2005) Two-step self-assembly of nanodisks into plate-built cylinders through oriented aggregation, *Journal of Physical Chemistry B 109*(23), 11548–11551.

Chen, Y., Williams, J. S., Campbell, S. J., and Wang, G. M. (1999) Increased dissolution of ilmenite induced by high-energy ball milling, *Materials Science and Engineering: A 271*(1–2), 485–490.

Chen, Z. H., Belharouak, I., Sun, Y. K., and Amine, K. (2013) Titanium-based anode materials for safe lithium-ion batteries, *Advanced Functional Materials 23*(8), 959–969.

Deng, D., Kim, M. G., Lee, J. Y., and Cho, J. (2009) Green energy storage materials: Nanostructured TiO_2 and Sn-based anodes for lithium-ion batteries, *Energy & Environmental Science 2*(8), 818–837.

Djenizian, T., Hanzu, I., and Knauth, P. (2011) Nanostructured negative electrodes based on titania for Li-ion microbatteries, *Journal of Materials Chemistry 21*(27), 9925–9937.

Etacheri, V., Marom, R., Elazari, R., Salitra, G., and Aurbach, D. (2011) Challenges in the development of advanced Li-ion batteries: A review, *Energy & Environmental Science 4*(9), 3243–3262.

Glushenkov, A. M., Zhang, H. Z., and Chen, Y. (2008) Reactive ball milling to produce nanocrystalline ZnO, *Materials Letter 62*(24), 715–718.

Gong, D., Grimes, C. A., Varghese, O. K., Hu, W. C., Singh, R. S., Chen, Z., and Dickey, E. C. (2001) Titanium oxide nanotube arrays prepared by anodic oxidation, *Journal of Materials Research 16*(12), 3331–3334.

Han, X. G., Kuang, Q., Jin, M. S., Xie, Z. X., and Zheng, L. S. (2009) Synthesis of titania nanosheets with a high percentage of exposed (001) facets and related photocatalytic properties, *Journal of the American Chemical Society 131*(9), 3152–3153.

Hou, S. Y., Zou, Y. C., Liu, X. C., Yu, X. D., Liu, B., Sun, X. J., and Xing, Y. (2011) CaF_2 and CaF_2:Ln^{3+}(Ln = Er, Nd, Yb) hierarchical nanoflowers: Hydrothermal synthesis and luminescent properties, *Crystal Engineering Communication 13*(3), 835–840.

Hu, Y. S., Kienle, L., Guo, Y. G., and Maier, J. (2006) High lithium electroactivity of nanometer-sized rutile TiO_2, *Advanced Materials 18*(11), 1421–1426.

Jin, R. C., Chen, G., Wang, Q., Pei, J., Sun, J. X., and Wang, Y. (2011) PbTe Hierarchical nanostructures: Solvothermal synthesis, growth mechanism and their electrical conductivities, *Crystal Engineering Communication 13*(6), 2106–2113.

Jung, J. H., Kobayashi, H., Van Bommel, K. J. C., Shinkai, S., and Shimizu, T. (2002) Creation of novel helical ribbon and double-layered nanotube TiO_2 structures using an organogel template, *Chemistry of Materials 14*(4), 1445–1447.

Kasuga, T., Hiramatsu, M., Hoson, A., Sekino, T., and Niihara, K. (1999) Titania nanotubes prepared by chemical processing, *Advanced Materials 11*(15), 1307–1311.

Kim, Y. J., Gao, B. F., Han, S. Y., Jung, M. H., Chakraborty, A. K., Ko, T., Lee, C., and Wan, I. L. (2009) Heterojunction of $FeTiO_3$ nanodisc and TiO_2 nanoparticle for a novel visible light photocatalyst, *Journal of Physical Chemistry C 113*(44), 19179–19184.

Lee, K. T. and Cho, J. (2011) Roles of nanosize in lithium reactive nanomaterials for lithium ion batteries, *Nano Today 6*(1), 28–41.

Léonard, F. and Talin, A. A. (2011) Electrical contacts to one-and two-dimensional nanomaterials, *Nature Nanotechnology 6*(12), 773–783.

Li, X. L., Ge, J. P., and Li, Y. D. (2004) Atmospheric pressure chemical vapor deposition: An alternative route to large-scale MoS_2 and WS_2 inorganic fullerene-like nanostructures and nanoflowers, *Chemistry: A European Journal 10*(23), 6163–6171.

Liu, J. H. and Liu, X. W. (2012) Two-dimensional nanoarchitectures for lithium storage, *Advanced Materials 24*(30), 4097–4111.

Liu, S. H., Jia, H. P., Han, L., Wang, J. L., Gao, P. F., Xu, D. D., Yang, J., and Che, S. N. (2012) Nanosheet-constructed porous TiO_2-B for advanced lithium ion batteries, *Advanced Materials 24*(24), 3201–3204.

Lu, X. H., Wang, G. M., Zhai, T., Yu, M. H., Gan, J. Y., Tong, Y. X., and Li, Y. (2012) Hydrogenated TiO_2 nanotube arrays for supercapacitors, *Nano Letters 12*(3), 1690–1696.

Luo, Y. S., Zhang, W. D., Dai, X. J., Yang, Y., and Fu, S. Y. (2009) Facile synthesis and luminescent properties of novel flower like $BaMoO_4$ nanostructures by a simple hydrothermal route, *Journal of Physical Chemistry C 113*(12), 4856–4861.

Marom, R., Amalraj, S. F., Leifer, N., Jacob, D., and Aurbach, D. (2011) A review of advanced and practical lithium battery materials, *Journal of Materials Chemistry 21*(27), 9938–9954.

Ohara, S., Sato, K., Tan, Z. Q., Shimoda, H., Ueda, M., and Fukui, T. (2010) Novel mechanochemical synthesis of fine $FeTiO_3$ nanoparticles by a high-speed ball-milling process, *Journal of Alloys and Compounds 504*(1), L17–L19.

Ortiz, G. F., Hanzu, I., Djenizian, T., Lavela, P., Tirado, J. L., and Knauth, P. (2009) Alternative Li-ion battery electrode based on self-organized titania nanotubes, *Chemistry of Materials 21*(1), 63–67.

Ovenstone, J. and Yanagisawa, K. (1999) Effect of hydrothermal treatment of amorphous titania on the phase change from anatase to rutile during calcination, *Chemistry of Materials 11*(10), 2770–2774.

Park, J., Privman, V., and Matijevic, E. (2001) Model of formation of monodispersed colloids, *The Journal of Physical Chemistry B 105*(47), 11630–11635.

Pfanzelt, M., Kubiak, P., Fleischhammer, M., and Wohlfahrt-Mehrens, M. (2011) TiO_2 rutile—An alternative anode material for safe lithium-ion batteries, *Journal of Power Sources 196*(15), 6815–6821.

Pradhan, S. K., Reucroft, P. J., Yang, F. Q., and Dozier, A. (2003) Growth of TiO_2 nanorods by metalorganic chemical vapor deposition, *Journal of Crystal Growth 256*(1–2), 83–88.

Qiu, Y. C., Yan, K. Y., Yang, S. H., Jin, L. M., Deng, H., and Li, W. S. (2010) Synthesis of size-tunable anatase TiO_2 nanospindles and their assembly into anatase @ titanium oxynitride/titanium nitride graphene nanocomposites for rechargeable lithium ion batteries with high cycling performance, *ACS Nano 4*(11), 6515–6526.

Ren, Y. and Gao, L. (2010) From three-dimensional flower-like α-$Ni(OH)_2$ nanostructures to hierarchical porous NiO nanoflowers: Microwave-assisted fabrication and supercapacitor properties, *Journal of the American Ceramic Society 93*(11), 3560–3564.

Roy, P., Berger, S., and Schmuki, P. (2011) TiO_2 nanotubes: Synthesis and applications, *Angewandte Chemie International Edition 50*(13), 2904–2939.

Sarkar, S., Pradhan, M., Sinha, A. K., Basu, M., Negishi, Y. C., and Pal, T. (2010) An aminolytic approach toward hierarchical β-$Ni(OH)_2$ nanoporous architectures: A bimodal forum for photocatalytic and surface-enhanced Raman scattering activity, *Inorganic Chemistry 49*(19), 8813–8827.

Shim, H. W., Lee, D. K., Cho, I. S., Hong, K. S., and Kim, D. W. (2010) Facile hydrothermal synthesis of porous TiO_2 nanowire electrodes with high-rate capability for Li ion batteries, *Nanotechnology 21*(25), 255706–255715.

Simon, P. and Gogotsi, Y. (2008) Materials for electrochemical capacitors, *Nature Materials 7*(11), 845–854.

Sun, X. J., Wang, J. W., Xing, Y., Zhao, Y., Liu, X. C., Liu, B., and Hou, S. Y. (2011) Surfactant-assisted hydrothermal synthesis and electrochemical properties of nanoplate-assembled 3D flower-like $Cu_3V_2O_7(OH)_2\cdot 2H_2O$ microstructures, *Crystal Engineering Communication 13*(1), 367–370.

Tan, A. W., Pingguan-Murphy, B., Ahmad, B. R., and Akbarc, S. A. (2012) Review of titania nanotubes: Fabrication and cellular response, *Ceramics International 38*(6), 4421–4435.

Tao, T. and Chen, Y. (2013) Direct synthesis of rutile TiO_2 nanorods with improved electrochemical lithium ion storage properties, *Materials Letters 98*(1), 112–115.

Tao, T., Chen, Y., Zhou, D., Zhang, H. Z., Liu, S., Amal, R., Sharma, N., and Glushenkov, A. M. (2013) Expanding the applications of the ilmenite mineral to the preparation of nanostructures: TiO_2 nanorods and their photocatalytic properties in the degradation of oxalic acid, *Chemistry: A European Journal 19*(3), 1091–1096.

Tao, T., Glushenkov, A. M., Chen, Q. Y., and Chen, Y. (2010) Air-assisted growth of tin dioxide nanoribbons, *Journal of Nanoscience and Nanotechnology 10*(8), 5015–5019.

Tao, T., Glushenkov, A. M., Liu, H. W., Liu, Z. W., Dai, X. J. J., Chen, H., Ringer, S. P., and Chen, Y. (2011) Ilmenite $FeTiO_3$ nanoflowers and their pseudocapacitance, *The Journal of Physical Chemistry C 115*(35), 17297–17302.

Tsai, M. C., Chang, J. C., Sheu, H. S., Chiu, H. T., and Lee, C. Y. (2009) Lithium ion intercalation performance of porous laminal titanium dioxides synthesized by sol-gel process, *Chemistry of Materials 21*(3), 499–505.

Wang, G. P., Zhang, L., and Zhang, J. J. (2012) A review of electrode materials for electrochemical supercapacitors, *Chemical Society Reviews 41*(2), 797–828.

Yang, Z. G., Choi, D., Kerisit, S., Rosso, K. M., Wang, D. H., Zhang, J., Graff, G., and Liu, J. (2009) Nanostructures and lithium electrochemical reactivity of lithium titanites and titanium oxides: A review, *Journal of Power Sources 192*(2), 588–598.

Yu, J. and Chen, Y. (2010) One-dimensional growth of TiO_2 nanorods from ilmenite sands, *Journal of Alloys and Compounds 504*(1), S364–S367.

Yu, J., Chen, Y., and Glushenkov, A. M. (2009) Titanium oxide nanorods extracted from ilmenite sands, *Crystal Growth & Design 9*(2), 1240–1244.

Yue, W. B., Randorn, C., Attidekou, P. S., Su, Z. X., Irvine, J.T. S., and Zhou, W. Z. (2009) Syntheses, Li insertion, and photoactivity of mesoporous crystalline TiO_2, *Advanced Functional Materials 19*(17), 2826–2833.

Zhang, Q. F., Uchaker, E., Candelaria, S. L., and Cao, G. Z. (2013) Nanomaterials for energy conversion and storage, *Chemical Society Reviews 42*(7), 3127–3171.

Zhang, Q. H. and Gao, L. (2003) Preparation of oxide nanocrystals with tunable morphologies by the moderate hydrothermal method: Insights from rutile TiO_2, *Langmuir 19*(3), 967–971.

Zhou, W. J., Liu, H., Boughton, R. I., Du, G. J., Lin, J. J., Wang, J. Y., and Liu, D. (2010) One-dimensional single-crystalline Ti–O based nanostructures: Properties, synthesis, modifications and applications, *Journal of Materials Chemistry 20*(29), 5993–6008.

Zhu, G. N., Wang, Y. G., and Xia, Y. Y. (2012) Ti-based compounds as anode materials for Li-ion batteries, *Energy & Environmental Science 5*(5), 6652–6667.

Index

A

Adatom-induced magnetism, 264, 266–267
AFM images, nanosheets, 124, 126–127
Aluminum–BNNT system
    formability, 516, 518–519
    interfacial reaction
        AlB_2 crystals, 511–513
        crystallographic orientation, 514
        diffusion kinetics, 513
        distinct globular crystals, 511–512
        fabrication, 515
        fracture toughness, 514
        macroscale composite structure, 513
    strengthening, 515, 517
    synthesis, 515–516
Angle-resolved photo-emission spectroscopy
    (ARPES), 152, 154
Annealing, 143, 590
Arc discharge, 5
    BCN direct synthesis, 144–145
    BNNTs fabrications, 93
Assembling process, 2D metal oxide nanosheets
    flocculation, 238–240
    freeze-drying method, 237
    layer-by-layer fabrication
        electrostatic sequential deposition, 240–241
        Langmuir–Blodgett method, 241–242
    schematic illustration, 237–238
Atomic structure
    nanosheets
        experimental aspects, 22–23
        theoretical aspects, 22
    nanotubes
        experimental aspects, 16–17
        polygonization (*see* Polygonization)
        theoretical aspects, 15–16
Au-BNNTs
    carbon-doped BNNTs, 533
    emission current density, 533–534
    emitter surface, 534
    field-emission test, 535
    field enhancement factor, 532

Fowler–Nordheim theory, 533–534
    Keithley 485 picoammeter, 533
    work function, 535
Automated nanotube builders, 310
Azobenzene polymer, 132

B

Ball milling method
    BNNTs (*see* Boron nitride nanotubes (BNNTs))
    nanoribbons (*see* Boron nitride nanoribbons
        (BNNRs))
    titanium-based nanorods and nanosheets,
        589–590
B-doped graphene
    DOS and PDOS, 148, 150
    output and transfer characteristics, 150–151
    valence band spectra dispersion, 150–151
Bioimaging, two-dimensional layered nanosheets,
    353–354
Biosensors, two-dimensional layered nanosheets
    definition, 350
    field-effect transistor, 351
    graphene-based FRET biosensors, 352–353
    impedimetric sensors, 351
BN, *see* Boron nitride (BN)
BNNRs, *see* Boron nitride nanoribbons (BNNRs)
BNNSs, *see* Boron nitride nanosheets (BNNSs)
BNNTs, *see* Boron nitride nanotubes (BNNTs)
Bonded interactions, 309
Boron carbonitride (BCN) nanotube/nanosheet
    arc discharge and laser ablation, 144–145
    CVD growth (*see* Chemical vapor
        deposition (CVD))
    definition, 141
    electronic property
        B-doped graphene, 148–150
        boron nitrogen Co-doped graphene, 156–158
        N-doped graphene, 150–156
        ternary nanotubes, 158–159
    luminescence
        band gap, 162
        characterization, 158

609